Advances in Intelligent Systems and Computing

Volume 440

About this Series

The series "Advances in Intelligent Systems and Computing" contains publications on theory, applications, and design methods of Intelligent Systems and Intelligent Computing. Virtually all disciplines such as engineering, natural sciences, computer and information science, ICT, economics, business, e-commerce, environment, healthcare, life science are covered. The list of topics spans all the areas of modern intelligent systems and computing.

The publications within "Advances in Intelligent Systems and Computing" are primarily textbooks and proceedings of important conferences, symposia and congresses. They cover significant recent developments in the field, both of a foundational and applicable character. An important characteristic feature of the series is the short publication time and world-wide distribution. This permits a rapid and broad dissemination of research results.

More information about this series at http://www.springer.com/series/11156

Roman Szewczyk · Cezary Zieliński
Małgorzata Kaliczyńska
Editors

Challenges in Automation, Robotics and Measurement Techniques

Proceedings of AUTOMATION-2016, March 2–4, 2016, Warsaw, Poland

Editors
Roman Szewczyk
Industrial Research Institute for Automation and Measurements PIAP
Warsaw
Poland

Małgorzata Kaliczyńska
Industrial Research Institute for Automation and Measurements PIAP
Warsaw
Poland

Cezary Zieliński
Industrial Research Institute for Automation and Measurements PIAP
Warsaw
Poland

ISSN 2194-5357 ISSN 2194-5365 (electronic)
Advances in Intelligent Systems and Computing
ISBN 978-3-319-29356-1 ISBN 978-3-319-29357-8 (eBook)
DOI 10.1007/978-3-319-29357-8

Library of Congress Control Number: 2015961030

Printed on acid-free paper

This Springer imprint is published by SpringerNature
The registered company is Springer International Publishing AG Switzerland

Preface

Innovation, especially attained through advanced technologies, is one of the key factors influencing the quality of life of both individual people and whole nations. Among technologies currently having a major impact on innovation are: control, industrial automation, robotics and measurement techniques. For this reason problems associated with these technologies are at the centre of interest of researchers, both from the point of view of theoretical challenges and practical applications.

This book presents the set of papers accepted for presentation at the international conference named Automation, held in Warsaw during 2–4 March of 2016. It presents the research results presented by top experts in the fields of industrial automation, control, robotics and measurement techniques. Each chapter presents a thorough analysis of a specific technical problem which is usually followed by numerical analysis, simulation, and description of results of implementation of the solution of a real-world problem.

We are convinced that the presented theoretical results, practical solutions and guidelines will be of great value for both researchers working in the area of engineering sciences and for practitioners solving industrial problems.

Warsaw
December 2015

Roman Szewczyk
Cezary Zieliński
Małgorzata Kaliczyńska

Contents

About the Editors

Prof. Roman Szewczyk received both his Ph.D. and D.Sc. in the field of mechatronics. He specializes in the modelling of properties of magnetic materials as well as in sensors and sensor interfacing, in particular magnetic sensors for security applications. He leads the development of a sensing unit for a mobile robot developed for the Polish Police Central Forensic Laboratory and of methods of non-destructive testing based on the magnetoelastic effect. Professor Szewczyk was involved in over 10 European Union funded research projects within the FP6 and FP7 as well as projects financed by the European Defence Organization. Moreover, he was leading two regional and national scale technological foresight projects and was active in the organization and implementation of technological transfer between companies and research institutes. Roman Szewczyk is Secretary for Scientific Affairs in the Industrial Research Institute for Automation and Measurements (PIAP). He is also Associate Professor at the Faculty of Mechatronics, Warsaw University of Technology and a Vice-chairman of the Academy of Young Researchers of the Polish Academy of Sciences.

Prof. Cezary Zieliński received his M.Sc. Eng. degree in control in 1982, Ph.D. degree in control and robotics in 1988, the D.Sc. (habilitation) degree in control and robotics in 1996, all from the Faculty of Electronics and Information Technology, Warsaw University of Technology, Warsaw, Poland, and Full Professorship in 2012. Currently he is Full Professor both in the Industrial Research Institute for Automation and Measurement (PIAP) and the Warsaw University of Technology, where he is Director of the Institute of Control and Computation Engineering. Since 2007 he has been a member of the Committee for Automatic Control and Robotics, the Polish Academy of Sciences. Professor Zieliński is Head of the Robotics Group in the Institute of Control and Computation Engineering working on robot control and programming methods. His research interests focus on robotics in general and in particular include: robot programming methods, formal approach to the specification of architectures of multi-effector and multi-receptor systems, robot kinematics, robot position-force control, visual servo control and design of digital circuits. He is the author/co-author of over 180 conference and journal papers as well as books concerned with the above-mentioned research subjects.

Dr. Małgorzata Kaliczyńska received her M.Sc. Eng. degree in cybernetics from the Faculty of Electronics, Wrocław University of Technology, and her Ph.D. degree in the field of fluid mechanics from the Faculty of Mechanical and Power Engineering at the same university. Now she is Assistant Professor in the Industrial Research Institute for Automation and Measurement (PIAP) and editor of the scientific and technological magazine "Measurements, Automation, Robotics". Her areas of research interest include distributed control systems, Internet of Things, information retrieval and webometrics.

Part I
Control and Automation

Hierarchical Models in Modern Control Theory

Dmitry A. Novikov

Abstract Historical and modern approaches towards heterogeneous models and hierarchical modeling in control theory are listed. A unified approach towards a design of hierarchical models of multiagent systems is described and illustrated by the example of diffuse bomb model. Problems of "hierarchical automation" in organization-technical and production systems are described. General problems of hierarchical modeling are underlined.

Keywords Heterogeneous model · Hierarchical modeling · Multiagent system · Diffuse bomb problem · Production systems

1 Introduction: Heterogeneous Models and Hierarchical Modeling

In recent years, control theory more and more addresses the term of system "*heterogeneity*" comprehended, in the first place, as the multiplicity of its mathematical description (e.g., descriptive dissimilarity of separate subsystems: the type and scale of time/space of subsystems functioning, multi-type descriptive languages for certain regularities of a studied object, etc.) [1]. "Heterogeneity" also means complexity appearing in (qualitative, temporal and functional) *dissimilarity*, (spatial and temporal) *distribution* and the *hierarchical/networked structure* of a controlled object and an associated control system.

An adequate technology for design and joint analysis of a certain set of heterogeneous systems models is the so-called *hierarchical modeling*. According to this technology, models describing different parts of a studied system or its different properties (perhaps, with different levels of detail) are ordered on the basis of some logic, thereby forming a hierarchy or a sequence (a horizontal chain). Generally, lower hierarchical levels correspond to higher levels of detail in modeled systems

D.A. Novikov (✉)
Institute of Control Sciences, Moscow, Russia
e-mail: novikov@ipu.ru

R. Szewczyk et al. (eds.), *Challenges in Automation, Robotics and Measurement Techniques*, Advances in Intelligent Systems and Computing 440, DOI 10.1007/978-3-319-29357-8_1

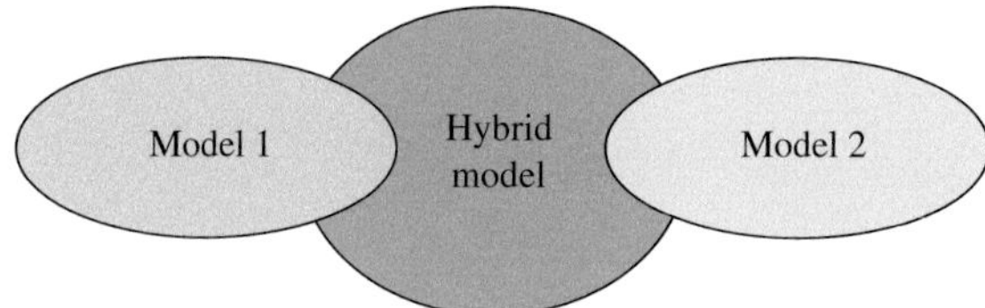

Fig. 1 The narrow interpretation of a hybrid model

description. Each element of a sequence possesses almost same level of detail, and the results (outputs) of a current model represent input data for a next model. Such approach to modeling was born and further developed in the 1960–1970s, e.g. see [2].

In some sense, hierarchical models are a wider category than hybrid models and the multi-model approach. A *hybrid model* is a model combining elements of two or more models reflecting different aspects of a studied phenomenon or process and/or employing different apparatuses (languages) of modeling—see Fig. 1. For instance, a hybrid model can include discrete and continuous submodels, digital and analog submodels, and so on.

In the wider interpretation, a hybrid model represents a complex of models each chosen under well-defined conditions, see Fig. 2. As an example, consider hybrid dynamic systems (HDS, also known as switching systems). The expression in the right-hand side of the HDS differential equation is chosen from a given set of options depending on the current state of the system and/or time and/or auxiliary conditions.

Within the *multi-model approach*, several models are used sequentially or simultaneously with further or current analysis and selection of "best" results.

Hierarchical (sequential) models may have a more complex structure, see Fig. 3. At each level, a model can be hybrid or follow the multi-model approach.

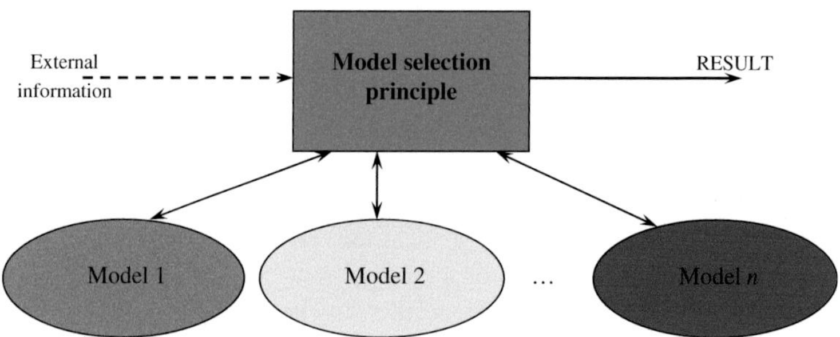

Fig. 2 The modern interpretation of a hybrid model. The multi-model approach

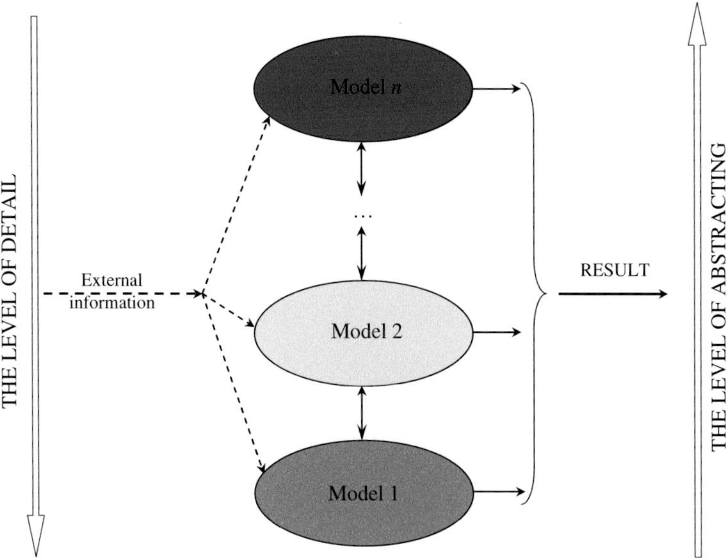

Fig. 3 A hierarchical (sequential) model

Hierarchical models lead to the problems of *aggregation* and *decomposition* well-known in mathematical modeling.

The next subsections give several examples of hierarchical models and the discussion of their drawbacks and advantages.

2 Hierarchical Models of Multiagent Systems

For the recent 15–20 years, a modern tendency in control theory has been seeking towards "*miniaturization*," "*decentralization*" and "*intellectualization*" in systems of very many interacting autonomous *agents* having social, technical or informational nature. Inherent properties of multi-agent systems (MAS) such as *decentralized* interaction and agents' multiplicity induce fundamentally new emergent properties (*autonomy*, smaller *vulnerability* to unfavorable factors, etc.) crucial in several applications [3–5].

In MAS the hierarchy of models is *inter alia* generated by the functional structure of the agent. The latter may have several hierarchical levels, see Fig. 4 [6]. The lowest (operational) level is responsible for implementation of actions (e.g., motion stabilization with respect to a preset path). Tactical level corresponds to actions' choice (including interaction with other agents). Strategic level is in charge of *decision-making*, *learning* and *adaptivity of behavior*. And finally, the highest

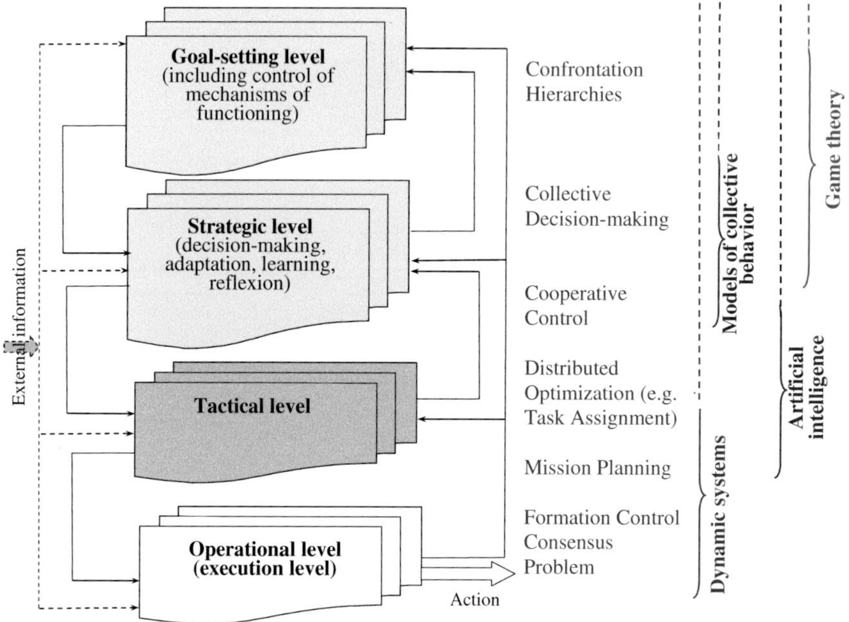

Fig. 4 The hierarchical structure of an agent in MAS

level (goal-setting) answers the principles of goal-setting and choice of the mechanisms of functioning for agents. The diffuse bomb model in Sect. 3 implements the general structure described by Fig. 4.

The structure presented by Fig. 4 seems rather universal. However, most realizations of multi-agent systems involve merely two lower levels and the framework of *dynamic systems theory*.

In *mission planning* problems, one can use different means of *artificial intelligence*, e.g., neural networks, evolutionary and logical methods, etc.

Also, let us mention *distributed optimization* (*agent-based computing*, see [7]) as a direction of modern optimization widespread in MAS. Its key idea consists in the following. An optimization problem of a multivariable function is decomposed into several subproblems solved by separate agents under limited information. For instance, each agent is “responsible” for a certain variable; at a current step, it chooses the value of this variable, being aware of the previous choice of some its “neighbors” and seeking to maximize its own local “goal function.” Given an initial (global) goal function, is it possible to find the “goal functions” of agents and their interaction rules so that the autonomous behavior of agents implements a centralized optimum? (in *algorithmic/computational game theory* [8, 9], this optimum can correspond to a Nash equilibrium or a Pareto efficient state of agents’ game).

Consider the strategic level of agent's architecture, which answers for adaptation, learning, reflexion and other aspects of strategic decision-making. *Game theory* and *theory of collective behavior* analyze interaction models for rational agents. In game theory, a common scheme consists in (1) describing the "model of a game," (2) choosing an equilibrium concept defining the stable outcome of the game and (3) stating a certain *control problem*–find the values of controlled "game parameters" implementing a required equilibrium [10].

Nowadays investigators gradually shift their efforts towards higher levels of agents' architecture, i.e., from consensus and communications problems to cooperative control and strategic behavior models of agents [11]. To justify this statement, let's analyze the networked control topics on main worldwide conferences on control theory and applications. Being subjective and not pretending to a complete overview, the author emphasizes triennial world congresses conducted by *International Federation of Automatic Control* (IFAC) and annual *Conferences on Decisions and Control* (CDC) under the auspices of *Institute of Electrical and Electronics Engineers* (IEEE). Alongside with these major events (or even jointly with CDC), there exist regular "national" conferences (*a*ctually, these conferences gather researchers from many other countries): *American Control Conference* (ACC) and *European Control Conference* (ECC). In the USSR, the role of such national conferences belonged to *AMCP*—All-Union Meetings on Regulation Theory (later, on Automatic Control and, then, on Control Problems).

Figures 5 and 6 specify the topics of networked control by the levels of agents architecture in MAS and problems treated at these levels. The following groups of topics have been identified via expertise: MAS and *consensus* problems, communications in MAS, *cooperative control*, upper levels of control (*strategic behavior* of agents), "others" (mostly, information and communication networks with a slight emphasis on control problems).

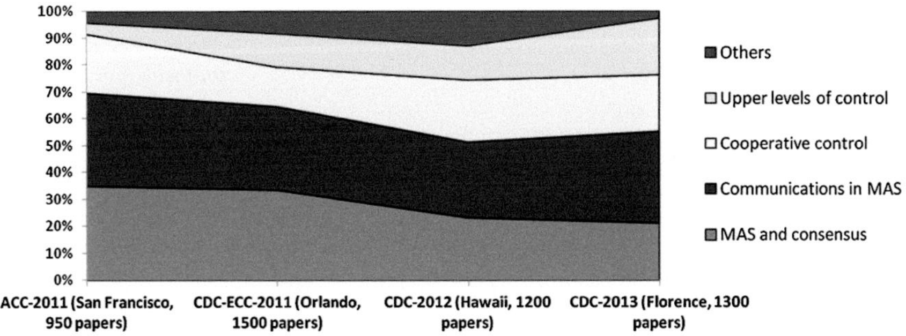

Fig. 5 Specification of networked control topics at ACC and CDC

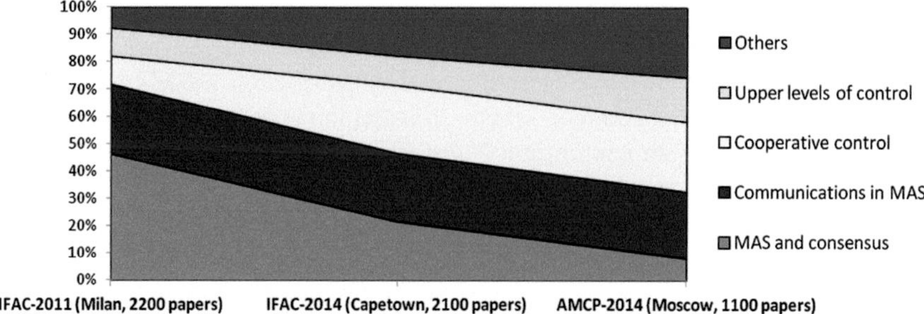

Fig. 6 Specification of networked control topics at IFAC Congresses and AMCP-2014

3 The Diffuse Bomb Model

An example of the hierarchical model of a MAS is the model of distributed penetration through a defense system (the so called diffuse bomb model [12]).

Consider a group of autonomous moving agents must hit a target with given coordinates. At each time step, any agent can be detected and destroyed by a defense system (with a certain probability). Detection/annihilation probability depends on agent's coordinates and speed, as well as on the relative arrangement of all objects in the group.

The problem is synthesizing algorithms of decentralized interaction among agents and their decision-making (the choice of direction and speed of their motion) to maximize the number of agents reaching the target. Agents appear "intelligent" in the following sense. Some agents (reconnaissance) can acquire on-line information on the parameters of the defense system. By observing the behavior of the reconnaissance agents, the rest ones perform "reflexion," assess the limits of dangerous areas and solve the posed problem.

The following hierarchical model defined by Table 1 serves for appraising and choosing most efficient algorithms of behavior in [12].

Table 1 The diffuse bomb model

Hierarchical level	Modeled phenomena/processes	Modeling tools
6	Choosing the set of agents and their properties	Discrete optimization methods
5	Choosing the paths and speeds of agents	Optimal control
4	Agent's forecast of the behavior of other agents	Reflexive games. The reflexive partitions method
3	Detection probability minimization based on current information	Algorithms of course choice
2	Collisions avoidance, obstacles avoidance	Algorithms of local paths choice
1	Object's movement towards a target	Dynamic motion equations

4 "Hierarchical Automation" in Organization-Technical Systems

Since the 1980s, *production systems* have followed a long path from flexible to holonic systems. In recent years, they attract the growing interest of researchers in connection with new market challenges: the efficiency of production specialization and decentralization, product and service differentiation, etc. There appear *networked productions* and "*cloud*" *productions*. Along with implementation of fundamentally new technologies of production (nanotechnologies, additive technologies, digital production, and so on), we observe gradual changes in its organization, i.e., the emphasis is shifted from operations automation to *control automation* at all life cycle stages.

Existing challenges such as:

- a huge number of product's customized configurations;
- integration of small- and large-scale production;
- lead-time reduction for an individual order;
- supply chains integration for stock optimization;
- and others call for solutions guaranteeing:
- the universality of production systems and their separate components;
- the capability of rapid and flexible adjustment with respect to new tasks;
- autonomous decision-making in production owing to high-level control automation;
- survivability, replicability and scalability owing to network-centric control and multi-agent technologies;
- decision-making in production with proper consideration of economic factors, etc.

Modern production systems have a hierarchical structure, as indicated by Fig. 7. And the complexity of control problems treated induces their decomposition into decision-making levels. Each level in control problems solution corresponds to its own goals, *models* and *tools* (Fig. 7) at each stage of control (organizing, planning, implementing, controlling and analyzing). Hence, in organizational-technical production systems it is possible (and necessary) to apply hierarchical modeling.

This possibility is implemented, but on an irregular and unsystematized basis. Obviously, one can solve real problems of automation, analysis and decision support for production systems only within appropriate computer-aided informational systems. As an illustration, consider the classes of such systems in the ascending order of their "hierarchical level":

- lower-level control systems (PLC, MicroPC, …);
- supervising and scheduling systems (SCADA, DCS, …);
- production planning and management systems (MRP, CRP, …, MRP2, …);
- integrated systems (MES, …, ERP., …);

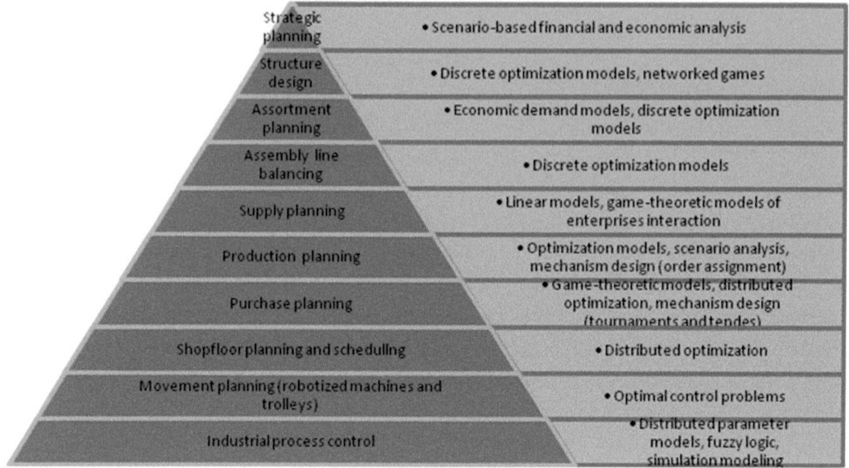

Fig. 7 Hierarchical models in production systems

- systems responsible for interaction with an external environment or development (SCM, CRM, PMS, …);
- upper-level analytic systems (OLAP, BSC, DSS, …).

These classes of systems use mathematical models, but very sparsely; as a rule, the higher is the level of hierarchy (this statement is true for separate informational systems and for integrated informational systems of product life cycle management (PLM) including computer-aided design systems, which realize the complex of the listed functions), the lesser is their usage. For instance, lower-level controllers employ in full automatic control theory; project management systems (PMS) incorporate classical algorithms for critical path search, Monte Carlo methods for project duration estimation, and heuristics for resources balancing; ERP systems and logistics systems (SCM) involve elementary results from stock management theory, and so on.

Nevertheless, full-fledged implementation of the so-called "*hard*" *models* and "quantitative science" (operations research, discrete optimization, data analysis and other branches of modern applied mathematics) in informational systems still waits in the wings.

Several global problems exist here. On the one hand, mathematical models require very accurate and actual information often associated with inadmissibly high organizational and other costs. On the other hand, in many cases "*soft*" *models* (putting things in order in production processes, implementation of typical solutions and standards in the form of qualitative best practices, etc.) yield an effect exceeding manyfold the outcomes of quantitative models, yet consume reasonable efforts. Therefore, it seems that quantitative models should be applied at the second stage, "extracting" the remainder of potential efficiency increase.

5 Conclusion

The forthcoming years will be remarkable for transition in control theory from the so-called *C^3 paradigm* (joint solution of *Control* + *Computations* + *Communications* problems) to *the C^5 concept* (*Control* + *Computations* + *Communications* + *Costs* + *Life Cycle*), when the above-mentioned problems are solved taking into account cost aspects (in the general sense) over the whole *life cycle* of a system including the joint design of a control system and its controlled object.

Speaking about "*networkism*," we have to touch "*network-centrism*" (network-centrism operates its own abbreviations differing from control theory (see above C^3 or C^5): C^3I—Command, Control, Communications and Intelligence, C^4I—Command, Control, Communications, Computers and Intelligence, and others) extremely fashionable nowadays (also called "*network-centric fever*"). It admits several interpretations covering organization and analysis principles of any networks in principle or temporary networks created for specific task or mission execution at a right place and right time (*networked organizations*, e.g., interaction of military units in a combat theater). This approach finds wide application in network-centric warfare problems for vertical and horizontal integration of all elements during a military operation (control, communication, reconnaissance and annihilation systems).

Another manifestation of "*networkism*" concerns the growing popularity of *distributed decision support systems*. The intensive development of information and communications technologies (ICT) increases the role of informational aspects of control in decentralized hierarchical systems (an example is decision-making support in distributed decision systems which integrate heterogeneous information on strategic planning and forecasting from different government authorities and industrial sectors). One of such aspects consists in *informational control* as a purposeful impact on the awareness of controlled subjects; therefore, a topical problem is to develop a mathematical apparatus providing an adequate description for an existing relationship between the behavior of system participants and their mutual awareness [10].

Design of intelligent analytic systems for informational and analytic support of goal-setting and control cycle represents another important informational aspect of control in decentralized hierarchical systems. Here it seems necessary to substantiate methodological approaches to control efficiency in decentralized control systems, including elaboration of principles and intelligent technologies for data acquisition, representation, storage and exchange.

We underline that an appreciable share of information required for situation assessment, goal-setting and control strategy choice in decentralized systems is ill-structured (mostly, in the form of text). And there arise the problems of relevant search and further analysis of such information [13]. The described circumstances bring to the need for suggesting new information retrieval methods (or even knowledge processing methods) based on proper consideration of its lexis and different quantitative characteristics and, moreover, on analysis of its semantics, separation of target data and situation parameters, assessment of their dynamics and scenario modeling of situation development in future periods.

Concluding this paper dedicated to heterogeneous models and hierarchical modeling, we underline a series of their common classes of problems. Modern controlled objects are complicated so that sometimes a researcher would hardly separate out purely hierarchical or purely networked components. In such cases, it is necessary to consider *networks of hierarchies* and *hierarchies of networks* [1].

First, at each level models have their own intricacies induced by a corresponding mathematical apparatus. Moreover, there arise "conceptual coupling" dilemmas and the *common language* problem among the representatives of different application domains.

Second, a complex of "*joined*" *models* inherits all negative properties of each component. Just imagine that, at least, one model in a "chain" admits no analytic treatment; then the whole chain is doomed to simulation modeling. The speed of computations in a chain is determined by the slowest component, and so on.

And third, it is necessary to assess the comparative efficiency of the solutions of aggregated problems, as well as to elaborate and disseminate typical solutions of corresponding control problems in order to transfer them to the engineering ground.

References

1. Novikov, D.: Cybernetics: From Past to Future, 107 p. Springer, Berlin (2016)
2. Mesarović, M., Mako, D., Takahara, Y.: Theory of Hierarchical Multilevel Systems, 294 p. Academic, New York (1970)
3. Ren, W., Yongcan, C.: Distributed Coordination of Multi-agent Networks, 307 p. Springer, London (2011)
4. Shoham, Y., Leyton-Brown, K.: Multiagent Systems: Algorithmic, Game-Theoretic, and Logical Foundations, 504 p. Cambridge University Press, Cambridge (2008)
5. Wooldridge, M.: An Introduction to Multi-Agent Systems, 376 p. Wiley, New York (2002)
6. Novikov, D.: Hierarchical models of warfare. Autom. Remote Control **74**(10), 1733–1752 (2013)
7. Boyd, S., Parikh, N., Chu, E., et al.: Distributed optimization and statistical learning via the alternating direction method of multipliers. Found. Trends Mach. Learn. **3**(1), 1–122 (2011)
8. Nisan, N., Roughgarden, T., Tardos, E., Vazirani, V. (eds.): Algorithmic Game Theory, 776 p. Cambridge University Press, New York (2009)
9. Mansour, Y.: Computational Game Theory, 150 p. Tel Aviv University, Tel Aviv (2003)
10. Novikov, D., Chkhartishvili, A.: Reflexion and Control: Mathematical Models, 298 p. CRC Press, London (2014)
11. Forrest, J., Novikov, D.: Modern trends in control theory: networks, hierarchies and interdisciplinarity. Adv. Syst. Sci. Appl. **12**(3), 1–13 (2012)
12. Korepanov, V., Novikov, D.: The diffuse bomb problem. Autom. Remote Control **74**(5), 863–874 (2013)
13. Novikov, D.: Big data and big control. Adv. Syst. Stud. Appl. **15**(1), 21–36 (2015)

Neural Modelling of a Yeast Fermentation Process Using Extreme Learning Machines

Maciej Ławryńczuk

Abstract This work details development of dynamic neural models of a yeast fermentation chemical reactor using Extreme Learning Machines (ELM). The ELM approach calculates very efficiently, without nonlinear optimisation, dynamic models, but only in the non-recurrent serial-parallel configuration. It is shown that in the case of the considered benchmark the ELM technique gives models which are also quite good recurrent long-range predictors, they work in the parallel configuration (simulation mode). Furthermore, properties of neural models obtained by the ELM and classical (optimisation-based) approaches are compared.

Keywords Neural networks · Extreme learning machines

1 Introduction

Neural networks [3], due to their excellent approximation ability, are used very frequently as models of nonlinear systems in many fields, e.g. in advanced control algorithms [6, 13], in pattern recognition [12], in interpretation of medical images [11], in fault diagnosis and fault-tolerant control [14] and in optimisation [15].

Typically, determination of parameters (weights) of neural networks (training) needs solving an optimisation problem [3]. Such a problem is nonlinear and it is likely to be non-convex and multi-modal. An alternative is to use Extreme Learning Machines [2]. In the ELM approach the weights of the second layer are determined explicitly, without nonlinear optimisation, while the weights of the first layer are chosen randomly. The ELM method yields non-recurrent serial-parallel models whereas in the case of dynamic systems the objective is to find recurrent models, which give good long-range prediction (the parallel configuration or the simulation mode) [10]. This work reports development of dynamic neural models of a yeast

M. Ławryńczuk (✉)
Institute of Control and Computation Engineering, Warsaw University of Technology, Ul. Nowowiejska 15/19, 00-665 Warsaw, Poland
e-mail: M.Lawrynczuk@ia.pw.edu.pl

R. Szewczyk et al. (eds.), *Challenges in Automation, Robotics and Measurement Techniques*, Advances in Intelligent Systems and Computing 440, DOI 10.1007/978-3-319-29357-8_2

fermentation reactor using the ELM approach and the classical (optimisation-based) method. Long-range prediction abilities and complexity of both model classes are compared.

2 Structures of Neural Model

Let u and y denote the input and output variables of a dynamic process, respectively. Figure 1 depicts two possible model configurations [10]. In the non-recurrent serial-parallel model the output signal for the sampling instant k is a function of the process input and output signal values from some previous instants

$$y_{\text{mod}}(k) = f(u(k-\tau), \ldots, u(k-n_{\text{B}}), y(k-1), \ldots, y(k-n_{\text{A}})) \tag{1}$$

where the integers $\tau, n_{\text{A}}, n_{\text{B}}$ determine the order of dynamics. In the recurrent parallel model (the simulation model), the past process outputs are replaced by the model outputs

$$y_{\text{mod}}(k) = f(u(k-\tau), \ldots, u(k-n_{\text{B}}), y_{\text{mod}}(k-1), \ldots, y_{\text{mod}}(k-n_{\text{A}})) \tag{2}$$

The serial-parallel model is a one-step-ahead predictor (the Autoregressive with eXogenous input (ARX) model), the parallel one is a multiple-steps-ahead predictor (the Output Error (OE) model).

The feedforward neural network with two layers is the most popular structure [3]. Taking into account Eq. (1) or Eq. (2), the network has $n_{\text{A}} + n_{\text{B}} - \tau + 1$ input nodes, K nonlinear hidden neurons with the nonlinear transfer function $\varphi\colon \mathbb{R} \rightarrow \mathbb{R}$

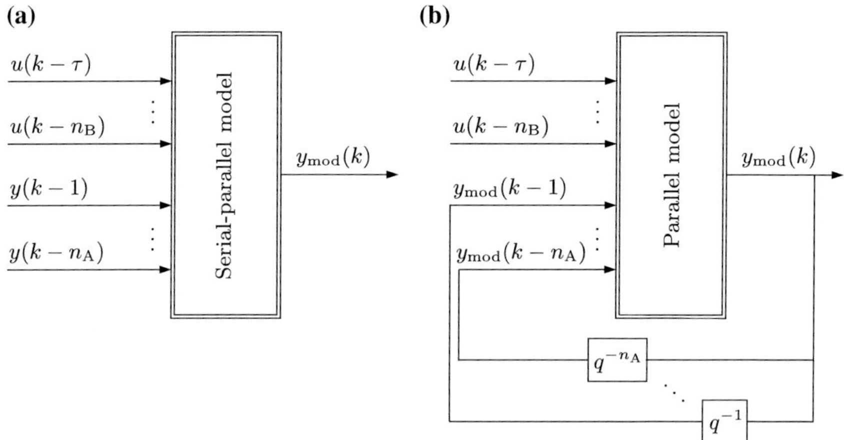

Fig. 1 Structures of dynamic models: **a** the serial-parallel model, **b** the parallel model

(e.g. $\varphi = \tanh$), one linear output element (sum) and one output $y_{\mathrm{mod}}(k)$. The weights of the first layer are denoted by $w^1_{i,j}$, where $i = 1, \ldots, K$, $j = 0, \ldots, n_{\mathrm{A}} + n_{\mathrm{B}} - \tau + 1$, the weights of the second layer are denoted by w^2_i, where $i = 0, \ldots, K$. The output signal of the neural ARX or OE model is

$$y_{\mathrm{mod}}(k) = w^2_0 + \sum_{i=1}^{K} w^2_i v_i(k)$$

where $v_i(k)$ denotes the output of the ith hidden node. Let $z_i(k)$ denote the sum of input signals of the ith hidden node. Hence, $v_i(k) = \varphi(z_i(k))$. For the ARX model defined by Eq. (1), one has

$$z_i(k) = w^1_{i,0} + \sum_{j=1}^{n_{\mathrm{B}}-\tau+1} w^1_{i,j} u(k - \tau + 1 - j) + \sum_{j=1}^{n_{\mathrm{A}}} w^1_{i,n_{\mathrm{B}}-\tau+1+j} y(k - j)$$

whereas for the OE model defined by Eq. (2), one has

$$z_i(k) = w^1_{i,0} + \sum_{j=1}^{n_{\mathrm{B}}-\tau+1} w^1_{i,j} u(k - \tau + 1 - j) + \sum_{j=1}^{n_{\mathrm{A}}} w^1_{i,n_{\mathrm{B}}-\tau+1+j} y_{\mathrm{mod}}(k - j)$$

3 Training of Neural Models

3.1 Classical Approach

Neural models are trained using a series of input-output samples. The objective is to find the values of the weights which result in good accuracy of the model. Accuracy is defined by the following Sum of Squared Errors (SSE) cost-function

$$\mathrm{SSE} = \sum_{k=k_{\mathrm{min}}}^{k_{\mathrm{max}}} (y_{\mathrm{mod}}(k) - y(k))^2 \tag{3}$$

where $y_{\mathrm{mod}}(k)$ is the output signal calculated by the model for the sampling instant k (in the serial-parallel or parallel configuration), $y(k)$ is the real value of the recorded process output, k_{min} and k_{max} define the training samples. Training consists in solving the unconstrained optimisation problem

$$\min_{w^1_{1,0},\ldots,w^1_{K,n_{\mathrm{A}}+n_{\mathrm{B}}-\tau+1},w^2_0,\ldots,w^2_K} \{\mathrm{SSE}\} \tag{4}$$

It is necessary to emphasise the fact that for training by means of the classical approach a nonlinear, possibly multi-modal, non-convex optimisation problem (4) must be solved. For this purpose a number of classical nonlinear optimisation algorithms may be used, e.g. the steepest descent method, the conjugated gradients methods, the quasi-Newton variable metrics methods, the Levenberg-Marquardt algorithm or heuristic optimisation approaches, e.g. evolutionary algorithms [3].

3.2 *Extreme Learning Machines*

In the ELM approach [2] the structure of the network the same as in the classical approach, but training does not need solving a computationally demanding nonlinear optimisation problem (4). The weights are calculated from the following procedure:

1. The weights of the first layer are assigned randomly.
2. The outputs of all hidden nodes (i.e. $v_1(k), \ldots, v_K(k)$) for all training data samples (i.e. for $k = k_{\min}, \ldots, k_{\max}$) are calculated.
3. The weights of the second layer are calculated analytically.

In order to simplify calculations it is assumed that there is a sufficient number of hidden nodes. Since the weights of the first layer are chosen randomly, the training optimisation problem (4) becomes

$$\min_{\boldsymbol{w}^2=\left[w_0^2 \ \ldots \ w_K^2\right]^{\mathrm{T}}} \{\mathrm{SSE}\} \tag{5}$$

Although the classical minimised objective function (3) may be used, it is more practical to use

$$\mathrm{SSE} = \sum_{k=k_{\min}}^{k_{\max}} (y_{\mathrm{mod}}(k) - y(k))^2 + \alpha \sum_{i=0}^{K} (w_i^2)^2 \tag{6}$$

where the regularisation term $\sum_{i=0}^{K}(w_i^2)^2$ minimises values of the weights of the second layer, $\alpha > 0$. The cost-function (6) may be expressed in a vector notation

$$\mathrm{SSE} = (\boldsymbol{y}_{\mathrm{mod}} - \boldsymbol{y})^{\mathrm{T}}(\boldsymbol{y}_{\mathrm{mod}} - \boldsymbol{y}) + \alpha(\boldsymbol{w}^2)^{\mathrm{T}}\boldsymbol{w}^2 = \left\|\boldsymbol{y}_{\mathrm{mod}} - \boldsymbol{y}\right\|^2 + \alpha \left\|\boldsymbol{w}^2\right\|^2$$

where $\boldsymbol{y}_{\mathrm{mod}} = \left[y_{\mathrm{mod}}(k_{\min}) \ \ldots \ y_{\mathrm{mod}}(k_{\max})\right]^{\mathrm{T}}$, $\boldsymbol{y} = \left[y(k_{\min}) \ \ldots \ y(k_{\max})\right]^{\mathrm{T}}$. The outputs of the hidden nodes for all training samples give a matrix of dimensionality $(k_{\max} - k_{\min} + 1) \times (K + 1)$

$$\boldsymbol{v} = \begin{bmatrix} v_0(k_{\min}) & \ldots & v_K(k_{\min}) \\ \vdots & \ddots & \vdots \\ v_0(k_{\max}) & \ldots & v_K(k_{\max}) \end{bmatrix}$$

Hence, in the case of the ARX configuration, the model output vector is $\boldsymbol{y}_{\mathrm{mod}} = \boldsymbol{v}\boldsymbol{w}^2$ and the minimised cost-function (6) becomes

$$\mathrm{SSE} = \left\| \boldsymbol{v}\boldsymbol{w}^2 - \boldsymbol{y} \right\|^2 + \alpha \left\| \boldsymbol{w}^2 \right\|^2 \tag{7}$$

As the minimised cost-function (7) is a second-order polynomial of the weights of the second layer, $\boldsymbol{w}^2$, they may be determined analytically, without nonlinear optimisation, by zeroing the derivative vector $\frac{\mathrm{dSSE}}{\mathrm{d}\boldsymbol{w}^2} = 2\boldsymbol{v}^{\mathrm{T}}(\boldsymbol{v}\boldsymbol{w}^2 - \boldsymbol{y}) + 2\alpha\boldsymbol{w}^2$, which gives

$$\boldsymbol{w}^2 = (\boldsymbol{v}^{\mathrm{T}}\boldsymbol{v} + \alpha \boldsymbol{I}_{(K+1)\times(K+1)})^{-1}\boldsymbol{v}^{\mathrm{T}}\boldsymbol{y} \tag{8}$$

4 Simulation Results

The considered process is a yeast fermentation reactor (*Saccharomyces cerevisiae*). Yeast is commonly used in many branches of the food industry, in particular in: bakeries, breweries, wineries and distilleries. The reactor manifests significantly nonlinear behaviour. It cannot be modelled precisely by means of linear models with constant parameters and it cannot be controlled efficiently by the classical linear control schemes [6, 8]. Neural networks may be successfully used to approximate behaviour of the process as described in [6–9]. Different nonlinear controllers may be used for the process, including a fuzzy-PI controller [1], an inverse neural-network controller [4], a reinforcement learning algorithm [5] and nonlinear Model Predictive Control (MPC) strategies [6–9].

During fermentation the reactor temperature must be maintained within a narrow range because temperature greatly influences process operation. Imprecise temperature control is likely to result in a reduction of fermentation yield [1]. Hence, the problem is to find a sufficiently precise model of the reactor temperature, which may be next used for developing a control system [6–9]. From the perspective of a control algorithm the reactor is a single-input single-output process: the coolant flow rate (F_{ag}) is the input (the manipulated variable), the reactor temperature (T_{r}) is the output (the controlled variable).

The first-principle model consists of a set of nonlinear differential equations [6–9]. It is treated as the "real" process. The first-principle model is simulated open-loop (without any controller) in order to obtain data sets necessary for model identification. Figure 2 depicts training and validation data sets. Each set has 4000 samples. The sampling period is 30 min [9]. The output signal contains small measurement noise. The training data set is used only for model training, i.e. the training error is minimised for that set. The validation data set is used only to calculate the validation error after training of different model structures. The validation error indicates generalisation abilities of the models. The models are finally compared taking into account the validation error.

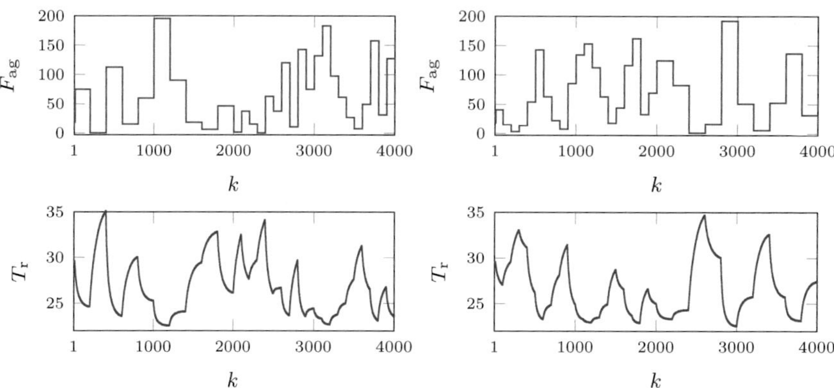

Fig. 2 The training data set (*left*) and the validation data set (*right*)

The hyperbolic tangent transfer function $\varphi = \tanh$ is used in the hidden layer. Because input and output process variables have different orders of magnitude, they are scaled: $u = 0.01(F_{ag} - F_{ag,nom})$, $y = 0.1(T_r - T_{r,nom})$, where $F_{ag,nom} = 18$ l/h, $T_{r,nom} = 29.573212$ °C correspond to the nominal operating point.

4.1 Training of Classical Neural Models

Accuracy of models is assessed taking into account the SSE index (3). For model training the efficient Broyden-Fletcher-Goldfarb-Shanno (BFGS) nonlinear optimisation algorithm is used, training is carried out in the parallel (recurrent) configuration. The second-order dynamics is assumed (i.e. $\tau = 1$, $n_A = n_B = 2$) as in [6, 8]. In order to find a neural model with good accuracy and generalisation abilities the networks with $K = 1, \ldots, 7$ hidden nodes are trained and compared. For each structure training is repeated 10 times (because of possible shallow local minima), all weights are initialised randomly. Table 1 presents properties of the best obtained models. Increasing the number of model parameters leads to reducing the training error ($\mathrm{SSE}_{\mathrm{OE}}^{\mathrm{train}}$). On the other hand, when $K > 3$, the models have too many weights and the validation error ($\mathrm{SSE}_{\mathrm{OE}}^{\mathrm{val}}$) increases, which means that the generalisation ability deteriorates. Hence, the model with as few as 3 hidden nodes is finally chosen. The model has only 19 weights. Figure 3 compares the validation data set and the output of the chosen neural model. The model works fine both in serial-parallel and parallel configurations.

Table 1 Comparison of the best neural models trained in the parallel configuration and in the classical approach; NP–the number of parameters

K	NP	$\mathrm{SSE}_{\mathrm{OE}}^{\mathrm{train}}$	$\mathrm{SSE}_{\mathrm{OE}}^{\mathrm{val}}$
1	7	1.1649×10^{1}	1.3895×10^{1}
2	13	3.2821×10^{-1}	3.2568×10^{-1}
3	19	2.0137×10^{-1}	*1.8273×10^{-1}*
4	25	1.9868×10^{-1}	1.9063×10^{-1}
5	31	1.3642×10^{-1}	1.9712×10^{-1}
6	37	1.3404×10^{-1}	2.0440×10^{-1}
7	43	1.2801×10^{-1}	2.9391×10^{-1}

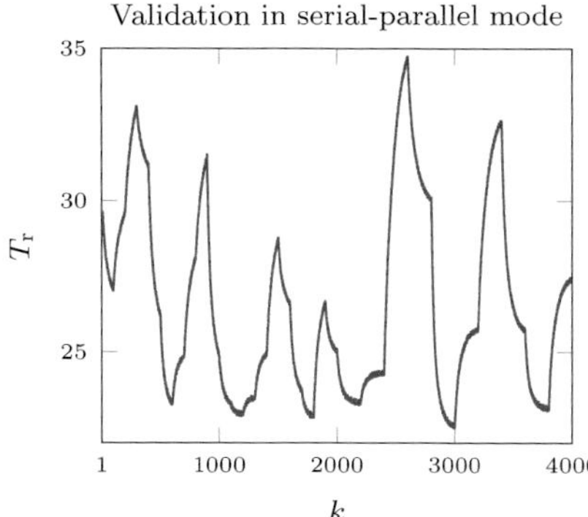

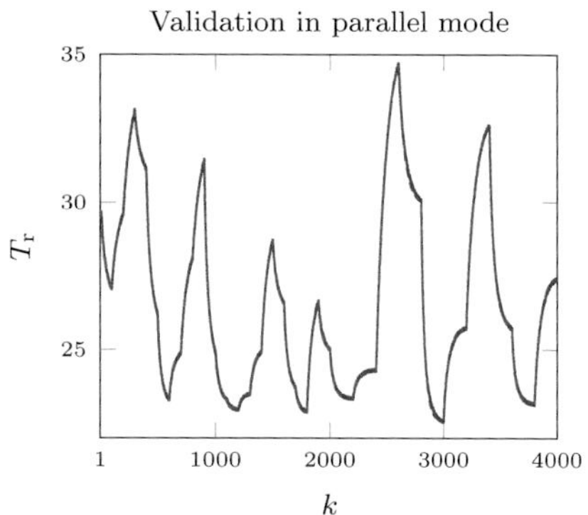

Fig. 3 The validation data set (*solid line*) versus the output of the best neural model with 3 hidden nodes trained in the parallel configuration and in the classical approach (*dashed line*), $\mathrm{SSE}_{\mathrm{ARX}}^{\mathrm{val}} = 1.2320 \times 10^{-1}$, $\mathrm{SSE}_{\mathrm{OE}}^{\mathrm{val}} = 1.8273 \times 10^{-1}$

4.2 Training of Extreme Learning Machines

In the classical approach to training, i.e. when the cost-function is minimised by means of a nonlinear optimisation algorithm, the model may be trained in both serial-parallel and parallel configurations. When the model is trained in the ELM manner, training is possible only in the first mode. Nevertheless, since the objective is to obtain good dynamic models, the ELM neural models are validated in the parallel configuration after training. The models with $K = 1, 2, \ldots, 100$ hidden nodes are considered, for each model structure the weights are calculated 20 times. The SSE error with the regularisation term (6) is minimised during training, the models are evaluated using the error (3), i.e. without that term. Validation errors of all determined models are shown in Fig. 4, both serial-parallel and parallel configurations are considered. Additionally, validation errors of the best selected neural models are given in Table 2. When $\alpha = 0$ (no regularisation during training), the models with too few and too many hidden nodes frequently give huge errors, particularly in the

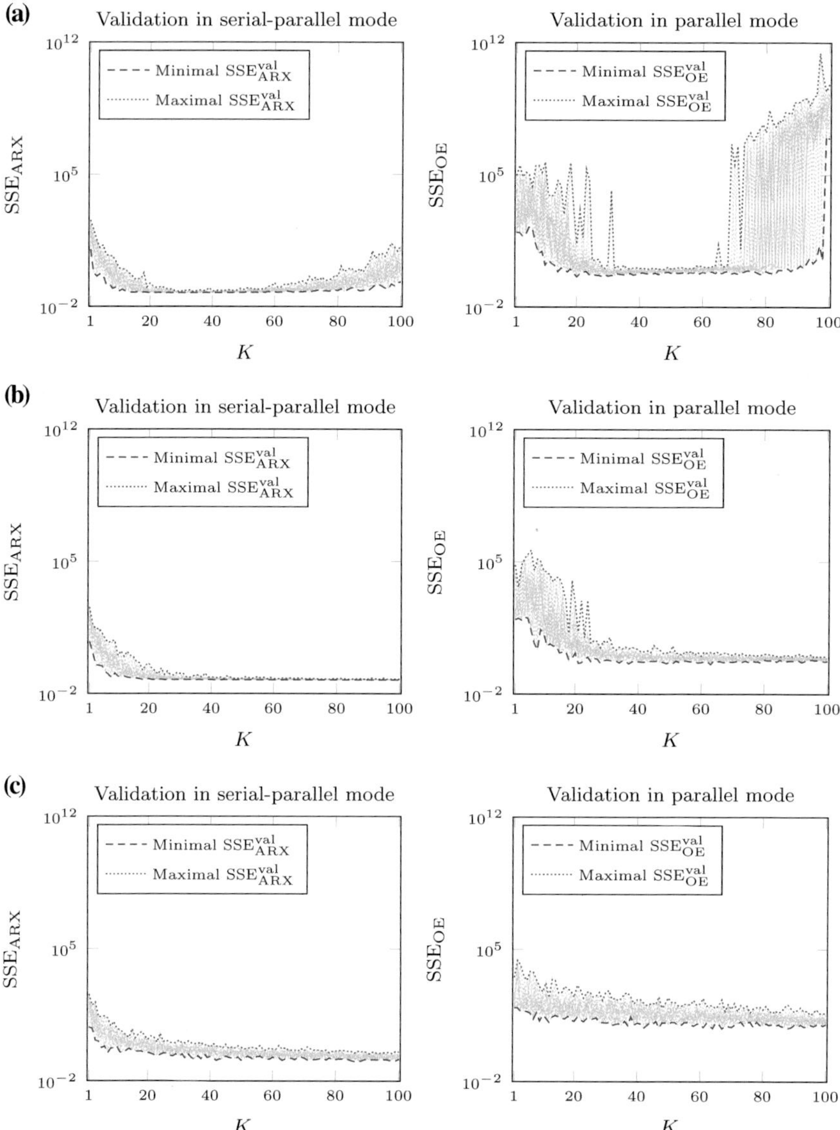

Fig. 4 Validation errors of all models trained in the serial-parallel configuration and in the ELM approach for different values of the regularisation parameter α: **a** $\alpha = 0$, **b** $\alpha = 0.01$, **c** $\alpha = 10$

parallel configuration, which is illustrated in Fig. 4a. Moreover, there are huge differences of accuracy between the best and the worst models, even for the same number of hidden nodes. As the regularisation parameter α increases, the differences between models of the same structure become smaller and smaller and the effect of

Table 2 Comparison of the best neural models trained in the serial-parallel configuration and in the ELM approach; NP–the number of parameters

		SSE^{val}_{OE}				
K	NP	$\alpha = 0$	$\alpha = 0.01$	$\alpha = 0.1$	$\alpha = 1$	$\alpha = 10$
10	61	1.6836×10^{0}	1.6927×10^{1}	1.7547×10^{1}	2.9732×10^{1}	3.2173×10^{1}
20	121	*3.8370×10^{-1}*	1.4811×10^{0}	4.6646×10^{0}	8.3131×10^{0}	1.5030×10^{1}
30	181	4.2560×10^{-1}	6.5626×10^{-1}	1.2046×10^{0}	4.0492×10^{0}	1.2573×10^{1}
40	241	6.2061×10^{-1}	5.3584×10^{-1}	1.0773×10^{0}	4.6231×10^{0}	1.2441×10^{1}
50	301	6.2207×10^{-1}	5.1635×10^{-1}	7.5840×10^{-1}	3.3643×10^{0}	1.0516×10^{1}
60	361	6.9239×10^{-1}	4.9123×10^{-1}	7.8591×10^{-1}	3.5661×10^{0}	1.1595×10^{1}
70	421	6.7860×10^{-1}	5.1162×10^{-1}	8.8329×10^{-1}	2.8592×10^{0}	7.3480×10^{0}
80	481	8.0274×10^{-1}	6.6283×10^{-1}	7.9178×10^{-1}	2.3020×10^{0}	9.0120×10^{0}
90	541	1.1602×10^{-1}	5.7329×10^{-1}	6.8762×10^{-1}	2.3021×10^{0}	9.7736×10^{0}
100	601	8.3138×10^{-1}	5.5840×10^{-1}	5.7272×10^{-1}	1.9188×10^{0}	1.0406×10^{1}

overparameterisation is not present (for $K \leq 100$). Unfortunately, the bigger the parameter α, the bigger the errors.

The best model trained using the ELM approach is obtained for 20 hidden nodes (it has 121 weights). Its validation error is $SSE^{val}_{OE} = 3.8370 \times 10^{-1}$ whereas the model trained by means of the classical approach needs only 3 hidden nodes (19 weights) and it is characterised by the error $SSE^{val}_{OE} = 1.8273 \times 10^{-1}$.

Figure 5 compares the validation data set and the output of three neural models with 20 hidden nodes trained in the serial-parallel configuration and in the ELM approach. The best model (minimal SSE^{val}_{OE} for $\alpha = 0$) works fine both in serial-parallel and parallel configurations. Conversely, the worst neural model (maximal SSE^{val}_{OE} for $\alpha = 0$) fails to give good long-range prediction in the parallel configuration. When the regularisation factor is big ($\alpha = 10$), the best model (minimal SSE^{val}_{OE}) also gives a significant error in the parallel configuration.

5 Conclusions

The ELM approach makes it possible to very efficiently find values of the weights of neural networks since they are calculated explicitly, without nonlinear optimisation. Although the ELM technique trains networks only in the non-recurrent serial-parallel configuration, for the considered yeast fermentation reactor they perform well also in the recurrent parallel mode. As the input weights are chosen randomly, the ELM approach gives the best network with 20 hidden nodes whereas in the classical approach only 3 nodes are sufficient.

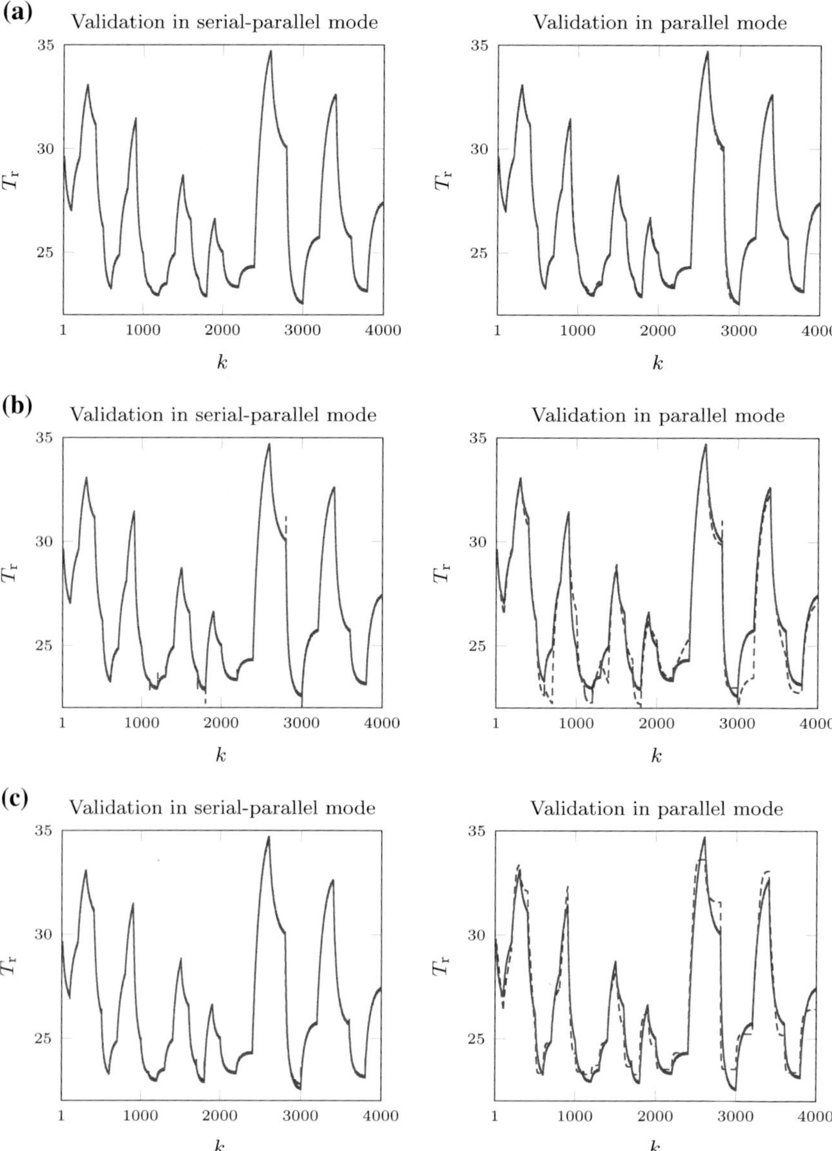

Fig. 5 The validation data set (*solid line*) versus the output of the neural model with 20 hidden nodes trained in the serial-parallel configuration and in the ELM approach (*dashed line*): **a** the best neural model (minimal SSE_{OE}^{val} for $\alpha = 0$), $SSE_{ARX}^{val} = 5.2808 \times 10^{-2}$, $SSE_{OE}^{val} = 3.8370 \times 10^{-1}$; **b** the worst neural model (maximal SSE_{OE}^{val} for $\alpha = 0$), $SSE_{ARX}^{val} = 1.6595 \times 10^{-1}$, $SSE_{OE}^{val} = 2.2227 \times 10^{1}$, **c** the best neural model (minimal SSE_{OE}^{val} for $\alpha = 10$), $SSE_{ARX}^{val} = 3.3633 \times 10^{-1}$, $SSE_{OE}^{val} = 1.5030 \times 10^{1}$

References

1. Fonseca, R.R., Schmitz, J.E., Fileti, A.M.F., da Silva, F.V.: A fuzzy-split range control system applied to a fermentation process. Bioresour. Technol. **142**, 475–482 (2013)
2. Huang, G.-B., Zhou, H., Ding, X., Zhang, R.: Extreme learning machine for regression and multiclass classification. IEEE Trans. Syst. Man Cybern. Part B Cybern. **42**, 513–529 (2012)
3. Haykin, S.: Neural Networks and Learning Machines. Prentice Hall, Englewood Cliffs (2009)
4. Imtiaz, U., Assadzadeh, A., Jamuar, S.S., Sahu, J.N.: Bioreactor temperature profile controller using inverse neural network (INN) for production of ethanol. J. Process Control **23**, 731–742 (2013)
5. Li, D., Qian, L., Jin, Q., Tan, T.: Reinforcement learning control with adaptive gain for a Saccharomyces cerevisiae fermentation process. Appl. Soft Comput. **11**, 4488–4495 (2011)
6. Ławryńczuk, M.: Computationally efficient model predictive control algorithms. In: A Neural Network Approach, Studies in Systems, Decision and Control, vol. 3, Springer, Heidelberg (2014)
7. Ławryńczuk, M.: Online set-point optimisation cooperating with predictive control of a yeast fermentation process: a neural network approach. Eng. Appl. Artif. Intell. **24**, 968–982 (2011)
8. Ławryńczuk, M.: Modelling and nonlinear predictive control of a yeast fermentation biochemical reactor using neural networks. Chem. Eng. J. **145**, 290–307 (2008)
9. Nagy, Z.K.: Model based control of a yeast fermentation bioreactors using optimally designed artificial neural networks. Chem. Eng. J. **127**, 95–109 (2007)
10. Nelles, O.: Nonlinear system identification. In: From Classical Approaches to Neural Networks and Fuzzy Models. Springer, Berlin (2001)
11. Ogiela, M., Tadeusiewicz, R.: Modern computational intelligence methods for the interpretation of medical images. In: Studies in Computational Intelligence, vol. 84. Springer, Heidelberg (2008)
12. Ripley, B.D.: Pattern Recognition and Neural Networks. Cambridge University Press, Cambridge (1996)
13. Tatjewski, P.: Advanced control of industrial processes. In: Structures and Algorithms. Springer, London (2007)
14. Witczak, M.: Fault diagnosis and fault-tolerant control strategies for non-linear systems: analytical and soft computing approaches. In: Lecture Notes in Electrical Engineering, vol. 266. Springer, Heidelberg (2014)
15. Yan, Z., Wang, J.: Robust model predictive control of nonlinear systems with unmodeled dynamics and bounded uncertainties based on neural networks. IEEE Trans. Neural Networks Learn. Syst. **25**, 457–469 (2014)

Project and Simulation of a Portable Device for Measuring Bioelectrical Signals from the Brain for States Consciousness Verification with Visualization on LEDs

Szczepan Paszkiel, Wojciech Hunek and Andrew Shylenko

Abstract This paper introduces a design of low cost, portable and simple brain computer interface device. Proposed prototype is dedicated to creating a control signal depending on the selected states of users consciousness. RGB LED is used to visualize mind states of a human subject based on generated control signals. Paper consists of documentation and simulation for EEG analog frond-end together with key concepts and structure of the Arduino-based digital side. Different aspects of BCI process flow are considered, such as: electrodes types and designs, EEG signal noise and filtering approaches, brain patterns classification problem.

Keywords Brain computer interface · EEG · LTSpice simulation · Prototype development · Wireless · Portable · States of consciousness · RGB LED

1 Introduction

With advances in technology the issue of interaction between man and machine becomes more urgent. The usual methods of human-computer interaction are limited by the capabilities of the human body. Communication occurs through a biological bridge of nerves, muscles and sensory organs. Therefore, tactile, auditory, voice, visual interfaces are limited in speed of response, accuracy and volume of transmitted data. Technologies, such as augmented reality (AR), which allows virtual reality mixing with real world, require new approaches in input methods, and information representation, like the Brain Computer Interfaces (BCI) [1]. BCI allows to get rid of slow biological layer between brain and machine. It replaces the biological

PhD. Eng. Szczepan Paszkiel, Assistant Professor; PhD. Eng. Wojciech Hunek, Professor; M. Sc., Eng. Andrew Shylenko.

S. Paszkiel (✉) · W. Hunek · A. Shylenko
Opole University of Technology, Faculty of Electrical Engineering,
Automatic Control and Informatics, Institute of Control and Computer Engineering,
Gen. Kazimierza Sosnkowskiego 31, 45-271 Opole, Poland
e-mail: s.paszkiel@po.opole.pl

R. Szewczyk et al. (eds.), *Challenges in Automation, Robotics and Measurement Techniques*, Advances in Intelligent Systems and Computing 440, DOI 10.1007/978-3-319-29357-8_3

communication channel with fully hardware one. This leads to fast, reliable, universal direct data interface. The most popular and easiest way of implement BCI is to use a electroencephalography (EEG) measurement. Low price, portability and usability of the BCI together with the improvement of processing algorithms of the brain signals working on microcontrollers, open the possibility for a widespread use of such kinds of interfaces. Compact BCI can be efficient on short distances with simple tasks, as an input device for smartphones, PCs, mobile robots and even household appliance. Close range radio interfaces such as bluetooth or Wi-Fi are ideal for bridging between BCI and controlled devices. At the same time BCI should be also useful in complex applications, such as: (a) diagnosing the condition of the human organism, e.g. preventing machinist falling asleep on the duty [2]; (b) in areas that require a rapid response, e.g. shooting from different types of weapons, trigger control; (c) and the classical task of managing movement control for people with disabilities [3]. Based on the wide range of applications, the researches in the field of cheap, handy, portable BCIs are important priority.

2 Brain Computer Interface Technology

A brain computer interface is a direct communication link between the brain and an external machine, which do not rely on peripheral nerves and muscles [4]. Many modern BCI implementations are based on discovery of the electrical activity of the human brain and the development of electroencephalography (EEG) in 1924 by German neurologist Hans Berger. The main recognized applications of the BCI are: human machine interaction (HMI): interface devices between humans and machines; bioengineering: assisting equipment for people with disabilities; human subject observation: diagnosis and research of neurological and sleep disorders, attention and states of consciousness monitoring, health tracking; neuroscience studies: approaches to found interconnections between observable behavior of the subject with recorded brain data. The idea of this paper lies within range of HMI problem. The typical HMI process flow consists of the following steps: brain signal must be measured, filtered, amplified and classified to generate control sequence.

There are several common features regarding BCI data analysis [5]. EEG signal have poor signal-to-noise ratio, it is distorted, noisy and outlined due to amplification process and conditions of measurements. Data density of the BCI measurements are very high. It must be extracted and analyzed from several channels at the same time, before it can be classified. Brain activity is cyclic and time related. Therefore time-variation information is crucial to BCI features classification. Additionally brain signals is non-stationary. It is very varieties over time and from session to session. The problem is compounded due to small data training sets. The HMI BCI structure begins with electrodes mounted on the head of the subject (Fig. 1). EEG brain waves through electrodes are transmitting to analog front-end, where signals amplification and filtering from noise taking place. The next stage is digital signal analysis. Analog signal is converted to digital form and passed to future extraction

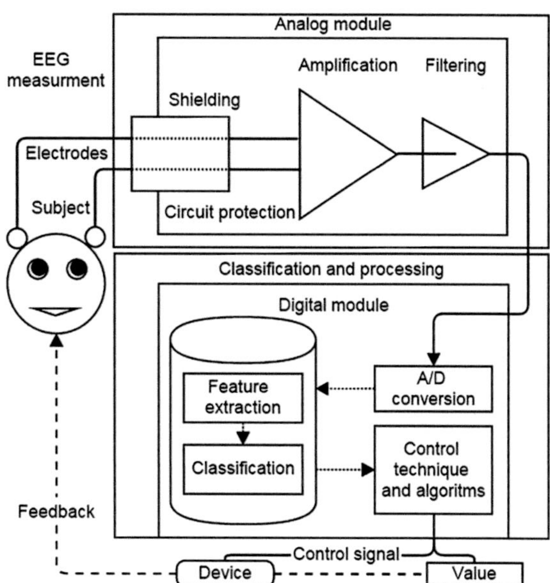

Fig. 1 Typical EEG-based BCI

algorithms. By using the mentioned algorithms the extracted features are obtained for classification process. Control algorithms use classified patterns to generate control commands. Subject gives the possibilities to observing the result of brainwave activity and adapts the brain activity using feedback loop.

The first stage between EEG signal amplification task and subject is choosing of electrodes. Electrodes can be: active electrodes have an amplifier very close to the electrode on the scalp, the signals from active electrodes are stronger and less noisy; passive electrodes are small precious metal pads or cups. In general, electrodes should be chemically inactive, must provide good electrical contact with scalp. They are made of highly conductive chemically passive materials (for example Ag, Ag Cl, Au and Cu). Impedance of the passive electrodes should be matched. Otherwise, unequal impedance and bad contact lead to the poor signal-to-noise ratio of the measurements. In order to reduce the impedance and maintain good contact, the skin must be abraded and moistened with electrode gel before placing electrodes. So-called dry electrodes do not require electrode above solutions and skin preparation for operation. EEG circuit must provide protection against dangerous currents, which can damage an equipment and harm the human subject: isolation should be done with optocuplers, isolation transformers and wireless interfaces of battery powered devices etc.; components inputs should be protected against high current and ESD spikes. Amplifier circuit is expected to be low noise with high common-mode rejection ratio and ability to handle high impedance sources. Gain must be at least 10000 before A/D conversion, as signal is in uV ranges. In EEG amplifiers usually several stages of amplifications are used: instrumentation preamplifier (INA118, AD620, etc.); op-amp second stage amplification to required level.

One of the main parts in EEG signal processing is noise filtering presented in the poor EEG signal. The following techniques are usually used to reduce the noise: offset correction between electrodes; Driven Right Leg (DRL) and other methods for commonmode rejection improvement; 50/60 Hz analog and digital notch filters; low-pass/high-pass/band-pass filters; shielding (cable shielding, Faraday cage); ground loops avoiding; low electrode impedance (use of conductive gel, skin cleaning, high-quality electrodes); anti-aliasing filters (Low-pass). Sampling rate must be twice as maximum measured frequency to avoid aliasing. For maximum frequency of 120 Hz, ADC sampling rate should be near 400 Hz and more. At least 12 bits of resolution is recommended to obtain the good EEG readings. Better resolution leads to greater dynamic range and better signal recognition.

Electrical brain activity is much more complex then EEG measurement data. In general, we see a signals summation of billions neurons. Thus EEG is heuristic process and therefore not suitable to read minds. Complex analysis and classification algorithms must be applied to EEG signal to obtain any useful information. After signal discretization there are few methods to extract BCI features for further classification. One of the most used is Discrete Fourier Transform (DFT) and its fast implementation FFT. But there are also other methods for future extraction like power spectral density approach with autoregressive coefficients approach and time frequency transformations. There are five recognized categories that cover the most used algorithms in BCI classification systems: linear classifiers, where a classification decision based on the value of linear combinations of the extracted future characteristics. It is one of the fastest methods; nearest-neighbor classifiers, which classify any feature based on the closest training pattern; Bayes classifiers, based on Bayes theorem; neural networks and a combination of different classifiers. Because of the microcontrollers (MCU) limitations it is hard to implement complex analysis of brain wave data on it. To generate some control signal with MCU we are limited in fast data analysis and simple linear classifiers techniques [6].

3 Prototype Design and Analog Frontend LTSpice Simulation

According to the theoretical background, described before, the following hardware requirements were formulated: analog Front-End frequency range: 0.1–100 MHz; measurement voltage amplitude: 0.5–100 V; signal must be amplified 1000–20000 times before ADC conversion; protection against noise from external sources (e.g. 50/60 Hz electrical network noise); easy to assemble and making process. Circuit should be based on widespread components; modularity, flexibility and portability; design must be safe for human subjects; lowest price and availability of components without sacrificing functionality. The human subject is connected to the front end through three EEG electrodes. The front end is the analogue electronics circuit, which consists of an instrumentation amplifier followed by signal amplification

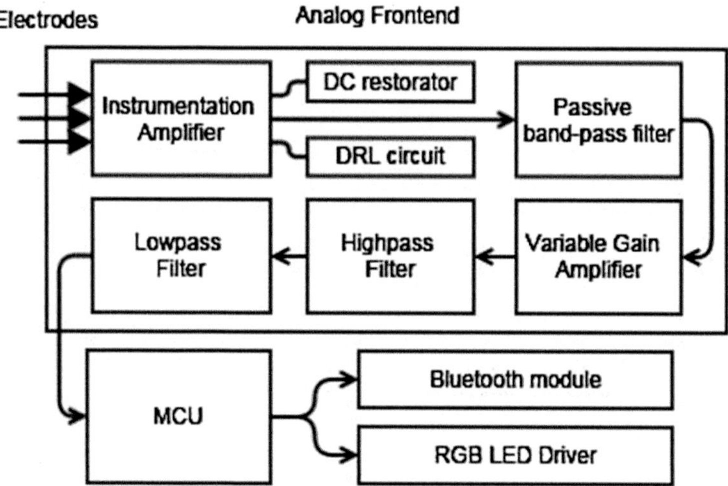

Fig. 2 Analog Frontend processing stages

and the following processing stages (Fig. 2): Driven Right Leg (DRL) circuit; direct current (DC) restorator; variable gain amplifier; highpass and low-pass filters; signal level shifter (optional); voltage buffer is used to create virtual ground from 5 V power source. Preamplifier used in the first stage is AD620 instrumentation amplifier [7].

The AD620 is a low cost, high accuracy instrumentation amplifier that requires only one external resistor to set gains of 1–10000. The gain of the AD620 in the prototype is set using three resistors and equals: G = 24.58. Furthermore, the AD620 offers only 1.3 mA max supply current and accuracy of maximum 40 ppm nonlinearity, low offset voltage of 50 V max and offset drift of 0.6 V/C max. The AD620 suits well as a preamplifier due to low input voltage noise of 20 nV at 1 Hz for gain more than 10 and ability to handle high impendent sources. The AD620 provides high Common Mode Rejection Ratio (CMRR). With low CMRR, commonmode voltages, which are typically 1 V, is amplified and prevent the EEG signal being recovered and recognized. The CMRR ratio of AD620 is in the range of 110–120 dB for gain G = 25 in the frequency range from DC to 50 Hz. The AD620 can be replaced with AD623, INA118 or other low cost, low power amplifier. As operational amplifiers for filters, level shifter, DC restorator and DRL circuit two CLC4011 quad amplifiers were considered. CLC4011 is Low Power, Low Cost, Rail-to-Rail I/O Amplifier from EXAR suitable for portable medical equipment. Main characteristics of the CLC4011 are: 136 A current per channel; 2.5 V min power supply; 4.9 MHz bandwidth; output swings to within 20 mV on either rail; 5.3 V/s slew rate; 21 nV input voltage noise; 16 mA output current. CLC4011 can be replaced with LMV774 or other low cost, low power, Rail to-Rail amplifier. The Driven Right Leg (DRL) is added in the amplifier design to minimize common-mode interference (e.g. 50/60 Hz noise). Common-mode interference is inevitably conjugate into the subject and electrodes from environment. The Driven Right Leg is incorporated in electrocardiography (ECG) equipment and can efficiently increase CMRR and avoid use of analog notch filters, which will cut useful EEG signal in other side. DRL circuit is built according AD620 datasheet. As shown in Fig. 6 the DRL is a feedback circuit that inverts common-mode voltage and superimpose it to the input signal. As a result, the common-mode noise observed at the electrodes can be significantly reduced. The output of the AD620 is highly sensitive to deviation in electrodes resistance, causing baseline drift of the DC voltage. This phenomenon is referred to as baseline wander effect. The DC clamper (DC restorator) is used to compensate DC offset common-mode voltage and to cooperate with the AD620 op-amps, finally to reduce CMR. The DC restorator build upon inverting op-amp integrator circuit in the feedback loop is connected to reference pin of the AD620. It provides injecting a precise offset to the preamplifiers saturated output signal with constant DC level, regardless of the change in skin contact resistance. By the use of DC clamper circuit as AD620 reference input we also reduce the number of required electrodes to 3 for one channel device. The high-pass and low-pass filters are used to cutoff frequency components of the EEG signal below 0.3 Hz and above 100 Hz. There are three stages of filtering incorporated: passive band-pass filter after AD620 preamplifier, lower passband Fh = 0.16 Hz, upper passband Fl = 120 Hz; the high pass active filter is implemented by using a 3rd-order multiple feedback architecture with Butterworth response and cut-off frequency Fh = 0.15 Hz, its function is blocking the DC potential presents in processing signal; the low pass active filter is implemented by using two 2nd-order filter stages with overall gains of 25 and cut-off frequency Fl = 120 Hz, therefore, it is effectively a 4thorder filter, the main purpose of high pass filter is to avoid aliasing, first stage is Sallen-Key architecture and the second is multiple feedback. Due to 2.5 V virtual ground implementation of the power supply circuit the signal level is suitable for Arduino single supply ADC. Inverted amplifier level shifter circuit is useful for conversion 5 V signal level into suitable range of the 3.3 V of microcontroller. Circuit shifts amplified signal down through 0.85 V (Fig. 3), so that it can be sampled by an ADC of more powerful microprocessor, e.g. the STM32F4.

Portable BCI device is meant to be powered by 6 AAA batteries. Arduino onboard 5 V linear regulator provides a current for Analog Front End circuit and also for microcontroller digital side. Simple voltage buffer circuit is implemented to provide stable Virtual Ground for amplifiers instead of a split power supply. To verify designed circuit the LTSpice model was built and tested. Dummy subject circuit is using as signal source with simulation of four main brain wave bands (Alpha, Beta, Gamma, Delta) and common-mode voltage source. Worst case electrode impedance for dry electrodes in the model is 150 kΩ with 12 kΩ impedance mismatch. Frequency response of the simulation circuit represents lowest (−3 dB) cut-off frequency Fl = 0.3 Hz and highest (−3 dB) cut-off frequency Fh = 100 Hz for the ideal components tolerances. More precision the Monte Carlo simulation considers resistors and capacitors tolerances. Figure 5 shows Monte Carlo simulation results for 1, 5, and 10 % passive components tolerances.

The 5 and 10 % components are almost unusable as they produce high drift in filter characteristics (Fig. 4). The compromising solution between cost and accuracy

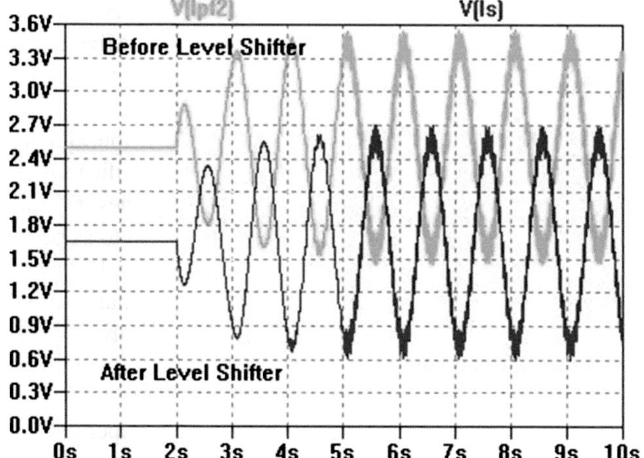

Fig. 3 Model sine wave response before and after level shifting

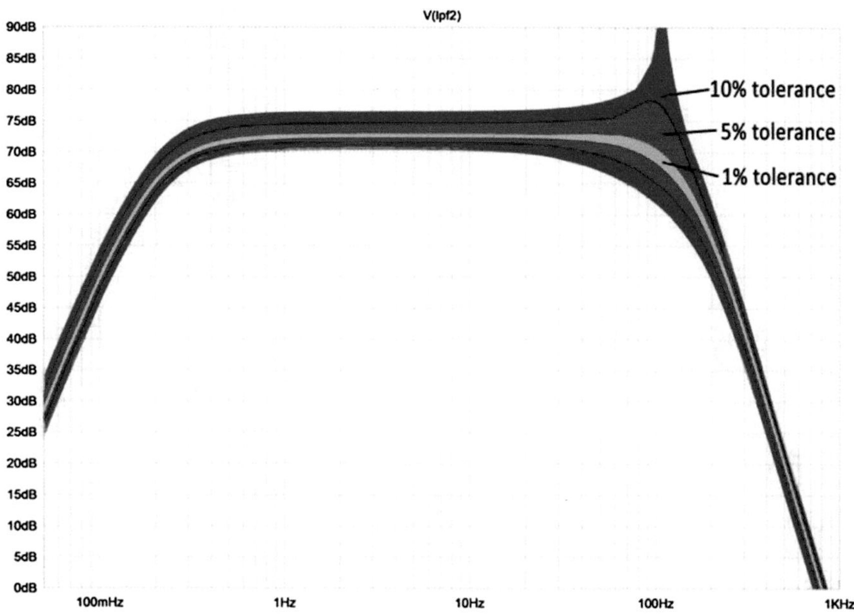

Fig. 4 Frequency responses for different components tolerances

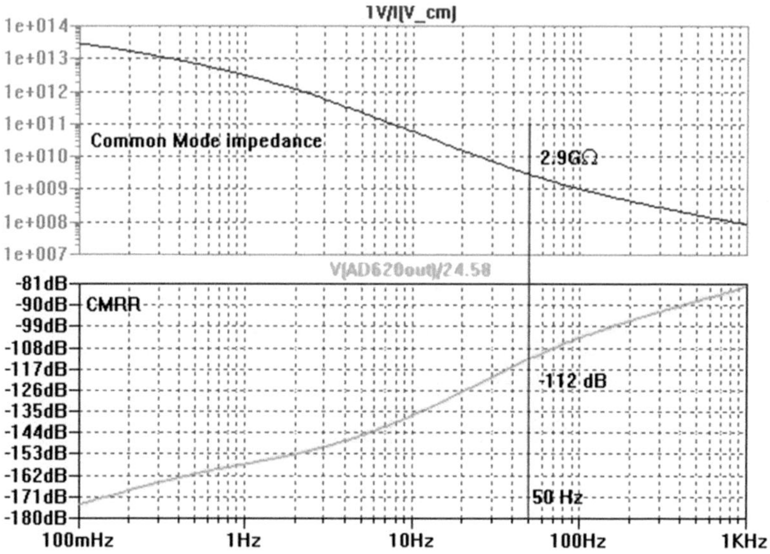

Fig. 5 CMRR and impedance of AD620 output

is to use 1 % tolerance resistors and 5 % tolerance capacitors. CMR ratio of DRL and AD620 circuit combination is also heavily dependent on components tolerances. Even 1 % mismatch in R5, R4, R8 resistors values results in CMRR decreases about a 40 dB in comparison to ideal case CMMR of 112 dB with common-mode impedance value near 2.9 MΩ for 50 Hz common-mode noise (Fig. 5).

4 Signal Processing

After amplification the signal is fed to the Arduino MCU board ADC pin. Arduino nano is wide-spread, low-cost, 8-bit ATmega 328p development board with 10-bits built-in ADC (maximum sampling frequency is 9615 Hz), UART and hardware PWM generator. HC-06 UART Bluetooth module connected to Arduino through UART provides the wireless communication ability. With wireless capability the device becomes portable and safe for subject as it is physically disconnected from dangerous current sources. Bluetooth link can be used to further EEG signal analysis on the PC and for transmission of the control signal. EEG signal is processed through ATmega 328p ADC. Sampling rate depends on available microcontroller resources but, considering analog frontend cut-off frequencies, minimum of 400 samples per second is required to avoid the aliasing artifacts. The connection of the Fourier Transformation and spectral power extraction is one of the most applied method for signal processing and analysis. The FFT converts time-domain signals to the frequency domain. Algorithm is based on every-known Discrete Fourier transform (DFT) as in

Table 1 Mind states with associated LED colors

Mind condition	Wave patterns	Frequency component (Hz)	LED color
Active thinking, focusing, anxious	Gamma	30–100	Red LED
Normal waking state of consciousness	Beta	12–30	Green LED
Relaxation and meditation	Alpha/Theta	3–12	Blue LED

Eq. (1) and by applying that to the EEG signal the EEG frequencies can be separated.

$$X_k = \sum_{k=0}^{N-1} x_n e^{-jwk\frac{n}{N}}, \ k = 0, 1, \ldots, N-1 \tag{1}$$

More efficient algorithm called Fast Fourier Transform (FFT) can compute the same result with only O(NlogN), FFT is widely used in many engineering and science applications. For 8-bit AVR microcontroller used in Arduino even FFT can be slow and demanding because of it operates on complex data. There is even faster solution than Fourier transformation: The Fast Hartley Transform (FHT). It is doing exactly what the FFT is, but it is specifically designed for real data (2):

$$H_k = \frac{1}{N} \sum_{n=0}^{N-1} x_n \left(\cos\left(\frac{2\pi nk}{N}\right) + \sin\left(\frac{2\pi nk}{N}\right) \right), \quad k = 0, 1, \ldots, N-1 \tag{2}$$

The algorithm of EEG data processing consists of the following steps: (1) Sampling the EEG signal at the rate of 400 Hz. (2) Applying the FHT to n-gathered samples. (3) Getting the amplitude spectrum from the result of the FHT at the 128 frequency. (4) Averaging the k-FHT vector. (5) Analyzing the obtained vector basis of the frequency magnitude and generating the PWM signal. Classification of obtained frequency is shown in Table 1.

As a the LED drivers the three general purpose widely available NPN Transistors 2N3904 are used. Transistors are connected in switching configuration. Resistors R7, R8 and R9 limiting the red, blue and green LEDs current to 30 mA.

5 Electrodes Design and Housing

The best results in EEG measurement can be obtained by the use of factory-made electrodes. Their impendency have matched and have optimized for EEG recordings. The cost of copper EEG electrodes is near 30 dolars for 10 pcs. The minimum price for silver plating (Ag/AgCl) electrodes is near 45 dolars for 10 pcs, (for cheapest

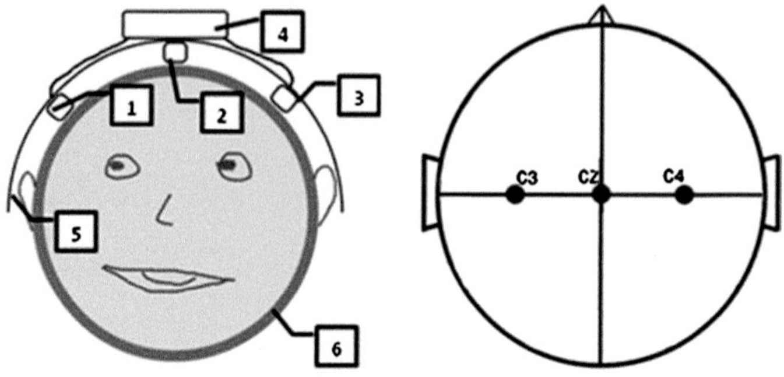

Fig. 6 General design concept and 10–20 system electrodes location

models). The advantage of using the Ag/AgCl electrodes is their low frequency noise performance. Other option is to use self-made electrodes. One of the most popular designs is hair band design (Fig. 6). It consists of head-band, sponge plugs and coaxial plug as electrodes. Based on the Hair band design the following configuration of the BCI with electrodes is proposed (Fig. 6).

Connection between parts and the general design concept is self-explanatory. Coaxial connector with sponge pads (1, 2, 3) are mounted on the head-band (5) through the drilled holes by nuts. Holes must be drilled according to 10–20 system (Fig. 6). DRL electrode is placed on Cz reference point. Minus channel electrode is located on C3 point and plus electrode is placed symmetrically on the C4 location. Device housing (Fig. 7) holds battery compartment, analog front circuit board, MCU board and RGB LED drivers [8].

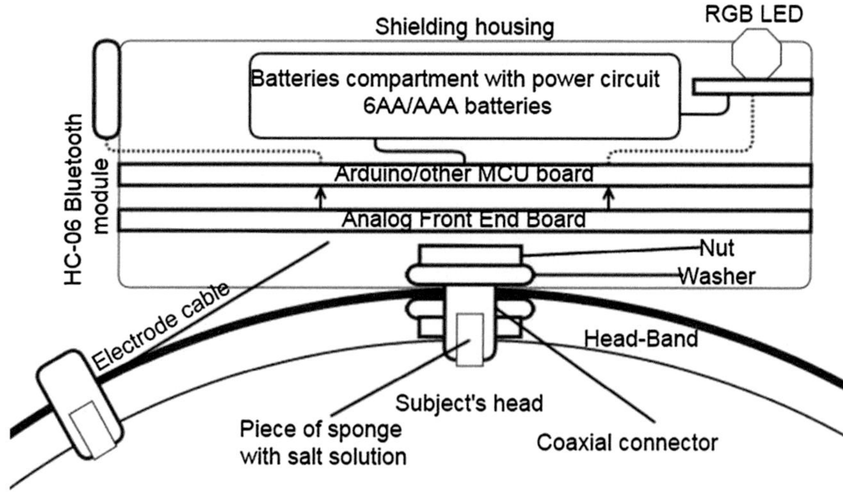

Fig. 7 Prototype BCI arrangement into housing

To maintain good contact with scalp, conductive solution should be applied to electrodes. The simplest solution recipe is to drop a tiny pinch of table salt (NaCl) into a little water. Salt/water ration is near 5 g to 0.1 l of water. Solution should be stirred and applied to the electrodes on the skin locations. Of course, any special conductive gels and solutions for (ECG/EEG) electrodes are pursuit.

6 Discussion and Future Work

This paper describes design and simulation of EEG based simple brain computer interface. The BCI has been designed for the use of brainwaves to predict the states of the subjects consciousness and generate the control signal. To classify different mental states the Fast Hartley Transform (FHT) with simple linear classifiers, filtering and averaging techniques was used. Due to the low computational power and available memory on the Arduino platform, complex algorithms are hard to implement there. After using more powerful microcontrollers, like STM32F401, we can apply the more complex data classification techniques with machine learning and neural feedback, for example the nearest neighbor algorithms, neural networks and even Bayesian classifiers. Another big improvement will be implementation of more than one the EEG channels to acquire data from bigger area of the scull. Moreover, the designing of the device has to agree with the safety requirements for medical electric equipment according to the IEC 60601-1-11 standard. The next step of our work will be the implementation of whole platform in the real world.

References

1. Lotte, F., Faller J., Guger, Ch., Renard, Y., Pfurtscheller, G., et al.: Combining BCI with Virtual Reality: Towards New Applications and Improved BCI. Springer (2013)
2. Liang, S.F., Lin, C.T., Wu, R.C., Chen, Y.C., Huang, T.Y., Jung, T.P.: Monitoring drivers alertness based on the driving performance estimation and the eeg power spectrum analysis, engineering in medicine and biology. In: Proceedings of the 27th Annual Conference Shanghai, China, September 1–4 (2005)
3. Blatt, R., Ceriani, S., Dal Seno, B., Fontana, G., Matteucci, M., Milgliore, D.: Brain control of a smart wheelchair. In: Paper Presented at the 10th International Conference on Intelligent Autonomous Systems (2008)
4. Wolpaw, J.R., Birbaumer, N., McFarland, D.J., Pfurtscheller, G., Vaughan, T.M.: Braincomputer interfaces for communication and control. Clin. Neurophysiol. **113**(6), 767–791 (2002)
5. Calvo, R.A., D'Mello, S.: Affect detection: an interdisciplinary review of models, methods, and their applications. IEEE Trans. Affect. Comput. **1**(1), 18–37 (2010)
6. Niedermeyer, E., Da Silva, F.L.: Electroencephalography: Basic Principles, Clinical Applications and Related Fields. Lippincot Williams and Wilkins. ISBN 0-7817-5126-8 (2004)
7. Nicolas-Alonso, L.F., Gomez-Gil, J.: Brain computer interfaces, a review. Sensors **12**(2), 1211–1279 (2012)
8. Klem, G.H., et al.: The ten-twenty electrode system of the International Federation. Electroencephalogr. Clin. Neurophysiol. **52** (suppl.) (1999)

Descriptor Fractional Discrete-Time Linear System and Its Solution—Comparison of Three Different Methods

Łukasz Sajewski

Abstract Descriptor fractional discrete-time linear systems are addressed. Three different methods for finding the solution to the state equation of the descriptor fractional linear system are considered. The methods are based on: Shuffle algorithm, Drazin inverse of the matrices and Weierstrass-Kronecker decomposition theorem. Effectiveness of the methods is demonstrated on simple numerical example.

Keywords Descriptor · Fractional · Solution · Method

1 Introduction

Descriptor (singular) linear systems have been considered in many papers and books [1–8]. First definition of the fractional derivative was introduced by Liouville and Riemann at the end of the 19th century [9, 10], another on was proposed in 20th century by Caputo [11] and next one in present times by Caputo-Fabrizio [12]. This idea has been used by engineers for modeling different processes [13, 14]. Mathematical fundamentals of fractional calculus are given in the monographs [9–11, 15]. Solution of the state equations of descriptor fractional discrete-time linear systems with regular pencils have been given in [7, 16] and for continuous-time in [5, 6]. Reduction and decomposition of descriptor fractional discrete-time linear systems has been considered in [17]. Application of the Drazin inverse method to analysis of descriptor fractional discrete-time and continuous-time linear systems have been given in [18, 19]. Solution of the state equation of descriptor fractional continuous-time linear systems with two different fractional has been introduced in [8].

Ł. Sajewski (✉)
Faculty of Electrical Engineering, Bialystok University of Technology,
Wiejska 45A, Bialystok 15-351, Poland
e-mail: l.sajewski@pb.edu.pl

R. Szewczyk et al. (eds.), *Challenges in Automation, Robotics and Measurement Techniques*, Advances in Intelligent Systems and Computing 440, DOI 10.1007/978-3-319-29357-8_4

In this paper three different methods for finding the solution to descriptor fractional discrete-time linear systems will be considered and illustrated on single example.

The paper is organized as follows. In Sect. 2 the basic informations on the descriptor fractional discrete-time linear systems are recalled. Shuffle algorithm method is described in Sect. 3. Drazin inverse method is given in Sect. 4. Section 5 reccals Weierstrass-Kronecker decomposition method. In Sect. 6 single numerical example, illustrating three methods is presented. Concluding remarks are given in Sect. 7.

The following notation will be used: $\mathfrak{R}$—the set of real numbers, $\mathfrak{R}^{n\times m}$—the set of $n\times m$ real matrices, Z_+—the set of nonnegative integers, I_n—the $n\times n$ identity matrix.

2 Preliminaries

Consider the descriptor fractional discrete-time linear system described by the state equation

$$E\Delta^\alpha x_{i+1} = Ax_i + Bu_i, i\in Z_+ = \{0, 1, \ldots\}, \tag{2.1}$$

where, $x_i\in\mathfrak{R}^n$, $u_i\in\mathfrak{R}^m$ are the state and input vectors, $A\in\mathfrak{R}^{n\times n}$, $E\in\mathfrak{R}^{n\times n}$, $B\in\mathfrak{R}^{n\times m}$, and the fractional difference of the order α is defined by

$$\Delta^\alpha x_i = \sum_{k=0}^{i} c_k x_{i-k}, \quad 0<\alpha<1, \tag{2.2a}$$

where

$$c_k = (-1)^k\binom{\alpha}{k} = (-1)^k\begin{cases} 1 & \text{for } k=0 \\ \frac{\alpha(\alpha-1)\ldots(\alpha-k+1)}{k!} & \text{for } k=1,2,\ldots \end{cases} \tag{2.2b}$$

It is assumed that $\det E=0$ and the pencil of the system (2.1) is regular, that is $\det[Ez-A]\neq 0$ for some $z\in C$ (the field of complex numbers). To find the solution of the system (2.1) at least three different methods can be used. These methods are the Shuffle algorithm method [17], the Drazin inverse method [18] and the Weierstrass-Kronecker decomposition method [7]. These methods was previously used to find the solution of the descriptor standard discrete-time linear systems and was extended to fractional systems. The question arise, does the order α has influence on the solution computed by the use of these methods?

In the next section, three different approaches to finding the solution to the state Eq. (2.1) of the descriptor fractional discrete-time linear systems will be given.

3 Shuffle Algorithm Method

First method is based on row and column elementary operations [20] and use the Shuffle algorithm to determine the solution [17].

By substituting (2.2a) into (2.1) we can write the state equation in the form

$$\sum_{k=0}^{i+1} Ec_k x_{i-k+1} = Ax_i + Bu_i, i \in Z_+, \tag{3.1}$$

where c_k is given by (2.2b). Applying the row elementary operations to (3.1) we obtain

$$\sum_{k=0}^{i+1} \begin{bmatrix} E_1 \\ 0 \end{bmatrix} c_k x_{i-k+1} = \begin{bmatrix} A_1 \\ A_2 \end{bmatrix} x_i + \begin{bmatrix} B_1 \\ B_2 \end{bmatrix} u_i, i \in Z_+, \tag{3.2}$$

where $E_1 \in \Re^{n_1 \times n}$ is full row rank and $A_1 \in \Re^{n_1 \times n}$, $A_2 \in \Re^{(n-n_1) \times n}$, $B_1 \in \Re^{n_1 \times m}$, $B_2 \in \Re^{(n-n_1) \times m}$. The Eq. (3.2) can be rewritten as

$$\sum_{k=0}^{i+1} E_1 c_k x_{i-k+1} = A_1 x_i + B_1 u_i \tag{3.3a}$$

and

$$0 = A_2 x_i + B_2 u_i. \tag{3.3b}$$

Substituting in (3.3b) i by $i + 1$ we obtain

$$A_2 x_{i+1} = -B_2 u_{i+1}. \tag{3.4}$$

The Eqs. (3.3a) and (3.4) can be written in the form

$$\begin{bmatrix} E_1 \\ A_2 \end{bmatrix} x_{i+1} = \begin{bmatrix} A_1 - c_1 E_1 \\ 0 \end{bmatrix} x_i - \begin{bmatrix} c_2 E_1 \\ 0 \end{bmatrix} x_{i-1} - \cdots - \begin{bmatrix} c_{i+1} E_1 \\ 0 \end{bmatrix} x_0 + \begin{bmatrix} B_1 \\ 0 \end{bmatrix} u_i + \begin{bmatrix} 0 \\ -B_2 \end{bmatrix} u_{i+1}. \tag{3.5}$$

If the matrix

$$[E_1^T \quad A_2^T]^T \tag{3.6}$$

is singular then applying the row operations to (3.5) we obtain

$$\begin{bmatrix} E_2 \\ 0 \end{bmatrix} x_{i+1} = \begin{bmatrix} A_{20} \\ \bar{A}_{20} \end{bmatrix} x_i + \begin{bmatrix} A_{21} \\ \bar{A}_{21} \end{bmatrix} x_{i-1} + \cdots + \begin{bmatrix} A_{2,i} \\ \bar{A}_{2,i} \end{bmatrix} x_0 + \begin{bmatrix} B_{20} \\ \bar{B}_{20} \end{bmatrix} u_i + \begin{bmatrix} B_{21} \\ \bar{B}_{21} \end{bmatrix} u_{i+1}, \tag{3.7}$$

where $E_2 \in \Re^{n_2 \times n}$ is full row rank with $n_2 \geq n_1$ and $A_{2,j} \in \Re^{n_2 \times n}$, $\bar{A}_{2,j} \in \Re^{(n-n_2)\times n}$, $j=0,1,\ldots,i$ $B_{2,k} \in \Re^{n_2 \times m}$, $\bar{B}_{2,k} \in \Re^{(n-n_2)\times m}$, $k=0,1$. From (3.7) we have

$$0=\bar{A}_{20}x_i+\bar{A}_{21}x_{i-1}+\cdots+\bar{A}_{2,i}x_0+\bar{B}_{20}u_i+\bar{B}_{21}u_{i+1}. \tag{3.8}$$

Substituting in (3.8) i by $i+1$ (in state vector x and in input u) we obtain

$$\bar{A}_{20}x_{i+1}=-\bar{A}_{21}x_i-\cdots-\bar{A}_{2,i}x_1-\bar{B}_{20}u_{i+1}-\bar{B}_{21}u_{i+2}. \tag{3.9}$$

From (3.7) and (3.9) we have

$$\begin{bmatrix} E_2 \\ \bar{A}_{20} \end{bmatrix} x_{i+1}=\begin{bmatrix} A_{20} \\ -\bar{A}_{21} \end{bmatrix} x_i+\begin{bmatrix} A_{21} \\ -\bar{A}_{22} \end{bmatrix} x_{i-1}+\cdots+\begin{bmatrix} A_{2,i} \\ 0 \end{bmatrix} x_0+\begin{bmatrix} B_{20} \\ 0 \end{bmatrix} u_i+\begin{bmatrix} B_{21} \\ -\bar{B}_{20} \end{bmatrix} u_{i+1}+\begin{bmatrix} 0 \\ -\bar{B}_{21} \end{bmatrix} u_{i+2} \tag{3.10}$$

If the matrix

$$[E_2^T \quad \bar{A}_{20}^T]^T \tag{3.11}$$

is singular then we repeat the procedure.

Continuing this procedure after finite number of steps p we obtain

$$\begin{bmatrix} E_p \\ \bar{A}_{p,0} \end{bmatrix} x_{i+1}=\begin{bmatrix} A_{p,0} \\ -\bar{A}_{p,1} \end{bmatrix} x_i+\begin{bmatrix} A_{p,1} \\ -\bar{A}_{p,2} \end{bmatrix} x_{i-1}+\cdots+\begin{bmatrix} A_{pi} \\ 0 \end{bmatrix} x_0+\begin{bmatrix} B_{p,0} \\ 0 \end{bmatrix} u_i+\begin{bmatrix} B_{p,1} \\ -\bar{B}_{p,0} \end{bmatrix} u_{i+1}+\cdots+\begin{bmatrix} 0 \\ -\bar{B}_{p,p-1} \end{bmatrix} u_{i+p-1} \tag{3.12}$$

where $E_p \in \Re^{n_p \times n}$ is full row rank, $A_{pj} \in \Re^{n_p \times n}$, $\bar{A}_{pj} \in \Re^{(n-n_p)\times n}$, $j=0,1,\ldots,p$ and $B_{pk} \in \Re^{n_p \times m}$, $\bar{B}_{pk} \in \Re^{(n-n_p)\times m}$, $k=0,1,\ldots,p-1$ with nonsingular matrix

$$[E_p^T \quad \bar{A}_{p.0}^T]^T \in \Re^{n\times n}. \tag{3.13}$$

In this, case premultiplying Eq. (3.12) by $\begin{bmatrix} E_p \\ \bar{A}_{p,0} \end{bmatrix}^{-1}$, we obtain the standard system

$$x_{i+1}=\hat{A}_0x_i+\hat{A}_1x_{i-1}+\cdots+\hat{A}_ix_0+\hat{B}_0u_i+\hat{B}_1u_{i+1}+\cdots+\hat{B}_{p-1}u_{i+p-1} \tag{3.14}$$

with the matrices

$$\begin{gathered}
\hat{A}_0=\begin{bmatrix} E_p \\ \bar{A}_{p,0} \end{bmatrix}^{-1}\begin{bmatrix} A_{p,0} \\ -\bar{A}_{p,1} \end{bmatrix},\quad \hat{A}_1=\begin{bmatrix} E_p \\ \bar{A}_{p,0} \end{bmatrix}^{-1}\begin{bmatrix} A_{p,1} \\ -\bar{A}_{p,2} \end{bmatrix},\ldots,\quad \hat{A}_i=\begin{bmatrix} E_p \\ \bar{A}_{p,0} \end{bmatrix}^{-1}\begin{bmatrix} A_{pi} \\ 0 \end{bmatrix},\\
\hat{B}_0=\begin{bmatrix} E_p \\ \bar{A}_{p,0} \end{bmatrix}^{-1}\begin{bmatrix} B_{p,0} \\ 0 \end{bmatrix},\quad \hat{B}_1=\begin{bmatrix} E_p \\ \bar{A}_{p,0} \end{bmatrix}^{-1}\begin{bmatrix} B_{p,1} \\ -\bar{B}_{p,0} \end{bmatrix},\ldots,\quad \hat{B}_{p-1}=\begin{bmatrix} E_p \\ \bar{A}_{p,0} \end{bmatrix}^{-1}\begin{bmatrix} 0 \\ -\bar{B}_{p,p-1} \end{bmatrix}.
\end{gathered} \tag{3.15}$$

Eventually, we reduce the descriptor system to standard system with delays. To compute the solution x_i of (3.14), now we can use methods given for standard discrete-time linear systems with delays, e.g. iterative approach (initial conditions are needed).

4 Drazin Inverse Method

Second method use the Drazin inverses of the matrices $\bar{E}$ and $\bar{F}$ [18].

Definition 4.1 [18] A matrix $\bar{E}^D$ is called the Drazin inverse of $\bar{E}\in\Re^{n\times n}$ if it satisfies the conditions

$$\bar{E}\bar{E}^D=\bar{E}^D\bar{E},\bar{E}^D\bar{E}\bar{E}^D=\bar{E}^D,\bar{E}^D\bar{E}^{q+1}=\bar{E}^q, \tag{4.1}$$

where q is the smallest nonnegative integer, satisfying condition $\operatorname{rank}\bar{E}^q=\operatorname{rank}\bar{E}^{q+1}$ and it is called the index of $\bar{E}$.

The Drazin inverse $\bar{E}^D$ of a square matrix $\bar{E}$ always exist and is unique [1]. If $\det\bar{E}\neq 0$ then $\bar{E}^D=\bar{E}^{-1}$. Some methods for computation of the Drazin inverse are given in [20].

Lemma 4.1 [18] The matrices $\bar{E}$ and $\bar{F}$ satisfy the following equalities:

1. $$\bar{F}\bar{E}=\bar{E}\bar{F}\text{ and }\bar{F}^D\bar{E}=\bar{E}\bar{F}^D,\bar{E}^D\bar{F}=\bar{F}\bar{E}^D,\bar{F}^D\bar{E}^D=\bar{E}^D\bar{F}^D, \tag{4.2a}$$

2. $$\ker\bar{F}_1\cap\ker\bar{E}=\{0\}, \tag{4.2b}$$

3. $$\bar{E}=T\begin{bmatrix}J&0\\0&N\end{bmatrix}T^{-1},\bar{F}=T\begin{bmatrix}A_1&0\\0&A_2\end{bmatrix}T^{-1},\bar{E}^D=T\begin{bmatrix}J^{-1}&0\\0&0\end{bmatrix}T^{-1},\ \det T\neq 0, \tag{4.2c}$$

   $J\in\Re^{n_1\times n_1}$, is nonsingular, $N\in\Re^{n_2\times n_2}$ is nilpotent, $A_1\in\Re^{n_1\times n_1}$, $A_2\in\Re^{n_2\times n_2}$, $n_1+n_2=n$,

4. $$(I_n-\bar{E}\bar{E}^D)\bar{F}\bar{F}^D=I_n-\bar{E}\bar{E}^D\text{ and }(I_n-\bar{E}\bar{E}^D)(\bar{E}\bar{F}^D)^q=0. \tag{4.2d}$$

Similar as in previous case, substitution of (2.2a) into (2.1) yields

$$Ex_{i+1}=Fx_i-\sum_{k=1}^{i}Ec_{k+1}x_{i-k}+Bu_i,i\in Z_+, \tag{4.3a}$$

where

$$F = A - Ec_1 \text{ and } \det[Ec - F] \neq 0 \text{ for some } c \in \mathrm{C}. \tag{4.3b}$$

Premultiplying (4.3a) by $[Ec - F]^{-1}$ we obtain

$$\bar{E}x_{i+1} = \bar{F}x_i - \sum_{k=1}^{i} \bar{E}c_{k+1}x_{i-k} + \bar{B}u_i, \tag{4.4a}$$

where

$$\bar{E} = [Ec - F]^{-1}E, \bar{F} = [Ec - F]^{-1}F, \bar{B} = [Ec - F]^{-1}B. \tag{4.4b}$$

Theorem 4.1 The solution to the Eq. (4.4a) with an admissible initial condition x_0, is given by

$$x_i = (\bar{E}^D\bar{F})^i\bar{E}^D\bar{E}x_0 + \sum_{k=0}^{i-1} \bar{E}^D(\bar{E}^D\bar{F})^{i-k-1}[\bar{B}u_k - \sum_{j=1}^{k} \bar{E}c_{j+1}x_{k-j}] + (\bar{E}\bar{E}^D - I_n)\sum_{k=0}^{q-1} (\bar{E}\bar{F}^D)^k\bar{F}^D\bar{B}u_{i+k} \tag{4.5}$$

where q is the index of $\bar{E}$. Proof is given in [18].

From (4.5) for $i = 0$ we have

$$x_0 = \bar{E}^D\bar{E}x_0 + (\bar{E}\bar{E}^D - I_n)\sum_{k=0}^{q-1} (\bar{E}\bar{F}^D)^k\bar{F}^D\bar{B}u_k. \tag{4.6}$$

In practical case, for $u_i = 0$, $i \in Z_+$ we have $x_0 = \bar{E}^D\bar{E}x_0$. Thus, the equation $\bar{E}x_{i+1} = Ax_i$ has a unique solution if and only if $x_0 \in \mathrm{Im}\bar{E}\bar{E}^D$, where Im denotes the image.

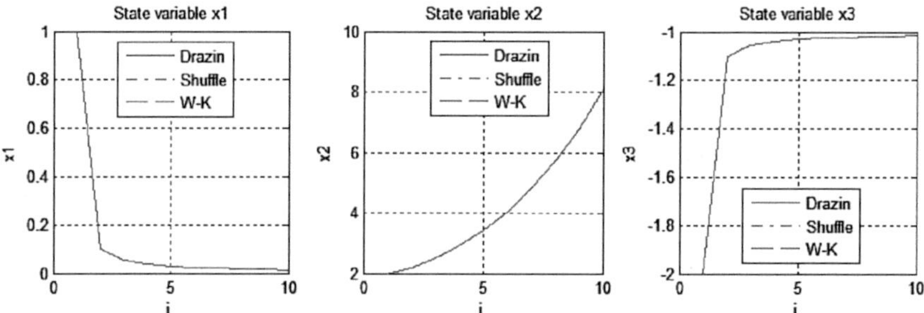

Fig. 1 Solution for $\alpha = 0.1$

5 Weierstrass-Kronecker Decomposition Method

Third method use the following Lemma, upon which the solution to the state equation will be derived.

Lemma 5.1 [7, 20, 21] If (2.3) holds, then there exist nonsingular matrices $P, Q \in \mathfrak{R}^{n \times n}$ such that

$$PEQ = \mathrm{diag}(I_{n_1}, N), \;\; PAQ = \mathrm{diag}(A_1, I_{n_2}), \tag{5.1}$$

where $N \in \mathfrak{R}^{n_2 \times n_2}$ is nilpotent matrix with the index μ (i.e. $N^{\mu} = 0$ and $N^{\mu-1} \neq 0$), $A_1 \in \mathfrak{R}^{n_1 \times n_1}$ and n_1 is equal to degree of the polynomial

$$\det[Es - A] = a_{n_1} z^{n_1} + \cdots + a_1 z + a_0, n_1 + n_2 = n. \tag{5.2}$$

A method for computation of the matrices P and Q has been given in [22].

Premultiplying the Eq. (2.1) by the matrix $P \in \mathfrak{R}^{n \times n}$ and introducing new state vector

$$\bar{x}_i = \begin{bmatrix} \bar{x}_i^{(1)} \\ \bar{x}_i^{(2)} \end{bmatrix} = Q^{-1} x_i, \bar{x}_i^{(1)} \in \mathfrak{R}^{n_1}, \bar{x}_i^{(2)} \in \mathfrak{R}^{n_2}, i \in Z_+, \tag{5.3}$$

we obtain

$$PEQQ^{-1} \Delta^{\alpha} x_{i+1} = PEQ \Delta^{\alpha} Q^{-1} x_{i+1} = PAQQ^{-1} x_i + PBu_i. \tag{5.4}$$

Applying (5.1) and (5.3) to (5.4) we have

$$\begin{bmatrix} I_{n_1} & 0 \\ 0 & N \end{bmatrix} \Delta^{\alpha} \begin{bmatrix} \bar{x}_{i+1}^{(1)} \\ \bar{x}_{i+1}^{(2)} \end{bmatrix} = \begin{bmatrix} A_1 & 0 \\ 0 & I_{n_2} \end{bmatrix} \begin{bmatrix} \bar{x}_i^{(1)} \\ \bar{x}_i^{(2)} \end{bmatrix} + \begin{bmatrix} B_1 \\ B_2 \end{bmatrix} u_i, i \in Z_+, \tag{5.5}$$

where

$$\begin{bmatrix} B_1 \\ B_2 \end{bmatrix} = PB, B_1 \in \mathfrak{R}^{n_1 \times m}, B_2 \in \mathfrak{R}^{n_2 \times m}. \tag{5.6}$$

Taking into account (2.2a), from (5.5) we obtain

$$\bar{x}_{i+1}^{(1)} = -\sum_{k=1}^{i+1} (-1)^k \binom{\alpha}{k} \bar{x}_{i-k+1}^{(1)} + A_1 \bar{x}_i^{(1)} + B_1 u_i = A_{1\alpha} \bar{x}_i^{(1)} + \sum_{k=2}^{i+1} (-1)^{k-1} \binom{\alpha}{k} \bar{x}_{i-k+1}^{(1)} + B_1 u_i \tag{5.7}$$

and

$$N\left[\bar{x}_{i+1}^{(2)}+\sum_{k=1}^{i+1}(-1)^k\binom{\alpha}{k}\bar{x}_{i-k+1}^{(2)}\right]=\bar{x}_i^{(2)}+B_2u_i, A_{1\alpha}=A_1+I_{n_1}\alpha. \quad (5.8)$$

The solution $\bar{x}_i^{(1)}$ to the Eq. (5.7) is well-known [20] and it is given by the following theorem.

Theorem 5.1 [7, 20] The solution $\bar{x}_i^{(1)}$ of the Eq. (5.7) is given by the formula

$$\bar{x}_i^{(1)}=\Phi_i\bar{x}_0^{(1)}+\sum_{k=0}^{i-1}\Phi_{i-k-1}B_1u_k, i\in Z_+, \quad (5.9)$$

where the matrices Φ_i are determined by the equation

$$\Phi_{i+1}=\Phi_iA_{1\alpha}+\sum_{k=2}^{i+1}(-1)^{k-1}\binom{\alpha}{k}\Phi_{i-k+1} \quad \Phi_0=I_{n_1}. \quad (5.10)$$

To find the solution $\bar{x}_i^{(2)}$ of the Eq. (5.8) for $N\neq 0$ nilpotent (e.g. for $N=\begin{bmatrix}0 & 1\\ 0 & 0\end{bmatrix}$ we have two equations with two unknown elements) we simple start with solving the equation related to zero row and then continue solving the rest of the equations, see e.g. [7, 20].

If $N=0$ then from (5.8) we have

$$\bar{x}_i^{(2)}=-B_2u_i, i\in Z_+. \quad (5.11)$$

From (5.3), for known $\bar{x}_i^{(1)}$ and $\bar{x}_i^{(2)}$, we can find the desired solution of the Eq. (2.1).

6 Example

Main goal of this chapter as well as whole paper, is to show, how to use presented methods, for computation of the solution of the fractional discrete-time linear system described by the Eq. (2.1). The following example will be used to describe the procedure for computation of the solution.

Find the solution x_i of the descriptor fractional discrete-time linear system (2.1) with the matrices

$$E=\begin{bmatrix}1 & 0 & 0\\ 0 & 1 & 0\\ 0 & 0 & 0\end{bmatrix}, \quad A=\begin{bmatrix}1 & 0 & 1\\ 0 & 1 & 0\\ -1 & 0 & -1\end{bmatrix}, \quad B=\begin{bmatrix}1\\ 0\\ -1\end{bmatrix} \quad (6.1)$$

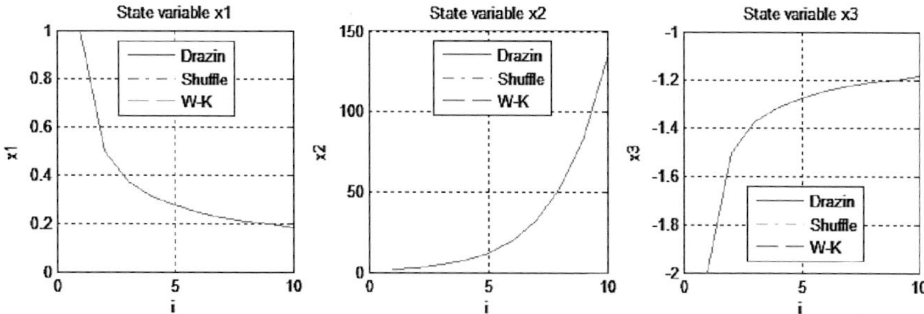

Fig. 2 Solution for $\alpha = 0.5$

for $\alpha = 0.5$, $u_i = u = 1$, $i \in Z_+$ and $x_0 = \begin{bmatrix} 1 & 2 & -2 \end{bmatrix}^T$ (T denotes the transpose).

In this case, $\det E = 0$ and the pencil of the system (2.1) witch (6.1) is regular since

$$\det[Ez - A] = \begin{vmatrix} z-1 & 0 & -1 \\ 0 & z-1 & 0 \\ 1 & 0 & 1 \end{vmatrix} = z(z-1). \tag{6.2}$$

6.1 Case of Shuffle Algorithm Method

Following Chap. 3, we compute

$$\begin{bmatrix} E & A & B \end{bmatrix} = \begin{bmatrix} 1 & 0 & 0 & 1 & 0 & 1 & 1 \\ 0 & 1 & 0 & 0 & 1 & 0 & 0 \\ 0 & 0 & 0 & -1 & 0 & -1 & -1 \end{bmatrix} = \begin{bmatrix} E_1 & A_1 & B_1 \\ 0 & A_2 & B_2 \end{bmatrix} \tag{6.3}$$

and the Eqs. (3.3a) and (3.3b) has the form

$$\sum_{k=0}^{i+1} c_k \begin{bmatrix} 1 & 0 & 0 \\ 0 & 1 & 0 \end{bmatrix} x_{i-k+1} = \begin{bmatrix} 1 & 0 & 1 \\ 0 & 1 & 0 \end{bmatrix} x_i + \begin{bmatrix} 1 \\ 0 \end{bmatrix} u_i, \tag{6.4a}$$

$$0 = \begin{bmatrix} -1 & 0 & -1 \end{bmatrix} x_i - u_i. \tag{6.4b}$$

Using (2.2b) we obtain $c_1 = -0.5$, $c_2 = 1/8$, …, $c_{i+1} = (-1)^{i+1} \left. \frac{\alpha(\alpha-1)\ldots(\alpha-i)}{(i+1)!} \right|_{\alpha=0.5}$ and the Eq. (3.5) has the form

$$\begin{bmatrix} 1 & 0 & 0 \\ 0 & 1 & 0 \\ -1 & 0 & -1 \end{bmatrix} x_{i+1} = \begin{bmatrix} 1.5 & 0 & 1 \\ 0 & 1.5 & 0 \\ 0 & 0 & 0 \end{bmatrix} x_i - \frac{1}{8}\begin{bmatrix} 1 & 0 & 1 \\ 0 & 1 & 0 \\ 0 & 0 & 0 \end{bmatrix} x_{i-1} - \cdots - c_{i+1}\begin{bmatrix} 1 & 0 & 1 \\ 0 & 1 & 0 \\ 0 & 0 & 0 \end{bmatrix} x_0 + \begin{bmatrix} 1 \\ 0 \\ 0 \end{bmatrix} u_i + \begin{bmatrix} 0 \\ 0 \\ 1 \end{bmatrix} u_{i+1}. \tag{6.5}$$

The matrix $\begin{bmatrix} E_1 \\ \bar{A}_{10} \end{bmatrix} = \begin{bmatrix} 1 & 0 & 0 \\ 0 & 1 & 0 \\ -1 & 0 & -1 \end{bmatrix} = \begin{bmatrix} E_1 \\ \bar{A}_{10} \end{bmatrix}^{-1}$ is nonsingular and the solution to the state Eq. (2.1) has the form

$$x_{i+1} = \hat{A}_0 x_i + \hat{A}_1 x_{i-1} + \cdots + \hat{A}_i x_0 + \hat{B}_0 u_i + \hat{B}_1 u_{i+1}, \tag{6.6}$$

where

$$\hat{A}_0 = \begin{bmatrix} E_1 \\ \bar{A}_{10} \end{bmatrix}^{-1} \begin{bmatrix} 1.5 & 0 & 1 \\ 0 & 1.5 & 0 \\ 0 & 0 & 0 \end{bmatrix}, \quad \hat{A}_1 = \frac{1}{8}\begin{bmatrix} E_1 \\ \bar{A}_{10} \end{bmatrix}^{-1} \begin{bmatrix} 1 & 0 & 1 \\ 0 & 1 & 0 \\ 0 & 0 & 0 \end{bmatrix}, \ldots,$$
$$\hat{A}_i = (-1)^{i-1}\frac{0.5(-0.5)\ldots(0.5-i)}{(i+1)!}\begin{bmatrix} E_1 \\ 0 \end{bmatrix}^{-1} \begin{bmatrix} 1 & 0 & 1 \\ 0 & 1 & 0 \\ 0 & 0 & 0 \end{bmatrix}, \quad \hat{B}_0 = \begin{bmatrix} E_1 \\ \bar{A}_{10} \end{bmatrix}^{-1} \begin{bmatrix} 1 \\ 0 \\ 0 \end{bmatrix}, \quad \hat{B}_1 = \begin{bmatrix} E_1 \\ \bar{A}_{10} \end{bmatrix}^{-1} \begin{bmatrix} 0 \\ 0 \\ 1 \end{bmatrix}. \tag{6.7}$$

The desired solution of the descriptor fractional system (2.1) with (6.1) has the form

$$x_i = \sum_{k=0}^{i-1} \hat{A}_k x_{i-k-1} + \sum_{k=0}^{p} \hat{B}_k u_{i-k-1}. \tag{6.8}$$

6.2 Case of Drazin Inverse Method

Following this chapter, we compute

$$F = A - Ec_1 = A + E\alpha = \begin{bmatrix} 1+\alpha & 0 & 1 \\ 0 & 1+\alpha & 0 \\ -1 & 0 & -1 \end{bmatrix} \text{ and } q = 1. \tag{6.9}$$

For $c = 1$ the matrices (4.4b) have the form

$$\bar{E} = \begin{bmatrix} -2 & 0 & 0 \\ 0 & -0.667 & 0 \\ 2 & 0 & 0 \end{bmatrix}, \quad \bar{F} = \begin{bmatrix} -1 & 0 & 0 \\ 0 & -1 & 0 \\ 0 & 0 & -1 \end{bmatrix}, \quad \bar{B} = \begin{bmatrix} 0 \\ 0 \\ -1 \end{bmatrix}. \tag{6.10}$$

Using e.g. formula $\bar{E}^D = V[W\bar{E}V]^{-1}W$ where $\bar{E} = VW = \begin{bmatrix} -2 & 0 \\ 0 & -0.667 \\ 2 & 0 \end{bmatrix} \begin{bmatrix} 1 & 0 & 0 \\ 0 & 1 & 0 \end{bmatrix}$, we compute

$$\bar{E}^D = \begin{bmatrix} -0.5 & 0 & 0 \\ 0 & -1.5 & 0 \\ 0.5 & 0 & 0 \end{bmatrix} \text{ and } \bar{F}^D = \bar{F}^{-1} = \begin{bmatrix} -1 & 0 & 0 \\ 0 & -1 & 0 \\ 0 & 0 & -1 \end{bmatrix}, \tag{6.11}$$

since $\det \bar{F} = 0.187 \neq 0$. Taking into account that

$$\bar{E}^D \bar{F} = \begin{bmatrix} 0.5 & 0 & 0 \\ 0 & 1.5 & 0 \\ -0.5 & 0 & 0 \end{bmatrix}, \quad \bar{E}\bar{E}^D = \begin{bmatrix} 1 & 0 & 0 \\ 0 & 1 & 0 \\ -1 & 0 & 0 \end{bmatrix}, \tag{6.12}$$

the desired solution for the descriptor fractional system (2.1) with (6.1) has the form

$$x_i = (\bar{E}^D\bar{F})^i \bar{E}^D \bar{E} x_0 + \sum_{k=0}^{i-1} \bar{E}^D (\bar{E}^D \bar{F})^{i-k-1} [\bar{B}u_k - \sum_{j=1}^{k} \bar{E}c_{j+1}x_{k-j}] + (\bar{E}\bar{E}^D - I_n) \sum_{k=0}^{q-1} (\bar{E}\bar{F}^D)^k \bar{F}^D \bar{B} u_{i+k} \tag{6.13}$$

where the coefficients c_j are defined by (2.2b). From (6.13) for $i = 0$ we have

$$x_0 = \bar{E}^D \bar{E} x_0 + (\bar{E}\bar{E}^D - I_3)\bar{F}^D \bar{B} u_0 = \begin{bmatrix} 1 & 0 & 0 \\ 0 & 1 & 0 \\ -1 & 0 & 0 \end{bmatrix} x_0 + \begin{bmatrix} 0 \\ 0 \\ -1 \end{bmatrix} u_0. \tag{6.14}$$

Hence, for given $u_0 = u = 1$, the initial condition $x_0 = [1 \quad 2 \quad -2]^T$ satisfy (6.14) and their are admissible.

6.3 Case of Weierstrass-Kronecker Decomposition Method

In this case the for (6.1) matrices P and Q have the form

$$P = \begin{bmatrix} 0 & 1 & 0 \\ 1 & 0 & 1 \\ 0 & 0 & -1 \end{bmatrix}, \quad Q = \begin{bmatrix} 0 & 1 & 0 \\ 1 & 0 & 0 \\ 0 & -1 & 1 \end{bmatrix} \tag{6.15}$$

and

$$PEQ = \begin{bmatrix} 1 & 0 & 0 \\ 0 & 1 & 0 \\ 0 & 0 & 0 \end{bmatrix}, \quad PAQ = \begin{bmatrix} 1 & 0 & 0 \\ 0 & 0 & 0 \\ 0 & 0 & 1 \end{bmatrix}, \quad PB = \begin{bmatrix} 0 \\ 0 \\ 1 \end{bmatrix}, \quad A_{1\alpha} = \begin{bmatrix} 1.5 & 0 \\ 0 & 0.5 \end{bmatrix},$$
$$(n_1 = 2, n_2 = 1). \tag{6.16}$$

The Eqs. (3.5) and (3.6) have the form

$$\bar{x}_{i+1}^{(1)} = \begin{bmatrix} 1.5 & 0 \\ 0 & 0.5 \end{bmatrix} \bar{x}_i^{(1)} + \sum_{k=2}^{i+1} (-1)^{k-1} \binom{0.5}{k} \bar{x}_{i-k+1}^{(1)}, i \in Z_+, \tag{6.17a}$$

$$\bar{x}_i^{(2)} = -B_2 u_i = -u_i, i \in Z_+. \tag{6.17b}$$

The solution $\bar{x}_i^{(1)}$ of the Eq. (6.17a) has the form

$$\bar{x}_i^{(1)} = \Phi_i \bar{x}_0^{(1)} + \sum_{k=0}^{i-1} \Phi_{i-k-1} B_1 u_k, i \in Z_+, \tag{6.18}$$

where

$$\Phi_0 = \begin{bmatrix} 1 & 0 \\ 0 & 1 \end{bmatrix}, \quad \Phi_1 = \begin{bmatrix} 1.5 & 0 \\ 0 & 0.5 \end{bmatrix}, \quad \Phi_2 = \begin{bmatrix} 2.125 & 0 \\ 0 & 0.125 \end{bmatrix}, \ldots. \tag{6.19}$$

From (5.3) for $i = 0$ we have

$$\bar{x}_0 = Q^{-1} x_0 = \begin{bmatrix} 0 & 1 & 0 \\ 1 & 0 & 0 \\ 1 & 0 & 1 \end{bmatrix} \begin{bmatrix} 1 \\ 2 \\ -1 \end{bmatrix} = \begin{bmatrix} 2 \\ 1 \\ 0 \end{bmatrix}, \quad \bar{x}_0^{(1)} = \begin{bmatrix} 2 \\ 1 \end{bmatrix}, \quad \bar{x}_0^{(2)} = 0. \tag{6.20}$$

The desired solution of the descriptor fractional system (2.1) with (6.1) is given by

$$x_i = Q \bar{x}_i = \begin{bmatrix} 0 & 1 & 0 \\ 1 & 0 & 0 \\ 0 & -1 & 1 \end{bmatrix} \begin{bmatrix} \bar{x}_i^{(1)} \\ \bar{x}_i^{(2)} \end{bmatrix}, \tag{6.21}$$

where $\bar{x}_i^{(1)}$ and $\bar{x}_i^{(2)}$ are determined by (6.18) and (6.17b), respectively.

6.4 Comparison of the Results

Using Matlab/Simulink computing environment, the solution for 10 first steps have been calculated and shown on the Figs. 1, 2 and 3, where Fig. 1 represent solution

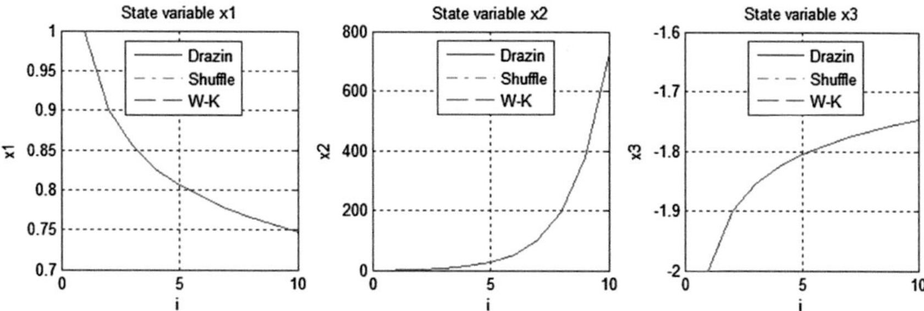

Fig. 3 Solution for $\alpha = 0.9$

for order $\alpha = 0.1$, Fig. 2 represent solution for order $\alpha = 0.5$ and Fig. 3 represent solution for order $\alpha = 0.9$. Additionally *solid line (blue)* represent solution obtained by Drazin inverse method, *dash-dot line (green)* represent solution obtained by Shuffle algorithm method and *dash-dash line (red)* represent solution obtained by Weierstrass-Kronecker decomposition method.

All three methods gives coherent result. Smaller order α, results in faster response stabilization (see state variable x_1, x_3). The greatest disadvantage of the Weierstrass-Kronecker decomposition method is its first step, that is decomposition, which is difficult for numeric implementation. Similar problem occurs in Shuffle algorithm method, where elementary row and column operation need to be applied. Finally, the Drazin inverse method, where most difficult part is computation of the Drazin inverse of the matrix E. In author opinion, this method suits best for numerical implementation, since computation of the Drazin inverse is easy for numerical implementation.

7 Concluding Remarks

The descriptor fractional discrete-time linear systems have been recalled. Three different methods for finding the solution to the state equation of the descriptor fractional discrete-time linear system have been considered. Comparison of computation efforts of the methods has been demonstrated on single numerical example. Iterative approach have been used to compute the desired solution of the systems.

In Drazin inverse method admissible initial conditions should be applied. In Shuffle algorithm method admissible initial conditions as well as future inputs should be known. The weak point of Weierstrass-Kronecker decomposition approach is computation of the P and Q matrices, where elementary row and column operations method is recommended. The same method is used for Shuffle algorithm. In summary, the Drazin inverse method seems to be most suitable for numerical implementation. An open problem is extension of these considerations to the system with different fractional orders.

Acknowledgment This work was supported by National Science Centre in Poland under work No. 2014/13/B/ST7/03467.

References

1. Campbell, S.L., Meyer, C.D., Rose, N.J.: Applications of the Drazin inverse to linear systems of differential equations with singular constant coefficients. SIAMJ Appl. Math. **31**(3), 411–425 (1976)
2. Dai, L.: Singular Control Systems, Lectures Notes in Control and Information Sciences. Springer, Berlin (1989)
3. Dodig, M., Stosic, M.: Singular systems state feedbacks problems. Linear Algebra Appl. **431** (8), 1267–1292 (2009)
4. Guang-Ren, D.: Analysis and Design of Descriptor Linear Systems. Springer, New York (2010)
5. Kaczorek, T.: Descriptor fractional linear systems with regular pencils. Int. J. Appl. Math. Comput. Sci. **23**(2), 309–315 (2013)
6. Kaczorek, T.: Singular fractional continuous-time and discrete-time linear systems. Acta Mechanica et Automatica **7**(1), 26–33 (2013)
7. Kaczorek, T.: Singular fractional discrete-time linear systems. Control Cybern. **40**(3), 1–8 (2011)
8. Sajewski, Ł.: Solution of the state equation of descriptor fractional continuous-time linear systems with two different fractional. Adv. Intell. Syst. Comput. **350**, 233–242 (2015)
9. Nishimoto, K.: Fractional Calculus. Decartess Press, Koriama (1984)
10. Oldham, K.B., Spanier, J.: The Fractional Calculus. Academmic Press, New York (1974)
11. Podlubny, I.: Fractional Differential Equations. Academic Press, San Diego (1999)
12. Losada, J., Nieto, J.: Properties of a new fractional derivative without singular kernel. Prog. Fract. Differ. Appl. **1**(2), 87–92 (2015)
13. Dzieliński, A., Sierociuk, D., Sarwas, G.: Ultracapacitor parameters identification based on fractional order model. Proc. ECC, Budapest (2009)
14. Ferreira, N.M.F., Machado, J.A.T.: Fractional-order hybrid control of robotic manipulators. In: Proceedings of the 11th International Conference on Advanced Robotics, ICAR, pp. 393–398. Coimbra, Portugal (2003)
15. Miller, K.S., Ross, B.: An Introduction to the Fractional Calculus and Fractional Differenctial Equations. Willey, New York (1993)
16. Kaczorek, T.: Solution of the state equations of descriptor fractional discrete-time linear systems with regular pencils. Tech. Transp. Szyn. 10, 415–422 (2013)
17. Kaczorek, T.: Reduction and decomposition of singular fractional discrete-time linear systems. Acta Mechanica et Automatica **5**(4), 1–5 (2011)
18. Kaczorek, T.: Application of Drazin inverse to analysis of descriptor fractional discrete-time linear systems with regular pencils. Int. J. Appl. Math. Comput. Sci. **23**(1), 29–33 (2013)
19. Kaczorek, T.: Drazin inverse matrix method for fractional descriptor continuous-time linear systems. Bull. Pol. Ac.: Tech. **62**(3), 409–412 (2014)
20. Kaczorek, T.: Selected Problems in Fractional Systems Theory. Springer, Berlin (2011)
21. Kaczorek, T.: Vectors and Matrices in Automation and Electrotechnics. WNT, Warszawa (1998)
22. Van Dooren, P.: The computation of Kronecker's canonical form of a singular pencil. Linear Algebra Appl. **27**, 103–140 (1979)

Implementation of Dynamic Matrix Control Algorithm Using a Microcontroller with Fixed-Point Arithmetic

Patryk Chaber

Abstract The aim of this paper is to describe software implementation of the Dynamic Matrix Control (DMC) algorithm using a microcontroller with fixed-point arithmetic. A 32-bit RISC ARM platform is used, which is cheap, but quite a powerful hardware system. To prevent register overflow and drastic loss of precision, a partial shifting of values technique is performed. The DMC algorithm with fixed-point arithmetic is applied to a laboratory thermal process and the obtained results are compared with those of the DMC algorithm implemented in floating-point arithmetic.

Keywords Model predictive control · Dynamic model control · Fixed-point arithmetic · Microcontroller

1 Introduction

Model Predictive Control (MPC) algorithms are commonly known to offer much better control quality comparing to the classical Proportional-Integral-Derivative (PID) approach [6, 9]. Moreover, they have some additional advantages, i.e. the ability to take the constraints into account, to efficiently control multiple-input multiple-output processes and a very flexible, universal formulation. There are many variants of MPC algorithms, which are currently used in many areas, i.e. for control of fluid level in multiple tank plants [1], for precision tracking control and constraints handling in mechatronic systems [4], for control of autonomous multi-jointed system movement [3], for control of distillation columns [10], for control of a HVAC system [8]. One of the most popular MPC algorithms is Dynamic Matrix Control (DMC). It may be used for controlling stable process (or stabilised by an additional technique) and when steady-state and dynamic properties of the controlled process are

P. Chaber (✉)
Institute of Control and Computation Engineering, Warsaw University of Technology, ul. Nowowiejska 15/19, 00-665 Warsaw, Poland
e-mail: P.Chaber@stud.elka.pw.edu.pl

R. Szewczyk et al. (eds.), *Challenges in Automation, Robotics and Measurement Techniques*, Advances in Intelligent Systems and Computing 440, DOI 10.1007/978-3-319-29357-8_5

(approximately) linear. In such a case it is possible to use a step-response model of the process which is obtained from the process in an exceptionally simple way. Furthermore, the DMC algorithm is charecterised by low computation burden, which makes it suitable for implementation on embedded devices (e.g. microcontrollers).

One of many factors that hamper a significant increase in the number of applications of the MPC algorithms for embedded devices is a very precisely limited time during which all the calculations must be completed. It may be difficult to meet and guarantee such a restriction. The need of floating-point arithmetic utilisation often lengthens the computations of the algorithm [2], which is not acceptable. One way to overcome the problem is to trade the precision of results for a faster calculation time [5]. Despite still growing popularity of devices equipped with Floating-Point Unit (FPU), there are many other hardware platforms which only offer fixed-point or integer variables representation. On the other hand, calculation in fixed-point arithmetic is usually simpler and therefore faster than in the case of using the FPU or software floating-point implementation. The use of an algorithm that is suited for fixed-point arithmetic allows to lower the costs of the processing unit and it increases the range of available hardware. The most popular technique developed with the aim of coping with a lack of floating-point arithmetic is to modify the Quadratic Programming (QP) solver [2, 7].

The objective of this work is to describe software implementation of the DMC algorithm on the microcontroller that does not support floating-point arithmetic. A 32-bit RISC ARM STM32F100RB unit is used. To prevent register overflow or precision loss which would occur during multiplication, one of the multiplicands is partially shifted. The number of bits to be shifted is chosen based on the values of the precalculated matrices—the other factor of mentioned multiplications. This allows to keep as many most significant bits of the result as possible, without the risk of overflowing the register.

2 Dynamic Matrix Control

The DMC algorithm is one of the most popular MPC algorithms. The process has n_{u} input signals (manipulated variables) and n_{y} output signals (controlled variables). The vector notation is used in this paper: $u(k) = [u_1 \ \dots \ u_{n_{\mathrm{u}}}]^{\mathrm{T}}$, $y(k) = [y_1 \ \dots \ y_{n_{\mathrm{y}}}]^{\mathrm{T}}$. In the DMC algorithm [9] at each consecutive sampling instant k, $k = 0, 1, 2, \dots$, a set of future control increments

$$\triangle U(k) = \begin{bmatrix} \triangle u(k|k) \\ \vdots \\ \triangle u(k + N_{\mathrm{u}} - 1|k) \end{bmatrix} \tag{1}$$

is calculated. It is assumed that $\triangle u(k + p|k) = 0$ for $p \geq N_{\mathrm{u}}$, where N_{u} is the control horizon. The aim of the algorithm is to minimise differences between the set-point trajectory $y^{\mathrm{sp}}(k + p|k)$ and the predicted values of the outputs $\hat{y}(k + p|k)$ (i.e.

predicted control errors) over the prediction horizon N. The unconstrained MPC optimisation task is

$$\min_{\Delta U(k)} \left\{ \sum_{p=1}^{N} \left\| y^{\mathrm{sp}}(k+p|k) - \hat{y}(k+p|k) \right\|_{\boldsymbol{M}_p}^2 + \sum_{p=0}^{N_\mathrm{u}-1} \left\| \triangle u(k+p|k) \right\|_{\boldsymbol{\Lambda}_p}^2 \right\} \tag{2}$$

where $\boldsymbol{M}_p \geq 0$ and $\boldsymbol{\Lambda}_p > 0$ are tuning matrices of dimensionality $n_\mathrm{y} \times n_\mathrm{y}$ and $n_\mathrm{u} \times n_\mathrm{u}$, respectively. Although the whole optimal future control policy (1) over the control horizon is calculated, only its first n_u elements (current control increments) are actually applied to the process, i.e. $u(k) = \triangle u(k|k) + u(k-1)$. At the next sampling instant, $k+1$, output measurements are updated, the prediction is shifted one step forward and the whole procedure is repeated.

Predicted values of process outputs, $\hat{y}(k+p|k)$, over the prediction horizon are calculated using a dynamic model of the controlled process. The model used for prediction in DMC is represented as step responses of the output variables to unit changes of manipulated variables. The model may be obtained in a simple way. Furthermore, as the model is linear, it is possible to formulate an analytic unconstrained control law. When the constraints on the manipulated variables must be taken into account, the calculated values are projected on the set of feasible solutions. For the considered Multiple Input Multiple Output (MIMO) process the step responses (for each control signal and each output signal) can be written as a set of matrices

$$\boldsymbol{S}_l = \begin{bmatrix} s_l^{1,1} & s_l^{1,2} & \cdots & s_l^{1,n_\mathrm{u}} \\ s_l^{2,1} & s_l^{2,2} & \cdots & s_l^{2,n_\mathrm{u}} \\ \vdots & \vdots & \ddots & \vdots \\ s_l^{n_\mathrm{y},1} & s_l^{n_\mathrm{y},2} & \cdots & s_l^{n_\mathrm{y},n_\mathrm{u}} \end{bmatrix}, \quad l = 1, 2, \ldots, D$$

where the coefficient $s_l^{i,j}$ denotes the value of the output signal i measured at iteration l after a step change of the input signal j. It is assumed for stable processes that after D sampling instants the step response coefficients are approximately constant i.e. $s_l^{i,j} = s_D^{i,j}$ for $l \geq D$ (D denotes the horizon of dynamics, the same value for all input-output channels is used).

When there are no constraints imposed on the manipulated and predicted controlled variables, optimisation of the cost-function (2) leads to the MPC control law

$$\triangle U(k) = \overline{\boldsymbol{K}}_1 \left[Y^{\mathrm{sp}}(k) - Y^0(k) \right] \tag{3}$$

where the free trajectory vector is $Y^0(k) = Y(k) + \boldsymbol{M}^{\mathrm{P}} \Delta U^{\mathrm{P}}(k)$, the matrix

$$\boldsymbol{K} = \left[\boldsymbol{M}^{\mathrm{T}} \boldsymbol{M} + \lambda \boldsymbol{I} \right]^{-1} \boldsymbol{M}^{\mathrm{T}} = \begin{bmatrix} \overline{\boldsymbol{K}}_1 \\ \overline{\boldsymbol{K}}_2 \\ \vdots \\ \overline{\boldsymbol{K}}_{N_\mathrm{u}} \end{bmatrix} = \begin{bmatrix} \boldsymbol{K}_{1,1} & \boldsymbol{K}_{1,2} & \cdots & \boldsymbol{K}_{1,N} \\ \boldsymbol{K}_{2,1} & \boldsymbol{K}_{2,2} & \cdots & \boldsymbol{K}_{2,N} \\ \vdots & \vdots & \ddots & \vdots \\ \boldsymbol{K}_{N_\mathrm{u},1} & \boldsymbol{K}_{N_\mathrm{u},2} & \cdots & \boldsymbol{K}_{N_\mathrm{u},N} \end{bmatrix}$$

is of dimensionality $n_u N_u \times n_y N$ and it is calculated once off-line using the step-response model of the process. The so called dynamic matrix consists of the step-response coefficients

$$M = \begin{bmatrix} S_1 & 0 & \cdots & 0 \\ S_2 & S_1 & \cdots & 0 \\ \vdots & \vdots & \ddots & \vdots \\ S_N & S_{N-1} & \cdots & S_{N-N_u+1} \end{bmatrix}$$

and is of dimensionality $n_y N \times n_u N_u$. The set-point vector and the vector of the output signals measured at the current sampling instants

$$Y^{sp}(k) = \begin{bmatrix} y^{sp}(k) \\ \vdots \\ y^{sp}(k) \end{bmatrix}, \; Y(k) = \begin{bmatrix} y(k) \\ \vdots \\ y(k) \end{bmatrix}$$

are of length $n_y \times N$, an additional matrix M^P, of dimensionality $n_y N \times n_u(D-1)$, containing the step-response coefficients, and the vector $\triangle U^P(k)$ of length $n_u(D-1)$, containing increments of some past values of the manipulated variables have the following structure

$$M^P = \begin{bmatrix} S_2 - S_1 & \cdots & S_D - S_{D-1} \\ S_3 - S_1 & \cdots & S_{D+1} - S_{D-1} \\ \vdots & \ddots & \vdots \\ S_{N+1} - S_1 & \cdots & S_{N+D-1} - S_{D-1} \end{bmatrix}, \; \triangle U^P(k) = \begin{bmatrix} \triangle u(k-1) \\ \vdots \\ \triangle u(k-(D-1)) \end{bmatrix}$$

3 Hardware Platform

The hardware platform used in this work is the STM32F100RB microcontroller mounted on the STM32VLDISCOVERY development board. The use of high-performance ARM®Cortex®-M3 32-bit RISC core operating at a frequency of 24 MHz (1.25 DMIPS), combined with single-cycle multiplication and hardware division makes this microcontroller suitable to be used for MPC algorithms. Its advantages include also large high-speed memory resources, that is 128 kbytes of flash memory and 8 kbytes of SRAM and fast communication possibilities i.e. 400 kHz I^2C, up to 12 Mbit/s master and slave SPI or up to 3 Mbit/s USART. Furthermore, the built-in programmer and debugger help to quickly develop error-free code.

Despite the fact that this microcontroller has many timers that allows to have up to 12 PWM outputs, it is not connected directly to the controlled process. An another microcontroller is used as an interface of the controlled process, assuring its safe work at the same time. The connection schemes for fixed-point and floating-point versions of DMC algorithm are shown in Fig. 1a and b, respectively. For commu-

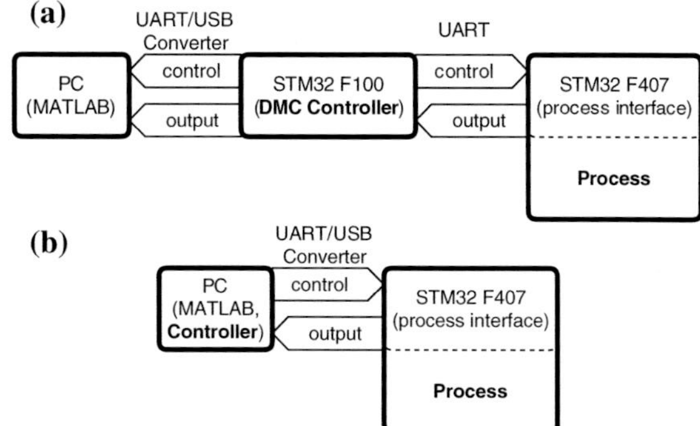

Fig. 1 Two schemes of hardware connections used **a** the DMC algorithm with fixed-point arithmetic, **b** the DMC algorithm with floating-point arithmetic

nication an UART with an USB-UART converter is used. Information is sent using custom simple protocol that allows to send control values and receive output values. Such a division of tasks between two separate microcontrollers is made to preserve simplicity and modularity of the DMC structure.

4 Process Description

The considered controlled process is a laboratory stand designed and built in the Faculty of Electronics and Information Technology of Warsaw University of Technology (Fig. 2a). The input (manipulated) signals of the process are voltage values at the pins of 4 fans and 2 power resistors used as heaters. The output (controlled) signals are temperature values measured using 4 temperature sensors communicating using One Wire standard. The control task is to change the values of PWM signal

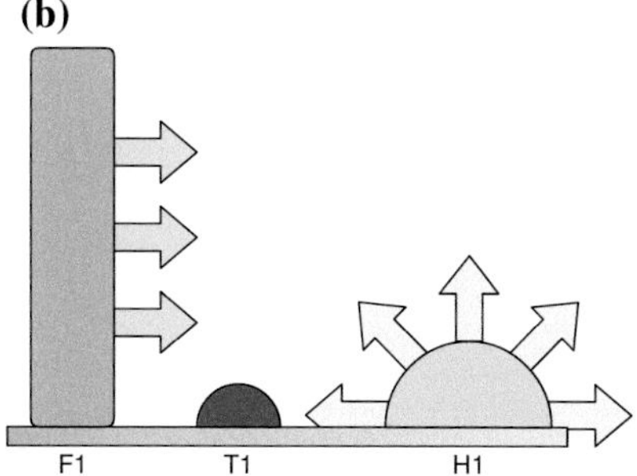

Fig. 2 The controlled process **a** the photo of full set of elements; 4 temperature sensors, 4 fans and 2 power resistors (heaters), **b** the scheme of subset of elements used during the experiments: F_1—a fan, H_1—a power resistor, T_1—a temperature sensor

which drive the fans and heaters, so that the measured temperature values are as close as possible to their given set-point values. The considered laboratory stand is highly configurable (including the possibility of setting input/output signals delay), which allows to use only a subset of available elements. This paper considers the following configuration of the process: the fan F_1, the heater H_1 and the temperature sensor T_1. In such a case the number of process input signals is $n_u = 2$ and the number of output signals is $n_y = 1$. The scheme depicting chosen elements of the process is shown in Fig. 2b. The manipulated variables are: the Pulse-Width Modulation (PWM) signal connected to the heater H_1 (the u_1 signal) and to the fan F_1 (the u_2 signal). The constraints imposed on the control signals values are $u_1^{\min} \geq u_1 \geq u_1^{\max}$ and $u_2^{\min} \geq u_2 \geq u_2^{\max}$, $u_1^{\min} = u_2^{\min} = u^{\min} = 0$, $u_1^{\max} = u_2^{\max} = u^{\max} = 100$. The value of $u^{\min}$ represents the PWM signal of zero width, and the value of $u^{\max}$ represents the PWM signal of a maximum width. The measurements are made each $T_p = 5$ s, which is also the maximum time a single iteration of DMC algorithm can take. In the nominal operating point of the process: $\bar{u}_1 = 50$, $\bar{u}_2 = 50$, $\bar{y} = 36\,^{\circ}\mathrm{C}$.

The step values used for measurement of the step-response model are: $\delta u_1 = \delta u_2 = 25$. After obtaining the step-response values $\tilde{s}_l^{1,1}$ (for the channel u_1 and y) and $\tilde{s}_l^{1,2}$ (for the channel u_2 and y) where $l = 1, 2, \ldots$, they are scaled as $s_l^{1,1} =$

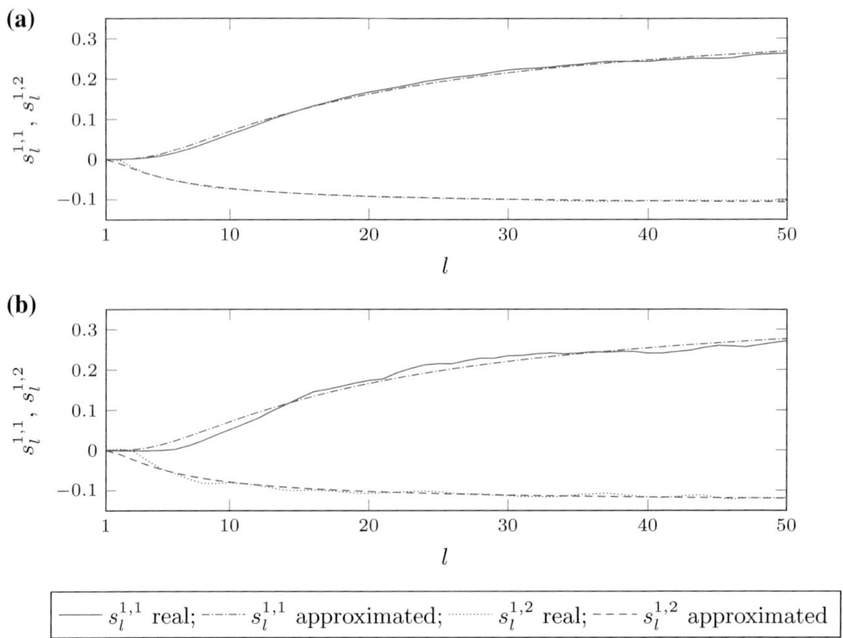

Fig. 3 Real values of the scaled step response values $s_l^{1,1}$ (*solid line*), $s_l^{1,2}$ (*dotted line*) and their approximations (*dash-dotted line* and *dashed line* respectively) **a** the process is not affected by external disturbances, **b** the process is affected by external disturbances (room air conditioning is turned on)

$(\tilde{s}_l^{1,1} - \bar{y})/\delta u_1$ and $s_l^{1,2} = (\tilde{s}_l^{1,2} - \bar{y})/\delta u_2$ for each $l = 1, 2, \ldots$. The scaled step response coefficients are shown in Fig. 3a. Despite small disturbances, step response values are almost constant for $l \geq 50$, therefore the horizon of dynamics is $D = 50$. It is worth underlining that experiments have been also performed in a slightly disturbed environment, where the step-response is much harder to obtain, which is shown in Fig. 3b. The approximation of each step response is calculated as $A \exp(-B/l)$, where $l = 1, 2, \ldots, D$ denotes a time instant. Parameters A and B are obtained as a results of minimisation of sum of squared errors between the approximation and the step response values.

5 Implementation

The DMC algorithm is implemented in C programming language using Standard Peripheral Library for STM32 microcontrollers. The initialisation code allocates memory for the matrices $\overline{\boldsymbol{K}}_1, \boldsymbol{M}^{\mathrm{P}}$, the vectors $Y(k)$, $Y^{\mathrm{sp}}(k)$, $Y^0(k)$, $\triangle U^{\mathrm{P}}(k)$ and the matrices for current measurements. All variables are of type `int32_t` which means that they are stored on 32 bits with one bit representing a sign. Instead of implementing floating-point arithmetic, which would require a lot of resources, a simple fixed-point notation is used. The idea is to use a Q15.16 notation, which represents the integer part of the variable on 15 bits with one additional bit for a sign and the fractional part on the other 16 bits. Thanks to that, it is possible to work on values from range from −32768 to 32767.9999847412109375 with a resolution of 2^{16} on 32 bits. The initialisation also defines the $\overline{\boldsymbol{K}}_1, \boldsymbol{M}^{\mathrm{P}}$ matrices which are calculated off-line, i.e. before the program starts. These matrices can be calculated using floating-point arithmetic and then converted to the Q15.16 notation, which is far more precise than calculating it in the Q15.16 notation and fixed-point arithmetic. Finally, the initial default values of all the other variables are set and the UART communication is turned on.

The main part of the program consists of the infinitely repeated steps:

1. The current value of the process output signal y is measured.
2. The future values of increments of both manipulated variables, u_1 and u_2, i.e. the decision vector (1) are calculated using the DMC algorithm.
3. The values of the manipulated variables for the current sampling instant, i.e. $u_1(k)$ and u_2(k), are applied to the process.
4. The program waits for new measurements and goes to step 1.

Calculation of the control values is the crucial part, where the problems with fixed-point arithmetic may occur. While addition or subtraction does not require any exceptional treatment, multiplication needs to be done carefully (division is done off-line). Taking into consideration that the free trajectory is $Y^0(k) = Y(k) + \boldsymbol{M}^{\mathrm{P}} \Delta U^{\mathrm{P}}(k)$, it is clear that the entries of the matrices $\boldsymbol{M}^{\mathrm{P}}$ and $\triangle U^{\mathrm{P}}(k)$ are shifted to the right by 16 bits (which is the same as multiplying these values by 2^{16}), their multiplication results in double shifted value. Assuming that the result is put into an another 32 bit register, an integer part of the multiplication is lost. Knowing that the output of this multiplication needs to be shifted back (divided by 2^{16}), it is possible to shift back one of the multiplication factors which would cause the result to be in the correct notation. The drawback of this process is that the fractional part of the factor shifted back is lost. The second solution is however better than the first one, because it causes only a reduction of precision instead of overflowing the buffer.

It is worth noting that there is no simple solution that does not involve loss of precision. Besides, it is more important to minimise the possibility of occurring the overflow problem than the rounding. The most reasonable solution is to partially shift the values that are multiplied. During that process some of the most significant bits of the integer part and some of the least significant bits of the fractional part are lost, which results in a lower risk of overflow and a lower precision loss. Fortunately, having the matrices $\overline{\boldsymbol{K}}_1$ and $\boldsymbol{M}^{\mathrm{P}}$ calculated off-line (before the algorithm starts), it can be estimated how their values are represented in the Q15.16 notation. This helps to estimate the number of bits of the shift that is required for the best accuracy of the multiplication results. The smaller the absolute values of the entries of the mentioned matrices are, the lower the number of bits shifted can be. For the considered process, the number of 8 bits shifted is chosen.

An another issue is to choose the operation order. Each shift causes a loss of the least significant bits of the fractional part. While the multiplication result must be shifted by 16 bits in total, the multiplicands can be shifted in various manner. In the considered implementation the absolute value of the entries of the precalculated matrices are so low that they are not shifted before multiplication. That helps to establish the order of operation, which in this case is as follows:

1. The elements of the vector $\triangle U^{\mathrm{P}}(k)$ are shifted by 8 bits to the right.
2. Multiplication of the shifted vector and the matrix $\boldsymbol{M}^{\mathrm{P}}$ is performed (loosing 8 of the most significant bits which should be 0).
3. The result is shifted by another 8 bits to the right, to fit it in the Q15.16 notation.

Analogously, multiplication used in the control law given by Eq. (3) is performed. The reasoning is as follows: denoting the number of bits of the first number to be multiplied as a and the number of bits of the second one as b, it is known that the maximum number of bits of the result is $a + b$. Therefore, to fit the result into the Q15.16 notation (that is a 31 bit variable) and knowing that one of the multiplicands is stored using a bits, the second one has to be stored on $31 - a$ bits. To achieve this, the second multiplicand should be rounded to the $31 - a$ most significant bits.

Right before sending the new values of the manipulated variables to the process, the calculated values are projected on the feasible set determined by constraints.

6 Experimental Results

The implemented DMC algorithm uses the prediction horizon $N = 10$, the control horizon $N_u = 3$, the weighting matrices are: $\boldsymbol{M}_p = 1$ and $\boldsymbol{\Lambda}_p = \boldsymbol{I}_{2\times2}$. This set of parameters gives good control quality as shown in Fig. 4. The output signal of the process stabilises at the set point value in approximately 300 s. There is some visible overshooting increasing with the distance from the working point, caused by the nonlinearity of the process. Nevertheless, despite using in the DMC algorithm the linear model, there are no oscillations of the output signal and no steady-state error.

Comparing experimental results obtained in fixed-point and floating-point DMC implementations, there are no significant differences. It has to be underlined that a single experiment takes as much as 50 min, thus the restoration of the object to

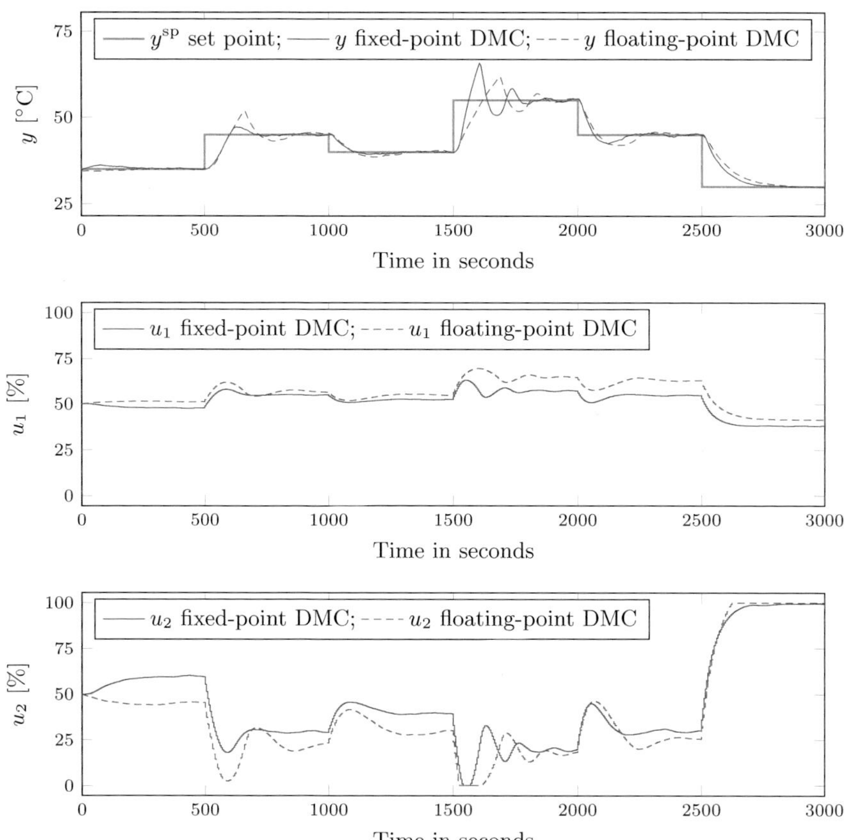

Fig. 4 The output (y) and input (u_1, u_2) signals of the process controlled using the DMC algorithm implemented in fixed-point (*solid line*) and floating-point (*dashed line*) arithmetic, the set-point (y^{sp}) is shown as a thick *gray line*

exactly the same initial point as the previous experiment is almost impossible in a real environment, in which time-varying disturbances affect the process. Two compared trajectories are not identical mainly because of different disturbances during both experiments.

Despite mostly smooth trajectories of the output signals, there are a few fragments where the temperature changes rapidly. They are best visible from the 1600th second of both experiments and result from the fact that the fans used in the laboratory stand are not able to move the propeller with too low voltage input. The exact amount of the PWM signal width at which the fan stops is 8 % and it is not starting until 19 % of width is reached. That is why the temperature value is still rising despite the control signal u_2 being greater than zero.

7 Conclusions

Popular DMC algorithm has been implemented on the microcontroller using fixed-point arithmetic. The developed implementation gives good and stable control of the considered MISO laboratory process using a low-cost hardware but still similar to the results obtained when the DMC algorithm is implemented using floating-point arithmetic in MATLAB running on PC. The constraints imposed on the manipulated variables are taken into account by projecting the calculated values onto the feasible set.

References

1. Bamimore, A., Taiwo, O., King, R.: Comparison of two nonlinear model predictive control methods and implementation on a laboratory three tank system. In: Proceedings of the 50th IEEE Conference on Decision and Control and European Control Conference, pp. 5242–5247 (2011)
2. Jerez, J.L., Constantinides, G.A., Kerrigan, E.C.: Towards a fixed point QP solver for predictive control. In: Proceedings of the 51st IEEE Annual Conference on Decision and Control, pp. 675–680 (2012)
3. Kayacan, E., Ramon, H., Saeys, W.: Learning in centralized nonlinear model predictive control: application to an autonomous tractor-trailer system. IEEE Trans. Control Syst. Technol. **23**(1), 197–205 (2015)
4. Lin, C.-Y., Liu, Y.-C.: Precision tracking control and constraint handling of mechatronic servo systems using model predictive control. IEEE/ASME Trans. Mechatron. **17**(4), 593–605 (2012)
5. Longo, S., Kerrigan, E.C., Constantinides, G.A.: A predictive control solver for low-precision data representation. In: Proceedings of the European Control Conference, pp. 3590–3595 (2013)
6. Ławryńczuk, M.: Computationally Efficient Model Predictive Control Algorithms: A Neural Network Approach. Stud. Syst. Decis. Control **3** (2014). Springer, Heidelberg
7. Patrinos, P., Guiggiani, A., Bemporad, A.: Fixed-point dual gradient projection for embedded model predictive control. In: Proceedings of the European Control Conference, pp. 3602–3607 (2013)

8. Shi, J., Jiang, Q., Cao, Z., Zhou, H., Yang, Y.: Design method of PID-type model predictive iterative learning control based on the two-dimensional generalized predictive control scheme. In: Proceedings of the 12th International Conference on Control Automation Robotics and Vision, pp. 452–457 (2012)
9. Tatjewski, P.: Advanced Control of Industrial Processes: Structures and Algorithms. Springer, London (2007)
10. Tousain, R.L., Bosgra, O.H.: Efficient dynamic optimization for nonlinear model predictive control-application to a high-density poly-ethylene grade change problem. In: Proceedings of the 39th IEEE Conference on Decision and Control, vol. 1, pp. 760–765 (2000)

Design of Control System for an Electrohydraulic Drive Based on the Valve with PMSM Motor

Dominik Rybarczyk, Piotr Owczarek and Arkadiusz Kubacki

Abstract The article describes design of control system and test stand for an electrohydraulic drive based on the new kind proportional valve with synchronous motor type PMSM (Permanent Magnets Synchronous Motor). In the second part, the chosen laboratory investigations of proposed valve with PMSM are presented. The study included the test of the basic characteristics of the drive such as step response and temperature of oil impact.

Keywords Electrohydraulic drive · Proportional valve · Permanent magnets synchronous motor

1 Introduction

Despite intensive development of the drive based on an electric motors, particularly visible in the last 30 years in associated primarily with the development of electronics and computer control systems, electrohydraulic drives are still used as actuators of multiple machines. Main advantage of electrohydraulic drives, with opposite to the electric drives, are easy implementation of linear displacement, ability to obtain very high power and high energy ratio: controlled to control—from 10^4 to 10^6. In commercial practice, electrohydraulic actuators capable of generating forces in order of the MN and obtaining speed of 0.1 m/s. Providing for such type of drives, high bandwidth (above 40 Hz) and positioning accuracy of better than ±0.01 mm, requires the use of the advanced control method and precised and fast valve.

D. Rybarczyk (✉) · P. Owczarek · A. Kubacki
Institute of Mechanical Technology, Poznan University of Technology,
Piotrowo 3, 60-965 Poznań, Poland
e-mail: Dominik.Rybarczyk@put.poznan.pl

P. Owczarek
e-mail: Piotr.Owczarek@put.poznan.pl

R. Szewczyk et al. (eds.), *Challenges in Automation, Robotics and Measurement Techniques*, Advances in Intelligent Systems and Computing 440, DOI 10.1007/978-3-319-29357-8_6

Many investigations and related publications about electrohydraulic drives focused on improvement them properties, by implementation of modern methods of control [1–3]. Murrenhoff [4] described the cross-cutting trends in the design and development of electrohydraulic valves. His article presented new type solution, by using of the direct drive in a proportional valve in hydraulic system. Research on build and properties of proportional valves were conducted by [5]. Paper [8] presented modelling and simulation of hydraulic spool valves by using simple mathematical expressions to describe the geometry of the sliding spool metering edge. Described here control stand is equipped with a new type of proportional valve with synchronous motor Author detailed described in: [6, 7].

2 Test Stand and Control System

2.1 *Assumptions*

During the design process, it was assumed that test stand should:

- allow the same valve tests, such as measuring the flow rate and recording the step response of the valve spool,
- allow the measurement of the actuator position and its changes over time,
- be able to performed drive test under load and measuring the force,
- have the possibility of quick and efficient testing of advanced control systems, such Model Following Control methods,
- have the possibility of quick reconfiguration of hardware electrohydraulic drive and its control system.

2.2 *Test Stand Structure*

Mechanical and measurement part consists of two blocks. The first one is allowed to measure the basis valve characteristics of the valve (Fig. 1) while the second is designed to test the entire electrohydraulic drive (Fig. 2). Control system was common for both parts. It consists of two main elements: the master controller like PLC with touch panel and the slave, which is a synchronous motor PMSM controller.

The flow meter type HySense QG100 is used, which parameters were: max. flow $Q = 30\ \mathrm{dm^3/min}$, max. pressure $p = 30$ MPa, non-linearity ±0.5 %, 1640 pulses per dm^3. It is connected to the counter module in the PLC. Hydraulic power supply characteristics were: motor power = 37 kW, maximum flow rate = 100 $\mathrm{dm^3/min}$, maximum pressure $p_0 = 40$ MPa, filtration at 6 μ.

Test stand was equipped with double acting hydraulic cylinder. The stroke of the hydraulic cylinder was 200 mm. The diameters of piston was $A = 40$ mm and the

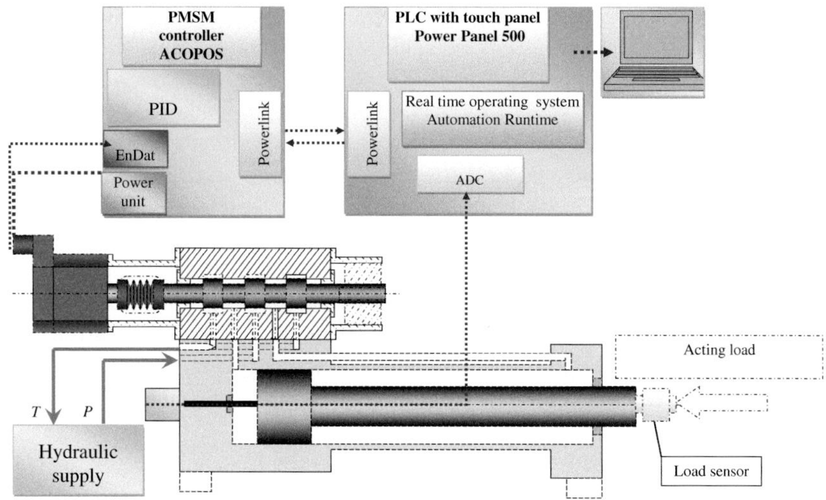

Fig. 1 Control stand for perform servo drive test

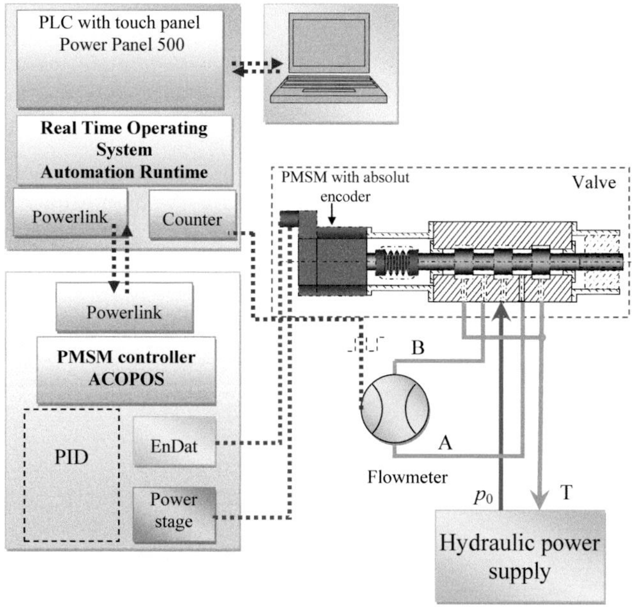

Fig. 2 Control stand for measure of the valve flow

piston rod $Aa = 63$ mm. The cylinder was equipped with internal magnetostrictive position sensor, providing the actual position of cylinder piston.

In the valve, Authors used PMSM motors type B&R 8LVA23. The motor is connected to the spool by flexible coupling bellow. Moving of the motor shaft caused rotation and simultaneously axial translation of the spool in the valve body. This movement was proportional to the angular motor displacement. Therefore, the described valve is defined as the "proportional". Direction of rotation determines the direction of spool translation and opening or closing of valve gaps. It results in the flow of oil to and from the actuator chambers and displacement of piston. The spool diameter was 10 mm. In the valve there were three rectangular gaps in the body. Dimension each of them were: 2.5 mm × 2 mm. Therefore valve size can be defined as 10 (below 64 dm^3/min) [6].

Basis parameters of the PMSM motor with is used in the valve were: the rated speed 3000 rev/min, rated current 2.9 A and stall torque 0.68 Nm. The motor is equipped with an absolute encoder type EnDat, providing a continuous information about the current position, even after a power failure. Also it assures high positioning accuracy (262144 pulses on revolution). Thanks to this the drive is able to assure the linear resolution of 0.5 μm [6, 9].

The test drive equipped with an auxiliary hydraulic system based on an outer cylinder for loading drive (Figs. 1 and 3). Additionally, there is mounted force sensor type HBMC9B, which allows measurement of the force up to 50 kN.

The PLC is equipped with the processor core type Intel Atom 1.6 GHz (Fig. 4). System was running under real-time operating system type Automation Runtime. The main control program is written in Structured Text and ANSI C. It has been divided into several tasks made of a certain determinism of time. The task responsible for the work of PID and MFC electrohydraulic servo drive controllers, synchronous motor control (homing, start, stop, setting the parameters of motion,

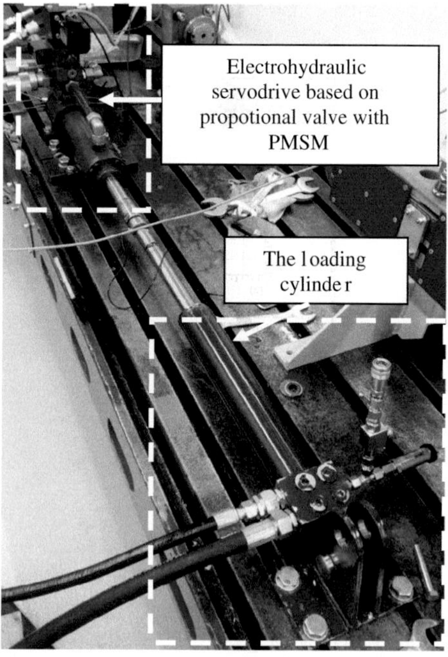

Fig. 3 View on the mechanical part of the control stand

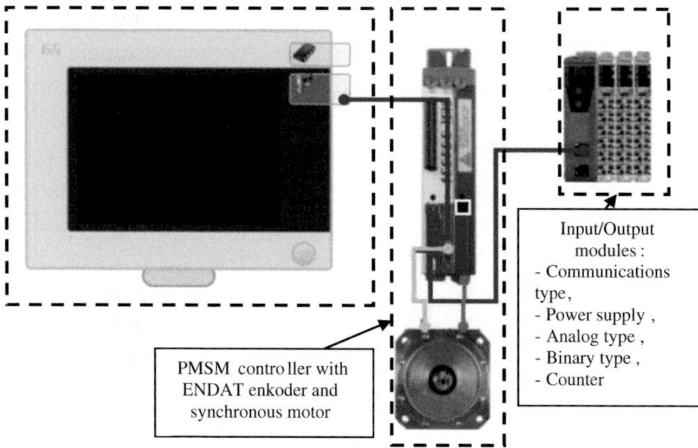

Fig. 4 Control systems schematic taken from Automation Studio software

acceleration, speed) was performed with a time base of 0.8 ms. With the same timebase, task responsible for communication between the inverter and the PLC are worked. Communication was carried out using the Powerlink interface [9]. The charge visualization was performed in steps of 12 ms, in order to not loading the CPU of PLC. There are also implements a thread and library which allowing user to generate code directly from Matlab Simulink software.

In order to performed easier communication between operator and service facilities Author designed the dedicated visualization on the touch panel, which significantly easier to perform and supervise tests (Fig. 5). On visualization,

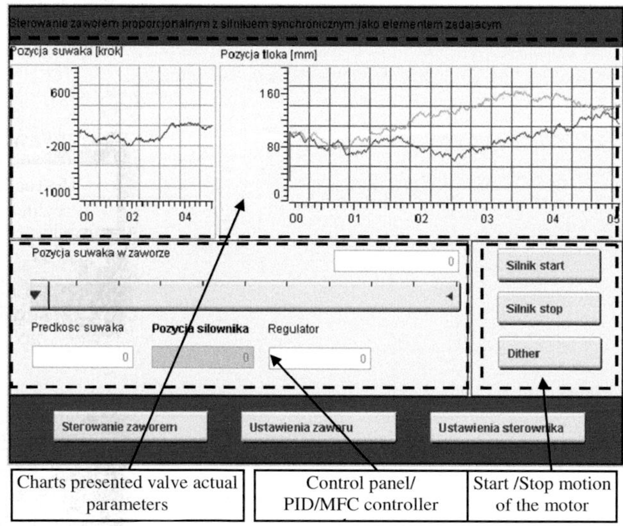

Fig. 5 Visualisation

immediately after the measurement, there are presented charts of the recorded signals on the screen. The application allowed the change the basic drive parameters such as: setting valve slider position, setting speed limits, setting the maximum range of drive motion. The graphs presented current parameters of the individual measured values such as fluid flow rate or motion velocity.

2.3 Testing

The electrohydraulic drive was tested by used of the step response signals, for values equal to: 20, 40, 60, 80 and 100 % of valve spool maximum displacement (in

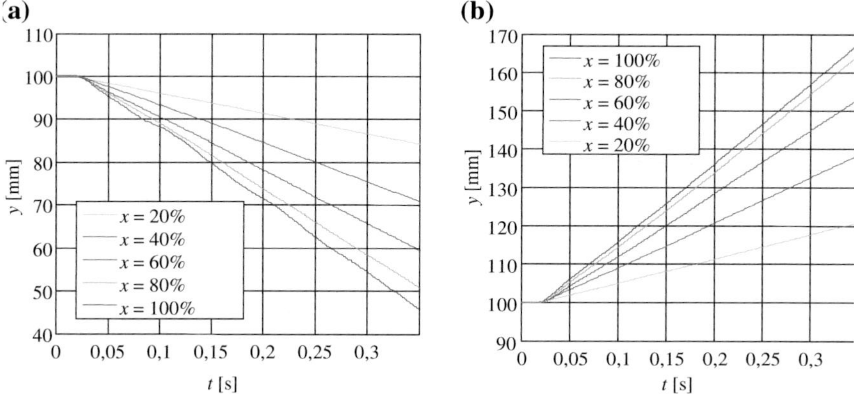

Fig. 6 Movement of electrohydraulic drive piston in open loop 15 MPa of hydraulic supply pressure: **a** valve open in "+" direction, **b** valve open in "–" direction

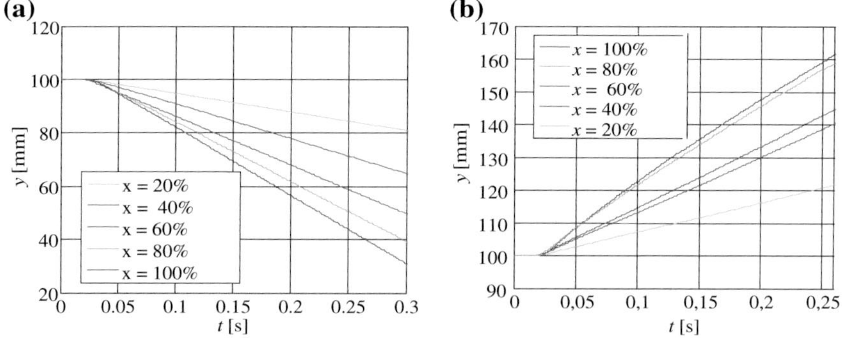

Fig. 7 Movement of electrohydraulic drive piston in open loop 10 MPa of hydraulic supply pressure: **a** valve open in "+" direction, **b** valve open in "–" direction

both directions). The experiment is performed for the supply pressure p_0 amounting to 5, 10 and 15 MPa. The results are shown in Figs. 6, 7 and 8. The recorded data indicate that the valve was characterized by a significant time delay, which is caused by PLC controller and communication interface, which is used to control the PMSM.

Collected charts show that the speed of movement of the piston is dependent on the supply pressure. Sample speed values for the pressure of 5 MPa are: the valve opening of 100 %–70 mm/s, the valve opening of 60 %–60 mm/s, the valve opening of 20 %–21 mm/s.

In test stand Authors used non-zinc hydraulic oil with high viscosity index type Draco HV 46 Premium Oil. During tests oil temperature is recorded. Collected values are compare in Table 1 and Fig. 9. Reflected differences in the speed of piston movements are caused of changes of oil viscosity [1, 2].

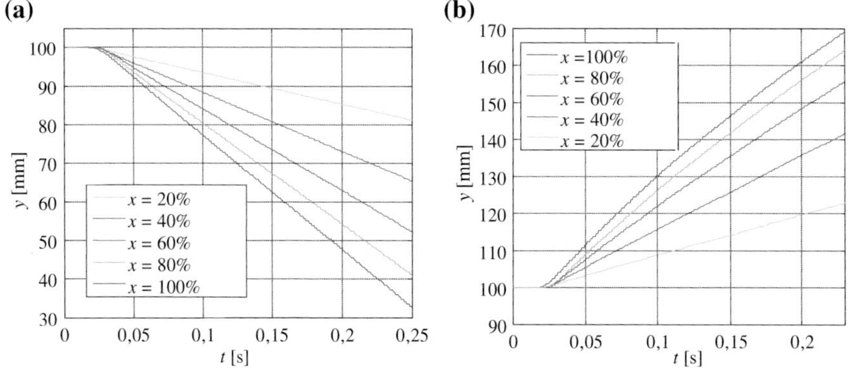

Fig. 8 Movement of electrohydraulic drive piston in open loop for 5 MPa of hydraulic supply pressure: **a** valve open in "+" direction, **b** valve open in "–" direction

Table 1 Oil temperature during drive tests

Oil temperature during drive tests (°C)			
x	5 (MPa)	10 (MPa)	15 (MPa)
100 % +	26.5	34.2	50.0
80 % +	25.3	32.7	48.0
60 % +	24.5	29.7	47.9
40 % +	24.5	27.7	47.5
20 % +	23.5	23.4	46.5
20 % –	27.3	37.3	42.6
40 % –	27.5	38.0	43.3
60 % –	27.8	38.9	47.9
80 % –	28.3	39.0	48.0
100 % –	28.3	39.6	45.7

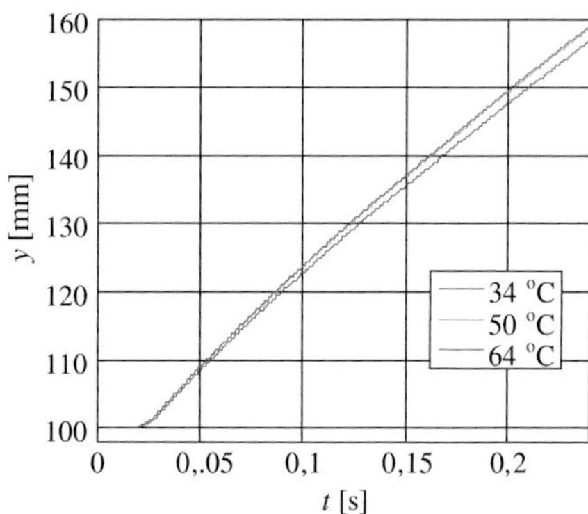

Fig. 9 Oil temperature impact on the electrohydraulic drive

3 Conclusion

In this paper the control stand of an electrohydraulic drive with proportional valve controlled by PM synchronous motor is presented. Mechanical and measurement part are consisted of the two blocks: to measure the basis valve characteristics and to test the entire electrohydraulic drive. Control system was common for both parts. It consists of the PLC with touch panel and a synchronous motor PMSM controller.

The article includes initial research—step response and the impact of oil temperature.

Performed test stand made allowed to conduct further comprehensive investigations of the described here valve and advanced control methods with used of whole drive.

Acknowledgments This paper was supported by the Polish Ministry of Science and Education, grant no. 02/23/DSPB/1208, Poland.

References

1. Chapple, P.: Principles of Hydraulic System Design. Coxmoor Publishing Company, Oxford (2003)
2. Cundiff, S.J.: Fluid Power Circuits and Control Fundamental and Applications. CRC Press, Boca Raton (2000)
3. Ming, X., Jin, B., Chen, G., Ni, J.: Speed-control of energy regulation based variable-speed electrohydraulic drive. Strojniški vestnik. J. Mech. Eng. **59**(7–8), 433–442 (2013). doi:10.5545/sv-jme.2012.911
4. Murrenhoff, H.: Trends in valve development. Ölhydraulik und Pneumatik, **46**, Nr. 4 (2003) (Germany 2003)
5. Schlegel, C., Bross, M., Beater, P.: HIL-simulation of the hydraulics and mechanics of an automatic gearbox. In: Proceedings of 2nd International Modelica Conference, pp. 67–75, Germany 2002

6. Milecki, A., Rybarczyk, D.: Modelling of electrohydraulic proportional valve with synchronous motor, Strojniški vestnik. J. Mech. Eng. **61**(9) (2015) (Slovenia 2015, SV—JME)
7. Rybarczyk, D., Sędziak, D., Owczarek, P., Owczarkowski, A.: Modelling of electrohydraulic drive with a valve controlled by synchronous motor. Advances in Intelligent Systems and Computing, ISBN 978-3-319-15796-2, pp. 215–222. Springer International Publishing Switzerland 2015
8. Šimic, M., Debevec, M., Herakovič, N.: Modelling of hydraulic spool-valves with specially designed metering edges. Strojniški vestnik. J. Mech. Eng. **60**(2), 77–83 (2014). doi:10.5545/sv-jme.2013.1104
9. B&R Automation. http://www.br-automation.com/. Accessed 23 Sept 2015

Electrohydraulic Valve with Two Stepping Motors

Andrzej Milecki and Jarosław Ortmann

Abstract The article describes the design of electrohydraulic valve with two stepping motors working differentially. The valve changes the flow proportionally to the sum of the motors steps. In the paper the application of this valve in a drive is presented and investigations results are shown. This results are compared with results obtained when only one motor is used, showing that the proposed new solution with two motors is significantly better.

Keywords Electrohydraulic valve · Stepping motor · Servo drive · Control

1 Introduction

Fluid power drives are able to generate very large forces measured in many kN, which are used to bend, crush, separate, and cut through thick, tough metal. Competition with electric drive systems has forced real progress towards new solutions in electrohydraulic drives. This is demonstrated in intensive introduction of advanced microelectronics. The introduction of microprocessor based control systems in hydraulic drives enables application of modern control theory. Thanks to this, today modern electrohydraulic drives can be regarded as both "strong" and "precise". Additionally, the drive connection to host controllers like CNC, PLC or industrial PC become easy. This facilitates application of electrohydraulic drives in different machines, presses, injection molding machines and many other devices. Nowadays, hydraulic drives enable moving of large masses, acting with large forces and simultaneously obtaining accuracy, comparable to electric drive systems, for example positioning accuracy of 1 μm and speed ratio of 1:10 000.

Electrohydraulic valves play a crucial role in every electrohydraulic drives. The first electrohydraulic servo drives were built in late forties of XX century. They

A. Milecki (✉) · J. Ortmann
Poznan University of Technology, Ul. Piotrowo 3, 60-965 Poznań, Poland
e-mail: Andrzej.milecki@put.poznan.pl

R. Szewczyk et al. (eds.), *Challenges in Automation, Robotics and Measurement Techniques*, Advances in Intelligent Systems and Computing 440, DOI 10.1007/978-3-319-29357-8_7

have used servo valves as control elements. Since the early eighties of XX century, the proportional valves have been used in hydraulic drives more and more widely. In these valves, as electromechanical transducers, proportional DC electromagnets are used. In comparison to servo valves, proportional valves are more cost effective, but their parameters, such as accuracy and frequency response, are a worse. Nowadays, the electrohydraulic valves are elements with a high grade of integration of the mechanical and electronic components, as well as computer based control elements and therefore such valves can be regarded as mechatronic devices. In electrohydraulic valve technology the design and development of high dynamic, low displacement electromechanical actuators is very important. In produced today servo valves torque motors are used and in proportional valves electromagnets are applied. However, in most cases these valves requires application of spool position measure element and special control valve. Moreover this valves does not enable to make small movements of the control spool. Therefore in some cases also stepping motors as electromechanical transformers are used [1–3]. Stepping motors are nowadays offered by various manufacturers in a wide range and with different control circuits. Electrohydraulic drives with the these motors used as electromechanical transducers have been produced for years by only a few companies like BW Hydraulic, Victory Controls, Curtiss Wright. Such a motor is able to make small steps with a frequency about a few kHz; thus to control valve gap cross section area with high accuracy. However, in order to fully open the valve, the stepping motor has to make thousands of steps, which takes several seconds. Therefore in order to improve the valve dynamics we proposed to use two stepping motors which work differentially.

2 Design of Proportional Valve with One and with Two Stepping Motors

At Poznan University of Technology investigations of stepping motors application in proportional valves has been conducted in the last few years [4].

Figure 1 shows physical model of one of the built there electrohydraulic servo drive in which proportional valve with stepping motor (1) is used as an initial element for spool displacement of a four edge hydraulic spool amplifier (3). The spool is connected with the stepping motor shaft by an elastic coupling element (2). On the other side of the valve, the spool is connected to the valve body by a thread (4), which turns the rotary displacement into linear and which is proportional to the number of stepping motor steps and. The shifting of the spool results in opening or closing of the gaps between spool and valve body. It results in the flow of oil, to and from the actuator chambers thus in displacement of the cylinder piston. Electrical position transducer (incremental encoder) mounted outside the cylinder is used to measure the current position of the mass moved by the hydraulic piston. The position is measured and compared with the input signal. According to the

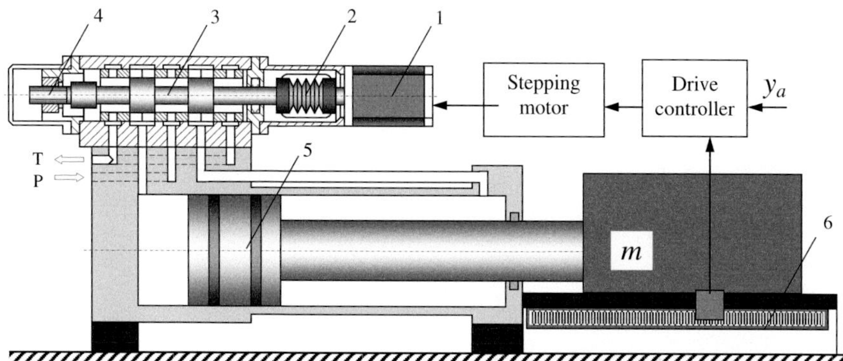

Fig. 1 Scheme diagram of the electrohydraulic servo drive with valve controlled by one stepping motor, where: *1*—stepping motor, *2*—elastic coupling, *3*—valve spool, *4*—thread, *5*—piston, *6*—incremental encoder

difference between those signals and to control algorithm, the controller produces an output signal, which is converted into a sequence of electric pulses given to the stepping motor coils.

The servo drive presented above was designed especially for low and very low velocities (less than 1 mm/s). In such case the phenomena that affect the drive behavior are different from those which affect normal work. One of the most important influence factor is contamination; their particles block the valve gaps. In order to assure the obtainment of low velocity control, the valve spool linear velocity was only 0.125 mm/s. As a result the servo drive was very accurate but not fast. Therefore the valve dynamics was very limited and it took to fully open the valve gaps as much as 5 s.

This was the main disadvantage of the drive and we decided to apply two stepping motors working differentially: one motor drives the screw and the second one the nut. If they are rotating in different directions the spool velocity was equal to the sum of both drives velocities. When they are rotating in the same directions with different velocities it was possible to obtain very low spool movement velocities. Additionally it is possible to apply stepping motors with different step angles and with different microsteps modes. The build and investigate out electrohydraulic servo drive with two stepping motors was both fast and accurate.

In Fig. 2a the scheme diagram of proposed by us four edge proportional hydraulic spool amplifier is shown. The first stepping motor (1) is used to rotate the valve spool (3), which ends with a screw (5). The valve spool is connected to the stepping motor shaft by an elastic coupling element (2), which enables the linear movement of the spool. This screw is placed in a nut, which is connected to the second stepping motor shaft (6). This nut can rotate because it is placed in bearings assembled to the valve housing. The photo of built out valve is shown in Fig. 2b. Both stepping motor drives can work in normal (full step), half-step or micro-stepping mode. In this solution the first stepping motor drives the spool and a

(a)

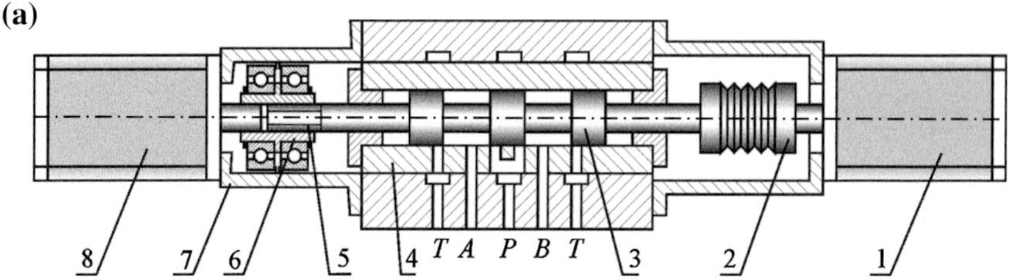

(b)

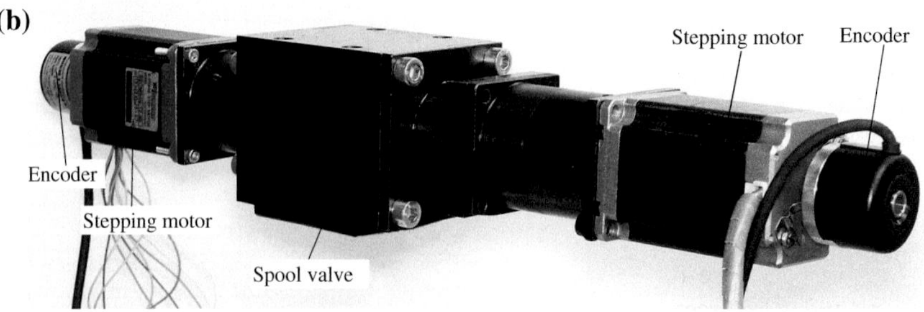

Fig. 2 Electrohydraulic spool valve with two stepping motors working differentially: **a** drawing, **b** photo

screw and a second stepping motor drives the nut. The rotation of both stepping motors caused the spool to linear movement. This linear movement is equal to the sum of angles θ_1 and θ_2, on which the motors rotors rotates. If these motors rotate with the same directions the linear spool movement is proportional to $\theta_1 - \theta_2$. In this case the linear spool movement can be controlled very accurately and its velocity can be very low. If the motors rotate in opposite directions this linear movement is proportional to $\theta_1 + \theta_2$. In this case the proposed valve is two times faster than the valve with single motor. In a normal mode both stepping motors worked with 0.36 degrees per full step or 1000 full steps per revolution. In order to assure this we used the microstepping mode, in which both motors can worked with 0.09 degrees per full step or 4000 full steps per revolution. The ballscrew pitch was equal to 2.5 mm.

3 Electrohydraulic Servo Drive with a New Valve

In Fig. 3 the scheme of the electrohydraulic servo drive with valve controlled by two stepping motors. Every stepping motor is controlled by motor control unit. The drive controller inputs are: assumed position y_a and measured position y_r.

The difference of this signals are the position error. The position error is the input signal to the controller. The described in this paper valve is completely new (far-ever found) and that's why for it a new, specially designed controller must be worked out. It should control simultaneously two stepping motors in order to achieve assumed spool position.

In Fig. 4 a block scheme of a drive is presented. Every bipolar 5-phases stepping motor type SECM569 T28 (holding torque 1.66 Nm, phase current 2.8 A) is governed by its own controller type SFC133, connected to a PC based main electrohydraulic drive controller, which calculates for each drive the number of steps to do. Two rotary encoders type MHK40 with resolutions of 3600 pulses/rev. measured the rotors positions which enable the stepping motor controllers to work in inner closed loops. The drives maximum frequency was equal 12 kHz.

The electrohydraulic cylinder position was measured by linear incremental encoder, which creates an outer feedback loop of the drive. At initial investigations we applied simple P-type controller (Fig. 4), which calculates the numbers of steps according to following equations

$$n_1 = k_{p1}e, n_2 = k_{p2}e. \quad (1)$$

Note, that because the motors are placed on opposite valve sides they in fact rotate in opposite directions when the signs of numbers n_1 and n_2 are the same, so they are working differentially. In the valve the backslash free ball screw type SEM-E-S 8 × 2.5R (Boshrexroth) was used. Its diameter was equal to 8 mm, pitch to 2.5 mm, nut length to 16 mm and screw length to 36 mm (accuracy T5). The

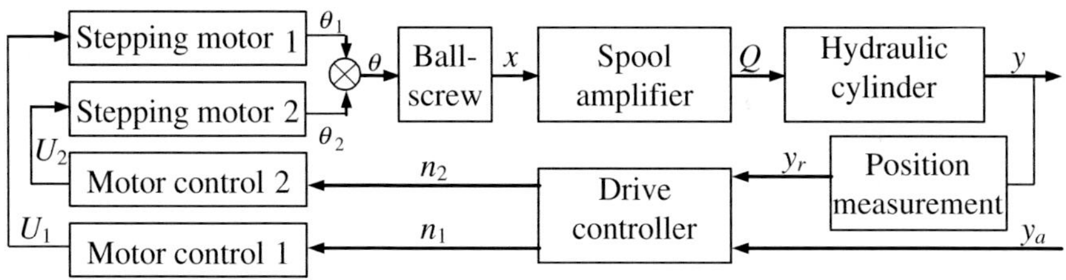

Fig. 3 Scheme of electrohydraulic servo drive with valve with two stepping motors

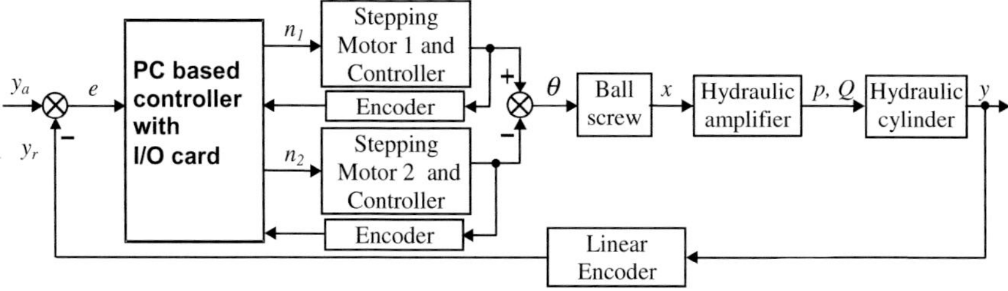

Fig. 4 Simulation results and errors; description see text

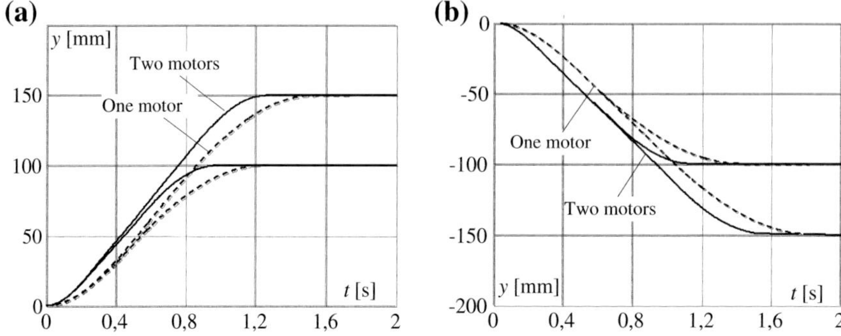

Fig. 5 Step responses of electrohydraulic servo drive controlled by the valve with one and two stepping motors for assumed displacements on 100 and 150 mm for a piston movement in: **a** forth direction (pushing), **b** back direction (pulling)

hydraulic amplifier spool diameter was equal to 10 mm. In amplifier body six gap windows with width of 2.5 mm and heights of 2.0 mm are made, which means that the gap length for every spool was equal to 6 mm. The hydraulic cylinder piston diameter was equal to 100 mm, piston rod diameter to 60 mm and stroke 400 mm. The electrohydraulic drive load was caused by the moved on the slideway mass equal to 100 kg. The supply pressure was 8 MPa, but it can be increased to 40 MPa; the resolution of the used incremental encoder was equal to 0.5 μm.

In order to compare the electrohydraulic servo drive with valve with single stepping motor with drive in which valve with two stepping motors are used the step responses are recorded, which are shown in Fig. 5. In this Figure step responses for two assumed distances 100 and 150 mm are shown. These step responses are recorded for two different piston movement directions. One can note that the step responses of servo drives with valve in which two stepping motors are used, are significantly faster. When one motor is used the settling time was equal to about 1.15 s for displacement 100 mm and 1.40 s for displacement 150 mm in one direction and accordingly 1.40 and 1.75 s for the second direction. When two motors are used the settling time was equal to about 0.95 s for displacement 100 mm and 1.25 s for displacement 150 mm in one direction and accordingly 1.05 and 1.45 s for the second direction. So, the settling time is about 20 % shorter when two stepping motors are used.

4 Conclusion

In the paper the new, electrohydraulic valve with two stepping motor is presented. In comparison to solutions with one stepping motor the new valve enables to obtain both low and relatively high valve spool velocity. A prototype electrohydraulic servo drive in which valve with two stepping motors was designed, constructed and investigated. The valve was used in electrohydraulic servo drive and the whole servo drive system was investigated. The application of a new valve improved the step responses settling time.

References

1. Murrenhoff, H.: Servohydraulik. Verlag Meinz, Aachen (1998)
2. Becker, M.: Schrittmotor als Aktuator fuer Hydraulik-Wegeventile, Oeilhydraulik und Pneumatic, (in German), No. 4, 2000
3. Milecki, A.: Elektrohydraulischer Servoantrieb mit Schrittmotor und Elektrische Rückführung, Ölhydraulik und Pneumatik, (in German), pp. 450–452, 6/1999
4. Milecki, A., Myszkowski, A.: Modelling of electrohydraulic servo drive used in very low velocity applications. Int. J. Model. Identif. Control **7**(3) (2009)

Controlling the Direction of Rotation of the Motor Using Brain Waves via Ethernet POWERLINK Protocol

Arkadiusz Kubacki, Arkadiusz Jakubowski, Dominik Rybarczyk and Piotr Owczarek

Abstract The paper presents preliminary results of the work with the Emotiv EPOC+™ system, which enabled in binary way control of objects. The system reads the brain waves from 14 plus 2 references electrodes. The controlled object was stepper motor with encoder released by the B&R company. For communication between the PC and engine control module was used Ethernet POWERLINK protocol, which allows data transfer with a minimum cycle time of 200 µs. Completely omitted PLC controller, which function was taken over the PC.

Keywords Emotiv EPOC+ · EEG · Ethernet POWERLINK · Stepper motor · Brain–computer interface

1 Introduction

Many studies regarding control of robots relates to their interaction with humans. Manual control gives way to verbal control or control using impulses from muscles (EMG) [1, 2]. Nowadays, EEG examination is normal and generally accessible therefore was taken the subject of object control using brain waves. The idea of control by "thinking" is nothing new especially in science-fiction movies, but in science occurs only a few years [3–16]. Currently, this form of control is confined

A. Kubacki (✉) · A. Jakubowski · D. Rybarczyk · P. Owczarek
Institute of Mechanical Technology, Poznan University of Technology,
Ul. Piotrowo 3, 60-965 Poznań, Poland
e-mail: arkadiusz.j.kubacki@doctorate.put.poznan.pl

A. Jakubowski
e-mail: arkadiusz.z.jakubowski@doctorate.put.poznan.pl

D. Rybarczyk
e-mail: dominik.rybarczyk@put.poznan.pl

P. Owczarek
e-mail: piotr.owczarek@put.poznan.pl

R. Szewczyk et al. (eds.), *Challenges in Automation, Robotics and Measurement Techniques*, Advances in Intelligent Systems and Computing 440, DOI 10.1007/978-3-319-29357-8_8

to moving the mouse cursor on the computer screen or characters through the virtual room [3, 5, 9]. Keep in mind that to control affect both psychological [17, 18] and neurological factors [17, 19]. A large impact on the results has also training. A person which first time has contact with this type of controller, in the first place must learn to repeat states of the brain, so as to allow recognize them. Another aspect is the identification of artefacts from the other parts of the body unrelated to the activity of the brain, e.g. heartbeat, eye movement or muscle activity.

2 Electroencephalography and Placement of Electrodes

Electroencephalography (EEG) is a noninvasive method of recording the electrical activity of the brain via electrodes placed on the skin of the skull [20].

Historically, the first registration of electroencephalogram from surface of the skull of a man was executed in 1925 by Hans Berger. Four years later he published them in the article "Uber das Elektrenkephalogramm des Menschen" [20]. Precisely in this article first time was used the name "electroencephalogram". Not to mention the Polish scientist Adolf Becku who discovered electroencephalogram as a side effect of their research in 1890 [20].

Today, the EEG uses a minimum of 10 evenly spaced electrodes on the scalp. Currently, the most widely used system placement of the electrodes is "10–20" system. It was proposed by Herbert H. Jasper in 1958 and shows the location of 21 electrodes. The name refers to the distance between the electrodes expressed as a percentage. This allows you to use the system regardless of the size of the head of the examined person [21] (Fig. 1).

The test can be performed on selected parts of the electrodes without some of them or their density in the area of interest to us [20].

In addition, during the EEG examination is used two additional reference electrode. Their location is not accidental and should be located in a place where

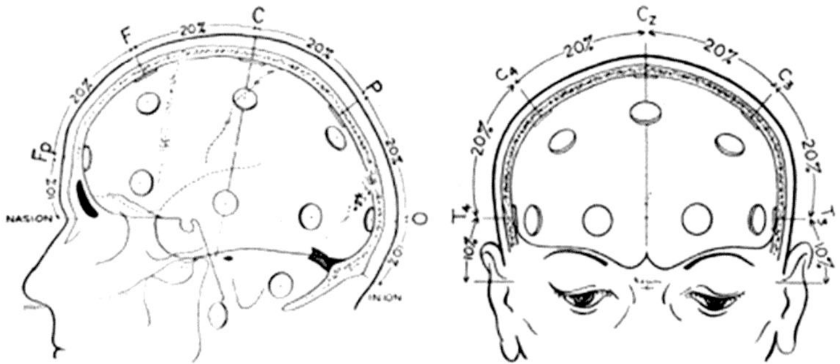

Fig. 1 Placement of electrodes in the "10–20" system [21]

bioelectric signals of the body will be similar to the one recorded by the electrodes. Most common are places located around the ears, forehead, nose or top of the head. In the last case we use only one reference electrode.

3 EEG Equipment Used in Research

The Emotiv EPOC+™ Headset is portable brain–computer interfaces based on electroencephalography with 14 plus 2 references electrodes situated in international system 10–20 comparable to a medical EEG system (Fig. 2).

It is equipped with electrodes: AF3, F7, F3, FC5, T7, P7, O1, O2, P8, T8, FC6, F4, F8, AF4. Placement of these electrodes is shown on picture below (Fig. 3).

Internal sampling rate of this system is 2048 Hz. Emotiv EPOC+™ operates at a resolution of 16 bit per channel and bandwidth is form 0.2 to 45 Hz. System has got additional digital notch filters at 50 and 60 Hz. It is equipped with built in digital 5th order Sinc filter. Communication between PC and Headset is wireless on 2.4 GHz band. It has LiPoly battery which is sufficient for 12 h [22]. In headset is mounted two axis gyroscope which can be use by programmer. With headset we get

Fig. 2 Emotiv EPOC™ Headset [22]

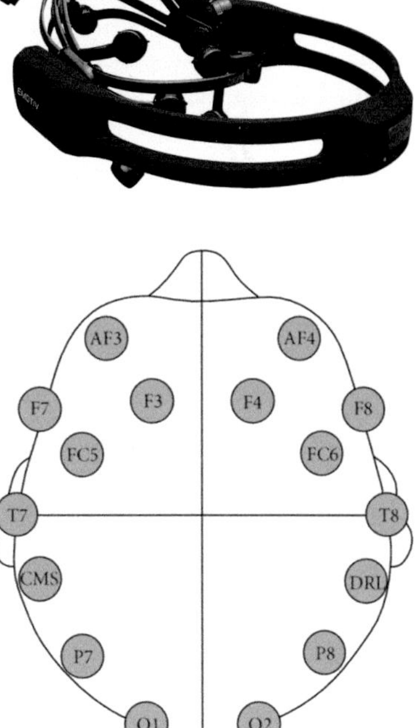

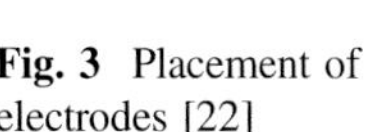

Fig. 3 Placement of electrodes [22]

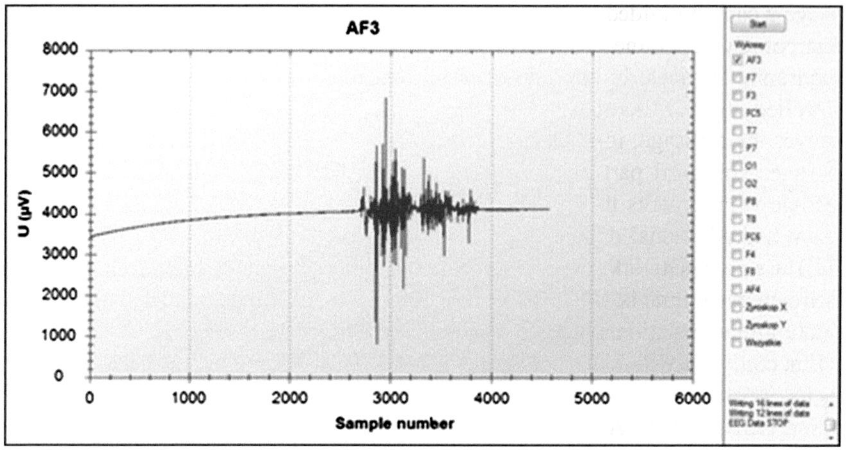

Fig. 4 Author's program to collect raw data from sensors

library which will allow to write application to collect raw data from sensors. Below is screen from program programmed by author (Fig. 4).

4 Sieć Ethernet POWERLINK

POWERLINK is deterministic and real-time variety of Ethernet protocol. Is compliant with IEE 802.3 standard. POWERLINK uses polling approach. Network always has one managing node which controls access. The others stations called controlled nodes can transmit data only upon request from the managing node. Thanks to all collisions have been eliminated [23] (Fig. 5).

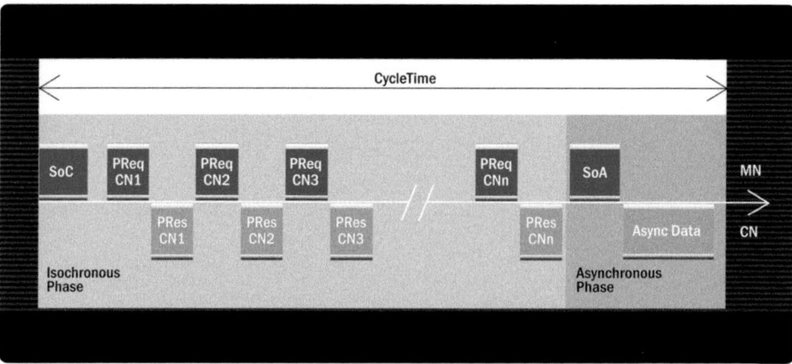

Fig. 5 Ethernet POWERLINK cycle [23]

Cycle can be divided into three periods. In the first period managing node send "Start of Cycle Frame" (SoC) to all controlled nodes. It does this in order to synchronize devices. In the second period managing node sends to the individual controlled node PollRequest (Preq) message. After that, when the controlled node receives the message, it sends data to all other stations using PollResponse (Presa) message. The third part of the cycle starts when managing node sends the SoA message which marks the beginning of the asynchronous phase. In this phase are transmitted additional data, such as parameters [23].

To a single network can be connected at the same time 240 stations and the minimum cycle time is 200 μs. During one cycle, it is possible to send a 1490 bytes of data into one station. Maximum asynchronous data transfer is 1500 bytes/cycle.

Ethernet POWERLINK is open protocol and can be implemented in any device. Open Source stack implementing the POWERLINK protocol called open-POWERLINK. This stack is both of managing node and of controlled node. Implementation of the POWERLINK protocol can be performed on operating systems like Windows, Linux or VxWorks and on controllers Altera and Xilinx FPGAs.

5 Controlling the Direction of Rotation of the Motor

The laboratory stand was designed to skip the PLC in the control of stepper motor. Below is an illustration which shows a scheme of stand (Fig. 6).

The test stand consists of a PC on which is running the openPowerlink library as managing node. As a POWERLINK interface using a standard network card integrated into the motherboard of the computer managing node. Managing node connects to the gateway X20BC0083 [24] released by the B&R company. In addition, it is plugged via X2X module to operate stepper motors X20SM1436 [25] and the digital output module X20DO9322 [26], which displays the connection

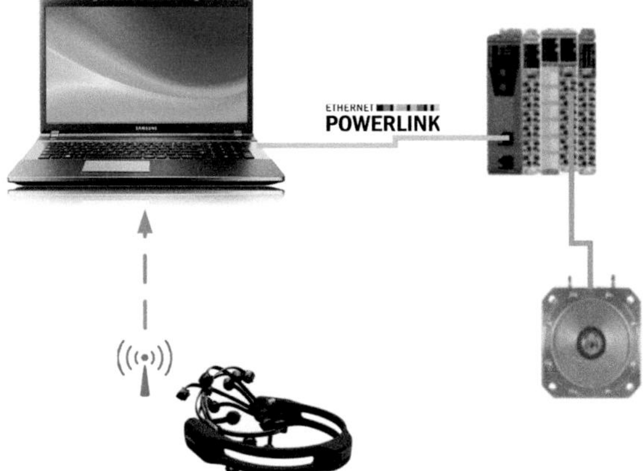

Fig. 6 Scheme of laboratory stand

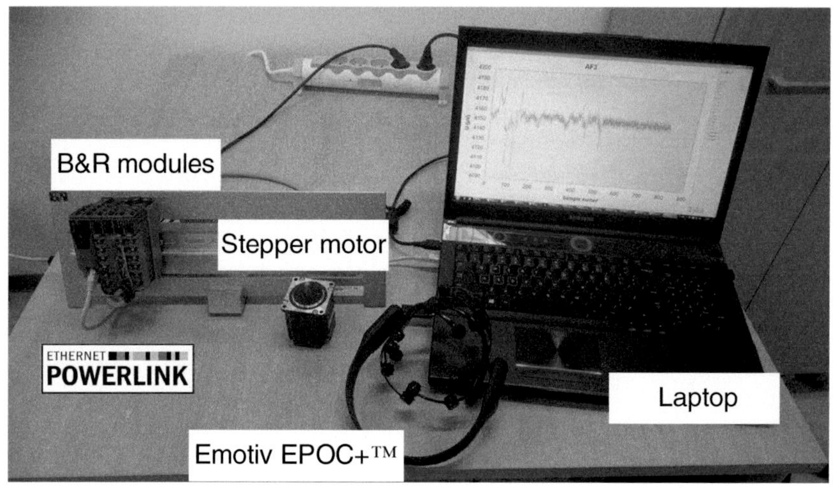

Fig. 7 Photography of the test stand

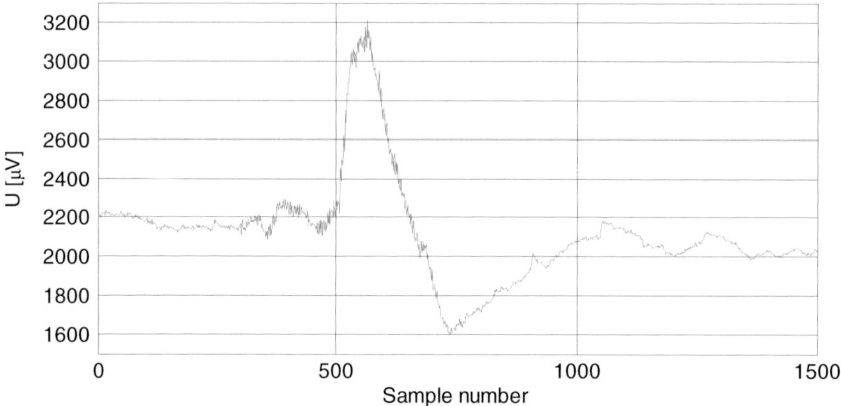

Fig. 8 Figure of AF3 electrode chart

status. To the module to control the stepper motor is attached a stepper motor with encoder 80MPD1.300S014–01 [27]. The engine features 200 steps per revolution. During the test the speed is set to 16 pulses per cycle. With a cycle time of 50 ms motor was rotated at 96 rpm. To receive brainwave was used earlier described Emotiv EPOC+™ Headset. Below is a photography of the final test stand (Fig. 7).

To control the direction of rotation is taken into consideration only one electrode AF3 located on the left side of the forehead of a patient. At this point, the research does not recognize individual thoughts of the focus or facial expressions. The condition for changing the direction was to cross the threshold of 2600 μV. The program activates the change on the rising edge of this signal. To achieve the required thresholds enough to properly focus the attention (Fig. 8).

6 Conclusion

In conclusion, it is possible to change the direction of rotation of the motor using brain waves. The test here presented is the basis of this further research that will have to find a way to recognize specific "thoughts" of man. It is inevitable to check the influence of artifacts on the same chart and possibly the opportunity to use them to control objects. In all studies should also take into account the possibility of creating a system that would learn the state of brain and add them to certain commands. It seems that the ideal for the above mentioned tasks would give artificial neural networks and fuzzy logic. In subsequent articles will be taken precisely this subject.

It is also worth mentioning POWERLINK protocol, which allows for very easy, and more importantly in a very fast way to transfer data and control elements of automation without a PLC which significantly reduces the response time to a given signal.

References

1. Bitzer, S., van der Smagt, P.: Learning EMG control of a robotic hand: towards active prostheses. In: Proceedings 2006 IEEE International Conference on Robotics and Automation. ICRA 2006, pp. 2819–2823 (2006)
2. Yang, D., Zhao, J., Y, Gu, Wang, X., Li, N., Jiang, L., Liu, H., Huang, H., Zhao, D.: An anthropomorphic robot hand developed based on underactuated mechanism and controlled by EMG signals. J. Bionic. Eng. **6**(3), 255–263 (2009)
3. Chapin, J.K., Moxon, K.A., Markowitz, R.S., Nicolelis, M.A.L.: Real-time control of a robot arm using simultaneously recorded neurons in the motor cortex. Nat. Neurosci. **2**(7), 664–670 (1999)
4. Nicolelis, M.A.L.: Brain–machine interfaces to restore motor function and probe neural circuits. Nat. Rev. Neurosci. **4**(5), 417–422 (Maj 2003)
5. Pfurtscheller, G., Neuper, C.: Motor imagery and direct brain-computer communication. Proc. IEEE **89**(7), 1123–1134 (Lipiec 2001)
6. Serruya, M.D., Hatsopoulos, N.G., Paninski, L., Fellows, M.R., Donoghue, J.P.: Brain-machine interface: instant neural control of a movement signal. Nature **416**(6877), 141–142 (2002)
7. Taylor, D.M., Tillery, S.I.H., Schwartz, A.B.: Direct cortical control of 3D neuroprosthetic devices. Science **296**(5574), 1829–1832 (2002)
8. Wessberg, J., Stambaugh, C.R., Kralik, J.D., Beck, P.D., Laubach, M., Chapin, J.K., Kim, J., Biggs, S.J., Srinivasan, M.A., Nicolelis, M.A.L.: Real-time prediction of hand trajectory by ensembles of cortical neurons in primates. Nature **408**(6810), 361–365 (2000)
9. Wolpaw, J.R., McFarland, D.J., Vaughan, T.M.: Brain-computer interface research at the wadsworth center. IEEE Trans. Rehabil. Eng. **8**(2), 222–226 (2000)

10. Lee, J.-H., Ryu, J., Jolesz, F.A., Cho, Z.-H., Yoo, S.-S.: Brain–machine interface via real-time FMRI: preliminary study on thought-controlled robotic arm. Neurosci. Lett. **450**(1), 1–6 (2009)
11. McFarland, D.J., Sarnacki, W.A., Wolpaw, J.R.: Electroencephalographic (EEG) control of three-dimensional movement. J. Neural Eng. **7**(3), 036007 (2010)
12. Millan, JdR, Renkens, F., Mourino, J., Gerstner, W.: Noninvasive brain-actuated control of a mobile robot by human EEG. IEEE Trans. Biomed. Eng. **51**(6), 1026–1033 (2004)
13. Oberman, L.M., McCleery, J.P., Ramachandran, V.S., Pineda, J.A.: eeg evidence for mirror neuron activity during the observation of human and robot actions: toward an analysis of the human qualities of interactive robots. In: Neurocomputing, Selected Papers from the 3rd International Conference on Development and Learning (ICDL 2004) Time Series Prediction Competition: The CATS benchmark 3rd International Conference on Development and Learning, vol. 70, no. 13–15, pp. 2194–2203 (2007)
14. Onose, G., Grozea, C., Anghelescu, A., Daia, C., Sinescu, C.J., Ciurea, A.V., Spircu, T., et al.: On the Feasibility of using motor imagery EEG-based brain–computer interface in chronic tetraplegics for Assistive robotic arm control: a clinical test and long-term post-trial follow-up. Spinal Cord **50**(8), 599–608 (Sierpie 2012)
15. Ranky, G.N., Adamovich, S.: Analysis of a commercial EEG device for the control of a robot arm. In: Proceedings of the 2010 IEEE 36th Annual Northeast Bioengineering Conference, pp. 1–2 (2010)
16. Wolpaw, J.R, Birbaumer, N., McFarland, D.J., Pfurtscheller, G., Vaughan, T.M.: Brain–computer interfaces for communication and control. Clin. Neurophys. **113**(6), 767–791 (2002)
17. Güler, İ., Übeyli, E.D.: Adaptive neuro-fuzzy inference system for classification of EEG signals using wavelet coefficients. J. Neurosci. Methods **148**(2), 113–121 (Październik 2005)
18. Ito, M., Tani, J.: On-Line imitative interaction with a humanoid robot using a dynamic neural network model of a mirror system. Adapt. Behav. **12**(2), 93–115 (2004)
19. Lindblom, J., Ziemke, T.: Social situatedness of natural and artificial intelligence: Vygotsky and beyond. Adapt. Behav. **11**(2), 79–96 (2003)
20. Klekowicz, H.: Opis i identyfikacja struktur przejściowych w sygnale EEG (2008)
21. Zyss, T.: Zastosowanie układu 10–20 w rozmieszczaniu elektrod do EEG. Przedsiębiorstwo Informatyki Medycznej ELMIKO (2007)
22. Emotiv EPOC Specifications. https://emotiv.com
23. POWERLINK Configuration and Diagnostics. https://www.br-automation.com
24. Datasheet X20(c)BC0083. https://www.br-automation.com
25. Datasheet X20SM1436. https://www.br-automation.com
26. Datasheet X20(c)DO9322. https://www.br-automation.com
27. Stepper motors user's manual. https://www.br-automation.com

System Responsive to ICT Security Incidents in the LAN

Marian Wrzesień and Piotr Ryszawa

Abstract The system responsive to ICT security incidents in the LAN is presented. The system is implemented in the Industrial Research Institute for Automation and Measurements PIAP, Poland. In everyday practice, while processing of information, it is necessary to deal with IT security incidents. They are single events or series of events related to the security of classified information. They threaten the confidentiality, availability and integrity of information. Implementation of the system required the following steps: choice of the sources of information based on which assessment of the level of threats will be carried out, definition of the method of analysis of data from the chosen sources of information, definition of a single, common storage of information about threats, realization of the feedback which will introduce modifications in the router in order to counteract threats. For the sources of information were selected: continuously recorded packet flows from CISCO router, information from RBL servers, web server logs and current behavior of LAN, observed by the system administrator. The methods of data analysis included: threat assessment based on the analysis of flows in the router, assessment of the level of threats based on a web server log analysis and assessment of risks in router and web server based on information from the RBL servers. For storage of data coming from sources of information, the MySQL database was used. The essence of feedback is a self-acting modification of Access Control List (ACL) on the CISCO router. As a result of the research work, the system was implemented which attempts to automatically remove LAN security threats.

Keywords Security · Netflow · ACL · Monitoring

M. Wrzesień (✉) · P. Ryszawa
Industrial Research Institute for Automation and Measurements PIAP, Warsaw, Poland
e-mail: Mwrzesien@piap.pl

P. Ryszawa
e-mail: pryszawa@piap.pl

R. Szewczyk et al. (eds.), *Challenges in Automation, Robotics and Measurement Techniques*, Advances in Intelligent Systems and Computing 440, DOI 10.1007/978-3-319-29357-8_9

1 Introduction

In everyday practice, while processing information, we have to deal with incidents of IT security, that is, single events or series of events related to the security of classified information, which threaten the confidentiality availability and integrity of the information. Due to the variety of security incidents, they were divided because of their nature and effects under the Regulations of the Prime Minister of Poland [1, 2]. The PIAP carried out actions that improve internal IT security. Introduction of restrictions on access from specific locations in the network, was preceded by an analysis of degree of impact of the WAN users on the systems in Industrial Research Institute for Automation and Measurements PIAP.

The aim of the work described in this article was to create a Responsive Software System for ICT Security Incidents in the LAN (RSSISI). The assumption was to build a system using existing network infrastructure without installing additional hardware devices, and to use existing information about flow of packets in the local network. Due to the functionality of the PIAP servers, the only ports vulnerable to incidents are network ports 22 and 80 of the web server, and network ports 587 and 995 of the mail server. The motivation of the study is associated with observed increased potentially unsafe network activity on those ports.

1.1 Review of Existing Solutions

On the market there are many solutions to protect IT resources from intruders. The most commonly used systems include: firewall, IDS (Intrusion Detection System), IPS (Intrusion Prevention System) [3]. Presented solution complements the commonly used security systems. The RSSISI system for risk assessment uses NetFlow data [4], pre-processed data from a web server and data from RBL servers (Real-time Blackhole List). The network security system based on NetFlow analysis was presented in article of Krmicek and Vykopal [5]. The "NetFlow Based Network Protection" system based solely on the analysis of NetFlow flows [6, 7]. The advantage of the system described in the present work is the use of threat information coming from different sources and assessment of risks using RBL servers.

1.2 Assumptions for the RSSISI System Design

During operation the system takes into account the following aspects:

1. Choice of sources of information on the basis of which assessment of the threat level will be carried out.
2. Method of analysis of data from the chosen sources of information.

3. Definition a single storage for information about threats.
4. Operation of the feedback that introduces modifications in the router in order to counteract the detected threats.

The system architecture is presented in Fig. 1.

1.3 Sources of Information

The system may have multiple sources of information about the risks. In the currently deployed version of the system, sources of information are:

- continuously recorded flows from the CISCO router [8],
- information from RBL servers,
- web server logs,
- current observations of LAN behavior by the system administrator.

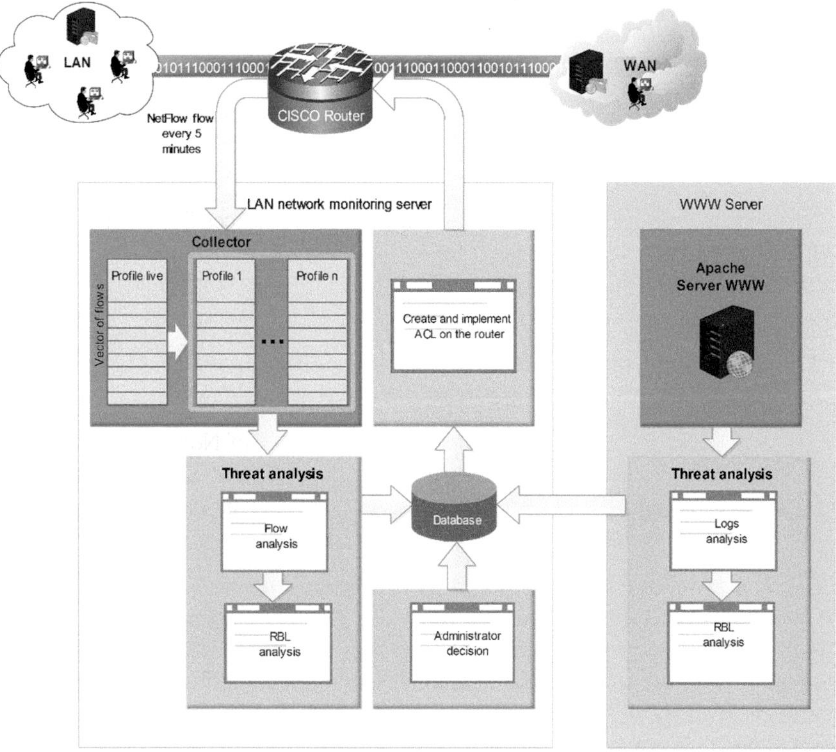

Fig. 1 System architecture

After analysis of the information from the mentioned sources, instructions for modification of the Access Control List (ACL) on the router are developed.

1.4 The Method of Data Analysis

Data processing consists in:

- Threat level assessment based on analysis of the flows in the router.
- Threat level assessment based on analysis of the web server logs.
- Assessment of risks in the router and the web server based on information from the RBL servers, which provide information about the IP addresses of servers that are sources of spam, viruses and other threats.

In addition, the data used to make decision may come from the network administrator.

1.5 Storage Location

A solution based on MySQL database located on a monitoring server implemented in the LAN was adopted as a place of storing information about the threats.

1.6 Feedback

The essence of feedback is a self-acting software modification of the Access Control List (ACL) on the CISCO router. To make it possible, a method of updating the ACL on the router without administrator intervention was developed.

2 Collection of the LAN Flows

The flow is an unidirectional stream of packets transmitted between the source and the target. Source and target are defined by the IP addresses (the network layer of the OSI model) and port numbers (transport layer of the OSI model). They can be assigned to both the LAN user and external sender. Flows in PIAP-LAN networks are collected in the LAN collector on the monitoring server, using the NetFlow protocol which is implemented in the CISCO router. The collector is equipped with OpenSource software—nfdump (version 1.6.8) and NfSen version (1.3.6). The NetFlow system purpose is gathering information about flows and their analysis (Fig. 2).

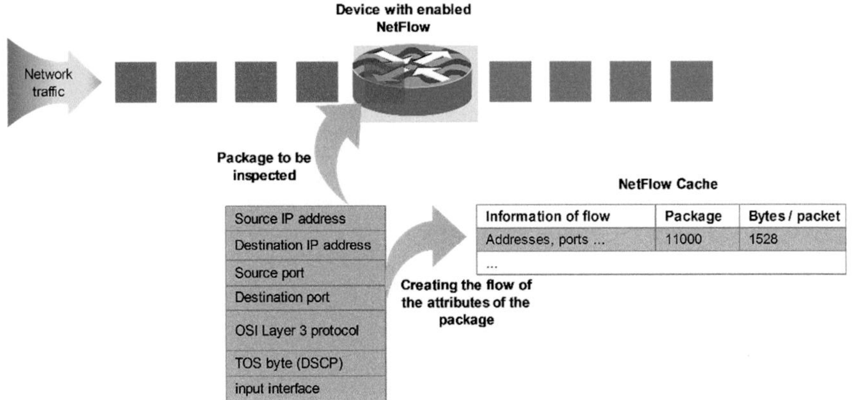

Fig. 2 Formation of NetFlow flow

Information about the packages analyzed in the CISCO router buffer are added to the router cache (called NetFlow Cache), and then sent to the collector. This operation is performed every 5 min [7]. The collector daemon nfcapd creates a vector flow (called Profile Live) by using nfprofile. The vector flow is divided into profiles defined by the network administrator. The purpose of the division into profiles is the ability to extract flows belonging to different recipients. Thus, for example, companies to which PIAP provides the Internet service, which fall within the profile 2 (NaskAC), are not protected, because PIAP does not interfere with internet access politics of the commercial customers. However, it is still possible to introduce protection against threats similar as in PIAP on individual request of the Internet service recipient [9, 10].

2.1 *Organization of the Collecting Cash*

Data from the live profile are selected based on the interface index (VLAN) or sub-interface of the router.

The nfdump software implemented in the collector contains the following tools:

- Nfcapd—receives Netflow packets from the router and saves them in a file in the collector,
- Nfdump—analyzes data stored by nfcapd in files,
- Nfprofile—creates a profile by selecting data stored in files by nfcapd and stores the results in new files,
- Nfreplay—for replication of profiles, reads files and sends them to the specified host on the network,
- Nfclean—cleans the profiles of outdated data (PIAP adopted 6-month data aging period).

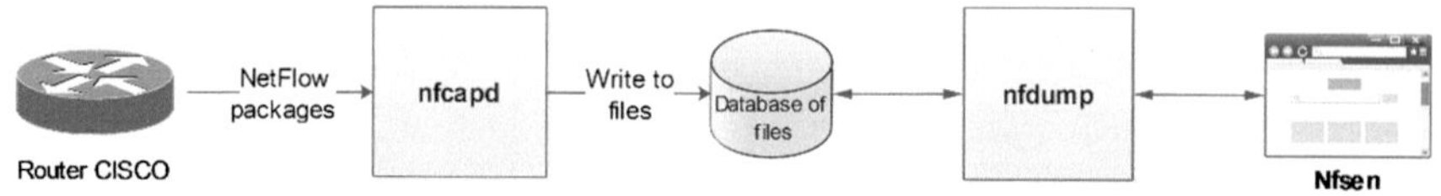

Fig. 3 The nfdump and NfSen architecture

Collector settings are stored in the nfsen.conf configuration file.

NfSen program does not provide user authentication mechanism. In order to ensure secure access to this data flow, authentication and restriction of IP addresses authorized to use the web user interface should be implemented at the level of the Apache configuration.

In Fig. 3 nfdump and NfSen architecture is presented.

3 Analysis of Flows and Automated Risk Prevention

Figure 4 presents the UML diagram of the operation of the RSSISI system, implemented in a LAN monitoring server [11].

According to the diagram, risk analysis is restricted to the profile 1, covering all devices in the PIAP-LAN network. The analysis is performed automatically at a time interval of 5 min.

In accordance with the accepted methods of flow analysis, apart from analyzing flows of individual profiles, it is also possible for the network administrator to perform manual analysis using nfdump tool directly in the collector via Linux system console. Both in manual and automatic method of analysis, filters available in the nfdump program can be used. In order to simplify the flow analysis process, the script analizaNetFlow.sh was developed (Fig. 5). This tool allows analysis of the amount of data downloaded ("top"), flow analysis ("flow") and flow analysis using appropriate formatting of results for the purposes of automatic risk analysis ("flow1").

For the purposes of automated threat analysis on the LAN, scripts analiza1.sh, rblcheck.sh, analiza2.sh, analiza3.sh, acl1.sh were developed. These scripts are launched every 5 min using cron. The cron launches run1.sh script (Fig. 6), which executes a sequence of other scripts.

The sequence of scripts creating and implementing an updated ACL on the router is run by the run2.sh script (Fig. 7).

During the execution of the analiza1.sh script (Fig. 8), the analysis of network traffic from external sources initiating transmission on ports: 21, 22, 23, 445, 587, 995, 1433, 3306, 3389 is performed. These ports were chosen because of the common interest of hackers. Network traffic on port 80 is analyzed in the WEB server.

IP addresses separated by flow analysis, are evaluated using RBL servers (rblcheck.sh script, Fig. 9). The RBL server list consists of 55 addresses. While

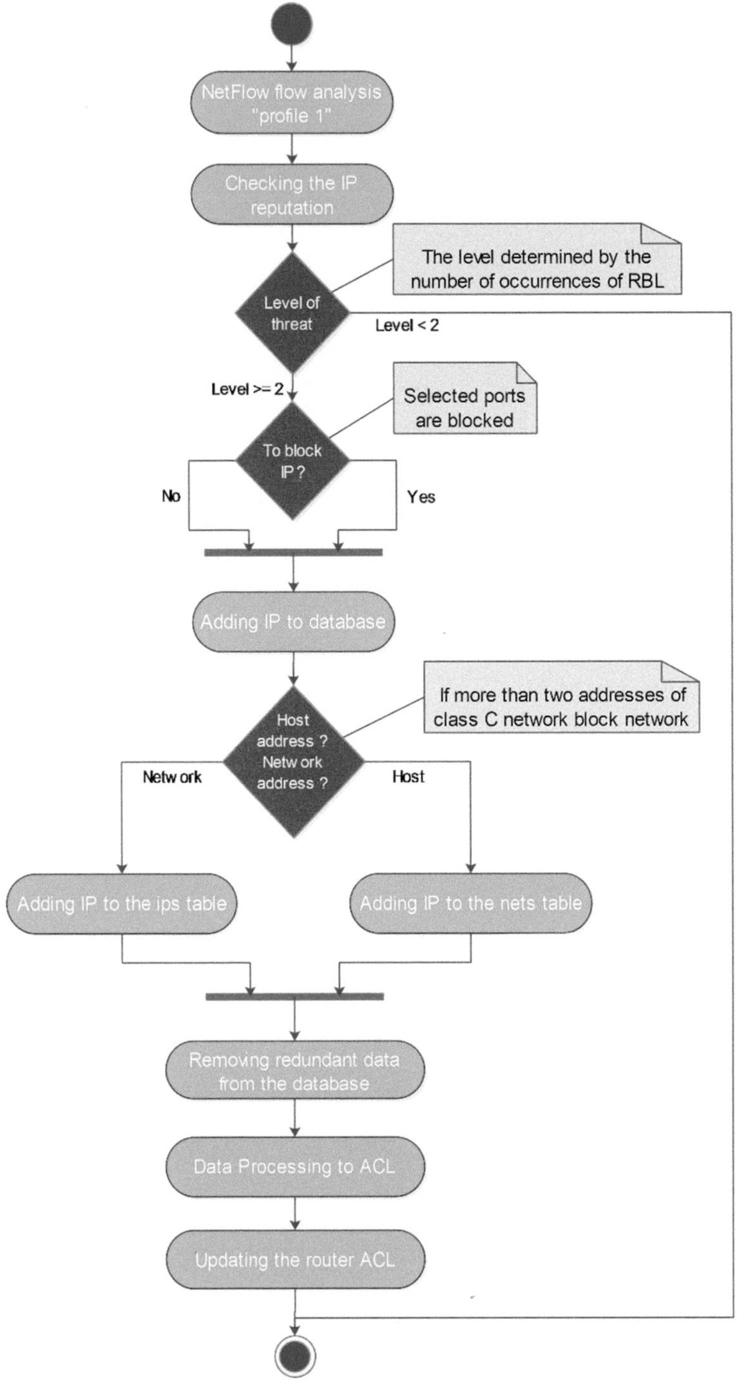

Fig. 4 The analysis and prevention of LAN security incidents

```
file1=$(ls -t /home/nfsen/profiles-data/$src/${src}_in/$dataRMD/ | head -1)
timepr=$(($d*1440+$m))
file2=$(find /home/nfsen/profiles-data/$src/${src}_in/$year/$month/$day/ -type f
-mmin -$timepr | sort | head -1 | awk -F"/" '{ print $10 }')
parameters=""
if [ "$parameters" == "" ]; then
      filter="$ip"
else
      filter="$parameters and $ip"
fi
filter="$parameters $ip"
if [ "$1" == "top" ]; then
        optionsnfdump="-M /home/nfsen/profiles-data/$src/${src}_out:${src}_in
-T  -R $year/$month/$day/$file2:$dataRMD/$file1 -n 30 -s ip/bytes"
elif [ "$1" == "flow" ]; then
        optionsnfdump="-M /home/nfsen/profiles-data/$src/${src}_out:${src}_in
-T  -R $year/$month/$day/$file2:$dataRMD/$file1 -c 30000"
elif [ "$1" == "flow1" ]; then
        optionsnfdump="-M /home/nfsen/profiles-data/$src/${src}_out:${src}_in
-T  -R $year/$month/$day/$file2:$dataRMD/$file1 -a  -A srcip -o fmt:%sa -c
30000"
```

Fig. 5 A snippet of analizaNetFlow.sh script

```
#!/bin/bash
/home/admin/analizaip/bin/analiza1.sh >> /home/admin/analizaip/Data/ip.txt
/home/admin/analizaip/bin/analiza2.sh
/home/admin/analizaip/bin/run2.sh
```

Fig. 6 The run1.sh script

```
#! /bin/bash
database="address"
DateN=`date +%Y-%m-%d:%H:%M:%S -d "-5 minutes"`
queryip="SELECT data FROM ips ORDER BY id DESC limit 1"
querynet="SELECT data FROM nets ORDER BY id DESC limit 1"
resultip=`mysql --login-path=address ${database} -Bse "${queryip}"`
resultnet=`mysql --login-path=address ${database} -Bse "${querynet}"`

if [[ $resultip > $DateN  ]] || [[ $resultnet > $DateN ]]; then
      /home/admin/analizaip/bin/analiza3.sh
        /home/admin/analizaip/bin/acl1.sh
        /home/admin/analizaip/bin/cisco1.sh
fi
```

Fig. 7 The run2.sh script

```
#!/bin/bash
ports=(21 22 23 445 587 995 1433 3306 3389)
pat="(^(192\.168\.)|^(10\.)|^(172\.1[6789]\.)|^(172\.2[0-9]\.)|^(172\.3[01]\.)|^(195\.187\.100\.)|^(195\.187\.148\.))"
for p in "${ports[@]}"
do
      address=($(/home/admin/analizaip/bin/analizaNetFlow.sh flow1 5 0 NaskAC 'dst port' $p | awk '{match($0,/[0-9]+\.
[0-9]+\.[0-9]+\.[0-9]+/); ip = substr($0,RSTART,RLENGTH); print ip}'))
      for i in "${address[@]}"
            do
            if [[ ! $i =~ $pat ]] ; then
                  /home/admin/analizaip/bin/rblcheck.sh $i 1 $p
            fi
      done
done
```

Fig. 8 The analiza1.sh script

designing the system, it was assumed that the presence of the tested IP address reported by any two RBL servers, is sufficient to add the tested IP to the list of dangerous IP addresses.

As a result of the analiza2.sh script execution (Fig. 10), dangerous IP addresses are entered into the database divided into individual IP addresses and Class C network. In order to reduce the number of entries in the ACL, IP addresses are automatically aggregated inside the Class C network. The policy is adopted that

```
#!/bin/bash
ip="$1"
RBL="bl.spamcop.net cbl.abuseat.org b.barracudacentral.org dnsbl.sorbs.net http.dnsbl.sorbs.net dul.dnsbl.sorbs.net
misc.dnsbl.sorbs.net smtp.dnsbl.sorbs.net socks.dnsbl.sorbs.net spam.dnsbl.sorbs.net web.dnsbl.sorbs.net
zombie.dnsbl.sorbs.net dnsbl-1.uceprotect.net dnsbl-2.uceprotect.net dnsbl-3.uceprotect.net pbl.spamhaus.org
sbl.spamhaus.org xbl.spamhaus.org zen.spamhaus.org bl.spamcannibal.org psbl.surriel.com ubl.unsubscore.com
rbl.spamlab.com dyna.spamrats.com noptr.spamrats.com spam.spamrats.com cbl.anti-spam.org.cn cdl.anti-spam.org.cn
dnsbl.inps.de drone.abuse.ch httpbl.abuse.ch dul.ru korea.services.net short.rbl.jp virus.rbl.jp spamrbl.imp.ch
wormrbl.imp.ch virbl.bit.nl rbl.suresupport.com dsn.rfc-ignorant.org ips.backscatterer.org spamguard.leadmon.net
opm.tornevall.org netblock.pedantic.org multi.surbl.org ix.dnsbl.manitu.net tor.dan.me.uk rbl.efnetrbl.org
dnsbl.dronebl.org access.redhawk.org db.wpbl.info rbl.interserver.net query.senderbase.org bogons.cymru.com
csi.cloudmark.com"
r_ip=$(echo $ip|awk -F"." '{for(i=NF;i>0;i--) printf i!=1?$i".":"%s",$i}')
rblcount=0
for rbl in $RBL
        do
#               echo $r_ip.$rbl
                result=$(dig +short $r_ip.$rbl)
                if [ ! -z "$result" ]
                then
                        #echo "$ip is in $rbl with code $result"
                        rblcount=$(($rblcount+1))
#               else
                        #echo "$rbl             clear";
                fi
done
if [ $rblcount -gt $2 ]
then
        echo $ip $rblcount
fi
```

Fig. 9 The rblcheck.sh script

```
#!/bin/bash
ipfile="/home/admin/analizaip/Data/ip.txt"
ipfiles="/home/admin/analizaip/Data/ips.txt"
from="monitor"
blockPort=(22 80 587 995)
i=0
database="address"
if [[ " ${blockPort[*]} " == *" $port "* ]]; then
    echo "blockPort contains $port"
fi
cat $ipfile | awk '{ print $1 " " $3 }' | sort -u > $ipfiles
while read ip
do
        IFS=" " read ipx port <<< "$ip"
        IFS=. read ip1 ip2 ip3 ip4 <<< "$ipx"
        net=$ip1.$ip2.$ip3
        if [[ " ${blockPort[*]} " == *" $port "* ]]; then
                block=1
        else
                block=0
        fi
        if [[ `grep $net $ipfiles | wc -l` -ge 2 ]]; then
                query="SELECT * FROM nets WHERE net LIKE '$net.0'"
                if [[ ! `mysql --login-path=address ${database} -Bse "${query}"` ]]; then
                        location=`geoiplookup $net | sed 's/GeoIP //g' | sed 's/ Edition//g' | sed 's/'\''/ /g'`
                        DATE=`date +%Y-%m-%d:%H:%M:%S`
                        query="insert into nets (net,location,source,port,block,data) values ('$net.0', '$location', '$from', '$port',
'$block', '$DATE')"
                        mysql --login-path=address ${database} -e "${query}"
                fi
        networks1+=($net)
        else
                queryi="SELECT * FROM ips WHERE ip LIKE '$ipx'"
                queryn="SELECT * FROM nets WHERE net LIKE '$net.0'"
                if [[ ! `mysql --login-path=address ${database} -Bse "${queryi}"` ]] && [[ ! `mysql --login-path=address ${database} -Bse
"${queryn}"` ]]; then
                        location=`geoiplookup $ipx | sed 's/GeoIP //g' | sed 's/ Edition//g' | sed 's/'\''/ /g'`
                        DATE=`date +%Y-%m-%d:%H:%M:%S`
                        query="insert into ips (ip,location,source,port,block,data) values ('$ipx', '$location', '$from', '$port', '$block',
'$DATE')"
                        mysql --login-path=address ${database} -e "${query}"
                fi
        ips1+=($ipx)
        fi
done <$ipfiles
```

Fig. 10 The analiza2.sh script

only two dangerous network IP addresses are enough to assume that the whole Class C network is potentially dangerous.

The blocked IP addresses (recommended to be blocked) are those, that are found to communicate on ports 22, 80, 587 and 995. The choice of ports 22, 80, 587 and 995 is associated with their use by the services available from the Internet.

Before creating the ACL list, redundant information must be removed from the database. For this purpose the analiza3.sh script is launched (Fig. 11). All addresses that already exist in the table of network addresses (nets) are removed from the table that contains IP addresses (ips).

After placing the information in the database, acl1.sh script is launched (Fig. 12), which creates the ACL in a form acceptable to the CISCO router.

In order to automatically update the ACL which contains dangerous addresses, the script cisco1.sh (Fig. 13) is launched. This script uses a scripting language called Expect. This language was created in order to automate activities performed through interactive programs such as telnet or ftp. In the case presented in the article, the system uses the secure ssh connection to execute the commands and scp command to upload ACL from the server to router memory.

```
#!/bin/bash
database="address"
query="SELECT net FROM nets WHERE block = 1"
while read net1
do
      IFS=. read ip1 ip2 ip3 ip4 <<< "$net1"
        net=$ip1.$ip2.$ip3
      queryd="DELETE FROM ips WHERE ip LIKE '$net.%'"
      mysql --login-path=addressA ${database} -Bse "${queryd}"
      echo $queryd
done < <(mysql --login-path=addressA ${database} -Bse "${query}")
```

Fig. 11 The analiza3.sh script

```
#!/bin/bash
AccessList=/home/admin/analizaip/Data/zabronioneNaskAC
database="address"
query="SELECT net FROM nets WHERE block = 1"
echo -n > $AccessList
echo "ip access-list extended zabronioneNaskAC" > $AccessList
while read net
do
      echo -e " deny   ip $net 0.0.0.255 any" >> $AccessList
done < <(mysql --login-path=address ${database} -Bse "${query}")
query="SELECT ip FROM ips WHERE block = 1"
while read ip
do
      echo -e " deny   ip host $ip any" >> $AccessList
done < <(mysql --login-path=address ${database} -Bse "${query}")
echo " permit ip any any" >> $AccessList
```

Fig. 12 The acl1.sh script

Fig. 13 The snippet of cisco1.sh script

```
#!/usr/bin/expect -f
set hostname "router.piap.pl"
set username "**********"
set password "**********"
set enablepassword "**********"
set timeout 30
spawn /usr/bin/scp /home/admin/analizaip/Data/zabronioneNaskAC
$username@$hostname:zabronioneNaskAC
                    ⋮
expect "*#"
send "copy flash:zabronioneNaskAC running-config\n\n"
expect "running-config"
send "conf t\n"
expect "(config)#"
send "interface vlan 50\n"
expect "(config-if)#"
send "ip access-group zabronioneNaskAC out\n"
expect "(config-if)#"
send "ip access-group zabronioneNaskAC in\n"
expect "(config-if)#"
send "end\n"
```

4 Hazard Data Storage

MySQL database is used for storing processed data about threats. The database uses a mechanism that allows scaling the system. This feature allows the use of additional sources of information about the risks. The database contains two tables that store information about hazardous IP addresses and networks from which the IP addresses come from. The tables contain information such as:

- IP address of the host or network,
- threat geographic location,
- the source from which comes the threat,
- port on which the network communication occurred in LAN,
- decision on whether to block the IP address as a danger,
- date when the IP address was added to the database.

In order to ensure appropriate level of security, in the database were created two users with different access to the database. To communicate with different sources of information about threats, the user who has rights only to add new records to the database is designated. The second user is granted full rights to the database subject to log on locally.

5 Analysis of Risks in the WEB Server

In Fig. 14, a draft of the system in the WEB server is presented.

A Python script developed by Romain Gaucher has been applied to analyze and detect threats for the PIAP Web site Server. The script is used to analyze the Apache server logs. In the system described in the present article, the fifth version of the script (scalp-05.py) is used. The script is a form of IDS (Intrusion Detection System) application, which is running on user computer. It uses the access log to selected web server pages in PIAP. Scalp-05.py is used in the analiza0.sh script (Fig. 15). This script contains tables of the analyzed websites logs. It analyzes the information at a frequency of 10 min, defined by the system administrator.

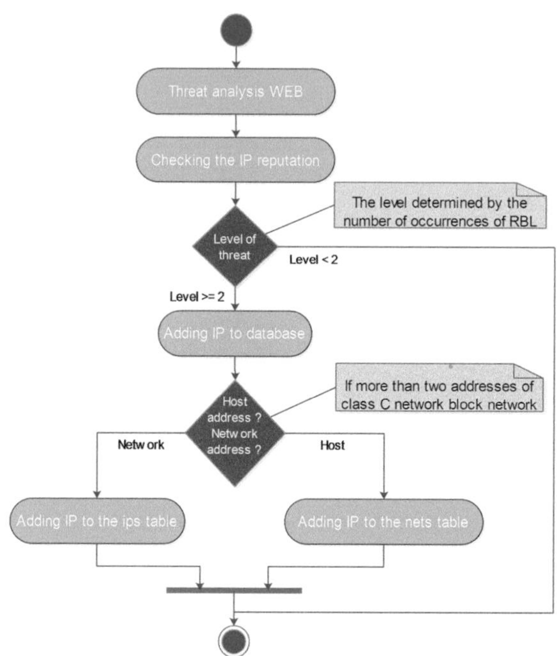

Fig. 14 Projekt systemu serwer WEB

```
#!/bin/bash
LANG=en_EN
startD=$(date +%d/%b/%Y:%H:%M -d "-10 minutes")
endD=$(date +%d/%b/%Y)
p="$startD;$endD"
echo $p
logs=(
      "/var/log/httpd/www.automaticon.pl-access_log"
      "/var/log/httpdPIAP/www.piap.pl-access_log"
      "/var/log/httpd/www.par.pl-access_log"
      "/var/log/httpd/www.design.piap.pl-access_log"
)
rm -rf /home/admin/Security/Data/scalp-out/*.txt
for i in "${logs[@]}"
do
/home/admin/Security/bin/scalp-05.py -l $i -f
/home/admin/Security/Data/default_filter.xml -o
/home/admin/Security/Data/scalp-out --period "$p"
        echo $i
done
```

Fig. 15 The analiza0.sh script

```
#!/bin/bash
address=($(cat /home/admin/Security/Data/scalp-out/*.txt | awk '{match($0,/
[0-9]+\.[0-9]+\.[0-9]+\.[0-9]+/); ip = substr($0,RSTART,RLENGTH); print ip}' |
sort -u ))
for i in "${address[@]}"
do
      /home/admin/Security/bin/rblcheck.sh $i 1
done
```

Fig. 16 The analiza1.sh script

IP addresses are extracted from the log files generated by the script analiza0.sh (Fig. 15) and evaluated in the same manner as during analysis of flows in the LAN (Fig. 16).

6 Summary

The study used a router and web server running CISCO and Open Source software. As a result of the research, System responsive to ICT security incidents in LAN was implemented. During tests lasting 146 days, 585 Class C network addresses and 2342 individual IPs have been blocked. The analysis of the information collected in the database, reveals that the majority of LAN security breach attempts took place outside working hours of the network administrator. Between 8 am and 4 pm hours the system has blocked 790 individual IP addresses and 178 Class C networks, and between 4 pm and 8 am hours, the system has blocked 1552 individual IP addresses and 612 class C networks. The result of the above addresses being blocked was reduction in severity of harmful effects of attacks on the LAN resources. Therefore, in order to enhance security during absence of network administrator, it is necessary to automate both the analysis process and application of adequate protection measures against intruders. The distinguishing feature of the described solution is building a scalable LAN security system using RBL servers commonly used in e-mail security systems.

References

1. Maj, M., Silicki, K.: Classification and terminology of the network security incidents. CERT POLSKA, Warsaw (1999). http://www.cert.pl/PDF/SECURE99_referatCP_klasyf.doc
2. Regulation of the Prime Minister of Poland from 20th July 2011 on the basic safety requirements of ICT, Warsaw (2011)
3. Wrzesień, M., Olejnik, Ł., Ryszawa, P.: IDS/IPS: Systems Detection and Prevention Against the Intrusion to Computer Networks. PAR, Warsaw (2013)
4. RFC 3954 (2004). http://www.ietf.org/rfc/rfc3954
5. Krmicek, V., Vykopal, J.: NetFlow Based Network Protection, Security and Privacy in Communication Networks. Springer (2012)

6. NetFlow Version 9 Flow-Record Format (2011). http://www.cisco.com/en/US/technologies/tk648/tk362/technologies_white_paper09186a00800a3db9_ps6601_Products_White_Paper.html
7. Introduction to Cisco IOS NetFlow—A Technical Overview (2012). http://www.cisco.com/en/US/prod/collateral/iosswrel/ps6537/ps6555/ps6601/prod_white_paper0900aecd80406232.html
8. Dooley, K., Brown, I.J.: Cisco. Cookbook. O'Reilly Media (2003)
9. Fry, Ch., Nystrom, M.: Security Monitoring. O'Reilly Media (2009)
10. Santos, O.: Network Security with NetFlow and IPFIX: Big Data Analytics for Information Security. Cisco Press (2015)
11. Lucas, M.W.: Network Flow Analysis. No Starch Press, Inc (2010)

Practical Aspects of Hammerstein Models for Nonstationary and Nonlinear Processes

Karol Koniuszewski and Paweł D. Domański

Abstract Practical experience shows that in real modeling situations an engineer has to cope with several issues associated with non-stationary behavior, nonlinearities, data deficiencies and identification process limitations caused by technology. The paper presents observations and practical solutions developed in such a situation. The goal of the modeling is to derive nonlinear model that captures most of the process characteristics and can be a base for process simulations and further design of alternative control structures. The model under consideration is of Hammerstein type. Proposed modified structure covers process nonlinearities and has embedded self-adaptation mechanism enabling satisfactory operation in non-stationary environment.

1 Introduction

This paper covers two interesting aspects of process modeling: identification in nonlinear, non-stationary environment and limitations (or specifics) of project realization in the real industrial world. In fact, both issues pervade each other. Additionally experience shows that classical original theoretical stipulations rarely meet reality. Practice is much more rich. Thus engineer performing modeling in industry has to be always prepared for strange situations and limitations revealing openness, flexibility and real engineering art. On the other hand theory cannot be neglected as it delivers tools that allow to solve upcoming problems. The ultimate goal has to be finally reached using the customized set of methods and algorithms.

K. Koniuszewski (✉) · P.D. Domański
Institute of Control and Computational Engineering, Warsaw University of Technology, Nowowiejska 15/19, 00-665 Warszawa, Poland
e-mail: k.koniuszewski@stud.elka.pw.edu.pl

P.D. Domański
e-mail: p.domanski@ia.pw.edu.pl

R. Szewczyk et al. (eds.), *Challenges in Automation, Robotics and Measurement Techniques*, Advances in Intelligent Systems and Computing 440, DOI 10.1007/978-3-319-29357-8_10

The goal is simple. There is an installation and its owner wants to improve performance. There are several ways of the improvement: mechanical reconstruction, instrumentation rehabilitation or control system upgrade. We focused on the last option. Decision was based on the evaluation of control system quality called Process Control Assessment (PCA) [1, 2].

Most often design and evaluation of the new control scheme requires simulations. To perform those we need to have an appropriate and relevant model. That model has to catch the most of the process dynamics and constraints, should not be too simplified and it should not be over-sized nor over-complicated as its calibration may be painful and identification time is limited.

The first principle models are the most probable candidates. However, they have two shortcuts. They are idealized and their calibration or fitting into industrial reality might be too difficult or sensitive. Empirical models are the second option. One may choose between different classes [3], like linear or nonlinear regression, Wiener-Hammerstein, neural networks, fuzzy, hybrid models, etc.

Hammerstein [4] structure was selected in the presented case, although other approaches were also considered and rejected.

The next design decision was associated with the model structure [5]. One have to select the appropriate signals that will be used as model inputs and outputs. The outputs selection is relatively easy as they should reflect control goals. Input selection [6] is rather unclear depending to different extents on signals availability and quality, process knowledge, statistical analysis, applied benchmarking indexes, etc. Concluding, model structure identification is somehow like an art.

Once the structure identification phase is achieved, we may progress towards estimation of model parameters. In our case we are focusing on Hammerstein structure. This model consists of two parts the procedure often consists of two separate phases: identification of the nonlinear static part and linear dynamic one. One may find many methods, starting from classical one [7], where parameters of the non-linear part are determined by any non-linear optimization, while the linear part may be determined by least squares approach. In the literature we may find many more papers considering different identification approaches, i.e. blind approach [8], iterative [9], multivariable methods [10] and aspects as for instance identification in closed loop [11].

In the presented case, authors had to find a model in the situation of the MISO process with non-stationary behavior. Different approaches were tested. Results of the adaptive model operation form an original result.

The paper is organized as follows. In the starting Sect. 2 the theoretical basis for the considered subject are presented. Then Sect. 3 presents and discusses practical observations, features and limitations found in course of paper. Those two background sections are followed by presentation of the practical modeling process example (Sect. 4). The paper concludes in Sect. 5 with the discussion of obtained results and presentation of open issue for future research.

2 Theoretical Aspects

In this section main theoretical background for the paper is presented. There are three main subjects that are included in the presented approach and require closer attention: model structure identification, Hammerstein multi-variable models, and aspects of model adaptation as the solution to non-stationarity.

2.1 *Model Structure Identification*

Model structure identification is quite a tedious task, not well defined and very often customized to the problem. Frequently it consists of several steps, while each of them gives fragmentary contribution to the overall methodology. The methods used in the considered example are shortly sketched below.

Cross correlation is a measure of similarity or dependence. A strong correlation between signals x and y means that the growth of x is accompanied by the growth of y and opposite.

$$\sigma_{xy} = \int (x(t) - \bar{x})\,(y(t) - \bar{y})\,dt \tag{1}$$

To get a value between $[-1, 1]$ standardization as a factor in the variance of the signal—called the covariance correlation is used.

Autocorrelation is a measure of the correlation signal $s(t)$ with its copy delayed by time shift τ [12].

$$\sigma_{s,s}(\tau) = \int s(t)s(t+\tau)dt \tag{2}$$

Function reaches maximum at zero and the existence of other peaks is associated with the presence of periodically recurring phenomena in the signal.

Histogram is the graphic way to present empirical distribution characteristics. Boxes are on the one hand determined by the class intervals' features, and their amount is determined by the number elements falling into a particular class. If the histogram shows only numbers, not the probability density, the width of the compartments should be equal. Histogram enables to get the feeling about probabilistic function distribution of the variable.

Butterworth low-pass filter is designed to have maximally flat frequency response in defined passband against close zero in the stop-band. Filter frequency is determined by 3 dB drop in passband with response roll of $n \cdot 6$ dB per octave, where n is the filter order [13].

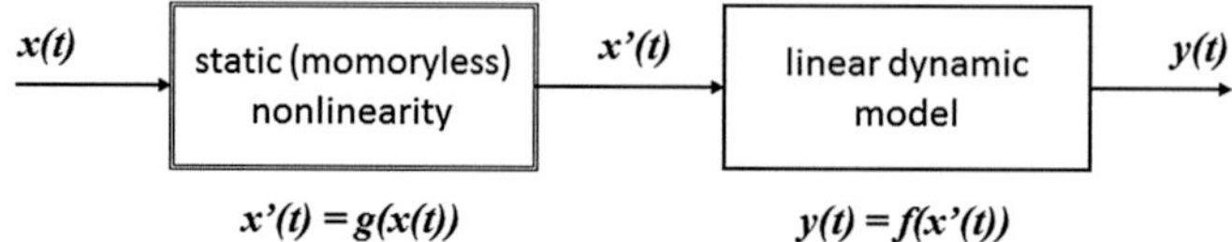

Fig. 1 Hammerstein model scheme

2.2 *Hammerstein Models*

They belong to the class of nonlinear dynamical models called block-oriented ones. It consists of a nonlinear memoryless static element followed by linear dynamics (see Fig. 1). Static nonlinear function $g(.)$ and linear dynamic one $f(.)$ can be generally of different origin [7]. In our case we mostly use static SISO and MISO polynomials as $g(.)$ and ARX empirical regression models as $f(.)$.

2.3 *Model Adaptation*

Adaptation of the models is needed when the assumption that process parameters are constant in time fails. For the sake of simplicity most of the modeling approaches, starting from the first principle described by stationary differential equations up to complex empirical models like neural networks or other heuristics, assume no variations in parameters. However, practice is not so ideal. Real process always vary in time due to the mechanical aging, technology/production variations or human interventions. Thus we should embed into the model adaptation mechanisms.

In literature the subject was historically of the great interest and we may find several different approaches [3, 14, 15]. We may find two main scenarios, which are analogous to the adaptive control schemes. One of them is called Gain Scheduling and means switching of the several models according to the known and varying external variables. The second one (Model Reference Adaptive Control) refers to the constant or recurrent model updates, like RLS (Recursive Least Squares). It happened that both approaches were tested.

3 Practical Considerations

The modeling approach described in the paper was mostly driven by industry requirements constrained by technology and I and C (Instrumentation and Control) limitations. There are two main common issues that should be always remembered. Historical data access, collection and preprocessing embedded in plant data historian is the first one. Aspects of identification in the mixed mode with some process loops operating in MANUAL and unfortunately the most of them in AUTO (resulting in close loop identification). Both those aspects are discussed below.

3.1 Data Collection and Preprocessing

Working with the real object is associated with a number of complications and problems. Considerations of economy or security limit or even block testing experiments. Complexity and the multitude of influencing factors often prevent model development based on physico-chemical equations. The only available solution is to work with historical data. Plant is normally working with the set of control loops. The operation generates a further complication for the identification process, in particular when operators switch to the manual mode.

The consequence of this action is the presence of constant value with no change despite process variations. This prevents the use of a standard approach to the identification based on the assumption of dynamic, not periodic signal. Process often has non-stationarity, resulting in variations depending on external unknown factors, like weather conditions, variable product quality and installation throughput. It makes simple determination of a steady state impossible.

The scale and interconnection with other installations make it difficult to determine the parameters affecting the object, because its function is affected by factors occurring in the other areas of installation. Thus close cooperation and consultancy with plant personnel is crucial. Additionally data collection is frequently not an easy part, as the historian may be hardly accessed and it often embeds strange modifications into the stored data. Nevertheless, an engineer beginning the modeling process must be aware of traps and problems associated with working in real industrial facility.

3.2 Close Loop Identification

Identification of the dynamics, which is inside of the close loop is often a challenge. In industry it is often impossible to open the loop, apply external identification input and collect data. Apart from the cost of stopping regular production and adding of external disturbance, everybody have concerns about installation safety and associated risks. However, in closed loop case standard identification policies fail due to the convergence criteria. There are two options:

- Indirect Process Identification: An overall model of the closed-loop is identified. The controller has to be known and than process model is evaluated.
- Direct Process Identification: The process is identified as it is, without intermediate step of closed-loop model identification.

In reality it is difficult to have knowledge about real controller. It is often complex structure with several blocks, like nonlinear functions $f(x)$, feedforwards, override controls, interlocking, normalizations, interactions with other loops and human interventions (biases, setpoint changes, etc.). Thus we are left to the Direct Process Identification option despite all deficiencies.

4 Real Case

The process under consideration originates from chemical facility performing natural gas processing. Required natural gas heating temperature is obtained in the natural gas heater. It is vertical apparatus consisting of three heat exchange sections, where natural gas gets heat from the hot flue gases resulting from combustion of the heating gas in the bottom radiation heater element. In this heater steam is added to natural gas. The process is considered to be MISO. Initially the all available process data was collected with 14 possible input signals and 1 output—natural gas temperature after the heater.

Three datasets of different type were collected: dynamic 5 s data and static 5 and 10 min. Combination of those sets enables4 evaluation of static process curves and further dynamic models.

Selection of appropriate signals from the entire set of measured and collected data was essential determining further actions of the entire modeling process. The first step was to determine the value of cross-correlation of the signals collected and measured for installation. According to the performed calculations the cross-correlation maps were evaluated (Fig. 2) helping to find existing dependencies. Dark areas of the map are of particular concern. Those signals formed the first set for further consultation with technology expert to confirm the findings.

Calculation of cross-correlation values serves not only as a tool for selection of relevant data, but actually reveals information on the process nature giving premises

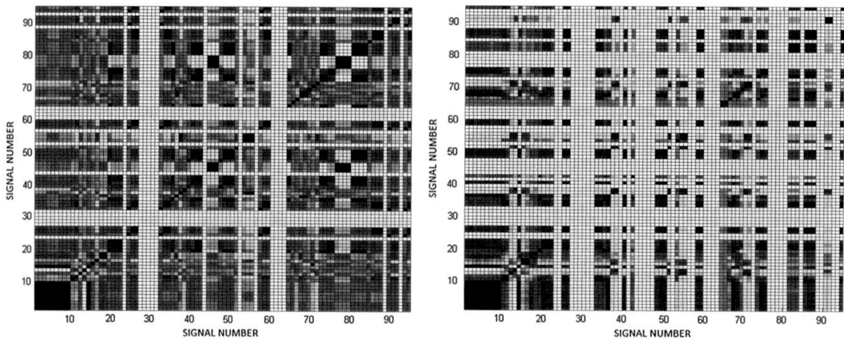

Fig. 2 Cross-correlation maps for static (*left*) and dynamic (*right*) data

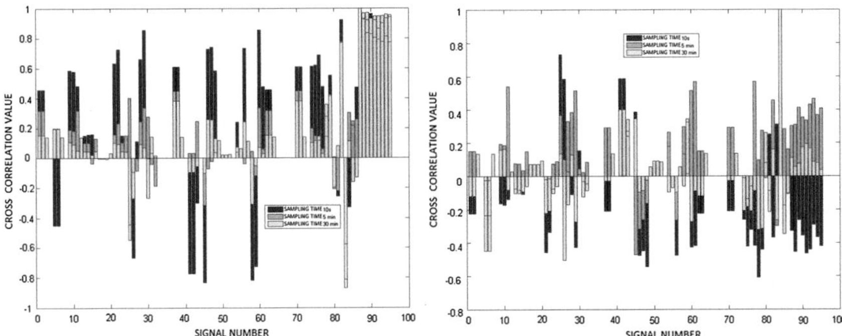

Fig. 3 Distribution of correlation values of the selected output for different datasets: dynamic (*blue*), static 5 min (*green*) and 10 min (*yellow*)

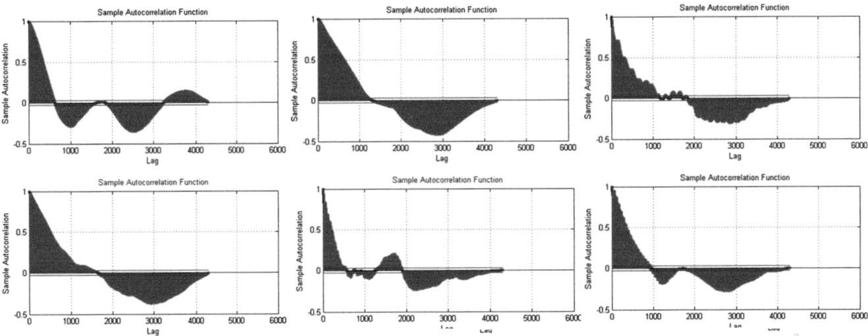

Fig. 4 Examplary autocorrelation delay identification

about model structure. Combination of the cross-correlation analysis for static and dynamic data helped to detect anomalies (Fig. 3).

Autocorrelation analysis was performed to determine the overlapping relationships and associated delays (Fig. 4). Determination of histograms allows for aggregation of interdependent signals. Confronting histogram with correlation analysis enables to remove undesired connections between signals.Further filtration of signals was performed in order to remove distortions contained therein. In particular this includes signal noises and removal of existing peaks of unknown origin (archivization process, control system errors, instrumentation malfunctioning, data collection, ...). Butterworth low-pass filter was customized for each signal and applied (Fig. 5).

Further structure simplification was obtained through data aggregation. It was mostly done manually through observation of alternative or interchangeable variables and operations like averaging. Examples are sketched on Fig. 6. Whole data conditioning and structure identification process reduced the number of inputs to 8 variables.

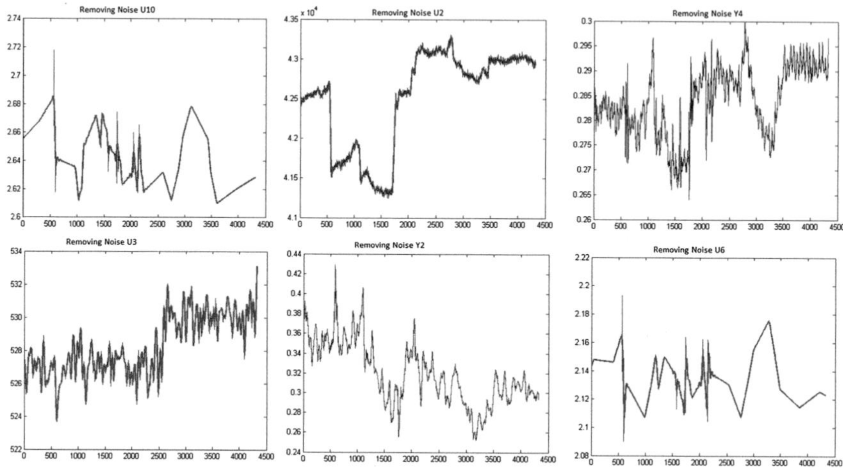

Fig. 5 Removing distortions of historical data sets using low-pass filter

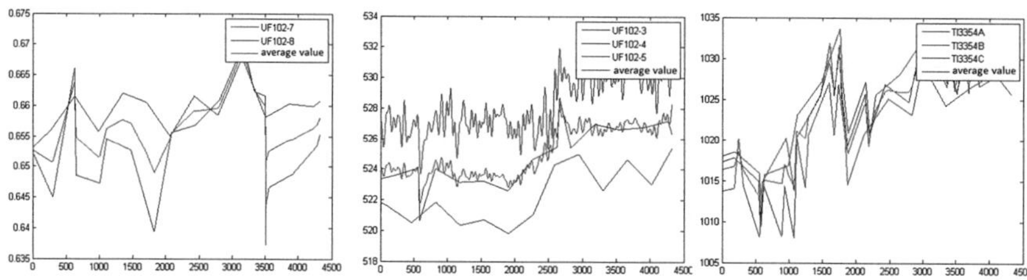

Fig. 6 Aggregation example for several signals relating to the same physical value (*blue*) and the output signal (*red*)

Once the model structure was defined the static element type was chosen. MISO polynomials were selected. The characteristics were approximated through first- or second-order polynomial curves. Their quality was measured by Minimum Square Error (Fig. 7). The case was modeled through a series of tests which, although did not give the fast result, guided towards better solution. Several different options for combination of SISO or MISO polynomials were checked, while regression ARX and ARIMA models were tested as the dynamic linear model.

Experiments demonstrated the weakness and unsatisfactory sensitivity of the static part against high quality and robustness of the dynamic element. This determined the next step, which was to improve the static block. Changes included adding of the adaptation.

At first the input data were clustered and several sub-models were identified. The model adapts through switching, similarly to Gain Scheduling. Unfortunately the approach did not work. There were not enough data to identify models for all subregions. Thus the overall adaptation was incomplete and could not be used.

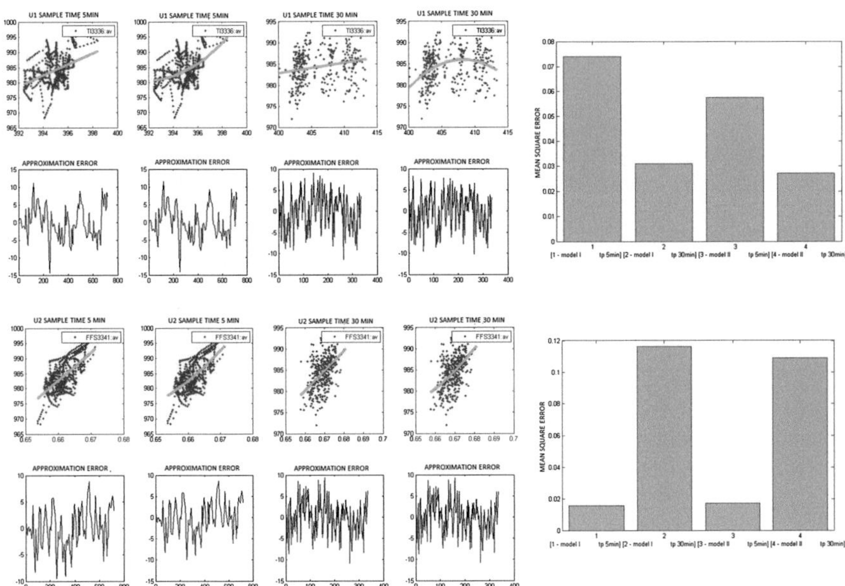

Fig. 7 Exemplary static characterstics

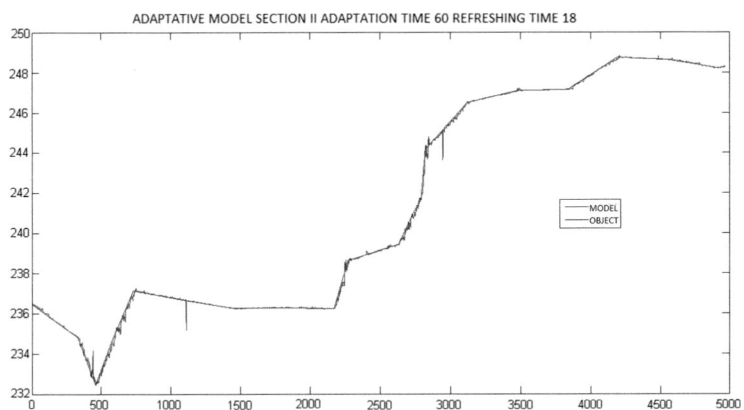

Fig. 8 Adaptive static model

Table 1 MSE comparison

	Hammerstein	Multi-regional	Adaptive
MSE	0.0118	4.1131	**0.0083**

The second option formed original approach of receding window relearning. The system keeps some memory window and with some sampling period relearns polynomial model according to the recent history stored in the window memory. This relatively simple and robust approach proved its effectiveness. Figure 8 shows its quality together with the MSE comparison—Table 1. Once the static part deficiency was improved, the overall MISO Hammerstein model could be put together.

5 Conclusions and Further Research

The paper presented the real process of the nonlinear multi-variable modeling for the real industrial installation of the natural gas heating facility. There are two main contributions of the paper: identification process in the real industrial environment and specific original approach to the non-stationary static characteristics identification based on the self-adaptation of polynomial-based multivariable model through recursive window-based memory.

Obtained models proved their efficiency tested on the several independent industrial data sets. Nonetheless, it should be stated that the model itself was not an ultimate goal. Derived model will be next used to design the APC control structure for its off-line testing validation (proof of concept). Once it is designed the new control scheme will be applied to the process and only then, formally, its efficient and robust operation will confirm identification quality. Thus presented developed model is only an intermediate milestone towards the final goal.

Acknowledgments The authors wish to thank JP Nawozy, GA ZAK S.A. for the support and access to process and data.

References

1. Thornhill, N.F.: Practical experiences in refinery control loop performance assessment. In: Proceedings of UKACC International Conference on Control, Control '96, pp. 175–180 (1996)
2. Jelali, M.: An overview of control performance assessment technology and industrial applications. Control Eng. Pract. **14**(5), 441–466 (2006)
3. Nelles, O.: Nonlinear System Identification: From Classical Approaches to Neural Networks and Fuzzy Models. Springer, Berlin (2001)
4. Hammerstein, A.: Nichtlineare Integralgleichungen nebst Anwendungen. Acta Math. **54**(1), 117–176 (1930)
5. Roffel, B., Betlem, B.: Process Dynamics and Control: Modeling for Control and Prediction. Wiley (2007)
6. Sindelar, R., Babuska, R.: Input selection for nonlinear regression models. IEEE Trans. Fuzzy Syst. **12**(5), 688–696 (2004)
7. Isermann, R., Münchhof, M.: Identification of Dynamic Systems: An Introduction with Applications. Springer, Berlin (2011)

8. Bai, E.-W.: A blind approach to Hammerstein model identification. IEEE Trans. Signal Process. **50**(7), 1610–1619 (2002)
9. Vörös, J.: An iterative method for Hammerstein-Wiener systems parameter identification. J. Electr. Eng. **55**(11), 328–331 (2004)
10. Wills, A., Schön, T.B., Ljung, L., Ninness, B.: Identification of Hammerstein-Wiener models. Automatica **49**(1), 7081 (2013)
11. Han, Y., de Callafon, R.: Closed-loop identification of Hammerstein systems using iterative instrumental variables. In: Proceedings of the 18th IFAC World Congress, vol. 18, part 1, pp. 13930–13935 (2011)
12. Dunn, P.F.: Measurement and Data Analysis for Engineering and Science. McGrawHill, New York (2005)
13. Binachi, G., Sorrentino, R.: Electronic Filter Simulation Design. McGraw-Hill (2007)
14. Bittanti, S., Picci, G. (eds.): Identification, Adaptation, Learning: The Science of Learning Models from Data. NATO Advanced Study Institute 'From Identification to Learning' (1996)
15. Bittanti, S.: Adaptation and Learning in Control and Signal Processing. Pergamon (2002)

Robust Scheduling Subject to Multi-project Environment Constraints

Grzegorz Bocewicz, Krzysztof Bzdyra and Zbigniew Banaszak

Abstract The main objective is to provide a declarative model enabling to state a constraint satisfaction problem aimed at multi project-like and mass customized oriented multi-product scheduling. Its formal representation enables to formulate the straight and reverse problem scheduling of a newly inserted orders portfolio subject to constraints imposed by an enterprise multi-project environment. Considered problems correspond to the routine questions faced by decision makers and are formulated either in the straight or reverse way, e.g. Is it possible to undertake a new portfolio under given resources availability (e.g. constrained in time) while guaranteeing disturbance-free execution of the already executed portfolios? What values and of what decision variables guarantee the production order (or orders portfolio) will completed while following assumed set of performance indexes? In that context a computationally effective approach aimed at simultaneous routing and allocation as well as batching and scheduling of a new portfolio subject to constraints imposed by a multi-project environment is sought. Provided examples illustrate both cases while taking into account robust scheduling requirements imposed by multi-project environment constraints.

Keywords Robust scheduling · Declarative modeling · Multimode resource-constrained multi-product scheduling problem

---

G. Bocewicz · K. Bzdyra (✉)
Department of Computer Science and Management, Koszalin University of Technology, Śniadeckich 3, 75-453 Koszalin, Poland
e-mail: krzysztof.bzdyra@tu.koszalin.pl

Z. Banaszak
Department of Business Informatics, Warsaw University of Technology, Narbutta 85, 02-524 Warsaw, Poland
e-mail: Z.Banaszak@wz.pw.edu.pl

R. Szewczyk et al. (eds.), *Challenges in Automation, Robotics and Measurement Techniques*, Advances in Intelligent Systems and Computing 440, DOI 10.1007/978-3-319-29357-8_11

1 Introduction

The resource-constrained multi-product scheduling problem where the activities of newly occurring production orders portfolio must be scheduled either as to minimize its duration or as to follow assumed deadline is quite well-known. It has been emerged and addressed by many researchers [1, 2]. However, its extensions, e.g. assuming each activity executed along production order under the presence of precedence and resource constraints in multiple production order environment can be performed in one of several modes representing a relationship between the resource requirements of the activity and its duration are rarely available.

Since the most frequently occurring questions are such as: Does a given production orders portfolio specified by its resources allocation guarantee the portfolio makespan will not exceed a given deadline? Does there exist a set of operation times guaranteeing the completion time of a given production orders portfolio will not exceed assumed deadline?, hence our contribution can be seen as an alternative approach to Decision Support Systems (DSSs) design, i.e. allowing one to take into account both: straight and reverse problems formulation. More precisely, allowing one to answer whether a given production order specified by its cost and completion time can be accepted in a manufacturing system specified by a given production capability and/or what manufacturing system capability guarantee completion of a given production orders portfolio under assumed cost and time constraints. In that context our contribution is dedicated to a computationally effective approach mostly aimed at simultaneous routing and allocation as well as batching and scheduling of a new portfolio subject to constraints imposed by a multi-project environment.

Existing methods assume that routing and allocation as well as batching and scheduling decisions are made independently, i.e. each production order is treated as an activity network and is assigned to processing units, and then divided into a number of batches (batching), and sequenced (scheduling). Several techniques have been proposed in the past 50 years, including MILP [3–5], Branch-and-Bound [6] or more recently Artificial Intelligence [7]. Most of them are oriented at multi-criteria planning problems formulated as direct ones focusing on questions such as: Is it possible to undertake the given project portfolio under a given resource availability while guaranteeing disturbance-free execution of already scheduled activities? Therefore, in order to consider an alternative, i.e. reverse planning problem (focusing on: Which values of the system parameters guarantee that the set of orders will be completed while giving a certain set of values for performance indexes?) this contribution proposes a declarative framework aimed at reference model of a multi project-like and mass customized oriented multi-product scheduling.

The considered model has to specify both the job shop capability and the production orders requirement in a unified way, i.e., through the description of determining them sets of variables and sets of constraints restricting domains of discrete variables. In other words, our reference model aimed at providing a decision problem encompassing equilibrium between possible expectations regarding potential orders completion and available production capabilities, has to focus on resources conflict resolution, i.e. conflicts arising in case different activities (operations) simultaneously request their access to common shared resources of limited quantity.

The remainder of this paper is organized as follows: Some related work are discussed in Sect. 2. Illustrative example of basic concepts are presented in Sect. 3. Section 4 introduces to our main contribution and provides problem formulation. Conditions sufficient for orders portfolio acceptance standing behind problem's solution are discussed in Sect. 5 and concluding remarks are presented in Sect. 6.

2 Related Work

A Resource Constrained Project Scheduling Problem (RCPSP) arises when the available resources are not enough to satisfy the requirements of the activities that can be performed concomitantly [8]. The RCPSP which involves scheduling project activities subjected to temporal and resource constraints, while minimizing the total project duration can be classified into single project scheduling problem and multi-project scheduling problem. Two main approaches to such strongly NP-hard problems [9] can be distinguish so far that is: mathematical programming and the artificial intelligence approach. Consequently, solutions to these problems can be obtained by many different methods (mainly aimed at tasks scheduling and resource allocation), widely covered and classified in the literature [10, 11]. Among them the methods providing exact solutions as mathematical programming, dynamic programming, etc., and approximately as evolutionary algorithms, tabu search, ant colony as well as dedicated heuristics [12] can be recalled. Moreover, many different advanced while problem oriented planning and scheduling software applications have been developed so far, e.g. SAP APS, Asprova by PMC, D-Opt by Greycon, etc. (www.capterra.com/production-scheduling-software).

Very limited work focuses on the joint technological processes, transportation routing and financial [8] as well as a new Decision Support System (DSS) oriented modeling framework enabling to evaluate and prototype alternative flows of Production Orders Portfolio (POP) in the manufacturing systems disposing untapped potential in production capacity [13]. Furthermore, there is another aspect of the addressed class of problems, namely multi-criteria decision making under uncertain conditions. Studies conducted so far on declarative models implemented in fuzzy sets framework, show that the fuzzy model of POP online control can be supported by constraint programming techniques [1]. Declarative approaches to systems and/or process modeling promise a high degree of flexibility. In that context, Constraint Programming (CP) framework can be considered as a well-suited framework for development of decision making software supporting small and medium size enterprises in the course of a variety, unique-product and mass customized oriented, i.e. multi-order-like, production [14].

Limited work concentrates on finding a computationally effective approach aimed at simultaneous routing and allocation as well as batching and scheduling of a new production order subject to constraints imposed by a multi-project environment. More precisely, allowing one to answer whether a given production order specified by its cost and completion time can be accepted in a manufacturing system specified by a given production capability and/or what manufacturing systems capability guarantee completion of a given production orders portfolio under assumed cost and time constraints. In that context the paper can be seen as continuation of our former work aimed at POP production flow planning focused on coordination of processes and activities involved in course of work orders processing [14–16].

3 Illustrative Example

The paper focuses on a multimode resource-constrained POP scheduling problem, which contains mode selection problem and activities scheduling problem for multiple orders subject to constraints imposed by a multi-project environment. In that context, taking into account that a schedule is robust, if reasonable disruptions have no significant effects on the performance of the planned or so-called base-line schedule, the paper focuses on POP scheduling guaranteeing that its each activity could be executed on at least one among a given number of alternative resources, i.e. on POP robust scheduling.

The problem considered adheres to the following assumptions: the considered POP consists of n parallel production orders, each order consists of activities. Due to assumed precedence order each activity can be started if and only if all of its preceding activities have been completed. Each activity must be performed in only one of possible modes, with each activity mode possibly having different processing times and different resource requirements. Mode switching is not allowed when an activity is being executed. The shared resources that activities require are nonrenewable resources which are limited for the entire production order. The interrelationship among activities is executed in a certain order using an activity-on-node representation, in which the node represents an activity and path arrows demonstrate the logical sequencing from node to node through the network. The starting time of each POP is dependent upon the system resources availability, i.e. the guarantee the all activities will be executed along preserved orders, while not disturbing already executed, formerly scheduled POPs.

In order to illustrate a main idea standing behind of our approach let us consider the case where actually scheduled production flow realized in the Flexible Manufacturing System (FMS) composed of 4 Machining Centers (MC) is depicted by the Gantt's chart shown in Fig. 1a.

Let us assume a new POP composed of only one production order (see Fig. 1b) is waiting for acceptance at the moment 5. Conditions enabling acceptance of the order are following: the order completion time cannot increase 16 u.t., the resources allocation as well as operation times cannot differ from alternatives assumed, see Fig. 1a. The sought schedule has to follows its robustness constraint guaranteeing that in case of unforeseen disruptions, e.g. malfunction of planned to use resource, at least one alternative mode replacing disabled resource can be used not leading to extension of assumed deadline.

The illustration of feasible solution assuming $t^1_{1,1} = t^3_{1,1}$, $t^2_{1,2} = t^3_{1,2}$, $t^1_{1,3} = t^2_{1,3} = t^4_{1,3}$, and $t^3_{1,3} = t^2_{1,3}$ is shown in Fig. 2a. The cases following possible disturbances (malfunction of machining centers) and alternative rescheduled plans are shown in Fig. 2b–d. It should be noted however, that in case the j-th operation planned for execution in the i-th production order could be served only by M_4 or M_2, the only available schedules (in case of M_4 malfunction) will result in

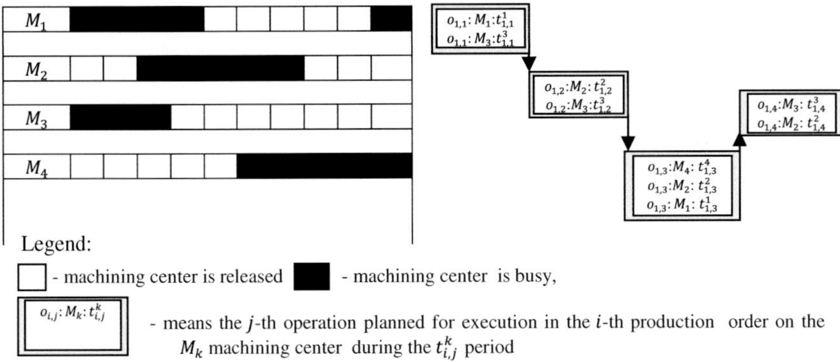

Fig. 1 Exemplary cases of: the Gantt's chart of planned production flow (**a**), activity-like network of acceptance awaiting production order (**b**)

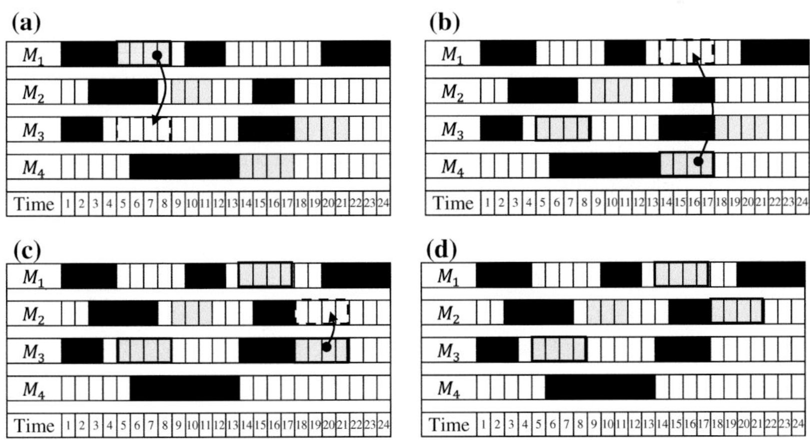

Fig. 2 Gantt's chart of the robust schedule: the base line schedule (**a**), exemplary cases of rescheduled solutions caused by machining centers malfunction (**b–d**)

overcoming of the assumed order completion deadline. Therefore, this remark motivates us to consider both the straight and reverse formulations of robust scheduling problem.

4 Problem Formulation

Well known problem of Multimode Resource-Constrained POP scheduling (MRCPOPS) subject to constraints imposed by a multi-project environment can be extended to robust MRCPOPS (R-MRCPOPS) problem. In such environment robust scheduling results in the POP schedule guaranteeing that each planned activity can be executed on at least one among a given set of alternative resources. In this paper we discuss due-date extension of R-MRCPOPS problem, i.e. cases assuming the all production orders have to be completed before a given deadline within assumed period of time.

Given FMS composed of the set of MCs and the base line scheduled describing their usage within a given time horizon. At a given moment within the time horizon H_{FMS} the new POP-like order requests its acceptance. POP is specified by the given set of activity networks, the deadline H_{POP}, and the time window within which all orders have to be completed. Since some of MCs being a multi-functional machine tools can replace each other, however with different operation times, hence sub-sets of MCs selected for alternative execution of distinguished operations are given too.

In this context our robust scheduling problem can be formulated either in:

- **a straight way** (i.e. corresponding to the question: What results from premises?), e.g. What POP schedule following a given set of operation times, the deadline H_{POP} and TW, as well as constraints imposed by a FMS base-line schedule and MC replaceability, guarantees assumed robustness of the POPs schedule?, or
- **a reverse way** (i.e. corresponding to the question: What implies conclusion?), e.g. What MC replaceability guarantees that the resultant schedule of a given POP will meet assumed robustness of the POP schedule?

In order to formulate robust scheduling problem in terms of a declarative framework the following assumptions regarding the sets of decision variables and corresponding domains as well as constrains are proposed.

Given the set of MCs $\mathbb{M} = \{M_1, \ldots, M_i, \ldots, M_{LM}\}$ (where M_i means the i-th MC and LM a number of MCs in the FMS) required for execution of operations belonging to a set of different production orders Z^*, in general case being a family of POPs. The set Z^* provides multi-project environment within which a newly introduced POP Z is scheduled.

Consider POP Z composed of $Z_i \in Z$ specified by the set of operations $O_i = \{o_{i,1}, \ldots, o_{i,j}, \ldots, o_{i,lo(i)}\}$, where $o_{i,j}$ means the j-th operation, while $lo(i)$ a number of production order Z_i operations. Operations execution order is specified by digraph $G_i = (O_i, E_i)$, vertices of which refer to operations from the set O_i and arcs $(o_{i,a}, o_{i,b}) \in E_i$ represent relation: $o_{i,a} \prec o_{i,b}$ ($\prec \subseteq O_i \times O_i$), which means: "$o_{i,a}$ *is executed before* $o_{i,b}$".

For each operation $o_{i,j}$ its execution (processing) time $t_{i,j} \in \mathbb{N}^+$ and the set of alternative modes of MCs is assigned $M_{i,j} \in \left\{mb^1_{i,j}, \ldots, mb^k_{i,j}, \ldots, mb^{la(i,j)}_{i,j}\right\}$, where $mb^k_{i,j} \in \mathbb{M}$, possibly having different processing times.

The following constrains have to be satisfied:

- each operation $o_{i,j}$ can be uniquely executed on the one MC from $M_{i,j}$,
- at each moment only one operation $o_{i,j}$ can be executed on the MC from $M_{i,j}$,
- the resources are non-preemptable, that means the resources of $M_{i,j}$ are reserved by operation $o^k_{i,j}$ until its completion, and cannot be used by other operations, (Note that usage of one, e.g. $mp^k_{i,j}$, among the set $M_{i,j}$ implies reservation of other resources from $M_{i,j}$, i.e. results in a schedule robustness.)
- resources from $M_{i,j}$ are released immediately after completion of actually executed operations.

Taking into account above mentioned assumptions the following problem can be considered: Given FMS, where simultaneously executed production orders $Z^* = \{Z^1, \ldots, Z^q\}$ are processed within assumed time horizon H_{FMS}, and the POP Z expecting to be completed within time horizon H_{POP} included in H_{FMS} ($H_{POP} \subseteq H_{FMS}$). Dose there exist robust schedule $X = (X_1, X_2, \ldots, X_n)$ for portfolio Z guaranteeing H_{POP} holds? Therefore the following Constraints Satisfaction Problem (CSP) [15, 16] can be considered:

$$CS = ((V, D), C) \tag{1}$$

Decision variables ***V***

- $X_i = \{x_{i,1}, \ldots, x_{i,j}, \ldots, x_{i,lo(i)}\}$ for $i = 1 \ldots n$—set of start times for the operation $o_{i,j}$ belonging to Z,
- $Y_i = \{y_{i,1}, \ldots, y_{i,j}, \ldots, y_{i,lo(i)}\}$ for $i = 1 \ldots n$—set of completion times for the operation $o_{i,j}$ belonging to Z, where $y_{i,j} = x_{i,j} + t_{i,j}$.

Fixed factors

- H_{FMS}—time horizon of considered base-line schedule,
- H_{POP}—time horizon of Z completion,
- $T_i = \{t_{i,1}, \ldots, t_{i,j}, \ldots, t_{i,lo(i)}\}$ for $i = 1 \ldots n$—set of the operation $o^k_{i,j}$ processing times,
- $\bar{X}_i = \{\bar{x}_{i,1}, \ldots, \bar{x}_{i,j}, \ldots, \bar{x}_{i,lo(i)}\}$—set of the start times for the operations $\bar{o}_{i,j}$ belonging to production orders $Z^1, \ldots, Z^q$ {see the base-line schedule},

- $\bar{Y}_i = \{\bar{y}_{i,1}, \ldots, \bar{y}_{i,j}, \ldots, \bar{y}_{i,lo(i)}\}$—set of the completion times for the operations $\bar{o}_{i,j}$ belonging to production orders $Z^1, \ldots, Z^q$. Note that over-lined terms, e.g. $\bar{x}_{i,j}$ and $\bar{y}_{i,j}$, differentiate specification of production orders belonging to Z^* from ones belonging to Z.

Domains *D*

$$x_{i,j} \in \mathbb{N}, y_{i,j} \in \mathbb{N} \quad \text{for } i = 1 \ldots n, j = 1 \ldots lo(i).$$

Constraints C

- precedence order $o_{i,a} \prec o_{i,b}$:

$$y_{i,a} \leq x_{i,a}, \forall (o_{i,a}, o_{i,b}) \in E_i, \tag{2}$$

- admissible allocation of the operation $o_{i,j}$ to the resource, e.g. $mb^k_{i,j}$, within H_{POP}:

$$x_{i,j} \in H, y_{i,j} \in H_{POP}, \quad \text{for } i = 1 \ldots n, j = 1 \ldots lo(i), \tag{3}$$

- admissible allocation of operation $o_{i,j}$ to the resource, e.g. $mb^k_{i,j}$, in available time windows (not occupied by the operation $\bar{o}_{a,b}$ from production orders $Z^1, \ldots, Z^q$ utilizing $mb^k_{i,j}$):

$$(y_{i,j} \leq \bar{x}_{a,b}) \vee (\bar{y}_{a,b} \leq x_{i,j})$$

for each $o_{i,j}$, $\bar{o}_{a,b}$, where

$$M_{i,j} \cap \bar{M}_{a,b} \neq \emptyset, \tag{4}$$

- mutual exclusion protocol synchronizing processing of operations, e.g. to $o_{i,j}$ and $o_{a,b}$, competing with access to the common shared $mp_{i,j}$:

$$(y_{i,j} \leq x_{a,b}) \vee (y_{a,b} \leq x_{i,j})$$

for each $o_{i,j}$, $o_{a,b}$, where

$$M_{i,j} \cap M_{a,b} \neq \emptyset. \tag{5}$$

5 Conditions Sufficient for Projects Portfolio Acceptance

In order to speed-up searching process, however at cost of losing of some admissible solutions let us consider, besides of the following (mentioned earlier) constraints:

- specifying an FMS current base-line schedule of time windows within which FMS workstations are not busy in assumed time horizon H_{FMS},
- determining the set of POP's production orders, the time horizon H_{POP}, sets of operation times and sets of alternative modes as well as technological orders (activities networks) for each $Z_i \in Z$,

the constraint assuming, that for each production order $Z_i \in Z$ a number of its alternative modes allocation variants match-up time windows available in currently observed base-line schedule (i.e., guaranteeing required robustness of the POP schedule). An idea standing behind of that constraint is to propose a sufficient condition guaranteeing there exists at least one solution even at cost of losing of some feasible solutions.

The notation introduced below allows us to propose the sought constraint through

^{n}—index of the n-th recursion step,
$\bar{X}^*$—set of start times of operations from Z^* (i.e., the base-line schedule of Z^*),
$Z_{LP} \in Z$—production order possessing the longest critical path,
$L(Z)$—function returning $Z_{LP} \in Z$,
X_{LP}—set of start times of operations from Z_{LP} (i.e., the robust schedule of Z_{LP}),
$\bar{X}^n$—set of start times of operations from $Z^* \cup Z_{LP}^{n-1}$,
$CS_{LP}(\bar{X}, Z_{LP})$—function returning robust schedule X_{LP} for Z_{LP} following extension of $\bar{X}$ taking into account newly allocated operations from Z_{LP}, i.e.:

$$CS_{LP}(\bar{X}, Z_{LP}) = \begin{cases} X_{LP} & if \quad for\, CS(1)\; there\; is\; a\; solution\; following\; Z_{LP}\; \text{and}\; \bar{X} \\ \varnothing & if \quad for\, CS(1)\; there\; is\; a\; lack\; of\; any\; solution \end{cases}$$

the following recurrence formula:

$$\boldsymbol{X_{LP}^n} = \boldsymbol{CS}\left(\boldsymbol{X^n}, \boldsymbol{Z_{LP}^n}\right), \boldsymbol{X}^n = \boldsymbol{X}^{n-1} \cup \boldsymbol{X_{LP}^{n-1}}; \boldsymbol{Z_{LP}^n} = \boldsymbol{L}(\boldsymbol{Z^n}); \boldsymbol{Z^n} = \boldsymbol{Z}^{n-1} \backslash \boldsymbol{Z_{LP}^{n-1}} \neq \varnothing \quad (6)$$

where: $\boldsymbol{X}^0 = \boldsymbol{X}^*$; $\boldsymbol{Z_{LP}^0} = \boldsymbol{L}\left(\boldsymbol{Z}^0\right)$; $\boldsymbol{Z}^0 = \boldsymbol{Z}$ are base cases.

In order to illustrate this condition let us consider the base-line schedule of the given Z^* shown in Fig. 3a and Z composed of two production orders specified by activity networks shown in Fig. 3b.

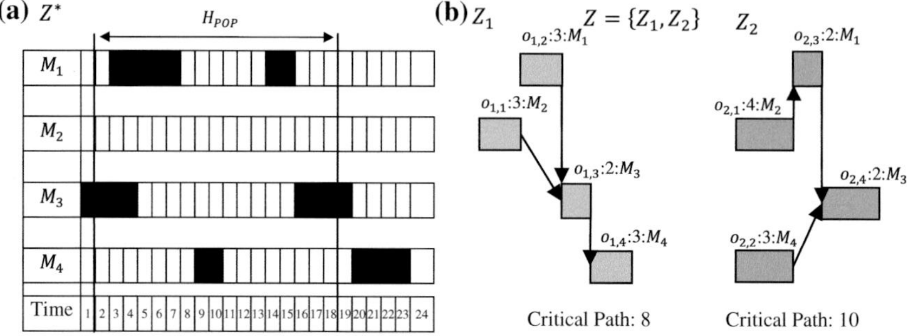

Fig. 3 Illustration of the sufficient condition: the base-line schedule (**a**), activity networks of POP considered for robust scheduling (**b**)

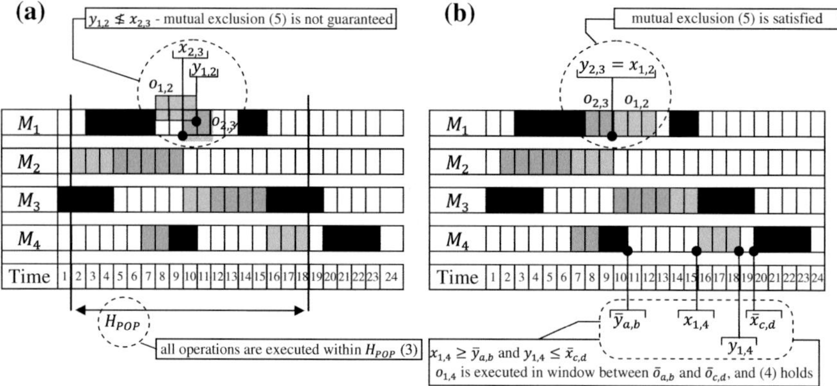

Fig. 4 Illustration of the sufficient condition: Gantt's chart following the case the sufficient condition (6) does not hold (**a**), the sufficient condition (6) holds (**b**)

Taking into account first two steps of the recurrence formula (6) (see below), the admissible solution following the sufficient condition is shown in Fig. 4b.

$$
\begin{aligned}
&X_{LP}^0 = CS\left(\bar{X}^0, Z_{LP}^0\right), \quad \bar{X}^0 = \bar{X}^*; && Z_{LP}^0 = L(Z^0); \quad Z^0 = Z \\
&X_{LP}^1 = CS\left(\bar{X}^1, Z_{LP}^1\right), \quad \bar{X}^1 = \bar{X}^0 \cup X_{LP}^0; && Z_{LP}^1 = L(Z^1); \quad Z^1 = Z^0 \backslash Z_{LP}^0 \\
&X_{LP}^2 = CS\left(\bar{X}^2, Z_{LP}^2\right), \quad \bar{X}^2 = \bar{X}^1 \cup X_{LP}^1; && Z_{LP}^2 = L(Z^2); \quad Z^2 = Z^1 \backslash Z_{LP}^1 \\
&\ldots \\
&X_{LP}^n = CS\left(\bar{X}^n, Z_{LP}^n\right), \quad \bar{X}^n = \bar{X}^{n-1} \cup X_{LP}^{n-1}; && Z_{LP}^n = L(Z^n); \quad Z^n = Z^{n-1} \backslash Z_{LP}^{n-1} \neq \varnothing
\end{aligned}
$$

6 Concluding Remarks

In this paper, the reference model of CSP was established to solve a due-date extension of robust multimode resource-constrained POP scheduling problem under constraints imposed by a multi-project environment. The model provides a formal framework enabling robust scheduling of POP, and resulting in a makespan following the given deadline and the time segment within which each production order should be completed. Subsequently, to solve this problem, the set of constraints was developed based on the particular nature of the model, which could be implemented in a constraint programming driven DSS enabling to deal with the selection of modes to activities, and with the scheduling of all activities. The results and analysis were illustrated through examples presenting introduced assumptions and following them sufficient condition supporting possible implementation of greedy algorithms.

It should be noted that our robust scheduling excepted value model was formulated with some assumptions, so it may not represent the precise production shop environment. Therefore, future research would be the consideration of more restrictions rather than assumptions, e.g. concerning the resources capacity. Another area of improvement would be the robustness metrics for due-time POPs scheduling. Moreover, more research needs to be devoted to more efficient heuristic methods and their implementation in problem dedicated DSSs.

References

1. Bouleimen, K., Lecocq, H.: A new efficient simulated annealing algorithm for the resource-constrained project scheduling problem and its multiple mode version. Eur. J. Oper. Res. **149**(2), 268–281 (2003)
2. Wang, L., Fang, C.: An effective shuffled frog-leaping algorithm for multi-mode resource-constrained project scheduling problem. Inf. Sci. **181**, 4804–4822 (2011)
3. Dang, Q.-V., Nielsen, I., Steger-Jensen, K., Madsen, O.: Scheduling a single mobile robot for part-feeding tasks of production lines. J. Intell. Manuf. **25**(6), 1271–1287
4. Khayat, G.E., Langevin, A., Riope, D.: Integrated production and material handling scheduling using mathematical programming and constraint programming. Eur. J. Oper. Res. **175**(3), 1818–1832 (2006)
5. Linderoth, J.T., Savelsbergh, M.W.P.: A computational study of branch and bound search strategies for mixed integer programming. INFORMS J. Comput. **11**, 173–187 (1999)
6. Beale, E.M.L.: Branch and bound methods for mathematical programming systems. Ann. Discret. Math. **5**, 201–219 (1979)
7. Sitek, P.: A hybrid approach to the two-echelon capacitated vehicle routing problem (2E-CVRP). Adv. Intell. Syst. Comput. **267**, 251–263 (2014)
8. Kanagasabapathi, B., Ananthanarayanan, K.: A simulation model for resource constrained scheduling of multiple projects. In: 21st Annual ARCOM Conference, SOAS, vol. 2, pp. 823–831. University of London. Association of Researchers in Construction Management (2005)
9. Browning, T.R., Yassine, A.A.: Resource-constrained multi-project scheduling: priority rule performance revisited. Int. J. Prod. Econ. **126**(2), 212–228 (2010)

10. Lombardi, M., Milano, M.: Optimal methods for resource allocation and scheduling: a cross-disciplinary survey. Constraints **17**, 51–85 (2012)
11. Schwindt, C., Zimmermann, J.: Handbook on Project Management and Scheduling (2015)
12. Leung, J.Y-T., Anderson J.H.: Handbook of Scheduling: Algorithms, Models, and Performance Analysis. Chapman & Hall/CRC (2004)
13. Bzdyra, K., Banaszak, Z., Bocewicz, G.: Multiple project portfolio scheduling subject to mass customized service. Progress in Automation, Robotics and Measuring Techniques. Series: Advances in Intelligent Systems and Computing, vol. 350, pp. 11–21. Springer (2015)
14. Bzdyra, K., Bocewicz, G., Banaszak, Z.: Mass customized projects portfolio scheduling—imprecise operations time approach. Appl. Mech. Mater. **791**, 70–80 (2015)
15. Banaszak, Z., Bocewicz, G.: Declarative modeling for production orders portfolio scheduling. Found. Manag. **6**(3), 7–24 (2014)
16. Bocewicz, G., Muszyński, W., Banaszak, Z.: Models of multimodal networks and transport processes. Bull. Pol. Acad. Sci. Tech. Sci. **63**(3), 636–650 (2015)

Selected Aspects of Automatic Maneuver Control to Avoid Moving Obstacles Resulting from the Simulation Analysis of the Course of Aircraft Movement

Jerzy Graffstein

Abstract The article is focused on the results of analysis aimed at selected variables, which are found to be important for the automatic flight control in case of passing by a moving obstacle. Desired values of state variables, significant for the run of automatically controlled evasive manoeuvre, are computed. The set of four control laws is proposed for selected object capable to perform manoeuvres from the considered class. The appropriate controller gain are computed to guarantee the fulfilment of criteria assumed for automatic control. Numerical flight simulations are completed to cover the aforementioned phases of evasive manoeuvre and selected scenarios of objects motion. The results obtained can be treated as the source of information opening the deeper insight into a behaviour of the controlled aircraft in case of known scenario of obstacle's motion.

Keywords Obstacle avoidance · Automatic control · Computer simulation · Flight dynamics

1 Introduction

The essential characteristics of collision avoidance systems will depend primarily on two of its components: the obstacle detection system [1] and the automatic control system in the collision avoidance maneuver. The following thesis was set up in the study: analysis of selected variables during the flight of the aircraft is of fundamental importance for developing algorithms and the structure of the automatic flight control of the aircraft having appropriate response to the threat of collision [2, 3]. The maneuver for the aircraft to avoid movable obstacles is a

J. Graffstein (✉)
Centre of Space Technologies Department, Institute of Aviation, Al. Krakowska, 110/114, 02-256 Warsaw, Poland
e-mail: jerzy.graffstein@ilot.edu.pl

R. Szewczyk et al. (eds.), *Challenges in Automation, Robotics and Measurement Techniques*, Advances in Intelligent Systems and Computing 440, DOI 10.1007/978-3-319-29357-8_12

complex issue due to, among others, a large number of variables that determine how to accomplish the required maneuver. These variables affect the conditions of risk, on the one hand, in various ways [2]. On the other hand, there are complex relationships between them determined by the characteristic properties of the dynamics for the specific flying object. The course of the avoidance maneuver was discussed for the selected object type and the scenario configuration of aircraft movement and obstacles was accepted for consideration. The object–to–obstacle distance at the moment of obstacle detection is determined by technical means used in obstacle detector, first of all by the type of sensor [4–6]. The knowledge of this distance and parameters of object's and obstacle's motion enables determination of the most convenient anti–collision manoeuvre and its parameters [7, 3, 8]. The study attempted to present the analysis method of selected variables from the simulation course of automatically controlled avoidance maneuvers of moving obstacles. The results presented in the analysis allowed conclusions on important aspects of automatic control to be formulated. Proposals for modification of the system resulting from the above proposals were presented.

2 The Trajectory of Avoidance Maneuvers

Defining the shape of the trajectory in order to avoid movable obstacles aims to achieve better control over the safety of the maneuver protecting against collision [9, 10]. It took into account, among other things, the conditions for minimizing the amount of elementary maneuvers as well their greatest possible unification. The last requirement allows for shaping the trajectory easily with a small number of parameters. The most important objective to achieve in forming the trajectory of avoiding moving obstacles is to avoid collisions safely and keep a minimum distance of not less than the value taking into account the adopted safety margin. The course of trajectory shown in Fig. 1 selected in the study allows a collision to be avoided, avoiding moving obstacles and returning to the previous pursued flight plan. In addition, it seeks to return the aircraft to fly along a trajectory segment

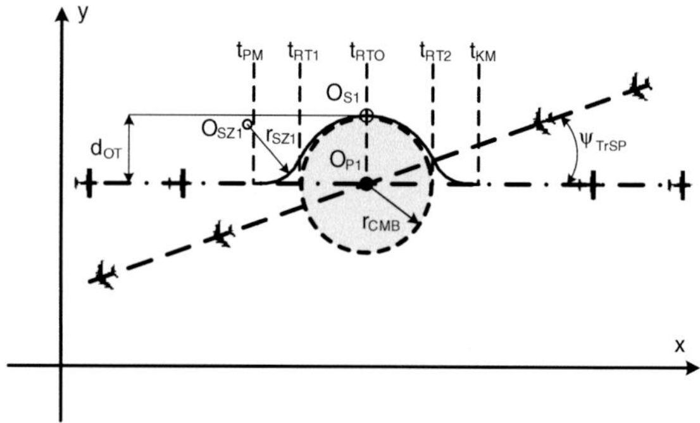

Fig. 1 The theoretical trajectory avoiding the moving obstacles

previously pursued. It was carried out at the shortest way in order to shape the course of the trajectory in question, primarily serving the following two parameters: the specified roll angle during the turn and the specified angle of yaw of the aircraft. It was necessary to complete three turns in order to obtain the complex maneuver illustrated. The first and third to the left and one to the right, each with the same roll angle adopted in the study with a value of 60°. The first turn made in a time interval of t_{PM} to t_{RT1} ensured that a collision was avoided and, as a result, the aircraft changed the yaw angle of approx. 40°. The second one undertaken in a time interval of t_{RT1} to t_{RT2} enabled the obstacles to be safely avoided. During its duration the aircraft changed the direction of flight of approx. 90°. The third one undertaken from t_{RT2} to t_{KM} ensured a return to the selected section of the trajectory which the aircraft moved along prior to commencing the first turn. During the last turn the aircraft changed its yaw angle by about 40°. While undertaking the complex maneuver, the maximum distance of the aircraft from the previously undertaken section of the flight plan was $d_{OT} = r_{CMB}$. Where the radius r_{CMB} is the total dimension of the aircraft, obstacles and the margin of safety [3, 11].

3 Control Laws and Desired State Variables

The specifics of the collision avoidance maneuver results in the need to establish high requirements for the operation of automatic control. Such a system should ensure a short time to adjust (especially when entering the turn), a small overshoot, and low oscillation level during the maneuver. Analysis of the duration of these phases of the maneuver with the selected situation was carried out in publication [12, 13]. The proposed automatic flight control of the aircraft carrying out the collision avoidance maneuver based on the control laws in four channels in the following form:

$$\begin{aligned}\delta_{HS} = {} & K^{H}_{1\Theta}(\Theta_z - \Theta) + K^{H}_{1Q}(Q_z - Q) + {}^{H}_{1W}(W_z - W) \\ & + K^{H}_{1z_1}(z_{1z} - z_1) + K^{H}_{1U}(U_z - U)\end{aligned} \tag{1}$$

$$\delta_{VS} = K^{V}_{1\Phi}(\Phi_z - \Phi) + K^{V}_{1P}(P_z - P) + K^{V}_{1V}(V) + K^{V}_{1R}(R_z - R) \tag{2}$$

$$\delta_{LS} = K^{L}_{1\Phi}(\Phi_z - \Phi) + K^{L}_{1P}(P_z - P) + K^{L}_{1V}(V) + K^{L}_{1R}(R_z - R) \tag{3}$$

$$\begin{aligned}\delta_{TS} = {} & K^{T}_{1\Theta}(\Theta_z - \Theta) + K^{T}_{1Q}(Q_z - Q) + K^{T}_{1W}(W_z - W) \\ & + K^{T}_{1z1}(z_{1z} - z_1) + K^{T}_{1U}(U_z - U)\end{aligned} \tag{4}$$

where the stated variables of the aircraft without index are measure in on-board of aircraft: $\mathbf{V_0} = [U \quad V \quad W]$, $\mathbf{\Omega} = [P \quad Q \quad R]$—components of linear and angular velocity, $[x_1 \quad y_1 \quad z_1]$, $\mathbf{\Lambda} = [\Phi \quad \Theta \quad \Psi]$—linear and angular position of the aircraft, the stated variables of the aircraft occurring in the control laws of the 'z' index is the desired value, calculated as shown in further parts of the study, $\mathbf{u} = [\delta_{HS} \quad \delta_{VS} \quad \delta_{LS} \quad \delta_{TS}]$—vector control (control surface deflection and throttle), K with double index—controller gains.

When selecting the appropriate controller gains, this formula ensures a turn with a specific roll angle which corresponds to the state of the predetermined radius turn. The values of controller gains have been calculated solving the algebraic Riccati equation (the method LQR) [14, 15] for steady state of aircraft manoeuvre. We are taking into account the aforementioned criteria including possible quick change of aircraft movement at the time of entry into the turn. In order to determine the value of controller gains, the integral quality index [2] was applied:

$$J_S = \int_{t=0}^{t=t_K} \left(\mathbf{x}^T\mathbf{Q}_1\mathbf{x} + \mathbf{u}^T\mathbf{R}_1\mathbf{u}\right)dt \tag{5}$$

where matrix elements by weight of $\mathbf{Q}_1$ and $\mathbf{R}_1$ were selected experimentally by way of digital movement simulation of the mathematical model of the aircraft.

The calculations were made initially only for flight conditions at a speed of 40 m/s (Table 1) being the average value of the interval velocity range within which the selected object can move, as shown by the results of simulation studies published later in the study for calculated ratios at lower speeds (Table 1) which did not meet the selected control criteria. Satisfactory behavior of automatic control is obtained after completing the calculation of controller gains (Table 2) for flight conditions at higher speed (50–60 m/s).

Avoiding the obstacle desired value of the yaw angle of the aircraft in the first stage:

$$\Psi_{Z1} = \Psi_{VS} - \beta_{ZNS} \tag{6}$$

The angle of the aircraft path determines it dependency (7):

$$\Psi_{VSZ} = \rho_2 + \arcsin[(V_P/V_S)\sin(\psi_{VP} - \rho_2)] \tag{7}$$

Table 1 Controller gains of the laws of control for the speed of 40 m/s

$K_{1U}^{H}\left[\frac{°s}{m}\right]$	$K_{1W}^{H}\left[\frac{°s}{m}\right]$	$K_{1Q}^{H}\left[\frac{°s}{rad}\right]$	$K_{1z1}^{H}\left[\frac{°}{m}\right]$	$K_{1\Theta}^{H}\left[\frac{°}{rad}\right]$
3.7103	0.7051	−12.869	0.15	−55.738
$K_{1V}^{V}\left[\frac{°s}{m}\right]$	$K_{1P}^{V}\left[\frac{°s}{rad}\right]$	$K_{1R}^{V}\left[\frac{°s}{rad}\right]$	$K_{1\Phi}^{V}\left[\frac{°}{rad}\right]$	$K_{\Psi}^{V}\left[\frac{°}{rad}\right]$
1.6837	2.7048	−36.336	−5.4	−3.6013
$K_{1V}^{L}\left[\frac{°s}{m}\right]$	$K_{1P}^{L}\left[\frac{°s}{rad}\right]$	$K_{1R}^{L}\left[\frac{°s}{rad}\right]$	$K_{1\Phi}^{L}\left[\frac{°}{rad}\right]$	$K_{1\Psi}^{L}\left[\frac{°}{rad}\right]$
2.146	9.7582	−19.728	92.1	6.0853
$K_{1U}^{T}\left[\frac{°s}{m}\right]$	$K_{1W}^{T}\left[\frac{°s}{m}\right]$	$K_{1Q}^{T}\left[\frac{°s}{rad}\right]$	$K_{1z1}^{T}\left[\frac{°}{m}\right]$	$K_{1\Theta}^{T}\left[\frac{°}{rad}\right]$
76.132	11.099	−41.396	−4.3036	−277.64

Table 2 Controller gains of the laws of control for the speed of 50 and 60 m/s

$K_{2U}^{H}\left[\frac{°s}{m}\right]$	$K_{2W}^{H}\left[\frac{°s}{m}\right]$	$K_{2Q}^{H}\left[\frac{°s}{rad}\right]$	$K_{2z1}^{H}\left[\frac{°}{m}\right]$	$K_{2\Theta}^{H}\left[\frac{°}{rad}\right]$
0.4477	0.3579	–9.8324	0.1032	−58.4015
$K_{2V}^{V}\left[\frac{°s}{m}\right]$	$K_{2P}^{V}\left[\frac{°s}{rad}\right]$	$K_{2R}^{V}\left[\frac{°s}{rad}\right]$	$K_{2\Phi}^{V}\left[\frac{°}{rad}\right]$	$K_{2\Psi}^{V}\left[\frac{°}{rad}\right]$
2.1645	1.9199	–41.286	–0.9571	0.0
$K_{2V}^{L}\left[\frac{°s}{m}\right]$	$K_{2P}^{L}\left[\frac{°s}{rad}\right]$	$K_{2R}^{L}\left[\frac{°s}{rad}\right]$	$K_{2\Phi}^{L}\left[\frac{°}{rad}\right]$	$K_{2\Psi}^{L}\left[\frac{°}{rad}\right]$
1.4796	6.2199	−16.125	120.594	0.0
$K_{2U}^{T}\left[\frac{°s}{m}\right]$	$K_{2W}^{T}\left[\frac{°s}{m}\right]$	$K_{2Q}^{T}\left[\frac{°s}{rad}\right]$	$K_{2z1}^{T}\left[\frac{°}{m}\right]$	$K_{2\Theta}^{T}\left[\frac{°}{rad}\right]$
452.1	23.5	−177.1	–3.9	−140.1

where angle line tangents to the circle have a diameter of r_{CMB} and the centre O_P:

$$\rho_2 = \beta_{SP} + \arcsin(r_{CMB}/r_{SP}), \tag{8}$$

$$\text{dla} \quad r_{SP} = r_{CMB}, \quad \arcsin(1) = \pi/2 \quad \rho_2 = \beta_{SP} + \pi/2, \tag{9}$$

and the line of sight angle [16]:

$$\beta_{SP} = arctg\left(\frac{y_{1P} - y_{1S}}{x_{1P} - x_{1S}}\right) \tag{10}$$

Taking into account the angle of drift of the aircraft β_{ZNS} expressed as a dependency:

$$\beta_{ZNS} = \arcsin\frac{\dot{y}_{1S}}{\sqrt{\dot{x}_{1S}^2 + \dot{y}_{1S}^2}} \tag{11}$$

In the adopted method the absolute value of a given roll angle has a constant value. For the case under consideration it is $\Phi_{Zl} = +60°$.

After reaching the predetermined value of the yaw angle, taking into account the specified experimental safety margin, changes in the value occur. The new desired value of the angle corresponds to the angle of the last path (of that angle) on that section of path:

$$\Psi_{Z2} = \Psi_{Dk} - \beta_{ZNS} \tag{12}$$

The example considered in this phase requires a turn from the predetermined value of the roll angle $\Phi_{Z2} = -60°$. After obtaining Ψ_{Z2} the yaw angle is still maintained according to the dependency of (12), however the predetermined roll angle is $\Phi_{Z3} = 0°$.

The total allowable duration of the collision avoidance maneuver in avoiding a collision with an obstacle should be no longer than:

$$t_{ZC} = (r_{SP} - r_{CMB})/V_{SP} \tag{13}$$

The specified speed of turn (in the terrestrial system around the axis Oz_g) taking into account the time to achieve [2] and the certain safety margin should not be less than:

$$\omega_{ZZ} \geq 2(\Psi_Z - \Psi_{S0})/(t_{ZC}) \tag{14}$$

where Ψ_{S0} is the initial yaw angle of the aircraft. Taking (14) into account and on the basis of kinematic dependencies of the aircraft [17, 13] the aircraft angular velocity components in the system associated with the aircraft is described in the following formula:

$$\begin{bmatrix} P_Z \\ Q_Z \\ R_Z \end{bmatrix} = \begin{bmatrix} 0 \\ 0 \\ \omega_{ZZ} \end{bmatrix} \begin{bmatrix} -\sin\Theta_Z \\ \sin\Phi_Z \cos\Theta_Z \\ \cos\Phi_Z \cos\Theta_Z \end{bmatrix} \tag{15}$$

where Θ_Z is the desired angle of pitch of the aircraft under equilibrium conditions carrying out a turn of a predetermined angle of roll.

4 Analysis of the Results of Simulation Tests

The adopted mathematical simulation model (among other, publications [17, 13]) was with reference to I-23 Manager type aircraft. This included a simplified representation of the actuator in the form of a first-order inertial term. A speed limit of deflection was implemented on all the control surfaces. The system of differential equations representing the aforementioned model was solved using MatLab software using the Runge-Kutta 4 th order integration step procedure amounting to 0.01 s. The object was flying at a constant speed at a constant level altitude of $H_S = 200$ m. At the same altitude, an obstacle was moving at a speed of $V_P = 72$ m/s, along a trajectory intersecting the trajectory of the aircraft at an angle of $\Psi_{TrSP} = 50°$. All elementary maneuvers (turns), included in the avoidance maneuver were carried out with a given angle roll of a value of $\Phi_{ZS} = 60°$. It is the maximum of safe value for stabilty flight of aircraft (for chosen type I-23 Manager) during evasive manoeuvre (experience on the simulation). The adopted radius of the circle (Fig. 2) $r_{CMB} = 80$ m resulted from the assumed safety margin of for example 50 m and a half of the sum of the size of the aircraft and obstacles. The value of safety margin depends on number of factors for example accuracy of data characterizing the obstacle, disturbance of flighte etc. Simulation of aircraft flight was carried out for three of its speeds amounting to 40, 50 and 60 m/s.

A comparison of the selected state variables and control during avoidance maneuvers in flight at speeds of $V_S = 40$ m/s shown in the graphs from (3) to (5).

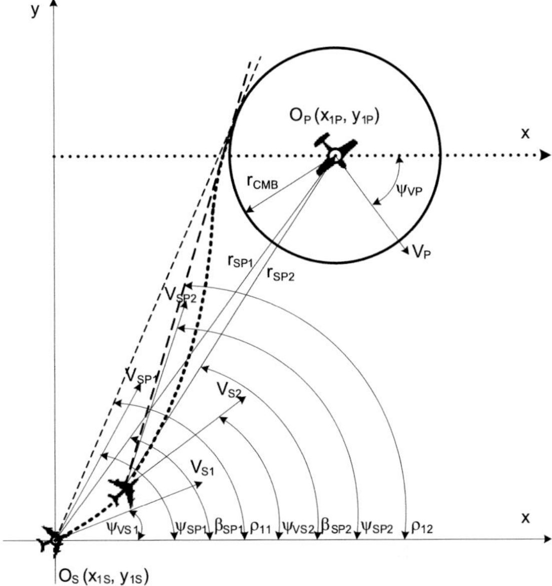

Fig. 2 The physical size of the aircraft and moving obstacles

Computer simulations were made under the same conditions, but with two different sets of controller gains of control laws. Meeting the criteria adopted to control controller gains were obtained (Table 1) calculated for the conditions at $V_S = 40$ m/s. For such a flight, the above coefficient criteria is not fulfilled, calculated using $V_S = 50$ m/s. In the latter case, the waveform angles (Fig. 4) and the angular velocity (Fig. 3) is characteristic, elongated by the regulation time—in the illustrated time period, the airplane does not reach the predetermined state. The

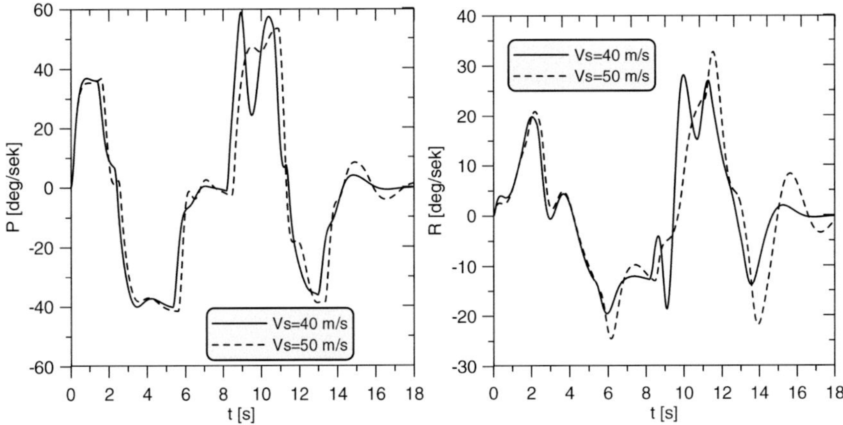

Fig. 3 A comparison of the angular velocity of roll and yaw of the plane obtained for automatic control of the controller gains from Tables 1 and 2

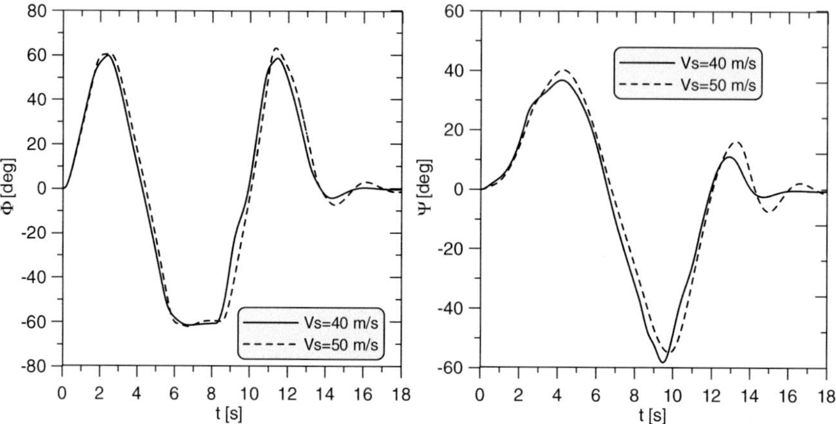

Fig. 4 A comparison of the angles of roll and yaw of the plane obtained for automatic control of the controller gains from Tables 1 and 2

differences are in the shape of trajectory, wherein for the coefficients in Table 2 a straight line is not achieved in the final stage, and the end portion of the trajectory is wavy. There was no evidence of unfavorable features of stated variables described above the trajectory in flight with $V_S = 60$ m/s for the controller gains of Table 2.

Overall, graphs from (3) to (10) refer to variable waveforms obtained from the avoidance maneuver simulation (along the trajectory with shape shown on Figs. 1, 5, 6, 7 and 8) started at different times in relation to that assumed. Starting the maneuver at an appropriate time results in the plane obtaining a minimum distance from the obstacle r_{SPmin}, different from that assumed. When this distance is greater than the assumed from the point of view of safety, it is more beneficial. But such a

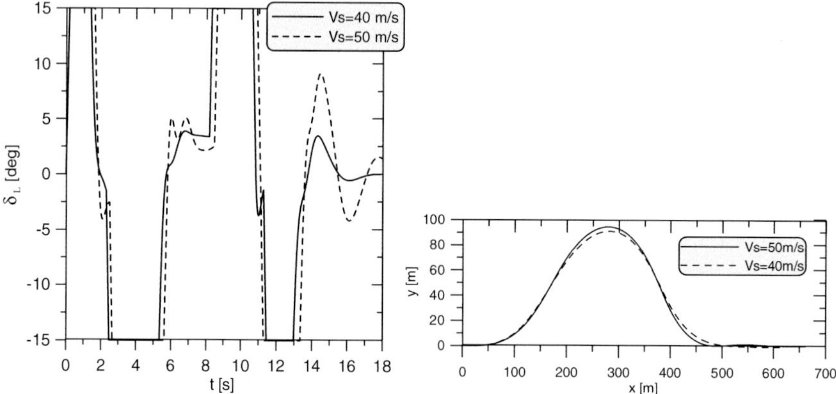

Fig. 5 Comparison of the aileron deflections and trajectory of the aircraft obtained for automatic control of the controller gains from Tables 1 and 2

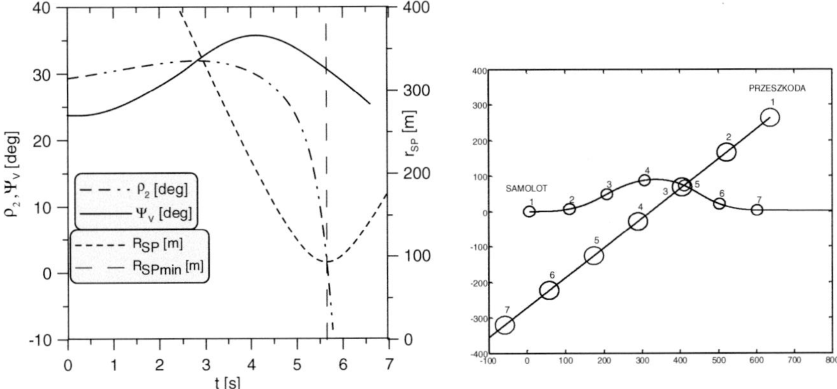

Fig. 6 The course of tangent angle, vector angle of relative velocity, the distance of the plane from obstacles and trajectory for a maneuver commencing earlier

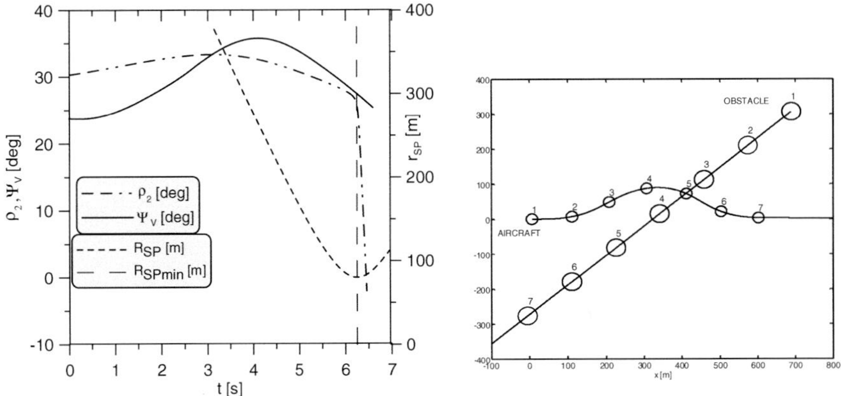

Fig. 7 The course of tangent angle, vector angle of relative velocity and the distance of the plane from obstacles and trajectory for a maneuver commencing in a time consistent with the assumptions

case requires sufficiently early detection of obstacles which, as a result of hardware limitations, it often occurs that it is not possible to obtain.

For a correct start time of the avoidance maneuver (Fig. 7) initially both angles ρ_2 and Ψ_V grow faster for the latter so that after about 3.0 s. angles are equal. After this ρ_2 slowly begins to decrease while Ψ_V still increases so that after a short time it starts to decline at a faster pace than ρ_2. From the moment of achieving r_{SPmin}, in which Ψ_V remained slightly larger than ρ_2 a sharp decline begins in the value of the angle ρ_2.

An example of the earlier ($\Delta t = +0.95$ s) start of the avoidance maneuver (Fig. 6) results in a higher r_{SPmin} value than the assumed by 13.2 m. It is

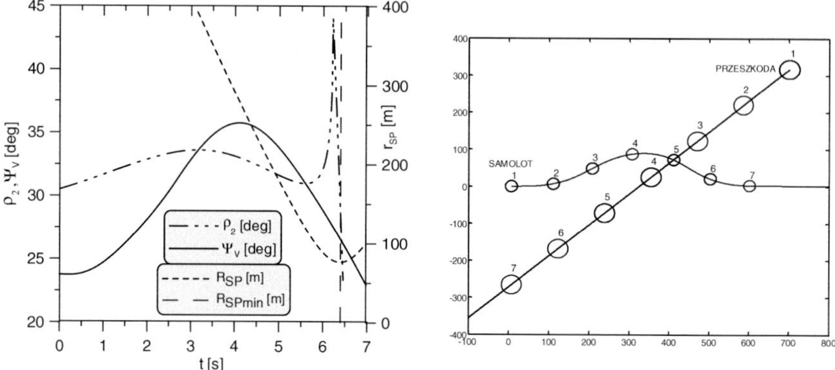

Fig. 8 The course of tangent angle, vector angle of relative velocity and the distance of the plane from obstacles and trajectory for a maneuver commencing later

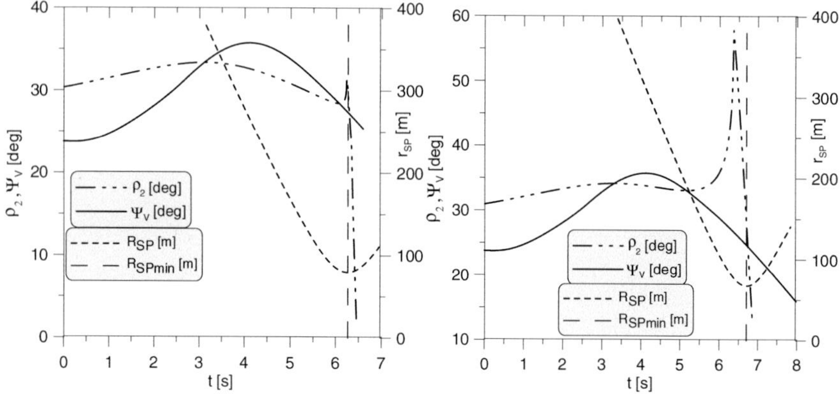

Fig. 9 The course of tangent angle, vector angle of relative velocity and the distance of the plane from obstacles for a two maneuvers commencing later

characteristic that after a slight and similarly continuing increase in the ρ_2 angle as in the previous cases and this value decreases at an accelerating pace. There is no sharp single increase of this speed as was the case for the conditions of Figs. 8 and 9. Consequently there is an increase in the difference of these angles even though the 1.5 s before reaching r_{SPmin} the downward trend in the value of the Ψ_V angle begins.

Most of all, however, there is shortage of time, which must necessarily be included for further consideration. Its occurrence increases the likelihood of a start which is too late for the required maneuver. This results in obtaining a smaller, minimum distance from the assumed and at the same time worsening the safety of the flight. In the graphs relating to this undesirable situation, three phases can be distinguished where the first angle Ψ_V increases and finally exceeds the ρ_2 angle,

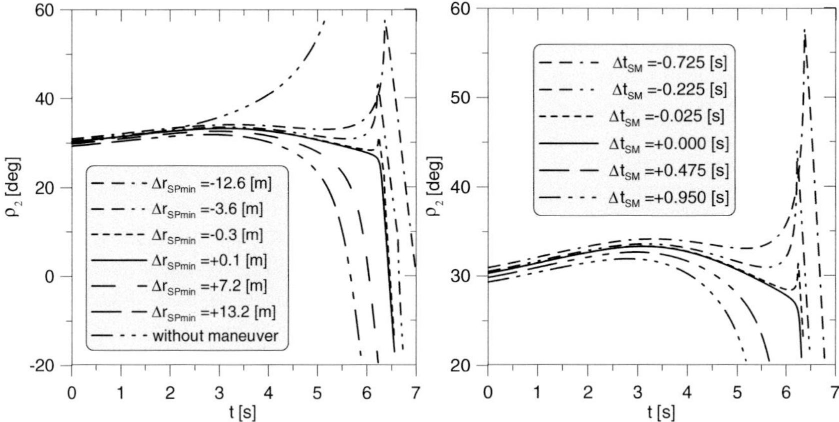

Fig. 10 The course of tangential angle for different start times of the avoidance maneuver

then the difference between them increases. In the next phase the Ψ_V angle decreases and at a certain point it becomes smaller than the ρ_2 angle. Even a short decrease of the Ψ_V angle below ρ_2 causes a minimum distance of r_{SPmin} to be achieved below that assumed. Almost immediately after achieving the condition of $\Psi_V < \rho_2$ a rapid increase in the ρ_2 angle follows. In the third phase, in close proximity to the minimum r_{SPmin} value there is a sufficiently rapid decline of the aforementioned ρ_2 angle, that despite the continuing trend of decrease in the Ψ_V angle, in a short time the condition is fulfilled $\Psi_V > \rho_2$.

On Fig. 10 a summary of the ρ_2 angle is presented for different time moments of starting the maneuver and in cases when no maneuver was made. The legend of that chart contains the value of the difference between the established and the resulting value of the minimum distance r_{SPmin}. In order to better understand the comparison of the transitional phases of the ρ_2 angle a significant portion of the graph is presented in Fig. 10. Increasing the delay of the start of the maneuver increases the maximum value of rapid change of the ρ_2 angle, for a delay of$\Delta t = -0.725$ s an irregular increase of the ρ_2 angle amounts to 25°.

The aforementioned increase in value ρ_2 starts, depending on the delay, from 0.3 to 1.1 s prior to achieving the r_{SPmin} value and ends approximately shortly after this value is exceeded.

5 Conclusions

The results of the analysis of flight simulation at the time of avoiding the moving obstacles allowed for patterns in the behavior of the relevant variable in terms of safety of aircraft to be determined. As a result of action, automatic control was achieved, the trajectory shape was adopted and the safety margin distances between the obstacle and the aircraft were established.

The relationship between the angle of the resultant velocity vector Ψ_V. was decisive in the case tested and the angle of the ρ_2 tangent. a greater value of Ψ_V angle was achieved for a certain period, the size ρ_2 does not ensure a lasting elimination of the danger of a collision or achieve the desired safety margin. An indication of permanently eliminating the threat of collision is a sufficiently long period of t_{SW} time, wherein the following condition is satisfied (13). Value t_{SW}. depends on many factors and developing a corresponding function describing its course is difficult and may not be necessary. On the basis of work analyzing the results of simulation studies, it can be concluded that the t_{SW} value is sufficient since the r_{SP} distance after achieving r_{SPmin} value begins to increase.

Flight simulation results confirm the need to check the controller gains calculated for the full range of selected state variables. The given control laws (1)–(4), together with the proposed method of calculating controller gains should function properly for small disruptions of flight. However, at a higher level, significant errors were expected, among other things, on the implementation of the trajectory of the aircraft thereby increasing the risk of collision. The late start of the avoidance maneuver, according to the analysis conducted earlier, will achieve less than the assumed distance of the aircraft from the obstacle which impairs flight safety. In order to eliminate these adverse effects, it is proposed to include additional elements in the laws of control on the basis of waveforms shown in graphs (3)–(5). They arise from taking into account the three previously existing variables: the difference between the selected tangential angle of the vector angle of the velocity of the aircraft in the Oxy plane, the rate of change of the difference and the trends of the tangent angle under consideration. The proposed modification will concern two control laws which obtain the following form in this way:

$$\begin{aligned}\delta_{VS} = {} & K^V_{1\Phi}(\Phi_z - \Phi) + K^V_{1P}(P_z - P) + K^V_{1V}(V) + K^V_{1R}(R_z - R) \\ & + K^V_{1\Psi_\rho}(\Psi_V - \rho_2) + K^V_{1\Delta\Psi_\rho}\Delta\psi_\rho + K^V_{1\rho}\dot{\rho}_2\end{aligned} \tag{16}$$

$$\begin{aligned}\delta_{LS} = {} & K^L_{1\Phi}(\Phi_z - \Phi) + K^L_{1P}(P_z - P) + K^L_{1V}(V) + K^L_{1R}(R_z - R) \\ & + K^L_{1\Psi_\rho}(\Psi_V - \rho_2) + K^L_{1\Delta\Psi_\rho}\Delta\psi_\rho + K^L_{1\rho}\dot{\rho}\end{aligned} \tag{17}$$

The final shape of modification of control laws and identifying additional controller gains require the need to carry out a number of simulation tests. The material presented in the study can be used as a guideline making it easier to select the parameters characterizing the chosen moving obstacle avoidance maneuvers.

References

1. Graffstein, J.: Anti-collision system with radar obstacle detector. Measur. Autom. Robot. **2**, 171–175 (2013) (Warsaw)
2. Curtis, J.W., Cloutier, JR.: Optimal obstacle avoidance using expanding trees and waypoint guidance. In: AIAA Guidance, Navigation, and Control Conference, pp. 1–8. Chicago (2009)
3. Graffstein, J.: An Influence of Selected State Variables on Accuracy of Aircraft Trajectory During Automatically Controlled Manoeuvre (in Polish), vol. 202, pp. 51–64. Transactions of the Institute of Aviation, Warsaw (2009)
4. Fasano, G., Forlenza, L., Accardo D., Moccia A.: Integrated obstacle detection system based on radar and optical sensors. In: AIAA Infotech and Aerospace Conference, pp. 1–17. Atlanta (2010)
5. Freeman, P., Moobrugger, P.: A low cost phased array solution for UAV collision avoidance. In: AIAA Infotech and Aerospace Conference, pp. 1–4. Atlanta (2010)
6. Higuchi, T., Toratani, D., Ueno S.: Double tetrahedron hexa-rotorcraft collision avoidance of indoor flying. In: International Congress of the Aeronautical Sciences ICAS, pp. 1–7 (2012)
7. Becker, M., Dantas, C., Macedo, W.P.: Obstacle avoidance procedure for mobile robots. In: International Congress of Mechanical Engineering, pp. 250–257. Ouro Preto (2006)
8. Smith, A.L., Harmon,F.G.: UAS Collision avoidance algorithm minimizing impact on route surveillance. In: AIAA Guidance, Navigation, and Control Conference, pp. 1–20. Chicago (2009)
9. Hashim, M.S.M., Tien-Fu, Lu.: Real-time control of time-critical trajectory planning for a car-like. Int. J. Eng. Res. Technol. **2**, 1–9 (2013)
10. Shim, D.H., Sastry, S.: An evasive maneuvering algorithm for UAVs in see–and–avoid situations. In: American Control Conference, pp. 3886–3891. New York (2007)
11. Graffstein, J.: Elements of collision threat detection process and automatically controlled emergency manoeuvre (in Polish). Measur. Autom. Robot. **2**, 383–387 (2012) (Warsaw)
12. Park, J-W., Kim, J.-H., Tahk M.-J.: UAV collision avoidance via optimal trajectory generation method. In: International Congress of the Aeronautical Sciences (ICAS), Brisbane (2012)
13. Phillips, W.F.: Mechanics of Flight. Willey, New Jersey (2010)
14. Athans, M., Falb, P.: Optimal control; an introduction to the theory and its applications, (translated into Polish). WNT, Warsaw (1969)
15. Kaczorek, T.: Theory of control, (in Polish), vols. 1, 2,. PWN, Warsaw (1981)
16. Lalish, E., Morgansen, K.A., Tsukamaki, T.: Decentralized reactive collision avoidance for multiple unicycle-type vehicles. In: American Control Conference, pp. 5055–5061. St. Louis (2009)
17. Maryniak, J.: General mathematical model of a controlled aircraft (in Polish). Mechanics in Flight, PTMTiS. pp. 575–592. Warsaw (1992)

Existence of Metzler Matrices with Given Spectra of Positive Stable Linear Systems and Electrical Circuits

Tadeusz Kaczorek

Abstract The problem of existence of Metzler matrices with given spectra is addressed for positive stable continuous-time linear systems and electrical circuits. Necessary and sufficient conditions for the existence of the Metzler matrices are established. It is shown that positive and stable electrical circuits have real eigenvalues. An open problem for positive electrical circuits is formulated. The considerations are illustrated by numerical examples of linear systems and positive electrical circuits.

Keywords Existence · Metzler matrix · Positive · Stable · System · Electrical circuit

1 Introduction

In positive electrical circuits the state variables and outputs take only nonnegative values for any nonnegative initial conditions and inputs. The positive standard and fractional order electrical circuits have been investigated in many papers and books [1–7]. A new class of normal electrical circuits has been introduced in [8]. The minimum energy control of electrical circuits has been investigated in [9]. Positive linear systems consisting of n subsystems with different fractional orders have been addressed in [10, 11]. Decoupling zeros of positive linear systems have been introduced in [12].

Determination of the state space equations for given transfer matrices is a classical problem, called the realization problem, which has been addressed in many papers and books [13–17]. An overview of the positive realization problem is given in [13, 16, 18]. The realization problem for positive continuous-time and

T. Kaczorek (✉)
Faculty of Electrical Engineering, Białystok University of Technology, Wiejska 45D, 15-351 Białystok, Poland
e-mail: kaczorek@isep.pw.edu.pl

R. Szewczyk et al. (eds.), *Challenges in Automation, Robotics and Measurement Techniques*, Advances in Intelligent Systems and Computing 440, DOI 10.1007/978-3-319-29357-8_13

discrete-time linear system has been considered in [16, 19–28] and for linear systems with delays in [16, 20, 25, 28–30]. The realization problem for fractional linear systems has been analyzed in [16, 31–33] and for positive 2D hybrid linear systems in [30]. A new modified state variable diagram method for determination of positive realizations with reduced number of delays for given proper transfer matrices has been proposed in [19]. The problem of determination of the set of Metzler matrices for given polynomials has been analyzed in [34].

In this paper the existence of Metzler matrices with given spectra of positive asymptotically stable linear systems and electrical circuits is investigated.

The paper is organized as follows. In Sect. 2 some basic definitions and theorems concerning positive and stable continuous-time linear systems are recalled and the problem is formulated. The solution of the problem is presented in Sect. 3. The problem for positive electrical circuits is analyzed in Sect. 4, where an open problem is formulated. Concluding remarks are given in Sect. 5.

The following notation will be used: $\Re$—the set of real numbers, $\Re^{n\times m}$—the set of $n\times m$ real matrices, $\Re_+^{n\times m}$—the set of $n\times m$ real matrices with nonnegative entries, $\Re^{n\times m}(s)$—the set of $n\times m$ rational matrices in s with real coefficients, I_n—the $n\times n$ identity matrix.

2 Continuous-Time Linear Systems and Problem Formulation

Consider the linear continuous-time autonomous system

$$\dot{x}(t)=Ax(t), x_0=x(0), \tag{1}$$

where $x(t)\in\Re^n$ is the state vector and $A\in\Re^{n\times n}$.

Definition 1 The system (1) is called (internally) positive if $x(t)\in\Re_+^n$ for all initial conditions $x_0\in\Re_+^n$.

Theorem 1 *[18] The system (1) is positive if and only if*

$$A\in M_n, \tag{2}$$

where M_n is the set of $n\times n$ Metzler matrices (i.e. real matrices with nonnegative off-diagonal entries).

Definition 2 The positive system (1) is called asymptotically stable if $\lim_{t\to\infty} x(t)=0$ for all $x_0\in\Re_+^n$.

Theorem 2 *[18] The positive system (1) is asymptotically stable if and only if one of the following conditions is satisfied:*

1. *All coefficients of the characteristic polynomial*

$$p(s) = \det[I_n s - A] = s^n + a_{n-1}s^{n-1} + \cdots + a_1 s + a_0 \tag{3}$$

are positive, i.e. $a_k > 0$ *for* $k = 0, 1, \ldots, n-1$.

2. *All principal minors of the matrix* $\hat{A} = -A = [\hat{a}_{ij}]$ *are positive, i.e.*

$$\hat{M}_1 = |\hat{a}_{11}| > 0, \hat{M}_2 = \begin{vmatrix} \hat{a}_{11} & \hat{a}_{12} \\ \hat{a}_{21} & \hat{a}_{22} \end{vmatrix} > 0, \ldots, \hat{M}_n = \det\hat{A} > 0 \tag{4}$$

The problem under the considerations for continuous-time positive and asymptotically stable systems can be stated as follows:

Given a set of n_1 *real and* n_2 *pairs of complex conjugate eigenvalues*

$$\lambda_{k_1} = -\alpha_{k_1}, k_1 = 1, \ldots, n_1, \lambda_{k_2} = -\beta_{k_2} \pm j\gamma_{k_2}, k_2 = 1, \ldots, n_2 \tag{5}$$

find asymptotically stable Metzler matrix $A \in M_n$.

3 Problem Solution

It is well-known that the symmetric matrix $A \in M_n$ has only real eigenvalues. Therefore, if $n_2 > 0$ the desired matrix is nonsymmetrical.

Lemma 1 *The Metzler matrix for* $n = 2$ *has only real eigenvalues.*

Proof The zeros of the polynomial

$$\det[I_2 s - A_2] = \begin{vmatrix} s - a_{11} & -a_{12} \\ -a_{21} & s - a_{22} \end{vmatrix} = s^2 - (a_{11} + a_{22})s + a_{11}a_{22} - a_{12}a_{21} \tag{6}$$

i.e. $s_1 = -\alpha_1$, $s_2 = -\alpha_2$ are real since

$$\begin{aligned} (a_{11} + a_{22})^2 - 4(a_{11}a_{22} - a_{12}a_{21}) &= a_{11}^2 + a_{22}^2 - 2a_{11}a_{22} + 4a_{12}a_{21} \\ &= (a_{11} - a_{22})^2 + 4a_{12}a_{21} \geq 0 \end{aligned} \tag{7}$$

for all $a_{ij} \geq 0$, $i, j = 1, 2$. □

From Lemma 1 we have the following theorem.

Theorem 3 *For* $n = 2$ *does not exist a Metzler matrix with a given pair of complex conjugate eigenvalues* $s_1 = -\beta_1 + j\gamma_1$, $s_2 = -\beta_1 - j\gamma_1$.

For $n = 3$ *we shall prove the following theorem.*

Theorem 4 There exists an asymptotically stable Metzler matrix

$$A_3 = \begin{bmatrix} -a_{11} & a_{12} & a_{13} \\ a_{21} & -a_{22} & a_{23} \\ a_{31} & a_{32} & -a_{33} \end{bmatrix} \in M_3 \tag{8}$$

with given eigenvalues

$$s_1 = -\alpha, s_2 = -\beta + j\gamma, s_3 = -\beta - j\gamma, 0 < \alpha < \beta, \gamma > 0 \tag{9}$$

if and only if

$$(\beta - \alpha)^2 > 3\gamma^2. \tag{10}$$

Proof The eigenvalues (9) of the matrix (8) satisfy the following well-known equalities

$$-(s_1 + s_2 + s_3) = \alpha + 2\beta = a_{11} + a_{22} + a_{33}, \tag{11}$$

$$\begin{aligned} s_1(s_2 + s_3) + s_2 s_3 &= 2\alpha\beta + \beta^2 + \gamma^2 \\ &= a_{11}(a_{22} + a_{33}) + a_{22}a_{33} - a_{12}a_{21} - a_{13}a_{31} - a_{23}a_{32}, \end{aligned} \tag{12}$$

$$\begin{aligned} -\det A = -s_1 s_2 s_3 &= \alpha(\beta^2 + \gamma^2) \\ &= a_{11}a_{22}a_{33} - a_{11}a_{23}a_{32} - a_{22}a_{13}a_{31} - a_{33}a_{12}a_{21} - a_{12}a_{31}a_{23} - a_{13}a_{21}a_{32}. \end{aligned} \tag{13}$$

Note that the sum $a_{11} + a_{22} + a_{33}$ takes its maximal value for $a_{11} = a_{22} = a_{33} = \frac{\alpha + 2\beta}{3}$ and the condition (12) can be satisfied for $a_{ij} \geq 0$, $i \neq j$ if and only if

$$a_{11}(a_{22} + a_{33}) + a_{22}a_{33} = \frac{(\alpha + 2\beta)^2}{3} > 2\alpha\beta + \beta^2 + \gamma^2 \tag{14}$$

It is easy to check that the condition (14) is equivalent to the condition (10) and if (10) holds then there exist $a_{ij} \geq 0$, $i \neq j$ of the matrix (8) such that (13) is satisfied. □

Example 1 Find the matrix (8) with the eigenvalues

$$s_1 = -1, s_2 = -2 + j0.1, s_3 = -2 - j0.1. \tag{15}$$

The characteristic polynomial of the matrix with the eigenvalues (15) has the form

$$(s+1)(s+2-j0.1)(s+2+j0.1)=s^3+5s^2+8.01s+4.01. \tag{16}$$

In this case the condition (10) is satisfied since $\alpha=1$, $\beta=2$, $\gamma=0.1$ and

$$(\beta-\alpha)^2=(2-1)^2>3\gamma^2=0.03. \tag{17}$$

Assuming $a_{11}=a_{22}=a_{33}=\frac{5}{3}$, $a_{13}=a_{21}=1$, $a_{31}=a_{23}=0$ we are looking for the values of a_{12} and a_{32} such that the polynomial (16) is the characteristic polynomial of the desired matrix

$$\begin{aligned}\det[I_3 s-A_3]&=\begin{vmatrix} s+\frac{5}{3} & -a_{12} & -1 \\ -1 & s+\frac{5}{3} & 0 \\ 0 & -a_{32} & s+\frac{5}{3}\end{vmatrix}\\ &=s^3+5s^2+\left(\frac{75}{9}-a_{12}\right)s+\left(\frac{5}{3}\right)^3-\frac{5}{3}a_{12}-a_{32}\\ &=s^3+5s^2+8.01s+4.01.\end{aligned} \tag{18}$$

From comparison of the coefficients of (18) we have

$$a_{12}=\frac{75}{9}-8.01=0.3233,\, a_{32}=\left(\frac{5}{3}\right)^3-\frac{5}{3}a_{12}-4.01=0.0808.$$

It is easy to check that the matrix

$$A_3=\begin{bmatrix} -1.6667 & 0.3233 & 1 \\ 1 & -1.6667 & 0 \\ 0 & 0.0808 & -1.6667\end{bmatrix} \tag{19}$$

has the desired eigenvalues.

Note that the matrix

$$\bar{A}_3=\begin{bmatrix} -1.6667 & 0 & 0.0808 \\ 1 & -1.6667 & 0.3233 \\ 0 & 1 & -1.6667\end{bmatrix} \tag{20}$$

is also a solution to the problem, i.e. its characteristic polynomial has also the desired form (18). The matrix (20) has been obtained from (19) by similarity transformation $\bar{A}_3=PA_3P^{-1}$ for the permutation matrix

$$P=\begin{bmatrix} 0 & 0 & 1 \\ 1 & 0 & 0 \\ 0 & 1 & 0\end{bmatrix} \tag{21}$$

Therefore, we have the following important conclusion.

Conclusion 1 The problem under considerations has many solutions. Any matrix $A = PAP^{-1}$ is also the desired solution to the problem for any permutation matrix $P \in \Re_{+}^{3\times 3}$ since

$$\det[I_3 s - \bar{A}_3] = \det P \det[I_3 s - A_3] \det P^{-1} = \det[I_3 s - A_3]$$

and $\bar{A}_3 \in M_3$ if and only if $A_3 \in M_3$.

It is well-known that for $n = 4$ the matrix $A_4 \in M_4$ has at least two real eigenvalues. For $n = 4$ the problem can be stated as follows. Given eigenvalues

$$s_1 = -\alpha_1,\ s_2 = -\alpha_2,\ s_3 = -\beta + j\gamma,\ s_4 = -\beta - j\gamma,\ \alpha_1 < \alpha_2 < \beta \tag{22}$$

find a Metzler matrix of the form

$$A_4 = \begin{bmatrix} -a_{11} & 0 & 0 & 0 \\ 0 & -a_{22} & a_{23} & a_{24} \\ 0 & a_{32} & -a_{33} & a_{34} \\ 0 & a_{42} & a_{43} & -a_{44} \end{bmatrix} = \begin{bmatrix} -a_{11} & 0 \\ 0 & A_3 \end{bmatrix},\ A_3 = \begin{bmatrix} -a_{22} & a_{23} & a_{24} \\ a_{32} & -a_{33} & a_{34} \\ a_{42} & a_{43} & -a_{44} \end{bmatrix} \tag{23}$$

with the desired eigenvalues (22).

Theorem 5 *For $n = 4$ there exists an asymptotically stable Metzler matrix (23) with given eigenvalues if and only if*

$$(\beta - \alpha_1)^2 > 3\gamma^2. \tag{24}$$

Proof We choose $a_{11} = \alpha_2$ and to find the Metzler matrix $A_3 \in M_3$ we apply the method given for $n = 3$. By Theorem 4 the problem for $n = 4$ has a solution if and only if the condition (24) is satisfied. □

Example 2 Find the Metzler matrix (23) with the eigenvalues

$$s_1 = -1,\ s_2 = -1.5,\ s_3 = -2 + j0.1,\ s_4 = -2 - j0.1. \tag{25}$$

To find the desired Metzler matrix we can use the results of Example 1. Note that the condition (24) is satisfied since (17) holds. In this case we have $a_{11} = \alpha_2$ and the matrix A_3 has the form (19). The desired matrix with eigenvalues (25) has the form

$$A_4 = \begin{bmatrix} -1.5 & 0 & 0 & 0 \\ 0 & -1.6667 & 0.3233 & 1 \\ 0 & 1 & -1.6667 & 0 \\ 0 & 0 & 0.0808 & -1.6667 \end{bmatrix}. \tag{26}$$

Note that any matrix $\bar{A}_4 = PA_4P^{-1}$ is also the desired solution for any permutation matrix $P_4 \in \Re_+^{4\times 4}$.

Now let us consider the problem for $n = 5$. Given eigenvalues

$$s_1 = -\alpha_1,\ s_2 = -\alpha_2,\ s_3 = -\alpha_3,\ s_4 = -\beta + j\gamma, s_5 = -\beta - j\gamma \tag{27}$$

find a Metzler matrix of the form

$$A_5 = \begin{bmatrix} -a_{11} & 0 & 0 \\ 0 & -a_{22} & 0 \\ 0 & 0 & A_3 \end{bmatrix} \in M_5,\ A_3 = \begin{bmatrix} -a_{33} & a_{34} & a_{35} \\ a_{43} & -a_{44} & a_{45} \\ a_{53} & a_{54} & -a_{55} \end{bmatrix} \in M_3 \tag{28}$$

with the desired eigenvalues (27).

Theorem 6 *For $n = 5$ there exists an asymptotically stable Metzler matrix (28) with given eigenvalues (27) if and only if the condition (24) is satisfied.*

Proof We choose $a_{11} = \alpha_2$, $a_{22} = \alpha_3$ and to find the Metzler matrix A_3 we apply the method given for $n = 3$. By Theorem 4 the problem for $n = 5$ has a solution if and only if the condition (24) is satisfied. □

Example 3 Find the Metzler matrix (28) with the eigenvalues

$$s_1 = -1,\ s_2 = -1.2,\ s_3 = -1.6,\ s_4 = -2 + j0.1,\ s_5 = -2 - j0.1. \tag{29}$$

To find the desired Metzler matrix we can use the results of Example 1. In this case we choose $a_{11} = -s_2 = 1.2$, $a_{22} = -s_3 = 1.6$ and the remaining eigenvalues as the eigenvalues of the matrix A_3. The eigenvalues s_1, s_4, s_5 satisfy the condition (24). Therefore, using the results of Example 1 we obtain the desired matrix in the form

$$A_5 = \begin{bmatrix} -1.2 & 0 & 0 & 0 & 0 \\ 0 & -1.6 & 0 & 0 & 0 \\ 0 & 0 & -1.6667 & 0.3233 & 1 \\ 0 & 0 & 1 & -1.6667 & 0 \\ 0 & 0 & 0 & 0.0808 & -1.6667 \end{bmatrix}. \tag{30}$$

Note that this procedure can be applied for any $n > 4$ if the desired Metzler matrix has only one pair of complex conjugate eigenvalues and for $n = 3k$, $k = 2, 3, \ldots$ if the desired Metzler matrix has k pairs of complex conjugate eigenvalues.

4 Positive Electrical Circuits

Consider the linear electrical circuits described by the state equation

$$\dot{x} = Ax + Bu, \; x_0 = x(0), \tag{31}$$

where $x = x(t) \in \Re^n$, $u = u(t) \in \Re^m$ are the state and input vectors and $A \in \Re^{n\times n}$, $B \in \Re^{n\times m}$.

The electrical circuit is called positive if $x(t) \in \Re^n_+$ for any $x_0 \in \Re^n_+$ and all $u(t) \in \Re^m_+$, $t \geq 0$.

Theorem 7 *[7] The electrical circuit is positive if and only if*

$$A \in M_n \text{ and } B \in \Re^{n\times m}_+, \tag{32}$$

where M_n is the set of $n\times n$ Metzler matrices.

Example 4 Consider the electrical circuit shown in Fig. 1 with given resistances R_k, $k = 1, \ldots, 8$, inductances L_1, L_2 and source voltages e_1, e_2.

Using the mesh method we obtain the equations

$$\begin{aligned} L_1\frac{di_1}{dt} &= -R_{11} + R_3 i_3 + R_5 i_4, \\ L_2\frac{di_2}{dt} &= -R_{22} i_2 + R_4 i_3 + R_7 i_4, \end{aligned} \tag{33}$$

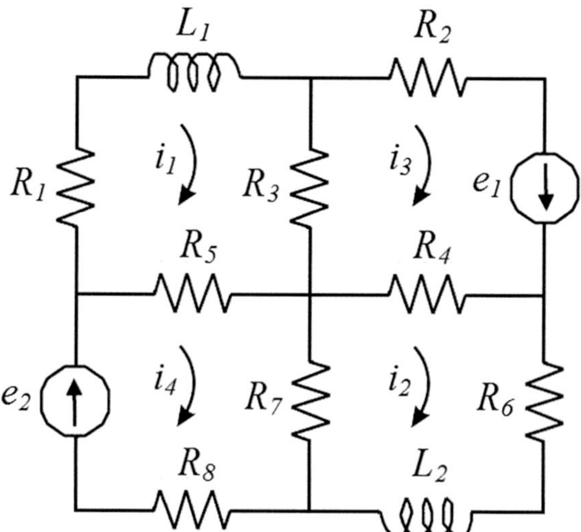

Fig. 1 Positive electrical circuit of Example 4

$$\begin{aligned} 0 &= R_3 i_1 + R_4 i_2 - R_{33} i_3 + e_1, \\ 0 &= R_5 i_1 + R_7 i_2 - R_{44} i_4 + e_2, \end{aligned} \tag{34}$$

where $R_{11} = R_1 + R_3 + R_5$, $R_{22} = R_4 + R_6 + R_7$, $R_{33} = R_2 + R_3 + R_4$, $R_{44} = R_5 + R_7 + R_8$ and i_k, $k = 1, \ldots, 4$ are the mesh currents.

From (34) we have

$$\begin{bmatrix} i_3 \\ i_4 \end{bmatrix} = \begin{bmatrix} \frac{R_3}{R_{33}} & \frac{R_4}{R_{33}} \\ \frac{R_5}{R_{44}} & \frac{R_7}{R_{44}} \end{bmatrix} \begin{bmatrix} i_1 \\ i_2 \end{bmatrix} + \begin{bmatrix} \frac{1}{R_{33}} & 0 \\ 0 & \frac{1}{R_{44}} \end{bmatrix} \begin{bmatrix} e_1 \\ e_2 \end{bmatrix}. \tag{35}$$

Substitution of (35) into (33) yields

$$\frac{d}{dt} \begin{bmatrix} i_1 \\ i_2 \end{bmatrix} = A \begin{bmatrix} i_1 \\ i_2 \end{bmatrix} + B \begin{bmatrix} e_1 \\ e_2 \end{bmatrix}, \tag{36}$$

where

$$\begin{aligned} A &= \begin{bmatrix} \frac{-R_{11}R_{33}R_{44} + R_3^2 R_{44} + R_5^2 R_{33}}{L_1 R_{33} R_{44}} & \frac{R_3 R_4 R_{44} + R_5 R_7 R_{33}}{L_1 R_{33} R_{44}} \\ \frac{R_3 R_4 R_{44} + R_5 R_7 R_{33}}{L_2 R_{33} R_{44}} & \frac{-R_{22}R_{33}R_{44} + R_4^2 R_{44} + R_7^2 R_{33}}{L_2 R_{33} R_{44}} \end{bmatrix}, \\ B &= \begin{bmatrix} \frac{R_3}{L_1 R_{33}} & \frac{R_5}{L_1 R_{44}} \\ \frac{R_4}{L_2 R_{33}} & \frac{R_7}{L_2 R_{44}} \end{bmatrix}. \end{aligned} \tag{37}$$

Note that the matrix A is a Metzler matrix and the matrix B has positive entries. Therefore, the electrical circuit is a positive one with real negative eigenvalues. The electrical circuit is positive and asymptotically stable and its matrix A satisfies the condition [7, 18]

$$-A^{-1} \in \mathfrak{R}_+^{n \times n}. \tag{38}$$

In general case we have the following theorem [7].

Theorem 8 *The electrical circuit composed of resistances, inductances and source voltages is positive and asymptotically stable for positive values of the resistances and inductances if and only if the number of the inductances is less or equal to the number of linearly independent meshes and each independent mesh contains at least one positive resistance. The matrix $A \in M_n$ of the asymptotically stable positive electrical circuit satisfies the condition (38).*

Dual results hold for positive asymptotically stable electrical circuits with given resistances, capacitances and source voltages [7].

Theorem 9 *[18] The Metzler matrix $A \in M_n$ of positive electrical circuit is asymptotically stable if and only if all coefficients of the characteristic polynomial*

$$\det[I_n\lambda - A] = \lambda^n + a_{n-1}\lambda^{n-1} + \cdots + a_1\lambda + a_0 = (\lambda - \lambda_1)(\lambda - \lambda_2)\ldots(\lambda - \lambda_n)$$

are positive, i.e. $a_k > 0$ for $k = 0, 1, \ldots, n-1$.

Theorem 10 *The Metzler matrix $A \in M_n$ of the positive electrical circuit composed of resistances and inductances or of resistances and capacitances has non-positive real eigenvalues. The eigenvalues are real negative if the electrical circuit has not independent meshes containing at least one positive resistance or at least one node with branches containing only capacitors and current sources.*

Proof Proof follows immediately from Theorem 8 and dual results hold for positive asymptotically stable electrical circuits composed of resistances, capacitances and source currents [7]. □

Remark 1 The Metzler matrix $A \in M_n$ may have a pair of complex conjugate eigenvalues only if the positive electrical circuit is composed of resistances, inductances and capacitances.

Remark 2 [18] The Metzler matrix $A \in M_n$ of positive electrical circuit has at least one real eigenvalue $\lambda_1 = \alpha$ satisfying the condition

$$\mathrm{Re}\,\lambda_k < \alpha \quad \text{for}\, k = 2, \ldots, n. \tag{39}$$

Example 5 Consider the positive electrical circuit shown in Fig. 2 with given positive resistances R_1, R_2, R, inductance L, capacitances C_1, C_2 and source voltage e.

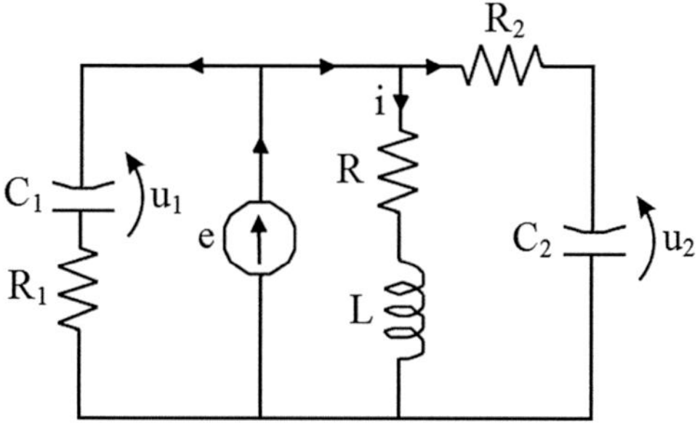

Fig. 2 Positive electrical circuit of Example 5

Using Kirchhoff's laws we may write the equations

$$\begin{aligned} e &= R_1 C_1 \frac{du_1}{dt} + u_1, \\ e &= Ri + L\frac{di}{dt}, \\ e &= R_2 C_2 \frac{du_2}{dt} + u_2, \end{aligned} \tag{40}$$

which can be written in the form

$$\frac{d}{dt}\begin{bmatrix} u_1 \\ u_2 \\ i \end{bmatrix} = A \begin{bmatrix} u_1 \\ u_2 \\ i \end{bmatrix} + Be, \tag{41}$$

where

$$A = \begin{bmatrix} -\frac{1}{R_1 C_1} & 0 & 0 \\ 0 & -\frac{1}{R_2 C_2} & 0 \\ 0 & 0 & -\frac{R}{L} \end{bmatrix}, \quad B = \begin{bmatrix} \frac{1}{R_1 C_1} \\ \frac{1}{R_2 C_2} \\ \frac{1}{L} \end{bmatrix}. \tag{42}$$

The electrical circuit is positive and asymptotically stable since $A \in M_3$, $B \in \Re_+^3$ and all diagonal entries of the matrix A are negative.

Example 6 Consider the positive electrical circuit shown in Fig. 3 for $n_1 = 3$, $n_2 = 4$ with given positive resistances R_1, R_2, R_3, R_4, inductances L_2, L_4, capacitances C_1, C_3 and source voltages e_0, e_2, e_4.

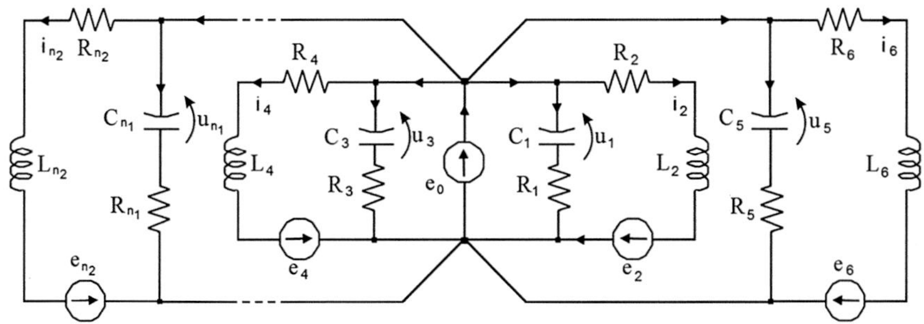

Fig. 3 Positive electrical circuit of Example 6

In this case the state equations have the form [7]

$$\frac{d}{dt}\begin{bmatrix} u_1 \\ u_3 \\ i_2 \\ i_4 \end{bmatrix} = A\begin{bmatrix} u_1 \\ u_3 \\ i_2 \\ i_4 \end{bmatrix} + B\begin{bmatrix} e_0 \\ e_2 \\ e_4 \end{bmatrix}, \tag{43}$$

where

$$A = \operatorname{diag}\begin{bmatrix} -\frac{1}{R_1C_1} & -\frac{1}{R_3C_3} & -\frac{R_2}{L_2} & -\frac{R_4}{L_4} \end{bmatrix}, \; B = \begin{bmatrix} \frac{1}{R_1C_1} & 0 & 0 \\ \frac{1}{R_3C_3} & 0 & 0 \\ \frac{1}{L_2} & \frac{1}{L_2} & 0 \\ \frac{1}{L_4} & 0 & \frac{1}{L_4} \end{bmatrix}. \tag{44}$$

The electrical circuit is positive and asymptotically stable since $A \in M_4$, $B \in \Re_+^{4\times 3}$ and all diagonal entries of the matrix A are negative.

Remark 3 The eigenvalues of the matrices A of presented positive asymptotically stable electrical circuits are real.

Therefore, there arises the following open problem:

Give an example of positive asymptotically stable electrical circuit with at least one pair of complex conjugate eigenvalues of the matrix A.

5 Concluding Remarks

The problem of existence of Metzler matrices with given spectra has been addressed. Necessary and sufficient conditions for a large class of positive and stable continuous-time linear systems and positive electrical circuits have been established (Theorems 3–6). It has been shown that positive and stable electrical circuits have usually real eigenvalues. An open problem for the positive electrical circuits has been formulated. The considerations have been illustrated by numerical examples of continuous-time linear systems and positive electrical circuits. The considerations can be easily extended to discrete-time linear systems.

Acknowledgment This work was supported by National Science Centre in Poland under work No. 2014/13/B/ST7/03467.

References

1. Kaczorek, T.: A class of positive and stable time-varying electrical circuits. Electr. Rev. **91**(5), 121–124 (2015)
2. Kaczorek, T.: Constructability and observability of standard and positive electrical circuits. Electr. Rev. **89**(7), 132–136 (2013)
3. Kaczorek, T.: Positive electrical circuits and their reachability. Arch. Electr. Eng. **60**(3), 283–301 (2011)
4. Kaczorek T.: Positive fractional linear electrical circuits. In: Proceedings of SPIE, vol. 8903. Bellingham WA, USA, Art. No 3903-35
5. Kaczorek, T.: Positive unstable electrical circuits. Electr. Rev. **88**(5a), 187–192 (2012)
6. Kaczorek, T.: Zeroing of state variables in descriptor electrical circuits by state-feedbacks. Electr. Rev. **89**(10), 200–203 (2013)
7. Kaczorek T., Rogowski K.: Fractional Linear Systems and Electrical Circuits. Studies in Systems, Decision and Control, vol. 13. Springer (2015)
8. Kaczorek, T.: Normal positive electrical circuits. IET Circ. Theory Appl. **9**(5), 691–699 (2015)
9. Kaczorek T.: Minimum energy control of positive electrical circuits. In: Proceedings of Conference MMAR, Miedzyzdroje, pp. 2–9, 2–5 Aug 2014
10. Kaczorek, T.: Positive linear systems with different fractional orders. Bull. Pol. Acad. Sci. Techn. **58**(3), 453–458 (2010)
11. Kaczorek, T.: Positive systems consisting of n subsystems with different fractional orders. IEEE Trans. Circuits Syst. Regul. Paper **58**(6), 1203–1210 (2011)
12. Kaczorek T.: Decoupling zeros of positive continuous-time linear systems and electrical circuits. Advances in Systems Science. Advances in Intelligent Systems and Computing, vol. 240. pp. 1–15. Springer (2014)
13. Benvenuti, L., Farina, L.: A tutorial on the positive realization problem. IEEE Trans. Autom. Control **49**(5), 651–664 (2004)
14. Farina, L., Rinaldi, S.: Positive Linear Systems: Theory and Applications. Wiley, New York (2000)
15. Kaczorek, T.: Linear Control Systems, vol. 1. Research Studies Press, Wiley, New York (1992)
16. Kaczorek T., Sajewski Ł.: Realization Problem for Positive and Fractional Systems. Springer (2014)
17. Shaked, U., Dixon, M.: Generalized minimal realization of transfer-function matrices. Int. J. Contr. **25**(5), 785–803 (1977)
18. Kaczorek, T.: Positive 1D and 2D Systems. Springer, London (2002)
19. Kaczorek, T.: A modified state variables diagram method for determination of positive realizations of linear continuous-time systems with delays. Int. J. Appl. Math. Comput. Sci. **22** (4), 897–905 (2012)
20. Kaczorek, T.: A realization problem for positive continuous-time linear systems with reduced numbers of delays. Int. J. Appl. Math. Comput. Sci. **16**(3), 325–331 (2006)
21. Kaczorek, T.: Computation of positive stable realizations for discrete-time linear systems. Comput. Prob. Electr. Eng. **2**(1), 41–48 (2012)
22. Kaczorek, T.: Computation of positive stable realizations for linear continuous-time systems. Bull. Pol. Acad. Techn. Sci. **59**(3), 273–281 (2011)
23. Kaczorek, T.: Computation of realizations of discrete-time cone systems. Bull. Pol. Acad. Sci. Techn. **54**(3), 347–350 (2006)
24. Kaczorek, T.: Positive and asymptotically stable realizations for descriptor discrete-time linear systems. Bull. Pol. Acad. Sci. Techn. **61**(1), 229–237 (2013)
25. Kaczorek, T.: Positive minimal realizations for singular discrete-time systems with delays in state and delays in control. Bull. Pol. Acad. Sci. Techn. **53**(3), 293–298 (2005)
26. Kaczorek, T.: Positive stable realizations of discrete-time linear systems. Bull. Pol. Acad. Sci. Techn. **60**(3), 605–616 (2012)

27. Kaczorek, T.: Positive stable realizations with system Metzler matrices. Arch. Control Sci. **21** (2), 167–188 (2011)
28. Kaczorek, T.: Realization problem for positive multivariable discrete-time linear systems with delays in the state vector and inputs. Int. J. Appl. Math. Comput. Sci. **16**(2), 101–106 (2006)
29. Kaczorek, T.: Determination of positive realizations with reduced numbers of delays or without delays for discrete-time linear systems. Arch. Control Sci. **22**(4), 371–384 (2012)
30. Kaczorek, T.: Positive realizations with reduced numbers of delays for 2-D continuous-discrete linear systems. Bull. Pol. Acad. Sci. Techn. **60**(4), 835–840 (2012)
31. Kaczorek, T.: Positive stable realizations for fractional descriptor continuous-time linear systems. Arch. Control Sci. **22**(3), 255–265 (2012)
32. Kaczorek, T.: Positive stable realizations of fractional continuous-time linear systems. Int. J. Appl. Math. Comput. Sci. **21**(4), 697–702 (2011)
33. Kaczorek, T.: Realization problem for fractional continuous-time systems. Arch. Control Sci. **18**(1), 43–58 (2008)
34. Kaczorek, T.: Determination of the set of Metzler matrices for given stable polynomials. Meas. Autom. Monit. **58**(5), 407–412 (2012)

Existence of System Matrices with Given Spectra of Positive Stable Discrete-Time Linear Systems

Tadeusz Kaczorek and Kamil Borawski

Abstract The problem of existence of system matrices with given spectra is addressed for positive stable discrete-time linear systems. Necessary and sufficient conditions for the existence of the system matrices are established. The considerations are illustrated by numerical examples of discrete-time linear systems.

Keywords Existence · System matrix · Positive · Stable · Discrete-time system

1 Introduction

In positive systems the state variables and outputs take only nonnegative values for any nonnegative initial conditions and inputs. An overview of state of the art in positive systems theory is given in [1, 2]. The positive standard and fractional order electrical circuits have been investigated in many papers and books [3–9]. Positive linear systems consisting of n subsystems with different fractional orders have been addressed in [10, 11]. Decoupling zeros of positive linear systems have been introduced in [12].

Determination of the state space equations for given transfer matrices is a classical problem, called the realization problem, which has been addressed in many papers and books [1, 13–16]. An overview of the positive realization problem is given in [2, 13, 15]. The realization problem for positive continuous-time and discrete-time linear system has been considered in [5, 15, 17–26] and for linear systems with delays in [15, 18, 22, 25, 27, 28]. The realization problem for frac-

T. Kaczorek (✉) · K. Borawski
Faculty of Electrical Engineering, Białystok University of Technology, Wiejska 45D, 15-351 Białystok, Poland
e-mail: kaczorek@isep.pw.edu.pl

K. Borawski
e-mail: kam.borawski@gmail.com

R. Szewczyk et al. (eds.), *Challenges in Automation, Robotics and Measurement Techniques*, Advances in Intelligent Systems and Computing 440, DOI 10.1007/978-3-319-29357-8_14

tional linear systems has been analyzed in [15, 29–31] and for positive 2D hybrid linear systems in [28]. A new modified state variable diagram method for determination of positive realizations with reduced number of delays for given proper transfer matrices has been proposed in [17]. The problem of determination of the set of Metzler matrices for given polynomials has been analyzed in [32].

In this paper the existence of system matrices with given spectra of positive asymptotically stable linear systems and electrical circuits is investigated.

The paper is organized as follows. In Sect. 2 some basic definitions and theorems concerning positive and stable discrete-time linear systems are recalled and the problem is formulated. The solution of the problem is presented in Sect. 3. Concluding remarks are given in Sect. 4.

The following notation will be used: $\Re$—the set of real numbers, $\Re^{n \times m}$—the set of $n \times m$ real matrices, $\Re_+^{n \times m}$—the set of $n \times m$ real matrices with nonnegative entries, I_n—the $n \times n$ identity matrix, Z_+—the set of nonnegative integers.

2 Discrete-Time Linear Systems and Problem Formulation

Consider the linear discrete-time autonomous system

$$x_{i+1} = Ax_i, \; i \in Z_+ = \{0, 1, \ldots\} \tag{1}$$

where $x_i \in \Re^n$ is the state vector and $A \in \Re^{n \times n}$ is the system matrix.

Definition 1 The system (1) is called (internally) positive if $x_i \in \Re_+^n$, $i \in Z_+$ for all initial conditions $x_0 \in \Re_+^n$.

Theorem 1 [2] *The system (1) is positive if and only if*

$$A \in \Re_+^{n \times n}. \tag{2}$$

Definition 2 The positive system (1) is called asymptotically stable if $\lim_{i \to \infty} x_i = 0$ for all $x_0 \in \Re_+^n$.

Theorem 2 [2] *The positive system (1) is asymptotically stable if and only if one of the following conditions is satisfied:*

1. All coefficients of the characteristic polynomial

$$p(z) = \det[I_n(z+1) - A] = z^n + a_{n-1}z^{n-1} + \cdots + a_1 z + a_0 \tag{3}$$

are positive, i.e. $a_k > 0$ for $k = 0, 1, \ldots, n-1$.

2. All principal minors of the matrix $\hat{A} = I_n - A = [\hat{a}_{ij}]$ are positive, i.e.

$$M_1 = |\hat{a}_{11}| > 0,\ M_2 = \begin{vmatrix} \hat{a}_{11} & \hat{a}_{12} \\ \hat{a}_{21} & \hat{a}_{22} \end{vmatrix} > 0,\ \ldots, M_n = \det \hat{A} > 0. \tag{4}$$

The problem under the considerations for discrete-time positive and asymptotically stable systems can be stated as follows:

Given a set of n_1 real and n_2 pairs of complex conjugate eigenvalues

$$\lambda_{k_1} = \alpha_{k_1}, k_1 = 1, \ldots, n_1, \lambda_{k_2} = \beta_{k_2} \pm j\gamma_{k_2}, k_2 = 1, \ldots, n_2 \tag{5}$$

find asymptotically stable system matrix $A \in \Re_+^{n \times n}$.

3 Problem Solution

It is well-known that the symmetric matrix $A \in \Re_+^{n \times n}$ has only real eigenvalues. Therefore, if $n_2 > 0$ the desired matrix is nonsymmetrical.

Lemma 1 *The system matrix for n = 2 has only real eigenvalues.*

Proof The zeros of the polynomial

$$\det[I_2 z - A_2] = \begin{vmatrix} z - a_{11} & -a_{12} \\ -a_{21} & z - a_{22} \end{vmatrix} = z^2 - (a_{11} + a_{22})z + a_{11}a_{22} - a_{12}a_{21}, \tag{6}$$

i.e. $z_1 = \alpha_1$, $z_2 = \alpha_2$ are real since

$$\begin{aligned} (a_{11} + a_{22})^2 - 4(a_{11}a_{22} - a_{12}a_{21}) &= a_{11}^2 + a_{22}^2 - 2a_{11}a_{22} + 4a_{12}a_{21} \\ &= (a_{11} - a_{22})^2 + 4a_{12}a_{21} \geq 0 \end{aligned} \tag{7}$$

for all $a_{ij} \geq 0$, $i, j = 1, 2$. □

From Lemma 1 we have the following theorem.

Theorem 3 *For n = 2 does not exist a system matrix with a given pair of complex conjugate eigenvalues* $z_1 = \beta_1 + j\gamma_1$, $z_2 = \beta_1 - j\gamma_1$.

For $n = 3$ we shall prove the following theorem.

Theorem 4 *There exists an asymptotically stable system matrix*

$$A_3 = \begin{bmatrix} a_{11} & a_{12} & a_{13} \\ a_{21} & a_{22} & a_{23} \\ a_{31} & a_{32} & a_{33} \end{bmatrix} \in \Re_{+}^{3\times 3} \tag{8}$$

with given eigenvalues

$$z_1 = \alpha, z_2 = \beta + j\gamma, z_3 = \beta - j\gamma, \alpha > \beta > 0, \gamma > 0 \tag{9}$$

if and only if

$$(\alpha - \beta)^2 > 3\gamma^2. \tag{10}$$

Proof The eigenvalues (9) of the matrix (8) satisfy the following well-known equalities

$$z_1 + z_2 + z_3 = \alpha + 2\beta = a_{11} + a_{22} + a_{33}, \tag{11}$$

$$\begin{aligned} z_1(z_2 + z_3) + z_2 z_3 &= 2\alpha\beta + \beta^2 + \gamma^2 \\ &= a_{11}a_{22} + a_{11}a_{33} + a_{22}a_{33} - a_{12}a_{21} - a_{13}a_{31} - a_{23}a_{32}, \end{aligned} \tag{12}$$

$$\begin{aligned} \det A &= z_1 z_2 z_3 = \alpha(\beta^2 + \gamma^2) \\ &= a_{11}a_{22}a_{33} - a_{11}a_{23}a_{32} - a_{22}a_{13}a_{31} - a_{33}a_{12}a_{21} + a_{12}a_{31}a_{23} + a_{13}a_{21}a_{32}. \end{aligned} \tag{13}$$

Note that the sum $a_{11} + a_{22} + a_{33}$ takes its maximal value for $a_{11} = a_{22} = a_{33} = \frac{\alpha + 2\beta}{3}$ and the condition (12) can be satisfied for $a_{ij} \geq 0$, $i \neq j$ if and only if

$$a_{11}(a_{22} + a_{33}) + a_{22}a_{33} = \frac{(\alpha + 2\beta)^2}{3} > 2\alpha\beta + \beta^2 + \gamma^2. \tag{14}$$

It is easy to check that the condition (14) is equivalent to the condition (10) and if (10) holds then there exist $a_{ij} \geq 0$, $i \neq j$ of the matrix (8) such that (13) is satisfied. □

Example 1 Find the matrix (8) with the eigenvalues

$$z_1 = 0.8,\ z_2 = 0.2 + j0.1,\ z_3 = 0.2 - j0.1. \tag{15}$$

The characteristic polynomial of the matrix with the eigenvalues (15) has the form

$$(z-0.8)(z-0.2-j0.1)(z-0.2+j0.1)=z^3-1.2z^2+0.37z-0.04. \tag{16}$$

In this case the condition (10) is satisfied since $\alpha=0.8$, $\beta=0.2$, $\gamma=0.1$ and

$$(\alpha-\beta)^2=(0.6)^2=0.36>3\gamma^2=0.03. \tag{17}$$

Assuming $a_{11}=a_{22}=a_{33}=0.4$, $a_{31}=a_{32}=1$, $a_{13}=a_{21}=0$ we are looking for the values of a_{12} and a_{23} such that the polynomial (16) is the characteristic polynomial of the desired matrix

$$\begin{aligned}\det[I_3z-A_3]&=\begin{vmatrix} z-0.4 & -a_{12} & 0 \\ 0 & z-0.4 & -a_{23} \\ -1 & -1 & z-0.4 \end{vmatrix}\\ &=z^3-1.2z^2+(0.48-a_{23})z-(0.4)^3+0.4a_{23}-a_{12}a_{23}\\ &=z^3-1.2z^2+0.37z-0.04.\end{aligned} \tag{18}$$

From comparison of the coefficients of (18) we have

$$a_{23}=0.48-0.37=0.11, a_{12}=\frac{-(0.4)^3+0.4\cdot 0.11+0.04}{0.11}=0.1818.$$

It is easy to check that the matrix

$$A_3=\begin{bmatrix} 0.4 & 0.1818 & 0 \\ 0 & 0.4 & 0.11 \\ 1 & 1 & 0.4 \end{bmatrix} \tag{19}$$

has the desired eigenvalues.

Note that the matrix

$$\bar{A}_3=\begin{bmatrix} 0.4 & 1 & 1 \\ 0 & 0.4 & 0.1818 \\ 0.11 & 0 & 0.4 \end{bmatrix} \tag{20}$$

is also a solution to the problem, i.e. its characteristic polynomial has also the desired form (18). The matrix (20) has been obtained from (19) by similarity transformation $\bar{A}_3=PA_3P^{-1}$ for the permutation matrix

$$P=\begin{bmatrix} 0 & 0 & 1 \\ 1 & 0 & 0 \\ 0 & 1 & 0 \end{bmatrix}. \tag{21}$$

Therefore, we have the following important conclusion.

Conclusion 1 The problem under considerations has many solutions. Any matrix $A = PAP^{-1}$ is also the desired solution to the problem for any permutation matrix $P \in \Re_+^{3\times 3}$ since

$$\det[I_3 z - \bar{A}_3] = \det P \det[I_3 z - A_3] \det P^{-1} = \det[I_3 z - A_3]$$

and $\bar{A}_3 \in \Re_+^{3\times 3}$ if and only if $A_3 \in \Re_+^{3\times 3}$.

It is well-known that for $n = 4$ the matrix $A_4 \in \Re_+^{4\times 4}$ has at least two real eigenvalues. For $n = 4$ the problem can be stated as follows. Given eigenvalues

$$z_1 = \alpha_1, z_2 = \alpha_2, z_3 = \beta + j\gamma, z_4 = \beta - j\gamma, \alpha_1 > \alpha_2 > \beta \tag{22}$$

find a system matrix of the form

$$A_4 = \begin{bmatrix} a_{11} & 0 & 0 & 0 \\ 0 & a_{22} & a_{23} & a_{24} \\ 0 & a_{32} & a_{33} & a_{34} \\ 0 & a_{42} & a_{43} & a_{44} \end{bmatrix} = \begin{bmatrix} a_{11} & 0 \\ 0 & A_3 \end{bmatrix}, A_3 = \begin{bmatrix} a_{22} & a_{23} & a_{24} \\ a_{32} & a_{33} & a_{34} \\ a_{42} & a_{43} & a_{44} \end{bmatrix} \tag{23}$$

with the desired eigenvalues (22).

Theorem 5 *For $n = 4$ there exists an asymptotically stable system matrix (23) with given eigenvalues if and only if*

$$(\alpha_1 - \beta)^2 > 3\gamma^2. \tag{24}$$

Proof We choose $a_{11} = \alpha_2$ and to find the system matrix $A_3 \in \Re_+^{3\times 3}$ we apply the method given for $n = 3$. By Theorem 4 the problem for $n = 4$ has a solution if and only if the condition (24) is satisfied. □

Example 2 Find the system matrix (23) with the eigenvalues

$$z_1 = 0.8, z_2 = 0.5, z_3 = 0.2 + j0.1, z_4 = 0.2 - j0.1. \tag{25}$$

To find the desired system matrix we can use the results of Example 1. Note that the condition (24) is satisfied since (17) holds. In this case we have $a_{11} = \alpha_2$ and the matrix A_3 has the form (19). The desired matrix with eigenvalues (25) has the form

$$A_4 = \begin{bmatrix} 0.5 & 0 & 0 & 0 \\ 0 & 0.4 & 0.1818 & 0 \\ 0 & 0 & 0.4 & 0.11 \\ 0 & 1 & 1 & 0.4 \end{bmatrix}. \tag{26}$$

Note that any matrix $\bar{A}_4 = PA_4P^{-1}$ is also the desired solution for any permutation matrix $P_4 \in \Re_+^{4\times 4}$.

Now let us consider the problem for $n = 5$. Given eigenvalues

$$z_1 = \alpha_1, z_2 = \alpha_2, z_3 = \alpha_3, z_4 = \beta + j\gamma z_5 = \beta - j\gamma \tag{27}$$

find a system matrix of the form

$$A_5 = \begin{bmatrix} a_{11} & 0 & 0 \\ 0 & a_{22} & 0 \\ 0 & 0 & A_3 \end{bmatrix} \in \Re_+^{5\times 5}, A_3 = \begin{bmatrix} a_{33} & a_{34} & a_{35} \\ a_{43} & a_{44} & a_{45} \\ a_{53} & a_{54} & a_{55} \end{bmatrix} \in \Re_+^{3\times 3} \tag{28}$$

with the desired eigenvalues (27).

Theorem 6 *For n = 5 there exists an asymptotically stable system matrix (28) with given eigenvalues (27) if and only if the condition (24) is satisfied.*

Proof We choose $a_{11} = \alpha_2$, $a_{22} = \alpha_3$ and to find the system matrix A_3 we apply the method given for $n = 3$. By Theorem 4 the problem for $n = 5$ has a solution if and only if the condition (24) is satisfied. □

Example 3 Find the system matrix (28) with the eigenvalues

$$z_1 = 0.8, z_2 = 0.6, z_3 = 0.3, z_4 = 0.2 + j0.1, z_5 = 0.2 - j0.1. \tag{29}$$

To find the desired system matrix we can use the results of Example 1. In this case we choose $a_{11} = z_2 = 0.6$, $a_{22} = z_3 = 0.3$ and the remaining eigenvalues as the eigenvalues of the matrix A_3. The eigenvalues z_1, z_4, z_5 satisfy the condition (24). Therefore, using the results of Example 1 we obtain the desired matrix in the form

$$A_5 = \begin{bmatrix} 0.6 & 0 & 0 & 0 & 0 \\ 0 & 0.3 & 0 & 0 & 0 \\ 0 & 0 & 0.4 & 0.1818 & 0 \\ 0 & 0 & 0 & 0.4 & 0.11 \\ 0 & 0 & 1 & 1 & 0.4 \end{bmatrix}. \tag{30}$$

Note that this procedure can be applied for any $n > 4$ if the desired system matrix has only one pair of complex conjugate eigenvalues and for $n = 3k$, $k = 2, 3, \ldots$ if the desired system matrix has k pairs of complex conjugate eigenvalues.

4 Concluding Remarks

The problem of existence of system matrices with given spectra has been addressed. Necessary and sufficient conditions for a large class of positive and stable discrete-time linear systems have been established (Theorems 3–6). The considerations have been illustrated by numerical examples of discrete-time linear systems. The considerations can be easily extended to fractional linear systems.

Acknowledgment This work was supported by National Science Centre in Poland under work No. 2014/13/B/ST7/03467.

References

1. Farina, L., Rinaldi, S.: Positive Linear Systems; Theory and Applications. Wiley, New York (2000)
2. Kaczorek, T.: Positive 1D and 2D Systems. Springer, London (2002)
3. Kaczorek, T.: A class of positive and stable time-varying electrical circuits. Electr. Rev. **91**(5), 121–124 (2015)
4. Kaczorek, T.: Constructability and observability of standard and positive electrical circuits. Electr. Rev. **89**(7), 132–136 (2013)
5. Kaczorek, T.: Positive electrical circuits and their reachability. Arch. Electr. Eng. **60**(3), 283–301 (2011)
6. Kaczorek T.: Positive fractional linear electrical circuits. In: Proceedings of SPIE, vol. 8903. Bellingham WA, USA, Art. No 3903–35
7. Kaczorek, T.: Positive unstable electrical circuits. Electr. Rev. **88**(5a), 187–192 (2012)
8. Kaczorek, T.: Zeroing of state variables in descriptor electrical circuits by state-feedbacks. Electr. Rev. **89**(10), 200–203 (2013)
9. Kaczorek, T., Rogowski, K.: Fractional Linear Systems and Electrical Circuits. Studies in Systems, Decision and Control, vol. 13. Springer (2015)
10. Kaczorek, T.: Positive linear systems with different fractional orders. Bull. Pol. Acad. Sci. Technol. **58**(3), 453–458 (2010)
11. Kaczorek, T.: Positive systems consisting of n subsystems with different fractional orders. IEEE Trans. Circ. Syst.—Regul. Pap. **58**(6), 1203–1210 (2011)
12. Kaczorek T.: Decoupling zeros of positive continuous-time linear systems and electrical circuits. Advances in Systems Science. Advances in Intelligent Systems and Computing, vol. 240, pp. 1–15. Springer (2014)
13. Benvenuti, L., Farina, L.: A tutorial on the positive realization problem. IEEE Trans. Auto. Control **49**(5), 651–664 (2004)
14. Kaczorek T.: Linear Control Systems, vol. 1. Research Studies Press, Wiley, New York (1992)
15. Kaczorek, T., Sajewski, Ł.: Realization Problem for Positive and Fractional Systems. Springer (2014)
16. Shaked, U., Dixon, M.: Generalized minimal realization of transfer-function matrices. Int. J. Control **25**(5), 785–803 (1977)
17. Kaczorek, T.: A modified state variables diagram method for determination of positive realizations of linear continuous-time systems with delays. Int. J. Appl. Math. Comput. Sci. **22** (4), 897–905 (2012)
18. Kaczorek, T.: A realization problem for positive continuous-time linear systems with reduced numbers of delays. Int. J. Appl. Math. Comput. Sci. **16**(3), 325–331 (2006)

19. Kaczorek, T.: Computation of positive stable realizations for discrete-time linear systems. Comput. Prob. Electr. Eng. **2**(1), 41–48 (2012)
20. Kaczorek, T.: Computation of positive stable realizations for linear continuous-time systems. Bull. Pol. Acad. Techn. Sci. **59**(3), 273–281 (2011)
21. Kaczorek, T.: Computation of realizations of discrete-time cone systems. Bull. Pol. Acad. Sci. Techn. **54**(3), 347–350 (2006)
22. Kaczorek, T.: Positive minimal realizations for singular discrete-time systems with delays in state and delays in control. Bull. Pol. Acad. Sci. Technol. **53**(3), 293–298 (2005)
23. Kaczorek, T.: Positive stable realizations of discrete-time linear systems. Bull. Pol. Acad. Sci. Techn. **60**(3), 605–616 (2012)
24. Kaczorek, T.: Positive stable realizations with system Metzler matrices. Arch. Control Sci. **21** (2), 167–188 (2011)
25. Kaczorek, T.: Realization problem for positive multivariable discrete-time linear systems with delays in the state vector and inputs. Int. J. Appl. Math. Comput. Sci. **16**(2), 101–106 (2006)
26. Kaczorek, T.: Positive and asymptotically stable realizations for descriptor discrete-time linear systems. Bull. Pol. Acad. Sci. Technol. **61**(1), 229–237 (2013)
27. Kaczorek, T.: Determination of positive realizations with reduced numbers of delays or without delays for discrete-time linear systems. Arch. Control Sci. **22**(4), 371–384 (2012)
28. Kaczorek, T.: Positive realizations with reduced numbers of delays for 2-D continuous-discrete linear systems. Bull. Pol. Acad. Sci. Technol. **60**(4), 835–840 (2012)
29. Kaczorek, T.: Positive stable realizations for fractional descriptor continuous-time linear systems. Arch. Control Sci. **22**(3), 255–265 (2012)
30. Kaczorek, T.: Positive stable realizations of fractional continuous-time linear systems. Int. J. Appl. Math. Comput. Sci. **21**(4), 697–702 (2011)
31. Kaczorek, T.: Realization problem for fractional continuous-time systems. Arch. Control Sci. **18**(1), 43–58 (2008)
32. Kaczorek, T.: Determination of the set of Metzler matrices for given stable polynomials. Measur. Aut. Monit. **58**(5), 407–412 (2012)

Two- and Three-Layer Recurrent Elman Neural Networks as Models of Dynamic Processes

Antoni Wysocki and Maciej Ławryńczuk

Abstract The goal of paper is to study and compare the effectiveness of two- and three-layer Elman recurrent neural networks used for modelling of dynamic processes. Training of such networks is discussed. For a neutralisation reactor benchmark system it is shown that the rudimentary Elman structure with two layers is much better in terms of accuracy and the number of parameters. Furthermore, its training is much easier.

1 Introduction

For modelling of dynamic processes neural networks of different structures are often used [3, 9]. In most cases, the perceptron networks with one hidden layer (Multi-Layer Perceptron, MLP) are used, the Radial Basis Function (RBF) networks are less popular. Because both MLP and RBF structures are static approximators, dynamics is introduced by feeding back some past model input and output signals as the inputs of the network. A conceptually more straightforward option is to use truly recurrent neural networks [3, 7], which are inherently dynamic systems. One well-known example of a recurrent structure is the Elman neural network [1, 3, 7]. It may be used in various applications: eg. in short-term temperature forecasting [4], in modelling of the flow of passengers in subway [5], in identification of the grammatical structure of literary works [6], in flight trajectory prediction [10], in air pressure control supplied to the disc drill subway tunnel under a river [12]. The classical Elman neural network has two layers. This paper studies effectiveness of two- and

A. Wysocki (✉) · M. Ławryńczuk
Institute of Control and Computation Engineering, Warsaw University of Technology, ul. Nowowiejska 15/19, 00-665 Warsaw, Poland
e-mail: A.T.Wysocki@stud.elka.pw.edu.pl

M. Ławryńczuk
e-mail: M.Lawrynczuk@ia.pw.edu.pl

R. Szewczyk et al. (eds.), *Challenges in Automation, Robotics and Measurement Techniques*, Advances in Intelligent Systems and Computing 440, DOI 10.1007/978-3-319-29357-8_15

three-layer Elman networks used for modelling of dynamic processes. The training algorithms is described. A neutralisation reactor benchmark system is considered in simulations.

2 Elman Neural Network with Two Layers

Figure 1 depicts the classical Elman neural network with two layers (with one hidden layer in different words). The Elman network is a structure with full recursion which means that the output signals of all neurons of the hidden layer are fed back through delay blocks to the input nodes. The network has one input $u(k-1)$, K hidden nodes with a transfer function φ and one output $y_{\mathrm{mod}}(k)$. The output of the network is

$$y_{\mathrm{mod}}(k) = w_0^{(2)} + \sum_{i=1}^{K} w_i^{(2)} v_i(k) \tag{1}$$

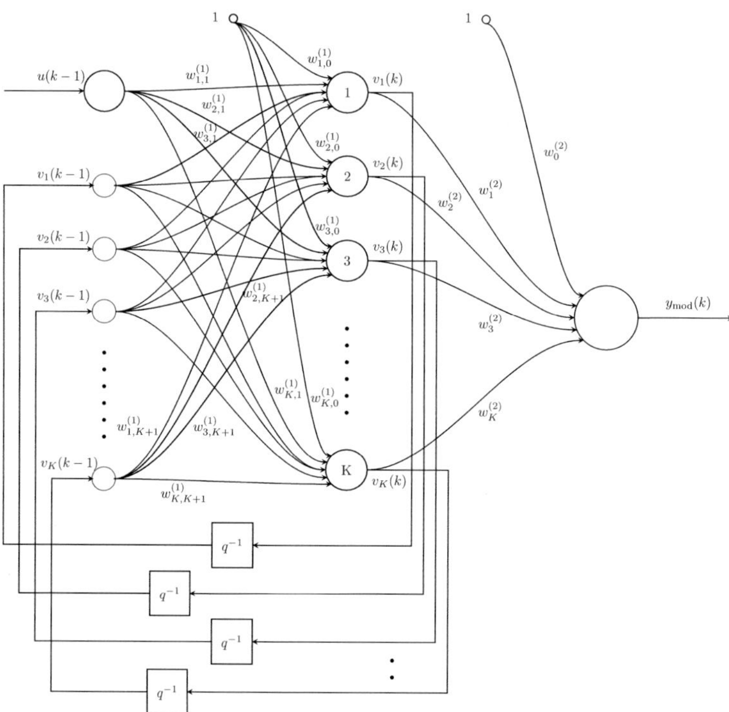

Fig. 1 The recurrent Elman neural network with two layers

where

$$v_i(k) = \varphi(z_i(k)) \tag{2}$$

is an output signal of the consecutive hidden neurons ($i = 1, \ldots, K$), $v_0(k) = 1$ is a bias and the sum of the input signals of the i-th neuron is

$$z_i(k) = w_{i,0}^{(1)} + w_{i,1}^{(1)} u(k-1) + \sum_{j=1}^{K} w_{i,j+1}^{(1)} v_j(k-1) \tag{3}$$

From Eqs. (1)–(3), one obtains

$$y_{\mathrm{mod}}(k) = w_0^{(2)} + \sum_{i=1}^{K} w_i^{(2)} \varphi\left(w_{i,0}^{(1)} + w_{i,1}^{(1)} u(k-1) + \sum_{j=1}^{K} w_{i,j+1}^{(1)} v_j(k-1) \right)$$

3 Elman Neural Network with Three Layers

Figure 2 shows the Elman neural network with three layers (with two hidden layers in different words). The output signals of all neurons of the first and the second hidden layers are fed back through delay blocks to the input nodes of the first and the second hidden layer, respectively. The network has one input $u(k-1)$, K and

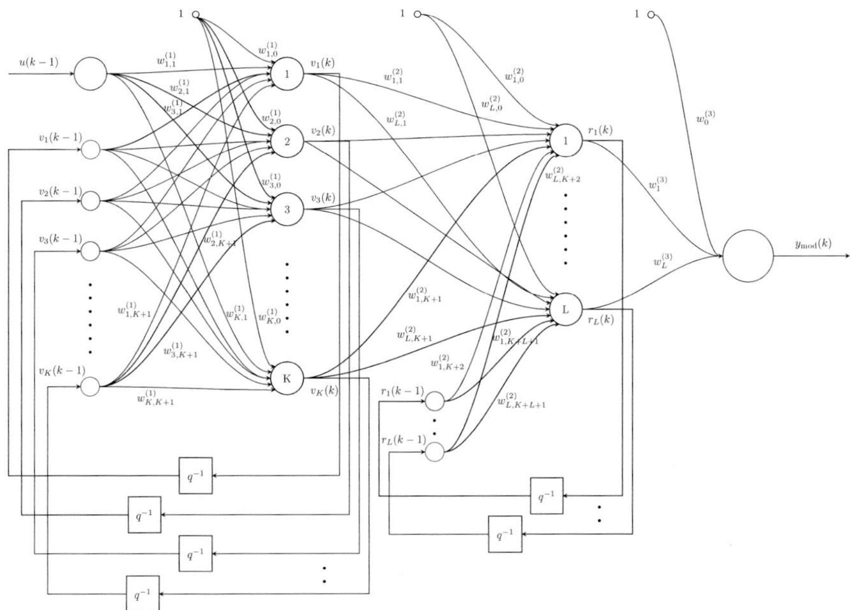

Fig. 2 The recurrent Elman neural network with three layers

L hidden nodes in the first and the second layers, respectively (the same transfer function φ is assumed) and one output $y_{\mathrm{mod}}(k)$. The output of the network is

$$y_{\mathrm{mod}}(k) = w_0^{(3)} + \sum_{i=1}^{L} w_i^{(3)} r_i(k) \tag{4}$$

where

$$r_i(k) = \varphi(t_i(k)) \tag{5}$$

is an output signal of the consecutive hidden neurons in the second layer ($i = 1, \dots, L$), $r_0(k) = 1$ is a bias and the sum of input signals of the i-th neuron of the second layer is calculated from

$$t_i(k) = w_{i,0}^{(2)} + \sum_{j=1}^{K} w_{i,j}^{(2)} v_j(k) + \sum_{j=1}^{L} w_{i,K+1+j}^{(2)} r_j(k-1) \tag{6}$$

where

$$v_j(k) = \varphi(z_j(k))$$

is an output signal of the consecutive hidden neurons in the first layer ($j = 1, \dots, K$), $v_0(k) = 1$ is a bias and the sum of input signals of j-th neuron of the first layer is

$$z_j(k) = w_{j,0}^{(1)} + w_{j,1}^{(1)} u(k-1) + \sum_{p=1}^{K} w_{j,p+1}^{(1)} v_p(k-1)$$

4 Training of Recurrent Elman Neural Networks

The objective of training of the neural network is to find the weight vector $\boldsymbol{w}$ in such a way that the approximation error is acceptable. The approximation error is usually defined as the sum of squared differences between the model output and the real output for some data set recorded during process operation

$$E(\boldsymbol{w}) = \sum_{k=1}^{P} (y_{\mathrm{mod}}(k) - y(k))^2 \tag{7}$$

where $y_{\mathrm{mod}}(k)$ is the output of the neural network, $y(k)$ is the training sample (i.e. a real value of process output), P is the number of training patterns. Training of the neural consists in minimising the error function (7). Due to the non-linear activation function φ, it is a nonlinear, unconstrained optimisation problem.

The general training algorithm may be summarised as follows ($it = 1, \ldots, it^{\max}$ denote the consecutive iterations):

0. Initialisation of the weight vector ($\boldsymbol{w}$), e.g. setting random initial values.
1. Calculation of the output $y_{\text{mod}}(k)$ of the neural network for all data samples $k = 1, \ldots, P$.
2. Calculation of the error function (7).
3. If a stopping criterion (e.g. imposed on the number of iterations or on the value of the error function or on the norm of its gradient) is satisfied, training is stopped.
4. The optimisation direction $\boldsymbol{p}_{it}$ is calculated (using e.g. the steepest descent method or a variable metric algorithm [8]).
5. The optimal step length η_{it} is found for the calculated direction $\boldsymbol{p}_{it}$.
6. The weights are updated from $\boldsymbol{w}_{it+1} = \boldsymbol{w}_{it} + \eta_{it}\boldsymbol{p}_{it}$ and the algorithm returns to the step 1.

The main issue is to calculate the gradients of the error function with respect to the weights of the network. For the Elman network with two layers they may be calculated analytically as discussed in [11]. In the case of the network with three layers the derivatives of the error function (7) with respect to the weight of the first, the second and the third layers are

$$\frac{\mathrm{d}E(\boldsymbol{w})}{\mathrm{d}w_{j,m}^{(1)}} = 2(y_{\text{mod}}(k) - y(k))\frac{\partial y_{\text{mod}}(k)}{\partial w_{j,m}^{(1)}}$$

for $j = 1 \ldots K,\ m = 0, \ldots, K+1$,

$$\frac{\mathrm{d}E(\boldsymbol{w})}{\mathrm{d}w_{i,j}^{(2)}} = 2(y_{\text{mod}}(k) - y(k))\frac{\partial y_{\text{mod}}(k)}{\partial w_{i,j}^{(2)}}$$

for $i = 1 \ldots L,\ j = 0, \ldots, K+1$ and

$$\frac{\mathrm{d}E(\boldsymbol{w})}{\mathrm{d}w_{i}^{(3)}} = 2(y_{\text{mod}}(k) - y(k))\frac{\partial y_{\text{mod}}(k)}{\partial w_{i}^{(3)}}$$

for $i = 0, \ldots, L$, respectively. Next, taking into account the output of the network (Eq. (4)), one obtains

$$\frac{\partial y_{\text{mod}}(k)}{\partial w_{j,m}^{(1)}} = \sum_{n=1}^{L} w_n^{(3)} \frac{\partial r_n(k)}{\partial w_{j,m}^{(1)}}$$

for $j = 1 \ldots K,\ m = 0, \ldots, K+1$,

$$\frac{\partial y_{\text{mod}}(k)}{\partial w_{i,j}^{(2)}} = \sum_{n=1}^{L} w_n^{(3)} \frac{\partial r_n(k)}{\partial w_{i,j}^{(2)}}$$

for $i = 0 \ldots K$, $j = 0 \ldots K+1$, $n = 1, \ldots, K$ and

$$\frac{\partial y_{\mathrm{mod}}(k)}{\partial w_i^{(3)}} = r_i(k)$$

for $i = 0, \ldots, K$. Using Eq. (5), the derivatives of the output signals of the second layer are

$$\frac{\partial r_n(k)}{\partial w_{j,m}^{(1)}} = \frac{\mathrm{d}\varphi(t_n(k))}{\mathrm{d}t_n(k)} \frac{\partial t_n(k)}{\partial w_{j,m}^{(1)}}$$

for $i = 1, \ldots, K$, $j = 0, \ldots, K+1$, $n = 1, \ldots, K$ and

$$\frac{\partial r_n(k)}{\partial w_{i,j}^{(2)}} = \frac{\mathrm{d}\varphi(t_n(k))}{\mathrm{d}t_n(k)} \frac{\partial t_n(k)}{\partial w_{i,j}^{(2)}}$$

for $i = 1, \ldots, K$, $j = 0, \ldots, K+1$, $n = 1, \ldots, K$. Finally, form Eq. (6), one calculates

$$\frac{\partial t_n(k)}{\partial w_{i,j}^{(2)}} = \begin{cases} X_j^{(1)}(k-1) + \sum_{p=1}^{L} w_{n,K+p}^{(2)} \frac{\partial r_p(k-1)}{\partial w_{i,j}^{(2)}} & \text{for } i = n \\ \sum_{p=1}^{L} w_{n,K+p}^{(2)} \frac{\partial r_p(k-1)}{\partial w_{i,j}^{(2)}} & \text{for } i \neq n \end{cases}$$

for $i = 0, \ldots, K$, $j = 0, \ldots, K+1$, $n = 1, \ldots, K$, where

$$X_j^{(1)}(k-1) = \begin{cases} 1 & \text{for } j = 0 \\ v_j(k) & \text{for } 0 < j \leqslant K \\ r_{j-K}(k-1) & \text{for } K < j \leqslant K+L \end{cases}$$

and

$$\frac{\partial t_n(k)}{\partial w_{j,m}^{(1)}} = \sum_{p=1}^{K} w_{n,p}^{(2)} \frac{\partial v_p(k)}{\partial w_{j,m}^{(1)}} + \sum_{q=1}^{L} w_{n,K+q}^{(2)} \frac{\partial r_q(k-1)}{\partial w_{j,m}^{(1)}}$$

for $i = 0, \ldots, K$, $j = 0, \ldots, K+1$, $n = 1, \ldots, K$, where

$$\frac{\partial v_p(k)}{\partial w_{j,m}^{(1)}} = \frac{\mathrm{d}\varphi(z_p(k))}{\mathrm{d}z_p(k)} \frac{\partial z_p(k)}{\partial w_{j,m}^{(1)}}$$

$$\frac{\partial z_p(k)}{\partial w_{j,p}^{(1)}} = \begin{cases} X_m^{(0)}(k-1) + \sum_{q=1}^{K} w_{p,q+1}^{(1)} \frac{\partial v_q(k-1)}{\partial w_{j,m}^{(1)}} & \text{for } m = p \\ \sum_{q=1}^{K} w_{p,q+1}^{(1)} \frac{\partial v_q(k-1)}{\partial w_{j,m}^{(1)}} & \text{for } m \neq p \end{cases}$$

for $i = 0, \ldots, K$, $j = 0, \ldots, K+1$, $n = 1, \ldots, K$, where

$$X_m^{(0)}(k-1) = \begin{cases} 1 & \text{for } m = 0 \\ u(k-1) & \text{for } m = 1 \\ v_{m-1}(k-1) & \text{for } 1 < m \leqslant K+1 \end{cases}$$

5 Simulation Results

The considered process is a neutralisation reactor (pH reactor), whose scheme and the static characteristics are shown in Fig. 3. The process is significantly nonlinear, it is frequently used as a benchmark to evaluate efficiency of different models. Acid HNO_3, base NaOH and buffer $NaHCO_3$ are mixed in the tank. From the modelling point of view, the process has one manipulated input (q_1) and one controlled outputs (pH). The fundamental model of the process consists of two differential equations [2]

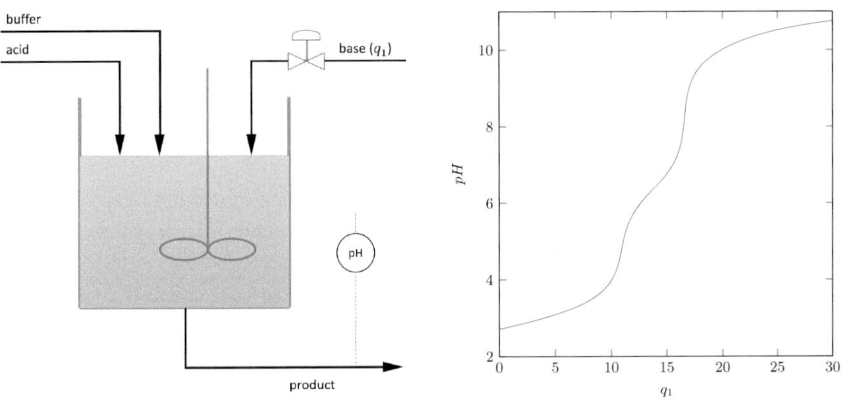

Fig. 3 The neutralisation reactor (*left*) and its static characteristics (*right*)

$$\frac{dW_a(t)}{dt} = \left(\frac{W_{a_1} - W_a(t)}{V}\right) q_1(t) + \left(\frac{W_{a_2} - W_a(t)}{V}\right) q_2 + \left(\frac{W_{a_3} - W_a(t)}{V}\right) q_3$$
$$\frac{dW_b(t)}{dt} = \left(\frac{W_{b_1} - W_b(t)}{V}\right) q_1(t) + \left(\frac{W_{b_2} - W_b(t)}{V}\right) q_2 + \left(\frac{W_{b_3} - W_b(t)}{V}\right) q_3$$

and the algebraic equation

$$W_a(t) + 10^{pH(t)-14} - 10^{-pH(t)} + W_b(t)\frac{1 + 2 \times 10^{pH(t)-pK_2}}{1 + 10^{pK_1-pH(t)} + 10^{pH(t)-pK2}} = 0$$

Parameters of the fundamental model are: $q_2 = 0.55$ ml/s, $q_3 = 16.60$ ml/s, $pK_1 = 6.35$, $pK_2 = 10.25$, $W_{a_1} = -3.05 \times 10^{-3}$ mol, $W_{b_1} = 5 \times 10^{-5}$ mol, $W_{a_2} = -3 \times 10^{-2}$ mol, $W_{b_2} = 3 \times 10^{-2}$ mol, $W_{a_3} = 3 \times 10^{-3}$ mol, $W_{b_3} = 0$ mol, $V = 2900$ ml. In the nominal operating point of the process $q_{1_0} = 15.55$ ml/s, $pH_0 = 7.0$, $W_{a_0} = -4.32 \times 10^{-4}$ mol, $W_{b_0} = 5.28 \times 10^{-4}$ mol.

Model identification usually needs three independent data sets: training, validation and test sets. Each of them should be large and representative enough so that the neural network after training would be able to correctly mimic behaviour of process for various operating points. The weights of the neural network are optimised taking into account the error for the first set (the training error). In consecutive iterations of training the training error is reduced, but too long training may lead to poor generalisation ability of the model. That is why during training the error for the validation data set is calculated (the validation error). When it grows, training is stopped. because further training may cause too strict dependence of the model on the data from the training set. Finally, the model is independently tested using the third data set.

The fundamental model of the reactor is used to generate three sets of data, each set has 1200 samples. For neural network training the process variables are scaled: $u = (q_1 - q_{1_0})/15$ and $y = (pH - pH_0)/4$. All the networks are trained using the Broyden-Fletcher-Goldfarb-Shanno (BFGS) variable-metric optimisation algorithm [8]. The maximal number of training steps is 500.

The classical Elman networks with $K = 3, \ldots, 7$ hidden nodes are evaluated. For each structure training is repeated 10 times (the weights are initialised randomly). Table 1 presents the errors of the best models. As the validation error for $K > 5$ grows, the network with 5 hidden neurons is chosen. Figure 4 compares the validation data set and the output of the chosen model, two fragments of the comparison are additionally enlarged. It can be noted that the network with two layers is able to precisely model the process.

Next, the Elman structure with three layers is considered. The networks with $K = 3, \ldots, 5$ and $L = 2, \ldots, 4$ hidden nodes are compared. For each structure training is repeated 10 times. Table 2 presents the errors of the best models. Taking into account the validation error, the structure with 4 neurons in the first layer and 3 neurons in the second layer is chosen. Figure 5 compares the validation data set and the output of the chosen model, two fragments of the comparison are additionally enlarged.

Table 1 The influence of the number of hidden nodes on the modelling quality of the Elman neural network with two layers, *NoP*—the number of parameters, $E_u(\boldsymbol{w})$—the training error, $E_w(\boldsymbol{w})$—the validation error, $E_t(\boldsymbol{w})$—the test error

Model	NoP	$E_u(\boldsymbol{w})$	$E_w(\boldsymbol{w})$	$E_t(\boldsymbol{w})$
$K = 3$	19	1.7881×10^{-3}	2.9912×10^{-3}	–
$K = 4$	29	1.5287×10^{-3}	2.1190×10^{-3}	–
$K = 5$	41	9.6537×10^{-4}	1.7456×10^{-3}	1.8953×10^{-3}
$K = 6$	55	7.2135×10^{-4}	1.8131×10^{-3}	–
$K = 7$	71	7.1601×10^{-4}	1.8042×10^{-3}	–

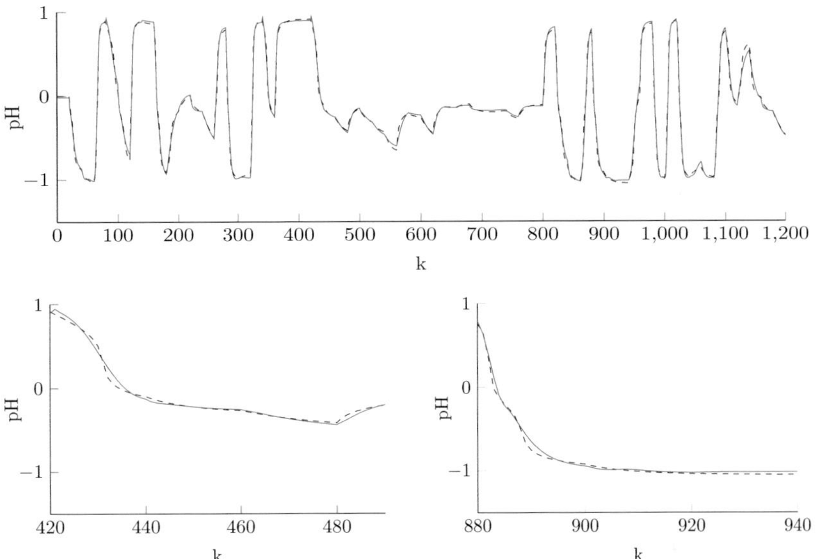

Fig. 4 Comparison of the validation data set (*dashed line*) and the output of the Elman neural network with two layers (*solid line*); the entire data set (*top*) and two enlarged fragments (*bottom*)

Table 2 The influence of the number of hidden nodes on the modelling quality of the Elman neural network with three layers, *NoP*—the number of parameters, $E_u(\boldsymbol{w})$—the training error, $E_w(\boldsymbol{w})$—the validation error, $E_t(\boldsymbol{w})$—the test error

Model	*NoP*	$E_u(\boldsymbol{w})$	$E_w(\boldsymbol{w})$	$E_t(\boldsymbol{w})$
$K = 3, L = 3$	40	3.3628×10^{-2}	6.9245×10^{-2}	–
$K = 4, L = 2$	41	5.8842×10^{-2}	6.3215×10^{-2}	–
$K = 4, L = 3$	52	2.8513×10^{-2}	4.9268×10^{-2}	9.2219×10^{-2}
$K = 5, L = 2$	54	2.3166×10^{-3}	6.2168×10^{-2}	–
$K = 4, L = 4$	65	3.1103×10^{-2}	5.3425×10^{-2}	–

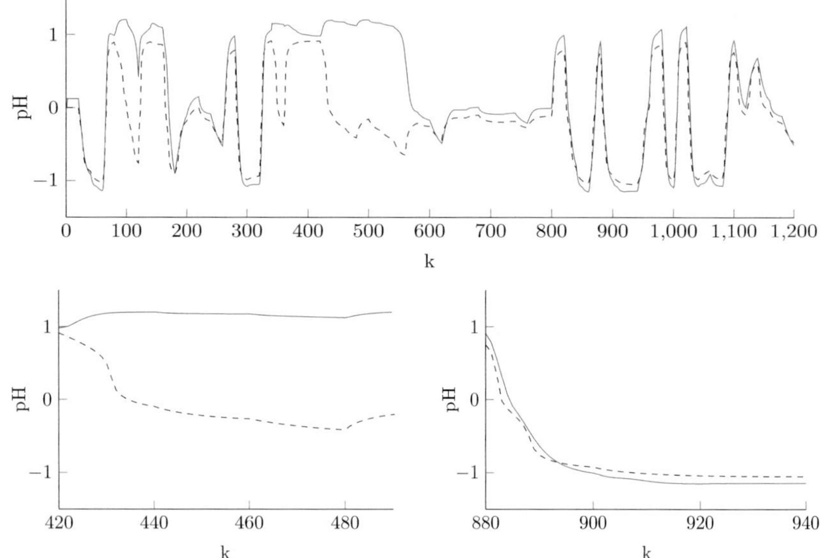

Fig. 5 Comparison of the validation data set (*dashed line*) and the output of the Elman neural network with three layers (*solid line*); the entire data set (*top*) and two enlarged fragments (*bottom*)

Unfortunately, when compared with Table 1 and Fig. 4, it is clear that the models with three layers, although they have more parameters, give much bigger errors (some 300 times) and the prediction is very imprecise.

It is interesting to considered not only model errors and generalisation, but training abilities of both consider model classes. Figure 6 depicts changes of the model errors (for the training data set) in consecutive training iterations in all 10 training attempts for the chosen networks. It is clear that the Elman network with two layers is trained much quicker than the one with three layers.

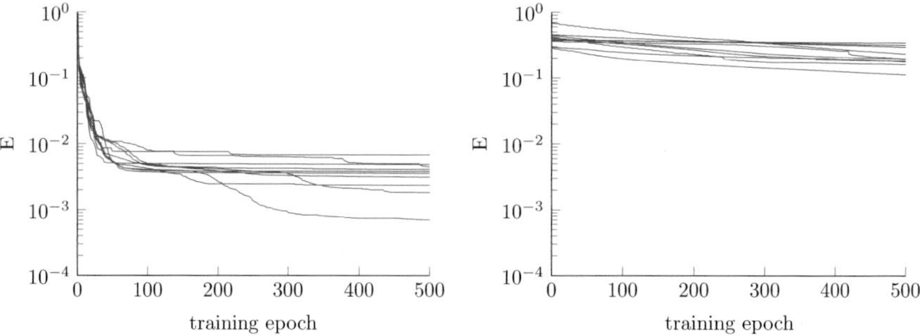

Fig. 6 Changes of the model errors (for the training data set) in the consecutive training iterations in all 10 training attempts: the Elman network with two layers, $K = 5$ (*left*); the Elman network with three layers, $K = 4, L = 3$ (*right*)

6 Summary

The paper compares the Elman recurrent neural network with two and three layers. The classical Elman network with two layers has a great ability to model dynamic systems with very good accuracy. Following this path an attempt has been made to verify if the Elman structure with three layers is able to give better or comparable modelling accuracy. This paper describes in details a gradient training algorithm for the Elman network with three layers. Next, for a benchmark neutralisation process, which is a is nonlinear dynamic system, a number of Elman networks with two and three layers are trained. It turns out that the Elman structure with three layers is very inefficient in comparison to the network with two layers because: (a) it gives unacceptable modelling errors, (b) it has many parameters, (c) training is difficult and needs a huge number of iterations. The reason for this may be that there are two separate recurrent layers which influence each other very much. Possibly both layers "fight each other", which has a very negative effect on approximation ability of the whole model.

References

1. Elman, J.L.: Finding structure in time. Cogn. Sci. **14**, 179–211 (1990)
2. Gómez, J.C., Jutan, A., Baeyens, E.: Wiener model identification and predictive control of a pH neutralisation process. IEEE Proc. Part D Control Theory Appl. **151**, 25–35 (2004)
3. Haykin, S.: Neural Networks-A Comprehensive Foundation. Prentice Hall, Upper Saddle River (2008)
4. Li, P.: Application of a hybrid quantized Elman neural network in short-term load forecasting. Int. J. Electr. Power Energy Syst. **55**, 749–759 (2014)
5. Li, Q.: Prediction of urban rail transit sectional passenger flow based on Elman neural network. Appl. Mech. Mater. **505–506**, 1023–1027 (2014)
6. Liou, J.-W.: Distributed representation of word by using Elman network. Int. J. Intell. Inf. Database Syst. **7**, 373–386 (2013)
7. Mandic, D.: Recurrent Neural Networks for Prediction. Wiley, Chichester (2001)
8. Nocedal, J., Wright, S.J.: Numerical Optimization. Springer, New York (2006)
9. Osowski, S.: Neural networks for information processing (in Polish). OWPW, Warsaw (2006)
10. Wang, J.-C.: Elman network based on EPSO-BP and its application in flight trajectory prediction. Control Decis. **28**, 1884–1888 (2013)
11. Wysocki, A.: Efficient training of Elman neural networks (in Polish). In: Malinowski, K., Józefczyk, J., Świątek, J. (eds.) Current Problems of Automation and Robotics, pp. 754–764, Exit, Wroclaw (2014)
12. Zhou, C., Ding, L.Y., He, R.: PSO-based Elman neural network model for predictive control of air chamber pressure in slurry shield tunneling under Yangtze river. Autom. Constr. **36**, 208–217 (2013)

Extension of Model Functionalities for Multi-echelon Distribution Systems Through the Introduction of Logical Constraints

Paweł Sitek and Jarosław Wikarek

Abstract Multi-echelon distribution systems are quite common in supply-chain and city logistic systems. The paper presents a concept of extending functionality of the multi-distribution models by introduction logical constraints. This is possible by using a hybrid approach to modeling and optimization the multi-echelon problems. In the hybrid approach, two environments of mathematical programming (MP) and constraint logic programming (CLP) were integrated. Logical constraints are associated with the transformation of the problem made by the CLP. The Two-Echelon Capacitated Vehicle Routing Problem (2E-CVRP) has been proposed as an illustrative example. The logical constraints on routes, cities etc. were introduced to the standard 2E-CVRP model. The presented approach will be compared with classical mathematical programming on the same data sets (known benchmarks).

Keywords Multi-echelon distribution systems · Constraint logic programming · Mathematical programming · Logical constraints

1 Introduction

The transportation of goods constitutes one of the main activities that influences economy and society, as it assures a vital link between suppliers and customers. Today, one of the most important aspects which takes place in freight transportation is the definition of different shipping strategies. There are several freight strategies, but they derive from the three main aspects [1]:

P. Sitek (✉) · J. Wikarek
Department of Information Systems, Kielce University of Technology, Kielce, Poland
e-mail: sitek@tu.kielce.pl

J. Wikarek
e-mail: j.wikarek@tu.kielce.pl

R. Szewczyk et al. (eds.), *Challenges in Automation, Robotics and Measurement Techniques*, Advances in Intelligent Systems and Computing 440, DOI 10.1007/978-3-319-29357-8_16

- **Hierarchical level**: This aspect can be defined using two groups of strategies (direct shipping and multi-echelon distribution. In the direct shipping, vehicles, starting from a depot, bring their freight directly to the destination, while in the multi-echelon systems, freight is delivered from the depot to the customers through an intermediate point.)
- **Transportation mode**: Due to different motivations (characteristics and quantity of freight to transport, distance, costs, etc.) and the offer in terms of transportation modes (road, railway, sea, air), a transportation solution will be defined using one or a combination of vehicles. If the transportation from origin to destination is realized using more than one of these modes, the transport is called multimodal or intermodal; otherwise, the strategies are monomodal.
- **Vehicle usage**: In some road transportation strategies, vehicles are loaded to capacity. This policy is known as truckload (TL). Instead, in other real applications, like in city logistics, most of the vehicles are not full-loaded, so the applied policy is known as Less-Than-Truckload (LTL).

In recent years multi-echelon systems have been introduced in different areas: logistics enterprises and express delivery service companies under competitions; hypermarkets and supermarkets products distribution; multimodal freight transportation; postal consignments and courier services, supply chains; E-commerce and home delivery services; city and public logistics. The multi-echelon distribution systems are the subject of intense study in recent years in connection with applications in city logistic [1, 2], multimodal transport and supply chains. The vast majority of models of optimization in freight transportation have been formulated as the mixed integer programming (MIP) or mixed integer linear programming (MILP) problems and solved using the operations research (OR) methods [3]. Their structures are similar and proceed from the principles and requirements of mathematical programming (MP).

An approach based on mathematical programming has weaknesses. Firstly, for the real size problems, it is time consuming and requires a lot of hardware resources. Secondly, it only allows modeling of linear and integer constraints [3]. The motivation undertake research, the results of which are presented in this paper was the need to develop a methodology that extends the capabilities of modeling and solving multi-echelon problems with logical constraints and better efficiency than mathematical programming methods. The need to introduce logical constraints for multi-echelon distribution models results from business, technological or legal conditions. In this context, it seems that better results will be obtained by the use of the constraint programming (CP) environments especially in modeling phase. The CP-based environments have the advantage over traditional methods of mathematical modeling in that they work with a much broader variety of interrelated constraints and allow producing "natural" solutions for highly combinatorial problems. The CP environments have declarative nature [4, 5]. The main contribution of this paper is extension of functionality of the multi-echelon models using hybrid approach (mixed CP with MILP). This approach has been shown for illustrative example (2E-CVRP—Two-Echelon Capacitated Vehicle Routing Problem) both in terms of modeling capability and efficiency solutions.

The paper is organized as follows. In Sect. 2 the Multi-Echelon Vehicle Routing Problems has been reviewed. In Sect. 3 the concept of hybrid approach to modeling and solving, and the hybrid solution environment have been presented. Then, the general description of illustrative example (2E-CVRP with logical constraints) has been discussed. Finally test instances for 2E-CVRP and some computational results were presented in Sect. 5. Conclusions and further works have been presented in Sect. 6.

2 Multi-echelon Systems

The hierarchical level in terms of transport strategies is the way the freight goes to the final destination. When the freight arrives to final destination without changing vehicle, a direct shipping or single-echelon strategy is applied, whereas when freight is derived from its origin to its final destination passing through intermediate points (satellites), where the freight is unloaded, then loaded into the same or into a different vehicle, we speak of a multi-echelon system. Especially in transportation, it is not always possible or convenient to deliver the goods directly to the destination. In fact, some transportation systems use intermediary points where some operations take place. The different vehicles that belong to these systems stop at some of these points, and in some cases the freight changes vehicle or even mode of transport. Moreover, some additional services, like labeling, packaging, re-packing etc., can be realized at these intermediary points. One of the basic problems in a multi-echelon systems is VRP. The Vehicle Routing Problem (VRP) is used to design an optimal route for a fleet of vehicles to service a set of customers' orders (known in advance), given a set of constraints. The VRP is of the NP-hard type. There are several variants and classes of VRP like the capacitated VRP (CVRP), VRP with Time Windows (VRPTW) and Dynamic Vehicle Routing Problems (DVRP), sometimes referred to as On-line Vehicle Routing Problems, 2E-CVRP (Two-Echelon Capacitated Vehicle Routing Problem) is a multi-echelon variant of CVRP etc. [1, 2, 6].

3 Hybrid Approach

Based on [4, 5, 7] and our previous work [8, 9] we observed some advantages and disadvantages of both CP-based and MP-based environments. An integrated approach of constraint programming (CP) and mixed integer/linear integer programming (MIP/MILP) can help to solve optimization problems that are intractable with either of the two methods alone [10–12]. Although Operations Research (OR) and Constraint Programming (CP) have different roots, the links between the two environments have grown stronger in recent years. Both environments involve variables and constraints. However, the types of the variables and constraints that are used, and the way the constraints are solved, are different in the two approaches [10]. MP-based environments relies completely on linear equations and inequalities in integer variables, i.e., there are only two types of constraints: linear arithmetic (linear equations or inequalities) and integrity (stating that the variables have to take their values in the integer numbers). In CP-based environments in addition to linear equations and inequalities, there are various other constraints: disequalities, non-linear, symbolic (*alldifferent, disjunctive, cumulative* etc.). In both MP-based and CP-based environments, there is a group of constraints that can be solved with ease and a group of constraints that are difficult to solve. The easily solved constraints in MP are linear equations and inequalities over rational numbers.

Integrity constraints are difficult to solve using mathematical programming methods and often the real problems make them NP-hard. In CP, domain constraints with integers and equations between two variables are easy to solve. The inequalities and general linear constraints (more than two variables), and symbolic constraints are difficult to solve, which makes real problems in CP NP-hard. What is important in the context of the presented research, CP-based environments easily allow you to model the logical constraints and integrate them with the others

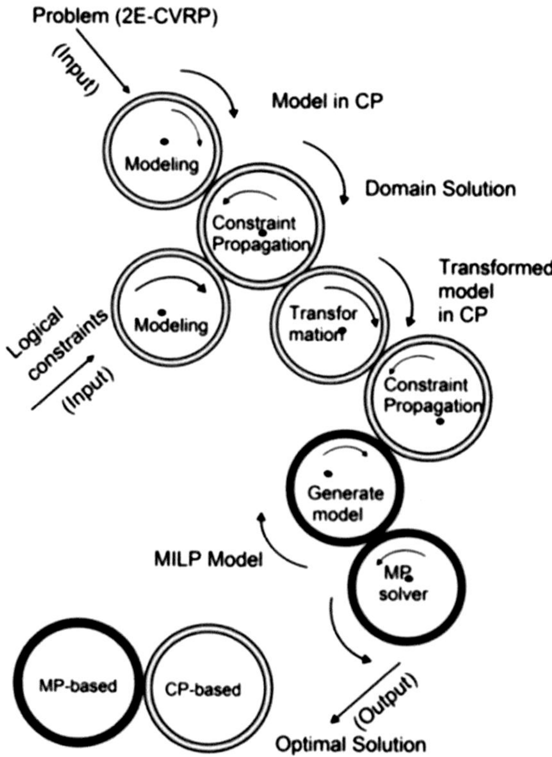

Fig. 1 Method of using hybrid approach for illustrative example (2E-CVRP)

constraints. Problems in CP-based environments on finite domains are typically solved using a form of search (backtracking, constraint propagation, and local search). Constraint propagation roughly amounts to choosing a small subset of a constraint set and to making some of its local constraints stronger by forbidding and eliminating some value assignment that would make the subset over constrained. The domain solution requires further solving process by backtracking search [5]. Backtracking search is a time-consuming procedure. Therefore we propose the use of authors' hybrid approach [8] to modeling and optimization the multi-echelon problems with logical constraints, which eliminates the backtracking mechanism and replaces it with the MP methods. The method of using hybrid approach to implementation and solutions for illustrative example is shown in Fig. 1.

4 Illustrative Example—Two-Echelon Capacitated Vehicle Routing Problem

The Two-Echelon Capacitated Vehicle Routing Problem (2E-CVRP) is an extension of the classical Capacitated Vehicle Routing Problem (CVRP) where the delivery depot-customers pass through intermediate depots (called satellites). As in CVRP, the goal is to deliver goods to customers with known demands, minimizing the total delivery cost in the respect of vehicle capacity constraints. Multi-echelon systems presented in the literature usually explicitly consider the routing problem at the last level of the transportation system, while a simplified routing problem is considered at higher levels [1, 6]. In 2E-CVRP, the freight delivery from the depot to the customers is managed by shipping the freight through intermediate depots. Thus, the transportation network is decomposed into two levels (Fig. 2): the 1st level connecting the depot (d) to intermediate depots (s) and the 2nd one connecting the intermediate depots (s) to the customers (c). The objective is to minimize the total transportation cost of the vehicles involved in both levels.

The mathematical model in the form of MILP was taken from [6, 13]. This is an arc-based model, i.e. the routes are composed of arcs between the depot and satellite, and satellite and individual customers. Constraints, decision variables and model parameters were discussed in detail in [6]. The model was implemented

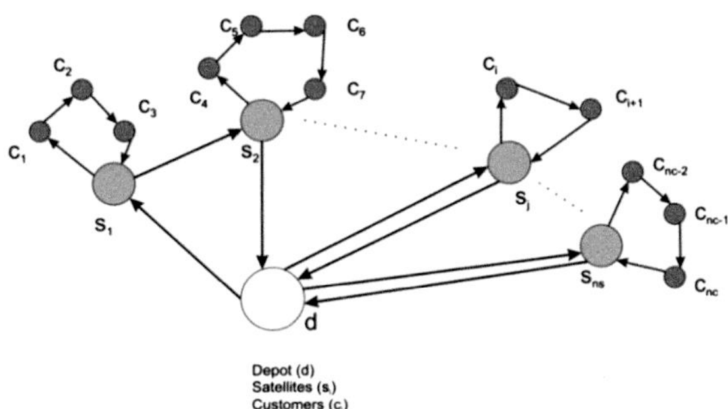

Fig. 2 Example of 2E-CVRP transportation network

using a hybrid approach wherein the numerous logical constraints were introduced (Fig. 1). Transformation model of illustration is to change the representation of the problem of arc-based on a route-based. In the route-based model routes are not assembled from individual arcs, but generated entirely on the basis of a set of data (facts) before you start problem solving. Therefore, it is kind of problem presolving. In the process of transformation, decision variables are changed (Tables 5, 6, 7 and 8 Appendix A). The model after transformation is shown in Appendix A.

4.1 Logical Constraints in 2E-CVRP

Logical constraints in 2E-CVRP could arise from the technological, business or law purposes. This kind of constraint is difficult to implement in the form of linear equations and inequalities. In the hybrid approach (Fig. 1), the logical constraints are implemented during generation of routes in the transformation process problem. This is done by generating only the acceptable routes, i.e. those that meet, among others, the logical constraints. The first of the logic constraint (LC_1) applies to the so-called time windows, that is not to exceed the operating time of the route. Such a constraint may result, e.g. with limited time drivers. Constraint LC_2 applies to the exclusion of customers on the route, i.e. that the two selected customers cannot get together to find a on the route. Such a restriction may result from trade and marketing agreements. On the other hand constraint LC_3 is contrary to the LC_2 (two selected customers must be on a given route). Last constraint LC_4 ensures that the selected customer would be no more than the n-th on the route.

5 Computational Examples (2E-CVRP)

For the final validation of the proposed hybrid approach, the benchmark data for 2E-CVRP was selected. 2E-CVRP, a well described and widely discussed problem, corresponded to the issues to which our hybrid approach was applied.

The instances for computational examples were built from the existing instances for CVRP denoted as E-n13-k4. All the instance sets can be downloaded from the website [14]. The instance set was composed of 5 small-sized instances with 1 depot, 12 customers and 2 satellites. The full instance consisted of 66 small-sized instances because the two satellites were placed over twelve customers in all 66 possible ways (number of combinations: 2 out of 12). All the instances had the same position for depot and customers, whose coordinates were the same as those of instance E-n13-k4. Small-sized instances differed in the choice of two customers who were also satellites (En13-k4-1, En13-k4-5, En13-k4-9, En13-k4-12, etc.). In the first stage the numerical experiments were conducted for the same data in two runs. The first run was a classical implementation of model [6] and its solution in

Table 1 The results of numerical examples for 2E-CVRP

E-n13-k4	MILP				Hybrid			
	Fc	T	C	V (int V)	Fc	T	C	V(int V)
E-n13-k4-18	246	79	1262	744(368)	246	10,28	21	1082(1079)
E-n13-k4-19	254	126	1262	744(368)	254	11,30	21	1082(1079)
E-n13-k4-20	276	487	1262	744(368)	276	12,14	21	1082(1079)
E-n13-k4-21	286	600*	1262	744(368)	286	16,11	21	1082(1079)
E-n13-k4-22	312	600*	1262	744(368)	312	9,97	21	1082(1079)
E-n13-k4-23	242	50	1262	744(368)	242	15,36	21	1082(1079)

Fc The optimal value of the objective function
T Time of finding solution
V(int V)/C The number of variables (integer variables)/constraints
*Calculations stopped after 600 s, the feasible value of the objective function

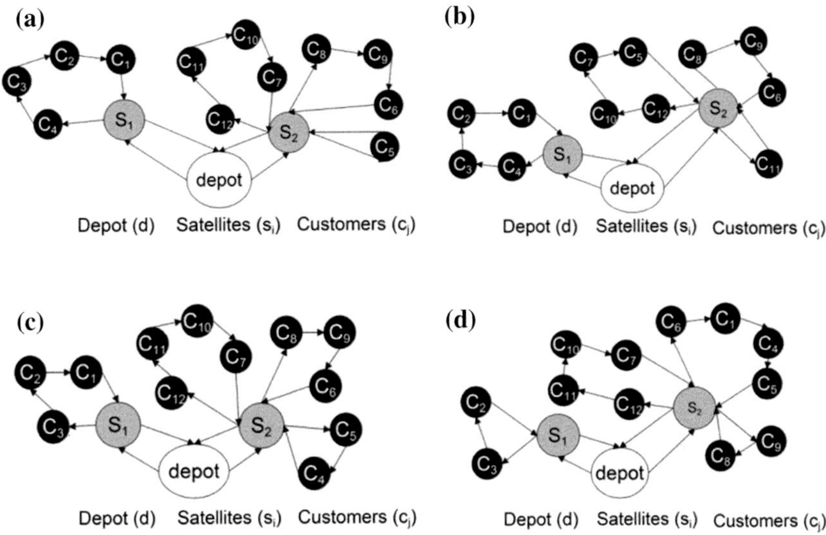

Fig. 3 **a** Example of 2E-CVRP transportation network for E-n13-k4-23 instance. **b** Example of 2E-CVRP transportation network for E-n13-k4-23 instance with logic constraint LC_1. **c** Example of 2E-CVRP transportation network for E-n13-k4-23 instance with logic constraint LC_2. **d** Example of 2E-CVRP transportation network for E-n13-k4-23 instance with logic constraint LC_4

the MILP environment (LINGO solver by LINDO Inc. [15]). In the second run the model was transformed and solved by the proposed hybrid approach.

The calculations were performed using a computer with the following specifications: Intel(R) Core(TM) 2 QQuad CPU Q6600 @ 2 × 2.40 GHz, 2.4 GHz, RAM 1.98 GB. The analysis of the results for the benchmark instances demonstrates that the hybrid approach may be a superior approach to the classical mathematical programming. For all examples, the solutions were found 6–40 times faster than

Table 2 The results of numerical examples for 2E-CVRP with logical constraints LC_1

E-n13-k4	T					
	50	60	70	80	90	100
E-n13-k4-13	–	–	–	292	288	288
E-n13-k4-23	–	290	242	242	242	242
E-n13-k4-24	–	258	242	242	242	242
E-n13-k4-25	254	254	252	252	252	252
E-n13-k4-47	–	–	300	274	274	274
E-n13-k4-48	–	–	308	280	280	280
E-n13-k4-49	–	–	312	280	280	280
E-n13-k4-64	–	–	–	–	362	358

T-the length of the time window

Table 3 The results of numerical examples for 2E-CVRP with logical constraints LC_2 (ex. customers)

E-n13-k4	Fc	T
E-n13-k4-13	290	1
E-n13-k4-23	260	3
E-n13-k4-24	246	2
E-n13-k4-25	252	3
E-n13-k4-47	300	3
E-n13-k4-48	304	3
E-n13-k4-49	304	2
E-n13-k4-64	368	4

Exclusion 3-4, 3-5, 3-6, 7-8, 7-9

Table 4 The results of numerical examples for 2E-CVRP with logical constraints LC_4

E-n13-k4	T	Fc	C	V(int V)	E-n13-k4	T	Fc	C	V(int V)
E-n13-k4-01	3	290	21	680(677)	E-n13-k4-33	3	288	21	680(677)
E-n13-k4-07	4	226	21	680(677)	E-n13-k4-40	4	306	21	680(677)
E-n13-k4-26	3	286	21	680(677)	E-n13-k4-13	3	288	21	680(677)
E-n13-k4-32	3	282	21	680(677)	E-n13-k4-23	3	286	21	680(677)

they are in the classical approach (Table 1 and Fig. 3a). In many cases the calculations ended after 600 s as they failed to indicate that the solution was optimal in MILP environment. As the presented example was formulated as a MILP problem, the hybrid approach was tested for the solution efficiency. Owing to the hybrid approach the 2E-CVRP models can be extended over logical, nonlinear, and other constraints. At the next stage logical constraints (LC_1, LC_2, LC_4) were introduced into the model. Table 2 and Fig. 3b present the results of the numerical experiments conducted for 2E-CVRPs with logical constraints LC_1. As you can see time windows affect the value of the objective function. For some of the T there is no solution. Table 3 and Fig. 3c present the results of the numerical experiments conducted for 2E-CVRPs with logical constraints relating to the situation where two delivery points (customers) can be handled separately but not together in one route. Table 4 and Fig. 3d show the results of the numerical experiments conducted for 2E-CVRPs with logical constraints LC_4. This constraint ensures that the selected customer 1, 2, 3 would be no more than the second on the route.

6 Conclusions

The possibility of extending models for multi-echelon systems of logical constraints significantly increase their functionality. Obtaining similar functionality using linear constraints is impossible or requires a very large number of constraints. This results in a significant increase in the time search for a solution. Design and solving a new class of models, complemented by logical constraints is possible by using a hybrid approach. The research in the context of the implementation illustrative example (2E-CVRP) also shows the high efficiency of hybrid approach (Sect. 5). 2E-CVRP is a known benchmark for the problem of routing with the available data instances [14]. The efficiency of the proposed approach is based on the reduction of the combinatorial problem and using the best properties of both environments. Further work will focus on running the optimization models with non-linear and other logical constraints, multi-objective, uncertainty, fuzzy-logic [16] etc., implementation in cloud-based environments [17] and implementation for logistic problems [18].

Table 5 Decision variables for MILP model [6]

Symbol	Description
Decision variables	
$X_{i,j}$	An integer variable of the 1st-level routing is equal to the number of 1st-level vehicles using *arc (i, j)*
$Y_{k,i,j}$	A binary variable of the 2nd-level routing is equal to 1 if a 2nd-level vehicle makes a route starting from satellite k and goes from node i to node j and 0 otherwise
$Q^1_{i,j}$	The freight flow *arc (i, j)* for the 1st-level
$Q^2_{k,i,j}$	The freight *arc (i, j)* where k represents the satellite where the freight is passing through.
$Z_{k,j}$	A binary variable that is equal to 1 if the freight to be delivered to customer *j* is consolidated in satellite *k* and 0 otherwise

Table 6 Summary indices, parameters and decision variables for transformed model

Symbol	Description
Indices	
n_s	Number of satellites
n_c	Number of customers
Ts	Number of possible routes from depot to satellites (CLP—determined)
Tc	Number of possible routes from satellites to customers (CLP—determined)
i	Satellite index
l	Depot-satellite route index
j	Customer index
k	Satellite-customer route index
M_1	Number of the 1st-level vehicles
M_2	Number of the 2nd-level vehicles
Input parameters	
s_s	Cost of loading/unloading operations of a unit of freight in satellite *s*
D_j	Demand required by customer *j*
Pc_k	Total demand for route *k* (CLP—determined)
Ks_l	Route 1 cost (CLP—determined)
Kc_k	Routek cost (CLP—determined)
$U_{l,i}$	If *i* is located on route *l* $U_{l,i} = 1$, otherwise $U_{l,i} = 0$
$W_{k,s}$	If satellite or receipient *s* is located on route k $W_{k,s} = 1$, otherwise $W_{k,s} = 0$
K_1	Capacity of the vehicles for the 1st level
Decision variables	
Y_l	If the tour takes place along the route l from the route set generated for level 1, then$Y_l = 1$, otherwise $Y_l = 0$
X_k	If the tour takes place along the route k from the route set generated for level 2, then $X_k = 1$, otherwise $X_k = 0$
Computed quantities	
Ps_l	Total demand for route l
$\min \sum_{k=1}^{Tc} (Y_k^T \cdot Kc_k) + \sum_{l}^{Ts} X_l^T \cdot Ks_l \quad (1)$	$\sum_{k=1}^{Tc} Y_k^T \leq M2 \quad (2)$
$\sum_{k=1}^{Tc} Y_k^T \cdot W_{k,j} = 1 \text{ for } j = 1 \ldots n_c \quad (3)$	$\sum_{k=1}^{Tc} X_k \cdot W_{k,i} \cdot Pc_k = \sum_{l}^{Ts} Ps_l \cdot U_{l,i} \text{ for } i = 1 \ldots n_s \quad (4)$
$\sum_{i=1}^{n_s} \sum_{k=1}^{Tc} Y_k^T \cdot W_{k,i} \cdot Pc_k = \sum_{l}^{Ts} Ps_l \quad (5)$	$X_l^T \cdot K1 \geq Ps_l \text{ for } l = 1 \ldots Ts \quad (6)$
$\sum_{l=1}^{Ts} X_;^T \leq M_1 \quad (7)$	$Y_k^T \in \{0,1\} \text{ for } k = 1 \ldots Tc \quad (8)$
$X_l^T \in C \text{ for } l = 1 \ldots Ts \quad (9)$	

Table 7 Constraints after transformation

Constraint	Description
T1	The objective function minimizes the sum of the routing and handling operations costs
T2/T7	The number of the routes in each level must not exceed the number of vehicles for that level (first/second)
T3	No overlapping deliveries to customers
T4/T5	The supply balance for satellites
T6	The number of tours for level 1 resulting from the capacity of vehicles.
T8/T9	The integrity and binarity

Table 8 Decision variables and constraints before and after transformation

Before	After	Description
Decision variables		
$X_{i,j}$	X_l^T	Transformation of decision variables 1-level from the arc model *arc (i, j)* to the route model *(l)*
$Q_{i,j}^1$		
$Y_{k.i.j}$	Y_k^T	Transformation of decision variables 2-level from the arc model *arc (i, j)* to the route model *(k)*

Appendix A

See Tables 5, 6, 7 and 8

References

1. Gonzalez-Feliu, J.: Models and Methods for the City Logistics: The Two-Echelon Capacitated Vehicle Routing Problem, Operations Research. Politecnico di Torino (2008)
2. Crainic, T., Ricciardi, N., Storchi, G.: Advanced freight transportation systems for congested urban areas. Transp. Res. Part C **12**, 119–137 (2004)
3. Schrijver, A.: Theory of Linear and Integer Programming. Wiley, New York (1998)
4. Apt, K., Wallace, M.: Constraint Logic Programming using Eclipse. Cambridge University Press, Cambridge (2006)
5. Rossi, F., Van Beek, P., Walsh, T.: Handbook of Constraint Programming (Foundations of Artificial Intelligence). Elsevier Science Inc, New York (2006)
6. Perboli, G., Tadei, R., Vigo, D.: The two-echelon capacitated vehicle routing problem: models and math-based heuristics. Transp. Sci. **45**, 364–380 (2012)
7. Bocewicz, G., Banaszak, Z.: Declarative approach to cyclic steady states space refinement: periodic processes scheduling. Int. J. Adv. Manuf. Technol. **67**(1–4), 137–155 (2013)
8. Sitek, P., Nielsen I.E., Wikarek, J.: A Hybrid multi-agent approach to the solving supply chain problems. Procedia Comput. Sci. **35**, KES, 1557–1566 (2014)
9. Sitek, P., Wikarek J.: A hybrid framework for the modelling and optimisation of decision problems in sustainable supply chain management. Int. J. Prod. Res. **53**(21), 6339–6343 (2015). doi:10.1080/00207543.2015.1005762

10. Jain, V., Grossmann, I.E.: Algorithms for hybrid MILP/CP models for a class of optimization problems. INFORMS J. Comput. **13**(4), 258–276 (2001)
11. Achterberg, T., Berthold, T., Koch, T., Wolter K.: Constraint Integer Programming: A New Approach to Integrate CP and MIP, Lecture Notes in Computer Science, vol. 5015, pp. 6–20 (2008)
12. Milano, M., Wallace, M.: Integrating operations research in constraint programming. Ann. Oper. Res. **175**(1), 37–76 (2010)
13. Sitek, P.: A hybrid approach to the two-echelon capacitated vehicle routing problem (2E-CVRP). Recent Adv. Autom. Robot. Measuring Tech. Adv. Intell. Syst. Comput. **267**, 251–263 (2014)
14. ORO Group Web-page. http://www.orgroup.polito.it/. Accessed 4 Oct 2015
15. Lindo Systems INC, LINDO™. www.lindo.com. Accessed 4 Oct 2015
16. Relich, M., Muszyński, W.: The use of intelligent systems for planning and scheduling of product development projects. Procedia Comput. Sci. **35**, 1586–1595 (2014)
17. Bąk, S., Czarnecki, R., Deniziak, S.: Synthesis of real-time cloud applications for internet of things. Turkish J. Electr. Eng. Comput. Sci. (2013). doi:10.3906/elk-1302-178
18. Grzybowska, K., Kovács, G.: Logistics Process Modelling in Supply Chain—algorithm of coordination in the supply chain—contracting, International Joint Conference SOCO'14-CISIS'14-ICEUTE'14. Adv. Intell. Syst. Comput. **299**, 311–320 (2014)

Shooting Methods to Solve Optimal Control Problems with State and Mixed Control-State Constraints

Andrzej Karbowski

Abstract The paper presents two important approaches to solve numerically general optimal control problems with state and mixed control-state constraints. They may be attractive in the case, when the simple time discretization of the state equations and expressing the optimal control problem as a nonlinear mathematical programming problem is not sufficient. At the beginning an extension of the optimal control theory to problems with constraints on current state and on current state and control simultaneously is presented. Then, two approaches to solve numerically the emerging boundary value problems: indirect and direct shooting method are described and applied to an example problem.

Keywords Optimal control · Numerical methods · State constraints · Mixed control-state constraints · Lagrange functionals · Shooting method · Multiple shooting method · Boundary value problem

1 Introduction

Shooting method is the basic numerical method used to solve ordinary differential equations, when instead of initial conditions for the state trajectory, as in Cauchy problem (aka initial value problem—IVP), we have terminal conditions. Such a problem is called boundary value problem (BVP) [13]. This approach can be easily adapted to solve these optimal control problems, which need a higher precision

A. Karbowski (✉)
Institute of Control and Computation Engineering, Warsaw University of Technology, Warsaw, Poland
e-mail: A.Karbowski@elka.pw.edu.pl

A. Karbowski
NASK, Research and Academic Computer Network, Warsaw, Poland

R. Szewczyk et al. (eds.), *Challenges in Automation, Robotics and Measurement Techniques*, Advances in Intelligent Systems and Computing 440, DOI 10.1007/978-3-319-29357-8_17

of the solution, than obtained from the standard numerical approach, based on time discretization. In the latter, state and control trajectories are represented by vectors and from the differential state equations difference equations are obtained, which are treated as a set of equality constraints in a static nonlinear programming problem.

There are two possible approaches to transform an optimal control problem to BVP: direct and indirect.

In the direct approach the control interval is divided into a certain number of subintervals, on which Cauchy problem is solved by an ordinary differential equation (ODE) solver. The initial conditions are generated iteratively by an optimizer, the constraints on state and mixed are checked in the discretization points of the time interval.

In the indirect approach BVP concerns not only state equations, but also the equations describing adjoint variables. It means, that for an optimal control problem, before using a solver we have to make a kind of preprocessing on the paper, based on the appropriate theory. In particular, we have to determine the number of switching points, where the state trajectory enters and leaves the constraint boundary. Moreover, in this approach we have to provide the function connecting the optimal control at a given time instant with current value of state and adjoint variables (i.e., the control law). Only general formulas should be given, their parameters: Lagrange multipliers, initial values of adjoint variables and the concrete values of switching points (i.e., times) will be the subject of optimization. Actually, in the fundamental book of Pontryagin et al. [11] one may found a suitable version of the maximum principle to formulate BVP for problems with the constraints function of order one (in Chap. 6). The generalization to problems of higher orders was first given by Bryson et al. [1, 2].

In the author's opinion, the BVPs corresponding to necessary conditions of optimality for quite a big class of optimal control problems with state and mixed control-state constraints can be derived in a simpler way, using local optimality theory. We adapt here the approach proposed by Wierzbicki [15], based on looking for stationary points of Lagrange functionals in abstract spaces. A similar reasoning, using generalized Kuhn-Tucker conditions in a Banach space, was presented by Jacobson et al. [5], but the derivation presented here is more straightforward.

In this paper we consider a Mayer-type problem with one state constraint of the order one with one boundary arc. The approach can be easily adapted to solve other problems with state and mixed constraints. After the derivation of the necessary conditions of optimality, a numerical approach where it can be used—the indirect shooting method—and an alternative to it—the direct shooting method—are shortly described and compared on a test problem taken from the literature.

2 Necessary Optimality Conditions for a General Optimal Control Problem with State and Mixed State-Control Constraints

For simplicity of formulas our presentation concerns problems with scalar control and constraint functions. The passage to the multidimensional case is obvious.

We want to determine a piecewise continuous control function $u(t) \in \mathbb{R}$, $t_0 \leq t \leq t_f$, which minimizes the Mayer functional

$$J(u) = g(x(t_f)) \tag{1}$$

subject to the constraints

$$\dot{x}(t) = f(x(t), u(t), t), \quad t_0 \leq t \leq t_f \tag{2}$$

$$x(t_0) = x_0 \tag{3}$$

$$r(x(t_f), t_f) = 0 \in \mathbb{R}^m \tag{4}$$

$$S(x(t), t) \leq 0, \quad t_0 \leq t \leq t_f \tag{5}$$

$$C(x(t), u(t), t) \leq 0, \quad t_0 \leq t \leq t_f \tag{6}$$

Here, $x(t) \in \mathbb{R}^n$ denotes a vector of the state variables, the constraint (4) describes boundary conditions, (5), (6) are nonstationary inequality constraints on current values of, respectively, state and state and control simultaneously. For simplicity, it is assumed, that the functions S, C are sufficiently continuously differentiable. The function $f(x(t), u(t), t)$ is allowed to be merely piecewise continuously differentiable with respect to time variable, for $t \in [t_0, t_f]$. The final time t_f is fixed. Problems with free final time or problems with integral terms in the performance index (Bolza or Lagrange) can be easily transformed into a problem of the type (1) by means of additional state variables. The constraints on control of the type:

$$u(t) \in U \subset \mathbb{R} \tag{7}$$

can be expressed as a specific case of the mixed functional constraints (6).

The Lagrangian for the problem (1)–(6) will contain the standard components:

- from the objective function

$$g(x(t_f)) \tag{8}$$

- from the initial conditions

$$\langle \rho, x(t_0) - x_0 \rangle \tag{9}$$

- from the state equation

$$\int_{t_0}^{t_f} \langle \eta(t), \dot{x}(t) - f(x(t), u(t), t) \rangle \, dt, \tag{10}$$

where $\rho \in \mathbb{R}^n$, η is the adjoint function with values in $\mathbb{R}^n$ and $\langle .,. \rangle$ denotes the scalar product, and some additional terms coming from the terminal, state and mixed constraints. These will be:

- the product

$$\nu^T r(x(t_f), t_f), \quad \nu \in \mathbb{R}^m \tag{11}$$

  from the terminal constraint (4)
- the integral

$$\int_{t_0}^{t_f} \mu_C(t) \, C(x(t), u(t), t) \, dt \tag{12}$$

  where

$$\mu_C(t) = \begin{cases} = 0, \; C(x(t), u(t), t) < 0 \\ \geq 0, \; C(x(t), u(t), t) = 0, \end{cases} \tag{13}$$

  from mixed control-state constraint (6).

The situation with the state constraint (5) is more complicated. When this constraint is active, that is $S(x(t), t) = 0$, e.g. on some interval $[t_1, t_2] \subseteq [t_0, t_f]$, its time derivative along the path must vanish. That is, we must have[1] [1, 2]

$$\frac{dS}{dt} = \frac{\partial S}{\partial t} + \left(\frac{\partial S}{\partial x}\right)^T \dot{x} = \frac{\partial S}{\partial t} + \left(\frac{\partial S}{\partial x}\right)^T f(x, u, t) = 0 \tag{14}$$

The expression (14) may or may not have explicit dependence on u. If it does, the expression (14) for $\frac{dS}{dt}$ plays on the boundary arc the same role as the function C defining mixed state-control variable constraint of the type (6). If it does not, we may take the next time derivative. We may repeat this procedure until, finally, some explicit dependence on u does occur [1, 2].

If this occurs on the qth time derivative, we will tell, that we have a qth-order state variable inequality constraint. In this case, the qth total time derivative of $S(x(t), t)$ will be the component representing state inequality constraint (5) in the Lagrangian. More precisely, let $S^{(k)}$ denotes the kth total time derivative of the state constraint function S, i.e.,

$$S^{(0)}(x(t), u(t), t) := S(x(t), t), \tag{15}$$

[1] In our convention a gradient of a scalar function is a column vector.

$$S^{(k+1)}(x(t),u(t),t) = \frac{d^{k+1}S}{dt^{k+1}}(x(t),u(t),t)$$

$$= \frac{\partial S^{(k)}}{\partial t}(x(t),u(t),t) + \left[\frac{\partial S^{(k)}}{\partial x}(x(t),u(t),t)\right]^T f(x(t),u(t),t), \quad k \geq 0 \tag{16}$$

The order q is the lowest order of derivative such, that $S^{(q)}$ contains the control explicitly:

$$\frac{\partial S^{(k)}(x,u,t)}{\partial u} \equiv 0, \quad k = 0,\ldots,q-1, \quad \frac{\partial S^{(q)}}{\partial u} \neq 0 \tag{17}$$

Hence, we may write in the following:

$$S^{(k+1)}(x(t),u(t),t) \equiv S^{(k+1)}(x(t),t), \quad k = 0,\ldots,q-1 \tag{18}$$

and the actual constraint will be:

$$S^{(q)}(x(t),u(t),t) = 0, \quad \text{for } t \in [t_1,t_2] \tag{19}$$

The corresponding component of the Lagrangian will have the form

$$\int_{t_0}^{t_f} \mu_S(t)\, S^{(q)}(x(t),u(t),t)\, dt \tag{20}$$

where

$$\mu_S(t) = 0, \quad \text{for } t \notin [t_1,t_2] \tag{21}$$

Since control of $S(x(t),t)$ is obtained only by changing its qth time derivative, to keep the system on the constraint boundary, the path entering onto the constraint boundary has to meet the following "tangency" constraints at time t_1 (or, equivalently at t_2) [1, 2]:

$$N(x(t_1),t_1) = \begin{bmatrix} S(x(t_1),t_1) \\ S^{(1)}(x(t_1),t_1) \\ \vdots \\ S^{(q-1)}(x(t_1),t_1) \end{bmatrix} = 0 \tag{22}$$

The corresponding component of the Lagrangian will be the scalar product:

$$\gamma^T N(x(t_1),t_1) \tag{23}$$

As an example we will consider the optimal control problem (1)–(5) (i.e. without mixed state-control constraint (6)) with the state constraint active on the interval $[t_1,t_2]$:

$$J(u) = g(x(t_f)) \tag{24}$$

subject to the constraints

$$\dot{x}(t) = f(x(t), u(t), t), \quad t_0 \le t \le t_f \tag{25}$$

$$x(t_0) = x_0 \tag{26}$$

$$r(x(t_f), t_f) = 0 \in \mathbb{R} \tag{27}$$

$$S(x(t), t) \le 0, \quad t_0 \le t \le t_f \tag{28}$$

We assume, that the state-constraint (28) is of order one, that is $q = 1$, $\gamma \in \mathbb{R}$.

The Lagrangian for this problem will be as follows:

$$L(x, u, \eta, \rho, \mu_S, \gamma, v) = g(x(t_f)) + \langle \rho, x(t_0) - x_0 \rangle + \int_{t_0}^{t_f} \langle \eta(t), \dot{x}(t) - f(x(t), u(t), t) \rangle \, dt +$$

$$+ \int_{t_0}^{t_f} \mu_S(t) S^{(1)}(x(t), u(t), t) \, dt + \gamma S(x(t_1), t_1) + v r(x(t_f), t_f) \tag{29}$$

Because of the possible discontinuity of the adjoint function $\eta(t)$ at time t_1, we will partition the integral stemming from the state equation, that is $\int_{t_0}^{t_f} \langle \eta(t), \dot{x}(t) - f(x(t), u(t), t) \rangle \, dt$ into two components concerning two subintervals: $[t_0, t_1)$ and $(t_1, t_f]$. It is not necessary to partition the second integral stemming from the state constraint, that is $\int_{t_0}^{t_f} \mu_S(t) S^{(1)}(x(t), u(t), t) \, dt$, although its Lagrange multiplier function $\mu_S(t)$ can be discontinuous at t_1 too, because this discontinuity has no possibility to transform itself further into independent, different components of the Lagrangian, including those being functions of the state vector at time t_1, that is $x(t_1)$. The reason is, that to the first integral the integration by parts can be applied, while to the second it cannot.

After the mentioned partition, the Lagrangian (29) will take the form:

$$L(x, u, \eta, \rho, \mu_S, \gamma, v) = g(x(t_f)) + \langle \rho, x(t_0) - x_0 \rangle + \int_{t_0}^{t_1^-} \langle \eta(t), \dot{x}(t) - f(x(t), u(t), t) \rangle \, dt +$$

$$+ \int_{t_1^+}^{t_f} \langle \eta(t), \dot{x}(t) - f(x(t), u(t), t) \rangle \, dt + \int_{t_0}^{t_f} \mu_S(t) S^{(1)}(x(t), u(t), t) \, dt +$$

$$+ \gamma (S(x(t_1), t_1) + v r(x(t_f), t_f) = g(x(t_f)) + \langle \rho, x(t_0) - x_0 \rangle +$$

$$+ \int_{t_0}^{t_1^-} \langle \eta(t), \dot{x}(t) \rangle \, dt - \int_{t_0}^{t_1^-} \langle \eta(t), f(x(t), u(t), t) \rangle \, dt +$$

$$+ \int_{t_1^+}^{t_f} \langle \eta(t), \dot{x}(t) \rangle \, dt - \int_{t_1^+}^{t_f} \langle \eta(t), f(x(t), u(t), t) \rangle \, dt +$$

$$+\int_{t_0}^{t_f} \mu_S(t) S^{(1)}(x(t), u(t), t)\, dt + \gamma S(x(t_1), t_1) + \nu r(x(t_f), t_f) \tag{30}$$

Applying integration by parts to components with state velocity $\dot{x}(t)$ we will get:

$$L(x, u, \eta, \rho, \mu_S, \gamma, \nu) = g(x(t_f)) + \langle \rho, x(t_0) - x_0 \rangle + \langle \eta(t_1^-), x(t_1^-) \rangle - \langle \eta(t_0), x(t_0) \rangle +$$

$$-\int_{t_0}^{t_1^-} \langle \dot{\eta}(t), x(t) \rangle\, dt - \int_{t_0}^{t_1^-} \langle \eta(t), f(x(t), u(t), t) \rangle\, dt + \langle \eta(t_f), x(t_f) \rangle - \langle \eta(t_1^+), x(t_1^+) \rangle +$$

$$-\int_{t_1^+}^{t_f} \langle \dot{\eta}(t), x(t) \rangle\, dt - \int_{t_1^+}^{t_f} \langle \eta(t), f(x(t), u(t), t) \rangle\, dt + \int_{t_0}^{t_f} \mu_S(t) S^{(1)}(x(t), u(t), t)\, dt +$$

$$+\gamma(S(x(t_1), t_1)) + \nu r(x(t_f), t_f) \tag{31}$$

Grouping together similar terms, taking into account the continuity of the state variables, we will get:

$$L(x, u, \eta, \rho, \mu_S, \gamma, \nu) = \langle \rho, x(t_0) - x_0 \rangle - \langle \eta(t_0), x(t_0) \rangle +$$

$$+\langle \eta(t_1^-), x(t_1) \rangle + \gamma(S(x(t_1), t_1)) - \langle \eta(t_1^+), x(t_1) \rangle +$$

$$-\int_{t_0}^{t_1^-} \left[\langle \dot{\eta}(t), x(t) \rangle + \langle \eta(t), f(x(t), u(t), t) \rangle \right] dt +$$

$$-\int_{t_1^+}^{t_f} \left[\langle \dot{\eta}(t), x(t) \rangle + \langle \eta(t), f(x(t), u(t), t) \rangle \right] dt + \int_{t_0}^{t_f} \mu_S(t) S^{(1)}(x(t), u(t), t)\, dt +$$

$$+g(x(t_f)) + \langle \eta(t_f), x(t_f) \rangle + \nu r(x(t_f), t_f) = \left[\langle \rho, x(t_0) - x_0 \rangle - \langle \eta(t_0), x(t_0) \rangle \right] +$$

$$+\left[\langle \eta(t_1^-), x(t_1) \rangle + \gamma(S(x(t_1), t_1)) - \langle \eta(t_1^+), x(t_1) \rangle \right] +$$

$$-\int_{t_0}^{t_f} \left[\langle \dot{\eta}(t), x(t) \rangle + \langle \eta(t), f(x(t), u(t), t) \rangle - \mu_S(t) S^{(1)}(x(t), u(t), t) \right] dt +$$

$$+\left[g(x(t_f)) + \langle \eta(t_f), x(t_f) \rangle + \nu r(x(t_f), t_f) \right] \tag{32}$$

It will be convenient now to define the Hamiltonian function for a part of the integrand expression:

$$H(x, u, \eta, \mu_S, t) = \langle \eta, f(x, u, t) \rangle - \mu_S S^{(1)}(x, u, t) \tag{33}$$

According to the theory presented in [15], the optimal solution is a stationary point of the Lagrangian. Owing to that, we will get the following conditions of optimality:

State equation:

$$\dot{x}(t) = f(x(t), u(t), t), \quad t \in [t_0, t_f], \ x(t_0) = x_0 \tag{34}$$

Adjoint differential equation:

$$\dot{\eta} = -H_x(x, u, \eta, \mu_S, t), \quad t \in [t_0, t_f], \ \eta(t_0) = \eta_0 \tag{35}$$

Initial point multipliers vector:

$$\rho = \eta(t_0) \tag{36}$$

Natural boundary conditions:

$$\eta(t_f) = -\frac{\partial(g + \nu r)}{\partial x(t_f)} \tag{37}$$

Switching (tangency) condition:

$$S(x(t_1), t_1) = 0 \tag{38}$$

Boundary arc condition:

$$S^{(1)}(x, u, t) = 0, \quad t \in [t_1, t_2] \tag{39}$$

Complementarity (sign) conditions:

$$\mu_S(t) = \begin{cases} = 0, \ S(x(t), t) < 0 \\ \geq 0, \ S(x(t), t) = 0 \end{cases} \tag{40}$$

Junction conditions:

$$\eta(t_1^+) = \eta(t_1^-) + \gamma S_x(x(t_1), t_1) \tag{41}$$

Optimality condition: Assuming, that $\forall t \in [t_0, t_f]$ the Hamiltonian $H(x, u, \eta, \mu_S, t)$ is strictly convex with respect to u

$$\frac{\partial H}{\partial u}(x, u, \eta, \mu_S, t) = 0, \quad \forall t \in [t_0, t_f] \tag{42}$$

Otherwise (e.g., when the Hamiltonian is linear with respect to u), the modifications proposed by Maurer and Gillessen [7, 8] can be applied.

The (unknown) parameters in this problem are: $\eta_0, t_1, t_2, \nu, \gamma$.

3 Indirect Multiple Shooting Technique

This short presentation is adapted from the Oberle article [9].

In the previous section we saw, that the necessary conditions for the general optimal control problems lead to BVP with switching conditions for the state $x(t)$ and adjoint $\eta(t)$ trajectory. The basic idea of the numerical treatment of such problems by multiple shooting technique is to consider the switching conditions as boundary conditions to be satisfied at some interior multiple shooting nodes [9]. Thus, the problem is transformed into a classical multipoint BVP [9, 13]:

Determine a piecewise smooth vector function $y(t) = [x(t), \eta(t)]$, which satisfies

$$\dot{y}(t) = f(y, u, t), \quad t_0 \le t \le t_f, \tag{43}$$

$$u = u_k(t, y), \quad \tau_k \le t \le \tau_{k+1}, \quad k = 0, \dots, p, \tag{44}$$

$$y(t_0) = \begin{bmatrix} x_0 \\ \eta_0 \end{bmatrix} \tag{45}$$

$$y(\tau_k^+) = h_k(y(\tau_k^-), \gamma_k), \quad k = 1, \dots, p, \tag{46}$$

$$r_i(y(t_f), \nu) = 0, \quad i = 1, \dots, n_1, \tag{47}$$

$$\tilde{r}_k(\tau_k, y(\tau_k^-)) = 0, \quad k = 1, \dots, p. \tag{48}$$

In this formulation, $\eta_0, \gamma, \nu, \tau_k, k = 1, \dots, p$ are unknown parameters of the problem, where the latter satisfy

$$t_0 =: \tau_0 < \tau_1 < \tau_2 < \dots < \tau_p < \tau_{p+1} := t_f \tag{49}$$

The trajectory may possess jumps of size given by Eq. (46). If a coordinate of $y(t)$ is continuous, then its h_k is identity. The boundary conditions and the switching conditions are described by Eqs. (47) and (48), respectively.

In every time stage k the numerical integration over the interval $[\tau_k, \tau_{k+1}]$ is done by any conventional IVP solver with stepsize control. The resulting system of nonlinear equations (47)–(48) can be solved numerically by a quasinewton method, e.g. from the Broyden family.

The approach presented in this section, where the solution is sought basing on the necessary optimality conditions, is called the indirect method.

4 Direct Multiple Shooting Technique

As it was shown in Sect. 2, the necessary conditions of optimality for general optimal control problems lead to BVP for the set of ODEs describing the evolution of the state $x(t)$ and adjoint variables $\eta(t)$ trajectories. Unfortunately, while the state variables are continuous, the adjoint variables may have jumps in points, where there are changes in the activity of constraints. Due to these jumps in $\eta(t)$ trajectory, the only possible method to solve this problem is the multiple shooting algorithm presented in Sect. 3. The problem is, that rather a good initial approximation of the optimal trajectory is needed and rather a large amount of work has to be done by the user to derive the necessary conditions of optimality, in particular the adjoint differential equations [14]. Moreover, the user has to know a priori the switching structure of the constraints (the number and the sequence of the switching points), that is, he/she must have a deep insight into the physical and mathematical nature of the optimization problem [10, 14]. When the structure of the optimal control is more complicated and the solution consists of several arcs, it may lead to a very coarse approximation of the optimal control trajectory. Another drawback of the indirect approach is its sensitivity to parameters of the model, e.g., even small change of them, or an additional constraint, may lead to complete change of the switching structure [12].

In direct approaches, at the beginning the optimal control problem is transformed into a nonlinear programming problem [3, 6, 14]. In direct shooting method this is done through a parameterization of the controls $u(t)$ on the subintervals of the control interval. That is, we take:

$$t_0 < t_1 < t_2 < \cdots t_p < t_{p+1} = t_f \tag{50}$$

$$u(t) = u_j(t, \alpha_j), \quad t \in I_j = [t_j, t_{j+1}], \quad \text{for } j = 0, 1, \ldots, p \tag{51}$$

where $\alpha \in \mathbb{R}^{n_\alpha}$ is a vector of parameters. For example, $u(t)$ may be: piecewise constant, piecewise linear or higher order polynomials, linear combination of some basis functions, e.g. B-splines. The basic idea is to simultaneously integrate numerically the state equations (2) on the subintervals I_j for guess initial points

$$z_j = x(t_j) \tag{52}$$

Then the values obtained at the ends of subintervals—we will denote them by $x(t_{j+1}; z_j, \alpha_j)$—are compared with the guesses z_{j+1}.

The differential equations, initial and end points conditions and path constraints define the constraints of the nonlinear programming problem, that is the problem (1)–(6) is replaced with:

$$\min_{\alpha,z} g(z_{p+1}) \tag{53}$$

$$z_{j+1} - x(t_{j+1}; z_j, \alpha_j) = 0, \ \ j = 0, \ldots, p \tag{54}$$

$$r(z_{p+1}) = 0 \tag{55}$$

$$S(z_j, t_j) \leq 0, \;\; j = 0, \ldots, p+1 \tag{56}$$

$$C(z_j, \alpha_j, t_j) \leq 0, \;\; j = 0, \ldots, p+1 \tag{57}$$

where $x(t_{j+1}; z_j, \alpha_j)$ for $j = 0, \ldots, p$ is the solution of ODE:

$$\dot{x}(t) = f(x(t), u_j(t, \alpha_j), t), \;\; t_j \leq t \leq t_{j+1} \tag{58}$$

$$x(t_j) = z_j \tag{59}$$

at $t = t_{j+1}$.

This nonlinear programming problem can be solved by any constrained, continuous optimization solver.

5 A Case Study

Let us consider the following optimal control problem taken from Jacobson and Lele paper [4]:

$$\min_{u \in \mathbb{R}} \int_0^1 (x_1^2 + x_2^2 + 0.005u^2)dt \tag{60}$$

where

$$\dot{x}_1 = x_2, \qquad x_1(0) = 0 \tag{61}$$

$$\dot{x}_2 = -x_2 + u, \qquad x_2(0) = -1 \tag{62}$$

$$x_2(t) \leq 8(t - 0.5)^2 - 0.5 \tag{63}$$

To transform this problem into a Mayer problem form (1)–(6) we have to introduce an additional state variable x_3 governed by the state equation:

$$\dot{x}_3 = x_1^2 + x_2^2 + 0.005u^2, \qquad x_3(0) = 0 \tag{64}$$

and replace the objective function (60) with

$$g(x(t_f)) = x_3(1) \tag{65}$$

which will be also minimized.

Now we may rewrite the optimization problem over the time interval $[t_0, t_f] = [0, 1]$, introducing the vector notation from the Sect. 2, as the Mayer problem:

$$\min_{u \in \mathbb{R}} g(x(t_f)) = x_3(1) \tag{66}$$

$$\dot{x} = f(x,u) = \begin{bmatrix} x_2 \\ -x_2 + u \\ x_1^2 + x_2^2 + 0.005u^2 \end{bmatrix} \tag{67}$$

$$x(t_0) = \begin{bmatrix} 0 \\ -1 \\ 0 \end{bmatrix} \tag{68}$$

$$S(x,t) = x_2 - 8(t - 0.5)^2 + 0.5 \leq 0 \tag{69}$$

First, the problem (66)–(69) was solved by direct shooting method, described in Sect. 4, for $p = 20$ time subintervals of equal length with the help of two Matlab functions: `ode45` (ODE solver; medium order method), `fmincon` (constrained nonlinear multivariable optimization solver).

After 254.3 s we obtained the performance index value equal 0.2796. The resulting trajectories of the state variables x_1 and x_2 are presented in Fig. 1. Analyzing the resulting x_2 state variable trajectory we may see, that it contains one boundary arc. To find its precise course the indirect shooting method described in Sect. 3, based on the theory presented in Sect. 2, was used.

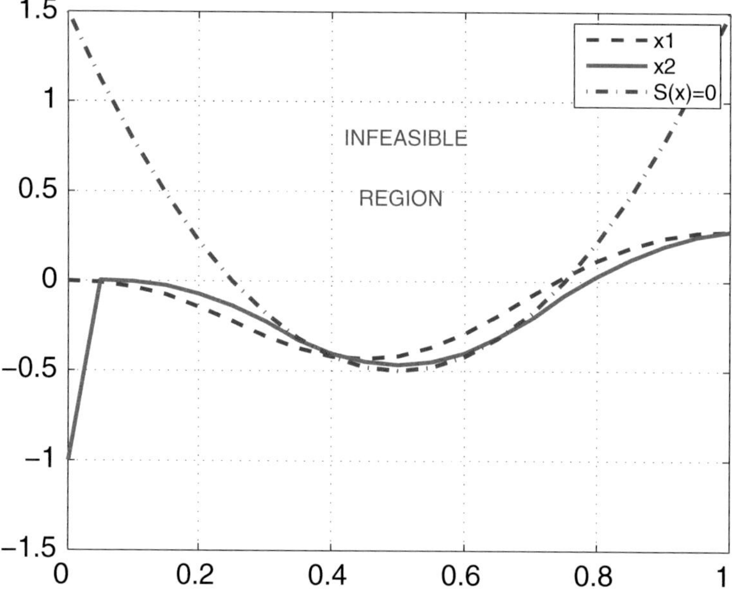

Fig. 1 Optimal state trajectories, obtained from the direct shooting method

To derive optimality conditions for our problem we will start from the determination of the order q of the state constraint (69).
We have from Eqs. (15), (16):

$$S^1(x,u,t) = \frac{\partial S}{\partial t} + \frac{\partial S^T}{\partial x} \cdot f(x,u) = -16(t-0.5) + [0\ \ 1\ \ 0] \begin{bmatrix} x_2 \\ -x_2 + u \\ x_1^2 + x_2^2 + 0.005u^2 \end{bmatrix} =$$

$$= -16(t-0.5) - x_2 + u \tag{70}$$

It means, that $q = 1$. Hence, according to Eq. (33) the Hamiltonian function will be as follows:

$$H(x,u,\eta,\mu_S,t) = \langle \eta, f(x,u) \rangle - \mu_S S^{(1)}(x,u,t) = \eta_1 x_2 + \eta_2(-x_2+u) +$$

$$+ \eta_3(x_1^2 + x_2^2 + 0.005u^2) - \mu_S\left[-16(t-0.5) - x_2 + u\right] \tag{71}$$

Applying the optimality condition (42) we will get:

$$\frac{\partial H}{\partial u} = \eta_2 + 0.01\eta_3 \cdot u - \mu_S = 0 \tag{72}$$

Taking into account the complementary condition (40), it means, that:

$$\hat{\mu}_S(t) = \begin{cases} \eta_2 + 0.01\eta_3 \cdot u, & t \in [t_1, t_2] \\ 0, & t \notin [t_1, t_2] \end{cases} \tag{73}$$

From Eq. (35) the adjoint equation will be:

$$\dot{\eta}_1 = -\frac{\partial H}{\partial x_1} = -2\eta_3 x_1 \tag{74}$$

$$\dot{\eta}_2 = -\frac{\partial H}{\partial x_2} = -\eta_1 + \eta_2 - 2\eta_3 x_2 - \mu_S \tag{75}$$

$$\dot{\eta}_3 = -\frac{\partial H}{\partial x_3} = 0 \tag{76}$$

Let us notice, that the Eq. (76) indicates, that

$$\eta_3(t) = const. \tag{77}$$

We may find its value from the natural boundary conditions (37). Using it, we will get:

$$\eta(1) = -\frac{\partial g}{\partial x(1)} = \begin{bmatrix} 0 \\ 0 \\ -1 \end{bmatrix} \tag{78}$$

Hence,

$$\eta_3(t) \equiv -1, \quad t \in [0, 1] \tag{79}$$

From (70) and (72), applying (39) and (79), the optimal control will be expressed as:

$$\hat{u}(t) = \begin{cases} 16(t - 0.5) + x_2(t), & t \in [t_1, t_2] \\ 100\eta_2(t), & t \notin [t_1, t_2] \end{cases} \tag{80}$$

Finally, the junction conditions (41) will give us the equation:

$$\eta(t_1^+) = \eta(t_1^-) + \gamma \begin{bmatrix} 0 \\ 1 \\ 0 \end{bmatrix} = \begin{bmatrix} \eta_1(t_1^-) \\ \eta_2(t_1^-) + \gamma \\ \eta_3(t_1^-) \end{bmatrix} \tag{81}$$

Now we may put all these things together and express them in the format required by the indirect shooting method, described in Sect. 3, that is a system of nonlinear equations, with unknowns: $\eta_{10}, \eta_{20}, t_1, t_2, \gamma$, to be satisfied at final and switching points, resulting from solution of ODEs with time functions: $x_1(t), x_2(t), x_3(t), \eta_1(t), \eta_2(t)$, defined on three intervals: $[0, t_1], [t_1, t_2], [t_2, 1]$, where starting points are defined by initial and junction conditions. Such a problem was solved under Matlab with the help of two Matlab functions: `ode45` (mentioned above) and `fsolve` (a solver of

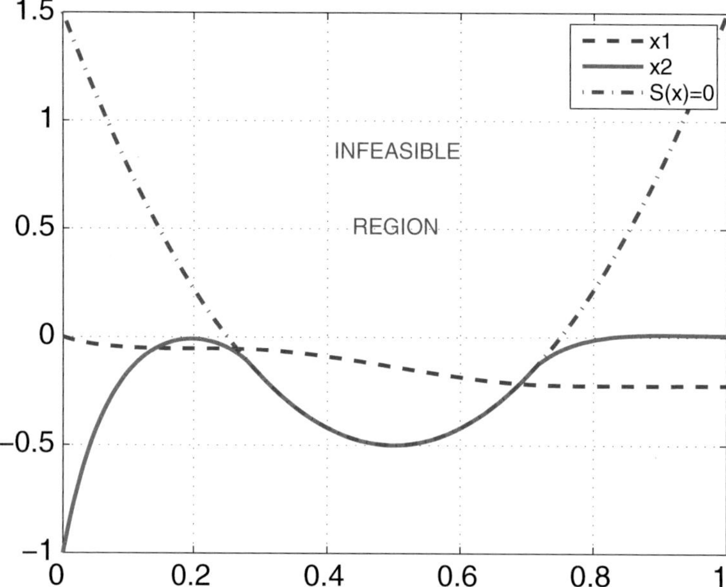

Fig. 2 Optimal trajectories of the state variables x_1 and x_2, obtained from the indirect shooting method

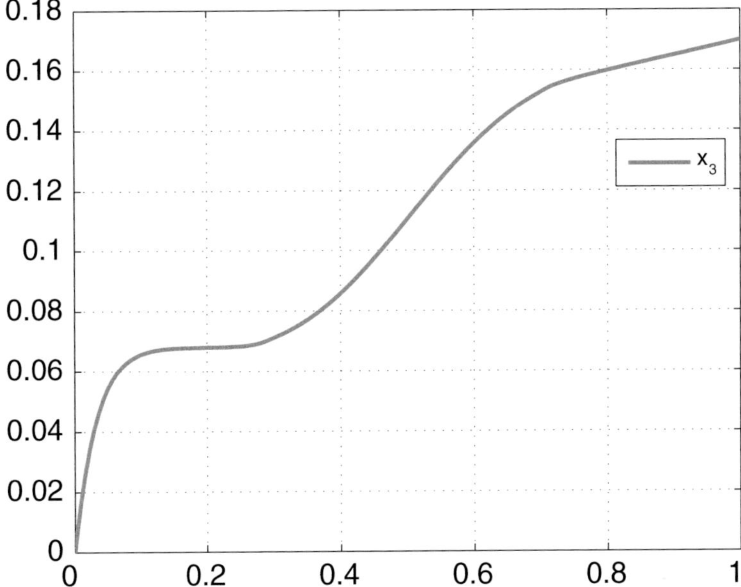

Fig. 3 Optimal trajectory of the state variable x_3, obtained from the indirect shooting method

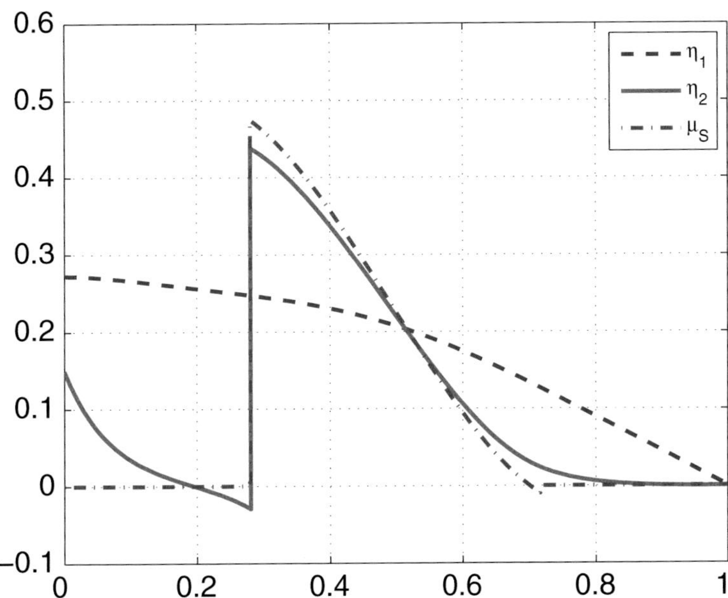

Fig. 4 Optimal trajectories of the adjoint variables η_1, η_2 and the Lagrange multiplier for state constraint μ_S, obtained from the indirect shooting method

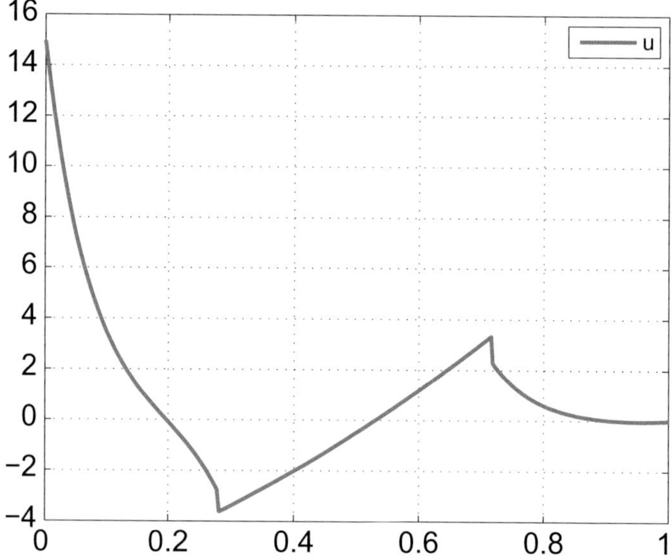

Fig. 5 Optimal control trajectory obtained from the indirect shooting method

systems of nonlinear equations of several variables). The calculations took 4.1 s, the obtained optimal value of the performance index was 0.1698. The results are presented in Figs. 2, 3, 4 and 5. One may see, that indeed, the solution is much more precise than that of the direct method, despite that the number of unknowns in the indirect method was much smaller than the dimension of the decision vector in the direct method (5 vs. 83) and the time of calculations was much shorter (4.1 s vs. 254.3 s on PC with Intel Core i7-2600K CPU@3.40 GHz processor).

6 Conclusions

The advantage of the direct shooting approach is, that the user does not have to be concerned with adjoint variables or switching structures. The main disadvantage of it is the lower accuracy of the obtained solution, than that of the indirect method, where the infinite-dimensional problem is solved. In particular, in the indirect method, in contrast to direct methods, no approximations of the controls have been undertaken [3, 6, 14]. However, as the number of states is large compared to the number of controls, direct methods may be more efficient.

Nowadays, indirect methods are most often applied, when high accuracy of the solution is crucial [12], e.g., in the aerospace, chemical and nuclear reactors, robot manipulators, medical apparatus domain. Typically, initial guesses of the optimal state and control trajectories are generated by applying direct methods.

In the paper this methodology was applied to a case study concerning a Lagrange problem with a single state constraint and fully confirmed its usefulness.

References

1. Bryson, A.E., Denham, W.F., Dreyfus, S.E.: Optimal programming problems with inequality constraints, i: necessary conditions for extremal solutions. AIAA J. **1**, 2544–2550 (1963)
2. Bryson, A.E., Ho, Y.-C.: Applied Optimal Control: Optimization, Estimation, and Control. Taylor & Francis (1975)
3. Gerdts, M.: Direct shooting method for the numerical solution of higher-index dae optimal control problems. J. Optim. Theory Appl. **117**, 267–294 (2003)
4. Jacobson, D.H., Lele, M.M.: A transformation technique for optimal control problems with a state variable inequality constraint. IEEE Trans. Autom. Control **14**, 457–464 (1969)
5. Jacobson, D.H., Lele, M.M., Speyer, J.L.: New necessary conditions of optimality for control problems with state-variable inequality constraints. J. Math. Anal. Appl. **35**, 255–284 (1971)
6. Leineweber, D.B., Bauer, I., Bock, H.G., Schlöder, J.P.: An efficient multiple shooting based reduced SQP strategy for largescale dynamic process optimization. Part 1: theoretical aspects. Comput. Chem. Eng **27**, 157–166 (2003)
7. Maurer, H., Gillessen, W.: Application of multiple shooting to the numerical solution of optimal control problems with bounded state variables. Computing **15**, 105–126 (1975)
8. Maurer, H.: Numerical solution of singular control problems using multiple shooting techniques. J. Optim. Theory Appl. **18**, 235–257 (1976)
9. Oberle, H.J.: Numerical solution of minimax optimal control problems by multiple shooting technique. J. Optim. Theory Appl. **50**, 331–357 (1986)
10. Oberle, H.J., Grimm, W.: BNDSCO -A Program for the Numerical Solution of Optimal Control Problems. Report No. 515 der DFVLR (German Test and Research Institute for Aviation and Space Flight) (1989)
11. Pontryagin, L.S., Boltyanskii, V.G., Gamkrelidze, R.V., Mishchenko, E.F.: The Mathematical Theory of Optimal Processes. Wiley/Intersciense (1962)
12. Sager, S.: Numerical methods for mixed-integer optimal control problems. Ph.D. Thesis, University of Heidelberg (2005)
13. Stoer, J., Bulirsch, R.: Introduction to Numerical Analysis. Springer, Berlin (1979)
14. von Stryk, O., Bulirsch, R.: Direct and indirect methods for trajectory optimization. Ann. Oper. Res. **37**, 357–373 (1992)
15. Wierzbicki, A.: Models and Sensitivity of Control Systems. Elsevier Science Publishers, Amsterdam (1984)

Power Factor Correction with ARM®*Cortex*®—*M*4 *Processor*

Kristian Gašparini, Wojciech Giernacki and Saša Sladić

Abstract ARM cortex-M4 processors family appeared in year 2006 and since than becomes a most common processor family used in mobile phone applications and other appliances including personal computers, televisions, photocopiers and similar. In this article it has been used in order to obtain efficient power factor correction (PFC). It was tested modified average current method with modification in sense of histeresis, in sense of current sensor and in sense of driver in order to decrease operating voltage and to increase efficiency for low power, low speed wind power generation systems. A simulation and practical results were presented.

Keywords ARM cortex-M4 processor · Power factor correction (PFC) · Average current method · Current sensor modification · Low power applications

1 Introduction

Increasing the quality of electric energy becomes more important especially with increasing share of photovoltaic and wind energy. This type of energy sources are highly time dependent so their managment is crucial for stability in electrical networks. Some important merits of electric energy quality are power factor (pf), total harmonic distortion (THD), distortion factor kd, etc. Power factor (pf) of electric system has been defined as a ratio of active power P and apparent power S [1]:

$$(pf) = \frac{P}{S} \tag{1}$$

K. Gašparini · S. Sladić
Faculty of Engineering, University of Rijeka, Vukovarska 58, 51000 Rijeka, Croatia
e-mail: kristian.gasparini@riteh.hr

S. Sladić
e-mail: sladics@riteh.hr

W. Giernacki (✉)
Faculty of Control and Information Engineering, Poznan University of Technology,
3A Piotrowo St., 60-965 Poznań, Poland

R. Szewczyk et al. (eds.), *Challenges in Automation, Robotics and Measurement Techniques*, Advances in Intelligent Systems and Computing 440, DOI 10.1007/978-3-319-29357-8_18

or

$$(pf) = k_d k_\theta = \frac{I_{1rms}}{I_{rms}} * cos\Theta, \tag{2}$$

where:

k_d—*distortion factor*,
k_Θ—*displacement factor*,
I_{rms}—*rms value of complete current signal*,
I_{1rms}—*first harmonic component of load current.*

It is desirable that power factor should have a value close to 1, which means voltage and current in electric system have to be in the same phase (without phase-shift, $k_\theta = cos\theta = 1$). However, harmonic content of current could decrease the power factor, also [1]. So, waveform of current should be defined by sine function as a grid voltage. It means that first harmonic component of load current equals to rms value of complete current signal (including higher harmonic components).

Power factor correction becomes a mature technology in last two decades [1, 2]. New microcontrollers with different programming tools adds a new possibilities in algorithm improvement. It seems that PIC microcontrollers are used extensively [3, 4], but also an other approaches including power factor correction with support may be found on the web [5]. Concerning algorithms development, technology improvements in microcontrollers and capacitor technology a new impuls was given to development of PFC approaches [6]. Number of scientific papers concerning the power factor correction is lower comparing to begining of the century. However, the main concerns remained the same: increasing efficiency, decreasing the number of sensors (especially current sensors) and developing the new algorithms which will follow that trend ensuring the high quality of electric energy [1, 7, 8]. In recent work [7] a modified topology with implicit current control was tested avoiding the usage of current sensor. Similar approach was chosen in [8]. In that case an current sensorless PFC was implemented. In this paper current sensor (usually ordinary resistor) was avoided but voltage drop on 1 cm converter connection line which was connected to operational amplifier input (LM358), so the sensor voltage output (up to 3V) could be achieved without additional joule losses. In this case an efficiency will not be decreased and power converter dynamics will be improved compared to recent papers [7, 8]. However, this approach was not common in older papers where few ohm resistors (Fig. 2) were used [2]. Besides that, an step-up transformer (forward topology) was used instead of classic industrial driver in order to increase efficiency and avoid implementation of integrated circuits with 15 V DC supply.

2 Average Current Method

Average current method involves adopting of rectified voltage waveform as the reference waveform. However, it needs to be scaled in order to obtain the same value of average load current. Average current method may be applied on different circuits [1, 2].

According to literature more then several different topologies may be used for power factor correction [9, 10], however boost topology is the most common. Its input characteristic is not perfectly linear (Fig. 1, left) but the IGBT (insulated gate bipolar transistor) could be switched on easier than in other topologies (e.g. buck topology) since its emiter is connected to the circuit referent point (ground) as shown in Figs. 1 and 2.

In this article a boost converter was chosen, because it has quite linear output characteristic enabling a low harmonic distortion [9]. Another reason of choosing this topology is the possibility to *switch-on* and *switch-off* a IGBT power electronic switch using stepup transformer (Forward topology). Its only function is to increase a control signal level from 3,3 V (used in STM32F407 microcontroller) to 15 V used by power IGBT (designed in classic silicon technology). That means a both driver and current sensor are modified (using 3,3 V instead of 15 V supply) so the increased efficiency could be achieved [7, 8] with current sensor. Furthermore, comparing to classic algorithm [2], in this paper sampling frequency could be chosen separately from switching frequency which was not quite defined in older algorithm description when the analog approach was more common.

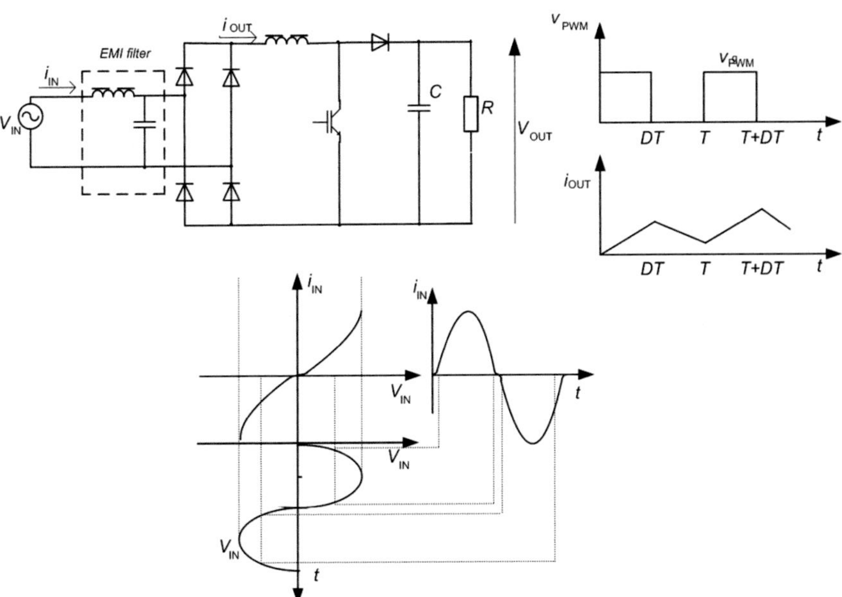

Fig. 1 Single-phase rectifier with Boost converter power factor correction module with switching sequence *Vpwm(t)* and idealized waveform of load current *Iout* and input characteristic (relation between input voltage *Vs* and input current *is*)

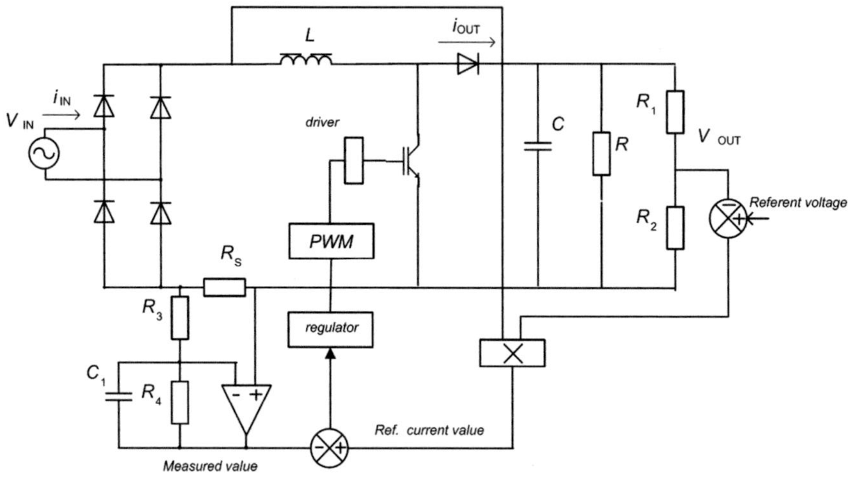

Fig. 2 Block diagram of power factor correction according to average current value [2]

3 Simulation Results

Simulation could be achieved in different software tools. For this purpose a Simplorer program has been chosen. It has been shown (Table 1) that power factor could be increased from 0,48 to 0,99. In the same time total distortion factor (THD) may be decreased from 166 to 12 %. However, a switching frequency of 2 kHz has been chosen in order to obtain visible current ripple and recognisable switching in the frame of few periods (Fig. 3). According to simulations choosing a higher switching frequency, an even better results may be obtained.

Table 1 Comparison of results obtained for the case without PFC (first column) and with PFC (second column); in second case a higher inductance where chosen

Without PFC	With PFC
Simulation parameters: 0,5 mH, 230 V, 1 mF, 10 Ohm	Simulation parameters: 3 mH, 230 V, 1 mF, 10 Ohm
P = 2,349 kW	P = 7,906 kW
S = 4,9168 kVA	S = 7,9987305 kVA
Q = 4,31922 kVAr	Q = 1,2120589 kVAr
(pf) = 0,477789	(pf) = 0,988452
THD = 165,764889 %	THD = 12,18 %

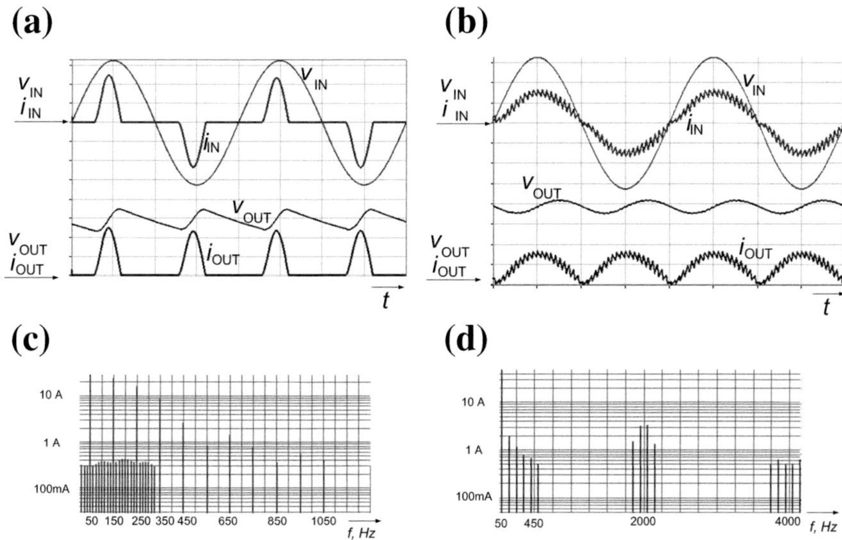

Fig. 3 Comparison of simulated results between rectified D class current (RC load) with waveform obtained with power factor correction (*right*) circuit accompanied by their frequency domain (100 V/div, 5A/div, 2,5 ms/div) obtained with switching frequency 2kHz. **a** Voltages and currents without PFC. **b** Voltages and currents with PFC. **c** Current in frequency domain without PFC. **d** Current in frequency domain with PFC

4 Power Factor Correction with STM32F407

The same algorithm which has been applied in simulation program Simplorer, could be applied in microcontroller using a microcontroller programming tool. Many programs may be applied with ARM cortex (compatible) microcontrollers [11]. Some of them are Keil, IAR Embedded Workbench for ARM, Gnu Compiler (gcc), Code Composer studio by Attolic, GNU Tools for ARM Embedded Processors by ARM, IAR Embedded Workbench for ARM by IAR, Keil MDK-ARM by Keil, MikroC by Mikroelektronika [12], Visual Studio by Microsoft and others [13]. In this case a MikroC was used because of extensive materials on web [12]. For research purposes 32-bit STM32F407 microcontroller with hardware floating point unit was used. It has Cortex M4F core, as well as 168 MHz chip clock, 1 MB Flash and 192 kB RAM memory.

According to connection from Fig. 2 a simple code could be written:

```
Iref=600;
Iin3=15*IinAC;
ISUM=Iref-Iin3;
DUTY SET=ISUM;
```

So, a desired duty cycle has been defined as a difference between input voltage which has been used as referent current value (Iref) (which amplitude has been adapted according to output voltage—Fig. 2) and measured current (current through resistor Rs in Fig. 2 which is converter line strip without additional resistor from Fig. 2). Program is written in CMSIS (Cortex Microcontroller Software Interface Standard) and it is not compatible with earlier (not cortex) ARM processors [14].

Period = PWM TIM1 Init(10000);
PWM TIM1 Start(PWM CHANNEL1,
GPIO MODULE TIM1 CH1 PE9);
PWM TIM1 Set Duty(DUTY SET,
PWM NON INVERTED,
PWM CHANNEL1);

where a switching frequency was defined by: *Period = PWM TIM1 Init(10000);* In this way a numerous PWM outputs could be defined. Applying the PWM signal to power circuit via step-up transformer, input current follow the voltage waveform (wave carrier PWM) and that could be noticed on practical results where measured waveforms coincide with simulated ones (Fig. 4).

In order to compare results obtained without PFC (Fig. 5) and after turning it on (Fig. 6), a both cases are shown.

Applying an PWM signal to PFC transistor, it could be noticed that maximum value of load current decreases.

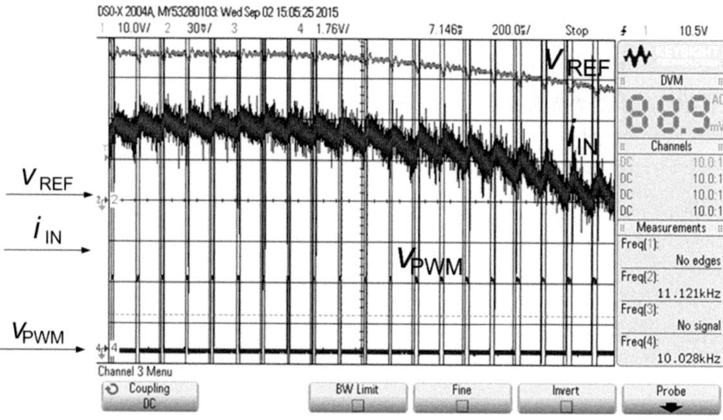

Fig. 4 Measured waveforms of input voltage, accompanied with input current and PWM signal for switching the PFC circuit transistor (10 kHz switching frequency) (10 V/div., 2 A/div., 100 us/div.)

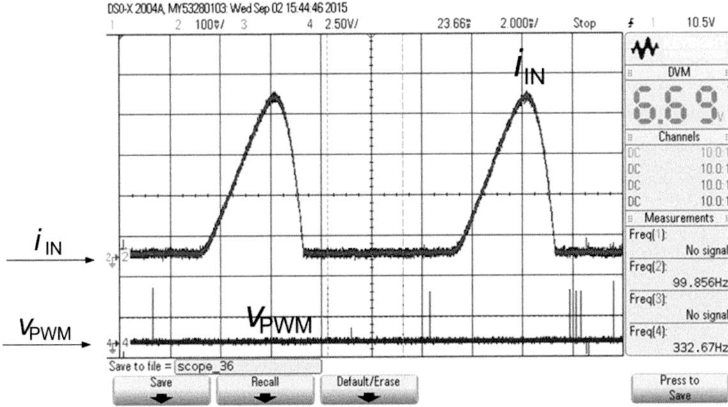

Fig. 5 Measured D class load current (RC load) with no switching of additional (PFC) transistor (10 V/div., 2 A/div., 2 ms/div.)

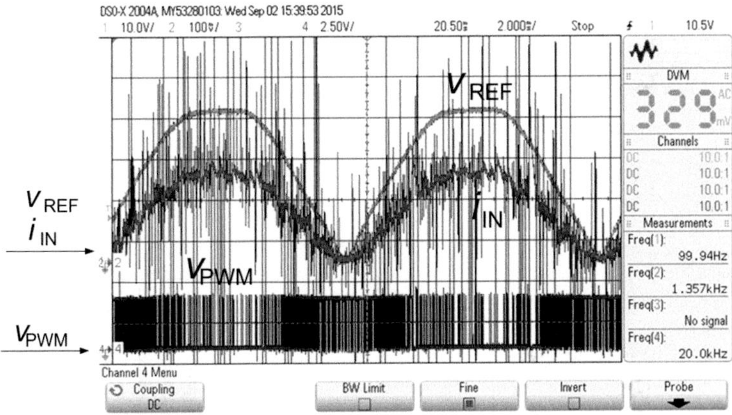

Fig. 6 Measured waveforms of input voltage, accompanied with input current and PWM signal for switching the PFC circuit transistor (20 kHz switching frequency)

5 Conclusions

Similarly as other modern electronic devices, power converters become more sophisticated. Compared to earlier ones they could involve sleep mode for night charging of batteries (automatic switching-off), additional measurement of power factor or even harmonic contient of rectifier input current as well as internet connection for data acquistion and monitoring. STM32F407 with rich peripheral enables more sophisticated approach to power quality control in electric networks in general. In tested prototipe, at switching frequencies above 20 kHz an electromagnetic interference could be noted which was not present at simulations, however during the tests upper frequency limit was not reached implying that this system is faster than earlier similar systems and high level of EMC was obtained in processor inspite of EM disturbances. Concerning a PFC algorithm implementation, current sensor is built without additional resistor, by measuring voltage drop on converter prototipe and complete converter operates on 3,3 V instead on older industrial 15 V standard supply which makes it appropriate for low power wind power applications at low wind speeds in frame of distributed energy production systems and smart grids.

References

1. Kumar, P., Sharma, P.R., Kumar, A.: Power factor correction on RISC controller. In: Advances in Power Electronics and Instrumentation Engineering Communications in Computer and Information Science, vol. 148, pp. 83–87. Springer (2011)
2. Erickson, R., Maksimović, R.: Fundamentals of Power Electronics. Kulwer Academic Press (2001)
3. Sonje, P., Soman, A.: Power factor correction using PIC microcontroller. Int. J.Eng. Inovative Technol. (IJEIT), **3**(4) (2013)
4. Murad, A.: Design and Implementation of Microcontroller-Based Controling of Power Factor Using Capacitor Banks with Load Monitoring, Global Journal of Researches in Engineering, Global Journals Inc. (USA), Electrical and Electronics Engineering, vol. 13, Issue 2, Version 1.0 (2013)
5. Sladić, S., Franković, D., Mužić, I.: Mechatronic tools for active power factor correction. Eng. Rev. **27**(2), 59–69 (2007) (Accessed 27.10.2015). http://hrcak.srce.hr/index.php?show=clanakŹid_clanak_jezik=41656
6. Heger, C.A., Sen, P.K., Morroni, A.: Power factor correction a fresh look into today's electrical systems. In: Cement Industry Technical Conference, 2012 IEEE-IAS/PCA 53rd, pp. 1–13, 14–17 May 2012
7. Angulo, M., Ruiz-Caballero, D.A., Lago, J., Heldwein, M.L., Mussa, S.A.: Active power filter control strategy with implicit closed-loop current control and resonant controller. In: IEEE Trans. Ind. Electron. **60**(7) (2013)
8. El Moniem, M.S.A., Azazi, H.Z., Mahmoud, S.A.: Alexandria Eng. J. **53**(1), 69–79 (2014)
9. Wei, H., Batarseh, I.: Comparison of Basic Converter Topologies for Power Factor Correction, Southeastcon '98. Proceedings, pp. 348–353. IEEE, 24–26 April 1998, Orlando, FL
10. Mishra, A.K., Pathak, M.K., Das, S.: Isolated converter topologies for power factor correction a comparison. In: 2011 International conference on Energy, Automation and Signal (ICEAS), 28–30 Dec 2011
11. Yiu, J.: The Definitive Guide to ARM cortex-M3 and Cortex-M4 Processors, Newnes (2013)
12. Mikroelektronika on web, http://www.mikroe.com/arm/compilers/
13. List of ARM Cortex-M development tools, https://en.wikipedia.org/wiki/. List of ARM Cortex-M development tools (Accessed 27.10.2015)
14. Dulos tutorial, Getting started with CMSIS (Accessed 27.10.2015). https://www.doulos.com/knowhow/arm/CMSIS/CMSIS_Doulos_Tutorial.pdf

Concept and Design of New Type Valve with Helix Type Spool

Dominik Rybarczyk and Czesław Jermak

Abstract The paper presents the design of new kind of electrohydraulic valve with helix type rotary spool. First part, provides a literature overview concerning on the construction of the hydraulic valves. Then, Authors present a pair of spool works principle of a new type valve. Finally, basic equations describing the valve are formulated and its simulation model is implemented in MATLAB-Simulink environment. The valve flow characteristics are collected from the model.

Keywords Electrohydraulic valve · Helix type spool · Modelling

1 Introduction

Electrohydraulic servo drives can be controlled by two types of electrohydraulic valves: servovalves and proportional valves. The servovalves are used in applications which require high-accuracy and high dynamics parameters. Proportional valves have worst dynamic parameters, but they are much cheaper. Therefore these elements are commonly used in many industrial devices. The basics of proportional valves design have been established about 30 years ago and from this time there is no significant progress in this area [1, 2].

The main elements of electrohydraulic proportional valve is a pair of spool, which consists of a spool and bushing. Their shape defines the number of connected ways and control positions. The spool of the valve has the form of different diameters rolls. The spool moves in both direction and its control by the proportional magnets. The shape of the spool edges determines the characteristics of the fluid flow changes

D. Rybarczyk (✉) · C. Jermak
Institute of Mechanical Technology, Poznan University of Technology,
Piotrowo 3, 60-965 Poznań, Poland
e-mail: Dominik.Rybarczyk@put.poznan.pl

C. Jermak
e-mail: Czeslaw.Jermak@put.poznan.pl

R. Szewczyk et al. (eds.), *Challenges in Automation, Robotics and Measurement Techniques*, Advances in Intelligent Systems and Computing 440, DOI 10.1007/978-3-319-29357-8_19

as a function of the spool position. The ports in the valve body may have the different shapes. Mostly, they are triangular, rarely rectangular, semi-circular, sinusoidal like. The choice of the shape of the edges determines the size of the distributor, the maximum operating pressure and the limit of leaks. The springs used in valves require moving the spool to the center position (zero) after a power failure or power cut-off. The valves may have overlap control edges: positive, negative or zero (physically not possible due to technology performance) [1, 2].

In the literature from the last 20 years, only a few papers focused on the applications of different types of spool shapes in hydraulic proportional valves. Murrenhoff [3] described the cross-cutting trends in the design and development of electrohydraulic valves. He presented an interesting solution, using the direct drive in a proportional valve. In another solution the use of the mechanism with four independent flow spools, was proposed. In his proposals, there is no changing of spool design (he focus only on the setpoint element).

U.S. Patent [4] describes the construction of the slider, which profile shaped balance hydrodynamic forces. Geometry of the profile is a combination of several rays of a circle. Proposed in patent force balanced hydraulic spool valve reduces the forces required for moving the spool, thereby allowing potentially greater metering control flexibility for hydraulic remote and electro-proportional flow systems. The lower forces also reduce physical strength necessary for the introduction in manually operated systems, thus tending to reduce operator fatigue as well. Shape of the spool offers improved linearity in flow rates.

U.S. Patent [5] is presented multi-way valve provided with a lever mechanism and the arrangement of two sliders. The displacement of one of them causes synchronous movement of the second one. Described here valve is dedicated to control the direction of fluid flow under pressure.

In the article [6] described electro-valve, which uses a spool made of asymmetric control blocks. They are two cylinders with different diameters chosen for the different flow characteristics. The result is an unbalanced effect of mechanical load compensation valve.

In the patent [7] presents the valve slider in the form of a sleeve with openings control. As a result, the hydraulic oil can pool slider, but also provides a flow path for him. The spool valve assembly includes a pilot operated spool valve that is moveable by differential pressure across the valve. The spool valve assembly includes a spool placed in a chamber. A first volume of fluid at a control pressure is disposed between a first spool and a first chamber ends. A second volume of fluid at a reference pressure is disposed between a second spool end and a second chamber end. The reference pressure varies as the position of the spool valve varies, due to a concurrent throttling effect of the spool valve [7].

In the paper [8] Šimic presented a new approach to modelling and simulation of hydraulic spool valves by using simple mathematical expressions to describe the geometry of the sliding spool metering edge. Thanks to this, different shapes of spool metering edges in combination with other functional elements could be used in design of hydraulic valves.

Myszkowski and Milecki described in [9] the use of a stepping motor in the valve to obtain a very low velocity of the electrohydraulic servodrive. Described in article drive was able to move with a very low constant speed equal to 1 μm/s. However, the drive maximum speed was only 0.125 mm/s, which was rather low and which reduced the number of possible applications. In disseretation [10] Author described different shape of holes (ports) in valves and and their effect on the flow ratio and characteristics.

2 Valve Construction

2.1 *Concept and Design*

The essence of the present invention is the use of a multi-way hydraulic rotary spool with a double cylindrical helix and a sleeve in a cylindrical shape with notched holes. Advantage of these construction is no need for use the gearbox to convert rotary motion into linear motion when using rotary motors in valve. The shape of the holes determines the type of flow characteristics. In the case of use of oval-shaped holes (ports) the characteristics become non-linear. Nonlinearities can be compensated by the control system.

The scheme of the valve proposed in this paper is presented in Figs. 1 and 2. The valve spool is actuated by a stepper motor. Thanks to use microstepping mode the drive is able to assure the linear resolution of 0.5 μm. This motor is connected to the spool by flexible bellows coupling. The rotation of the motor causes rotation of spool. This movement is proportional to the angular motor displacement. Direction of rotation determines the direction of spool translation and opening or closing of valve gaps. It results in the flow of oil to and from the actuator chambers and displacement of piston.

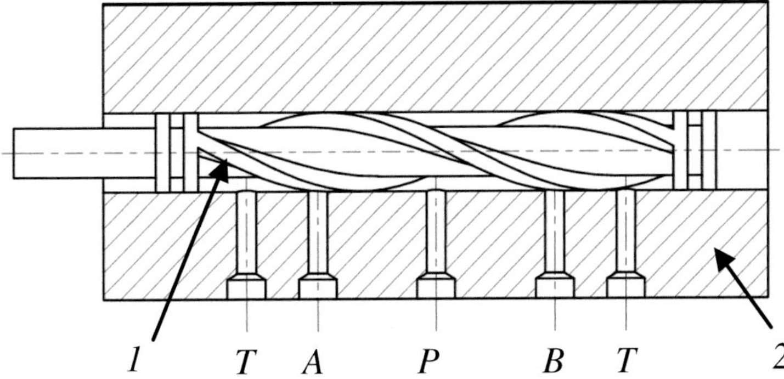

Fig. 1 Scheme of valve with helix type spool

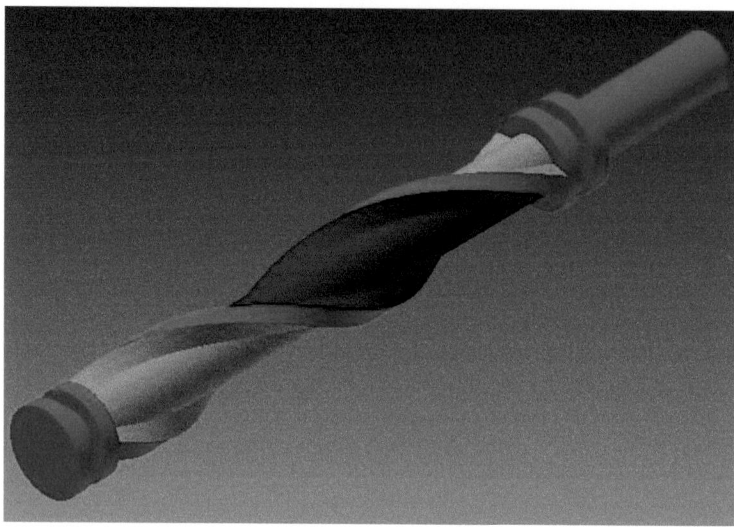

Fig. 2 View of helix type spool

2.2 Operation Modes

Depending on the direction of rotation of the slider, the valve can be one of three states (Fig. 3):

(a) default position—the ports A and B are closed,
(b) turn the slider to the right position, opening the port A and B, the connection port B port P and port A to port T
(c) turn the slider to the left position, opening the port A and B, the connection port A to port P and the port B port T.

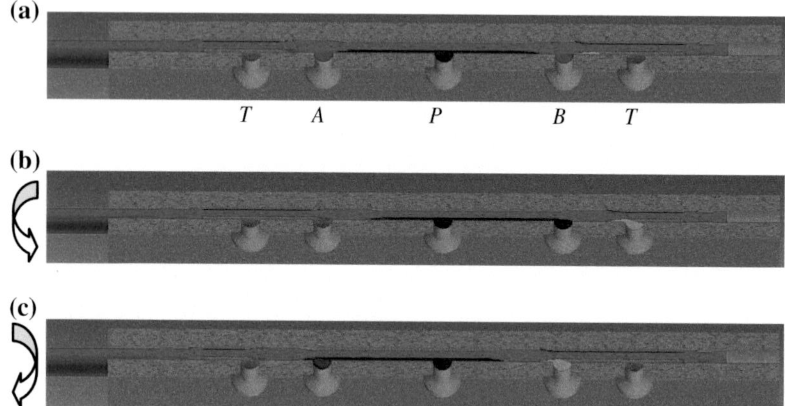

Fig. 3 Operating modes of the valve

By changing the motor shaft angle, it is possible to change the flow rate.

2.3 Modelling

The flow through the valve is turbulent type [1, 2], because of the sharp type edges. The flow can be described using following equation:

$$Q = K_Q \sqrt{\Delta p} \cdot A, \tag{1}$$

where: K_Q is a flow coefficient, Δp pressure drop on a valve (Fig. 4).

The flow coefficient K_Q depends on the following parameters:

$$K_Q = \mu_d \cdot \sqrt{2/\rho}, \tag{2}$$

$$A = \frac{1}{2} \cdot \left(\frac{\pi \alpha}{180^\circ} - \sin(\alpha) \right) \cdot r^2, \tag{3}$$

where μ_d is the discharge coefficient, ρ—oil density: 900 dm^3/min, r—dimension of the valve gap (6 mm).

The value of the discharge coefficient μ_d depends mainly on the geometry of the slot. In the case of valve considered in this paper, this value is usually taken as 0.64.

Its static behaviors can be characterized using so called square root equations, which describe the oil flow through the hydraulic nozzles. These flows can be described by following equations:

$$\text{for:}\, x > 0: \begin{array}{l} Q_a(t) = K_Q \sqrt{p_0 - p_a(t)} \cdot x(t) \\ Q_b(t) = K_Q \sqrt{p_b(t)} \cdot x(t) \end{array}, \tag{4}$$

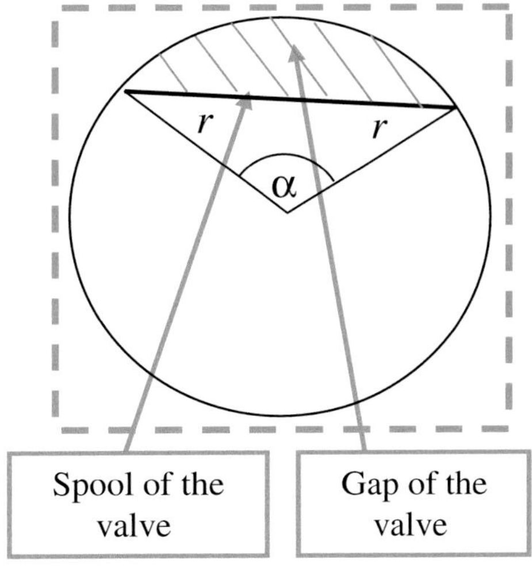

Fig. 4 Geometry of the valve gap

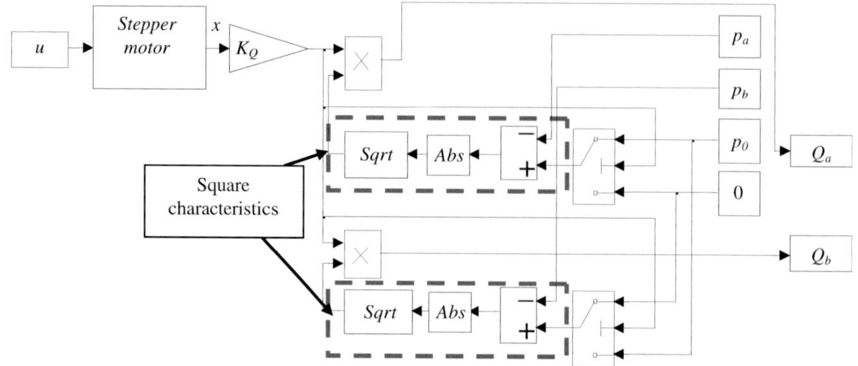

Fig. 5 Model of the valve with helix type spool in MATLAB/Simulink

$$\text{for: } x < 0: \begin{array}{l} Q_a(t) = K_Q\sqrt{p_a(t)} \cdot x(t) \\ Q_b(t) = K_Q\sqrt{p_0 - p_b(t)} \cdot x(t) \end{array}, \tag{5}$$

where x is a spool displacement (mm), Q_a flow through the gap A (dm^3/min), Q_b flow through the gap B (dm^3/min), K_Q flow coefficient, p_0 supply pressure (Pa), and p_a, p_b pressures in chambers A and B (Pa).

Model of the valve with helix type spool is performed in MATLAB/Simulink software (Fig. 5).

Figure 6 shows measured flow characteristics from the model in relation to motor position, obtained for supply pressures 6, 8, 10, 12, 14, 16, 18 and 20 MPa.

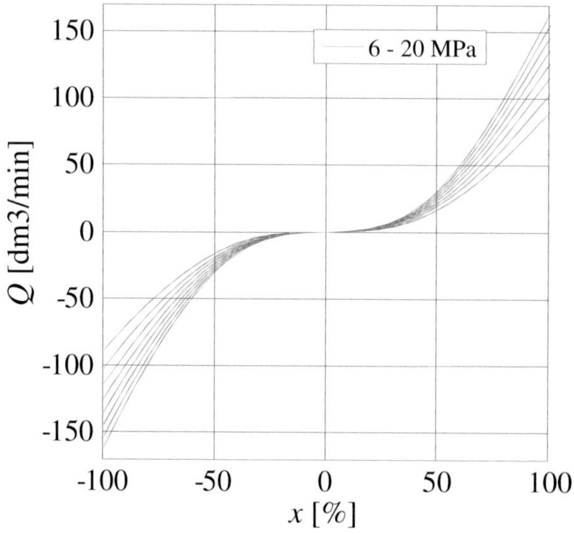

Fig. 6 Static flow characteristic of the valve

The flow during test was measured between output ports A and B of the valve. In the given conditions a maximum flow for 6 MPa was about 80 dm^3/min, for 20 MPa—about 162 dm^3/min. The flow characteristics was nonlinear due to the shape of the ports in valve body.

3 Conclusion

The paper presents the design of new kind of electrohydraulic valve built on base of the rotary spool with double helix, and a cylindrical sleeve with holes. The presented here concept and design valve and the helix type spool was patented by Authors in Pat/1738 P.412449. First part of the article described construction and gives some examples of different kind of spool in electrohydraulic valves. Next, the basic equations describing the valve are formulated and its simulation model is implemented in MATLAB/Simulink environment. The valve flow characteristics are collected from the model.

Advantage of these construction is no need for use the gearbox to convert rotary motion into linear motion when using rotary motors in valve. The shape of the holes determines the type of flow characteristics. In the case of use of oval-shaped holes (ports) the characteristics become non-linear. Nonlinearities can be compensated by the control system

Presented here valve is characterized by compact dimensions and high dynamics of movement. The described here invention can be used to control the direction and flow rate of the hydraulic fluid.

References

1. Chapple, P.: Principles of Hydraulic System Design. Coxmoor Publishing Company, Oxford (2003)
2. Cundiff, S.J.: Fluid Power Circuits and Control Fundamental and Applications. CRC Press, Boca Raton (2000)
3. Murrenhoff, H.: Trends in valve development. Ölhydraulik und Pneumatik **46**(4), 1–36 (2003)
4. Patent: Force balanced hydraulic spool valve US 4941508 A
5. Patent: Multiway valve for switching a flow of fluid under pressure with parallel disposition of valve bores, and valve assembly kit, US 6688332 B2, 28 Sty 2002
6. Rybarczyk, D., Milecki, A.: Modeling and control of proportional valve with synchronous motor. In: Valiulis, A.V., Černašėjus, O., Mokšin, V. (eds.) Solid State Phenomena, vols. 220–221, pp. 457–462 (2015) doi:10.4028/www.scientific.net/SSP.220-221.457
7. Patent: Pilot operated spool valve, US 8387659 B2, 28 Mar 2008
8. Šimic, M., Debevec, M., Herakovič, N.: Modelling of hydraulic spool-valves with specially designed metering edges. Strojniški vestnik—J. Mech. Eng. **60**(2), 77–83 (2014). doi:10.5545/sv-jme.2013.1104

9. Myszkowski, A., Milecki, A.: Modelling of electrohydraulic servo drive used in very low velocity applications. Int. J. Model. Ident. Control **7**(3), 246–254 (2009). doi:10.1504/IJMIC.2009.027211
10. Myszkowski, A.: Badanie możliwości uzyskiwania małych prędkości ruchu przez liniową serwojednostkę elektrohydrauliczną z silnikiem skokowym i elektrycznym sprzężeniem zwrotnym, rozprawa doktorska, Wydział Budowy Maszyn i Zarządzania, Ph.D. Disseretation, Poznań 2003

Structure and Functionalities of Ship Autopilot Simulator

Marcin Jamro, Dariusz Rzońca, Jan Sadolewski, Andrzej Stec, Zbigniew Świder, Bartosz Trybus and Leszek Trybus

Abstract Heading control, track control, rate of turn and turn radius are basic maneuvers executed by ship autopilots. The paper presents structure and functionalities of an autopilot simulator, written in ST language in CPDev environment. It runs both on PC and on ARM-based controller for anticipated real-time tests. Sample results of ship dynamics identification through zig-zag maneuver and of wave filtering by an observer are presented. The simulator has been developed in cooperation with a Dutch company.

Keywords Ship autopilot · Control loops · Identification · State observer

M. Jamro (✉) · D. Rzońca · J. Sadolewski · A. Stec · Z. Świder
B. Trybus · L. Trybus
Rzeszów University of Technology, Rzeszów, Poland
e-mail: mjamro@prz-rzeszow.pl

D. Rzońca
e-mail: drzonca@prz-rzeszow.pl

J. Sadolewski
e-mail: js@prz-rzeszow.pl

A. Stec
e-mail: astec@prz-rzeszow.pl

Z. Świder
e-mail: swiderzb@prz-rzeszow.pl

B. Trybus
e-mail: btrybus@prz-rzeszow.pl

L. Trybus
e-mail: ltrybus@prz-rzeszow.pl

R. Szewczyk et al. (eds.), *Challenges in Automation, Robotics and Measurement Techniques*, Advances in Intelligent Systems and Computing 440, DOI 10.1007/978-3-319-29357-8_20

1 Introduction

Autopilot steers the ship on a reference course by controlling the rudder. Actual course is measured by gyroscope. In track mode, the autopilot can also steer the ship along prescribed path, with position determined by GPS navigation. While cruising at rough sea, the rudder should not be affected by waves.

Since several years Rzeszów University of Technology has been cooperating with Praxis Automation Technology B.V. in the Netherlands on implementation of CPDev package [1] for programming ship monitoring and control systems manufactured by Praxis. A year and a half ago joint development of autopilot prototype has been undertaken, with Rzeszów responsible for high level software. Some preliminary results have already been reported [2]. Trial version of such software implemented on PC and ARM-based controller with 8.4" TFT panel has been completed recently. It involves 4DOF model of container ship for simulated steering and comparison of functionalities with commercial solutions, such as [3, 4]. Current operator interface is oriented towards learning and testing, so it is not a target solution.

The autopilot is designed to operate in the following modes: HC—Heading Control (course), TC—Track Control, RAD—Turn Radius, ROT—Rate-of-Turn, MANUAL—tiller or wheel steering. MAIN display (default) is shown in Fig. 1. Buttons on the left select corresponding modes. Buttons on the right select displays and adjust parameters.

Dynamics of the simulated ship is identified by standard maneuvers and adjusted to speed. Wave filtering makes the rudder almost not affected by wave-induced yaw motions. Sensitivity of autopilot controllers is determined by single parameter, adjusted automatically or manually. Automatic adjustment means high sensitivity at calm sea and low at rough sea, or at trailing wave. Essential details of such functionalities are explained in the following sections.

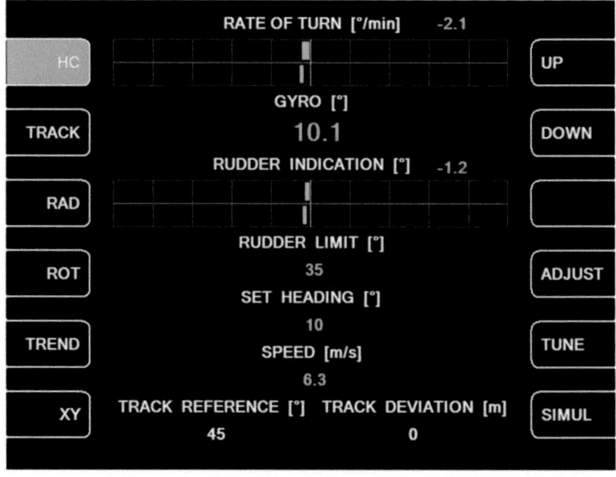

Fig. 1 MAIN display of the autopilot simulator

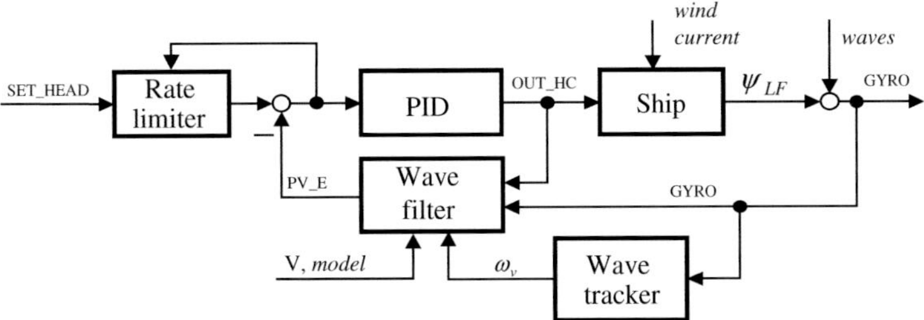

Fig. 2 Heading control system

2 Heading Control Structure

Architecture of the autopilot while operating in the HC mode is shown in Fig. 2. SET_HEAD denotes set heading, OUT_HC rudder angle and GYRO course measurement.

Rate limiter limits rate of set-point change for PID controller to avoid over-steering (overshoot). The limit depends on controller sensitivity [2]. PID settings are computed using ship dynamics data, sensitivity and speed. At rough sea, when PV_E course estimate is an input, OUT_HC is only slightly affected by waves.

State observer, based on ship and wave models [5, 6], generates filtered estimate PV_E of the course GYRO by removing wave-induced yaw component. Besides GYRO measurement the estimator uses rudder angle OUT_HC and frequency ω_v of yaw motions. Wave tracker identifies the frequency ω_v from GYRO. The frequency depends on sea state and relation between ship course and wind direction. The tracker automatically switches itself off at calm sea and switches on at normal or rough.

3 Track Control Structure

Track control system shown in Fig. 3 is of cascade structure with track controller as primary and heading controller as secondary.

PID track controller uses cross track error XTE computed from GPS measurements as the input. If XTE error occurs, the controller computes course correction FD to waypoint reference TRACK_REF (see also Fig. 1). Sensitivity of the track controller is directly related to sensitivity of the heading controller, so it can be a single tuning parameter. Therefore both controllers always behave similarly.

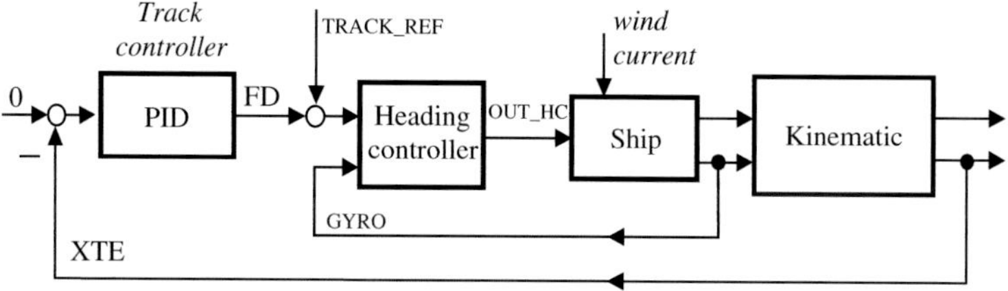

Fig. 3 Track control system

4 Adjustments

Basic parameters of the autopilot are set from ADJUSTMENTS display (Fig. 4). The first one, i.e. Minimum wave, denotes limit of wave-induced motions when the wave tracker switches itself on and begins to evaluate frequency. The tracker is dormant at calm sea. Trim means position of the rudder to counteract strong wind or sea current (convenient for MANUAL mode). If despite filtering the rudder is still affected by waves, small Dead zone can make it completely insensitive.

Track leg reference denotes a course for next section (leg) of the voyage. It becomes TRACK REFERENCE in the MAIN display (Fig. 1) after accepting Start for Track next leg.

Turn RAD and turn ROT maneuvers performed usually during sea trials are specified by Radius, Rate-of-turn, Angle and Direction. Given such data the autopilot generates corresponding time function fed as the reference SET_HEAD into the heading controller (Fig. 2).

Measurement filters are parameterized by time constants. The last line in the display, Rudder Activity, is not an adjusted parameter but indicator of average rudder motion (control activity). It helps to choose proper value of Minimum wave for a particular ship, decide when to turn on wave filter, etc.

Fig. 4 ADJUSTMENTS display for basic parameters

ADJUSTMENTS

Parameter	Value
Minimum wave [deg]	0.1
Trim [deg]	0.0
Rudder limit [deg]	35.0
Dead zone [deg]	0.0
TRACK leg reference [deg]	0
TRACK next leg	Start
RAD/ROT radius [km]	1.0
RAD/ROT rotation [deg/min]	20.0
RAD/ROT turn angle [deg]	360
RAD/ROT direction	<< PS
Measurement filter: Gyro [s]	1.0
Measurement filter: ROT [s]	1.0
Measurement filter: Speed [s]	1.0
Measurement filter: Track [s]	1.0
Rudder Activity [deg]	1.5

BACK ACCEPT MAIN TREND XY UP DOWN ADJUST TUNE SIMUL

5 Statuses and Tunings

STATUS/TUNINGS display (Fig. 5) indicates actual statuses of autopilot modes and sets remaining parameters. Among seven operating modes, i.e. from Adapt to MANUAL, Adapt + Filter means that identified ship model is used to adjust PID settings to actual speed and that the wave filter is active. If speed is too low, autopilot switches itself into MANUAL. Wave filter mode set to On + Slow means that controller sensitivity is automatically adjusted to frequency of yaw motions. This is recommended for really rough sea or for trailing waves.

Controller sensitivity is a single parameter affecting settings of heading and track controllers. It is defined as the ratio of the ship time constant (T below) to closed-loop time constant for heading control [2]. So for instance sensitivity 2.0 means that the closed-loop responses are twice faster than those of the ship itself.

A few lines in the display handle ship identification maneuver, zig-zag or sinusoid, with Rudder change amplitude. Standard first order Nomoto model [5, 7]

$$\frac{r(s)}{\delta(s)} = \frac{k}{Ts+1} \tag{1}$$

is identified, given measurements of rate-of-turn r and rudder angle δ. Zig-zag data are processed by least-squares (off-line) and the sinusoid by model reference (on-line, no data buffering). Inverses of the identified k and T, normalized with respect to ship length and speed, are presented as Adapt rudder ratio and Adapt counter rudder, respectively (notions familiar to marine community). The user may eventually change them. Sample trend of zig-zag maneuver involving two switchings is shown in Fig. 6.

We shall explain now how the wave filter is implemented. Following [5], state observer has been chosen since computation of its gains in this particular case is less demanding than those of the Kalman filter. This matters for on-line

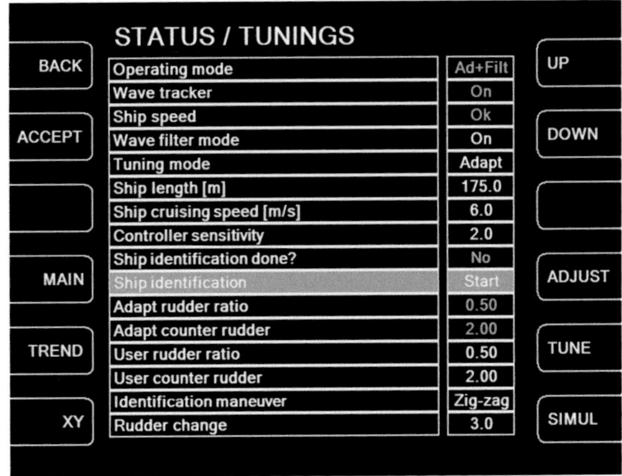

Fig. 5 STATUS/TUNING display

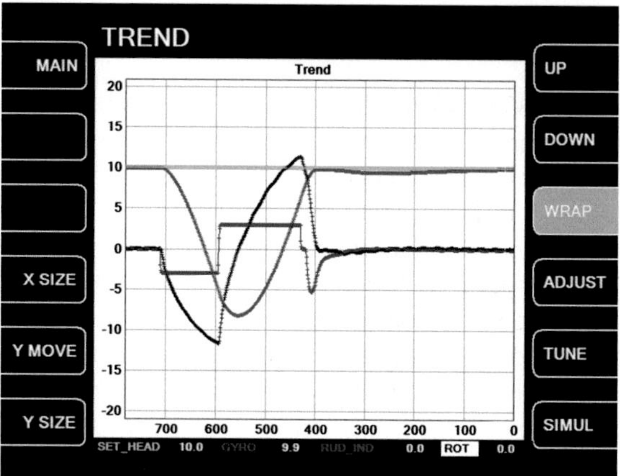

Fig. 6 TREND display for zig-zag identification maneuver

microcontroller implementation. Let the course ψ and rate-of-turn r be split into low and high frequency components, driven respectively by the rudder angle δ and a white noise. The high-frequency course component ψ_{HF}, which represents wave-induced yaw motions, is usually described by [5, 6]

$$\psi_{HF}(s) = \frac{s}{s^2 + 2\xi\omega_v s + \omega_v^2} w_v(s). \tag{2}$$

Introducing state vector $\left[\psi_{LF}, r_{LF}, \psi_{HF}, r_{HF}\right]^T$, using the ship model (1) and wave model (2), the overall state space representation consists of the following matrices

$$A = \begin{bmatrix} 0 & 1 & 0 & 0 \\ 0 & -\frac{k}{T} & 0 & 0 \\ 0 & 0 & 0 & 1 \\ 0 & 0 & -\omega_v^2 & -2\xi\omega_v \end{bmatrix}, \quad B = \begin{bmatrix} 0 \\ 1 \\ 0 \\ 0 \end{bmatrix}, \quad C^T = \begin{bmatrix} 1 \\ 0 \\ 1 \\ 0 \end{bmatrix} \tag{3}$$

with the control δ and the output $\psi = \psi_{LF} + \psi_{HF}$, so OUT_HC and GYRO in Fig. 2, respectively (course is an integral of rate-of-turn). It was demonstrated in [5], that for such sparse matrices analytic expressions for observer gains can be found, given some observer poles. This holds also for discrete case implemented in the autopilot.

Since the ratio k/T does not depend on ship speed [5, 7], the frequency ω_v evaluated by wave tracker is the only parameter which affects observer gains for a given ship. Therefore if ω_v changes sufficiently due to course change or wind, computation of the gains must be repeated. Gain-scheduling could be another option.

6 Simulations

SIMULATIONS display is shown in Fig. 7. Two ships can be simulated, a container from [8, 9] (with fixed parameters) and an abstract ship described by Nomoto model (1), whose Length, Cruising speed, Rudder ratio and Counter rudder can be adjusted in fairly large ranges to account for small and large ships. Sea state is determined by Beaufort number, wave and sea current Directions, Wave period and Current speed. X, Y coordinates determine position of the ship since beginning of simulated voyage, as it would be in case of real GPS data. Wave tracker period and Magnitude indications verify whether sea state (Beaufort) and Wave period set above are evaluated properly. Shaft revolutions are an input for steering machine simulator.

Effectiveness of wave filtering to eliminate rudder motions can be assessed at TREND display shown in Fig. 8. The display involves plots of SET HEADING

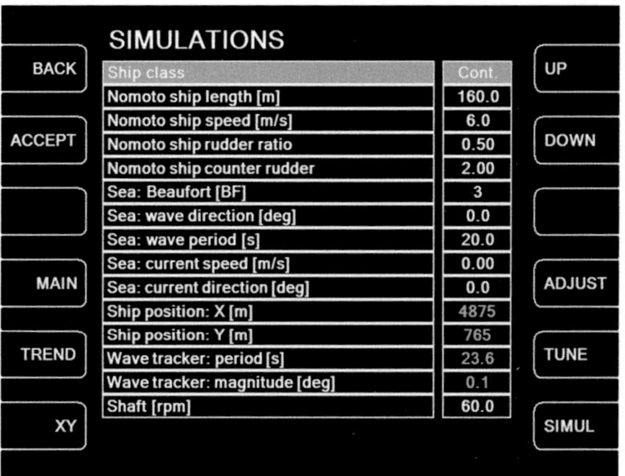

Fig. 7 SIMULATIONS display for choosing a ship and setting sea state

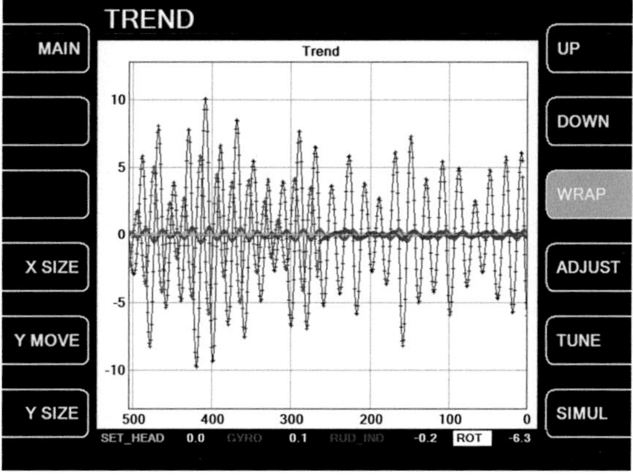

Fig. 8 Switching on the wave filter at rough sea

(green), GYRO (red), RUDDER INDICATION (blue), RATE_OF_TURN (black). At the left side of the display the wave filter is switched off, so the rudder moves in both directions trying to counteract portside and starboard motions of the ship. Later the filter is switched on, so after a while rudder motions become negligible. GYRO is almost the same as SET HEADING.

7 Conclusions

An overview of structure and functionalities of ship autopilot simulator, written in ST language in CPDev environment, has been presented. Heading control system, the basic one, involves PID controller, wave tracker, wave filter and set-point rate limiter. The tracker determines frequency of wave-induced ship motions, whereas wave filter removes them from gyrocompass measurements. Hence the rudder is not affected by waves.

The track control system, with additional GPS measurement of ship position, is of cascade structure with track controller as primary and heading controller as secondary. Sensitivity of the two controllers is adjusted by a single parameter. Ship dynamics data are identified by zig-zag or sinusoidal maneuvers and adjusted to ship speed.

The simulator involves models of two ships, real and abstract (Nomoto). Parameters of the abstract model can be adjusted in fairly broad ranges to account for small and large ships. Sea state is simulated as well, from calm sea to rough.

Displays of the simulator interface are designed to test and learn operation of the autopilot functionalities under different sea conditions. So ADJUSTMENTS, STATUS/TUNINGS and SIMULATIONS displays are in fact lists of indicators and parameter adjustments. It has been initially assumed that simulator interface components will be arranged in another, user-friendly way, in eventual target solution.

References

1. Jamro, M., Rzońca, D., Sadolewski, J., Stec, A., Świder, Z., Trybus, B., Trybus, L.: CPDev engineering environment for modeling, implementation, testing, and visualization of control software. In: Szewczyk, R., Zieliński, C., Kaliczyńska, M. (eds.) Recent Advances in Automation, Robotics and Measuring Techniques. AISC, vol. 267, pp. 81–90. Springer, Heidelberg (2014)
2. Trybus, L., Świder, Z., Stec, A.: Tuning rules of conventional and advanced ship autopilot controllers. In: Szewczyk, R., Zieliński, C., Kaliczyńska, M. (eds.) Progress in Automation, Robotics and Measuring Technics. AISC, vol. 350, pp. 303–311. Springer, Heidelberg (2015)
3. Autopilot AP2000 Track Pilot. Kongsberg Maritime Ship Systems AS (2001)
4. Simrad AP70/AP80 Operator Manual. Navico Holding AS (2012)
5. Fossen, T.I.: Guidance and Control of Ocean Vehicles, 4th edn. Wiley, Chichester (1999)

6. Fossen, T.I, Perez, T.: Kalman filtering for positioning and heading control of ships and offshore rigs. IEEE Control Syst. Mag. 32–46 (2009)
7. Lisowski, J.: Ship as Automatic Control Plant. Wyd. Morskie, Gdańsk (1981) (in Polish)
8. Stec, A.: Ship maneuvering model for autopilot simulator. In: Szewczyk, R., Zieliński, C., Kaliczyńska, M. (eds.) Progress in Automation, Robotics and Measuring Techniques. AISC, vol. 350, pp. 265–274. Springer, Heidelberg (2015)
9. Marine Systems Simulator. http://www.marinecontrol.org/Tutorial.html

Server Power Consumption: Measurements and Modeling with MSRs

Piotr Arabas and Michał Karpowicz

Abstract Precise model of power consumption is crucial for the design of controllers improving energy effectiveness of servers. This paper shows how the power usage statistics available via processor registers may be correlated to the total power consumption measurements of servers performing different types of operations. Identification of such relations may support development of power consumption models and application specific energy-aware server controllers.

Keywords Power saving in computer systems · Modeling · Identification

1 Introduction

The problem of reducing energy consumption of servers and devices used in ICT has been increasingly gaining importance [1, 2]. The main drivers are economic and technical, however, raising environmental awareness is of considerable significance as well. Since fossil fuels are still dominant source of energy, limiting demand on electricity is one of the basic methods to reduce emission. It should be noted that almost all energy provided to electric devices changes into heat which in turn must be discharged through cooling system at additional costs.

Hardware manufacturers have been improving energy efficiency of devices in recent years. In a straightforward step new technologies were introduced at the physical layer—e.g. passive optical networks, or new generation processors [3, 4]. As a consequence, energy-aware controllers and protocols were designed and implemented [1, 5]. Examples of novel solutions include CPU clock frequency scaling

P. Arabas (✉) · M. Karpowicz
Research and Academic Computer Network NASK, ul. Wawozowa 18, 02-796 Warszawa, Poland
e-mail: parabas@ia.pw.edu.pl

M. Karpowicz
e-mail: Michal.Karpowicz@nask.pl

R. Szewczyk et al. (eds.), *Challenges in Automation, Robotics and Measurement Techniques*, Advances in Intelligent Systems and Computing 440, DOI 10.1007/978-3-319-29357-8_21

mechanisms [6] and network-wide controllers coordinating interconnected devices [7–12] such as servers, firewalls and software routers.

In order to construct an energy-efficient control algorithm adequate power consumption model is clearly a necessity. Identification of such a model involves complicated measurements and dedicated equipment. As a solution to this problem one may view the use of power usage measurements provided by recent Intel® processors through model specific registers (MSRs) [13]. MSRs allow for high frequency readings of processor power consumption. Clearly, total power consumed by the server is correlated to the power consumption of the processor. However, the nature of this relation may depend on particular application profile of the server engaging variable set of computing resources. The aim of this work is to identify the correlation model for a selected set of application scenarios.

2 Experimental Setting

Experimental network consisted of five desktop PCs equipped Intel i7 processor, 8 GB of RAM, a quad 1 Gb/s Ethernet cards (Broadcom BCM5719), and a typical household electricity meter with remote access. This way it was possible to measure the power consumed by the one of computers while the remaining four could be used as traffic generators. Another computer was set up to record measurements in the database. The electricity meter used allows to read active power at the sampling rate of approximately 8 s with resolution of 1 W, so to collect the sufficient amount of data the experiments were designed to last at least several minutes. MSRs were read with a modified version of the `power_gov`[1] program. The modifications covered sampling rate adjustment. Namely, in the course of experiments the sampling rate MSRs was set to 10 ms, while the resolution of power measurement was in the order of $\frac{1}{100}$ W.

In order to determine the relationship between the power measurements read from the CPU MSRs and the overall server power consumption measurements a large number of experiments must be carried out. To replicate real world applications the following scenarios were proposed (Fig. 1).

Computation Server Carrying Intensive Arithmetic Operations

A modified version of the `stress`[2] benchmark was used to generate computational workload. To vary the intensity of operations `sleep()` instruction was appropriately introduced into the inner benchmark loop to stop it for a specified time interval and, as a consequence, to reduce the mean processor load.

[1] https://software.intel.com/en-us/articles/intel-power-governor.

[2] http://people.seas.harvard.edu/~apw/stress/.

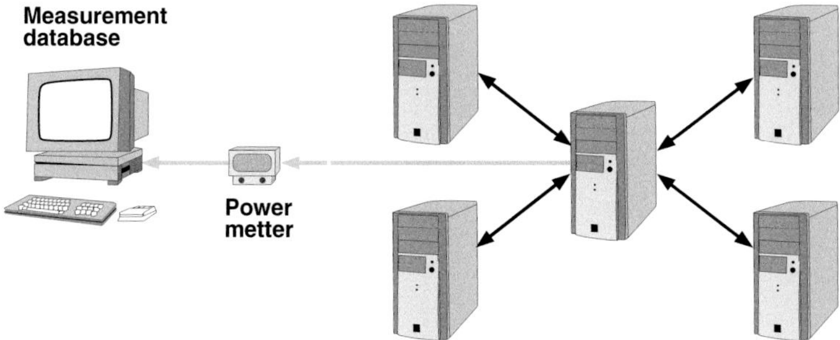

Fig. 1 Testbed topology

Software Router Forwarding Traffic Between Its Interfaces

Network traffic was generated by `iperf` benchmark, popular in standard Linux distributions. To simplify the analysis only UDP packets were sent, which allowed to easily keep constant bitrate regardless of the network load.

Video Transcoding Server

Finally, the video flows were streamed and then transcoded using `mencoder`[3] program. Forwarding the traffic was possible using `netcat` tunnel. As the result part of kernel level forwarding procedures was bypassed by the userspace program possibly reducing use of Ethernet board offloading mechanisms.

3 Results of Experiments

In order to identify the relationship between the power drawn by the processor and the total power consumed by the computer, in each experiment a series of measurements was conducted differing in workload levels imposed onto the system.

3.1 Scenario 1—Computation Server

Two parameters of the modified `stress` program were adjusted subsequently to perform measurements in varying conditions: the idle time between successive portions of calculations performed and the number of concurrently running threads (Table 1).

In the scenario considered the operating system was dynamically adjusting the frequency of CPU to the workload observed every 100 ms. The default `ondemand` governor was activated for this purpose. It should be noticed that the characteristics

[3]http://www.mplayerhq.hu.

Table 1 Parameters of modified `stress` in experiment 1

Idle time (μs)	96000	48000	24000	12000	6000	3000	2000	1200	700	160	10	0
Workload (%)	3	4	10	19	30	50	60	70	80	89	95	100

of the CPU governor imply relatively frequent switching of the CPU clock, up to the maximum value [14]. MSRs were sampled every 10 ms, whereas the total power consumed by the server every 8 s. Figure 2 illustrates the observed collections of samples. Despite different time scales close correlation between total power consumption and the processor power can be observed.

Figure 3 presents the relationship between the total power consumption and the CPU power consumption. Every point in this figure presents the results of one of 96 runs of the `stress` benchmark. For each of them the average is calculated with more than 70 measurements for total power and 60,000 samples for processor power.

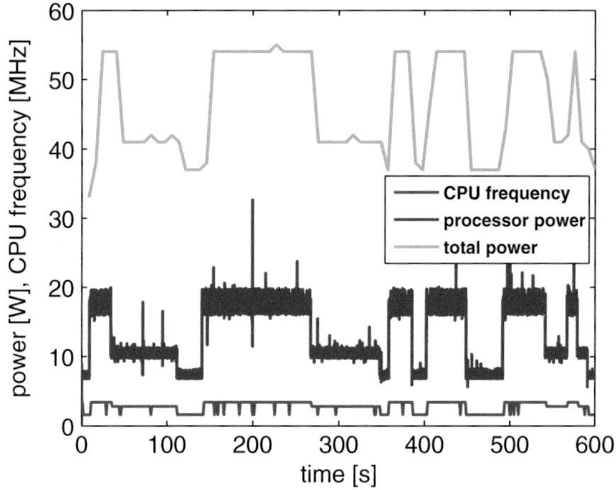

Fig. 2 Comparison of total power consumed by the server, processor power read from MSRs and cpu clock frequency

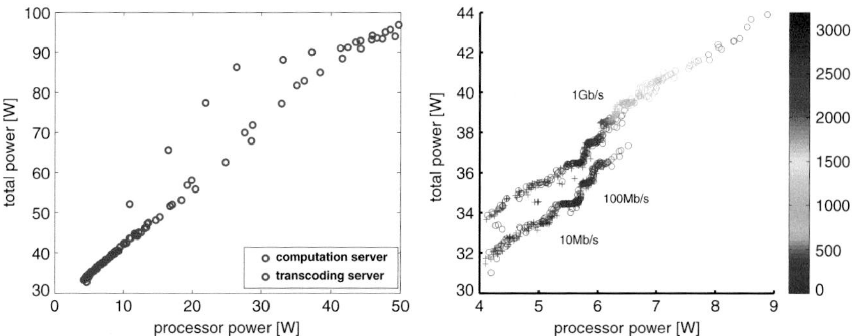

Fig. 3 Comparison of relationship between total power consumed by the computer and power consumed by the processor for computation server and streaming server scenarios (*left*) and software router scenario (*right*), *color scale* shows transfer rates in Mb/s

It may be easily seen that the shape of characteristic can be approximated by a linear function. This suggests that the main burden of arithmetic processing is carried by the processor itself so the overall power consumption follows that of the processor.

3.2 Scenario 2—Software Router

In this scenario three sets of experiments were performed to identify power consumption profile of a software router operating with different network adapter speeds, namely 10, 100 Mb/s and 1 Gb/s. In every case the network traffic flows were routed between all four ports of the server's network card. The rates were varied independently on two directions—there were 256 combinations covering range 0–800 Mb/s for 1 Gb/s mode. In lower modes traffic rates were limited to 144 combinations for 100 Mb/s and 49 when the adapters were switched to 10 Mb/s. For every combination two experiments were performed—with a single and bidirectional traffic. Maximum rate of traffic was limited to 800 Mb/s due to the hardware limitations related to transmission of 1500B UDP frames.

The measurements are summarized in Fig. 3—the lower lines contain samples gathered in 10 and 100 Mb/s modes, the upper one illustrates the measurements taken when interfaces worked with full speed, + markers point results obtained for single direction of traffic and ∘ for bidirectional traffic. The common feature is that the power consumption in software router scenario is much lower than in the case of computing server. It must be also noted that regardless of the traffic forwarding rate the processor frequency remained at its lower limit. These two facts show that the network traffic forwarding is a relatively simple task for the PC machine. Another important observation is a difference in power consumption that seems to depend on the network adapter speed applied. Clearly, the power consumption follows the same linear pattern in each case, but switching to 1 Gb/s mode results in approximately 2 W higher power consumption in the corresponding range of traffic forwarding rates. This may be explained by switching on additional circuits for full duplex mode. Contrary to the previous scenario, the power usage characteristics have much more undulating shape. This may be interpreted as an influence of measurements made with resolution limited to 1 W. Nevertheless, approximation with linear models seems to be possible, at least when extremely high precision in modeling is not required. Finally, it is worth noting that both single and two directional samples are grouped together in the graph. This seems to support the popular assumption that it is the total workload, not its distribution among the ports of a network card, that has a dominant influence on the power consumption profile (cf. [5, 15]).

3.3 Scenario 3—Video Stream Transcoding

Transcoding of video data is a processor-intensive task. In the course of experiments the server was capable of processing up to six streams only, with network ports working well bellow their limits (single stream produced approximately 20 Mb/s). Power consumption characteristic is shown in Fig. 3. Its shape significantly differs from the one analyzed in scenario 2. The collected measurements clearly form a concave pattern, which indicates the effect of saturation. Proper interpretation of the observed outcomes requires a somewhat deeper insight into the transconding process. Namely, the CPU intensive operations are performed subject to very sharp time regime constraints. Consequently, under sufficiently high workload delays begin to occur leading to saturation. Being relatively small and rare they do not cause transmission breaks (the receive buffers are large), however in the setting considered it was not possible to reliably transmit 7 or more streams. The coincidence with the number of cores (4) suggests that the source of the delays can be related to thread switching and concurrent memory access.

The slope of the initial part of the characteristic is higher than in previous scenario—it may be attributed to the higher processor workload caused by tunneling data through `netcat`. In this scenario some forwarding functions were also supported by a user space program instead of being offloaded to the network interface card. A shape of characteristic indicates that, for a reasonably accurate approximation in the full scope of workload levels a nonlinear function should be used.

4 Model Identification

The results of experiments demonstrated that power profiles may be successfully described by polynomial models of low degree.

For scenarios 1 and 2 application of linear function is reasonable. Clearly, linear regression cannot reproduce all features observed in Fig. 3, however it is expected that the accuracy of approximation shall be sufficient for most applications. Scenario 3 requires a bit more complex approach. It seems that it is worthwhile to consider two cases depending on the workload. For a smaller number of streams (in the range 1–4) linear model can be used. In the full range it is necessary to revert to nonlinear concave model. The advantage of both approaches is filtering of disturbances. Another important factor is that identification may be carried out using relatively small amount of data.

The following equation presents an example of cubic model for scenario 3:

$$p(w) = \alpha_0 + \alpha_1 w + \alpha_2 w^2, \tag{1}$$

where w is the power read from MSRs, $p(w)$ is the total power consumption and α_0, α_1 and α_2 are model coefficients. The linear model used for all other scenarios, as well as for the narrower range of the power consumed by the processor in scenario 3, can be treated as a special case of the model (1). The values of identified coefficients are presented in Table 2 (least squares method was used).

A comparison of models for three software router scenarios (modes 10, 100 Mb/s and 1 Gb/s) together with computation server are presented in Fig. 4. The graph is limited to the initial range due to the lower power consumption in the software router scenario than in the case of computation server. The increased power demand in 1 Gb/s mode is clearly visible. The line fitted for computation server starts at similar level to 1 Gb/s software router—it can be explained by the fact that the network interfaces are in this case configured in this particular mode (although they do not carry any traffic). The slope of the line for a computing server is lower, indicating the need of separate models.

The slope of the line fitted to the data for the video transcoding scenario is significantly higher than in the case of computation server (see Fig. 4). An important feature is in the range of 5–6 streams—it can not be modeled by linear function, however cubic approximation seems to be suitable.

Table 2 Power consumption model coefficients

Scenario	α_0	α_1	α_2
Computation server	27.33	1.46	0
Software router 10 Mb/s	23.22	2.04	0
Software router 100 Mb/s	22.76	2.19	0
Software router 1 Gb/s	23.37	2.41	0
Video transcoding—full range	10.6	4.43	−0.06
Video transoding—1–4 streams	28.49	2.21	0

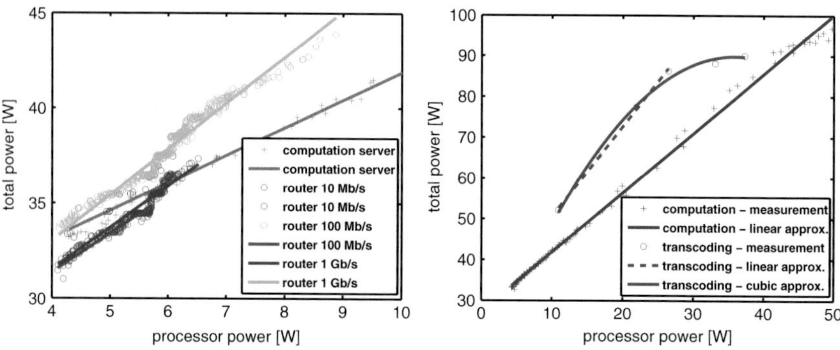

Fig. 4 Fitting models for: software router and computation server (*left*) and video transcoding (*right*), *points* and *lines* represent measurements and approximation respectively

5 Model Validation

Additional experiments were carried out to collect separate set of samples which could be used to validate models. To asses quality of fit the mean absolute value of error was calculated:

$$e_0 = \frac{1}{N} \sum_{i=1}^{N} | p(w_i) - p_i |, \tag{2}$$

where w_i is i-th sample of the processor power read from MSRs, p_i is corresponding value of the total power consumption, $p(w_i)$ is the total power computed with previously identified model and N is the number of samples.

As in the previous experiments the measurements were performed for three scenarios: computation server, software router and video streams transcoding. Parameters of first two groups of experiments are provided in Tables 3 and 4. The third experiment covered transcoding 1–6 video streams with the rate reduced to approximately 80 % of previous value.

Results of verification experiment for the computing server scenario are illustrated in Fig. 5. Lightly colored area indicates ±5 % deviation from the measurements. Except for a single sample, the line corresponding to the model falls in the marked area with the average value of absolute error (2) not greater than 3.1 %. This can be interpreted as a confirmation of the model linearity, whereby the adopted accuracy (5 %) should be sufficient for most control applications. A single sample showing greater deviation is most likely a measurement error or the effect of system process wake up e.g. file indexing, possibly using a hard disk and so significantly increasing the power consumption. Slight irregularities of the characteristic may, however, indicate more complex nature of the observed system.

Model verification for software router scenario also gave similar results, wherein a slightly higher accuracy may be observed. The areas marked in Fig. 5 with lighter color correspond to the 2 % deviation from the measured values. The average absolute value of error is 0.4 %, 0.8 % and 0.9 % for respectively modes of 10, 100 Mb/s and 1 Gb/s. Better accuracy is possible thanks to using more samples, while interval of function argument (i.e. the power consumed by the processor) is significantly narrower than in scenario 1.

Table 3 Parameters of modified `stress` in the validation experiment 1

Idle time (μs)	32000	10000	4500	1200	90	5
Load (%)	8	22	42	71	90	96

Table 4 Rates of traffic generated with `iperf` in validation experiment 2

No.	1	2	3	4	5	6	7	8
Rate (Mb/s)	0.5	3	7	15	50	90	300	700

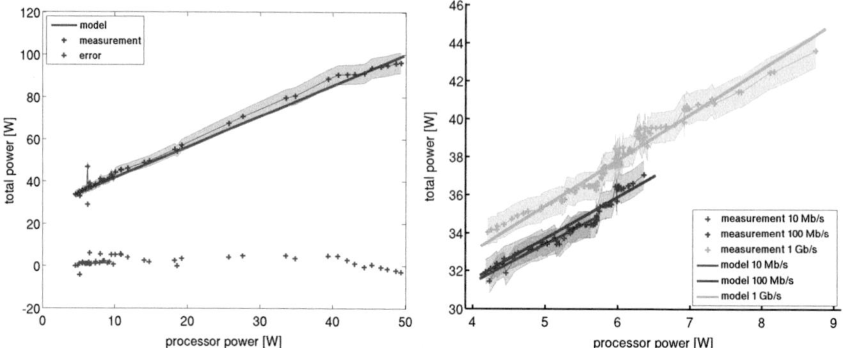

Fig. 5 Model validation for: scenario 1—computation server (*left*) and scenario 2—software router (*right*)

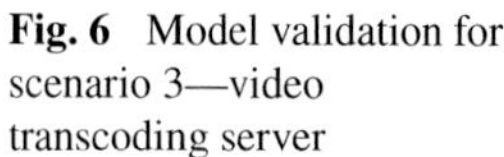

Fig. 6 Model validation for scenario 3—video transcoding server

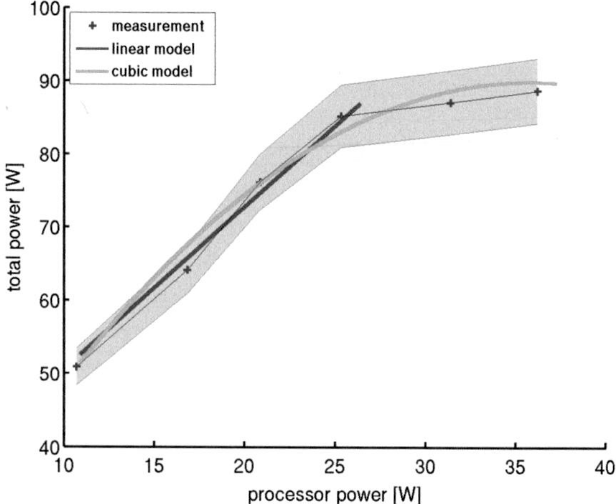

The case of video transcoding server is, as was mentioned earlier more complex and requires a nonlinear model to be applied. Figure 6 shows the results of the verification of previously tuned model, lightly blue margin is as in the scenario 1 set to 5 %. Both models fit within the imposed accuracy. The placement of the measurement points against model line indicates that the range, which was adopted for the linear model is appropriate. In other words the saturation point of the total power characteristic is well defined by the value of processor power.

6 Conclusions

The presented measurements allowed to identify models binding the total power consumed by a server to the processor power consumption monitored through MSRs. The most important observation is that the form of the model depends on the type of tasks performed by the server. In relatively simple cases, such as pure numerical calculations or basic traffic forwarding, it is permissible to use linear model. The case of video transcoding server is more complex, but approximation with cubic function seems to be sufficient. In both variants the number of model parameters is small—2 in the case of linear model and 3 for cubic polynomial. These can be relatively quickly identified, as for accurate determination of their values only few measurements are necessary. This suggests that it is possible to tune both models based on measurements taken by a simple power-meter or laboratory multi-meter.

The obtained results clearly show dependence of the power consumption profile on the type of server workload. Consequently, in order to design an energy-efficient device control algorithms it is necessary to construct and identify dedicated models suitable for specific application. This, however, suggests that application-specific controllers may be capable of increasing the efficiency of power usage above the level provided by the universal controllers, such as default Linux `ondemand` governor, cf. [6, 16].

The presented scenarios, though typical, do not exhaust the list of possibilities. In particular, a more complete set of model could be constructed with benchmarks involving storage and full set of processor instructions. Additional measurements available in the operating system, e.g. traffic statistics or instruction call counters, are also expected to improve the quality of modeling.

References

1. Nedevschi, S., Popa, I., Iannacone, G., Wetherall, D., Ratnasamy, S.: Reducing network energy consumption via sleeping and rate adaptation. In: Proceedings of the 5th USENIX Symposium on Networked Systems Design and Implementation, pp. 323–336 (2008)
2. Chabarek, J., Sommers, J., Barford, P., Estan, C., Tsiang, D., Wright, S.: Power awerness in network design and routing. In: Proceedings of the 27th Conference on Computer Communications (INFOCOM 2008), pp. 457–465 (2008)
3. Venkatachalam, V., Franz, M.: Power reduction techniques for microprocessor systems. ACM Comput. Surv. **37**(3), 195–237 (2005)
4. Tucker, R.S., Parthiban, R., Baliga, J., Hinton, K.: Evolution of WDM optical IP networks: a cost and energy perspective. J. Lightwave Technol. **27**(3), 243–252 (2009)
5. Bolla, R., Bruschi, R., Carrega, A., Davoli, F.: Theoretical and technological limitations of power scaling in network devices. In: Proceedings of the 2010 Australasian Telecommunication Networks and Applications Conference, pp. 37–42 (2010)
6. Karpowicz, M.: Energy-efficient CPU frequency control for the Linux system. Concurrency Comput. Pract. Experience (2015). doi:10.1002/cpe.3476
7. Chiaraviglio, L., Mellia, M., Neri, F.: Energy-aware backbone networks: a case study. Proc. IEEE Int. Conf. Commun. Workshops **2009**, 1–5 (2009)

8. Vasić, N., Kostić, D.: Energy-aware traffic engineering. In: Proceedings of the 1st International Conference on Energy-Efficient Computing and Networking E-ENERGY (2010)
9. Niewiadomska-Szynkiewicz, E., Sikora, A., Arabas, P., Kamola, M., Mincer, M., Kołodziej, J.: Dynamic power management in energy-aware computer networks and data intensive systems. Future Gener. Comput. Syst. **37**, 284–296 (2014)
10. Qureshi, A., Weber, R., Balakrishnan, H., Guttag, J., Maggs, B.: Cutting the electric bill for internet-scale systems. SIGCOMM Comput. Commun. Rev. **39**(4), 123–134 (2009)
11. Kozakiewicz, A., Malinowski, K.: Network traffic routing using effective bandwidth theory. Eur. Trans. Telecommun. **20**(7), 660–667 (2009)
12. Karpowicz, M., Arabas, P., Niewiadomska-Szynkiewicz, E.: Energy-aware multi-level control system for a network of Linux software routers: design and implementation. IEEE Syst. J. (2015). In press
13. Intel® 64 and IA-32 architectures software developers manual. http://www.intel.com/content/dam/www/public/us/en/documents/manuals/64-ia-32-architectures-software-developer-manual-325462.pdf (2015)
14. Pallipadi, V., Starikovskiy, A.: The ondemand governor: past, present, and future. Proc. Linux Symp. **2**, 215–230 (2006)
15. Bolla, R., Bruschi, R., Carrega, A., Davoli, F., Suino, D., Vassilakis, C., Zafeiropoulos, A.: Cutting the energy bills of internet service providers and telecoms through power management. Comput. Netw. **56**(10), 2320–2342 (2012)
16. Karpowicz, M., Arabas, P.: Preliminary results on the Linux libpcap model identification. In: Proceedings of the 20th IEEE International Conference on Methods and Models in Automation and Robotics (2015)

Multi-agent System for On-Line Game Matchmaking

Dan Latocha and Piotr Arabas

Abstract The task of choosing players to form optimal games is formulated as a binary mathematical programing problem. An efficient heuristic is proposed and decomposed to build a multi-agent system. The quality of such a solution is compared with the quality of the exact one, found with the MILP solver. The scalability of the system is verified by numerical experiments.

Keywords Matchmaking · MILP · Multi-agent system

1 Introduction

The popularity of on-line games has been rising rapidly in last years. It is estimated that approximately 700 millions people were playing on-line games in 2013 [1]. Many of them do it regularly—in 2015 more than half of players competed with others at least once a week [2]. As a result, in 2015, there were nearly 145 million active players each day [3]. Players in on-line games are usually matched automatically. Games facing such a tremendous load need a system which will match users into teams providing an entertaining game but also minimizing the cost (e.g. energy cost). High level optimization can lead to reduction of CO_2 pollution in a way similar to described in [4, 5].

D. Latocha (✉)
Institute of Control and Computation Engineering, Warsaw University of Technology, ul. Nowowiejska 15/19, 00-665 Warszawa, Poland
e-mail: research@danlatocha.com

P. Arabas
Research and Academic Computer Network NASK,
ul. Wawozowa 18, 02-796 Warszawa, Poland
e-mail: parabas@ia.pw.edu.pl

R. Szewczyk et al. (eds.), *Challenges in Automation, Robotics and Measurement Techniques*, Advances in Intelligent Systems and Computing 440, DOI 10.1007/978-3-319-29357-8_22

The paper is organized as follows: next Sect. 2 introduces the notion of matchmaking. Section 3 presents mathematical formulation of the problem together with a distributed heuristic development. Section 4 describes the design of a multi-agent system. In Sect. 5, the results of computational experiments are provided and finally Sect. 6 concludes on them.

2 Problem Description

The term "matchmaking" when referred to on-line computer games covers usually all aspects of finding appropriate combinations of users to form an entertaining game. In most multiplayer games users may join in teams making the process a two-stage one—first forming teams and then allocating them to games.

A typical approach is to use users rating to find a match for them. This leads to two problems: how to asses users rating to allow finding adequate teams and games and how to use this information to form games.

This paper concentrates on the second problem. Basing on [6–8], we use players skill and experience to describe their performance in every mode of the game. The problem is especially difficult when allocation of users to multiple servers is considered. A typical approach [9] is to first select the server based on player's location. In the solution proposed in this paper, the task is solved jointly in order to find better solutions. Another important advantage of our approach is higher scalability. In the case of raising load, new servers may be added dynamically, in any location.

Augmenting suggestions found in [10, 11], we propose the following requirements that a correctly constructed game should meet:

1. The game should be balanced i.e. the rating parameters (e.g. skill or experience) of the users should be similar, resulting in equal win probabilities.
2. Quality of service perceived by players should be maximized.
3. Overall cost of serving games should be minimized.

Solving such problem involves multiple criteria—we propose to tackle it by scalarization, allowing to tune the balance of the quality and cost. We formulate a mathematical programing task, and use it as a benchmark for much more effective distributed heuristic, implemented using multi-agent framework [12]. This way we obtain a scalable, distributed algorithm contrary to [9] or [10]. Another difference is that we try to optimize not only the game composition but also the quality of service and cost, while other authors usually focus only on one of these aspects (e.g. transmission delay [9] or detailed player characteristics [10]).

3 Modeling the Problem

3.1 Full Model

The model bellow covers all aspects of the matchmaking process in a basic form, i.e. as they are defined in the previous section. Such formulation involves nonlinear functions and may be solved by limited number of solvers. However, we believe that it allows to describe all requirements a balanced match should meet.

3.1.1 Objective Function

The objective function is a sum of six components related to all aims of the match-making process. The first of them aggregates transmission delays of all players, the next is a sum of absolute values of skill difference in teams. The third component is a sum of differences between the most and the least skilled players in subsequent matches. Then, analogically, there is a sum of differences between players with the highest and the lowest experience value. The cost of all servers is summarized in the fourth component, while the last is a penalty for leaving unmatched players.

$$
\begin{aligned}
J(x_{tp}^{sm}) = {} & c_\pi \sum_{\substack{s\in S,\\ m\in M}} \max_{\substack{t\in T\\ p\in P}} x_{tp}^{sm}\pi_{ps} + c_\tau \sum_{\substack{s\in S\\ m\in M}} \left| \sum_{p\in P} x_{1p}^{sm}\sigma_p - \sum_{p\in P} x_{2p}^{sm}\sigma_p \right| \\
& + c_\sigma \sum_{\substack{s\in S\\ m\in M}} \left(\max_{\substack{t\in T\\ p\in P}} x_{tp}^{sm}\sigma_p - \min_{\substack{\{t\in T,\ p\in P:\\ x_{tp}^{sm}=1\}}} \sigma_p \right) + c_\eta \sum_{\substack{s\in S\\ m\in M}} \left(\max_{\substack{t\in T\\ p\in P}} x_{tp}^{sm}\eta_p - \min_{\substack{\{t\in T,\ p\in P:\\ x_{tp}^{sm}=1\}}} \eta_p \right) \\
& + c_\kappa \sum_{\substack{s\in S\\ m\in M}} \kappa_s \max_{\substack{t\in T\\ p\in P}} x_{tp}^{sm} + c_\nu \left(N_p - \sum_{\substack{s\in S, m\in M\\ t\in T, p\in P}} x_{tp}^{sm} \right),
\end{aligned}
\tag{1}
$$

where: x_{tp}^{sm} is a binary variable equal 1 when player p is allocated to server s, match m and team t, c_π, c_τ, c_σ, c_η, c_κ, c_ν are weight coefficients. S, M, T and P denotes sets of all servers, matches, teams and players respectively, π_{ps} is player p to server s delay, σ_p and η_p are the skill and experience of player p, κ_s is the cost of server s, N_p is the total number of players.

3.1.2 Constraints

The following inequalities describe limitations imposed on matches.

$$\forall s \in S, \; m \in M, \; t \in T \qquad \sum_{p \in P} x_{tp}^{sm} = 3 \; \cup \; \sum_{p \in P} x_{tp}^{sm} = 0 \tag{2}$$

$$\forall s \in S, \; m \in M \qquad \sum_{t \in T, \, p \in P} x_{tp}^{sm} = 6 \; \cup \; \sum_{t \in T, \, p \in P} x_{tp}^{sm} = 0 \tag{3}$$

$$\forall p \in P \qquad \sum_{s \in S, \, m \in M, \, t \in T} x_{tp}^{sm} \leq 1 \tag{4}$$

$$\forall s \in S, \; m \in M, \; t \in T, \; p \in P \qquad x_{tp}^{sm} \pi_{ps} \leq 200 \tag{5}$$

$$\forall s \in S, \; m \in M \qquad \max_{\substack{t \in T \\ p \in P}} x_{tp}^{sm} \eta_p - \min_{\substack{\{t \in T, \; p \in P: \\ x_{tp}^{sm} = 1\}}} \eta_p \leq 2000 \tag{6}$$

$$\forall s \in S, \; m \in M \qquad \max_{\substack{t \in T \\ p \in P}} x_{tp}^{sm} \sigma_p - \min_{\substack{\{t \in T, \; p \in P: \\ x_{tp}^{sm} = 1\}}} \sigma_p \leq 1200 \tag{7}$$

$$\forall s \in S, \; m \in 1 \ldots N_m - 1 \qquad \sum_{t \in T, \, p \in P} x_{tp}^{sm} \geq \sum_{t \in T, \, p \in P} x_{tp}^{s,m+1} \tag{8}$$

Constraints (2) and (3) ensure appropriate number of players in teams and matches (3) and (4), (6) allocates every player to only one team and match, (5) limits transmission delay between users and servers—200 ms was used. Maximal difference in experience and skill among players in the match is limited by (6) and (7). Finally (8) orders matches (N_m being the number of matches). Constraints (5)–(7) resemble objective function and may be considered superfluous, however they were included to impose all requirements the match should meet.

3.1.3 Optimization Task

The optimization task is defined as a minimization of the objective function (1) with respect to binary decision variable x_{tp}^{sm}:

$$\min_{x_{tp}^{sm}} J(x_{tp}^{sm}), \tag{9}$$

subject to:

$$\begin{gathered} \text{inequalities } (2-8), \\ x_{tp}^{sm} \in \{0, 1\} \end{gathered} \tag{10}$$

Solving this task may be difficult as the number of binary variables grows quickly with rising number of players and servers. The reduction to the standard form will result in an even larger problem. With this in mind, we decided not to develop formulation beyond presented above, but to seek appropriate heuristics.

3.2 Decomposition

The task (9) and (10) is difficult to solve, however it can be decomposed to build a distributed algorithm as the performance index (1) is additive with respect to s:

$$J(x_{tp}^{sm}) = c_v N_p + \sum_{s \in S} J_s(x_{tp}^{sm}), \tag{11}$$

where $J_s(x_{tp}^{sm})$ is part of (1) with index s omitted under subsequent sums. Most of the constraints may be decomposed in a similar way—e.g. for (2):

$$\forall m \in M,\ t \in T \sum_{p \in P} x_{tp}^{sm} = 3 \ \cup \ \sum_{p \in P} x_{tp}^{sm} = 0, \tag{12}$$

The same may be done for (3), (5)–(8). The constraint (4) can not be decomposed as it couples allocation of players to teams among servers.

3.3 Greedy Algorithm

Server Task, ST: generates matches (i.e. forms two teams of selected players) and sorts them with respect to match evaluation index:

$$\begin{aligned} Q_m = {} & c_\pi \sum_{p \in P_m} \pi_p + c_T \left| \sum_{p \in C_{m,1}} \sigma_p - \sum_{p \in C_{m,2}} \sigma_p \right| + c_\sigma \left(\max_{p \in P_m} \sigma_p - \min_{p \in P_m} \sigma_p \right) \\ & + c_\eta \left(\max_{p \in P_m} \eta_p - \min_{p \in P_m} \eta_p \right) + c_\kappa \kappa_s, \end{aligned} \tag{13}$$

where: P_m denotes users in the match m and $P_{m,i}$, $i \in \{1, 2\}$ these in the match m and team i. The match evaluation index is part of $J_s(x_{sm}^{tp})$ related to the single match. The substitution of maximal delay with a sum shortens computations.

The potential teams may be selected randomly, but have to be feasible in the sense of partitioned constraints (2), (3), (5)–(8). As the number of combinations may be very high, the output of ST is limited to $n_3 > N_p$ best matches.

Player Task, PT: PTs receive matches from servers and select the best one using the same index (13) so the same match should be chosen. Automatically, the global constraint (4) is satisfied. The algorithm is suboptimal, with every player acting egoistically. Typically, however similar users will have similar values of (13), so they will be grouped making it possible for PTs to choose sensible matches. The advantage of the ST-PT mechanism is that new matches may be generated on the arrival of new players and sent to PT for evaluation. The algorithm may react dynamically to requests contrary to the task (9) and (10), which, in such case, must be laboriously reformulated and solved again.

4 Multi-agent System

Multi-agent system is based on the greedy algorithm. The procedure is repetitive allowing to allocate some users in the first round and then to wait for new ones. It is also possible to change the number of servers during system operation adding new ones in periods of high load or shutting down unused ones.

4.1 User Agent

User agent (UA) implements the idea of PT (see Sect. 3.3) in a multi-agent framework. It starts with registering with server agents (SA). As UA knows the values of delay to all servers, it may decide to register only with SAs satisfying constraint (5) limiting number of SAs involved. After that, UA enters negotiation phase and waits for SAs to propose matches. Answers are collected for time t_a, then one of them, minimizing (13) is selected and requested. The UAs wait for time t_n for the SA to confirm selected match. However, if UA receives better proposition within time t_n it may cancel previous selection and request a new one. On reception of a match confirmation from the SA the process is ended. Otherwise UA repeats all steps, beginning with the registration as the list of servers may change.

The important point is registering with server agents. In the base version of the algorithm, all server agents with adequate transmission delay are contacted. It is, however, possible to limit the number of SAs selected to e.g. 5 with best delays. Involving all servers provides a solution closer to the optimal one while limiting their number reduces combinations to generate and evaluate.

4.2 Server Agent

Server agent implements ST by managing two lists: the list of registered users and the list of all possible matches. As user agents register only with selected servers the number of combinations is limited.

The operation of the server agent is triggered by user registration or acceptance/cancellation of previously proposed match. On user arrival, new matches are generated according to the match generation algorithm. If some of them are better than the ones already stored, or include the new player, they are added to the list and proposed to the relevant users.

A match is accepted when all user agents involved confirm it. Then, server agent sends the final match proposition message to all of them and after receiving confirmations removes the match and players from it's lists. Additionally, server agent checks if there are any matches containing any of removed users—if so, it sends cancellation message to all user agents involved, and removes these matches from the list.

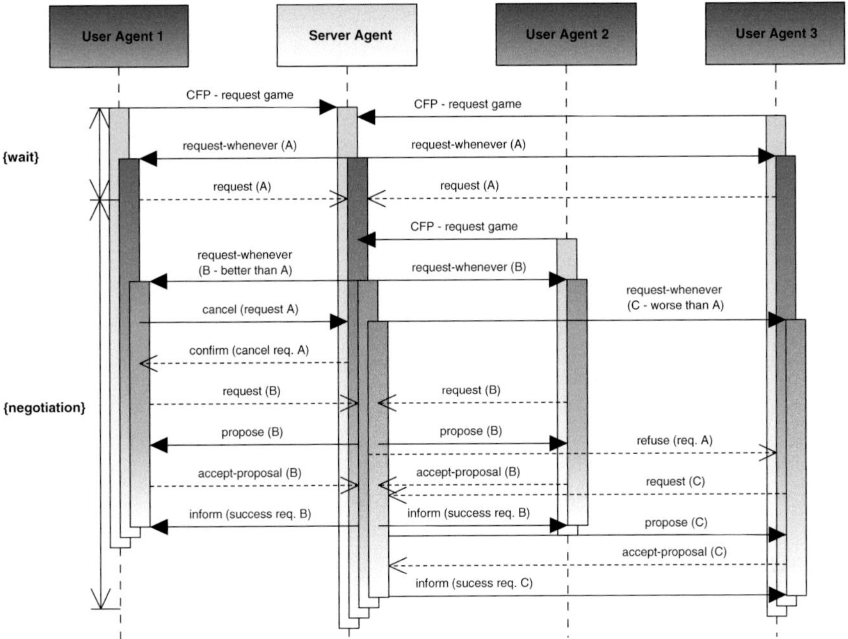

Fig. 1 Communication between agents

A cancellation request may be received by server agent when new (i.e. better) match was accepted by the user agent. In this case, the match referenced should be removed from the list. The player however will remain on player list until acceptance of newly proposed match. Figure 1 presents the communication where user agent 1 cancels match A in favor of match B.

4.3 Match Generation Algorithm

As number of possible combinations may be huge, it is necessary to propose some means to effectively constrain them (e.g. for teams of 3–6 players in a match and 100 players there is $\binom{100}{6} \approx 10^9$ combinations). In order to mitigate this, a two stage algorithm is proposed. In stage 1 n_1 teams are randomly selected and evaluated using the following index:

$$Q_t = \sum_{p \in P_t} \pi_p + \max_{p \in P_t} \sigma_p - \min_{p \in P_t} \sigma_p + \max_{p \in P_t} \eta_p - \min_{p \in P_t} \eta_p, \tag{14}$$

where P_t are players of team t. Obviously, (14) is a part of overall match evaluation index (13) related to the single team with weighting coefficients set to 1.

In stage 2 $n_2 < n_1$ of the best teams are combined into matches. As the skipped teams are inferior to selected ones it is unlikely they could form valuable matches—thanks to this the quality is not significantly impaired.

Matches conforming to quality constraints are sorted according to (13). Going from the top (best) match, a new match may be added to the match list if it is better than the one for any of players included in the new combination. Finally, the match list is truncated to n_3 best matches consisting of different players.

5 Implementation and Experiments

The multi-agent algorithm was implemented using JADE (Java Agent DEvelepment Framework)[1] conformant with Foundation for Intelligent Physical Agents (FIPA) standards. FIPA architecture provides predefined framework to construct efficient and reliable messaging, agent management and location.

To find an exact solution, IBM ILOG CPLEX Optimization Studio[2] was used. A great elasticity of its OPL language allowed using some, non-standard for MIP solvers, functions (eg. max) existing in the problem definition.

5.1 Player Data Generation

As the real data was unavailable, it was necessary to generate a set of players. Following information on leagues parameters in League of Legends game [11] the normal distribution was adopted for skill. The experience was modeled with lognormal distribution to conform with ESA statistics [2]. For modeling player to server delays a normal distribution with 60 ms mean was chosen. The server costs were modeled using normal distribution. The details are provided in Table 1.

5.2 Exact Solution

As previously mentioned, an exact solution of the mathematical programing task (9) and (10) was possible only for very small sizes—24 users and 3 servers being the maximum—it was used throughout further experiments as a benchmark.

[1] http://jade.tilab.com.

[2] https://www.ibm.com/developerworks/downloads/ws/ilogcplex/.

Table 1 Details of distributions used to generate test data

Parameter	Distribution	μ	σ	Remarks
σ_p	Normal	1200	500	[11]
η_p	Log-normal	0	1.25	[2], mean 262
π_{ps}	Normal	60	25	
κ_s	Normal	90	30	

5.3 *Multi-agent System Solution*

First, solutions obtained by solving mathematical programming task and using multi-agent system were compared. Five sets of 24 users were generated. The number of servers was set to 3. Table 2 presents computation times. Exact solution is valuable as benchmark for the agent system. Table 3 presents values of performance index components averaged for 15 runs of the system together with the exact solution. When comparing them, one can see that the agent system performs visibly worse, however they should be scaled with respect to the system requirements. Taking this into account, one must notice that in the worse case—for team skill difference both values are extremely small—far below 100, while average value is 1200. Similarly, transmission delay is bellow average, which suggests that most users got well suited servers (i.e. with short delay).

The scalability was tested using datasets with 24–966 players and 3–15 servers. The system operated with the following parameters: waiting time $t_a = 10$ s, negotiation time $t_n = 30$ s, match generation constants $n_1 = 5625$, $n_2 = 750$ and $n_3 = 25000$. As the process is iterative, it was allowed that last 6 players remain unallocated if not matched within 5 s. Additionally, 2.5 s margin was provided for communica-

Table 2 Comparison of computation times for 24 players example

Solution method	Set 1	Set 2	Set 3	Set 4	Set 5
Exact	4 h 52 min	7 h 21 min	10 h 10 min	11 h 54 min	3 h 5 min
Agent	13.2 s	17.6 s	12.3 s	26.7 s	20.8 s

Table 3 Comparison of solution quality for 24 players example

	Exact	Agent	% difference
Avg. delay (s)	48	53.8	12.2
Avg. team skill diff.	42	71.8	70.9
Avg. player skill diff.	383.4	427.1	11.4
Avg. player experience diff.	400.3	482.3	20.5
Avg. cost per game	73.5	83.6	13.7

Table 4 Comparison of computation times for varying number of players and servers

No. of players	24	36	60	120	240	396	966
No. of servers	3	3	3	5	7	8	9
Exact	4 h 52 min	–	–	–	–	–	–
Agent	13.2 s	31.4 s	47.3 s	44.3 s	46.4 s	52 s	67.9 s

tion. Summing up one round of algorithm should last no longer than 47.5 s. Table 4 presents computation times averaged from 15 runs—it may be seen that for two largest cases they are longer than 47.5 s. Such a situation suggests some scaling problems. However, it is worth noting that the execution time is still acceptable. The quality of the solutions is depicted in Fig. 2—it shows clearly that all parts of performance index becomes lower with the rising number of players. This behavior is not a surprise, as there should exist more valuable combinations in the larger set, however it is not obvious that it would be easier to find them using relatively simple, random based algorithm. The results testify that choice of parameters n_1-n_3 provides reasonable effectiveness of the algorithm.

The third set of experiments was devoted to the testing of how the algorithm performs in the conditions of growing load. The number of servers was set to 5 and the numbers of players was changed from 24 to 966. The results are presented in Fig. 3. It may be seen that the quality of the solution is very similar to the previous case and the greatest difference is the growing server costs—it is necessary to rise expenses by 35 % to keep the quality.

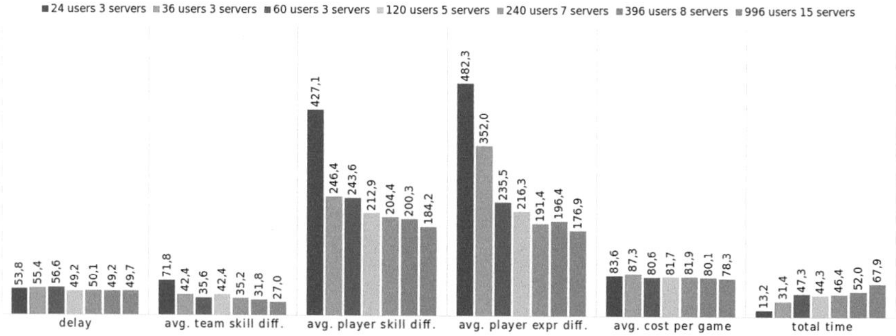

Fig. 2 Performance index components and computation times for number of users varying from 24 to 966 and number of servers in the range 3–15

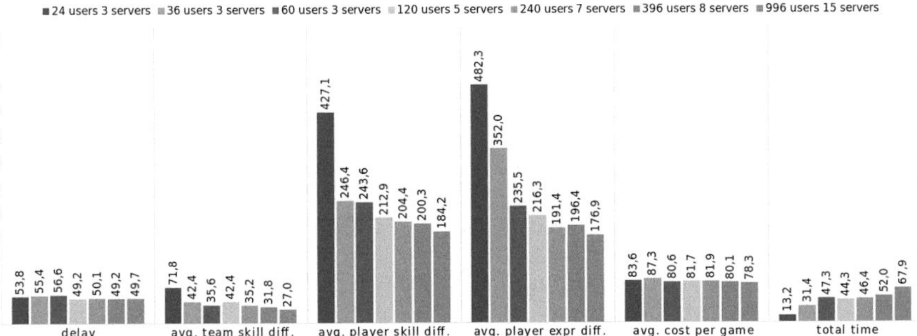

Fig. 3 Performance index components and computation times for number of users varying from 24 to 966 and 5 servers

6 Conclusions

The paper presented a development of a mathematical model and a multi-agent system for matchmaking in on-line games. Although the model is relatively complicated and the resulting mathematical programing task cannot be solved effectively, it can be used to develop an efficient heuristic. This heuristic is by design distributed and may be easily implemented within a multi-agent framework. The greatest advantage of the proposed system is that it may react dynamically to new requests. The experiments proved positive properties of the solution. They also showed that the system can serve up to thousand simultaneous requests. For more popular games, limiting the number of server agents contacted by user agents may allow to better spread the load and eliminate excessive combinations.

References

1. Spil Games: State of online gaming report. http://www.academia.edu/5499175/State_of_Gaming_2013_UK_FINAL (2013)
2. Ipsos MediaCT for Entertainment Software Association (ESA): The 2015 essential facts about the computer and video game industry. http://www.theesa.com/wp-content/uploads/2015/04/ESA-Essential-Facts-2015.pdf (2015)
3. Statista.com: Number of daily active online gamers worldwide as of April 2013, by region (in millions). http://www.statista.com/statistics/261262/number-of-online-gamers-worldwide-by-region/ (2015)
4. Karbowski, A., Jaskóła, P.: Two approaches to dynamic power management in energy-aware computer networks methodological considerations. In: Proceedings of Federated Conference on Computer Science and Information Systems (FedCSIS 2015), pp. 1189–1193 (2015)
5. Niewiadomska-Szynkiewicz, E., Sikora, A., Arabas, P., Kamola, M., Mincer, M., Kołodziej, J.: Dynamic power management in energy-aware computer networks and data intensive computing systems. Future Gener. Comput. Syst. **37**, 284–296 (2014)
6. Herbrich, R., Minka, T., Graepel, T.: True skill^{TM}: a Bayesian skill rating system. Adv. Neural Inf. Proc. Syst. (NIPS 06) **19**, 569–576 (2007). MIT

7. Lanzi, P.L., Loiacono, D., Stucchi, R.: Evolving maps for match balancing in first person shooters. In: Proceedings of IEEE Conference on Computational Intelligence and Games (CIG), pp. 1–8. IEEE (2014)
8. Tarlow, D., Graepel, Th., Minka, T.: Knowing what we don't know in NCAA football ratings: understanding and using structured uncertainty. In: Proceedings of the 2014 MIT Sloan Sports Analytics Conference (SSAC 2014), pp. 1–8 (2014)
9. Agarwal, Sh., Lorch, J.R.: Matchmaking for online games and other latency-sensitive P2P systems. In Proceedings of ACM SIGCOMM 2009 Conference on Data Communication, pp. 315–326. ACM, New York (2009)
10. Delalleau, O., Comtal, E., Thibodeau-Laufer, E., Ferrari, R., Bengio, Y., Zhang, F.: Beyond skill rating: advanced matchmaking in ghost recon online. IEEE Trans. Comput. Intell. AI Games **4**(3), 167–177 (2012). IEEE
11. LOL matchmaking explained. http://forums.na.leagueoflegends.com/board/showthread.php?t=12029
12. Wooldridge, M.: An Introduction to MultiAgent Systems. Wiley (2002)

Evaluation of High Efficiency Operating Rules for Grate-Fired Boilers

Tomasz Janiuk and Paweł D. Domański

Abstract Synthesis of control rules (extraction of knowledge) from historical data may be used for optimal control of the pulverized coal boiler, however it appears to be quite a difficult task. The best effects, allowing fast design and reliable tuning, might be obtained through comprehensive approach with use of all available sources of information. Presented methodology covers one aspect: knowledge extraction through analysis of process operational historical data. Proposed methodology is specifically developed for the task: synthesis of the control rules and conditions for improvement of the combustion process efficiency based on the historical. The approach is stepwise. It starts from the historical data acquisition, followed by the efficiency calculations, data processing. The methodology ends with evaluation of the control rules that may suggest boiler operators optimal operation. This approach copes with several practical problems met during daily boiler operation. Proposed algorithm uses original combination of heuristics, polynomial approximation and data clustering to find patterns in scattered data allowing to assign operational rules to meet the goals. The approach is validated on the real data from the industrial grate boiler.

Keywords Grate boilers · Combustion efficiency · Rule-based control · Data clustering

1 Introduction

The ultimate goal of the proposed approach is to develop methodology for synthesis of control rules that could improve combustion process efficiency using historical data that represent long-term process performance. Proposed approach is dedicated

T. Janiuk · P.D. Domański (✉)
Institute of Control and Computational Engineering, Warsaw University of Technology, Nowowiejska 15/19, 00-665 Warszawa, Poland
e-mail: p.domanski@ia.pw.edu.pl

T. Janiuk
e-mail: tomaszjaniuk@gmail.com

R. Szewczyk et al. (eds.), *Challenges in Automation, Robotics and Measurement Techniques*, Advances in Intelligent Systems and Computing 440, DOI 10.1007/978-3-319-29357-8_23

to the task of boiler combustion improvement. Following sections presents preliminary assumptions, expectations and motivation for this research.

1.1 Rationale

There are two main reasons that drive development of the proposed methodology. The first one is strictly economical. Power plant should be able to work with the highest achievable efficiency, which means spending less fuel to produce the same amount of energy. This implies costs reduction and boiler operational savings. The second reason is related with environment. Nowadays, every power plant can not exceed emission limitations for waste and pollution generation, which force plant management to introduce new solutions. Reduction in fuel used implies lower carbon dioxide (CO_2) emission and enables to embedded different target characteristics, like for instance nitrogen oxides (NOx) and sulfur oxides (SOx).

Analysis of the research on grate firing boilers shows that these subject is much less penetrated then PC boiler, gas/oil fired units or fluidized bed boilers [1, 2] and only few solutions may be found on that subject. On the other hand there are quite a lot of recent papers [3–5] describing research on the mathematical modeling of the grate firing of the biomass and waste.

One of the first discussions on the subject is [6]. The idea is to enhance standard single loop controllers with cascade control and the compensation system for the grate combustion power. In [7, 8] the authors researched the idea to do control using fuzzy models. Neural networks were also tested in [9, 10]. In [11] we can find MPC approach to the grate boiler control. One can notice that it is many years behind pulverized coal boilers which were MPC controlled already in late 1970s. This approach was further considered in [12, 13].

One have to notice that these newest application require more knowledge and measurements on the boiler. However in practice it is hardly met, thus qualitative approaches posses an interesting alternative and this way will be further investigated.

1.2 Grate Boiler Combustion

On picture below there is presented typical arrangement of the grate fired boiler (Fig. 1). Fuel is fed into the furnace pushed onto the moving grate by feeders. The fuel goes through the combustion processes transported on the moving grate. We can distinguish the following phases of this process: drying, pyrolysis, char burning and cooling. The process of combustion is substantially influenced by: fuel used, oxygen concentration, temperature, pressure, residence time and additives.

Residual ashes are removed onto the ash conveyor. Beneath the grate there are several inputs of primary air. Secondary air (and/or OFA) are fed into the combustion above the grate at the freeboard. The heat of the flue gases is transferred to the process steam in a boiler after combustion.

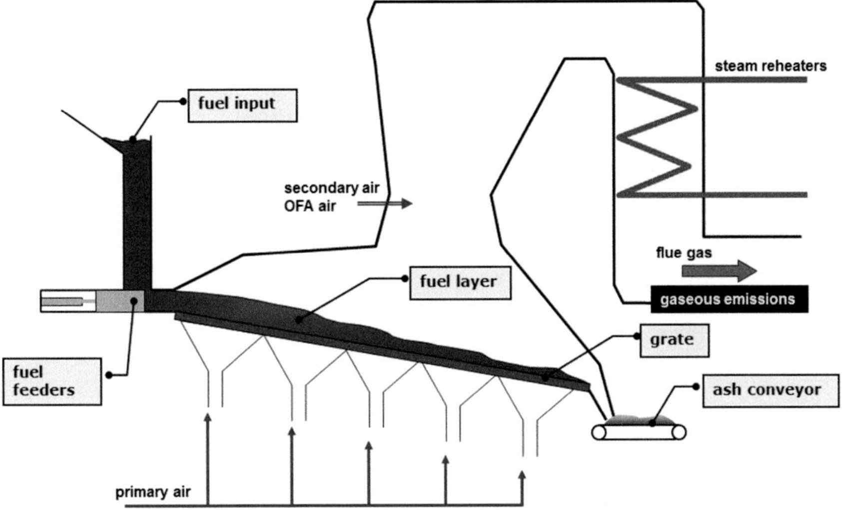

Fig. 1 Grate boiler combustion process

The flue gases also transport different pollutants, such CO_2, NOx, SOx, CO, C_xH_y, tar, HCl/Cl_2, PAH, PCDD/PCDF, heavy metals, particulate matter, and incompletely burned carbon particles.

Calculation of the boiler combustion efficiency is the entry data for proposed algorithm. Data selection for further project phases is based on it as only the operating points at high efficiency are taken into consideration. Boiler *heat efficiency* (η) is described as the ratio between heat embedded in the steam at the boiler outlet against the energy provided to the boiler with the fuel.

There are two main approaches to evaluate boiler efficiency: direct and measured heat losses method (indirect one). The direct one is not commonly used and it is not applied in our project. The indirect method was used [14].

2 Approach

Algorithm design process is divided into several phases:

1. **Examination of boiler efficiency plot** representing efficiency scattered data as a function of boiler load (steam flow) is used: $eff = f(steam\_flow)$.
2. **Extraction of maximum efficiency front** from the plot or minimum front in case of heat losses and its parametrization with n-th order polynomial (called *the best-achievable curve*).
3. Next **derive reasonable neighborhood of the best-achievable curve** and select points included in this set (*the best-achievable set*).

4. For the best-achievable data one may **find theirs associated MVs—Manipulated Variables**. In case of the boiler efficiency it could be excess air, fuel flow, mill load, secondary air flow, OFA flow, etc.
5. Next phase is to **analyze the plot of the best-achievable MVs**.
   At this point some discussion is required. In the ideal case it would be expected that these arguments would form some kind of the characteristics. If it is true, best-achievable argument curves would be easily evaluated that could be used for control. However in practice it is not so. These figures form highly scattered clouds without any clear pattern. To extract knowledge from that situation the following steps were used.
6. **Introduce fuzzy clustering for both input arguments and output** scattered data.
7. **Identification of the core fuzzy rules** using associations from original data, i.e. combining extracted fuzzy sets for inputs with fuzzy sets of outputs.

At the end it is expected to obtain algorithm, which analyses provided data and then extracts specific control rules. Obtained output, in textual and graphical form, may be applied as an advisory for the boiler operator to keep the best efficiency of the process. The reason for presenting output in two forms, textual and graphical, is that the first one will present detail data, and the second will represent the output in more legible way.

Proposed approach ends with selected control rules. They are in form of the fuzzy rules. Lotfi Zadeh proposed fuzzy logic [15] as a notion that allows to create systems directly from human intuition and experience. Fuzzy logic has become a commonly used technology and through the large number of successful applications different tools and methodologies have evolved. In recent years, fuzzy logic has proven well its broad potential in industrial automation applications.

One of the required features in the proposed methodology is to divide domain into subregions and data clustering is prosed to cover this aspect. Clustering is a process of finding in data groups that data within the same subset is as similar as possible and simultaneously it is different from data in other subsets. In fuzzy clustering similarity of the point with a cluster is represented by membership function whose values are between zero and one. For the purpose of this work the Fuzzy C-means Algorithm [16] was used. This clustering approach is very popular and was successfully applied to systems for automatic extraction of fuzzy rules from historical data [17]. Finally trapezoidal fuzzy set approximation is implemented using algorithm proposed in [18].

3 Methodology

The role of this chapter is to explain practical aspects of the proposed methodology.

Exemplary data used in this presentation originate from the grate boiler firing wood residues. Grate fired boilers are the most popular steam generators in the

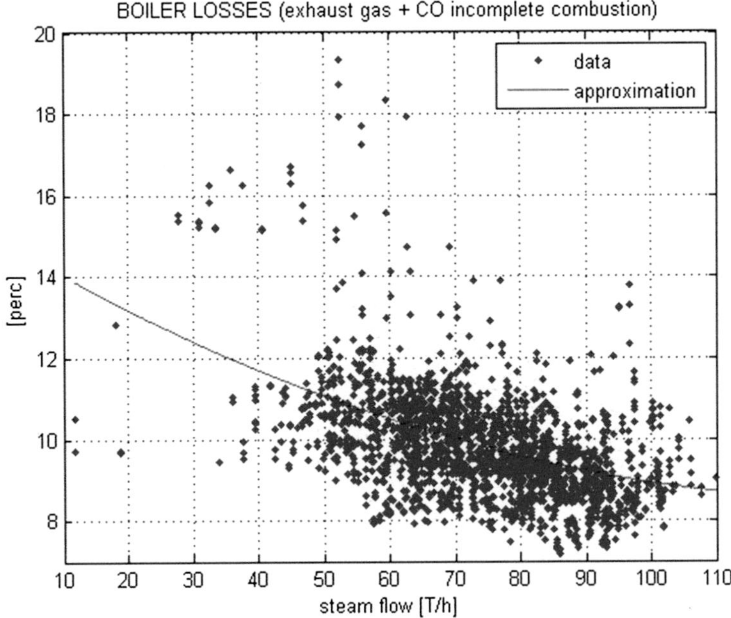

Fig. 2 Boiler losses scattered points and second order polynomial fitting curve

industrial applications. They are relatively simple enabling large flexibility especially with different fuels, including industrial and domestic waste. In our example we consider the following boiler data:

- Inputs: primary air flow, and four fuel feeders called Mill A, B, C and D,
- Output: boiler losses as the sum of stack and incomplete combustion loss.

Boiler losses are calculated using indirect method. Typical results are sketched on the Fig. 2. This form the starting point of the analysis.

The first phase is to find *the best-achievable curve* and associated points. The purpose is in need to choose data that is associated with the lowest losses, which are responsible for the highest efficiency. The first step is to divide the data into indivisible regions, which consists of points percentage given as a parameter of a function.

Next number of minima from every region is chosen to calculate their mean value. Subsequently we add two extreme points from the left and from the right, respectively to assure that the curve covers whole domain. Afterwards function approximates indicated points with a polynomial of a desired rank which is an expected result. Best heat engineering practice shows that second order polynomials are enough to capture heat efficiency curves and such orders are used in the paper.

Minimum approximation curve (*best-achievable curve*) is determined. Further, data points meeting the requirement of fitting between minimum curve and similar one shifted up by 10 % are selected (Fig. 3). Data that fall into the set are selected.

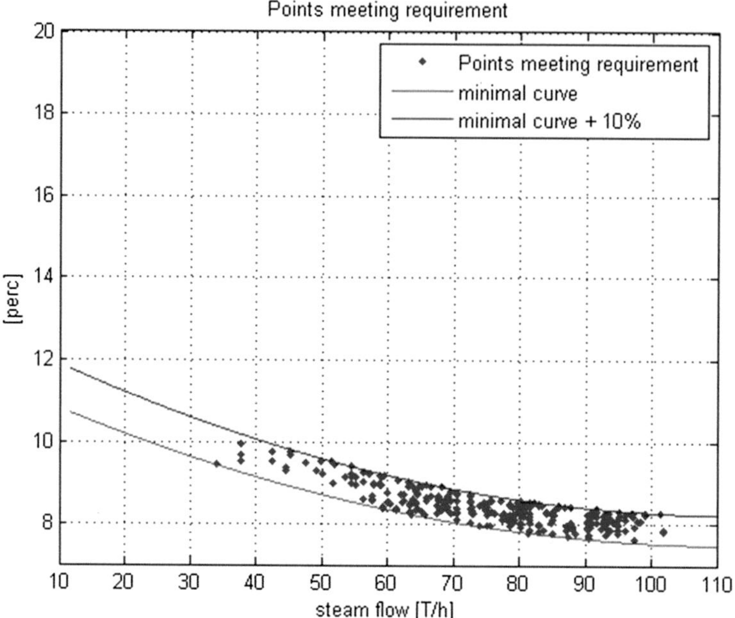

Fig. 3 Evaluation of the best achievable set with points belonging to it

Associated with them inputs are extracted. In the considered example there are five arguments—model inputs defined above.

Next, we need to look after any pattern that could be used in synthesis of the best efficiency dependencies. The following ones have been verified:

1. Dependency between primary air flow and individual mill flows.
2. Dependency between output point distance from the best-achievable curve.
3. Dependency between ratio of one mill flow against the sum of flows.
4. Two dimensional dependency between fuel and primary air against steam flow may be presented as a 3D graph.

Obtained characteristics are in-line with combustion theory and control practice. This confirms assumed hypotheses. It is very important because in case of data corruption any further steps would lead to false results. Decision about final approach was decided after the analysis of all possible scenarios and hypothesis. However, we do not get clear dependencies curves. Character of the plots is rather scattered. Thus there is a need to use help of practical expertise to select appropriate heuristics.

The first selection was about feeders signals. Two approaches were verified: direct feeder fuel flows and the ratio of single feeder against total fuel flow. The second one has an advantage as it keeps practical information where relatively more fuel is put, despite boiler load. For the further analysis the following inputs were selected: steam flow and fuel ratios for each feeder. Selected methodology is as follows:

1. Perform data fuzzy clustering for output and all inputs.
2. Identify control rules using dependencies between inputs and output and assigning rules with the most significant number of belonging points.

Data clustering process is split into two parts. Data are divided into two sets. First, algorithm is run and the rules with their number are determined. Than, it is validated with second data set, if result is reasonable and rules may be considered as reliable. Finally, the clusters are approximated by trapezoidal shape.

Once the fuzzy clusters are set, the rules may be identified through matching of the dependencies provided in analyzed data:

1. Cluster with the highest membership value is selected for every point.
2. All combinations are verified and the most frequent ones are selected.
3. Rules are checked against discrepancies and those with the highest average membership value are finally chosen.

4 Exemplary Results

Results presented below are derived using real process data from the grate boiler. Authors were considering simulation verification however, we have decided that it would not be relevant and results would be misleading (too optimistic). The process is very complex, thus the model should be simplified. Data obtained from such simulation would not represent real situation at all. Dependencies would be straightforward and the conclusions would not be realistic. Unfortunately (or fortunately) practice is more reach and unpredictable. As we had an access to industrial anonymous data, we have decided to present exemplary evaluation on those data. Confirmation for the approach can be only verified practically at site. Unfortunately, there was no such an occasion yet. Thus the validation could be done only using best practices and expertise.

First, the algorithm was run in simplified version assuming constant number of rules. The reason was simple. We wanted to verify practical quality of the obtained rules checking switching surfaces to verify correctness and completeness of obtained rules. Each input was clustered with two fuzzy sets while the output with three.

The obtained rules and resulting shape of the control surfaces was according to the expectations and in accordance with theoretical knowledge. At that point we cannot say that the algorithm is not incorrect.

Finally it was run with full flexibility and the following structure of the fuzzy clusters was obtained: steam flow: *3*, primary air flow: *5*, ratio for feeder A: *3*, ratio for feeder B: *7*, ratio for feeder C: *7* ratio for feeder D: *5*. Exemplary clusters for inputs and output are sketched on Fig. 4. One may notice strange covering of the domain by fuzzy cluster for Feeder B. It would be misleading if we would be interested to build and use the whole, complete, rule base. In our case we are interested only in the most dominant (efficient) rules for each operating point and thus such strange fuzzy clustering is not a limitation.

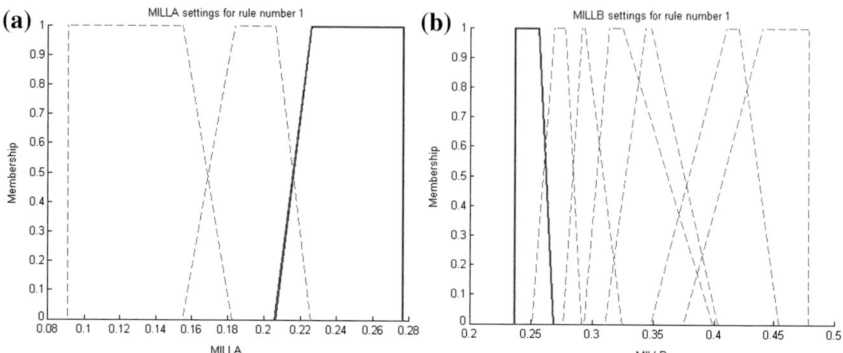

Fig. 4 Exemplary fuzzy clusters with highlighted rule. **a** Feeder A flow clustering. **b** Feeder B flow clustering

Table 1 Identified fuzzy rules

St_Flow	Prim_Air_Flow	M_A	M_B	M_C	M_D
1	5	3	1	2	4
2	5	3	1	3	5
3	2	1	6	1	5

At the end, the rules are defined. Our model has five inputs and one output. The fuzzy clustering algorithm for the output has found three dominating fuzzy sets (Table 1). It has to be noted that the rule base is not complete—these are only three rules. But they show direction in which operators should drive the process to reach the efficient operating point. Formally the other ones (worse rules) are not needed. They represent the best operation for different boiler operating points—three best rules for three steam flow fuzzy sets.

4.1 Results Discussion

Conclusions are derived from the results. First we may see the combustion dependencies existing in input and output data. For <u>low</u> steam flow (rule #1), between 34 and 70 T/h the best rule is as follows:

- Load for feeder A should be set at the highest (3).
- Load for feeder B should be set at the lowest (1).
- Load for feeder C should be kept at the middle (2).
- Load for feeder D should be set at the highest (4).
- Primary air flow is advised to be set in the range from 58 to 66.07 $\frac{\text{kNm}^3}{\text{h}}$.

Need for high level of primary air (lowering incomplete combustion loss) is caused by high amount of carbon monoxide in flue gases.

In case of mid steam flow (rule #2) setting of fuel feeders and primary air flow is almost the same. Only the load for feeders C and D was slightly increased to range *3* and *5*, respectively. It means uneven fuel delivery, which in fact results in even combustion. We notice that even combustion inside of the boiler is associated with the highest efficiency. This effect for the coal boilers is described in [19].

For high load the rules are opposite. The ideal case would be, when combustion is uniform in the whole volume of the chamber. In practical world it is never the case. Fuel feeders asymmetric operation is a way to bring back quasi homogeneous combustion within the boiler. This is visible especially for high power production, where dynamics of the combustion is supposed to be the fastest. Feeders uneven operation tames the effects of combustion turbulences and, as a result, provide homogeneous combustion.

5 Conclusions and Further Research

Proposed algorithm presents practical way to derive control rules (operators advisor). It may be used as a tool to improve efficiency of the power boiler. It was obtained combining practical heuristics, combustion experience, data processing and fuzzy logic methods. From the analysis of obtained results one may conclude that the assumptions and expectations stated in the initial phase of the project are fulfilled. Regarding the theoretical approach, results of an algorithm application on data occur to be correct and valuable.

There are no extraordinary outcomes or phenomena that could not be explained with scientific reasoning. However, it is difficult to state whether the algorithm is useful from practical point of view. Theoretical simulation will not be effective as it is extremely difficult or even impossible to get practical and reasonable simulator resembling real process with high accuracy. Due to the rules used in its development we will be always close to the curves and the cloud of points will not appear in such an extent as in reality. Thus the only way of methodology testing is application to the real process. Unfortunately it was not possible in the considered case.

If the algorithm is subject to the further development there would be several aspects that should be added or improved. The things that would be included to the algorithm are:

- Automated data collection and synchronization with I and C infrastructure.
- Graphical User Interface for visualization.
- Alternative approaches highlighted in paper might be validated.
- Automated parametrization for easier operation by inexperienced user.
- Validation of the methodology robustness testing it for different boilers, i.e. various structure, fuels etc.

References

1. Labbe, D., Roberts, D., Gordon, L.: Optimizing heat rate with model predictive control on riley turbo-furnace unit (2002)
2. Spinney, P., James, R.: The boiler optimization journey, from NOX box to diverse problem solver. Power Eng. (2011)
3. Van Kuijk, H.A.J.A., Bastiaans, R.J.M., Van Oijen, J.A., De Goey, L.P.H.: Grate furnace combustion: a submodel for the solid fuel layer. Int. J. Multiscale Comput. Eng. **6**(1), 103–111 (2008)
4. Ryu, C., Yang, Y.B., Khor, A., Yates, N.E., Sharifi, V.N., Swithenbank, J.: Effect of fuel properties on biomass combustion: part I. Experiments—fuel type, equivalence ratio and particle size. Fuel **85**(7–8), 1039–1046 (2006)
5. Yang, Y., Ryu, C., Khor, A., Yates, N., Sharifi, V., Swithenbank, J.: Effect of fuel properties on biomass combustion. Part II. Modelling approach—identification of the controlling factors. Fuel **84**(16), 2116–2130 (2005)
6. Kortela, U., Marttinen, A.: Modelling, identification and control of a grate boiler. In: American Control Conference, pp. 544–549 (1985)
7. Li, W., Chang, X., Wang, Y., Ma, C.: Design of fuzzy logic controller for combustion control of stoker-fired boilers based on behavior modeling. In: Proceedings of 6th International Fuzzy Systems Conference, vol. 1, pp. 453–458. IEEE (1997)
8. Li, W., Chang, X.: A neuro-fuzzy controller for a stoker-fired boiler, based on behavior modeling. Control Eng. Pract. **7**(4), 13 (1999)
9. Wilcox, S.: Monitoring and control of Stoker-fired boiler plant using neural networks - a jointly funded project with BCURA, Technical Report (1999)
10. Chong, A.: Neural network models of the combustion derivatives emanating from a chain grate stoker fired boiler plant. In: IEE Seminar on Advanced Sensors and Instrumentation Systems for Combustion Processes, vol. 2000, pp. 6–6. IEE (2000)
11. Smith, D.: Optimizing waste fuel boiler control with multivariable predictive control. In: Conference Record of 2006 Annual Pulp and Paper Industry Technical Conference, pp. 1–4. IEEE
12. Rismayasari, D., Joelianto, E., Chaerani, D.: The implementation of robust-optimization-based model predictive control to waste heat boiler. In: International Conference on Instrumentation, Control and Automation (2009)
13. Paces, N., Voigt, A., Jakubek, S., Schirrer, A., Kozek, M.: Combined control of combustion load and combustion position in a moving grate biomass furnace. In: 2011 19th Mediterranean Conference on Control and Automation (MED), pp. 1447–1452. IEEE (2011)
14. Kitto, J., Stultz, S.C.: Steam, Its Generation and Use, 42nd edn. Babcock & Wilcox Co. (2007)
15. Zadeh, L.A.: Fuzzy logic = computing with words. IEEE Trans. Fuzzy Syst. **4**(2), 103–111 (1996)
16. Bezdek, J.C., Ehrlich, R., Full, W.: FCM: the fuzzy c-means clustering algorithm. Comput. Geosci. **10**(2–3), 191–203 (1984)
17. Sugeno, M., Yasukawa, T.: A fuzzy-logic-based approach to qualitative modeling. IEEE Trans. Fuzzy Syst. (1993)
18. Domański, P.D.: Application of the qualitative methods for modelling and design of control systems (in polish). Ph.D. dissertation, Institute of Control, Warsaw University of Technology (1996)
19. Gao, H., Majeski, A., Runstedtler, A.: A method to target and correct sources of unburned carbon in coal-fired utility boilers. Fuel **108**, 484–489 (1996)

Detection of Artefacts from the Motion of the Eyelids Created During EEG Research Using Artificial Neural Network

Arkadiusz Kubacki, Arkadiusz Jakubowski and Łukasz Sawicki

Abstract This article shows the results of the work on the system to recognize artefacts during the EEG research. The focus is on recognizing only one but the most common artefact which is eyes blinking. Recognition was used six artificial neural networks with 1, 2, 5, 10, 100 and 1000 hidden layers. For its learn were used 16765 samples. This article is based on of Emotiv EPOC+™ system and the MATLAB environment.

Keywords Emotiv EPOC+ · EEG · Artificial neural network · Brain–computer interface

1 Introduction

Nowadays EEG research is common however, a problem associated with it not changed since the beginning [1–5]. The biggest problem is misinformation resulting from the movement of the head and the parts of face [6]. They are troublesome for physicians and interfere in the conduct automatic recognition, e.g. Epilepsy or sleep phase [7]. The artefact signal may be larger by an order of magnitude over what we lose to information from an EEG research [8]. There are many ways to detect such artefacts. Often used for the other devices as gyroscope [9], information from magnetic resonance imaging (fMRI) [10, 11] or uses sophisticated algorithms [12]. In the

A. Kubacki (✉) · A. Jakubowski · Ł. Sawicki
Institute of Mechanical Technology, Poznan University of Technology,
Ul. Piotrowo 3, 60-965 Poznań, Poland
e-mail: arkadiusz.j.kubacki@doctorate.put.poznan.pl

A. Jakubowski
e-mail: arkadiusz.z.jakubowski@doctorate.put.poznan.pl

Ł. Sawicki
e-mail: lukasz.w.sawicki@doctorate.put.poznan.pl

R. Szewczyk et al. (eds.), *Challenges in Automation, Robotics and Measurement Techniques*, Advances in Intelligent Systems and Computing 440, DOI 10.1007/978-3-319-29357-8_24

case of this study was used artificail neural networks to detect artefacts. Neural networks are currently used for face recognition [13], weather forecasting [14, 15], forecasting the possibility of cancer [16] or forecasting demand for electricity [17].

2 Placement of Electrodes and Equipment Used in Research

The most common method of placing the electrode is "10–20" system, which uses at least 10 electrodes which are equally spaced from each other [18] (Fig. 1).

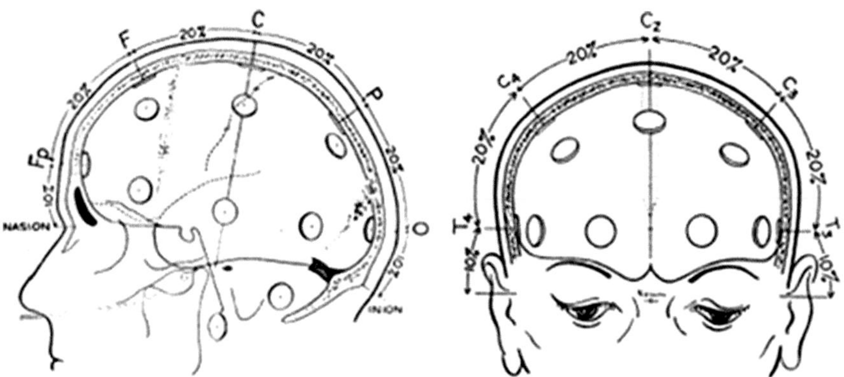

Fig. 1 Placement of electrodes in the "10–20" system [18]

Fig. 2 Placement of electrodes [20]

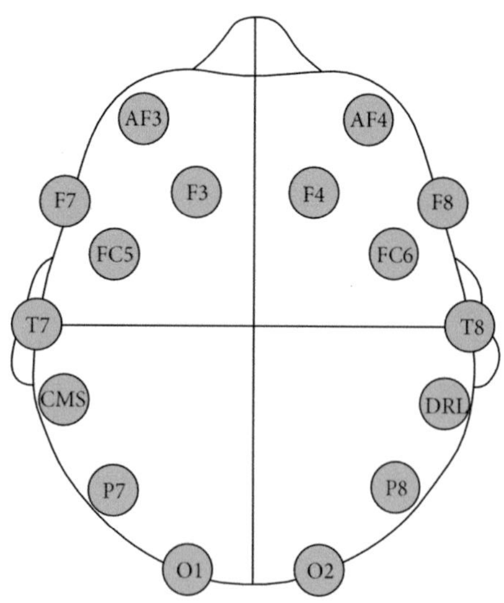

Carrying out tests does not oblige us to use all of the electrodes. We can skip some of them in areas not of interest to us or density in another area [19].

The Emotiv EPOC+™ Headset is a brain-computer interfaces based on electroencephalography. It has a built in 14 plus 2 reference electrodes, the arrangement of which agrees with the "10–20" system. Mounted electrodes are: AF3, F7, F3, FC5, T7, P7, O1, O2, P8, T8, FC6, F4, F8, AF4, whose location is shown above (Fig. 2).

3 Artefacts and Their Classification

During the EEG research are recorded not only the impulses coming from the brain but also bioelectric activity derived from the organism or even from the external environment. All signals not originating from the brain are called artefacts [7]. This phenomenon is associated EEG tests from the beginning and is very well known and are classified [21, 22]. The following is a the most popular classification of artefacts [18] (Fig. 3).

Over the years, a number of ways to detect and minimize them on the EEG graph. The most common method is the rejection of the waveform in place of the artifact. It rejected a fragment from all channels, even if the artifact occurred only

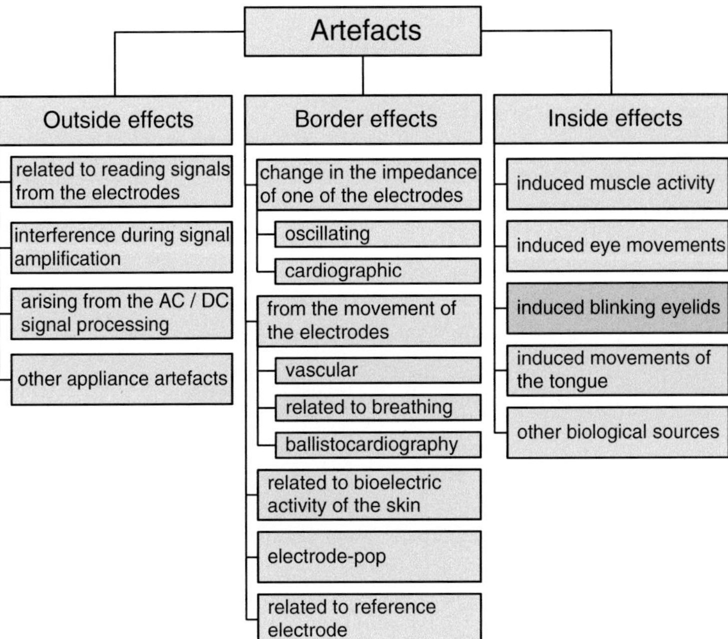

Fig. 3 Artefacts classification [18]

on one of them [23]. Artifacts can be detected in many ways. The oldest method is to observe the patient and marking its activity on the graph [7]. Another method is to use gyroscopes and collection of them information about head movement [9]. There are also other methods [10, 12]. This article will provide a new method based on artificial neural networks.

4 Detection of Artefacts by Artificial Neuron Networks

The test stand consists of a PC with MATLAB version R2013a and Emotiv EPOC +™ (Fig. 4).

In MATLAB implemented six artificial neural networks with different number of hidden layers. In all it was carried out the same training using the same input data. They were taken into consideration information from the three electrodes: AF3, AF4, FC6. Posted 16765 samples, which presented 40 charts with closed eyes and as many other face activities including closure of one eye or eyebrow lifting. The MATLAB research model shown below (Fig. 5).

The following is a confusion matrix of learning networks with one hidden layer (Fig. 6).

On the green fields shown compatibility of input values and the values provided by the neural network. On one hidden layer error is 13.2 %. The following table shows the error in percentage depending on the number of hidden layers (Table 1).

Classification quality very well represent the ROC charts. The closer the plot line the more uncertain the result of classification. Below are two charts ROC network with 1 hidden layer and network with 100 hidden layers obtained after the acquisition of all artificial neural networks (Fig. 7).

As you can see above, network with only one hidden layer classification result is uncertain. In the second case, the prediction result is correct at low coefficients. All implemented artificial neural networks have been tested with the same test data. The following is a test chart (Fig. 8).

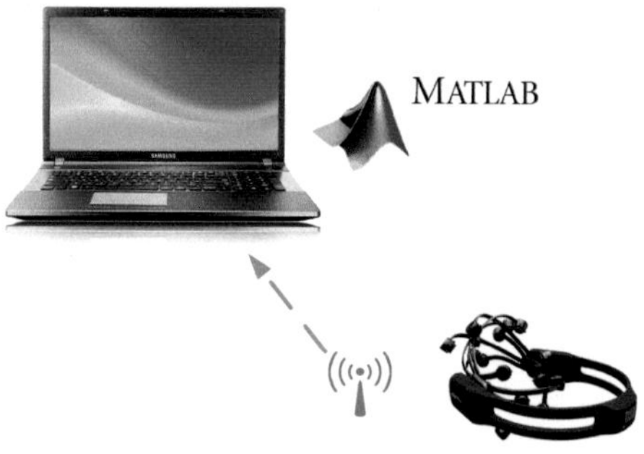

Fig. 4 Scheme of laboratory stand

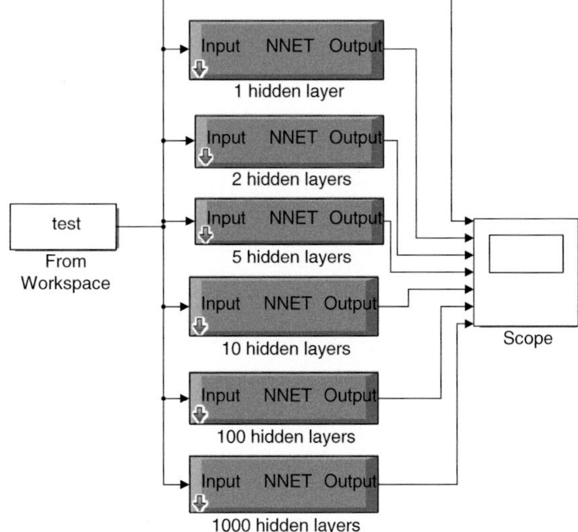

Fig. 5 Research model of artificial neural networks in MATLAB

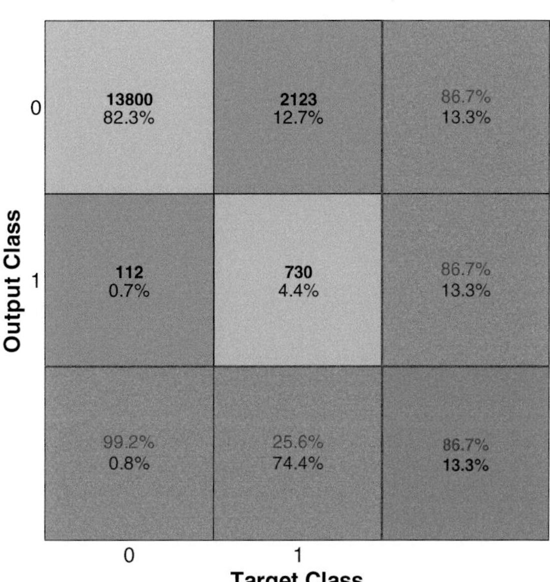

Fig. 6 Confusion matrix of learning networks with one hidden layer

The data is divided into three zones. The first part shows 10 blinking eyes. In the second 4 times has been raised eyebrows. In the last part blinks alternately. The graphs below is a response to artificial neural networks (Fig. 9).

It turns out that, despite the unfavorable effect of the learning best in this case a network with only one hidden layer. In other cases, the network also takes into account the noise. By setting the threshold for the value of 0.9 only if 5 hidden

Table 1 The prediction error depending on the number of hidden layers

The number of hidden layers	Error (%)
1	*13,3*
2	5,6
5	*5,7*
10	4,5
100	*5,7*
1000	5,6

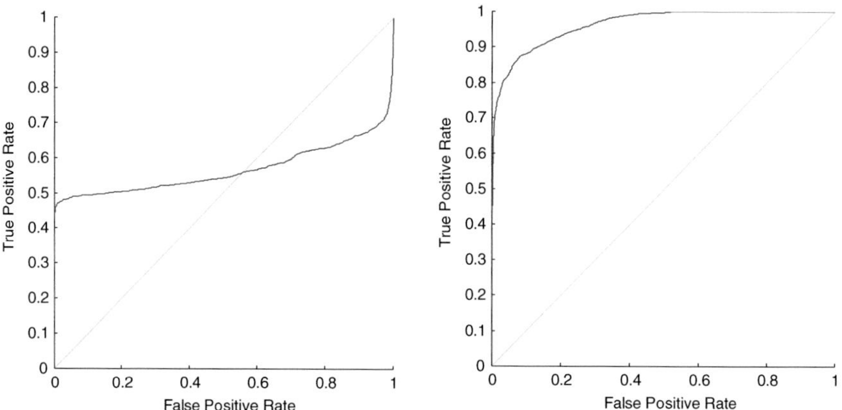

Fig. 7 ROC graphs of learning artificial neural networks: the 1 hidden layer (*left side*) and up to 100 layers of hidden (*right*)

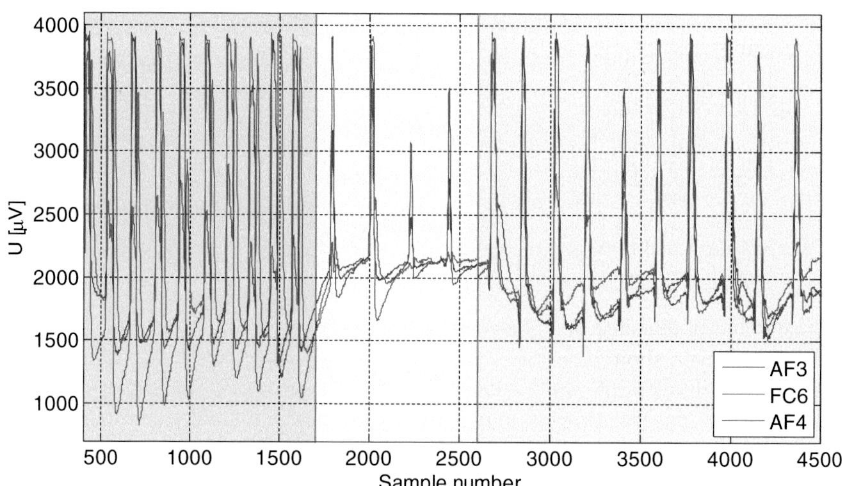

Fig. 8 The test data divided into zones: closing and opening eyes (*left*), lift the eyebrows (*middle*), the opening and closing only one eye (*right side*)

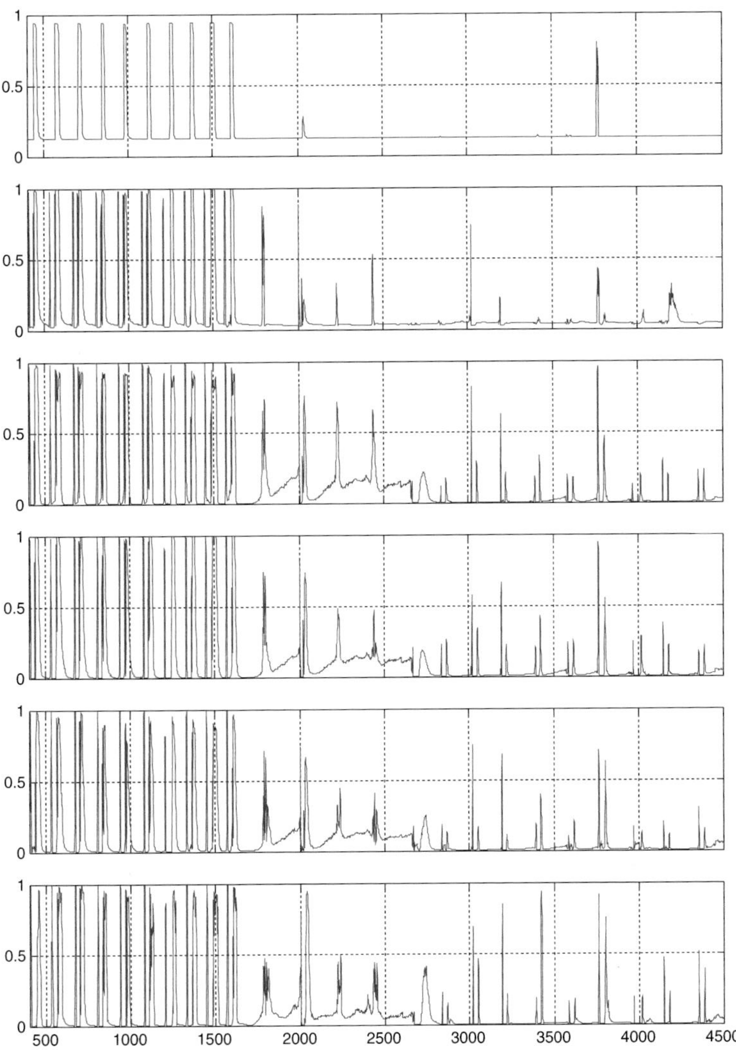

Fig. 9 Feedback neural network (top 1, 2, 5, 10, 100, 1000 hidden layers)

layers number of detected blink disagreed. In cases where a network with more than one hidden layer should be programmatically delete duplicate detection of the signal edges during blinking.

5 Conclusion

In conclusion, it is possible the use of artificial neural networks to detect movement of the eyelids. With information about the event we can filter them from EEG graph or use it to control the object. The next stage of research will be to create a model based on artificial neural networks, which recognize the entire facial expressions. To improve recognition of the events by the network with a larger number of hidden layers, enter the filtration of the input signal. We should also collect a greater amount of training data. Another aspect worth examining is the difference in the signal generated by different people. The control system based on reading brain waves must have the calibration process consisting in learning artificial neural networks.

References

1. Viola, F.C., Thorne, J., Edmonds, B., Schneider, T., Eichele, T., Debener, S.: Semi-Automatic identification of independent components representing EEG artifact. Clin. Neurophysiol. **120** (5), 868–877 (2009)
2. Gwin, J.T., Gramann, K., Makeig, S., Ferris, D.P.: Removal of movement artifact from high-density eeg recorded during walking and running. J. Neurophysiol. **103**(6), 3526–3534 (2010)
3. Makeig, S., Debener, S., Onton, J., Delorme, A.: Mining event-related brain dynamics. Trends Cogn. Sci. **8**(5), 204–210 (2004)
4. Nonclercq, A., Mathys, P.: Quantification of motion artifact rejection due to active electrodes and driven-right-leg circuit in spike detection algorithms. IEEE Trans. Biomed. Eng. **57**(11), 2746–2752 (2010)
5. Romo Vázquez, R., Vélez-Pérez, H., Ranta, R., Louis Dorr, V., Maquin, D., Maillard, L.: Blind source separation, wavelet denoising and discriminant analysis for EEG artefacts and noise cancelling. Biomed. Signal Process. Control **7**(4), 389–400 (2012)
6. Verleger, R.: Should we really use different estimates for correcting EEG artefacts produced by blinks and by Saccades? J. Psychophysiol. **14**(4), 204–206 (2000)
7. O'Regan, S., Faul, S., Marnane, W.: Automatic detection of EEG artefacts arising from head movements. In: 2010 Annual International Conference of the IEEE Engineering in Medicine and Biology Society (EMBC), pp. 6353–56 (2010)
8. Joyce, C.A., Gorodnitsky, I.F., Kutas, M.: Automatic removal of eye movement and blink artifacts from eeg data using blind component separation. Psychophysiology **41**(2), 313–325 (2004)
9. O'Regan, S., Faul, S., Marnane, W.: Automatic detection of EEG artefacts arising from head movements using EEG and gyroscope signals. Med. Eng. Phys. **35**(7), 867–874 (2013)
10. Niazy, R.K., Beckmann, C.F., Iannetti, G.D., Brady, J.M., Smith, S.M.: Removal of FMRI environment artifacts from EEG data using optimal basis sets. NeuroImage **28**(3), 720–737 (2005)
11. Allen, P.J., Josephs, O., Turner, R.: A method for removing imaging artifact from continuous EEG recorded during functional MRI. NeuroImage **12**(2), 230–239 (2000)
12. Delorme, A., Sejnowski, T., Makeig, S.: Enhanced detection of artifacts in EEG data using higher-order statistics and independent component analysis. NeuroImage **34**(4), 1443–1449 (2007)

13. Rowley, H.A., Baluja, S., Kanade, T.: Neural network-based face detection. IEEE Trans. Pattern Anal. Mach. Intell. **20**(1), 23–38 (1998)
14. Gardner, M.W., Dorling, S.R.: Artificial neural networks (the Multilayer Perceptron)—a review of applications in the atmospheric sciences. Atmos. Environ. **32**(14–15), 2627–2636 (1998)
15. Lee, K.Y., Cha, Y.T., Park, J.H.: Short-term load forecasting using an artificial neural network. IEEE Trans. Power Syst. **7**(1), 124–132 (1992)
16. Khan, J., Wei, J.S., Ringnér, M., Saal, L.H., Ladanyi, M., Westermann, F., Berthold, F., et al.: Classification and diagnostic prediction of cancers using gene expression profiling and artificial neural networks. Nat. Med. **7**(6), 673–679 (2001)
17. Park, D.C., El-Sharkawi, M.A., Marks, II R.J., Atlas, L.E., Damborg, M.J.: Electric load forecasting using an artificial neural network. IEEE Trans. Power Syst. **6**(2), 442–449 (1991)
18. Zyss, T.: Zastosowanie układu 10–20 w rozmieszczaniu elektrod do EEG. Przedsiębiorstwo Informatyki Medycznej ELMIKO (2007)
19. Klekowicz, H. Opis i identyfikacja struktur przejściowych w sygnale EEG (2008)
20. Emotiv EPOC Specifications. https://emotiv.com
21. Gevins, A.S., Yeager, C.L., Diamond, S.L., Spire, J., Zeitlin, G.M., Gevins, A.H.: Automated analysis of the electrical activity of the human brain (EEG): a progress report. Proc. IEEE **63** (10), 1382–1399 (1975)
22. Sanei, S., Chambers, J.A.: EEG Signal Processing: Sanei/EEG Signal Processing. Wiley, West Sussex (2007)
23. Vigário, R.N.: Extraction of ocular artefacts from EEG using independent component analysis. Electroencephalogr. Clin. Neurophysiol. **103**(3), 395–404 (1997)

Gradient Approach to Curve Fitting with Incomplete Data

Karol Koniuszewski and Paweł D. Domański

Abstract The paper deals with the problem of nonlinear curve fitting in situation of incomplete data. Research was motivated by the industrial identification of Hammerstein models. It was noticed that for the model robustness and quality the fitness of the static nonlinear element is much more crucial then efficiency of dynamic operation of linear part. Industrial data are incomplete, i.e. they do not cover the whole process domain. It is due to the operation around selected steady states, close loop operation, extensive manual model use, etc. In case of multi-regional approach we often get regions with no data. Proposed methodology is addressing that issue. Included results are for both simulation and real industrial case.

1 Introduction

Presented article covers an interesting practical aspect of Hammerstein model identification: process modeling in non-linear and non-stationary environment coupled with practical constraints of empirical identification frequently observed in real industrial world.

The class of Wiener-Hammerstein models forms an attractive approach for industrial modeling. The separation of the static nonlinearity block and dynamic linear model is well understood and has strong links with real process industry situation, where for instance nonlinearity associated with valve or damper (actuator) is closely coupled with the process itself, which can safely be linearized around operating point.

K. Koniuszewski (✉) · P.D. Domański
Institute of Control and Computational Engineering, Warsaw University of Technology, Nowowiejska 15/19, 00-665 Warszawa, Poland
e-mail: k.koniuszewski@stud.elka.pw.edu.pl

P.D. Domański
e-mail: p.domanski@ia.pw.edu.pl

R. Szewczyk et al. (eds.), *Challenges in Automation, Robotics and Measurement Techniques*, Advances in Intelligent Systems and Computing 440, DOI 10.1007/978-3-319-29357-8_25

Hammerstein models [1] consists of a non-linear memoryless static element followed by linear dynamics. Static nonlinear function $g(.)$ and linear dynamic one $f(.)$ can be generally of different origin [2]. In practice we mostly use static SISO and MISO polynomials as $g(.)$ and ARX empirical regression models as $f(.)$.

Identification of the Hammerstein model starts with structure identification. As it is not a main subject of that paper we may only mention that especially appropriate inputs selection [3] is rather unclear depending on signals availability and quality, process knowledge, statistical analysis, applied benchmarking indexes, etc. Afterwards we may progress towards estimation of model parameters. The model consists of two parts and the procedure often consists of two separate phases: identification of the nonlinear static part and linear dynamic one. One may find many methods, starting from classical one [2], where parameters of the non-linear part are determined by any non-linear optimization, while the linear part may be determined by least squares approach. In the literature we may find many more papers considering different identification approaches, i.e. blind approach [4], iterative [5], multi-variable methods [6] and aspects as for instance identification in closed loop [7].

In our case we are focusing on the first element—nonlinear curve identification. The main issue addressed here is its non-stationarity and incomplete data being available. Real process always vary in time due to the mechanical aging, technology/production variations or human interventions. Thus we should embed into the model adaptation mechanisms.

In literature the subject was historically of the great interest and we may find several different approaches [8–10]. We may find two main scenarios, which are analogous to the adaptive control schemes. One of them is called Gain Scheduling and means switching of the several models according to the known and varying external variables. The second one (Model Reference Adaptive Control) refers to the constant or recurrent model updates, like RLS (Recursive Least Squares). In our case we have selected approach similar to the Gain Scheduling obtaining original results using gradient maps to reconstruct unknown characteristics in empty regions with no information.

The paper is organized as follows. In the first Sect. 2 developed methodology is presented. Then Sect. 3 consists of two main sections: Sect. 3.1 presenting artificially generated simulation results for two 2D cases and Sect. 3.2 with real industrial case for the fan analysis. The paper concludes with Sect. 4 consisting of results discussion and presentation of possible open issues for future research.

2 Methodology

The subject of curve fitting is well known in literature with many methods. One of the most popular approaches is connected with polynomial fitting. Approximation and interpolation with polynomials can be done with different algorithms [11], however the least squares fitting seems to be most popular. The main limitation for

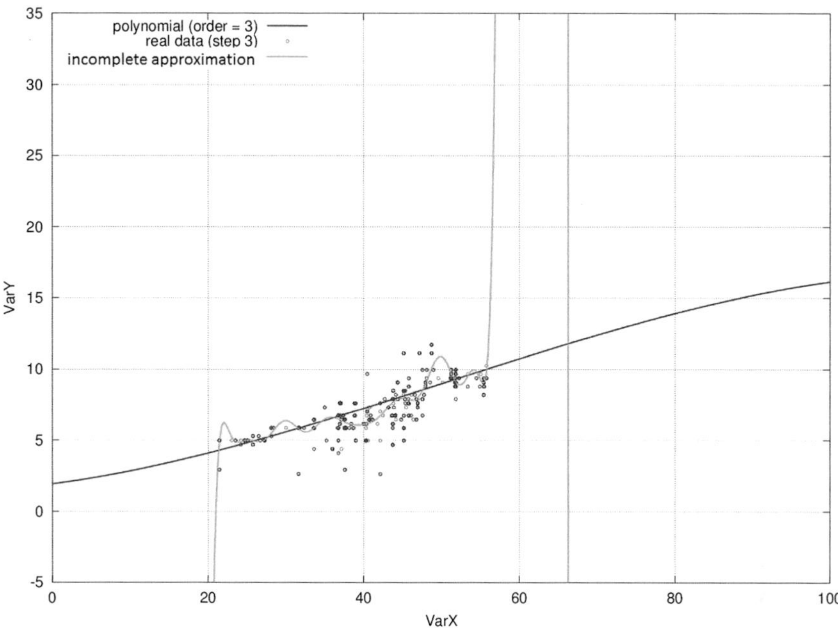

Fig. 1 Incomplete data for polynomial approximation: (*green*)—incomplete, (*magenta*) appropriate

good fitting is data. Ideally the data should cover the whole domain to enable perfect catching of the characteristics. Industrial data rarely meet this condition. Process data, gathered from normal installation operation, are mostly clustered around operating steady state points. Thus there are many wholes in data or data cover only small section of domain (see Fig. 1).

There are solutions to those features like for instance clustering of the domain into several regions (bins) and storing of data within those bins. But if there will be regions without any data we should develop something else. The proposed algorithm is as follows:

1. take historical multidimensional data,
2. perform independent classical polynomial fitting in each of the dimensions,
3. according to the obtained results generate multidimensional gradient map calculated as numerical central difference for interior data points as a matrix of discrete slopes in both directions,
4. take one point from the data and using gradient map regenerate the multidimensional surface point after point calculating it as the step response from the previous one,
5. shift the surface to fit it into the data using any selected performance index, like ISE (Integral Square Error) or IAE (Integral Absolute Error).

3 Results

Now modeling examples are presented. At first two-dimensional simulation examples are presented, while the third subsection shows results of the static nonlinearity identification for the industrial case.

3.1 Simulation Example

3.1.1 Example #1

We are considering the following two dimensional function (1).

$$Y_1 \left(u_1, u_2 \right) = u_1^2 + u_2^2 \tag{1}$$

where $1 < u_1, u_2 < 100$. Shape of the function is sketched below (Fig. 2a).

On the following figure (Fig. 2b) artificially generated points are shown. We clearly see that those points do not fully cover the whole domain in the uniform way. In such a way the classical polynomial interpolation will be biased. These data are used for analysis. At first each of the separate one-dimensional characteristics (Fig. 3) are verified, i.e. $Y_1 \left(u_1 \right)$ and $Y_1 \left(u_2 \right)$. Basing on that separate approximations are checked and selected. Thus second order polynomials are selected and used for the generation of the gradient map (Fig. 4).

Once the map is generated it may be fitted into the points. Starting from one selected point using the gradient map the shape is reconstructed and finally it's shifted to fit points according to some fitting index (see Fig. 5). Resulting approximation normalized error is equal to 0.7614. Percentage of good fitting is 99.24 %.

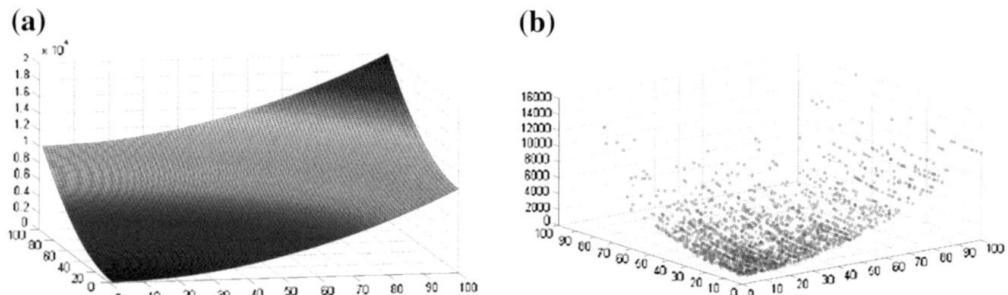

Fig. 2 Example #1: function shape and generated points. **a** Example #1: function shape. **b** Example #1: generated points

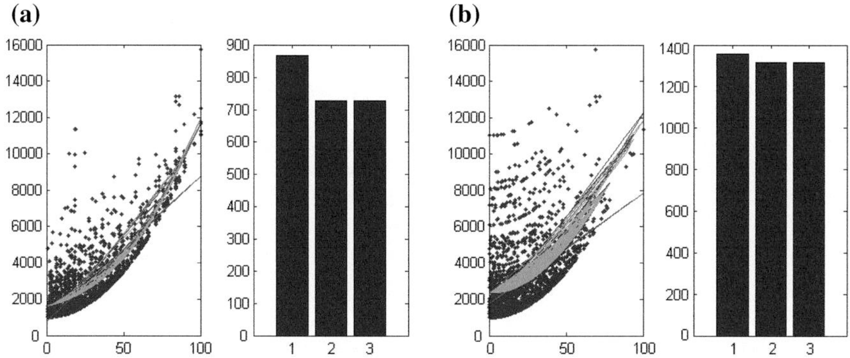

Fig. 3 Example #1: independent modeling. **a** Approximation $Y_1(u_1)$. **b** Approximation $Y_1(u_2)$

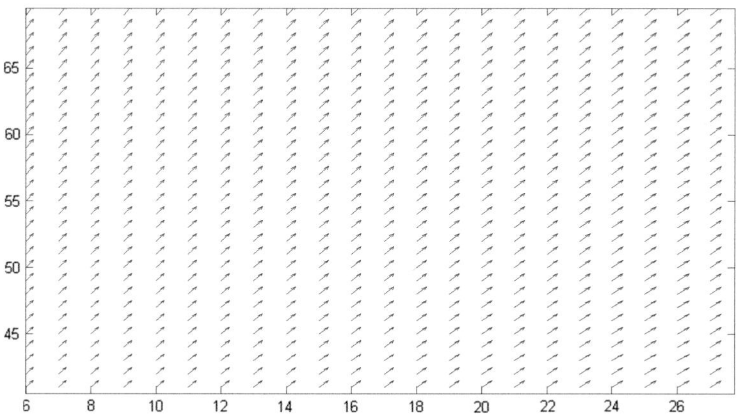

Fig. 4 Example #1: gradient map

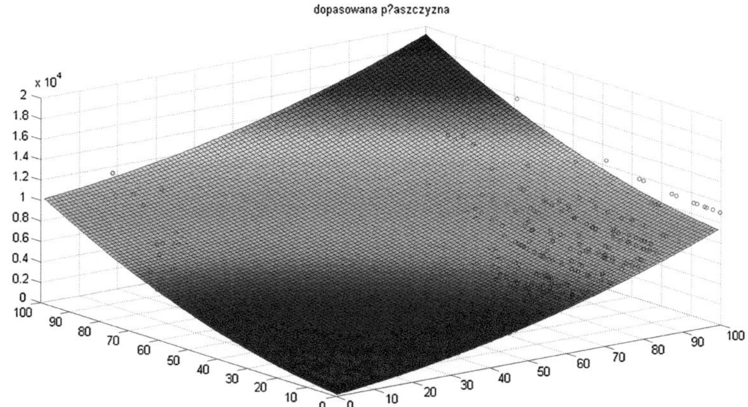

Fig. 5 Example #1: final result of curve fitting into incomplete data

3.1.2 Example #2

We are considering the following two dimensional function (2).

$$Y_2\left(u_1, u_2\right) = \frac{sin\left(u_1^2 + u_2^2\right)}{u_1^2 + u_2^2} \tag{2}$$

where $0 \leq u_1, u_2 \leq 1$. The shape of the function is sketched below (Fig. 6a).

On the following figure (Fig. 6b) artificially generated points are shown. We clearly see that those points do not fully cover the whole domain in the uniform way. In such a way the classical polynomial interpolation will be biased.

These data are used for analysis. At first each of the separate one-dimensional characteristics (Fig. 7) are verified, i.e. $Y_1\left(u_1\right)$ and $Y_1\left(u_2\right)$. Basing on that separate approximations are checked and selected.

Thus second order polynomials are selected and used for the generation of the gradient map (Fig. 8). Once the map is generated it may be fitted into the points. Starting from one selected point using the gradient map the shape is reconstructed and finally it's shifted to fit points (Fig. 9) according to some fitting index.

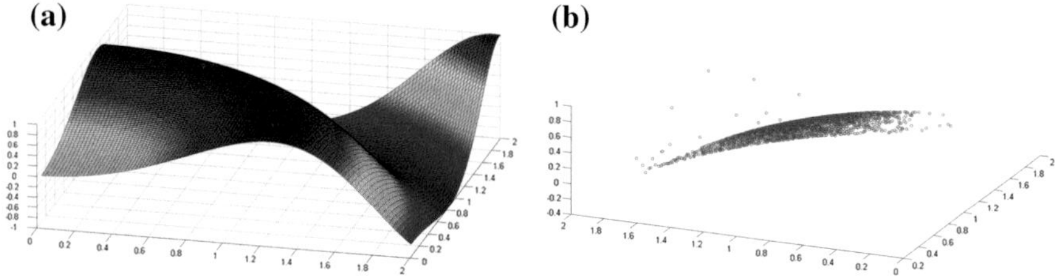

Fig. 6 Example #2: function shape and generated points. **a** Example #2: function shape. **b** Example #2: generated points

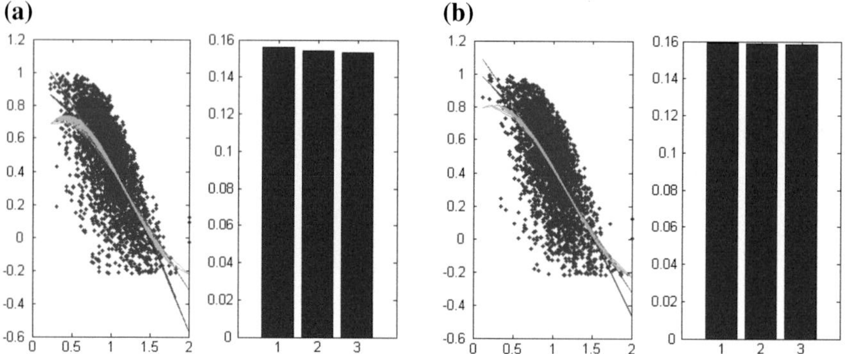

Fig. 7 Example #2: independent modeling. **a** Approximation $Y_1\left(u_1\right)$. **b** Approximation $Y_1\left(u_2\right)$

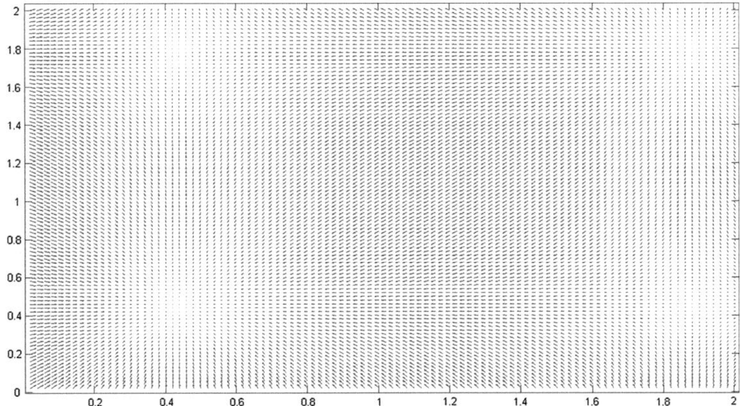

Fig. 8 Example #2: gradient map

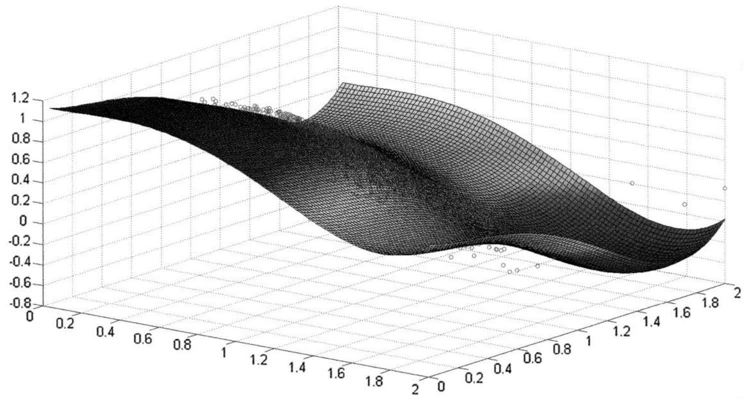

Fig. 9 Example #2: final result of curve fitting into incomplete data

Resulting approximation normalized error is equal to 15.8324. Percentage of good fitting is 84.17 %. We see that this function was more difficult to approximate as the local dimension functions were polynomials versus original sine function. Comparison of both the function and approximated models is sketched on Fig. 10.

3.2 Industrial Example

The industrial example is connected with the identification of the characteristics of the industrial fan. There are two separate data sets for that case. The data for a clean fan and a dirty one. Each data file consists of three columns: fan load (kW), fan speed

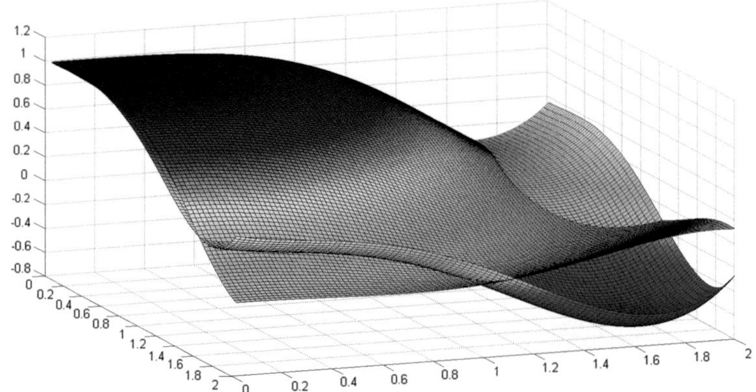

Fig. 10 Example #2: comparison of the model versus original function

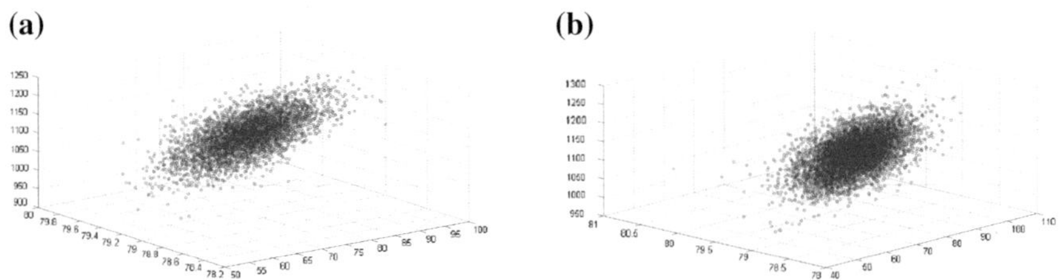

Fig. 11 Fan data points. **a** Clean fan data. **b** Dirty fan data

(rpm) and fan damper position (%). Both data sets are presented as cloud points on Fig. 11.

Both data sets are modeled according to presented algorithm separately. Modeling results are sketched on one common picture to show differences between both models (Fig. 12). According to the expectations there is visible difference between both models.

To better see the difference between the models the error curve is sketched on the following picture (Fig. 13) in for of the error map.

From the modeling standpoint one may finish the results analysis at that point. However, observing differences between both clean and dirty characteristics one may notice the following observations:

- dirty fan works in a very similar way to the dirty one in the region of low input parameters (darker region in comparison),
- Dirty fan has problems with high loads,
- energy consumption of a dirty fan increases.

Those facts are quite clear, once we know where the data originates. If not such an approach can play a support role in the process analysis and diagnostics.

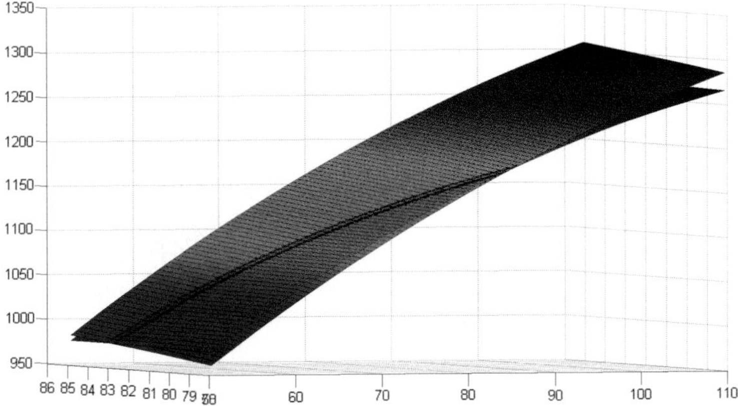

Fig. 12 Comparison of both fan models

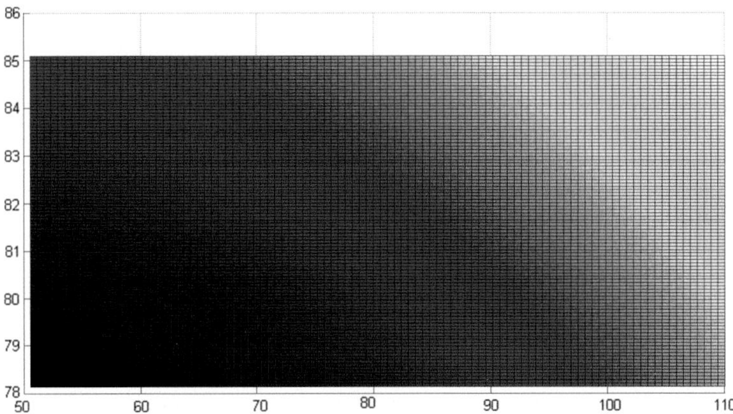

Fig. 13 Error map between clean and dirty curves

4 Conclusions and Further Research

The paper deals with the subject of the reconstruction of static nonlinear characteristics. The rationale of the paper originates from the issue of the nonlinearity identification for Hammerstein models. Nonetheless, the approach may be applied independently while any static nonlinear curve identification is needed. The algorithm is visualized with three examples.

The first example is very easy and the reconstruction quality is expectedly high. The second function has more complex shape however, fitting quality is quite good except areas without relevant data. There is required more research to improve performance in those conditions.

Finally the real example based on the real-time data from industrial fan was presented. The interesting issue of that example was that there were two characteristics covered by data: performance of clean and dirty fan. Comparison of those two models brought some interesting observations about the equipment under consideration.

As the future development one could consider three subjects:

- using of the more complex base functions in the process of the gradient map generation,
- extension of the approach to more dimensions,
- evaluation of some additional fitting mechanisms that could improve modeling in case of the insufficient data,
- and finally consider live model adaptation and tracking of non-stationary behaviour.

References

1. Hammerstein, A.: Nichtlineare Integralgleichungen nebst Anwendungen. Acta Math. **54**(1), 117–176 (1930)
2. Isermann, R., Münchhof, M.: Identification of Dynamic Systems. An Introduction with Applications. Springer, Berlin (2011)
3. Sindelar, R., Babuska, R.: Input selection for nonlinear regression models. IEEE Trans. Fuzzy Syst. **12**(5), 688–696 (2004)
4. Bai, E.-W.: A blind approach to Hammerstein model identification. IEEE Trans. Signal Process. **50**(7), 1610–1619 (2002)
5. Vörös, J.: An iterative method for Hammerstein-Wiener systems parameter identification. J. Electr. Eng. **55**(11), 328–331 (2004)
6. Wills, A., Schön, T.B., Ljung, L., Ninness, B.: Identification of Hammerstein-Wiener models. Automatica **49**(1), 70–81 (2013)
7. Han, Y., de Callafon, R.: Closed-loop identification of Hammerstein systems using iterative instrumental variables. In: Proceedings of the 18th IFAC World Congress, vol. 18, part 1, pp. 13930–13935 (2011)
8. Bittanti, S., Picci, G. (eds.): Identification, Adaptation, Learning The Science of Learning Models from Data. NATO Advanced Study Institute 'From Identification to Learning' (1996)
9. Nelles, O.: Nonlinear System Identification: From Classical Approaches to Neural Networks and Fuzzy Models. Springer, Berlin (2001)
10. Bittanti, S.: Adaptation and Learning in Control and Signal Processing. Pergamon (2002)
11. Phillips, G.M.: Interpolation and Approximation by Polynomials. Springer, New York (2003)

A Non Integer Order Model of Frequency Speed Control in AC Motor

Krzysztof Oprzędkiewicz and Tomasz Kołacz

Abstract In the paper new non integer order models for speed control in AC motor are proposed. The models have the form of hybrid transfer functions containing both integer order and non integer order parts. Parameters of models were assigned with the use of least square method. The proposed models were compared to integer order transfer function model with delay identified with the use of MATLAB. Results of experiments show that the proposed hybrid models containing fractional order part and integer order part are comparable in the sense of square cost function to reference integer order model and simultaneously they can be simpler to identify and their computational complexity is smaller than reference integer order model.

Keywords AC motor • Frequency speed control • Fractional order transfer function • Oustaloup approximation

1 An Introduction

Fractional order models are able to describe a number of physical phenomena from area of electrotechnics (heat transfer, diffusion etc.) properly and accurate. Fractional—order approach can be interpreted as generalization of well known integer-order models. Fractional order systems have been presented by many Authors [1–9]. An example of identification fractional order system can be found in

K. Oprzędkiewicz (✉)
Faculty of Electrotechnics, Automatics, Informatics and Biomedical Engineering, Department of Automatics and Biomedical Engineering, AGH University of Science and Technology, Kraków, Poland
e-mail: kop@agh.edu.pl

T. Kołacz
High Vocational School in Tarnów, Polytechnic Institute, Tarnów, Poland
e-mail: tmk@pwsztar.edu.pl

R. Szewczyk et al. (eds.), *Challenges in Automation, Robotics and Measurement Techniques*, Advances in Intelligent Systems and Computing 440, DOI 10.1007/978-3-319-29357-8_26

[4]. The proposal of generalization of the Strejc transfer function model into fractional area was given in [10].

In the paper a proposal of new, non integer order model for electric AC drive is presented. A motor speed control model in the shape of a simplified input-output transfer function is considered. The integer order model has the form of transfer function with delay. The proposed non integer order models are hybrids containing both integer order and non integer order parts.

The paper is organized as follows: at the beginning a speed control principle for induction AC motor is given, then simplified input-output transfer function models for considered plant are proposed. Furthermore, the Oustaloup approximation (the ORA approximation) is presented. Afterwards results of experiments used to order estimation and parameters identification for all proposed models are presented and discussed.

2 A Speed Control Rule for Induction AC Motor

As described in [11–13], when an induction machine with number of pole pairs p_p operates as a motor, the rotor speed, ω_M, is less than the synchronous velocity, ω_{syn}. Consequently, lines of magnetic flux intersect rotor conductors, inducing the EMFs and currents. The difference between these velocities is given by:

$$\omega_{sl} = \omega_{syn} - p_p \omega_M \tag{1}$$

Dividing the slip velocity ω_{sl} by ω_{syn}, we get the so-called slip s of the motor, defined as follows:

$$s = \frac{\omega_{sl}}{\omega_{syn}} = 1 - p_p \frac{\omega_M}{\omega_{syn}} \tag{2}$$

After transforming the above relation, we obtain a formula for speed of the induction motor as a function of the supply frequency, number of pole pairs of the magnetic field and slip of the motor. It has the following form:

$$n_M = 60 \frac{f}{p_p}(1 - s) \tag{3}$$

The simplest approach to open loop speed control of inductance motor is method known as Constant Volts/Hertz. In this strategy of control the stator voltage should be adjusted in proportion to the supply frequency to maintain the flux (and the same torque) at a constant level, what is written as:

$$\Lambda_S \approx \frac{V_S}{\omega} = 2\pi \frac{V_S}{f} \tag{4}$$

All the above relations describe the steady state dependence between frequency and rotor speed in considered motor. The dynamics of speed control process is more complicated and it is required to be described with the use of suitable models. These models are also known, but in the paper new, simplified input–output, non integer order transfer function models will be considered.

3 Transfer Function Models of Speed Control

The problem of model estimation for given experimental results has been considered for many years. Fundamental results for integer order models are given in [14–16]. Non integer order model parameter estimation typically consists in numerical minimization of selected cost function with respect to model parameters.

3.1 Cost Functions Used in Model Estimation

The performance of all considered models was estimated with the use of typical cost functions. Denote the error of approximation in k-step by $e^{+}(k)$:

$$e^{+}(k) = y^{+}(k) - y_{m}^{+}(k), \quad k = 1, \ldots, K_s \tag{5}$$

In (5) K_s is a number of samples, $y^{+}(k)$ and $y_{m}^{+}(k)$ denote the values of step responses of plant and model in k time moment respectively. Then the first cost function is Mean Squared Error (MSE):

$$MSE = \frac{1}{K_s} \sum_{k=1}^{K_s} \left(e^{+}(k)\right)^2 \tag{6}$$

The next applied cost function is Final Prediction Error (FPE) (see for example [14, 15]):

$$FPE = \frac{K_s + M_{FO}}{K_s - M_{FO}} \frac{1}{K_s} \sum_{k=1}^{K_s} \left(e^{+}(k)\right)^2 \tag{7}$$

In (7) K_s denotes the number of samples, M_{FO} denotes the number of free parameters of the model, applied to minimize a suitable cost function during its identification.

3.2 An Integer Order Transfer Function Model

The simplified, input–output, integer order model of rotor speed as a function of frequency f_z can be expressed as the following *M*th order transfer function with delay:

$$G_{IO}(s) = ke^{-\tau s}\frac{\sum_{m=0}^{M} b_m s^m}{\sum_{m=0}^{M} a_m s^m} \tag{8}$$

In (8) k is a steady-state gain of the system, τ is a dead time of model, $a_0 \ldots a_M$ and $b_0 \ldots b_M$ are coefficients of model, *M* denotes the order of model. All parameters of transfer function (8) can be assigned with the use of MALTAB via minimization of cost functions (6) and (7). The model order *M* can be estimated with the use of approach presented in [14], p. 572. It consists in calculating selected cost function (6) or (7) for increasing values of order *M*. If the significant improvement of cost function during crossing from *M* to *M + 1* is observed, then the estimated order of model is equal *M* + 1.

3.3 The Proposed Non Integer Order Models

The proposed non-integer order, hybrid models have the form close to hybrid models discussed in [4, 5, 16]:

$$G_{NO1}(s) = \frac{k}{(T_\alpha s^\alpha + 1)(T_{n1}s + 1)^{n_1}} \tag{9}$$

$$G_{NO2}(s) = \frac{k}{(T_\alpha s^\alpha + 1)(T_\beta s^\beta + 1)(T_{n2}s + 1)^{n_2}} \tag{10}$$

In (9), (10) $0 < \alpha,\beta < 2$ denote the non integer orders of the model, T_α, T_β and T_{n1}, T_{n2} denote time constants for non integer order and integer order parts respectively, n_1 and n_2 denote the orders of integer order part.

4 The Oustaloup Approximation

The approximation proposed by Oustaloup (so called ORA approximation) was presented in [17], see also for example [3]. It allows us to approximate an elementary non integer order transfer function s^α with the use of finite, integer order approximation expressed as underneath:

$$s^{\alpha} \cong k_f \prod_{n=1}^{N} \frac{1 + \frac{s}{\mu_n}}{1 + \frac{s}{\nu_n}} \tag{11}$$

In (11) N denotes the approximation order, μ_n and ν_n are coefficients of approximation calculated as follows:

$$\begin{aligned} \mu_1 &= \omega_l \sqrt{\eta} \\ \nu_n &= \mu_n \gamma, \quad n = 1, \ldots, N \\ \mu_{n+1} &= \nu_n \eta, \quad n = 1, \ldots, N-1 \end{aligned} \tag{12}$$

Where:

$$\begin{aligned} \gamma &= \left(\frac{\omega_h}{\omega_l}\right)^{\frac{\alpha}{N}} \\ \eta &= \left(\frac{\omega_h}{\omega_l}\right)^{\frac{1-\alpha}{N}} \end{aligned} \tag{13}$$

Parameters calculated with the use of approximation (11)–(13) are valid in the frequency range [ω_l; ω_h]. A steady-state gain k_f is calculated to assure the convergence the step response of approximation to step response of the real plant in a steady state.

5 Experimental Results

The experimental system is shown in Fig. 1. It contains frequency inverter Schneider Electric\Telemecanique Altivar 28, asynchronous 3-phase motor 3SG90L-4, 1.5 kW, 230/400 V, 1440/min and measuring card Advantech PCL818HD. The speed is measured with the use of tachogenerator connected via industrial card to virtual measuring panel DasyLab. At this platform, the signal is conditioned and can be written in the form, which is easy to further use.

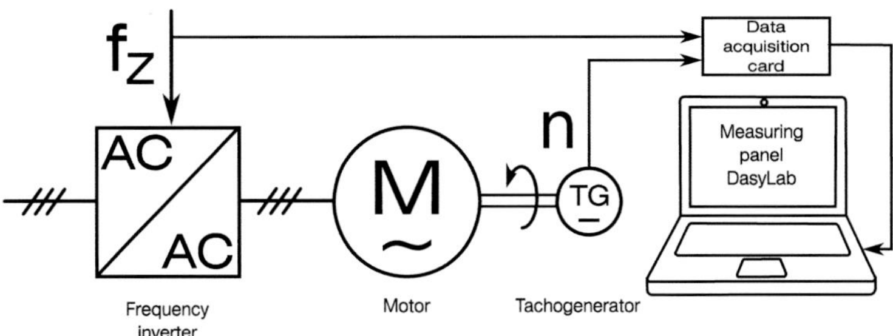

Fig. 1 The experimental system

The input signal was an input frequency f_z, the output was the angular speed of the rotor, measured with the use of tachogenerator.

The exemplary, normalized step response showing the dependence between step change of input frequency and rotor speed is shown in Fig. 2. This step response was tested for step change of input frequency f_z, from 20 to 50 Hz. The sample time during measurements was equal 0.2 ms. Number of collected samples was equal: $K_s = 5392$.

Estimation of model parameters started with estimation of the integer order model is expressed by (8). The first parameter was the order of estimation M. To calculate it, the cost functions (6) and (7) were applied. Calculations were done with the use of MATLAB function ***tfest***, the order M was estimated with the use of approach presented in Sect. 3.1. Results are given in the Table 1 and shown in Fig. 3.

The turning point in the run both cost functions describes model no11, containing 5 poles and 3 zeros. Further increasing complexity of model does not significantly improve the MSE and FPE functions. This allows us to select the order of model no 11 as proper order of integer order model (8). The parameters of this selected model are shown in Table 2.

The order estimation for both proposed hybrid models (9) and (10) is a little bit more complicated, because two orders (ORA approximation order N and integer orders n_1 or n_2) need to be tested. It can be done with the use of the above approach, but the plot of cost function (6) should show the dependence on both orders.

Firstly, the model (9) was investigated. Results of tests are shown in Table 3 and Fig. 4. Calculations were done for frequency range: $[\omega_l; \omega_h] = [0.1; 10]$.

From Table 3 and Fig. 4 we can conclude at once, that the optimal orders of the proposed hybrid model (9) are equal: $N = 2$ and $n_1 = 3$, what gives the summarized

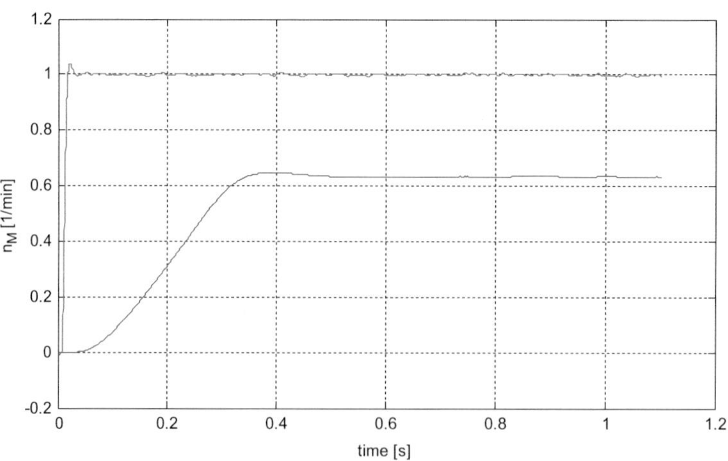

Fig. 2 The rotor speed as function of time during step change of input frequency f_z from 20 to 50 Hz

order of proposed model equal 5. The number of free coefficients during model optimization is equal 3. Optimal parameters of model (9) are given in Table 4.

Subsequently, the model (10) was tested. Calculations were done also for frequency range: $[\omega_l; \omega_h] = [0.1; 10]$. Results of tests are shown in Table 5 and Fig. 5.

Table 1 Estimation of model order M for integer order model (8)

No.	Parameters of estimated transfer functions		MSE (6)	FPE (7)
	Number of poles	Number of zeros		
1	3	0	4.591e-05	4.573e-05
2		1	5.066e-05	5.247e-05
3		2	4.387e-05	4.458e-05
4		3	4.026e-05	4.061e-05
5	4	0	5.201e-05	5.255e-05
6		1	3.693e-05	3.709e-05
7		2	1.879e-05	1.931e-05
8		3	1.127e-05	1.141e-05
9		4	9.661e-06	9.768e-06
10	5	2	3.625e-06	3.64e-06
11		3	1.561e-06	1.569e-06
12		4	1.239e-06	1.245e-06
13		5	1.136e-06	1.143e-06
14	6	2	7.325e-07	7.366e-07
15		3	7.195e-07	7.239e-07
16		4	7.264e-07	7.309e-07
17		5	6.649e-07	6.694e-07
18		6	9.027e-07	9.094e-07

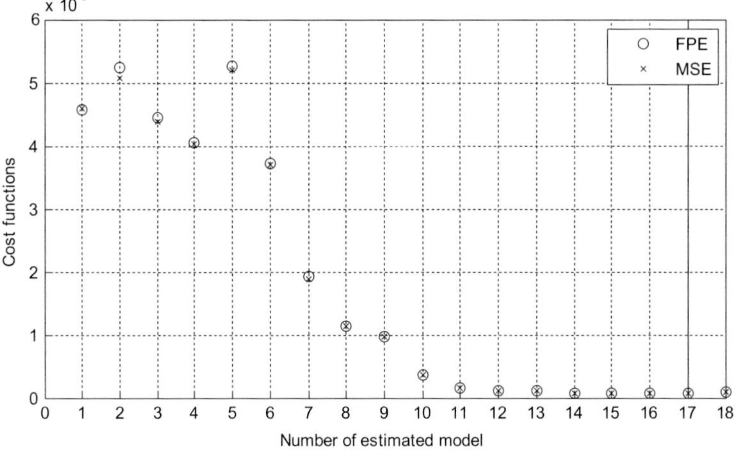

Fig. 3 Cost functions (6), (7) as functions of model order for integer order model (8)

Table 2 Parameters of integer order model (8) no 11

Number of poles	5
Number of zeros	3
Number of free coefficients M_{FO}	9
Numerator coefficients	$b_3 = -30.91$, $b_2 = 2657$, $b_1 = 2.232 * 10^4$, $b_0 = -1.552 * 10^6$
Denominator coefficients	$a_5 = 1.00$, $a_4 = 44.57$, $a_3 = 1692$, $a_2 = 3.367 * 10^4$, $a_1 = 4.233 * 10^5$, $a_0 = 2.444 * 10^6$
Dead time τ [s]	0.000612
MSE	$0.1561 * 10^{-5}$
FPE	$0.1569 * 10^{-5}$

Table 3 Values of cost function MSE (6) for hybrid model (9) and different orders N of ORA approximation and integer orders n_I

	$n_1 = 1$	$n_1 = 2$	$n_1 = 3$	$n_1 = 4$
$N = 2$	$0.8094 * 10^{-4}$	$0.1465 * 10^{-4}$	$\mathbf{0.1168 * 10^{-4}}$	$0.2164 * 10^{-4}$
$N = 4$	$0.8266 * 10^{-4}$	$0.1411 * 10^{-4}$	$0.1204 * 10^{-4}$	$0.2237 * 10^{-4}$
$N = 6$	$0.8183 * 10^{-4}$	$0.1388 * 10^{-4}$	$0.1217 * 10^{-4}$	$0.2259 * 10^{-4}$
$N = 8$	$0.8149 * 10^{-4}$	$0.1379 * 10^{-4}$	$0.1222 * 10^{-4}$	$0.2267 * 10^{-4}$

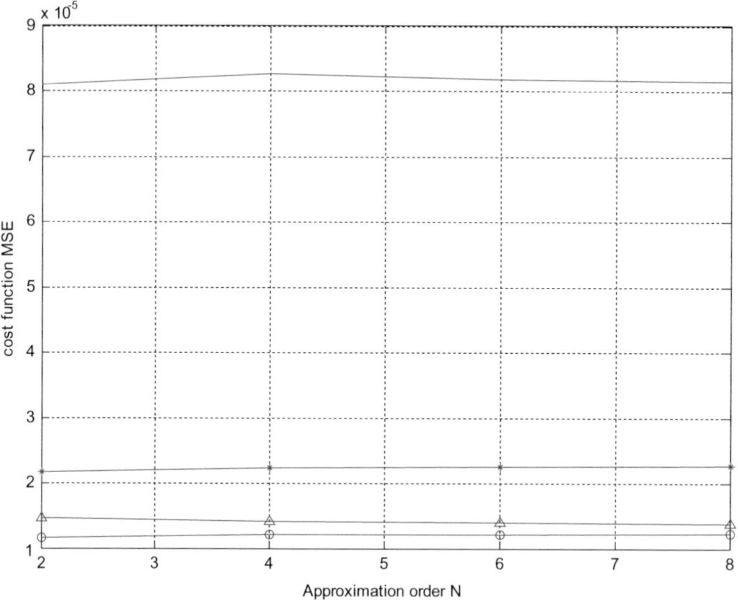

Fig. 4 Cost function MSE (6) as a function of model order for hybrid model (9) and orders n_1: $n_1 = 1$ ('.'), $n_1 = 2$ ('^'), $n_1 = 3$ ('o'), $n_1 = 4$ ('*')

From Table 5 and Fig. 5 we can conclude that the optimal orders of the model (10) are equal: $N = 10$ and $n_2 = 1$, what gives the summarized order of proposed model equal 21, because we need to use two different ORA approximants for both non integer order factors in transfer function (10). The number of free coefficients during model optimization is equal 5. Optimal parameters of model (10) are given in Table 6.

The comparison of both proposed hybrid models (9) and (10) to integer order model (8) in sense of MSE function (6), number of free coefficients M_{FO} and summarized model order is given in the Table 7. In Table 7 *Np* denotes the order of Pade approximation necessary to modeling the delay at MATLAB platform. This order is not directly given, but it always must be considered during use of transfer function model with delay. Typically, it should be set between 5 and 10.

From analysis of Table 7 we can conclude at the beginning, that the integer order model no 11 is the most accurate from all models presented in Table 1. However, it requires to identify 9 parameters and its real order is relatively high (between 10 and 15 with Pade approximant).

Next, the hybrid model (9) with number of free coefficients equal 3 and summarized order equal 5 is more accurate in the sense of MSE, than integer order models no 2 and 5 with 5 free coefficients and summarized order greater than 5.

Furthermore, the hybrid model (10) is comparable to integer order model no 11 in the sense of cost function and it contains only 5 free coefficients to identify, but its computational complexity is the highest from all discussed models, because it requires to use two different ORA approximants with relatively high order.

Finally, it can be concluded that the most sensible is to use the hybrid model (9) containing one non integer order part and integer order part. This model is able to assure the good accuracy (in the sense of MSE cost function), comparable to integer

Table 4 Optimal parameters of model (9)

N	n_1	α	T_α [s]	T_{n1} [s]	MSE	FPE	M_{FO}
2	3	1.729	0.0068	0.0656	$1.1679 * 10^{-5}$	$1.1692 * 10^{-5}$	3

Table 5 Values of cost function MSE (6) for hybrid model (10) and different orders *N* of ORA approximation and integer orders n_2

	$n_2 = 1$	$n_2 = 2$	$n_2 = 3$	$n_2 = 4$
$N = 2$	$0.7799 * 10^{-4}$	$0.1620 * 10^{-4}$	$0.1092 * 10^{-4}$	$0.1729 * 10^{-4}$
$N = 4$	$0.0807 * 10^{-4}$	$0.1631 * 10^{-4}$	$0.1138 * 10^{-4}$	$0.1711 * 10^{-4}$
$N = 6$	$0.0877 * 10^{-4}$	$0.1146 * 10^{-4}$	$0.1144 * 10^{-4}$	$0.1770 * 10^{-4}$
$N = 8$	$0.0778 * 10^{-4}$	$0.1427 * 10^{-4}$	$0.1146 * 10^{-4}$	$0.1770 * 10^{-4}$
$N = 10$	$\mathbf{0.0771 * 10^{-4}}$	$0.1590 * 10^{-4}$	$0.1149 * 10^{-4}$	$0.1803 * 10^{-4}$
$N = 12$	$0.0772 * 10^{-4}$	$0.1587 * 10^{-4}$	$0.1150 * 10^{-4}$	$0.1804 * 10^{-4}$
$N = 14$	$0.0779 * 10^{-4}$	$0.1585 * 10^{-4}$	$0.1150 * 10^{-4}$	$0.1585 * 10^{-4}$

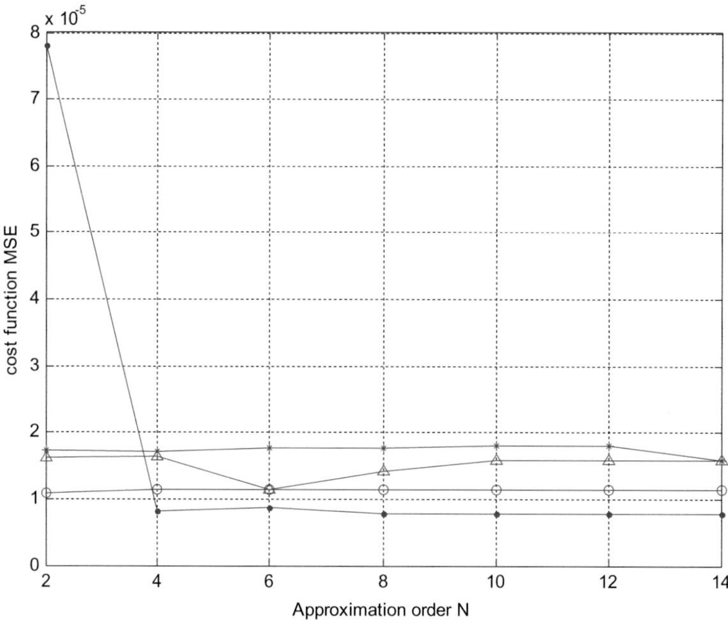

Fig. 5 Cost function MSE (6) as a function of model order for hybrid model (10) and orders n_2: $n_2 = 1$ ('.'), $n_2 = 2$ ('^'), $n_2 = 3$ ('o'), $n_2 = 4$ ('*')

Table 6 Optimal parameters of model (10)

N	n_2	α	β	T_α [s]	T_β [s]	T_{n2} [s]	MSE	FPE	M_{FO}
10	1	1.792	1.4266	0.0054	0.0266	0.1719	$7.7123 * 10^{-6}$	$7.7266 * 10^{-6}$	5

Table 7 Comparison of all discussed models

Model	MSE	Summarized order	Number of free coefficients M_{FO}
Integer order (8)—model no 2	$5.066 * 10^{-5}$	$3 + N_p$	5 (3 poles, 1 zero, dead time)
Integer order (8)—model no 5	$5.201 * 10^{-5}$	$4 + N_p$	5 (4 poles, dead time)
Integer order (8)—model no 11	$0.1561 * 10^{-5}$	$5 + N_p$	9 (5 poles, 3 zeros, dead time)
Fractional order (9)	$1.1679 * 10^{-5}$	5 ($N = 2$, $n_1 = 3$)	3 (α, T_α, T_{n1})
Fractional order (10)	$0.7712 * 10^{-5}$	21 ($2 * N = 20$, $n_2 = 1$)	5 (α, β, T_α, T_β, T_{n2})

order model with comparable order and number of free coefficients. However, this integer order model is more complicated to identify and more complex during calculations.

6 Final Conclusions

The final conclusions for the paper can be formulated as follows:

- In the paper the hybrid models for speed control in AC electric motor were proposed and discussed. The proposed models were compared to integer order transfer function model with delay, identified with the use of tools available at MATLAB platform.
- From simulations it can be concluded, that the most sensible is to apply the simple hybrid model with one non integer order part and integer order part, because this model assures good accuracy and it is simpler to identify and implementation. The second model with two non integer order parts is more accurate and also easy to identify, but its computational complexity is high.
- The proposed models are planned to be applied to construct the model-based control algorithms for the considered AC drive (for example ramp control algorithms).

Acknowledgments This paper was supported by the AGH (Poland)—project no 11.11.120.815.

References

1. Caponetto, R., Dongola, G., Fortuna, L., Petras, I.: Fractional Order Systems. Modeling and Control Applications. World Scientific Series on Nonlinear Science, Series A, vol. 72. World Scientific Publishing (2010)
2. Kaczorek, T.: Selected Problems in Fractional Systems Theory. Springer (2011)
3. Merikh-Bayat, F.: Rules for selecting the parameters of Oustaloup recursive approximation for the simulation of linear feedback systems containing $PI^{\lambda}D^{\mu}$ controller. Commun. Nonlinear Sci. Numer. Simulat. **17**, 1852–1861 (2012)
4. Mitkowski, W., Obrączka, A.: Simple identification of fractional differential equation. Solid State Phenom. **180**, 331–338 (2012)
5. Mitkowski, W. Oprzędkiewicz, K.: Application of fractional order transfer functions to modeling of high- order systems. In: 7 IFIP Conference, Klagenfurt, 9–12 Sept 2013
6. Mitkowski, W., Skruch, P.: Fractional-order models of the supercapacitors in the form of RC ladder networks. Bull. Pol. Acad. Sci. Tech. Sci. **61**(3), 581–587 (2013)
7. Mitkowski, W.: Finite-dimensional approximations of distributed RC networks. Bull. Pol. Acad. Sci. Tech. Sci. **62**(2), 263–269 (2014)
8. Vinagre, B.M., Podlubny, I., Hernandez, A., Feliu, V.: Some Approximations of fractional order operators used in control theory and applications. Fractional Calc. Appl. Anal. **3**(3), 231–248 (2000)

9. Weilbeer, M.: Efficient Numerical Methods for Fractional Differential Equations and their Analytical. Technischen Universität Braunschweig, Doktors Dissertation, pp. 1–224 (2005)
10. Oprzędkiewicz, K.: A Strejc model-based, semi- fractional (SSF) transfer function model, Automatyka/Automatics; AGH UST **16**(2), 145–154 (2012). http://journals.bg.agh.edu.pl/AUTOMAT/2012.16.2/automat.2012.16.2.145.pdf
11. Krishnan, R.: Electric Motor Drives. Modeling, Analysis and Control. Prentice Hall, Upper Saddle River (2001)
12. Leonhard, W.: Control of Electrical Drives. Springer, Berlin (2001)
13. Trzynadlowski, A.: Control of Induction Motors. Academic Press, San Diego (2000)
14. Isermann, R., Muenchhof, M.: Identification of Dynamic Systems. An Introduction with Applications. Springer (2011)
15. Ljung, L., Glad, T.: Modeling of Dynamic Systems. Prentice Hall (1994)
16. Mańczak, K. Nahorski, Z.: Computer Identification of Dynamic Systems, PWN 1983 (in Polish) (1983)
17. Oustaloup, A., Levron, F., Mathieu, B., Nanot, F.M.: Frequency-band complex noninteger differentiator: characterization and synthesis. IEEE Trans. Circuits Syst. I: Fundam. Theory Appl. I, **47**(1), 25–39 (2000)

An Estimation of Accuracy of Oustaloup Approximation

Krzysztof Oprzędkiewicz, Wojciech Mitkowski and Edyta Gawin

Abstract In the paper a new accuracy estimation method for Oustaloup approximation is presented. Oustaloup approximation is a fundamental tool to describe fractional-order systems with the use of integer-order, proper transfer function. The accuracy of approximation can be estimated via comparison of impulse responses for plant and Oustaloup approximation. The impulse response of the plant was calculated with the use of an accurate analytical formula and it can be interpreted as a standard. Approach presented in the paper can be applied to effective tuning of Oustaloup approximant for given application (for example in FO PID controller). The use of proposed method does not require us to know time response of a modeled controller. The proposed methodology can be easily generalized to another known approximations. Results of simulations show that the good performance of approximation is reached for low order and narrow angular frequency range.

Keywords Fractional order transfer function · Oustaloup approximation

---

K. Oprzędkiewicz (✉) · W. Mitkowski
Faculty of Electrotechnics, Automatics, Informatics and Biomedical Engineering,
Department of Automatics and Biomedical Engineering, AGH University
of Science and Technology, Kraków, Poland
e-mail: kop@agh.edu.pl

W. Mitkowski
e-mail: wojciech.mitkowski@agh.edu.pl

E. Gawin
Polytechnic Institute, High Vocational School in Tarnów, Tarnów, Poland
e-mail: e_gawin@pwsztar.edu.pl

R. Szewczyk et al. (eds.), *Challenges in Automation, Robotics and Measurement Techniques*, Advances in Intelligent Systems and Computing 440, DOI 10.1007/978-3-319-29357-8_27

1 An Introduction

Fractional order models are able to describe a number of physical phenomena from area of electrotechnics (heat transfer, diffusion, etc.) properly and accurate. Fractional–order approach can be interpreted as generalization of known integer-order models. Fractional order systems has been presented by many Authors [1–4]. An example of identification fractional order system can be found in [5, 6]. The proposal of generalization the Strejc transfer function model into fractional area was given in [7].

A modeling of fractional–order transfer function in MATLAB/Simulink requires us to apply integer order, finite dimensional, proper approximations. An important problem is to assign parameters of approximation correctly and to estimate its accuracy. The best known approximation presented by Oustaloup (see for example [8, 9]) bases on frequency approach. This is caused by a fact, that for fractional order systems the Bode magnitude plot can be drawn exactly and its parameters can be applied to approximants calculation.

Additionally, for elementary fractional-order elements an analytical form of step and impulse responses is known (see [1, 10]). These responses can be applied as reference to estimate a correctness of built approximant.

However, models obtained with the use of Oustaloup approximation are not always fully satisfying. This is caused by the fact that their accuracy is determined by proper selecting a frequency range and order of approximation.

The goal of this paper is to discuss an application of a method proposed by authors in [11] to accuracy estimation of Oustaloup approximation. The presented approach uses analytical formulas of impulse response of elementary non integer order integrator, cost functions and numerical calculations proposed by the authors, done with the use of MATLAB. The approach shown in the paper can be also applied to effective selecting parameters of the Oustaloup approximation during modeling FO PID controller and another systems containing FO integrators. Additionally, it does not require the use a step response of modeled element.

Particularly, in the paper the following problems will be presented:

- Non-integer order integrator,
- The Oustaloup approximation,
- Cost functions describing the accuracy of approximation,
- Simulation results.

2 Non-integer Order Integrator

Let us consider an elementary non-integer order integrator described with the use of transfer function (1). This transfer function can be applied for example to model integral part of FO PID controller.

$$G_{\alpha}(s) = \frac{1}{s^{\alpha}} \tag{1}$$

In (1) $\alpha \quad \boldsymbol{R}$ is a fractional-order of the plant.

The analytical form of the impulse response $y_a(t)$ for plant described with the use of (1) is as follows (see[1, pp. 8, 9]):

$$y_a(t) = L^{-1}\left\{\frac{1}{s^{\alpha+1}}\right\} = \cdot \frac{t^{\alpha-1}}{\Gamma(\alpha)} \tag{2}$$

where $\Gamma(..)$ denotes complete Gamma function:

$$\Gamma(\alpha) = \int_{0}^{\infty} e^{-x} x^{\alpha-1} dx \tag{3}$$

Let us assume, that the impulse response described by (2) and (3) is the accurate response. This implies, that it can be applied as a standard to estimate the accuracy of approximation.

3 The Oustaloup Approximation

The method proposed by Oustaloup (see for example [8, 9]) allows us to approximate an elementary non-integer order transfer function s^{α} with the use of a finite and integer-order approximation expressed as underneath:

$$s^{\alpha} \cong k_f \prod_{n=1}^{N} \frac{1 + \frac{s}{\mu_n}}{1 + \frac{s}{\nu_n}} = G_{ORA}(s) \tag{4}$$

In (4) N denotes the order of approximation, μ_n and ν_n denote coefficients calculated as underneath:

$$\begin{aligned} \mu_1 &= \omega_l \sqrt{\eta} \\ \nu_n &= \mu_n \gamma, \quad n = 1, \ldots, N \\ \mu_{n+1} &= \nu_n \eta, \quad n = 1, \ldots, N-1 \end{aligned} \tag{5}$$

where:

$$\begin{aligned} \gamma &= \left(\frac{\omega_h}{\omega_l}\right)^{\frac{\alpha}{N}} \\ \eta &= \left(\frac{\omega_h}{\omega_l}\right)^{\frac{1-\alpha}{N}} \end{aligned} \tag{6}$$

In (6) ω_l and ω_h describe the range of angular frequency, for which parameters are calculated.

A steady-state gain k_f is calculated to assure the convergence the step response of approximation to step response of the real plant in a steady state.

Denote the impulse response of approximation (4) by $y_{ORA}(t)$. It can be written as follows:

$$y_{ORA}(t) = L^{-1}\{G_{ORA}(s)\} \tag{7}$$

The general form of the impulse response (7) is determined by poles and zeros of transfer function $G_{ORA}(s)$ described by (4). They are real and different. This implies, that the general form of (7) can be easily expressed as follows:

$$y_{ORA}(t) = k_f \sum_{i=1}^{N} c_i e^{-p_i t} \tag{8}$$

In (8) k_f denotes the steady-state gain of the approximation, c_i denote coefficients of transfer function (4) factorization. The impulse response (7) or (8) can be evaluated numerically with the use of MATLAB/Simulink.

4 Cost Functions Describing the Accuracy of Approximation

Let us assume that the impulse response $y_a(t)$ described by (2) is the accurate response. Let $y_{ORA}(t)$ denotes the impulse response of approximation, described by (7) or (8). Then the approximation error $e_a(t)$ can be defined as follows:

$$e_a(t) = y_a(t) - y_{ORA}(t) \tag{9}$$

Furthermore, let us introduce the following cost functions, describing the accuracy of approximation:

$$I_{\max}(\Delta, N) = \max_t |e_a(t))| \tag{10}$$

$$I_2(\Delta, N) = \int_0^{\infty} e_a^2(t)dt \tag{11}$$

In (10) and (11) $e_a(t)$ denotes the approximation error described by (9). Both cost functions (10) and (11) for given plant (described by α) are functions of approximation parameters: order N and angular frequency range from ω_l to ω_h. It can be expected, that increasing N for constant frequency range should increase an approximant quality, described by cost functions (10) and (11). However, results of simulations point that too high value of N can cause bad conditioning of a model and consequently, make it useless.

The fastest method to check proper setting of the approximation parameters N and range of angular frequency described by values ω_l and ω_h is to calculate both proposed cost functions (10) and (11) as functions of approximation parameters: order and frequency range. An example of such a tuning of the Oustaloup approximant with the use of simulations will be shown in the next section.

5 Simulation Results

As an example let us consider the application of Oustaloup approximation to model the elementary fractional-order integrator described by (1). The calculations were run for time range from 0.05 to 0.5 s with step 0.02 s. Values of both cost functions (10) and (11) for different ranges of angular frequency from ω_l to ω_h, approximation order N and fractional order α are given in Table 1. There were considered the following values of approximation parameters: $N = 5, 10, 15, 25$, $\alpha = 0.2, 0.5, 0.9$.

Let $e_a(t)$ be an approximation error described by (9). Exemplary diagrams $e_a(t)$ as a function of a time t for selected values of parameters from Table 1: α, ω_l, ω_h and N are shown in Figs. 1 and 2.

From Table 1 and Figs. 1, 2 we can conclude at once, that the good approximant can be obtained with the use of approximation order N smaller than 20 and the narrower range of angular frequency between ω_l and ω_h. However, the more precise estimation of order N requires us to make next experiments. To estimate the

Table 1 Values of cost functions (10) and (11) for different α, ω_l, ω_h and N

Exp. No	Range of angular frequency		N	Fractional order α					
				0.2		0.5		0.9	
	ω_l	ω_h		I_{max}	I_2	I_{max}	I_2	I_{max}	I_2
1	0.1	10	5	0.3141	0.0031	0.1023	4.23e−004	0.0047	3.7088e−006
2	0.01	100	5	0.4581	0.0062	0.2800	0.0027	0.0255	2.9980e−005
3	0.001	1000	5	0.4543	0.0066	0.2815	0.0030	0.0271	3.6842e−005
4	0.1	10	10	0.3027	0.0030	0.0965	3.8646e−004	0.0047	3.7807e−006
5	0.01	100	10	0.4561	0.0062	0.2802	0.0027	0.0255	2.9964e−005
6	0.001	1000	10	0.4668	0.0065	0.2962	0.0030	0.0282	3.7121e−005
7	0.1	10	20	0.2997	0.0029	0.0951	3.7700e−004	0.0047	3.8009e−006
8	0.01	100	20	0.4560	0.0062	0.2800	0.0027	0.0255	2.9945e−005
9	0.001	1000	20	0.4665	0.0064	0.2961	0.0030	0.0282	3.7116e−005

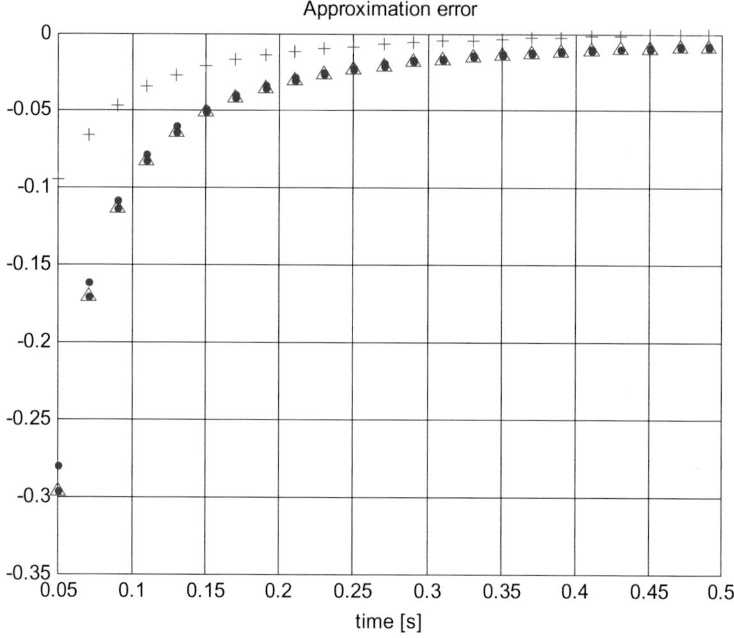

Fig. 1 Approximation error $e_a(t)$ described by (9) for: $\alpha = 0.5$, $N = 20$ and: experiments no: 7 (+), 8 (.), 9 (^), (variable range of angular frequency)

approximation order N assuring the reasonable value of cost function (11) can be applied an approach presented in [12, pp. 573, 574] to estimate a model order. A criterion to determine an approximation order is the rate of change the cost function (11) as a function of order N: if the increase of N causes firstly big and next small improvement of calculated cost function, then this "threshold" value of N is a sensible value of approximation order.

To estimate the order N the cost function (11) as a function of N was calculated. Calculations were done for ω_l = 0.1, ω_h =10, α = 0.2, α = 0.5 and N = 2.20. Results are shown in Fig. 3.

From Fig. 3 we can conclude that the approximation order N greater than 8 does not significantly improve the cost function (11) for both tested fractional orders α of Oustaloup approximation. This allows us to formulate conclusion that the order of approximation assuring the good accuracy and simultaneously lowest possible computational complexity of Oustaloup approximation is equal 8.

Additionally, it can be concluded that the accuracy of Oustaloup approximation improves with increasing fractional order α.

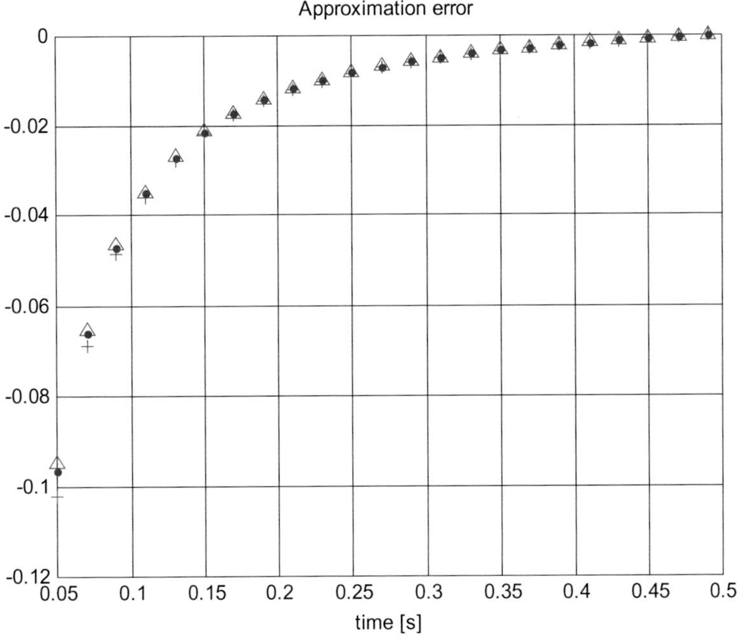

Fig. 2 Approximation error $e_a(t)$ described by (9) for: $\alpha = 0.5$, $N = 5$ (+), 10 (.), 20 (^), experiments no: 1, 4, 7 respectively

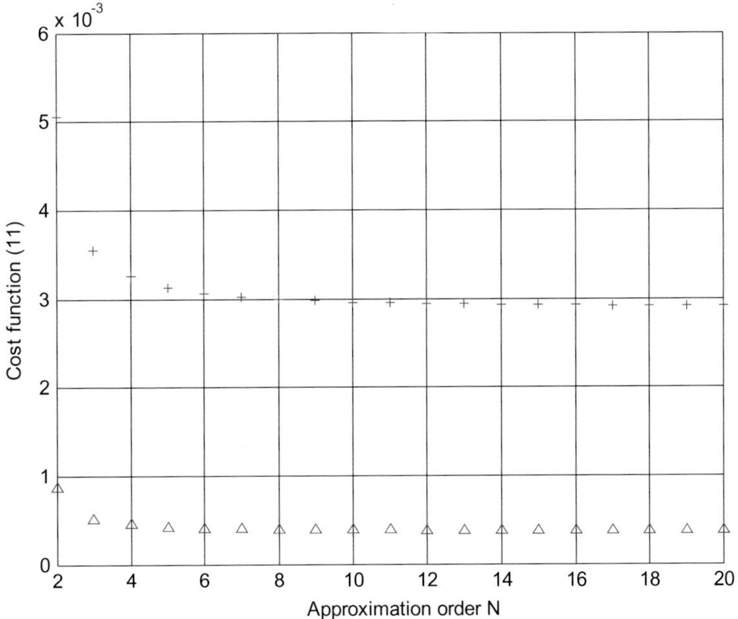

Fig. 3 Cost function (11) as a function of approximation order N for $\alpha = 0.2$ (+) and $\alpha = 0.5$ (+)

6 Final Conclusions

The final conclusions from the paper can be formulated as follows:

- In the paper the analysis of accuracy the Oustaloup approximation as a function of its parameters (order N and range of angular frequency from ω_l to ω_h) was presented. Different fractional-orders were also tested.
- The accuracy of Oustaloup approximation is stronger dependent on range of angular frequency between ω_l and ω_h, than approximation order, described by N.
- The order of approximation N, equal 8 assures the good accuracy of approximation. The improving of this order does not significantly improve this accuracy, but it increases a computational complexity of approximant.
- Decreasing the width range of angular frequency from ω_l to ω_h improves the accuracy of approximation in the sense of considered cost functions.
- Method presented in the paper can be applied to effective tuning of Oustaloup approximant for elementary non integer order integrator, independently on its gain (describing for example an integral and derivative actions in FO PID controller).
- The approach presented in this paper was applied to Charef approximation—see [11]. It can be also applied to discrete PSE and CFE approximations. This problem is planned to be considered by authors.
- As an another area of further investigations will be formulating analytical conditions directly associating with the cost functions (10) and (11) with integrator order α and approximation parameters: N, ω_l and ω_h.

Acknowledgements This paper was partially supported by the AGH (Poland)—project no 11.11.120.815 and partially supported by the AGH (Poland)—project no 11.11.120.817.

References

1. Caponetto, R., Dongola, G., Fortuna, L., Petras, I.: Fractional Order Systems. Modeling and Control Applications. World Scientific Series on Nonlinear Science, Series A, vol. 72. World Scientific Publishing (2010)
2. Kaczorek, T.: Selected Problems in Fractional Systems Theory. Springer (2011)
3. Mitkowski, W., Skruch, P.: Fractional-order models of the supercapacitors in the form of RC ladder networks. Bull. Polish Acad. Sci. Tech. Sci. **61**(3), 581–587 (2013)
4. Mitkowski, W.: Finite-dimensional approximations of distributed RC networks. Bull. Polish Acad. Sci. Tech. Sci. **62**(2), 263–269 (2014)
5. Mitkowski, W., Obrączka, A.: Simple identification of fractional differential equation. Solid State Phenom. **180**, 331–338 (2012)
6. Mitkowski, W., Oprzędkiewicz, K.: Application of fractional order transfer functions to modeling of high-order systems. 7 IFIP Conference, Klagenfurt (2013)

7. Oprzędkiewicz, K.: A Strejc model-based, semi- fractional (SSF) transfer function model. Automatyka/Automatics; AGH UST 2012 vol. 16 no. 2, pp. 145–154 (2012). Direct link to text: http://journals.bg.agh.edu.pl/AUTOMAT/2012.16.2/automat.2012.16.2.145.pdf
8. Merikh-Bayat, F.: Rules for selecting the parameters of Oustaloup recursive approximation for the simulation of linear feedback systems containing $PI^{\lambda}D^{\mu}$ controller. Commun. Nonlinear Sci. Numer. Simul. **17**, 1852–1861 (2012)
9. Oustaloup, A., Levron, F., Mathieu, B., Nanot, F.M.: Frequency-band complex noninteger differentiator: characterization and synthesis. IEEE Trans. Circuits Syst. I: Fundam. Theory Appl. I **47**(1), 25–39 (2000)
10. Petras, I.: Realization of fractional order controller based on PLC and its utilization to temperature control. Transfer inovaci nr **14**(2009), 34–38 (2009)
11. Mitkowski, W., Oprzędkiewicz, K.: An estimation of accuracy of Charef approximation. Theoretical developments and applications of non-integer order systems. In: Domek, S., Dworak, P. (eds.) 7th Conference on Non-Integer Order Calculus and its Applications (Lecture Notes in Electrical Engineering; ISSN 1876–1100; vol. 357). Szczecin, Poland, pp. 71–80. Springer (2016)
12. Isermann, R., Muenchhof, M.: Identification of Dynamic Systems. An Introduction with Applications. Springer (2011)

Low-Cost Multifunction Controller Prototype for Sawmill Wood Processing

Bartosz Trybus and Mateusz Wietecha

Abstract The paper presents concepts and implementation of a small multifunction controller. The device has been designed to control and monitor wood processing in a sawmill. It runs three control tasks simultaneously, namely climate control, drying control, surveillance system, and one HMI (Human-Machine Interface) task. CPDev environment has been used to develop programs in IEC 61131-3 LD, ST and SFC languages. Control algorithm of the drying process with SFC and ST is presented in more detail. CPDev virtual machine executes the tasks concurrently on the controller. Programmable graphical operator HMI created with the CPVis tool is also handled by the controller. The prototype employs the well-known Raspberry Pi 2 platform, reducing costs of the device.

Keywords Sawmill · IEC 61131-3 · SFC · LD · HMI · Raspberry Pi

1 Introduction

Compact programmable controllers become more and more versatile nowadays. A modern PLC can replace a set of classic devices, providing multiple control tasks, Ethernet connectivity and sometimes a built-in graphical HMI panel. Although such full-featured controllers from leading manufacturers are more expensive than single-function devices, costs of infrastructure and maintenance drop significantly (no need of extra wiring, signal converters etc.).

B. Trybus (✉) · M. Wietecha
Department of Computer and Control Engineering, Rzeszow University of Technology, al. Powstańców Warszawy 12, 35-959 Rzeszów, Poland
e-mail: btrybus@prz.edu.pl
URL: http://kia.prz.edu.pl

M. Wietecha
e-mail: d281@stud.prz.edu.pl

R. Szewczyk et al. (eds.), *Challenges in Automation, Robotics and Measurement Techniques*, Advances in Intelligent Systems and Computing 440, DOI 10.1007/978-3-319-29357-8_28

Limited expenses for control equipment are usually important for such branches of industry as wood processing or brick works. Therefore, in this paper we propose a low-cost multifunction control solution for a small sawmill. It involves a single all-in-one compact programmable controller. The prototype device has been built using a miniature computer, namely the popular Raspberry Pi 2 platform.[1] In the sawmill application it runs three control tasks concurrently, i.e. climate control in the production floor, wood drying control and machine equipment surveillance. The programs have been created according to the IEC 61131-3 standard in LD (Ladder Diagram), ST (Structured Text) and SFC (Sequential Function Chart) languages using CPDev programming environment [1]. Additionally, the controller involves graphical operator panel with multiple switchable pages (displays) for process monitoring and changing settings [2]. The operator panel can be accessed either locally, by using a touch screen, or remotely via a Web browser. Open and standardized structure allows easy extension or customization for particular mill of control programs and the operator panel HMI.

The paper is organized as follows. First, the main concepts of wood processing in a sawmill are given. Three automation areas are discussed: climate control, surveillance system and wood drying control. Section 3 presents the multifunction controller and its software structure. Then, two elements of sawmill application of the controller are covered, i.e. drying control algorithm in Sect. 4 and operator HMI in Sect. 5.

2 Sawmill Automation

Production management requires constant monitoring and control of running processes. Automation of this area may focus on particular problem, such as temperature control, or cover a wider range of functions. Here, a wood production plant is considered. This type of facility may involve such activities as product manufacturing, material workflow, internal logistics, organization of factory floor, etc. In this paper we focus on the sawmill hall and its integral elements, such as the wood-drying kiln or the sawing machinery.

Figure 1 presents an exemplary plan of the sawmill hall. Its main parts are: loading ramps, lumber production line, wood dryer with a kiln, storage and packaging yards. The plan also contains lumber and log buffers, log storage yard and an office. Other arrangements of sawmills can be found in [3] or [4].

The solution presented here covers the following automation areas: floor climate control, surveillance of machine equipment and drying control.

[1]Raspberry Pi Foundation, http://www.raspberrypi.org.

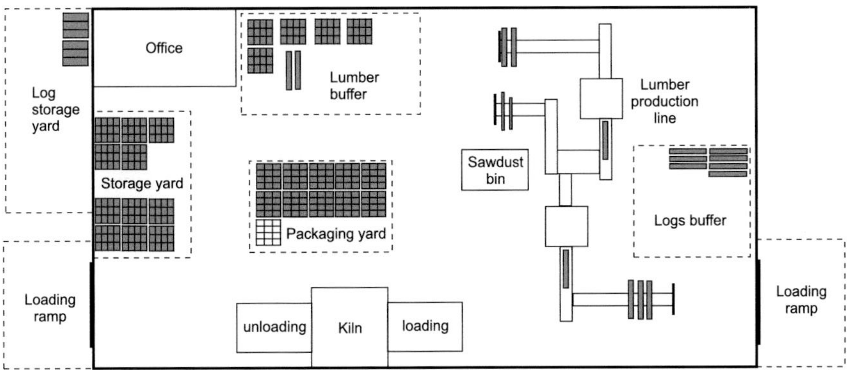

Fig. 1 Sample structure of a sawmill

2.1 Climate Control

The goal of the system is to keep required climate condition in the production floor. The operator should monitor air temperature during production, humidity and dustiness (too high degree may cause a fire hazard [5]). To achieve proper air conditions, an automated climate monitoring system may be introduced. Figure 2 shows key elements of such system in the sample sawmill hall.

The hall's roof is equipped with ventilation flaps and extraction ducts which are used for temperature and dustiness control. The interior is supplied with temperature, humidity and particle concentration sensors. The operator is able to set up temperature, humidity and should be alarmed when the maximum dust threshold is reached inside the hall.

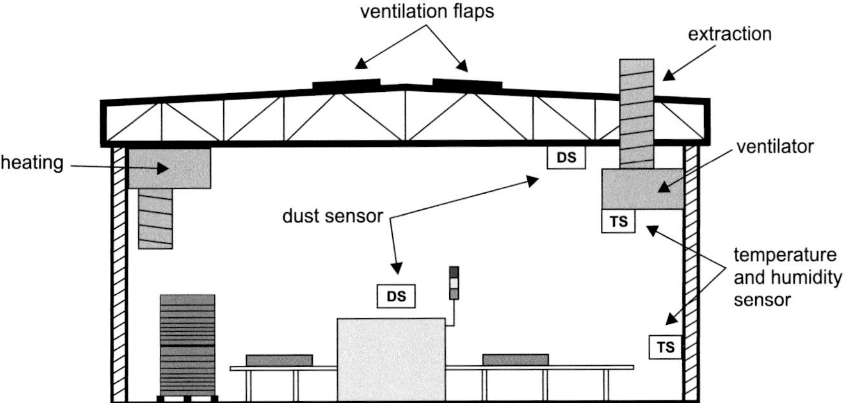

Fig. 2 Elements of climate control in the sawmill

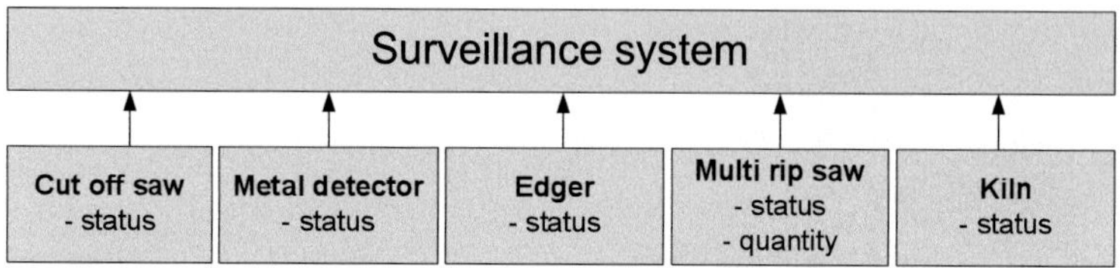

Fig. 3 Structure of the surveillance system

2.2 Surveillance System

The system allows supervision of machinery in the sawmill. In the proposed solution, each machine reports its status (running, stopped, error) and operation mode (automatic, manual). The quantity of manufactured lumber is also monitored. Therefore, the production manager is able to trace machine operation from the office room (Fig. 1). The information gathered in the system can be observed by the operator on the HMI display. Remote Web access to the display is also possible (see Sect. 5).

Figure 3 shows main components of the surveillance system. As seen, four machines are monitored in the sample application, namely cut off saw, metal detector, edger, multi rip saw and drying kiln.

2.3 Wood Drying Control

Drying reduces the moisture content of wood by heating it with hot air in a high temperature chamber (kiln) [6]. Duration of heating depends on the wood moisture and wood species and may take from a few to several days.

Figure 4 presents components of a drying kiln. The conveyors transport the wood material between the gates of the drying chamber. Automated drying proposed here will utilize signals from temperature sensor and optical sensors. The solution will be detailed later in Sect. 4.

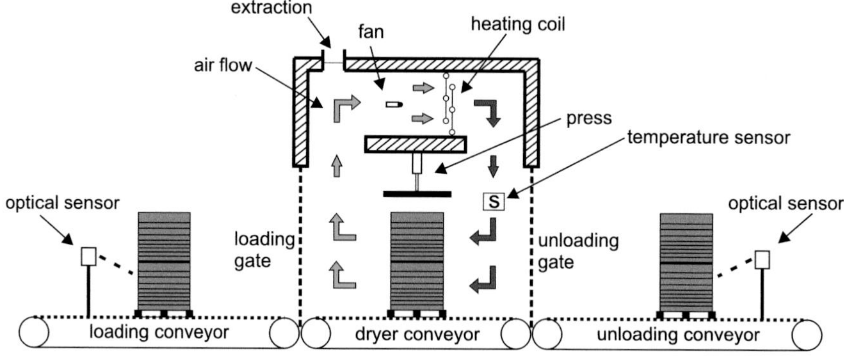

Fig. 4 Wood dryer

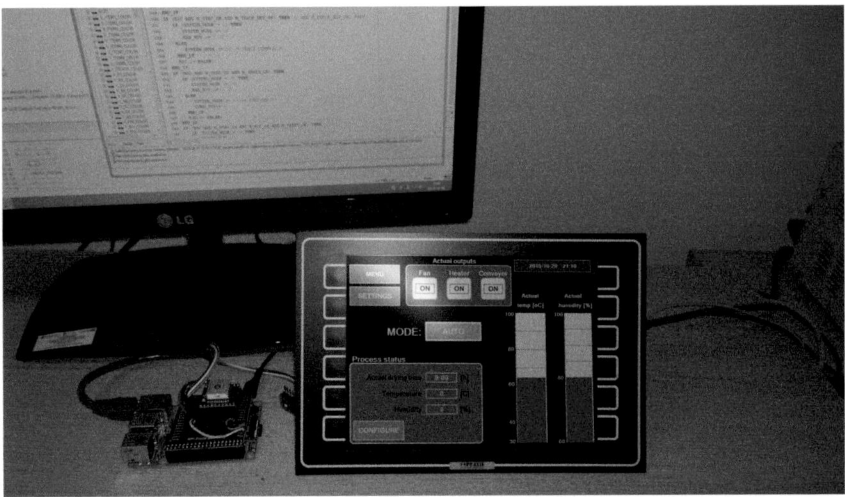

Fig. 5 Prototype of the multifunction controller

3 Multifunction Controller Prototype

Main concept of the prototype was to use a low-cost hardware solution, yet powerful enough to handle various functions in the sawmill, i.e. climate control, drying control and surveillance system. Another required feature was the ability to integrate a graphical operator panel. We have found out Raspberry Pi 2 miniature computer platform to be suitable, having efficient quad-core ARM7 CPU running at 900 MHz and 1 GB of RAM. It also provides Ethernet connectivity and HDMI display port.

The laboratory prototype of the controller is pictured in Fig. 5 with Raspberry Pi 2 on the left. The operator panel shown on the right has a resolution of 800×600 pixels.

General structure of controller software is presented in Fig. 6. CPDev virtual machine (VM) is run by Linux Raspbian operating system. The VM is the main component which executes control programs [7]. The VM has been ported to Raspberry Pi 2 by implementing some platform-specific functions. Here, the VM runs four tasks defined in the CPDev project for the sawmill, i.e. three control tasks (drying, climate, surveillance) and one HMI task. One can notice a resemblance of Fig. 6 and the software model structure defined in IEC 61131-3 [8]. Here, the sawmill *configuration* in formed out of the four tasks. It can be seen that LD, ST and SFC languages have been used to write control and HMI programs that constitute the tasks.

Since the tasks work in parallel, the VM utilizes multitasking at the operating system level to run concurrent threads. The tasks share global variables as defined in [8]. Passing data between tasks via globals, so tasks read variable set by another task, was first introduced to CPDev in hardware FPGA-based multiprocessor controller [9]. Here the same concept has been applied in a software solution.

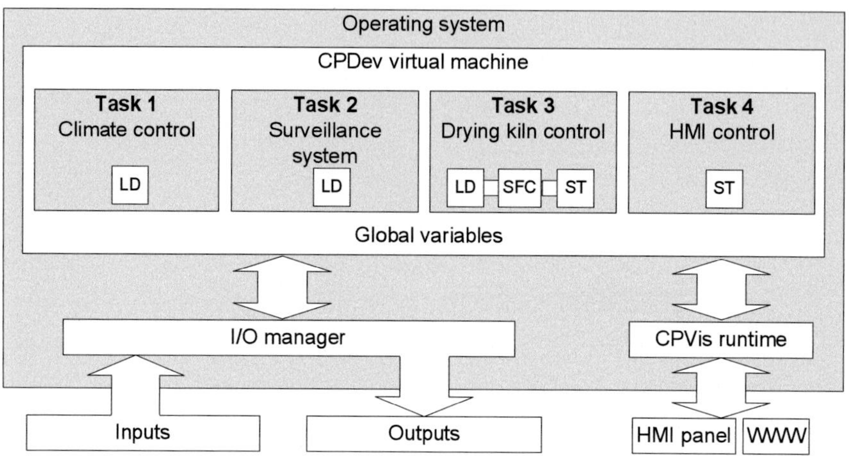

Fig. 6 Software structure of the controller

Global variables are mapped to controller inputs and outputs via I/O manager (Fig. 6). This module allows to bind particular variables to I/Os at design stage, making the mapping highly configurable. For example, a variable can be bound to various outputs (e.g. to a output signal pin or/and to a network). Various concepts of I/O handling in CPDev have been discussed in [10].

4 IEC 61131-3 Drying Program

As an example of IEC 61131-3 software run by the multifunction controller, we will now briefly characterize the drying control task. SFC, ST and LD languages have been used to program the task.

The drying process can be split into three stages: loading, drying and unloading. SFC provides a suitable notation for such sequential processes. The appropriate diagram for wood drying created in the CPDev environment is shown in Fig. 7. First the INITIALIZATION takes place. Then the system is waiting for a signal from the optical sensor (see Fig. 4) which indicates a wood material is ready to process. During the LOADING step, the wood material is loaded into the drying chamber. Therefore, the loading gate is opened and the belt conveyor is started. After the material has been loaded into the chamber, the conveyor is stopped and the gate is closed.

The action CONV1_2 associated with the LOADING step of the SFC diagram has been programmed in LD language. Its portion responsible for starting the belt conveyor is presented in Fig. 8. It can be seen that in case of automatic mode (AUTO_MODE), two input signals are used: end of previous drying (END_DRYING) and presence of wood material on the loading conveyor (LOADING_MATERIAL_PRESENT). Starting of the belt will not be possible if

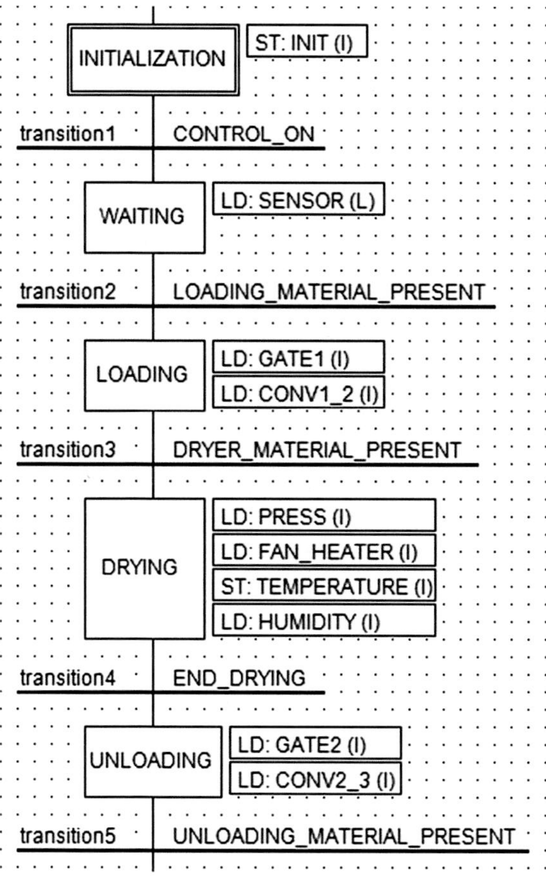

Fig. 7 SFC of the drying process

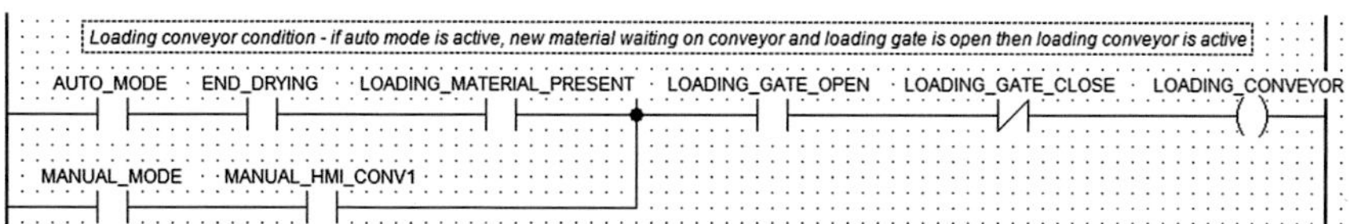

Fig. 8 LD portion for starting the loading conveyor

the loading gate is not opened to avoid damage. The manual control mode of loading conveyor (MANUAL_MODE) allows to start it by hand from the HMI panel. The open gate restriction is also verified here.

After the material is loaded, the DRYING step is started (Fig. 7). First, the wood is prepared for drying by squeezing it with a press. Then the air heating coil and fan are turned on. The fan forces air circulation run inside the chamber as shown is

Fig. 4. The heated air runs through the dried material and then returns back to the fan and the heater.

After the drying, the UNLOADING step begins. The press is released and the unloading gate opens. The wood is taken out of the chamber and put on the unloading conveyor. Then it is picked and moved to the storage yard.

The operator monitors the drying process and may set its parameters. The drying chamber is equipped with temperature and humidity sensors. This allows to adjust power of the fan and temperature of the heating coil accordingly. Furthermore, by controlling the air extraction, one can pull in dry air from the outside and remove humid air from the chamber.

5 HMI Operator Panel

The multifunction controller has been set up with an operator panel. The solution is based on the CPVis mechanism [2], which allows to integrate design of the HMI panel with IEC 61131-3 development. The display pages are composed of graphic elements (controls), such as bargraphs, text boxes with process values, push buttons, state indicators, etc. Global variables are bound to parameters of the controls. User interaction, such as switching display pages is handled by IEC 61131-3 code. ST language has been used for this purpose in the presented solution. In addition, the code is used to prepare some values for display (e.g. to adjust precision, change units, set colors dynamically, etc.).

Figure 9 shows two display pages of HMI for the sawmill project. The left display (Fig. 9a) is for the automated mode of drying control. It involves two push-buttons (upper left) to enter main menu and system settings page. The component group labeled "Actual outputs" (center) presents on/off state of three devices: fan, heater, conveyor. "Process status" group (lower left) displays important operating parameters, such as actual and configured drying time (hours), current temperature

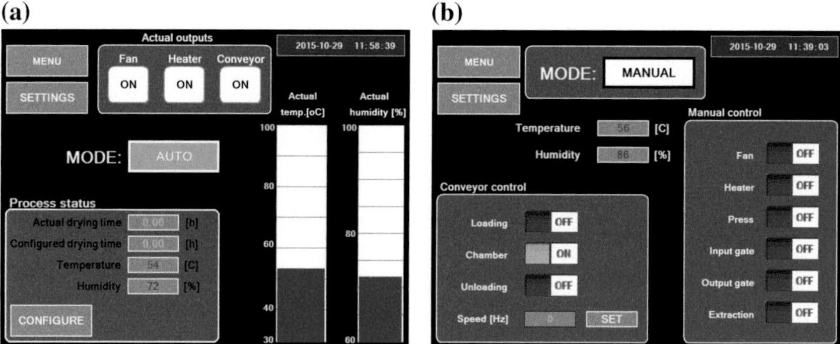

Fig. 9 HMI displays of **a** auto, **b** manual drying control modes

and humidity in the chamber. The push-button "CONFIGURE" is used to set up drying parameters. The two bargraphs on the right show temperature and humidity in a visual form. The bargraph color changes depending on the value range (normal, too high, too low), so the operator can easily observe whether the process runs properly and may react accordingly.

Figure 9b shows the drying display in the manual mode. By using the "Manual control" section (right) the operator is able to switch on/off all devices in the chamber. Temperature and humidity are displayed in the center. "Conveyor control" group (lower right) allows to set up three conveyors: loading, chamber and unloading. The belt speed can also be configured manually. As previously mentioned, the operator cannot start the conveyors when a gate is closed or the press is not released.

Since the controller hardware provides Ethernet connectivity, a Web browser can be used to access the HMI remotely. In such scenario, the CPVis runtime serves graphic screens to a WWW server (Fig. 6).

6 Summary

Sawmill automation requires executing multiple tasks at the same time. It has been shown that a versatile controller can be constructed for this purpose. The device integrates various functions in a single, relatively cheap device, handling two control activities (climate, drying), machine monitoring and graphical operator panel.

The solution is programmable in IEC 61131-3 languages. Control programs as well as HMI graphics can be easy adjusted for particular sawmill or reconfigured e.g. when a new machine is installed.

References

1. Jamro, M., Rzońca, D., Sadolewski, J., Stec, A., Świder, Z., Trybus, B., Trybus, L.: CPDev engineering environment for modeling, implementation, testing, and visualization of control software. In: Szewczyk, R., Zieliński, C., Kaliczyńska, M. (eds.) Recent Advances in Automation, Robotics and Measuring Techniques. Advances in Intelligent Systems and Computing, vol. 267, pp. 81–90. Springer, Heidelberg (2014)
2. Jamro, M., Trybus, B.: IEC 61131–3 programmable human machine interfaces for control devices. In: The 6th International Conference on Human System Interaction (HSI), 2003, pp. 48–55 (2013)
3. Generalized sawmill layout. http://www.afrc.uamont.edu/pattersond/Coursework/Undergrad/sawmills.htm
4. Sawmill layout. http://www.granttimbers.com.au/page.php?id=9
5. Injury prevention resources for wood products manufacturing - sawmills. http://www2.worksafebc.com/Portals/WoodProducts/Prevention-Sawmills.asp?ReportID=37458
6. Drying of wood products. http://www.afrc.uamont.edu/pattersond/Coursework/Undergrad/wood_&_water.htm
7. Trybus, B.: Development and implementation of IEC 61131–3 virtual machine. Theor. Appl. Inf. **23**(1), 21–35 (2011)

8. IEC 61131–3 - programmable controllers - part 3: programming languages (2013)
9. Hajduk, Z., Trybus, B., Sadolewski, J.: Architecture of FPGA embedded multiprocessor programmable controller. IEEE Trans. Ind. Electron. **62**(5), 2952–2961 (2015)
10. Rzońca, D., Sadolewski, J., Trybus, B.: Interfacing inputs and outputs with IEC 61131–3 control software. In: Szewczyk, R., Zieliński, C., Kaliczyńska, M. (eds.) Recent Advances in Automation, Robotics and Measuring Techniques. Advances in Intelligent Systems and Computing, vol. 267, pp. 229–238. Springer, Heidelberg (2014)

Application of SURF Algorithm for Real-Time Estimation of Angle and Central Point of a Tracked Object

Łukasz Sawicki, Arkadiusz Kubacki and Piotr Owczarek

Abstract In the paper a using of 2D color camera and a program based on the Speeded-Up Robust Features (SURF) algorithm was presented. The main aim of this article was to find influence of amount of tracked points on accuracy of position and angle tracking. Required data are obtained by finding corresponding key points between image captured from the camera and a reference image of the tracked object. The program has been written in C# with using of Emgu CV which is an image processing library.

Keywords Emgu CV · Camera · SURF · Speeded-Up robust features · Algorithm

1 Introduction

During the years many visual human-machine interfaces appeared based on inter alia color [1], depth or stereo cameras [2]. Controlling a machine by using a visual system is a complex issue with many variables and no solution appears to be ideal. Each device has some flaws, for example color cameras are subject to light variations [3], depth cameras obtain only contours of objects and are vulnerable for infrared light interference. Nevertheless the most common approach is to use color cameras because of their low price and availability. They are in almost every modern notebook.

Ł. Sawicki (✉) · A. Kubacki · P. Owczarek
Institute of Mechanical Technology, Poznan University of Technology, Poznan, Poland
e-mail: lukasz.w.sawicki@doctorate.put.poznan.pl

A. Kubacki
e-mail: arkadiusz.j.kubacki@doctorate.put.poznan.pl

P. Owczarek
e-mail: piotr.owczarek@put.poznan.pl

R. Szewczyk et al. (eds.), *Challenges in Automation, Robotics and Measurement Techniques*, Advances in Intelligent Systems and Computing 440, DOI 10.1007/978-3-319-29357-8_29

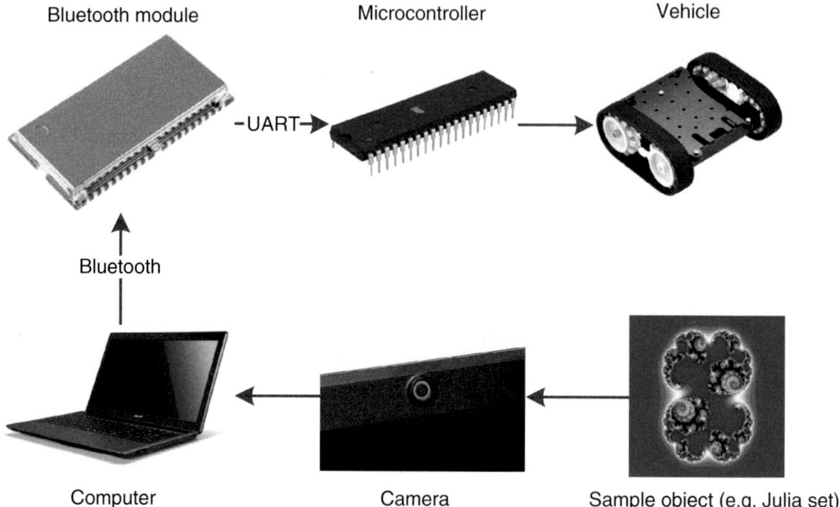

Fig. 1 Control scheme

The system consists of hardware and software. In purpose to process obtained images there are used some specific libraries. One of the most popular is OpenCV which is free for commercial and academic use and was designed mostly for real-time applications. It is written in C/C++ and supports Windows, Linux, Mac OS, iOS and Android [4]. However in this study the Emgu CV library is used which is based on OpenCV and has almost the same functions which can be called from .NET compatible languages inter alia C# [5]. These functions allow to capture and process images not only from an ordinary 2D camera but also from a stereo camera or depth camera like the Microsoft Kinect. By using the Emgu CV library a user friendly window application can be made.

This paper is about a method, which allows to track a specific object in two dimensions by using a color camera and a computer with special software based on the Speeded-Up Robust Features (SURF) algorithm. The main goal of this software is to control a vehicle by sending specific data via Bluetooth. The sent data contains information in which direction and at what speed the vehicle should move. Before each use the program needs calibration by taking a picture of the tracked object (Fig. 1).

2 Vision System and Mathematical Background

As it was mentioned to track an object in the presented study a special software was used. It was based on the SURF algorithm presented in 2006 by Bay [6] which is a faster version of the Scale-invariant feature transform (SIFT) method by Lowe [7, 8].

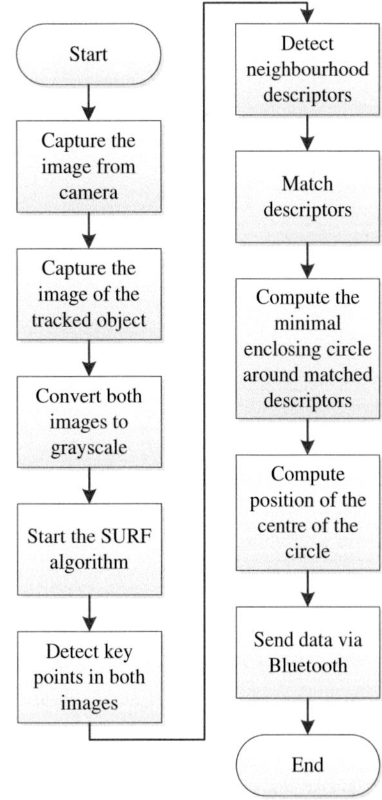

Fig. 2 Scheme of the algorithm

In the first two steps according to Fig. 2 two images are being captured. One is the current view from the camera, the second is the image of the tracked object which can be updated at any time by clicking the "Update image" button. The next step is converting both images into grayscale which is a basic requirement of this method. Afterwards the SURF algorithm is being started by using a function from the Emgu CV library which is called *SURFDetector()*. The main parameter of this function is the threshold for the determinant of a Hessian matrix, mostly it is between 300 and 500 in accordance with the OpenCV documentation. From this number depends how many key points are going to be found. The results of changing this parameter are shown in the third section.

In the further step key points are being detected by using an Emgu CV function *surfDetector.DetectKeyPointsRaw()*. The point detection algorithm uses an approximation of the determinant of the Hessian matrix H(z, σ) which is a square matrix of second order partial derivatives, e.g. $L_{xx}(z, \sigma)$. The point of using this matrix is to detect blob structures by searching for the determinant's maximum value at a certain location z = (x, y) and scale σ in an image *I*.

$$H(z,\sigma) = \begin{bmatrix} L_{xx}(z,\sigma) & L_{xy}(z,\sigma) \\ L_{xy}(z,\sigma) & L_{yy}(z,\sigma) \end{bmatrix} \tag{1}$$

For the approximation of the Hessian matrix box filters were used. In order to reduce their computation time integral images were applied also known as summed area tables. The integral image $I_{\sum}(z)$ at location z represents the sum of pixel intensities in the input image $I(x, y)$ within a rectangular region [9].

$$I_{\Sigma}(z) = \sum\nolimits_{x' \leq x, y' \leq y} I(x', y') \tag{2}$$

The main advantage of this approach is that it only needs three additions to compute pixel intensities in any rectangular area and the calculation time is independent of its size.

The next step according to Fig. 2 is computing descriptors by using an Emgu CV function called *surfDetector.ComputeDescriptorsRaw()*. A descriptor describes pixel intensities within the key point neighborhood [10]. The algorithm of obtaining them consists of two steps. First the orientation of each interest point needs to be calculated based on a circular region around it which size is dependent on the scale value σ. It was done by computing and later adding the Haar wavelet filters responses in x and y directions within the mentioned region. In order to decrease the time of filter computing an integral images were used. The outcome of the mentioned operations is a local orientation vector.

In the second step a square region around the key point is made and orientated along the local orientation vector. The region is split into regular sub-regions. For each sub-region two Haar wavelet responses are computed: one in horizontal d_x and other in vertical direction d_y. Then the obtained values d_x and d_y are summarized in each sub-region and also their absolute values are calculated. Thus the feature vector v (descriptor) consists of these 4 computed variables.

$$v = \left(\sum d_x, \sum d_y, \sum |d_x|, \sum |d_y|\right) \tag{3}$$

In the next step found descriptors need to be matched between the current view of the camera and the image of the tracked object. In order to do that the k-nearest neighbors algorithm was used which is built in the Emgu CV library and is called *bruteForceMatcher.KnnMatch()*. Simply it finds the k-nearest matches of the given descriptor which of course could lead to some mismatches [11]. Nevertheless there are some countermeasures, for example an Emgu CV function *VoteForSizeAndOrientation* which is checking if the matched descriptors are not rotated and scaled more than the fixed boundaries.

If everything is alright a homography matrix is build which is used to combine the image of the tracked object and the current view of the camera by using the corresponding key points. In order to do so is used a function called *GetHomographyMatrixFromMatchedFeatures()*.

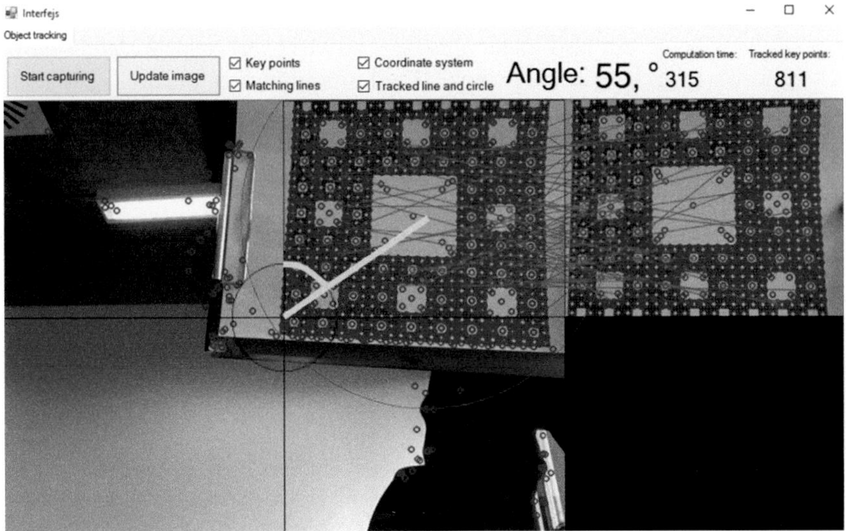

Fig. 3 A running software with SURF algorithm

In purpose to show an angle and position a circle is drawn around the matched points in the current view of the camera and then the center of the circle is being connected with the origin of the coordinate system by a line. Afterwards the angle α between the mentioned line and the vertical axis of the coordinate system is computed by using Eq. 4. In the equation arccosine's parameter is a ratio of the vertical coordinate y to the length of the line connecting the origin and the circle's center. The distance was computed by using a basic formula for the length between two points in the two dimensional Euclidian space.

$$a = \arccos\left(\frac{\frac{h}{2} - y}{\sqrt{\left(x - \frac{w}{3}\right)^2 + \left(y - \frac{h}{2}\right)^2}}\right) \tag{4}$$

where: h—height of the image; w—width of the image; x—current horizontal position of the circle's center; y—current vertical position of the circle's center.

In the Fig. 3 a working gesture recognition software is shown where the tracked object is an image of a fractal called the Sierpinski's carpet and where by using bolded lines is shown the angle α and the line connecting the circle's center with the origin of the coordinate system.

After obtaining the angle α its value is coded and transmitted via Bluetooth to the vehicle's Bluetooth module where it goes to the microcontroller and then on the motors. Before sending the computed angle it is being divided by 10 in order that the vehicle become less susceptible to software errors. Therefore the received angle value by the microcontroller varies from 0 to 18. The transmitted data consist not only of the α value but also of the direction in which the vehicle should move namely forward, backward, left or right. The last information which is sent is the velocity at which the vehicle should drive.

3 Experimental Results

In order to examine the Speeded-Up Robust Features algorithm for computation time and stability of the obtained angle value, a set of test images was created. This set consists of three fractals which are called the Mandelbrot set [12], the Julia set [12] and the Sierpinski's carpet [13]. They were chosen because of their diverse structures and color fill.

The set of pictures was tested with different preset Hessian threshold values which varied from 0 to 1000. From a relation seen in Fig. 5 one can conclude that there is more points for lower threshold. The parameter change should be made with great caution because if the value is too small there are many background points which could distort the result. From a relation seen in Fig. 6 one can conclude that the computation time for high threshold is lower. An advantage of a higher threshold is that if the tracked image has many significant key points then the chance that the outcome would be distorted is smaller. Each image was printed on a sheet of A4 paper and the software was tested on a notebook with Intel Core i7-3630QM 2.4 GHz 6 MB Cache and a camera with a resolution of 1280 × 720px. The results are shown in Figs. 4 and 5.

The Fig. 6 shows that the higher the threshold gets the lower computation time is required. The reason of this state is that there are fewer key points to be computed in the captured image. The largest amount of key points in the outcome has the Sierpinski's carpet because of its many lines which intersect and create corners, T-junctions etc.

In order to present how changes of the Hessian threshold affect the result, namely the stability of the calculation result of the angle, a further study was

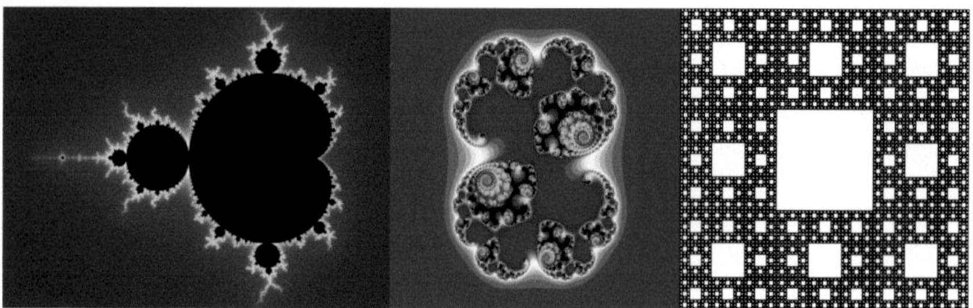

Fig. 4 Set of test images. From *left* to *right*: Mandelbrot set [14], Julia set [15], Sierpinski's carpet [16]

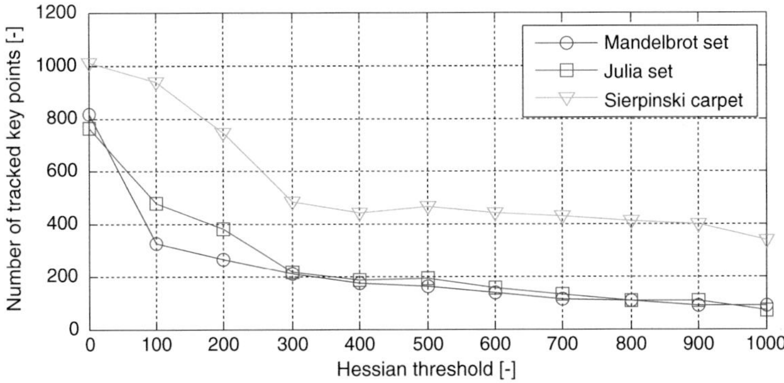

Fig. 5 Graph showing the relation between the Hessian threshold and number of found key points in the tracked image

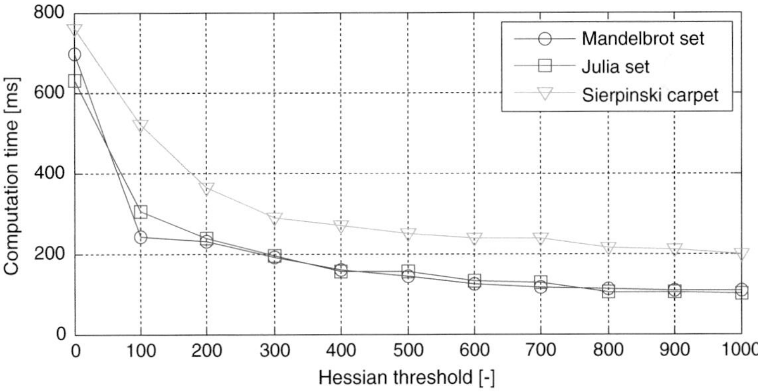

Fig. 6 Graph showing the relation between the Hessian threshold and software computation time

conducted. The results of this research were shown in Fig. 7. The study consisted of five steps: setting a value of the Hessian threshold (from 0 to 1000), taking a reference image of the tracked object (the Sierpinski's carpet), changing the object's location by 90°, recording the angle value for 20 s when the object is not moving, calculating the mean angle variations from the obtained data.

From Fig. 7 one can conclude that the smallest angle variations are for the Sierpinski's carpet when the threshold value is set on 600. When the Hessian threshold is low there are many background points which are the most likely reason for such big angle changes. On the other hand when the threshold value is high there are too few key points in the image to obtain a stable outcome.

Described tests had been done in similar light conditions and in the same environment but due to the fact that even slight light variations affect the outcome repeating the experiments with the same exact results are unlikely.

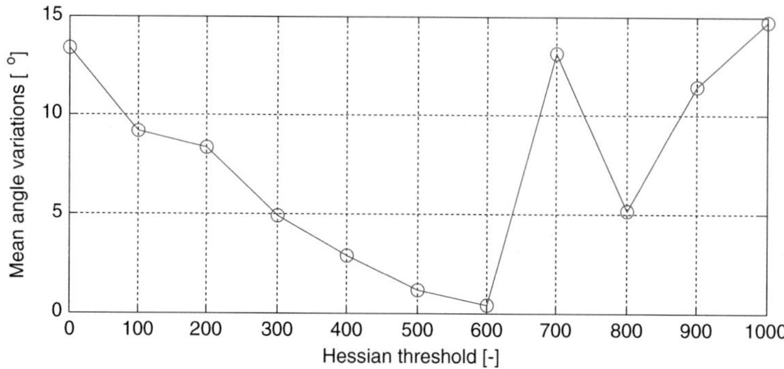

Fig. 7 Graph showing the relation between the Hessian threshold and the mean angle variations

4 Conclusion

In this article an application of the Speeded-Up Robust Features (SURF) algorithm was presented which is an upgraded version of the Scale-Invariant Feature Transform (SIFT) algorithm. In the second section the theoretical basis were described explaining how the software works. In the third section the experimental results were presented which consisted of testing the algorithm's computation time and stability by using a set of images namely three fractals. The main aim of the study was to examine the SURF algorithm for the stability of the obtained angle value in a real-time application and evaluate its suitability.

The main advantage of this software is the possibility to track various objects on the condition of course that they would have enough key points. In order to obtain a stable result the Hessian threshold cannot be too small or too big. The value of the threshold depends on the tracked object and the light conditions. There are some downsides of this software, for example before every use it needs calibration but therefore it could work in various circumstances. This software also requires a powerful computer in order to work in real time. In the future work the research of vehicle's control system will be described.

Acknowledgment The work described in this paper was funded from 02/23/DS-PB/120 (Nowe techniki w urządzeniach mechatronicznych).

References

1. Xu, Y. Gu, J., Tao, Z., Wu, D.: Bare Hand Gesture Recognition with a Single Color Camera. In: 2nd International Congress on Image and Signal Processing, 2009. CISP '09, pp. 1–4 (2009)
2. Li, X., Hong, K.-S.: Korean chess game implementation by hand gesture recognition using stereo camera. In: 2012 8th International Conference on Computing Technology and Information Management (ICCM), vol. 2, pp. 741–744 (2012)

3. Nakazawa, Y., Makino, H., Nishimori, K., Wakatsuki,. D., Komagata, H.: Indoor positioning using a high-speed, fish-eye lens-equipped camera in Visible Light Communication. In: 2013 International Conference on Indoor Positioning and Indoor Navigation (IPIN), pp. 1–8 (2013)
4. OpenCV documentation. http://opencv.org
5. Emgu CV documentation. http://emgu.com
6. Bay, H., Tuytelaars, T., Van Gool, L.: Surf: Speeded up robust features. In: Computer vision–ECCV, pp. 404–417. Springer, Berlin (2006)
7. Lowe, D.G.: Object recognition from local scale-invariant features. In: The Proceedings of the Seventh IEEE International Conference on Computer Vision, vol. 2, pp. 1150–1157. IEEE (1999)
8. Panchal, P.M., Panchal, S.R., Shah, S.K.: A comparison of SIFT and SURF. Int. J. Innovative Res. Comput. Commun. Eng. **1**(2), 323–327 (2013)
9. Viola, P., Jones, M.: Rapid object detection using a boosted cascade of simple features. In: Proceedings of the 2001 IEEE Computer Society Conference on Computer Vision and Pattern Recognition. CVPR 2001. vol. 1, pp. I-511. IEEE (2001)
10. Mikolajczyk, K., Schmid, C.: A performance evaluation of local descriptors. IEEE Trans. Pattern Anal. Mach. Intell. **27**(10), 1615–1630 (2005)
11. Kramer, O.: K-nearest neighbors. In: Dimensionality Reduction with Unsupervised Nearest Neighbors. pp. 13–23. Springer, Berlin (2013)
12. Zhang, X., Xu, Z.: Implementation of Mandelbrot set and Julia Set on SOPC platform. In: 2011 International Conference on Electronics, Communications and Control (ICECC), pp. 1494–1498 (2011)
13. Wang, Y., Luo, J., Li, Y.: Investigations on the K-Sierpinski carpet fractal antenna. In: Cross Strait Quad-Regional Radio Science and Wireless Technology Conference (CSQRWC), vol. 1, pp. 382–385 (2011)
14. Mandelbrot set. http://en.wikipedia.org/wiki/Mandelbrot_set
15. Julia set. http://en.wikipedia.org/wiki/Julia_set
16. Sierpinski carpet. http://en.wikipedia.org/wiki/Sierpinski_carpet

Classes of Digraph Structures Corresponding to Characteristic Polynomials

Krzysztof Hryniów and Konrad Andrzej Markowski

Abstract This paper introduces the first classification of digraph structures corresponding to characteristic polynomials. It was found that digraph structures created can be divided into three classes with different feasibility for different polynomials—only structures of one class are found to be independent from wages of polynomial's terms. In this paper classification of structures is described, along with method how to divide them and illustrated with examples.

Keywords Digraphs · Characteristic polynomial · Digraph structures · Digraph classes

1 Introduction

In last 2 years there was presented parallel algorithm for finding the determination of characteristic polynomial realisations of dynamic systems based on multi-dimensional digraphs theory [5]. It is an alternative to canonical forms of the system [8], i.e. constant matrix forms, which satisfy the system described by the transfer function. With the use of those forms, we are able to write only one realisation of the system, while algorithm presented in [5] allows for finding all possible sets of matrices which fit into the system transfer function [6, 7]. The digraph theory was applied to the analysis of dynamical systems. The use of multi-dimensional digraph

---

Research has been financed with the funds of the Statutory Research of 2016.

K. Hryniów (✉) · K.A. Markowski
Faculty of Electrical Engineering, Institute of Control and Industrial Electronics, Warsaw University of Technology, Koszykowa 75, 00-662 Warsaw, Poland
e-mail: Krzysztof.Hryniow@ee.pw.edu.pl
URL: http://nas.isep.pw.edu.pl/repo

K.A. Markowski
e-mail: Konrad.Markowski@ee.pw.edu.pl

R. Szewczyk et al. (eds.), *Challenges in Automation, Robotics and Measurement Techniques*, Advances in Intelligent Systems and Computing 440, DOI 10.1007/978-3-319-29357-8_30

theory was proposed for the first time in the paper [4] to the analysis of positive two-dimensional systems.

Still, multi-dimensional digraphs used for characteristic polynomial realisations are not fully defined and determined—and there is space for further research about properties of possible solutions obtained. It is known that some structures obtained won't satisfy the polynomial or there will be need for solving a system of polynomial equations to determine the coefficients of state matrices from them. The scope of this article is to examine structures that do not generate the solution instantly (i.e. just by examination of the digraph) to find all possible proper digraph structures for the characteristic polynomial.

Notion. In this paper the following notion will be used. The set $n \times m$ real matrices will be denoted by $\mathbb{R}^{n \times m}$ and $\mathbb{R}^n = \mathbb{R}^{n \times 1}$. If $\mathbf{G} = [g_{ij}]$ is a matrix, we write $\mathbf{G} \geqslant 0$ (matrix $\mathbf{G}$ is called non-negative), if $g_{ij} \geqslant 0$ for all i, j; $\mathbf{G} > 0$ (matrix $\mathbf{G}$ is called positive), if $\mathbf{G} \geqslant 0$ and any $g_{ij} > 0$; $\mathbf{G} \gg 0$ (matrix $\mathbf{G}$ is called strictly positive), if $g_{ij} > 0$ for all i, j. The set of $n \times m$ real matrices with non-negative entries will be denoted by $\mathbb{R}_+^{n \times m}$ and $\mathbb{R}_+^n = \mathbb{R}_+^{n \times 1}$. The $n \times n$ identity matrix will be denoted by $\mathbf{I}_n$.

1.1 Characteristic Polynomial

Let $\mathbb{F}$ be a field e.g., of the real number $\mathbb{R}$. The function $P(w_1, w_2, \dots, w_j)$ of the variable $w_1, w_2, \dots, w_j$, where $j = 1, 2, \dots, \infty$

$$p(w_1, w_2, \dots, w_j) = w_1^{n_1} w_2^{n_2} \dots w_j^{n_j} - \sum_{i_1=0}^{n_1} \sum_{i_2=0}^{n_2} \dots \sum_{i_j=0}^{n_j} a_{i_1, i_2, \dots i_j} w_1^{i_1} w_2^{i_2} \dots w_j^{i_j} \tag{1}$$

is called polynomial $p(w_1, w_2, \dots, w_j)$ in the variable $w_1, w_2, \dots, w_j$, over the field $\mathbb{F}$, where $a_{i_1, i_2, \dots, i_j} \in \mathbb{F}$ for $i = 0, 1, 2, \dots, n$ and for $j = 1, 2, \dots, \infty$ are called the coefficients of the polynomial.

The set of polynomial (1) over the field $\mathbb{F}$ will be denoted by $\mathbb{F}[w_1, w_2, \dots, w_j]$ where $j = 1, 2, \dots, \infty$.

If $a_{n_1, n_2, \dots, n_j} \neq 0$, then the non-negative integral $n = n_1 + n_2 + \dots + n_j$ is called the degree of a polynomial and is denoted $deg\, p(w_1, w_2, \dots, w_j)$, i.e., $n = deg\, p(w_1, w_2, \dots, w_j)$. The polynomial is called monic, if $a_{n_1, n_2, \dots, n_j} = 1$ and zero polynomial, if $a_{i_1, i_2, \dots, i_j} = 0$ for $i = 0, 1, \dots, n$ and for $j = 1, 2, \dots, \infty$.

Interested reader may find definition and properties of the characteristic polynomial in books on linear algebra, for example in [2, Chap. 9].

Remark 1 If we consider one-dimensional discrete-time system described by the equations: $x_{i+1} = \mathbf{A}x_i + \mathbf{B}u_i$; $y_i = \mathbf{C}x_i + \mathbf{D}u_i$ for $i \in \mathbb{Z}_+$ then, characteristic polynomial (1) consist from one variable z and have the form: $d(z) = z^n - \sum_{i=0}^{n} d_i z^i = z^n - d_{n-1} z^{n-1} - \dots - d_1 z - d_0$. Similarly, characteristic polynomial consist from variable s, if we consider one-dimensional continuous-time system.

Remark 2 If we consider two-dimensional discrete-time system described by the equations: $x_{i+1,j+1} = \mathbf{A}_1 x_{i+1,j} + \mathbf{A}_2 x_{i,j+1} + \mathbf{B}_1 u_{i+1,j} + \mathbf{B}_2 u_{i,j+1}$; $y_{ij} = \mathbf{C}x_{ij} + \mathbf{D}u_{ij}$ for $i \in \mathbb{Z}_+$ and $j \in \mathbb{Z}_+$ then, characteristic polynomial (1) consist from two variables z_1 and z_2 and have the form: $d(z_1, z_2) = z_1^n z_2^n - \sum_{i=0}^{n} \sum_{j=0}^{n} d_{ij} z_1^i z_2^j = z_1^n z_2^n - d_{n-1,n} z_1^{n-1} z_2^n - d_{n,n-1} z_1^n z_2^{n-1} - \cdots - d_{10} z_1 - d_{01} z_2 - d_{00}$. Similarly, characteristic polynomial consist from variables: s_1 and s_2 if we consider two-dimensional continuous-time system; s and z if we consider hybrid system. In an analogous way we proceed with higher dimensions systems.

1.2 Digraphs

A directed graph (called also digraph) $\mathfrak{D}$ consists of a non-empty finite set $\mathbb{V}(\mathfrak{D})$ of elements called vertices and a finite set $\mathbb{A}(\mathfrak{D})$ of ordered pairs of distinct vertices called arcs. We call $\mathbb{V}(\mathfrak{D})$ the vertex set and $\mathbb{A}(\mathfrak{D})$ the arc set of $\mathfrak{D}$. We will often write $\mathfrak{D} = (\mathbb{V}, \mathbb{A})$ which means that $\mathbb{V}$ and $\mathbb{A}$ are the vertex set and arc set of $\mathfrak{D}$, respectively. The order of $\mathfrak{D}$ is the number of vertices in $\mathfrak{D}$. The size of $\mathfrak{D}$ is the number of arc in $\mathfrak{D}$. For an arc (v_1, v_2), the first vertex v_1 is its tail and the second vertex v_2 is its head. More information about digraph theory is given in [1, 9]. A two-dimensional digraphs $\mathfrak{D}^{(2)}$ is a directed graph with two types of arcs and input flows. For the first time, this type of digraph was presented in [3, 4]. When we generalise this approach, we can define n-dimensional digraphs $\mathfrak{D}^n$ in the following form.

Definition 1 An n-dimensional digraphs $\mathfrak{D}^{(n)}$ is a directed graph with q types of arcs and input flows. In detail, it is $(\mathbb{S}, \mathbb{V}, \mathfrak{A}_1, \mathfrak{A}_2, \ldots, \mathfrak{A}_q, \mathfrak{B}_1, \mathfrak{B}_2, \ldots, \mathfrak{B}_q)$, where $\mathbb{S} = \{s_1, s_2, \ldots, s_m\}$ is the set of sources, $\mathbb{V} = \{v_1, v_2, \ldots, v_n\}$ is the set of vertices, $\mathfrak{A}_1$, $\mathfrak{A}_2$, …, $\mathfrak{A}_q$ are the subsets of $\mathbb{V} \times \mathbb{V}$ which elements are called $\mathfrak{A}_1$-arcs, $\mathfrak{A}_2$-arcs, …, $\mathfrak{A}_q$-arcs respectively, $\mathfrak{B}_1$, $\mathfrak{B}_2$, …, $\mathfrak{B}_q$ are the subsets of $\mathbb{S} \times \mathbb{V}$ which elements are called $\mathfrak{B}_1$-arcs, $\mathfrak{B}_2$-arcs, …, $\mathfrak{B}_q$-arcs respectively.

There exists $\mathfrak{A}_1$-arc ($\mathfrak{A}_2$-arc, …, $\mathfrak{A}_q$-arc) from vertex v_j to vertex v_i if and only if the (i,j)-th entry of the matrix $\mathbf{A}_1$ ($\mathbf{A}_2$, …, $\mathbf{A}_q$) is non-zero. There exists $\mathfrak{B}_1$-arc ($\mathfrak{B}_2$-arc, …, $\mathfrak{B}_q$-arc) from source s_l to vertex v_j if and only if the l-th entry of the matrix $\mathbf{B}_1$ ($\mathbf{B}_2$, …, $\mathbf{B}_q$) is non-zero.

Remark 3 $\mathfrak{A}_q$-arcs and $\mathfrak{B}_q$-arcs, are drawn by the other colour or line style.

Example 1 For the system described by the matrices

$$\mathbf{A}_1 = \begin{bmatrix} 0 & 0 & 1 \\ 1 & 0 & 0 \\ 0 & 1 & 0 \end{bmatrix}, \mathbf{A}_2 = \begin{bmatrix} 1 & 0 & 0 \\ 0 & 0 & 1 \\ 1 & 0 & 0 \end{bmatrix}, \mathbf{A}_3 = \begin{bmatrix} 0 & 0 & 0 \\ 1 & 0 & 1 \\ 0 & 0 & 1 \end{bmatrix}, \mathbf{B}_1 = \begin{bmatrix} 1 \\ 0 \\ 0 \end{bmatrix} \tag{2}$$

we can draw digraphs $\mathfrak{D}^{(3)}$ consisting from vertices v_1, v_2, v_3 and source s_1. Digraph corresponding to system (2) is presented on Fig. 1.

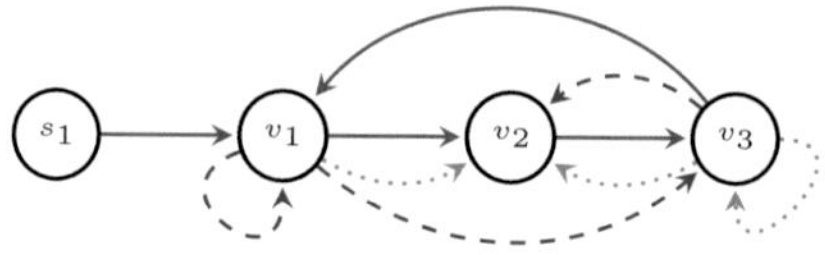

Fig. 1 Three-dimensional digraphs $\mathfrak{D}^{(3)}$ corresponding to matrices (2)

1.3 Digraph Creation

The digraph creation algorithm introduced in [5] starts with creating digraphs for all monomials in the characteristic polynomial, then joins them by the use of disjoint union to create all possible variants of digraphs representing the polynomial realisation. When multiplying the characteristic polynomial (1) by $w_1^{-n_1} w_2^{-n_2} \dots w_j^{-n_j}$ for $j = 1, 2, \dots \infty$, we obtain

$$p\left(w_1, w_2, \dots, w_j\right) = 1 - \sum_{i_1=0}^{n_1} \sum_{i_2=1}^{n_2} \dots \sum_{i_j=0}^{n_j} a_{i_1 i_2 \dots i_j} w_1^{i_1-n_1} w_2^{i_2-n_2} \dots w_j^{i_j-n_j} \tag{3}$$

The method finds state matrices $\mathbf{A}_k$, $k = 1, 2, \dots, \infty$ using decomposing characteristic polynomial (3). In the first step, we decompose polynomial (3) into a set of simple monomials $M_1, \dots, M_j$. The factor of 1 is a special case, as it is used in the topology to represent digraph vertices, so polynomial (3) can be represented as

$$p\left(w_1, w_2, \dots, w_j\right) = 1 - M_1 - \dots - M_j. \tag{4}$$

For each of monomials $M_1, \dots, M_j$ we create digraph representation that consists of n vertices and one n-arc cycle, where n is a sum of powers of all variables of the monomial. Each monomial digraph in fact represents a simple polynomial digraph $1 - M_i$, $i = 1, \dots, j$. After creation of all digraph representations of monomials in the polynomial, we can determine all possible characteristic polynomial realisations using all combinations of the digraph monomial representations. Finally, we combine received digraphs into one digraph which is corresponding to the characteristic polynomial (3) by disjoint union of monomial digraphs.

Example 2 Lets take polynomial $p_1(z_1, z_2) = 1 - z_1^{-2} z_2^{-1} - z_1^{-1} z_2^{-1}$. To create digraph representation for it we need first to create digraph representation for two monomials. Representation for the first monomial is presented on Fig. 2a and for the second on Fig. 2b. For simple polynomial $p_2(z_1, z_2) = 1 - z_1^{-2} z_2^{-1}$ digraph representation would be the same as for monomial (and presented on Fig. 2a). To achieve digraph representation for our polynomial p_1 we need to add digraph representations of monomials by means of disjoint union (without creation of multiarcs of the same colour). On Fig. 2c presented is one of possible digraph representations for polynomial p_1. Another, presented on Fig. 2d, can be obtained by adding second monomial to vertices 2 and 3 instead of vertices 1 and 2. As can be seen those aren't all the possible representations of polynomial p_1.

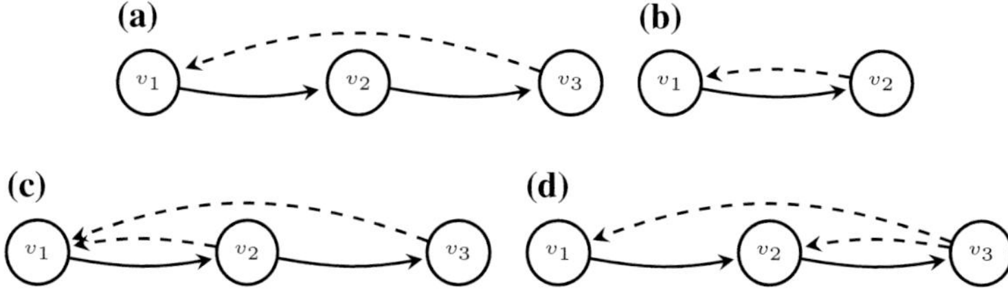

Fig. 2 **a** Digraph $\mathfrak{D}_1^{(2)}$ corresponding to polynomial $1 - z_1^{-2}z_2^{-1}$; **b** Digraph $\mathfrak{D}_2^{(2)}$ corresponding to polynomial $1 - z_1^{-1}z_2^{-1}$; Sample polynomial digraph: corresponding to union of digraphs **c** $\mathfrak{D}_3^{(2)} = \mathfrak{D}_1^{(2)} + \mathfrak{D}_2^{(2)}$; **d** $\mathfrak{D}_4^{(2)} = \mathfrak{D}_1^{(2)} + \mathfrak{D}_2^{(2)}$

2 Problem Formulation

The algorithm presented in [5, 7] is based on the multi-dimensional digraph theory to allow the creation of a complete set of solutions of characteristic polynomial realisations—this is what differs the method from other state-of-the-art solutions like canonical forms, as they are capable of finding only a few of existing realisations. As algorithm is able to find all the possible structures there is the need of checking the validity of them, as not all digraph structures created from monomial sub-graphs, according to the principles presented in Sect. 1.3, are a valid digraph representation of the characteristic polynomial. From some of them it is impossible to obtain state matrices that will satisfy the polynomial, while others generate solutions for which it is needed to get the coefficients of state matrices by solving a system of polynomial equations that in some cases can be under-determined. Those structures were in previous articles marked as invalid for reasons of different method of solving, slowing down the algorithm or removing the advantage of checking the matrix structure directly from digraph, but need to be examined in more detail to find all possible proper digraph structures for the characteristic polynomial and that is the scope of this article.

3 Classes of Digraph Structures

Extensive study and experimentation shoved that obtained digraph structures can be grouped into three classes. Some structures are valid for all possible coefficients of characteristic polynomial (given in symbolic form) and have minimal number of arcs needed. Those structures were examined in detail in [7] and here are denoted as class $\mathcal{K}_1$. Some structures give proper solution for given coefficients of characteristic polynomial—their structure can contain some additional arcs and there is need to solve a set of linear equations to get wages of digraph arcs. Those are denoted as class $\mathcal{K}_3$. And there are structures that cannot guarantee proper solution for given characteristic polynomial (or in some specific cases we are unable to determine if

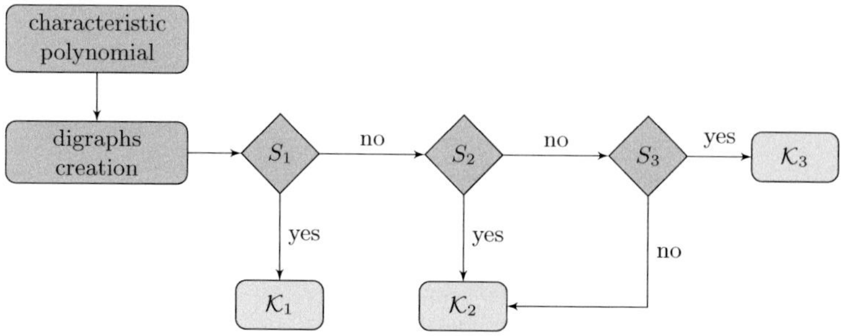

Fig. 3 Classes structure

the solution is possible) that are denoted as class $\mathcal{K}_2$. Figure 3 illustrates how we determine to which class given digraph structure belongs. Conditions S_1, S_2 and S_3 presented on the Fig. 3 are stated below.

Condition S_1: There exist positive state matrices of the discrete time linear system corresponding to the characteristic polynomial (3) if for digraph $\mathfrak{D}^{(n)} = \mathfrak{D}_1^{(n)} + \mathfrak{D}_2^{(n)} + \cdots + \mathfrak{D}_M^{(n)}$ all of the following conditions are met:

(S_{1a}): $\mathbb{V}_1 \cap \mathbb{V}_2 \cap \cdots \cap \mathbb{V}_M \neq \{\emptyset\}$,
(S_{1b}): the number of cycles in digraph $\mathfrak{D}^{(n)}$ equals M;

where M is a number of monomials in characteristic polynomial and $\mathbb{V}_k$ is a set of vertices of digraph $\mathfrak{D}_k^{(n)}$ of k-th monomial.

Condition S_2: Every digraph structure belongs to class $\mathcal{K}_2$ and cannot satisfy the given characteristic polynomial (3) if there exists a single cycle that is representing any of terms that is not existing in that polynomial (i.e. has its a_{i_n} wage equal to zero).

Condition S_3: If for every term not existing in characteristic polynomial (3) (i.e. with a_{i_n} wage equal to zero) there exist none or at least two cycles corresponding to that term and the resultant system of equations is not under-determined (i.e. the number of unknowns does not outnumber the number of equations) we can determine the wages for all arcs that satisfy given characteristic polynomial and digraph structure belongs to class $\mathcal{K}_3$.

Class $\mathcal{K}_1$: Digraph structures belonging to class $\mathcal{K}_1$ satisfy all characteristic polynomials of given type (with the same number and power of terms) for any $a_{i_1,i_2,\ldots,i_j}! = 0$ wages. Those are digraph structures that are the most thoroughly examined in previous papers and that can be computed quickly using digraph-based GPGPU (General-Purpose Computation on Graphics Processing Units) methods as there is no need of solving a system of polynomial equations.

Class $\mathcal{K}_2$: Digraph structures belonging to class $\mathcal{K}_2$ cannot satisfy the given characteristic polynomial (or we are unable to determine the solution due to problem with solving a system of under-determined equations) and are considered invalid for given characteristic polynomial. It is worth noting that in case of fulfilling S_2 condition such structures will be improper solutions for all characteristic polynomials with the same terms, no matter the $a_{i_1,i_2,\ldots,i_j}! = 0$ wages and in case of not fulfilling S_3 condition such structures are only improper for given wages of characteristic polynomial and possibly can be made proper with change of wages of characteristic polynomial's terms.

Class $\mathcal{K}_3$: Digraph structures belonging to class $\mathcal{K}_3$ satisfy given characteristic polynomial with specific $a_{i_1,i_2,\ldots,i_j}$ wages, but unlike class $\mathcal{K}_1$ structures cannot be computed directly using digraph-based method and solving a set of equations is also needed, which significantly slows down the algorithm of finding them.

4 Example

4.1 *Class* $\mathcal{K}_1$

Example 3 Let as consider the following example. For the given characteristic polynomial

$$d(z) = 1 - z^{-1} - z^{-2} - z^{-3} \tag{5}$$

determine entries of the state matrix **A** using digraph theory.

In the first step we write the following initial conditions: number of colours in digraph: *colour* := 1; monomials: $M_1 = z^{-1}$; $M_2 = z^{-2}$; $M_3 = z^{-3}$. For every simple monomial M_1, M_2 and M_3 we determine all possible realisations using digraph theory. On the Fig. 4a is presented digraph structure realisation of the monomial M_1; on Fig. 4b is presented digraph structure realisation of the monomial M_2 and on Fig. 4c is presented digraph structure realisation of the monomial M_3. In the next step using all combinations of the digraph monomial representations we determine all possible digraph structure, which satisfy characteristic polynomial (5). One of the possible digraph structures, which satisfies Condition S_1, is presented on Fig. 5.

Fig. 4 Monomials

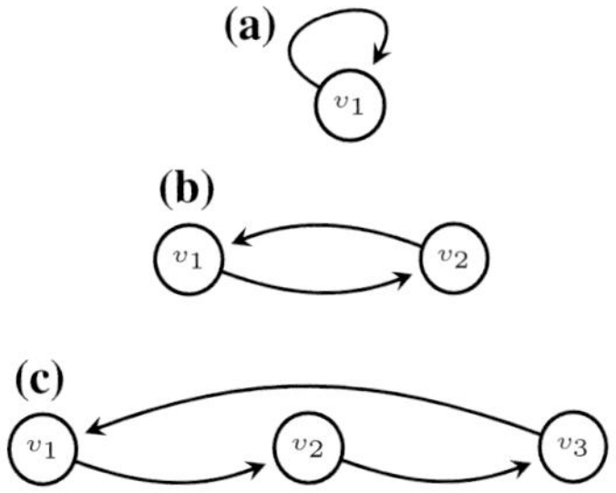

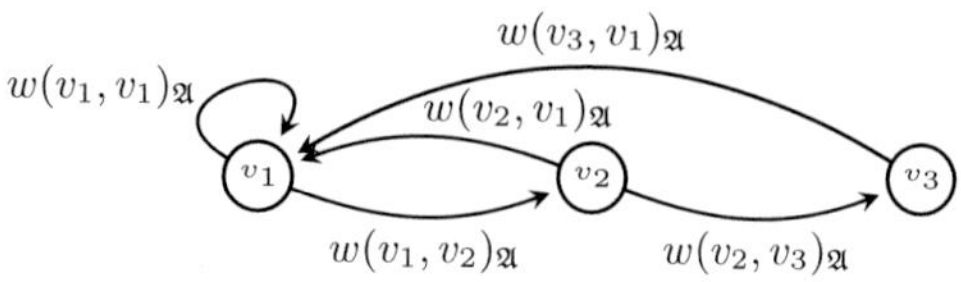

Fig. 5 One-dimensional digraph corresponding to characteristic polynomial (5)

Finally, we write state matrices in the following form:

$$\mathbf{A} = \begin{bmatrix} w(v_1, v_1)_{\mathfrak{A}} & w(v_2, v_1)_{\mathfrak{A}} & w(v_3, v_1)_{\mathfrak{A}} \\ w(v_1, v_2)_{\mathfrak{A}} & 0 & 0 \\ 0 & w(v_2, v_3)_{\mathfrak{A}} & 0 \end{bmatrix} \tag{6}$$

4.2 Class $\mathcal{K}_2$

Example 4 Let as consider the following example. For the given characteristic polynomial

$$d(z_1, z_2) = 1 - z_1^{-2} z_2^{-1} - z_1^{-2} - z_1^{-1} z_2^{-1} \tag{7}$$

determine entries of the state matrices $\mathbf{A}_1$ and $\mathbf{A}_2$ using digraph theory.

Solution. On Fig. 6 is presented digraph structure corresponding to the characteristic polynomial (7). Considered digraph consist from three digraphs (presented on Fig. 7), corresponding to monomials: $M_1 = z_1^{-2} z_2^{-1}$ (see Fig. 7a); $M_2 = z_1^{-2}$ (see Fig. 7b); $M_3 = z_1^{-1} z_2^{-1}$ (see Fig. 7c).

Therefore investigated digraph does not belong to the class S_1 (as the condition (S_{1_a}) is not met). In digraph structure presented on Fig. 6 appears additional cycle presented on Fig. 8 and digraph structure belong to the class S_2 (as it has exactly one redundant cycle in the digraph). Additional cycle in digraph makes that in the characteristic polynomial appears additional monomial that should not be represented. The characteristic polynomial will have the following structure: $\widetilde{d}(z_1, z_2) = d(z_1, z_2) + z_1^{-1} z_2^{-2}$. In this class of digraph we can not determine arcs weights fulfilling the characteristic polynomial (7).

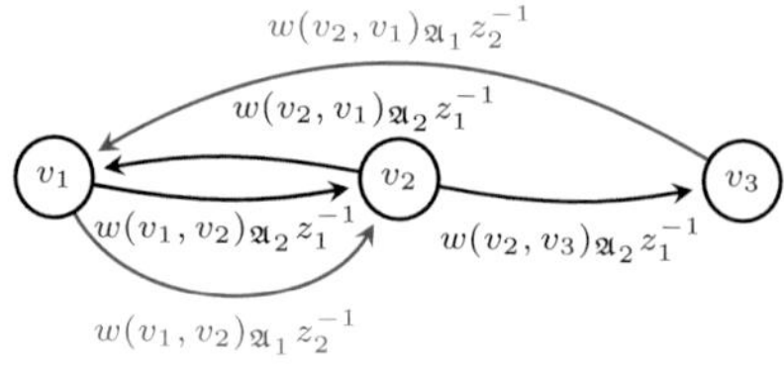

Fig. 6 A two-dimensional digraph $\mathfrak{D}^{(2)}$ structure corresponding to (7)

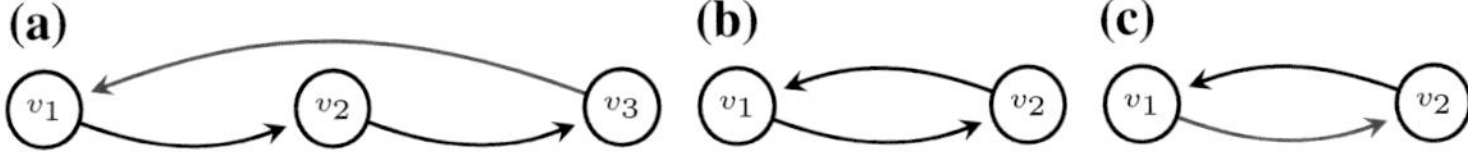

Fig. 7 Monomials

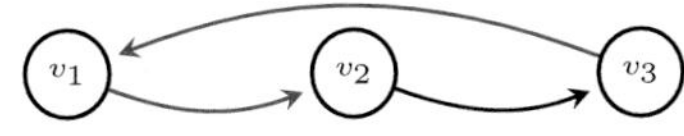

Fig. 8 Additional cycle in two-dimensional digraph structure

4.3 Class $\mathcal{K}_3$

Example 5 Let as consider the following example. For the given characteristic polynomial

$$d(z_1, z_2) = 1 - a_1 z_1^{-2} - a_2 z_2^{-2} \tag{8}$$

determine entries of the state matrices $\mathbf{A}_1$ and $\mathbf{A}_2$ using digraph theory.

Solution. On Fig. 9 is presented digraph structure corresponding to the characteristic polynomial (8). Considered digraph consist from four digraphs: two digraphs corresponding to the monomials M_1 (see Fig. 10a) and M_2 (see Fig. 10b) and two additional digraphs (see Fig. 10c, d) which generate additional monomials M_3 and M_4 that are not occurring in the characteristic polynomial (8). Therefore investigated digraph does not belong to the class S_1 (as the conditions (S_{1_a}) and (S_{1_b}) are not met) and does not belong to the class S_2 (as it has more than one redundant cycle in the graph).

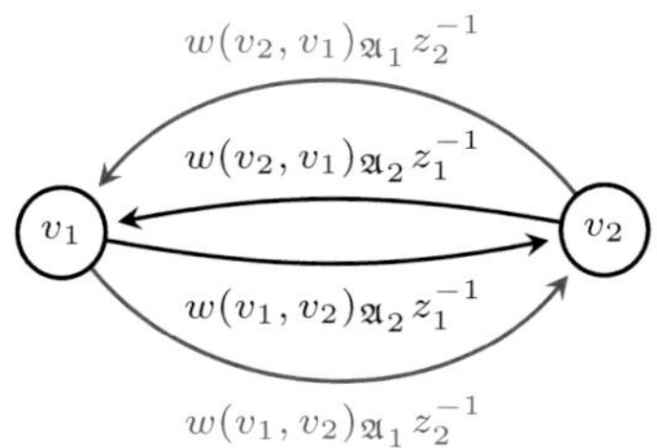

Fig. 9 A two-dimensional digraph $\mathfrak{D}^{(2)}$

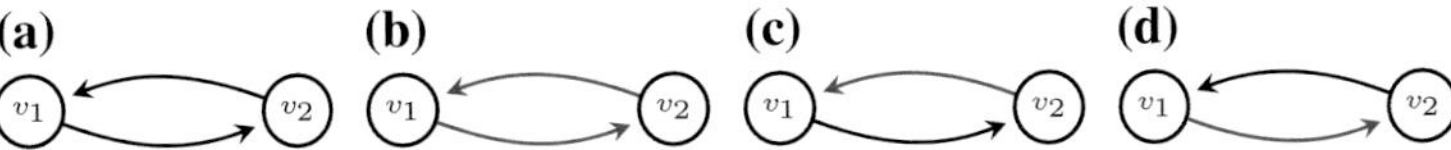

Fig. 10 Monomials

Using digraph structure we have the following characteristic polynomial:

$$d(z_1,z_2) = 1 - \overbrace{\underbrace{w(v_1,v_2)_{\mathfrak{A}_2} w(v_2,v_1)_{\mathfrak{A}_2} z_1^{-2}}_{M_1}}^{a_1} - \overbrace{\underbrace{w(v_1,v_2)_{\mathfrak{A}_1} w(v_2,v_1)_{\mathfrak{A}_1} z_2^{-2}}_{M_2}}^{a_2} - \underbrace{w(v_1,v_2)_{\mathfrak{A}_2} w(v_2,v_1)_{\mathfrak{A}_1} z_1^{-1} z_2^{-1}}_{M_3} - \underbrace{w(v_1,v_2)_{\mathfrak{A}_1} w(v_2,v_1)_{\mathfrak{A}_2} z_1^{-1} z_2^{-1}}_{M_4} \tag{9}$$

In this situation we have four possible variants:

V_1 : In the first variant the following relations must be satisfied:

$$\pm\, w(v_1,v_2)_{\mathfrak{A}_2} \cdot w(v_2,v_1)_{\mathfrak{A}_1} \mp w(v_1,v_2)_{\mathfrak{A}_1} \cdot w(v_2,v_1)_{\mathfrak{A}_2} = 0, \tag{10}$$

(a) The cycles cancel each other. In this case the following relations must be satisfied:

$$4 \cdot w(v_1,v_2)_{\mathfrak{A}_2} \cdot w(v_2,v_1)_{\mathfrak{A}_2} \cdot w(v_1,v_2)_{\mathfrak{A}_1} \cdot w(v_2,v_1)_{\mathfrak{A}_1} \leqslant 0. \tag{11}$$

If the conditions: (10) and (11) are met, then there exist a solution in the class $\mathcal{K}_3$.

(b) The cycles do not cancel each other. In this case the following relations must be satisfied:

$$4 \cdot w(v_1,v_2)_{\mathfrak{A}_2} \cdot w(v_2,v_1)_{\mathfrak{A}_2} \cdot w(v_1,v_2)_{\mathfrak{A}_1} \cdot w(v_2,v_1)_{\mathfrak{A}_1} \geqslant 0. \tag{12}$$

If the conditions: (10) and (12) are met, then there exist a solution in the class $\mathcal{K}_2$.

V_2 : In the second variant the following relations must be satisfied:

$$\pm\, w(v_1,v_2)_{\mathfrak{A}_2} \cdot w(v_2,v_1)_{\mathfrak{A}_1} \cdot w(v_1,v_2)_{\mathfrak{A}_1} \cdot w(v_2,v_1)_{\mathfrak{A}_2} \neq 0 \tag{13}$$

(a) The cycles cancel each other. In this case the following relations must be satisfied:

$$4 \cdot w(v_1,v_2)_{\mathfrak{A}_2} \cdot w(v_2,v_1)_{\mathfrak{A}_2} \cdot w(v_1,v_2)_{\mathfrak{A}_1} \cdot w(v_2,v_1)_{\mathfrak{A}_1} \leqslant \left(w(v_1,v_2)_{\mathfrak{A}_2} \cdot w(v_2,v_1)_{\mathfrak{A}_1} \cdot w(v_1,v_2)_{\mathfrak{A}_1} \cdot w(v_2,v_1)_{\mathfrak{A}_2}\right)^2 \tag{14}$$

If the conditions: (13) and (14) are met, then there exist a solution in the class $\mathcal{K}_3$.

(b) The cycles do not cancel each other. In this case the following relations must be satisfied:

$$4 \cdot w(v_1, v_2)_{\mathfrak{A}_2} \cdot w(v_2, v_1)_{\mathfrak{A}_2} \cdot w(v_1, v_2)_{\mathfrak{A}_1} \cdot w(v_2, v_1)_{\mathfrak{A}_1} \geqslant \left(w(v_1, v_2)_{\mathfrak{A}_2} \cdot w(v_2, v_1)_{\mathfrak{A}_1} \cdot w(v_1, v_2)_{\mathfrak{A}_1} \cdot w(v_2, v_1)_{\mathfrak{A}_2}\right)^2 \quad (15)$$

If the conditions: (13) and (15) are met, then there exist a solution in the class $\mathcal{K}_2$.

5 Concluding Remarks

In this paper there is introduced the first classification of digraph structures that are used to solve characteristic polynomials. Three classes of such structures are determined along with conditions how to classify digraph solutions obtained with parallel algorithm into each of classes. This allows to fully check the validity of solutions for given characteristic polynomial and determine if we want only solutions that can be obtained in fast and easier way using the algorithm (the ones in $\mathcal{K}_1$ class) or if we want to check all solutions, despite the need to solve a system of polynomial equations (adding solutions from $\mathcal{K}_3$ class). We can also determine that given digraph structure is invalid solution for given characteristic polynomial. Such basic classification is the first step for determination of properties of different digraph structures (like how can be check reachability and availability from digraph only, without need of system matrices) and introducing methods for finding best solutions fast.

References

1. Bang-Jensen, J., Gutin, G.: Digraphs: Theory, Algorithms and Applications, 2nd edn. Springer, London (2009)
2. Blyth, T., Robertson, E.: Basic Linear Algebra, 2nd edn. Springer, London (2002)
3. Fornasini, E., Valcher, M.E.: Directed graphs, 2D state models, and characteristic polynomials of irreducible matrix pairs. Linear Algebra Appl. **263**, 275–310 (1997)
4. Fornasini, E., Valcher, M.E.: On the positive reachability of 2D positive systems. LCNIS 297–304 (2003)
5. Hryniów, K., Markowski, K.A.: Parallel digraphs-building algorithm for polynomial realisations. In: Proceedings of 2014 15th International Carpathian Control Conference (ICCC), pp. 174–179 (2014). http://dx.doi.org/10.1109/CarpathianCC.2014.6843592
6. Hryniów, K., Markowski, K.A.: Digraphs minimal realisations of state matrices for fractional positive systems. In: R. Szewczyk, C. Zieliński, M. Kaliczyńska (eds.) Progress in Automation, Robotics and Measuring Techniques, Advances in Intelligent Systems and Computing, vol. 350, pp. 63–72. Springer (2015). doi:10.1007/978-3-319-15796-2_7. http://dx.doi.org/10.1007/978-3-319-15796-2_7
7. Hryniów, K., Markowski, K.A.: Parallel multi-dimensional digraphs-building algorithm for finding a complete set of positive characteristic polynomial realisations of dynamic system. Appl. Math. Comput. (Submitted to)
8. Kaczorek, T.: Polynomial and Rational Matrices. Springer, London (2007)
9. Wallis, W.D.: A Beginner's Guide to Graph Theory. Biiokhäuser (2007)

Reachability of Standard and Fractional Continuous-Time Systems with Piecewise Constant Inputs

Krzysztof Rogowski

Abstract The reachability of standard and fractional-order continuous-time systems with piecewise constant inputs is addressed. Necessary and sufficient conditions for the existence of such piecewise constant inputs that steers the system from zero initial conditions to the given final state in desired time are derived and proved. As examples of such systems the electrical circuits with DC switched voltage sources are presented.

Keywords Piecewise constant input · Reachability · Fractional system

1 Introduction

Reachability is one of the most important structural property of dynamical systems [1, 3, 9]. It appears in many different areas, such as computational models, celluar automata, Petri nets, decision procedures, game theory, etc. The problem of reachability of positive systems has been addressed in [4, 10]. Reachability of linear hybrid systems described by the general model have been considered in [8].

To the best knowledge of the author reachability of fractional continuous-time systems when the input is a piecewise constant function has not been considered yet.

The structure of the paper is following. In Sect. 2 the reachability of standard continuous-time systems with piecewise constant inputs is considered. Necessary and sufficient conditions for the existence of piecewise constant input that steers the systems from zero initial conditions to the given final states are given. Similar problem will be solved for fractional systems in Sect. 3. Concluding remarks are given in Sect. 4. The considerations are illustrated by examples of electrical circuits with DC switched voltage sources.

K. Rogowski (✉)
Faculty of Electrical Engineering, Bialystok University of Technology,
Wiejska 45D, 15-351 Bialystok, Poland
e-mail: k.rogowski@pb.edu.pl

R. Szewczyk et al. (eds.), *Challenges in Automation, Robotics and Measurement Techniques*, Advances in Intelligent Systems and Computing 440, DOI 10.1007/978-3-319-29357-8_31

The following notation will be used. $\mathbb{R}$ is the set of real numbers, $\mathbb{R}^{n\times m}$—the set of $n \times m$ matrices with real entries. The identity matrix of size n will be denoted by $\mathbb{I}_n$.

2 Reachability of Standard Continuous-Time Linear Systems with Piecewise Constant Inputs

Let us consider the standard continuous-time system described by the state equation [3, 9]

$$\frac{\mathrm{d}x(t)}{\mathrm{d}t} = Ax(t) + Bu(t), \tag{1}$$

where $x(t) \in \mathbb{R}^n$ is the state vector, $u(t) \in \mathbb{R}^m$ is the input vector of the system for $t \geq 0$ and matrices $A \in \mathbb{R}^{n\times n}$, $B \in \mathbb{R}^{n\times m}$.

The solution to the state equation (1) with initial condition $x_0 = x(0) \in \mathbb{R}^n$ and input vector $u(t) \in \mathbb{R}^m$ for $t \geq 0$ is given by [3, 9]

$$x(t) = e^{At}x_0 + \int_0^t e^{A(t-\tau)}Bu(\tau)\mathrm{d}\tau = e^{At}x_0 + \int_0^t e^{A\tau}Bu(t-\tau)\mathrm{d}\tau. \tag{2}$$

Now, we will consider the continuous-time system (1) with piecewise constant inputs vector $u(t)$, i.e.

$$u(t) = \begin{cases} U_1 \text{ for } 0 \leq t < t_1 \\ U_2 \text{ for } t_1 \leq t < t_2 \\ \quad\vdots \end{cases} \tag{3}$$

where $U_1, U_2, \ldots \in \mathbb{R}^m$ are input vectors with constant components.

Definition 1 The standard continuous-time linear system (1) is called reachable for piecewise constant inputs in time $t_f > 0$ if there exists such piecewise constant input vector (3), that steers the system from zero initial conditions $x_0 = x(0) = 0$ to arbitrary final state $x_f = x(t_f) \in \mathbb{R}^n$.

Let us assume, that $\det A \neq 0$ and t_f is divided into r periods in which the input vector takes constant values $U_1, U_2, \ldots, U_r$. Then, the following theorem holds.

Theorem 1 *The system (1) is reachable for piecewise constant input in time $t_f > 0$ if and only if the matrix*

$$\begin{aligned} R_c = &\left[e^{At_f} - e^{A(t_f-t_1)}\; e^{A(t_f-t_1)} - e^{A(t_f-t_2)}\; \ldots\; e^{A(t_f-t_{r-1})} - \mathbb{I}_n\right] \\ &\times \text{blockdiag}[A^{-1}B, \ldots, A^{-1}B] \in \mathbb{R}^{n\times rm} \end{aligned} \tag{4}$$

has n linearly independent columns, i.e.

$$\operatorname{rank} R_c = n. \tag{5}$$

Proof Substituting (3) into (2) for zero initial conditions and $t = t_f > 0$ we obtain

$$\begin{aligned}
x(t_f) = x_f &= \int_0^{t_f} e^{A(t_f-\tau)} B u(\tau) \mathrm{d}\tau \\
&= \int_0^{t_1} e^{A(t_f-\tau)} B U_1 \mathrm{d}\tau + \int_{t_1}^{t_2} e^{A(t_f-\tau)} B U_2 \mathrm{d}\tau + \cdots + \int_{t_{r-1}}^{t_f} e^{A(t_f-\tau)} B U_1 \mathrm{d}\tau \\
&= \left[e^{At_f} - e^{A(t_f-t_1)}\right] A^{-1} B U_1 + \left[e^{A(t_f-t_1)} - e^{A(t_f-t_2)}\right] A^{-1} B U_2 \\
&\quad + \left[e^{A(t_f-t_{r-1})} - \mathbb{I}_n\right] A^{-1} B U_r \\
&= \left[e^{At_f} - e^{A(t_f-t_1)} \;\; e^{A(t_f-t_1)} - e^{A(t_f-t_2)} \;\; \ldots \;\; e^{A(t_f-t_{r-1})} - \mathbb{I}_n\right] \\
&\quad \times \operatorname{blockdiag}[A^{-1}B, \ldots, A^{-1}B] \begin{bmatrix} U_1 \\ U_2 \\ \vdots \\ U_r \end{bmatrix} = R_c U,
\end{aligned} \tag{6}$$

since [6]

$$\int e^{A\tau} \mathrm{d}\tau = \left(e^{A\tau} - \mathbb{I}_n\right) A^{-1}.$$

The Eq. (6) has a piecewise constant solution U for arbitrary final state $x_f \in \mathbb{R}^n$ if and only if the matrix R_c satisfies the condition (5). □

The piecewise constant input that steers the system (1) from zero initial state to the final state x_f in time $t \in [0, t_f]$ is given by the formula

$$U = R_c^+ x_f, \tag{7}$$

where $R_c^+ \in \mathbb{R}^{rm \times n}$ is the right pseudoinverse of the rectangular matrix R_c given by [2, 6]

$$R_c^+ = R_c^T \left[R_c R_c^T\right]^{-1} + \left(\mathbb{I}_{rm} - R_c^T \left[R_c R_c^T\right]^{-1} R_c\right) K_1 \text{ for arbitrary } K_1 \in \mathbb{R}^{rm \times n}; \tag{8a}$$

$$R_c^+ = K_2 \left[R_c K_2\right]^{-1} \quad \text{for arbitrary} \quad K_2 \in \mathbb{R}^{rm \times n}, \ \det\left[R_c K_2\right] \neq 0. \tag{8b}$$

Example 1 Consider the electrical circuit shown in Fig. 1 with given resistances $R_1 = 0.1\,\Omega$, $R_2 = 1\,\Omega$, $R_3 = 2\,\Omega$, inductance $L = 0.5$ H and capacitance $C = 0.1$ F.

Using Kirchhoff's laws we may formulate the state equations (1) of the circuit shown in Fig. 1

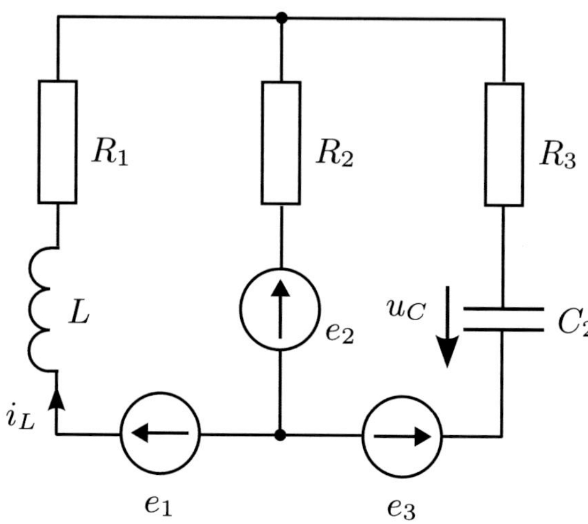

Fig. 1 Electrical circuit of Example 1

$$\frac{\mathrm{d}}{\mathrm{d}t}\begin{bmatrix} i_L(t) \\ u_C(t) \end{bmatrix} = A\begin{bmatrix} i_L(t) \\ u_C(t) \end{bmatrix} + B\begin{bmatrix} e_1(t) \\ e_2(t) \\ e_3(t) \end{bmatrix}, \tag{9a}$$

where

$$A = \begin{bmatrix} -\frac{R_1}{L} - \frac{R_2R_3}{L(R_2+R_3)} & \frac{R_2}{L(R_2+R_3)} \\ -\frac{R_2}{C(R_2+R_3)} & -\frac{1}{C(R_2+R_3)} \end{bmatrix} = \begin{bmatrix} -1.53 & 0.67 \\ -3.33 & -3.33 \end{bmatrix}, \tag{9b}$$

$$B = \begin{bmatrix} \frac{1}{L} & -\frac{R_3}{L(R_2+R_3)} & \frac{R_2}{L(R_2+R_3)} \\ 0 & -\frac{1}{C(R_2+R_3)} & \frac{1}{C(R_2+R_3)} \end{bmatrix} = \begin{bmatrix} 2 & -1.33 & -0.67 \\ 0 & -3.33 & 3.33 \end{bmatrix}. \tag{9c}$$

Let us assume that $r = 3$, and $t_1 = 1$ s, $t_2 = 2$ s and $t_f = 3$ s. Then the inputs takes three constant value components vectors E_1 for $t \in [0, 1)$, E_2 for $t \in [1, 2)$ and E_3 for $t \in [2, 3)$. We will show that the electrical circuit described by the Eq. (9a) with matrices (9b) and (9c) is reachable for piecewise constant inputs $U = \begin{bmatrix} E_1 & E_2 & E_3 \end{bmatrix}^T$ in time $t_f = 3$ s.

Using Theorem 1 we have

$$R_c = 10^{-2} \times \begin{bmatrix} -0.36 & 0.04 & 0.32 & 4.81 & -9.11 & 4.30 & 86.52 & -81.92 & -4.60 \\ -0.58 & 1.51 & -0.92 & -17.64 & 19.60 & -1.96 & -72.79 & -30.13 & 102.92 \end{bmatrix} \tag{10}$$

and the condition (5) is met, since

$$\operatorname{rank} R_c = n = 2. \tag{11}$$

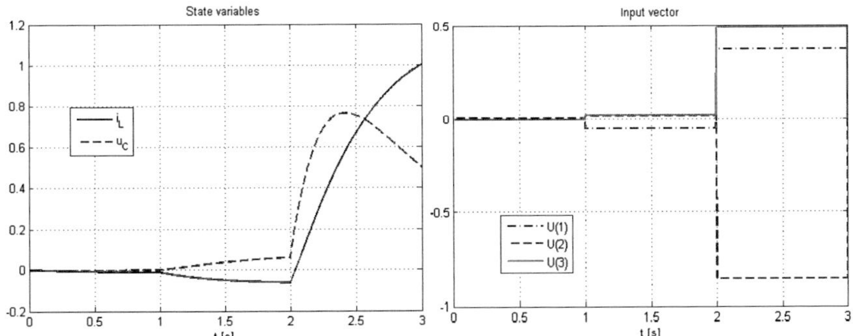

Fig. 2 State variables and inputs of Example 1

Therefore, there exists piecewise constant input U that steers the circuit from zero initial conditions $x_0 = \left[i_L(0)\ u_C(0)\right]^T = \left[0\ 0\right]^T$ to the arbitrary given final state $x_f = x(t_f) = \left[i_L(t_f)\ u_C(t_f)\right]^T$.

Let $i_L(t_f) = i_L(3) = 1$ A, $u_C(t_f) = u_C(3) = 0.5$ V. Then, the piecewise constant inputs can be computed using (7) and (8a) with $K_1 = 0$

$$U = \begin{bmatrix} E_1 \\ E_2 \\ E_3 \end{bmatrix} = R_c^T \left[R_c R_c^T\right]^{-1} x_f = 10^{-2} \times \begin{bmatrix} -0.61 \\ 0.81 \\ -0.20 \\ -4.87 \\ 2.17 \\ 2.70 \\ 37.25 \\ -85.84 \\ 48.59 \end{bmatrix}. \tag{12}$$

The state variables of the electrical circuit with piecewise constant inputs (12) are shown in Fig. 2.

3 Reachability of Fractional Continuous-Time Systems with Piecewise Constant Inputs

Let us consider the fractional continuous-time system described by the state equation [5, 7]

$$D^\alpha x(t) = Ax(t) + Bu(t), \tag{13}$$

where $x(t) \in \mathbb{R}^n$ is the state vector, $u(t) \in \mathbb{R}^m$ is the input vector of the system for $t \geq 0$, matrices $A \in \mathbb{R}^{n \times n}$, $B \in \mathbb{R}^{n \times m}$ and

$$D^{\alpha}f(t) = \frac{\mathrm{d}^{\alpha}}{\mathrm{d}t^{\alpha}}f(t) = \frac{1}{\Gamma(N-\alpha)}\int_0^t (t-\tau)^{N-\alpha-1}\frac{\mathrm{d}^N f(\tau)}{\mathrm{d}t^N}\mathrm{d}\tau \tag{14}$$

is the α-order ($\alpha \in \mathbb{R}$) fractional derivative described by the Caputo operator, where $N-1 \leq \alpha < N$, $N \in \mathbb{N}$ and $\Gamma(x)$ is the Euler gamma function.

The solution to the state equation (13) with initial condition $x_0 = x(0) \in \mathbb{R}^n$ and input vector $u(t) \in \mathbb{R}^m$ for $t \geq 0$ is given by [5, 7]

$$x(t) = \Phi_0(t)x_0 + \int_0^t \Phi(t-\tau)Bu(\tau)\mathrm{d}\tau = \Phi_0(t)x_0 + \int_0^t \Phi(\tau)Bu(t-\tau)\mathrm{d}\tau, \tag{15}$$

where

$$\Phi_0(t) = \sum_{k=0}^{\infty}\frac{A^k t^{k\alpha}}{\Gamma(k\alpha+1)} = E_{\alpha}(At^{\alpha}), \tag{16a}$$

$$\Phi(t) = \sum_{k=0}^{\infty}\frac{A^k t^{(k+1)\alpha-1}}{\Gamma[(k+1)\alpha]} = t^{\alpha-1}E_{\alpha,\alpha}(At^{\alpha}). \tag{16b}$$

and $E_{\alpha}(x)$, $E_{\alpha,\beta}(x)$ are the one and two parameters Mittag-Leffler functions, respectively [5, 7].

Now we define the notion of reachability of the system described by the state equation (13) with piecewise constant inputs (3).

Definition 2 The fractional continuous-time system (13) is called reachable in time $t_f > 0$ if there exists a piecewise constant input vector (3) that steers the system from zero initial conditions $x_0 = x(0) = 0$ to every given final state $x_f = x(t_f) \in \mathbb{R}^n$.

Similarly as in previous section, let us assume that t_f is divided into r periods in which the input vector takes constant values $U_1, U_2, \ldots, U_r$. Then, the following theorem holds.

Theorem 2 *The fractional continuous-time system (13) is reachable for piecewise constant inputs in time $t_f > 0$ if and only if the matrix*

$$R_f = \left[E(t_0) - E(t_1)\ E(t_1) - E(t_2)\ \cdots\ E(t_{r-1}) - E(t_r)\right] \in \mathbb{R}^{n\times rm}, \tag{17}$$

where

$$E(t_i) = (t_f - t_i)^{\alpha}E_{\alpha,\alpha+1}[A(t_f - t_i)^{\alpha}]B$$

for $i = 0, 1, \ldots, r$ and $t_0 = 0$, $t_r = t_f$ has n linearly independent columns, i.e.

$$\mathrm{rank} R_f = n. \tag{18}$$

Proof Substituting (3) and (16b) into (15) for zero initial conditions and $t = t_f > 0$ we obtain

$$x(t_f) = x_f = \int_0^{t_f} \Phi(t_f - \tau)Bu(\tau)\mathrm{d}\tau$$
$$= \int_0^{t_1} \Phi(t_f - \tau)\mathrm{d}\tau BU_1 + \int_{t_1}^{t_2} \Phi(t_f - \tau)\mathrm{d}\tau BU_2 + \cdots + \int_{t_{r-1}}^{t_f} \Phi(t_f - \tau)\mathrm{d}\tau BU_r. \tag{19}$$

Taking into account, that

$$\int \Phi(t - \tau)\mathrm{d}\tau = -(t - \tau)^{\alpha} E_{\alpha,\alpha+1}[A(t - \tau)^{\alpha}]$$

and from the properties of the gamma function $[(k+1)\alpha]\Gamma[(k+1)\alpha] = \Gamma[(k+1)\alpha + 1]$ we have

$$x_f = \begin{bmatrix} E(t_0) - E(t_1) & E(t_1) - E(t_2) & \cdots & E(t_{r-1}) - E(t_r) \end{bmatrix} \begin{bmatrix} U_1 \\ U_2 \\ \vdots \\ U_r \end{bmatrix} = R_f U. \tag{20}$$

From (20) it follows that this equation has a solution U for given final state x_f and given final time $t_f > 0$ if and only if the condition (18) is satisfied. □

The piecewise constant input that steers the system (13) from zero initial state to the final state x_f in time $t \in [0, t_f]$ is given by the formula

$$U = R_f^+ x_f, \tag{21}$$

where $R_f^+ \in \mathbb{R}^{rm\times n}$ is the right pseudoinverse of the rectangular matrix R_f given by one of the following equalities [2, 6]

$$R_f^+ = R_f^T \left[R_f R_f^T\right]^{-1} + \left(\mathbb{I}_{rm} - R_f^T \left[R_f R_f^T\right]^{-1} R_f\right) K_1 \text{ for arbitrary } K_1 \in \mathbb{R}^{rm\times n}; \tag{22a}$$

$$R_f^+ = K_2 \left[R_f K_2\right]^{-1} \quad \text{for arbitrary} \quad K_2 \in \mathbb{R}^{rm\times n}, \ \det\left[R_f K_2\right] \neq 0. \tag{22b}$$

Example 2 Consider the fractional electrical circuit from Example 1 shown in Fig. 1 with $\alpha = 0.8$.

Using Kirchhoff's laws for the fractional electrical circuit shown in Fig. 1, it can be easily shown, that the state equation of this circuit has the form

$$\frac{d^\alpha}{dt^\alpha}\begin{bmatrix} i_L(t) \\ u_C(t) \end{bmatrix} = A\begin{bmatrix} i_L(t) \\ u_C(t) \end{bmatrix} + B\begin{bmatrix} e_1(t) \\ e_2(t) \\ e_3(t) \end{bmatrix} \tag{23}$$

with the matrices given by (9b) and (9c).

We will show that the fractional electrical circuit is reachable for constant input vector $U = \left[E_1\ E_2\ E_3\right]^T$ in time $t_f = 3$ s.

Let us assume that $r = 3$, and $t_1 = 1$ s, $t_2 = 2$ s and $t_f = 3$ s. Then the inputs takes three constant value components vectors E_1 for $t \in [0, 1)$, E_2 for $t \in [1, 2)$ and E_3 for $t \in [2, 3)$. We will show that the fractional electrical circuit described by the Eq. (23) with matrices (9b) and (9c) is reachable for piecewise constant inputs $U = \left[E_1\ E_2\ E_3\right]^T$ in time $t_f = 3$ s.

Using Theorem 2 we have

$$\begin{aligned} R_f &= \left[E(t_0) - E(t_1)\ E(t_1) - E(t_2)\ E(t_2)\right] \\ &= 10^{-2} \times \begin{bmatrix} 1.69 & -2.40 & 0.72 & 5.98 & -8.33 & 2.35 & 79.76 & -75.49 & -4.27 \\ -3.83 & 3.18 & 0.64 & -13.00 & 10.39 & 2.61 & -67.02 & -27.91 & 94.93 \end{bmatrix} \end{aligned} \tag{24}$$

and the condition (18) is met, since

$$\operatorname{rank} R_f = n = 2. \tag{25}$$

Therefore, there exists piecewise constant input vector U that steers the fractional electrical circuit form zero initial conditions $x_0 = \left[i_L(0)\ u_C(0)\right]^T = \left[0\ 0\right]^T$ to the arbitrary given final state $x_f = x(t_f) = \left[i_L(t_f)\ u_C(t_f)\right]^T$.

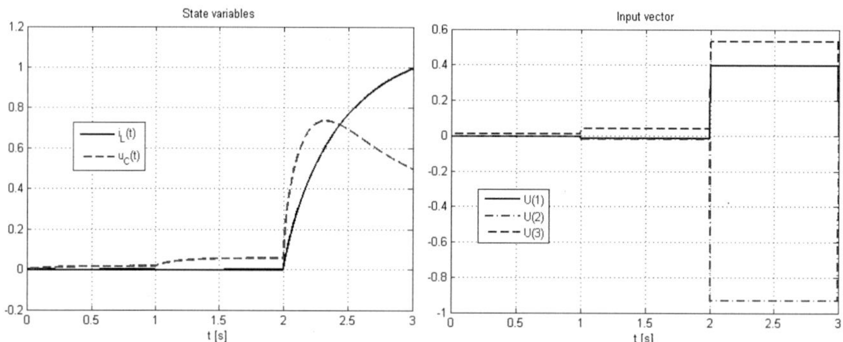

Fig. 3 State variables and inputs of Example 2

Let $i_L(t_f) = i_L(3) = 1$ A, $u_C(t_f) = u_C(3) = 0.5$ V. Then the piecewise constant input vector can be computed using (21) and (22a) for $K_1 = 0$

$$U = \begin{bmatrix} E_1 \\ E_2 \\ E_3 \end{bmatrix} = R_f^T \left[R_f R_f^T\right]^{-1} x_f \\ = 10^{-2} \times \left[-0.62\ -0.50\ 1.12\ -1.85\ -2.11\ 3.96\ 39.86\ -93.14\ 53.28\right]^T. \tag{26}$$

The state variables of the fractional electrical circuit with piecewise constant inputs (26) are shown in Fig. 3.

4 Concluding Remarks

The reachability of standard and fractional-order systems with piecewise constant inputs have been considered. Necessary and sufficient conditions for continuous-time systems have been established (Theorems 1 and 2). It has been shown that the conditions for reachability of systems with piecewise constant inputs are much more restrictive than for the systems with arbitrary (nonconstant) inputs (see [3, 5, 7]). The considerations are illustrated by example of electrical circuit with switched DC voltage sources.

An open problem is an extension of these considerations for descriptor (singular) standard and fractional-order systems and for two-dimensional systems.

Acknowledgments This work was supported by National Science Centre in Poland under work No. 2014/13/B/ST7/03467.

References

1. Antsaklis, P., Michel, A.N.: Linear Systems. Birkhäuser, Boston (2006)
2. Gantmacher, F.R.: The Theory of Matrices, vol. I and II. Chelsea Publishing Co., New York (1959) (Translated by K.A. Hirsch)
3. Kaczorek, T.: Linear Control Systems: Analysis of Multivariable Systems. Research Studies Press, New York (1992)
4. Kaczorek, T.: Positive 1D and 2D Systems. Communications and Control Engineering. Springer, London (2002)
5. Kaczorek, T.: Selected Problems of Fractional Systems Theory. Springer, Berlin (2011)
6. Kaczorek, T.: Vectors and Matrices in Automation and Electrotechnics. Scientific and Technical Publishing, Warsaw (1998). (in Polish)
7. Kaczorek, T., Rogowski, K.: Fractional Linear Systems and Electrical Circuits. Springer, Berlin (2015)
8. Kaczorek, T., Rogowski, K.: Reachability of linear hybrid systems described by the general model. Arch. Control Sci. **20**(2), 199–207 (2010)
9. Kailath, T.: Linear Systems. Prentice-Hall, New York (1980)
10. Valcher, M.E.: Reachability analysis for different classes of positive systems. In: Bru, R., Romero-Vivo, S. (eds.) Positive Systems. Lecture Notes in Control and Information Sciences, vol. 389, pp. 29–41. Springer, Heidelberg (2009)

Output Constraint Handling in Analytical MPC Algorithms Based on Hammerstein Models with Presumed Trajectory of Future Control Changes

Piotr M. Marusak

Abstract Novel numerically efficient, analytical MPC algorithms based on Hammerstein (nonlinear) models are proposed in the paper. They use the idea consisting in the assumption that the shape of a trajectory of future changes of a control signal is presumed in advance. In the proposed algorithms it is relatively easy to take output constraints into consideration on the entire prediction horizon. Thus, the constraint handling is very efficient because the control action can be appropriately modified many sampling instants before the potential constraint violation can occur.

Keywords Predictive control · Nonlinear control · Constrained control · Hammerstein models

1 Introduction

Novel analytical Model Predictive Control (MPC) algorithms based on Hammerstein models are proposed in the paper. In the algorithms trajectory of future control changes is presumed in advance. Therefore, despite the algorithms use a nonlinear model for prediction they need only very simple calculations to derive the control action (very simple control laws are obtained). The prediction can be used in the output constraint handling mechanism similar to the one described in [7]. This time, however, different prediction is used. Thus, the mechanism must have been reformulated. The new mechanism does not have the limitation that the control horizon must be of unity length.

The Hammerstein model consists of two blocks. The first one is a nonlinear static block. It precedes the second block—a linear dynamic one; see e.g. [2]. The Hammerstein model can be used to obtain prediction of process behavior in such a way

P.M. Marusak (✉)
Institute of Control and Computation Engineering, Warsaw University of Technology,
ul. Nowowiejska 15/19, 00-665 Warszawa, Poland
e-mail: P.Marusak@ia.pw.edu.pl

R. Szewczyk et al. (eds.), *Challenges in Automation, Robotics and Measurement Techniques*, Advances in Intelligent Systems and Computing 440, DOI 10.1007/978-3-319-29357-8_32

that numerically efficient algorithms can be obtained, see e.g. [5, 6, 8]. Assumption that future trajectory of changes of a control signal is presumed in advance can also contribute to simplification of the MPC algorithm, what will be demonstrated in the paper.

The presumed trajectory of changes of a control signal is the idea proposed first in the Predictive Functional Control (PFC) algorithm [1, 10, 11]. It consists in assuming that the trajectory of future control signal is described by a combination of some presumed functions (chosen during design of the algorithm). Application of this idea to analytical MPC algorithms based on linear models was described in [4], now it is used to modify MPC algorithms based on the Hammerstein (nonlinear) models. The proposed algorithms lack main disadvantages of the PFC algorithm listed in [10]. The algorithms are formulated in such a way that their features are preserved; they generate changes of the control signal and allow to use: long horizons, the control horizon, and the penalty for changes of the control signal. Thus, all tuning parameters typical for MPC algorithms are available in these algorithms, but it is also possible to formulate the analytical versions of the algorithms, and to choose the function describing future trajectory of changes of the control signal.

The output constraints often decide on safety and on economic efficiency of the process. Therefore, it is important to equip the analytical MPC controllers with an efficient mechanism of output constraint handling. Such a mechanism should not complicate the algorithm too much. In [7] the method of output constraint handling for analytical MPC algorithms based on Hammerstein model was proposed. Its idea will be also used in the method proposed in the paper. However, it is good to notice that in [7] a control horizon equal to 1 must have been used in order to obtain an efficient constraint handling mechanism. The solution proposed now lacks this limitation, at the same time advantages of the approach remain intact. Despite the constraint mechanism exploits the nonlinear model (the prediction is obtained using the Hammerstein model), it is relatively simple and efficient. It makes possible to take into consideration the values of the predicted output many sampling instants ahead.

In the next section, analytical MPC algorithms based on Hammerstein models and with presumed control signal trajectory are proposed. In Sect. 3 the output constraint handling mechanism based on prediction detailed in Sect. 2 is described. In Sect. 4 simulation experiments done in the example control system of a nonlinear control plant with delay, illustrating very good performance of the proposed approach, are presented. The paper is shortly summarized in the last section.

2 Analytical MPC Algorithms Based on Hammerstein Models

Control action is generated in the MPC algorithms using prediction of future behavior of the control plant many sampling instants ahead. To obtain the prediction a control plant model is used. The values of control variables are calculated in such a way that the prediction fulfills assumed criteria. Usually, the following optimization problem is formulated [1, 3, 10, 13] (i.e. minimization of future control errors when at the same time the control signal should not change too rapidly):

$$\arg\min_{\Delta u} \left\{ J_{\mathrm{MPC}} = \sum_{i=1}^{p} \left(\overline{y}_k - y_{k+i|k}\right)^2 + \sum_{i=0}^{s-1} \lambda_k \cdot \left(\Delta u_{k+i|k}\right)^2 \right\}, \tag{1}$$

where $\overline{y}_k$ is a set–point value, $y_{k+i|k}$ is a value of the output for the $(k+i)$th sampling instant, predicted at the kth sampling instant, $\Delta u_{k+i|k}$ are future changes in manipulated variable, $\lambda_k \geq 0$ is a weighting coefficient, p and s denote prediction and control horizons, respectively, $\boldsymbol{\Delta u} = \left[\Delta u_{k+1|k}, \ldots, \Delta u_{k+s-1|k}\right]^T$ is the vector of decision variables of the optimization problem (1). After minimization of the performance function the optimal vector of changes in the manipulated variable is obtained. From this vector, the $\Delta u_{k|k}$ element is applied in the control system and the algorithm passes to the next iteration.

The performance function from (1) can be expressed as:

$$J_{\mathrm{MPC}} = (\overline{\boldsymbol{y}} - \boldsymbol{y})^T \cdot (\overline{\boldsymbol{y}} - \boldsymbol{y}) + \boldsymbol{\Delta u}^T \cdot \boldsymbol{\Lambda}_k \cdot \boldsymbol{\Delta u}, \tag{2}$$

where $\overline{\boldsymbol{y}} = \left[\overline{y}_k, \ldots, \overline{y}_k\right]^T$ is the vector of length p, $\boldsymbol{\Lambda}_k = \lambda_k \cdot \boldsymbol{I}$ is the $s \times s$ matrix, $\boldsymbol{y} = \left[y_{k+1|k}, \ldots, y_{k+p|k}\right]^T$.

Assume that the predicted values of the output variable $y_{k+i|k}$ are derived using the Hammerstein model, as described in [5]; it will be now shortly reminded. In the Hammerstein model a linear dynamic block is preceded by a nonlinear static block in which a nonlinear function is used:

$$z_k = f(u_k), \tag{3}$$

where z_k is the output of the static block, and u_k is the input to the static block.

Assume that the dynamic part of the model has the form of the step response:

$$\widehat{y}_k = \sum_{n=1}^{p_d-1} a_n \cdot \Delta z_{k-n} + a_{p_d} \cdot z_{k-p_d}, \tag{4}$$

where $\widehat{y}_k$ is the output of the Hammerstein model, a_i are coefficients of the step response of the linear dynamic block, p_d is the horizon of the process dynamics, equal to the number of sampling instants after which the step response can be considered as settled.

The prediction is obtained in a way described in [5]. In this prediction method the Hammerstein model (4) is used to obtain the free response (it contains future values of the output variable calculated assuming that the control signal does not change in the future):

$$\widetilde{y}_{k+i|k} = \sum_{n=i+1}^{p_d-1} a_n \cdot \Delta z_{k-n+i} + a_{p_d} \cdot z_{k-p_d+i} + d_k, \tag{5}$$

where $\widetilde{y}_{k+i|k}$ is the element of the free response for the $(k+i)$th sampling instant, predicted at the kth sampling instant, $d_k = y_k - \widehat{y}_k$ is the DMC–type disturbance model (it is assumed the same for all future sampling instants). In order to simplify further calculations the influence of the future control changes is derived using a linear approximation of the Hammerstein model (4):

$$\widehat{y}_k^L = dz_k \cdot \left(\sum_{n=1}^{p_d-1} a_n \cdot \Delta u_{k-n} + a_{p_d} \cdot u_{k-p_d} \right), \tag{6}$$

where dz_k is a slope of the static characteristic near the z_k. It can be calculated analytically or numerically. In the latter case, which can be applied e.g. if non–differentiable membership functions are used in the fuzzy static model, the following formula can be used:

$$dz_k = \frac{f(u_k + du) - f(u_k)}{du}, \tag{7}$$

where du is a small number.

Finally, the following prediction is obtained [5]:

$$\boldsymbol{y} = \widetilde{\boldsymbol{y}} + \boldsymbol{A}_k \cdot \boldsymbol{\Delta u}, \tag{8}$$

where $\widetilde{\boldsymbol{y}} = \left[\widetilde{y}_{k+1|k}, \ldots, \widetilde{y}_{k+p|k}\right]^T$ is the free response trajectory;

$$\boldsymbol{A}_k = dz_k \cdot \boldsymbol{A} = dz_k \cdot \begin{bmatrix} a_1 & 0 & \ldots & 0 & 0 \\ a_2 & a_1 & \ldots & 0 & 0 \\ \vdots & \vdots & \ddots & \vdots & \vdots \\ a_p & a_{p-1} & \ldots & a_{p-s+2} & a_{p-s+1} \end{bmatrix}, \tag{9}$$

where $\boldsymbol{A}$ is the constant matrix.

After application of prediction (8) to the performance function (2) one obtains:

$$J_{\mathrm{HMPC}} = (\overline{\boldsymbol{y}} - \widetilde{\boldsymbol{y}} - \boldsymbol{A}_k \cdot \boldsymbol{\Delta u})^T \cdot (\overline{\boldsymbol{y}} - \widetilde{\boldsymbol{y}} - \boldsymbol{A}_k \cdot \boldsymbol{\Delta u}) + \boldsymbol{\Delta u}^T \cdot \boldsymbol{\Lambda}_k \cdot \boldsymbol{\Delta u}. \tag{10}$$

Note that the performance function (10) depends quadratically on decision variables $\boldsymbol{\Delta u}$. Thus, the minimization problem with (10) and without constraints has the following analytical solution:

$$\boldsymbol{\Delta u} = \left(\boldsymbol{A}_k^T \cdot \boldsymbol{A}_k + \lambda_k \cdot \boldsymbol{I}\right)^{-1} \cdot \boldsymbol{A}_k^T \cdot \left(\overline{\boldsymbol{y}} - \widetilde{\boldsymbol{y}}\right). \tag{11}$$

Unfortunately, the dynamic matrix A_k changes at each iteration, therefore calculation of (11) must be repeated at each iteration. Fortunately, it can be simplified.

The first method to simplify calculation of the control action, detailed e.g. in [6], relies on the fact that after assuming that $\lambda_k = dz_k^2 \cdot \lambda$ and using (9) one obtains:

$$\boldsymbol{\Delta u} = \frac{1}{dz_k} \cdot \boldsymbol{K} \cdot \left(\overline{\boldsymbol{y}} - \widetilde{\boldsymbol{y}}\right). \tag{12}$$

where the matrix $\boldsymbol{K} = \left(\boldsymbol{A}^T \cdot \boldsymbol{A} + \lambda \cdot \boldsymbol{I}\right)^{-1} \cdot \boldsymbol{A}^T$ does not change and thus the main part of calculations can be performed off–line.

The second method to simplify calculation of the control action will be now proposed. It is based on the idea of presumed trajectory of future control changes, thus it consists in assuming that:

$$\Delta u_{k+j-1|k} = \alpha_k \cdot \Delta u_B(j),\ j = 1, \ldots, s, \tag{13}$$

where α_k is a value which is changed at each iteration by the controller (it is a new decision variable). Therefore, the trajectory of future changes of the control signal will be given by:

$$\boldsymbol{\Delta u} = \alpha_k \cdot \boldsymbol{b}, \tag{14}$$

where $\boldsymbol{b} = \left[\Delta u_B(1) \Delta u_B(2) \ldots \Delta u_B(s)\right]^T$. After using (14) in (8) one obtains:

$$\boldsymbol{y} = \widetilde{\boldsymbol{y}} + \alpha_k \cdot \boldsymbol{A}_k \cdot \boldsymbol{b}. \tag{15}$$

Thus the minimization problem which must be solved by the controller is now as follows:

$$\arg\min_{\alpha_k} \left(\overline{\boldsymbol{y}} - \widetilde{\boldsymbol{y}} - \alpha_k \cdot \boldsymbol{A}_k \cdot \boldsymbol{b}\right)^T \cdot \left(\overline{\boldsymbol{y}} - \widetilde{\boldsymbol{y}} - \alpha_k \cdot \boldsymbol{A}_k \cdot \boldsymbol{b}\right) + \lambda_k \cdot (\alpha_k)^2 \cdot \boldsymbol{b}^T \cdot \boldsymbol{b}. \tag{16}$$

Analytical solution of the optimization problem is now given by the formula:

$$\alpha_k = \left(\boldsymbol{b}^T \cdot \boldsymbol{A}_k^T \cdot \boldsymbol{A}_k \cdot \boldsymbol{b} + \lambda_k \left(\boldsymbol{b}^T \cdot \boldsymbol{b}\right)\right)^{-1} \cdot \boldsymbol{b}^T \cdot \boldsymbol{A}_k^T \cdot \left(\overline{\boldsymbol{y}} - \widetilde{\boldsymbol{y}}\right). \tag{17}$$

Then future values of the control signal are given by:

$$\boldsymbol{\Delta u} = \left(\boldsymbol{b}^T \cdot \boldsymbol{A}_k^T \cdot \boldsymbol{A}_k \cdot \boldsymbol{b} + \lambda_k \left(\boldsymbol{b}^T \cdot \boldsymbol{b}\right)\right)^{-1} \cdot \boldsymbol{b}^T \cdot \boldsymbol{A}_k^T \cdot \left(\overline{\boldsymbol{y}} - \widetilde{\boldsymbol{y}}\right) \cdot \boldsymbol{b}. \tag{18}$$

Note that $\boldsymbol{b}$ is the vector thus $\left(\boldsymbol{b}^T \cdot \boldsymbol{A}_k^T \cdot \boldsymbol{A}_k \cdot \boldsymbol{b} + \lambda_k \left(\boldsymbol{b}^T \cdot \boldsymbol{b}\right)\right)$ is the scalar. Therefore the control action can be easily calculated using the formula:

$$\boldsymbol{\Delta u} = \frac{\boldsymbol{b}^T \cdot \boldsymbol{A}_k^T \cdot \left(\overline{\boldsymbol{y}} - \widetilde{\boldsymbol{y}}\right)}{\boldsymbol{b}^T \cdot \boldsymbol{A}_k^T \cdot \boldsymbol{A}_k \cdot \boldsymbol{b} + \lambda_k \left(\boldsymbol{b}^T \cdot \boldsymbol{b}\right)} \cdot \boldsymbol{b}. \tag{19}$$

Note that this time λ_k can be assumed any (also constant in time, e.g. $\lambda_k = \lambda$) and the solution is still easy to obtain. It can be also noticed that after assuming $\lambda_k = dz_k^2 \cdot \lambda$ the formula simplifies even more:

$$\boldsymbol{\Delta u} = \frac{1}{dz_k} \cdot \frac{\boldsymbol{b}^T \cdot \boldsymbol{A}^T \cdot \left(\overline{\boldsymbol{y}} - \widetilde{\boldsymbol{y}}\right)}{\boldsymbol{b}^T \cdot \boldsymbol{A}^T \cdot \boldsymbol{A} \cdot \boldsymbol{b} + \lambda \left(\boldsymbol{b}^T \cdot \boldsymbol{b}\right)} \cdot \boldsymbol{b}, \tag{20}$$

where $\frac{\boldsymbol{b}^T \cdot \boldsymbol{A}^T}{\boldsymbol{b}^T \cdot \boldsymbol{A}^T \cdot \boldsymbol{A} \cdot \boldsymbol{b} + \lambda(\boldsymbol{b}^T \cdot \boldsymbol{b})}$ can be calculated off–line.

Remark It is good to notice that in the proposed algorithm it is possible to constrain control signal values and control signal changes on the whole control horizon, like in the algorithm with presumed trajectory of control changes based on a linear model; for details see [4]. It is because this mechanism is independent on the model used.

3 Mechanism of Output Constraint Handling

Application of presumed trajectory of future control signal changes has also another advantage. It makes possible to constrain future output values on the whole prediction horizon. Moreover, it is a method which offers more flexibility than the one based on control horizon equal to 1, and proposed in [7].

Rewrite the prediction obtained in Sect. 2 using the assumption about presumed trajectory of future control signal changes. It is now given by:

$$y_{k+i|k} = \widetilde{y}_{k+i|k} + \alpha_k \cdot \sum_{j=1}^{s} a_{ij,k} \cdot b_j, \tag{21}$$

where $a_{ij,k}$ are elements from ith row and jth column of the dynamic matrix $\boldsymbol{A}_k$, and $b_j = \Delta u_B(j)$ is the jth element of the vector $\boldsymbol{b}$.

Assume that the following output constraints should be fulfilled:

$$y_{\min} \leq y_{k+i|k} \leq y_{\max}, \tag{22}$$

where $y_{\min}$ and $y_{\max}$ are lower and upper output limits, respectively. Application of the prediction (21) in the output constraints allows taking them into consideration many samplings instants ahead:

$$y_{\min} \leq \widetilde{y}_{k+i|k} + \alpha_k \cdot \sum_{j=1}^{s} a_{ij,k} \cdot b_j \leq y_{\max}. \tag{23}$$

Thus, the set of constraints grouping lower ones, put on current value of α_k, is described by:

$$\alpha_k \cdot \sum_{j=1}^{s} a_{ij,k} \cdot b_j \geq y_{\min} - \widetilde{y}_{k+i|k}, \tag{24}$$

and the second set of constraints grouping upper ones:

$$\alpha_k \cdot \sum_{j=1}^{s} a_{ij,k} \cdot b_j \leq y_{\max} - \widetilde{y}_{k+i|k}. \tag{25}$$

The following rules of α_k modifications leading to fulfillment of the output constraints can be formulated:

- for lower constraints:
  - if $\alpha_k \cdot \sum_{j=1}^{s} a_{ij,k} \cdot b_j < y_{\min} - \widetilde{y}_{k+i|k}$ then

$$u_k = \frac{y_{\min} - \widetilde{y}_{k+i|k}}{\sum_{j=1}^{s} a_{ij,k} \cdot b_j}; \tag{26}$$

- for upper constraints:
  - if $\alpha_k \cdot \sum_{j=1}^{s} a_{ij,k} \cdot b_j > y_{\max} - \widetilde{y}_{k+i|k}$ then

$$u_k = \frac{y_{\max} - \widetilde{y}_{k+i|k}}{\sum_{j=1}^{s} a_{ij,k} \cdot b_j}. \tag{27}$$

During prediction generation the DMC–type model of disturbance, containing also influence of modeling errors, was used. It is also present in the output constraint handling mechanism, in the free response, but if it is possible to assess the values of modeling errors, such knowledge can be easily used in the proposed mechanism.

Assume that the output prediction with uncertainty is described as follows:

$$\begin{aligned} \breve{y}_{k+i|k} &= y_{k+i|k} + r_{k+i|k} \\ &= \widetilde{y}_{k+i|k} + \alpha_k \cdot \sum_{j=1}^{s} a_{ij,k} \cdot b_j + r_{k+i|k}, \end{aligned} \tag{28}$$

where $r_{k+i|k}$ represent (usually unknown precisely) influence of modeling errors on the prediction. Assume that this uncertainty is bounded:

$$r_{k+i|k}^{\min} \leq r_{k+i|k} \leq r_{k+i|k}^{\max}, \tag{29}$$

where $r^{\min}_{k+i|k} \le 0$ and $r^{\max}_{k+i|k} \ge 0$ are minimum and maximum assessments of the $r_{k+i|k}$. Inclusion of this knowledge about modeling uncertainty in the rules (26) and (27) leads to the following, modified rules:

- for lower constraints:
  - if $\alpha_k \cdot \sum_{j=1}^{s} a_{ij,k} \cdot b_j < y_{\min} - \widetilde{y}_{k+i|k} - r^{\min}_{k+i|k}$ then

$$u_k = \frac{y_{\min} - \widetilde{y}_{k+i|k} - r^{\min}_{k+i|k}}{\sum_{j=1}^{s} a_{ij,k} \cdot b_j}; \tag{30}$$

- for upper constraints:
  - if $\alpha_k \cdot \sum_{j=1}^{s} a_{ij,k} \cdot b_j > y_{\max} - \widetilde{y}_{k+i|k} - r^{\max}_{k+i|k}$ then

$$u_k = \frac{y_{\max} - \widetilde{y}_{k+i|k} - r^{\max}_{k+i|k}}{\sum_{j=1}^{s} a_{ij,k} \cdot b_j}. \tag{31}$$

4 Simulation Experiments

The control plant under consideration is the ethylene distillation column used for tests also in [5–7]. It is a highly nonlinear control plant with significant time delay. The static part of the Hammerstein model used to obtain the prediction has the form of the fuzzy Takagi–Sugeno model [9, 12] detailed in [7]. The output variable y_k is the impurity of the product counted in ppm. It is assumed that it is constrained, and $y_k \le 285$ ppm. The control variable u_k is the reflux to product ratio; the higher it is the purer product is obtained. During the experiments it was assumed that the reflux is also constrained, and $4.05 \le u_k \le 4.4$.

The analytical MPC controller based on the Hammerstein model with future trajectory of changes of a control signal presumed in advance was designed for the control plant. In order to test the proposed output constraint handling mechanism, the controller was detuned. Thus, the values of the tuning parameters were assumed as follows: prediction horizon $p = 6$, control horizon $s = 3$ and $\lambda = 14 \cdot 10^5$, function describing the trajectory of future control signal changes $\Delta u_B(j) = 1/j$. Like in [7] during the experiments a Hammerstein model with polynomial static model was used to simulate the control plant, in order to test influence of modeling uncertainty.

The example responses obtained in the control system of distillation column are shown in Fig. 1. The set–point value $\overline{y} = 280$ ppm was assumed. When the mechanism of output constraint handling is not used in the controller the constraint is violated by around 20 ppm (dashed lines in Fig. 1). After application of the proposed mechanism the situation improves significantly. Despite modeling uncertainty is not explicitly taken into consideration in the output constraint handling mechanism (only

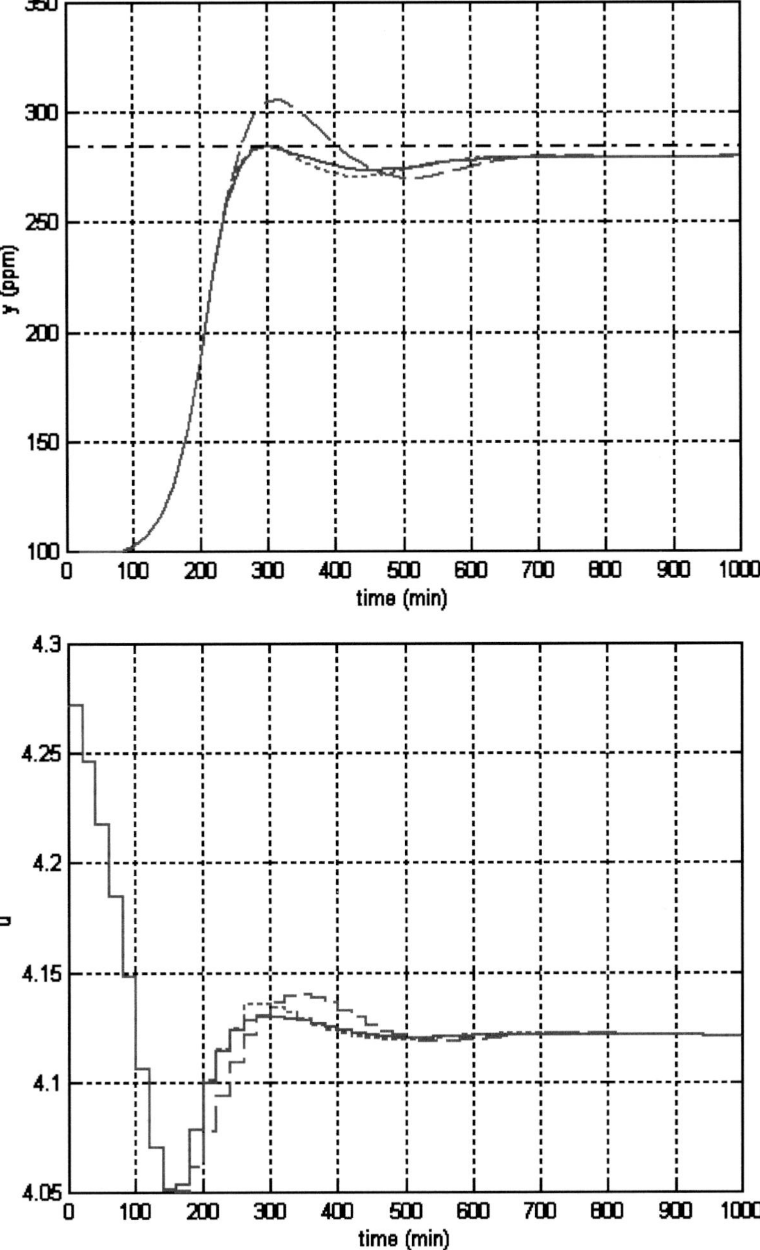

Fig. 1 Responses of the control system with analytical HMPC controller to the change of the set–point value to $\overline{y} = 280\,\text{ppm}$; output constraints: not taken into consideration (*dashed lines*), taken into consideration in on the whole prediction horizon (*dotted lines*), taken into consideration only in the last instant from prediction horizon (*solid lines*)

basic version of the mechanism is used) the result is satisfactory. It is the same case as in [7] and suggests that the method of prediction generation from [5] gives good results.

In the case when the constraints are taken into consideration in each instant from the prediction horizon (except first 4 steps due to the delay present in the control plant) the maximal output value achieves the constraint (dotted lines in Fig. 1). The similar situation takes place when only the last predicted output value from the prediction horizon is constrained (solid lines in Fig. 1). The maximal output value is slightly lower than in the previous case. Moreover, near the 400th minute of simulation the output values are closer to the set–point value than in the previous case. It illustrates that simplification of the mechanism by reduction of rules used to modify the control action can bring satisfactory results. It is also good to notice that in the algorithms with constraint handling mechanism employed, the control signal is modified in advance in order to fulfill the constraint and to improve operation of the controller.

5 Summary

The numerically efficient analytical HMPC algorithms were proposed. Thanks to the presumed trajectory of future control changes the proposed algorithms need very simple calculations to obtain the control action despite they are based on the Hammerstein (nonlinear) model. Moreover, application of the prediction method applied in the algorithms, in the output constraint handling mechanism, allows constraining the predicted output values many sampling instants ahead. It is done by appropriate modifications of the control action which can be done in advance, before potential constraint violation. As a result, fulfillment of constraints is relatively easy to assure and good performance is offered by the control systems with the proposed algorithms.

The proposed output constraint handling mechanism can be easily scaled to the given problem because it is often sufficient to constrain the predicted output values only in a few future instants from the prediction horizon, not in all of them. Thus, the algorithm can be simplified. The other feature useful for a designer of the control algorithm is possibility to use not only tuning parameters available in the MPC algorithms as a standard but also to choose the shape of the function describing the future control trajectory.

References

1. Camacho, E.F., Bordons, C.: Model Predictive Control. Springer (1999)
2. Janczak, A.: Identification of Nonlinear Systems Using Neural Networks and Polynomial Models: A Block-Oriented Approach. Springer, Heidelberg (2005)

3. Maciejowski, J.M.: Predictive Control with Constraints. Prentice Hall, Harlow (2002)
4. Marusak, P.: Predictive controllers with presumed trajectory of control changes and efficient mechanism of control signal constraints handling. Pomiary Automatyka Robotyka no. 2'2008 581–590 (2008). (in Polish)
5. Marusak, P.: On prediction generation in efficient MPC algorithms based on fuzzy Hammerstein models. Rutkowski, L., et al. (eds.) ICAISC 2010, Part I. LNAI, vol. 6113, pp. 136–143. Springer, Heidelberg (2010)
6. Marusak, P.: Numerically efficient analytical MPC algorithm based on fuzzy Hammerstein models. In: Dobnikar, A., Lotric, U., Ster, B. (eds.) ICANNGA 2011, Part II. LNCS, vol. 6594, pp. 177–185. Springer, Heidelberg (2011)
7. Marusak, P.: Efficient mechanism of output constraint handling for analytical predictive controllers based on Hammerstein models. In: Szewczyk, R., Zielinski, C., Kaliczynska, M. (eds.) Recent Advances in Automation, Robotics and Measuring Techniques. Advances in Intelligent Systems and Computing, vol. 267, pp. 137–146. Springer, Switzerland (2014)
8. Marusak, P.: Methods of prediction improvement in efficient MPC algorithms based on fuzzy Hammerstein models. In: Ngoc Thanh Nguyen (ed.) Transactions on Computational Collective Intelligence XIV. LNCS, vol. 8615, pp. 158–179. Springer, Heidelberg (2014)
9. Piegat, A.: Fuzzy Modeling and Control. Physica-Verlag, Berlin (2001)
10. Rossiter, J.A.: Model-Based Predictive Control. CRC Press, Boca Raton (2003)
11. Rossiter, J.A., Richalet, J.: Handling constraints with predictive functional control of unstable processes. In: American Control Conference, pp. 4746–4751. Anchorage (2003)
12. Takagi, T., Sugeno, M.: Fuzzy identification of systems and its application to modeling and control. IEEE Trans. Syst. Man Cybern. **15** 116–132 (1985)
13. Tatjewski, P.: Advanced Control of Industrial Processes. Structures and Algorithms. Springer, London (2007)

Realisation of Positive Continuous-Time Linear Systems Consisting of n Subsystems with Different Fractional Order

Konrad Andrzej Markowski

Abstract This paper presents a method for finding positive realisation of the continuous-time linear system consisting of n subsystems with different fractional orders. For the proposed method, a digraph-based algorithm was constructed. In this paper, we have shown how we can realise the transfer matrix which can be also stable. The proposed method was discussed and illustrated with some numerical examples.

Keywords Realisation · Fractional system · Digraphs · Subsystem · Algorithm · Stability

1 Introduction

In recent years, many researchers have been interested in positive linear systems. In this type of the system, state variables and outputs take only non-negative values. Analysis of positive one-dimensional (1D) systems is more difficult than of standard systems. Examples of positive systems include industrial processes involving chemical reactors, heat exchangers and distillation columns, storage systems, compartmental systems, water and atmospheric pollution models. An overview of the state of the art in the positive systems theory is given in [2, 5, 14, 18].

The first definition of the fractional derivative was introduced by Liouville and Riemann at the end of the 19th century. Mathematical fundamentals of fractional calculus are given in the monographs [4, 17, 23–26]. The notion of positive fractional discrete-time linear systems have been introduced in [17].

The realisation problem is a very difficult task. In many research studies, we can find the canonical form of the system, i.e. constant matrix form, which satisfies the

Research has been financed with the funds of the Statutory Research of 2016.

K.A. Markowski (✉)
Faculty of Electrical Engineering, Institute of Control and Industrial Electronics, Warsaw University of Technology, Koszykowa 75, 00-662 Warsaw, Poland
e-mail: Konrad.Markowski@ee.pw.edu.pl
URL: http://nas.isep.pw.edu.pl/repo

R. Szewczyk et al. (eds.), *Challenges in Automation, Robotics and Measurement Techniques*, Advances in Intelligent Systems and Computing 440, DOI 10.1007/978-3-319-29357-8_33

system described by the transfer function. With the use of this form, we are able to write only one realisation (or some by the transformation matrices) of the system, while there exist many sets of matrices which fit into the system transfer function. The realisation problem for positive discrete-time systems without and with delays was considered in [3, 5, 15, 16], while in [12] a solution for finding a set of possible realisations of the characteristic polynomial was proposed, that allows for finding many sets of matrices. In paper [19–22], the proposed method for finding minimal positive realisations is an extension of the method for finding a realisation of the characteristic polynomial. The optimisation of the proposed algorithm is presented in the paper [10, 13].

The digraphs theory was applied to the analysis of dynamical systems. The use of the multi-dimensional theory was proposed for the first time in the paper [6–8] for analysis of positive two-dimensional systems.

In this paper, a new method of determination positive minimal realisation for the fractional continuous one-dimensional system consisting of n subsystem with different fractional number will be given. The procedure will be illustrated with a numerical example.

Notion. In this paper the following notion will be used. The set $n \times m$ real matrices will be denoted by $\mathbb{R}^{n \times m}$ and $\mathbb{R}^n = \mathbb{R}^{n \times 1}$. If $\mathbf{G} = [g_{ij}]$ is a matrix, we write $\mathbf{G} \geqslant 0$ (matrix $\mathbf{G}$ is called non-negative), if $g_{ij} \geqslant 0$ for all i, j; $\mathbf{G} > 0$ (matrix $\mathbf{G}$ is called positive), if $\mathbf{G} \geqslant 0$ and any $g_{ij} > 0$; $\mathbf{G} \gg 0$ (matrix $\mathbf{G}$ is called strictly positive), if $g_{ij} > 0$ for all i,j. The set of $n \times m$ real matrices with non-negative entries will be denoted by $\mathbb{R}_+^{n \times m}$ and $\mathbb{R}_+^n = \mathbb{R}_+^{n \times 1}$. The $n \times n$ identity matrix will be denoted by $\mathbf{I}_n$.

Digraph. A directed graph (or just digraph) $\mathfrak{D}$ consists of a non-empty finite set $\mathbb{V}(\mathfrak{D})$ of elements called vertices and a finite set $\mathbb{A}(\mathfrak{D})$ of ordered pairs of distinct vertices called arcs [1, 9]. We call $\mathbb{V}(\mathfrak{D})$ the vertex set and $\mathbb{A}(\mathfrak{D})$ the arc set of $\mathfrak{D}$. We will often write $\mathfrak{D} = (\mathbb{V}, \mathbb{A})$ which means that $\mathbb{V}$ and $\mathbb{A}$ are the vertex set and arc set of $\mathfrak{D}$, respectively. The order of $\mathfrak{D}$ is the number of vertices in $\mathfrak{D}$. The size of $\mathfrak{D}$ is the number of arc in $\mathfrak{D}$. For an arc (v_1, v_2) the first vertex v_1 is its tail and the second vertex v_2 is its head. There exists an $\mathfrak{A}$-arc from vertex v_j to vertex v_i if and only if the (i,j)-th entry of the matrix $\mathbf{A}$ is non-zero. There exists a $\mathfrak{B}$-arc from source s to vertex v_j if and only if the l-th entry of the matrix $\mathbf{B}$ is non-zero. More information about use digraph theory in positive system is given in [6, 8].

2 Fractional Order Model

Let us consider the continuous-time fractional linear system described by state-space equations:

$$\begin{aligned} {}_0\mathfrak{D}_t^{\alpha} x(t) &= \mathbf{A}x(t) + \mathbf{B}u(t), \quad 0 < \alpha \leqslant 1 \\ y(t) &= \mathbf{C}x(t) + \mathbf{D}u(t) \end{aligned} \tag{1}$$

where $x(t) \in \mathbb{R}^n$, $u(t) \in \mathbb{R}^m$, $y(t) \in \mathbb{R}^p$ are the state, input and output vectors respectively and $\mathbf{A} \in \mathbb{R}^{n\times n}$, $\mathbf{B} \in \mathbb{R}^{n\times m}$, $\mathbf{C} \in \mathbb{R}^{p\times n}$ and $\mathbf{D} \in \mathbb{R}^{p\times m}$. The following Caputo definition of the fractional derivative will be used:

$$ {}_a^C\mathfrak{D}_t^\alpha = \frac{d^\alpha}{dt^\alpha} = \frac{1}{\Gamma(n-\alpha)} \int_a^t \frac{f^{(n)}(\tau)}{(t-\tau)^{\alpha+1-n}} d\tau \tag{2} $$

where $\alpha \in \mathbb{R}$ is the order of a fractional derivative, $f^{(n)}(\tau) = \frac{d^n f(\tau)}{d\tau^n}$ and $\Gamma(x) = \int_0^\infty e^{-t} t^{x-1} dt$ is the gamma function.

Theorem 1 *The Laplace transform of the derivative-integral* (2) *has the form*

$$ \mathcal{L}\left[{}_0^C\mathfrak{D}_t^\alpha\right] = s^\alpha F(s) - \sum_{k=1}^{n} s^{\alpha-k} f^{(k-1)}(0^+) \tag{3} $$

The proof of the Theorem 1 is given in [17].

Using the Laplace transform to (1), Theorem 1 and taking into account

$$ \begin{aligned} X(s) = \mathcal{L}\left[x(t)\right] &= \int_0^\infty x(t) e^{-st} dt, \\ \mathcal{L}\left[\mathfrak{D}^\alpha x(t)\right] &= s^\alpha X(s) - s^{\alpha-1} x_0 \end{aligned} \tag{4} $$

we obtain:

$$ \begin{aligned} X(s) &= \left[\mathbf{I}_n s^\alpha - \mathbf{A}\right]^{-1} \left[s^{\alpha-1} x_0 + \mathbf{B} U(s)\right] \\ Y(s) &= \mathbf{C} X(s) + \mathbf{D} U(s), \ \ U(s) = \mathcal{L}\left[u(t)\right]. \end{aligned} \tag{5} $$

Using (5) we can determine the transfer matrix of the system in the following form:

$$ \mathbf{T}(s) = \mathbf{C}\left[\mathbf{I}_n s^\alpha - \mathbf{A}\right]^{-1} \mathbf{B} + \mathbf{D}. \tag{6} $$

Let us consider single-input single-output multi-order fractional continuous-time linear system:

$$ \begin{bmatrix} \mathfrak{D}^{\alpha_1} x_1(t) \\ \vdots \\ \mathfrak{D}^{\alpha_n} x_n(t) \end{bmatrix} = \begin{bmatrix} \mathbf{A}_{11} & \dots & \mathbf{A}_{1n} \\ \vdots & \ddots & \vdots \\ \mathbf{A}_{n1} & \dots & \mathbf{A}_{nn} \end{bmatrix} \begin{bmatrix} x_1(t) \\ \vdots \\ x_n(t) \end{bmatrix} + \begin{bmatrix} \mathbf{B}_1 \\ \vdots \\ \mathbf{B}_n \end{bmatrix} u(t), \tag{7a} $$

$$ y(t) = \begin{bmatrix} \mathbf{C}_1 & \dots & \mathbf{C}_n \end{bmatrix} \begin{bmatrix} x_1(t) \\ \vdots \\ x_n(t) \end{bmatrix} + \mathbf{D} u(t), \tag{7b} $$

where $0 < \alpha_i < 1$, $i = 1, \ldots, n$ $x_i(t) \in \mathbb{R}^{n_i}$, $i = 1, \ldots, n$ is the state vector, $u(t) \in \mathbb{R}$ and $y(t) \in \mathbb{R}$ are the input and output of the system, $\mathbf{A}_{ij} \in \mathbb{R}^{n_i \times n_j}$, $\mathbf{B} \in \mathbb{R}^{n_i}$, $\mathbf{C} \in \mathbb{R}^{1 \times n_j}$, $i, j = 1, \ldots, n$ and $\mathbf{D} \in \mathbb{R}$. The transfer function of the system (7a) and (7b) has the form

$$\mathbf{T}(s) = \begin{bmatrix} \mathbf{C}_1 & \mathbf{C}_2 & \ldots & \mathbf{C}_n \end{bmatrix} \begin{bmatrix} \mathbf{I}_{n_1} s^{\alpha_1} - \mathbf{A}_{11} & -\mathbf{A}_{12} & \ldots & -\mathbf{A}_{1n} \\ -\mathbf{A}_{21} & \mathbf{I}_{n_2} s^{\alpha_2} - \mathbf{A}_{22} & \ldots & -\mathbf{A}_{2n} \\ \vdots & \vdots & \ddots & \vdots \\ -\mathbf{A}_{n1} & -\mathbf{A}_{n2} & \ldots & \mathbf{I}_{n_n} s^{\alpha_n} - \mathbf{A}_{nn} \end{bmatrix}^{-1} \begin{bmatrix} \mathbf{B}_1 \\ \mathbf{B}_2 \\ \vdots \\ \mathbf{B}_n \end{bmatrix} + \mathbf{D}, \tag{8}$$

where:

$$\mathbf{A} = \begin{bmatrix} \mathbf{A}_{11} & \mathbf{A}_{12} & \ldots & \mathbf{A}_{1n} \\ \mathbf{A}_{21} & \mathbf{A}_{22} & \ldots & \mathbf{A}_{2n} \\ \vdots & \vdots & \ddots & \vdots \\ \mathbf{A}_{n1} & \mathbf{A}_{n2} & \ldots & \mathbf{A}_{nn} \end{bmatrix}, \ \mathbf{B} = \begin{bmatrix} \mathbf{B}_1 \\ \mathbf{B}_2 \\ \vdots \\ \mathbf{B}_n \end{bmatrix}, \ \mathbf{C} = \begin{bmatrix} \mathbf{C}_1 & \mathbf{C}_2 & \ldots & \mathbf{C}_n \end{bmatrix}, \ \mathbf{D}. \tag{9}$$

Definition 1 The fractional system (7a) and (7b) is called the internally positive fractional system if and only if $x(t) \in \mathbb{R}_+^n$ and $y(t) \in \mathbb{R}_+^p$ for $t \geqslant 0$ for any initial conditions $x_0 \in \mathbb{R}_+^n$ and all inputs $u(t) \in \mathbb{R}_+^m$ for $t \geqslant 0$.

Definition 2 A square real matrix $\mathbf{A} = \left[a_{ij}\right]$ is called the Metzler matrix if its off-diagonal entries are non-negative, i.e. $a_{ij} \geqslant 0$ for $i \neq j$.

Definition 3 The fractional system (7a) and (7b) is positive if and only if

$$\mathbf{A} \in \mathbb{M}^{n \times n}, \ \mathbf{B} \in \mathbb{R}_+^{n \times m}, \ \mathbf{C} \in \mathbb{R}_+^{p \times n}, \ \mathbf{D} \in \mathbb{R}_+^{p \times m}. \tag{10}$$

Matrices (10) are called a positive realisation of the transfer function $\mathbf{T}(s)$ if they satisfy the equality (8). The realisation is called minimal if the dimension of the state matrix $\mathbf{A}$ is minimal among all possible realisation of (8).

Task: For the given transfer matrix (8), determine a minimal positive realisation (10) using the one-dimensional $\mathfrak{D}^{(1)}$ digraphs theory. The dimension of the system must be the minimal among possible.

3 Problem Solution

Let us be given the matrix $\mathbf{A}$ in the following form

$$\mathbf{A} = \begin{bmatrix} \mathbf{A}_{11} & 0 & \ldots & 0 \\ 0 & \mathbf{A}_{22} & \ldots & 0 \\ \vdots & \vdots & \ddots & \vdots \\ 0 & 0 & \ldots & \mathbf{A}_{nn} \end{bmatrix} \tag{11}$$

then the strictly proper transfer matrix can be written in the following form:

$$\mathbf{T}(s) = \begin{bmatrix} \mathbf{C}_1 & \mathbf{C}_2 & \dots & \mathbf{C}_n \end{bmatrix} \begin{bmatrix} \mathbf{I}_{n_1} s^{\alpha_1} - \mathbf{A}_{11} & 0 & \dots & 0 \\ 0 & \mathbf{I}_{n_2} s^{\alpha_2} - \mathbf{A}_{22} & \dots & 0 \\ \vdots & \vdots & \ddots & \vdots \\ 0 & 0 & \dots & \mathbf{I}_{n_k} s^{\alpha_k} - \mathbf{A}_{nn} \end{bmatrix}^{-1} \begin{bmatrix} \mathbf{B}_1 \\ \mathbf{B}_2 \\ \vdots \\ \mathbf{B}_n \end{bmatrix} + \mathbf{D}. \tag{12}$$

The transfer matrix (12) can be considered as a pseudo-rational function of the variable $\lambda_1 = s^{\alpha_1}, \lambda_2 = s^{\alpha_2}, \dots, \lambda_k = s^{\alpha_k}$ in the form:

$$\mathbf{T}(\lambda) = \tag{13}$$

$$\begin{bmatrix} \mathbf{C}_1 & \mathbf{C}_2 & \dots & \mathbf{C}_n \end{bmatrix} \begin{bmatrix} \mathbf{I}_{n_1} \lambda_1 - \mathbf{A}_{11} & 0 & \dots & 0 \\ 0 & \mathbf{I}_{n_2} \lambda_2 - \mathbf{A}_{22} & \dots & 0 \\ \vdots & \vdots & \ddots & \vdots \\ 0 & 0 & \dots & \mathbf{I}_{n_k} \lambda_k - \mathbf{A}_{nn} \end{bmatrix}^{-1} \begin{bmatrix} \mathbf{B}_1 \\ \mathbf{B}_2 \\ \vdots \\ \mathbf{B}_n \end{bmatrix} + \mathbf{D}. \tag{14}$$

From (15) we have $\mathbf{D} = \lim_{\lambda\to\infty} \mathbf{T}(\lambda)$ since $\lim_{\lambda\to\infty}[\mathbf{I}\lambda - \mathbf{A}] = 0$. The strictly proper transfer matrix is given by the equation:

$$\begin{aligned} \mathbf{T}_{sp}(\lambda) &= \mathbf{T}(\lambda) - \mathbf{D} \\ &= \begin{bmatrix} \mathbf{C}_1 & \mathbf{C}_2 & \dots & \mathbf{C}_n \end{bmatrix} \begin{bmatrix} \mathbf{I}_{n_1} \lambda_1 - \mathbf{A}_{11} & 0 & \dots & 0 \\ 0 & \mathbf{I}_{n_2} \lambda_2 - \mathbf{A}_{22} & \dots & 0 \\ \vdots & \vdots & \ddots & \vdots \\ 0 & 0 & \dots & \mathbf{I}_{n_k} \lambda_k - \mathbf{A}_{nn} \end{bmatrix}^{-1} \begin{bmatrix} \mathbf{B}_1 \\ \mathbf{B}_2 \\ \vdots \\ \mathbf{B}_n \end{bmatrix} \\ &= \frac{1}{\det\left[\left(\mathbf{I}_{n_1}\lambda_1 - \mathbf{A}_{11}\right)\left(\mathbf{I}_{n_2}\lambda_2 - \mathbf{A}_{22}\right)\dots\left(\mathbf{I}_{n_k}\lambda_k - \mathbf{A}_{nn}\right)\right]} \begin{bmatrix} \widetilde{n}_{11}(\lambda) & 0 & \dots & 0 \\ 0 & \widetilde{n}_{22}(\lambda) & \dots & 0 \\ \vdots & \vdots & \ddots & \vdots \\ 0 & 0 & \dots & \widetilde{n}_{nn}(\lambda) \end{bmatrix}, \end{aligned} \tag{15}$$

where

$$\begin{aligned} d(\lambda) &= \underbrace{\det\left(\mathbf{I}_{n_1}\lambda_1 - \mathbf{A}_{11}\right)}_{d_{11}(\lambda)} \underbrace{\det\left(\mathbf{I}_{n_2}\lambda_2 - \mathbf{A}_{22}\right)}_{d_{22}(\lambda)} \dots \underbrace{\det\left(\mathbf{I}_{n_k}\lambda_k - \mathbf{A}_{nn}\right)}_{d_{kk}(\lambda)} \\ &= d_{11}(\lambda)d_{22}(\lambda)\dots d_{kk}(\lambda), \qquad k = 1, 2 \dots, n. \end{aligned} \tag{16}$$

In the first step, we must find matrices $\mathbf{A}_{kk} \in \mathbb{R}_+^{n\times n}$, $k = 1, 2, \dots, n$; using decomposition characteristic polynomial (16). We decompose each simple polynomial into a set of simple monomials

$$d_{kk}(s) = \left(1 - d_{kk_{M_1}}(\lambda)\right) \cup \left(1 - d_{kk_{M_2}}(\lambda)\right) \cup \dots \cup \left(1 - d_{kk_{M_p}}(\lambda)\right), \tag{17}$$

where p is number of simple monomials $d_{kk_{M_p}}$ in the simple polynomial $d_{kk}(\lambda)$. For each simple monomial, we create digraphs representations. Then we can determine all possible simple polynomial realisations using all combinations of the digraphs monomial representations. Finally, we combine received digraphs in one digraph which is corresponding to a characteristic polynomial $d(\lambda)$.

Theorem 2 *There exists positive fractional continuous-time linear system described by the Eqs.* (7a) *and* (7b) *if a one-dimensional digraph corresponding to matrix* **A**, *satisfies the following conditions:*

- ***(C1)*** *the sets* $\mathbb{D}_{kk_{M_1}} \cap \mathbb{D}_{kk_{M_2}} \cap \cdots \cap \mathbb{D}_{kk_{M_p}}$, *where* $k = 1, 2, \ldots, n$; *and* p *is a number of simple monomials in simple polynomial* $d_{kk}(s)$ *corresponding to one-dimensional digraphs are not disjoint.*
- ***(C2)*** *the obtained one-dimensional digraph does not have additional cycles.*

The proof of the Theorem 2 is given in [11].

Remark 1 If we add in search positive realisation the stability condition then in Theorem 2 we must add the following condition: **(C3)** the poles of the characteristic polynomial are distinct real and negative. The proof of this condition results from the structure of the state matrix **A**, which must be a Metzler matrix.

Using the Theorem 2 we can construct the Algorithm 1.

Algorithm 1 *DetermineMinimalRealisation*()

```
monomial = 1;
Determine number of simple_polynomial in characteristic polynomial (16);
for simple_polynomial = 1 to k do
    Determine number of cycles in simple polynomial;
    for monomial = 1 to cycles do
        Determine one-dimensional digraph 𝔇^(1) for all monomial;
        MonomialRealisation(monomial);
    end for
    for monomial = 1 to cycles do
        Determine digraph as a combination of the digraph monomial representation;
        SimplePolynomialRealisation(monomial);
        if SimplePolynomialRealisation != cycles then
            Digraph contains additional cycles or digraph contains disjoint union;
            BREAK
        else if SimplePolynomialRealisation == cycles then
            Digraph satisfies characteristic polynomial;
            Determine weights of the arcs in digraph;
            Write state matrix A_kk|_{k=1,...,n};
            return (SimplePolynomialRealisation, A_kk|_{k=1,...,n});
        end if
    end for
    return (PolynomialRealisation, A = diag(A_kk|_{k=1,...,n}));
end for
```

Let us assume that the matrix $\mathbf{B}_k$, $k = 1, 2, \ldots, n$ and matrix **C** have the following form:

$$\mathbf{B}_k|_{k=1,2,\dots n} = \begin{bmatrix} b_{11}^k & b_{12}^k & \cdots & b_{1,m}^k \\ b_{21}^k & b_{22}^k & \cdots & b_{2,m}^k \\ \vdots & \vdots & \ddots & \vdots \\ b_{n,1}^k & b_{n,2}^k & \cdots & b_{n,m}^k \end{bmatrix}, \quad \mathbf{C} = \begin{bmatrix} c_{11} & c_{12} & \cdots & c_{1,n} \\ c_{21} & c_{22} & \cdots & c_{2,n} \\ \vdots & \vdots & \ddots & \vdots \\ c_{p,1} & c_{p,2} & \cdots & c_{p,n} \end{bmatrix}. \tag{18}$$

After determining the state matrix $\mathbf{A}_{kk} \in \mathbb{R}_+^{n\times n}$, $k = 1, 2, \dots, n$ from the Algorithm 1 and inserting matrices (18) to the Eq. (15) we obtain the polynomial $\bar{n}_{ij}(\lambda)$. After comparing variables with the same power of $\lambda_1, \lambda_2, \dots, \lambda_k$ polynomials $\bar{n}_{ij}(\lambda) = \widetilde{n}_{ij}(\lambda)$ we receive the set of equations. After solving the equation, we obtain matrices (10).

4 Numerical Example

For the given transfer matrix

$$\mathbf{T}_{sp}(s) = \frac{25s^{1.2} + 100s^{0.9} + 125s^{0.3} + 500}{s^{1.6} + 4s^{1.3} + 25s^{1.2} + 100s^{0.9} + 5s^{0.7} + 20s^{0.4} + 125s^{0.3} + 500}, \tag{19}$$

find a minimal positive stable realisation as an electrical circuit using the one-dimensional digraph theory for $0 < \alpha < 1$.
Solution. In the first step using pseudo-rational function of the variable $\lambda_1 = s^{0.3}$, $\lambda_2 = s^{0.9}$, $\lambda_3 = s^{0.4}$ we can write transfer matrix (19) in the following form:

$$\mathbf{T}_{sp}(\lambda) = \frac{25\lambda_1\lambda_2 + 100\lambda_2 + 125\lambda_1 + 500}{\lambda_1\lambda_2\lambda_3 + 4\lambda_2\lambda_3 + 25\lambda_1\lambda_2 + 5\lambda_1\lambda_3 + 100\lambda_2 + 20\lambda_3 + 125\lambda_1 + 500}. \tag{20}$$

After multiplying the nominator and denominator of the transfer function (20) by $\lambda_1^{-1}\lambda_2^{-1}\lambda_3^{-1}$, we obtain:

$$\begin{aligned} \mathbf{T}_{sp}(\lambda) &= \frac{\widetilde{n}(\lambda)}{d(\lambda)} \\ &= \frac{25\lambda_3^{-1} + 100\lambda_1^{-1}\lambda_3^{-1} + 125\lambda_2^{-1}\lambda_3^{-1} + 500\lambda_1^{-1}\lambda_2^{-1}\lambda_3^{-1}}{\begin{gathered}1 + 4\lambda_1^{-1} + 25\lambda_3^{-1} + 5\lambda_2^{-1} + 100\lambda_1^{-1}\lambda_3^{-1} + 20\lambda_1^{-1}\lambda_2^{-1} + \\ + 125\lambda_2^{-1}\lambda_3^{-1} + 500\lambda_1^{-1}\lambda_2^{-1}\lambda_3^{-1}\end{gathered}} \end{aligned} \tag{21}$$

The characteristic polynomial can be written in the following form:

$$\begin{aligned} d(\lambda) &= 1 + 4\lambda_1^{-1} + 25\lambda_3^{-1} + 5\lambda_2^{-1} + 100\lambda_1^{-1}\lambda_3^{-1} + 20\lambda_1^{-1}\lambda_2^{-1} + 125\lambda_2^{-1}\lambda_3^{-1} \\ &+ 500\lambda_1^{-1}\lambda_2^{-1}\lambda_3^{-1} = \underbrace{\left(1 + 4\lambda_1^{-1}\right)}_{d_{11}(\lambda)}\underbrace{\left(1 + 5\lambda_2^{-1}\right)}_{d_{22}(\lambda)}\underbrace{\left(1 + 25\lambda_3^{-1}\right)}_{d_{33}(\lambda)} \end{aligned} \tag{22}$$

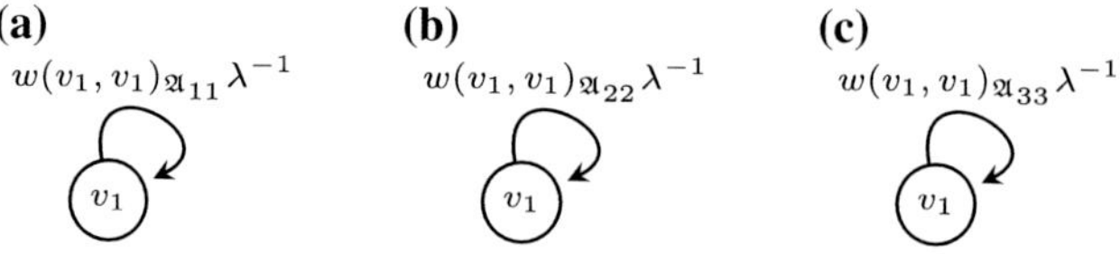

Fig. 1 Multi-dimensional digraphs corresponding to a simple polynomial: **a**—$d_{11}(\lambda)$; **b**—$d_{22}(\lambda)$; **c**—$d_{33}(\lambda)$

In the next step, we a decompose characteristic polynomial (22) into a set of simple polynomials $d_{11}(\lambda) = 1 + 4\lambda^{-1}$, $d_{22}(\lambda) = 1 + 5\lambda^{-1}$ and $d_{33}(\lambda) = 1 + 25\lambda^{-1}$. In the next step, for each simple monomial we write initial conditions. For a simple polynomial $d_{11}(\lambda)$, we have following conditions: number of vertices in digraph: *vertices* = 1; possible weights from which we will build digraphs: $(v_i, v_j)_{\mathfrak{A}_{11}}\lambda^{-1}$; monomials: $M_1 = 1 + 4\lambda^{-1}$.

Then, we determine all possible realisations of the simple polynomial $d_{11}(\lambda)$. In the considered example, we have only one realisation presented on Fig. 1a. The realisation meets conditions (C1) and (C2) of the Theorem 2. Finally, if we can determine positive stable realisation, we must verify the third condition. In the considered simple polynomial, the poles are real and negative. Described realisation satisfies the condition (C3). **The realisation does satisfy all conditions and is correct**. In this same way, we can determine realisations of the simple polynomial $d_{22}(\lambda)$ presented on Fig. 1b and $d_{33}(\lambda)$ presented on Fig. 1c. From the obtained digraphs, we can write a state matrix **A** in the form:

$$\mathbf{A} = \begin{bmatrix} \mathbf{A}_{11} & 0 & 0 \\ 0 & \mathbf{A}_{22} & 0 \\ 0 & 0 & \mathbf{A}_{33} \end{bmatrix} = \begin{bmatrix} w(v_1,v_1)_{\mathfrak{A}_{11}} & 0 & 0 \\ 0 & w(v_1,v_1)_{\mathfrak{A}_{22}} & 0 \\ 0 & 0 & w(v_1,v_1)_{\mathfrak{A}_{33}} \end{bmatrix} \tag{23}$$
$$= \begin{bmatrix} -4 & 0 & 0 \\ 0 & -5 & 0 \\ 0 & 0 & -25 \end{bmatrix}.$$

After inserting matrices (18) and (23) to the Eq. (15) we obtain the polynomial $\bar{n}(\lambda)$. After the comparison of the coefficients of the same power λ polynomials $\bar{n}(\lambda) = \widetilde{n}(\lambda)$, we receive the set of the equation. After solving them, we obtain the following matrices:

$$\mathbf{B} = \begin{bmatrix} 0 \\ 0 \\ 1 \end{bmatrix}, \quad \mathbf{C} = \begin{bmatrix} 0 & 0 & 25 \end{bmatrix}. \tag{24}$$

The desired positive realisation of the (19) is given by (23) and (24). The obtained realisation is stable, as can be seen on Fig. 2, showing the step response of the system

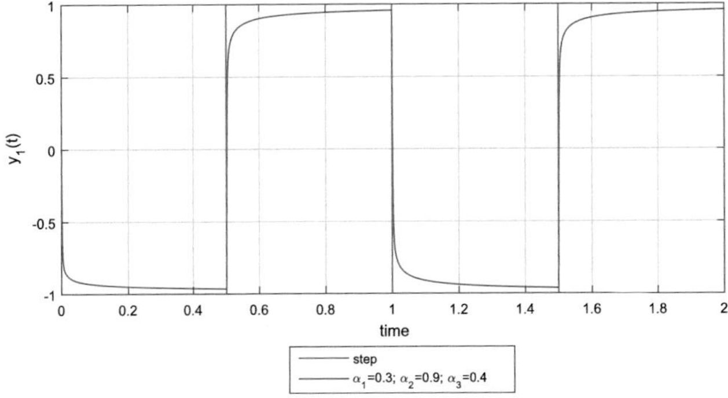

Fig. 2 Step response of the system (19) for $\alpha_1 = 0.3$; $\alpha_2 = 0.9$; $\alpha_3 = 0.4$

for $\alpha_1 = 0.3$; $\alpha_2 = 0.9$ and $\alpha_3 = 0.4$. Simulations were performed using the Matlab toolbox FOMCON (Fractional-order Modeling and Control) presented in the paper [27].

5 Concluding Remarks

The paper presents a method, based on the one-dimensional digraph theory, for finding the realisations consisting of n one-dimensional continuous-time fractional subsystems with different order. The difference between the proposed algorithm in this paper and currently used methods based on canonical forms of the system (i.e. constant matrix forms) is the creation of not one (or few) minimal realisations, but a set of every possible minimal realisation. Additionally using proposed method in easy way we can determine stable realization of the fractional system. Also determine electrical circuit realisation is possible in easy way.

Further work includes extension of the algorithm to find class of electrical circuits corresponding to the transfer matrix.

References

1. Bang-Jensen, J., Gutin, G.: Digraphs: Theory. Algorithms and Applications. Springer, London (2009)
2. Benvenuti, L., De Santis, A., Farina, L.: Positive Systems. Lecture Notes on Control and Information Sciences 294. Springer, Berlin (2003)
3. Benvenuti, L., Farina, L.: A tutorial on the positive realization problem. IEEE Trans. Autom. Control **49**(5), 651–664 (2004)
4. Das, S.: Functional Fractional Calculus. Springer, Heidelberg (2011). doi:10.1007/978-3-642-20545-3. http://dx.doi.org/10.1007/978-3-642-20545-3

5. Farina, L., Rinaldi, S.: Positive Linear Systems: Theory and Applications. Series on Pure and Applied Mathematics. Wiley, New York (2000)
6. Fornasini, E., Valcher, M.E.: Directed graphs, 2D state models, and characteristic polynomials of irreducible matrix pairs. Linear Algebra Appl. **263**, 275–310 (1997)
7. Fornasini, E., Valcher, M.E.: On the positive reachability of 2D positive systems. LCNIS 297–304 (2003)
8. Fornasini, E., Valcher, M.E.: Controllability and reachability of 2D positive systems: a graph theoretic approach. IEEE Trans. Circuits Syst. **I**(52), 576–585 (2005)
9. Foulds, L.R.: Graph Theory Applications. Springer (1991)
10. Hryniów, K., Markowski, K.: Optimisation of digraphs-based realisations for polynomials of one and two variables. In: Szewczyk, R., Zieliński, C., Kaliczyńska, M. (eds.) Progress in Automation, Robotics and Measuring Techniques. Advances in Intelligent Systems and Computing, vol. 350, pp. 73–83. Springer (2015). doi:10.1007/978-3-319-15796-2_8. http://dx.doi.org/10.1007/978-3-319-15796-2_8
11. Hryniów, K., Markowski, K.A.: Parallel digraphs-building algorithm for polynomial realisations. In: Proceedings of 2014 15th International Carpathian Control Conference (ICCC), pp. 174–179 (2014). http://dx.doi.org/10.1109/CarpathianCC.2014.6843592
12. Hryniów, K., Markowski, K.A.: Digraphs minimal realisations of state matrices for fractional positive systems. In: Szewczyk, R., Zieliski, C., Kaliczyska, M. (eds.) Progress in Automation, Robotics and Measuring Techniques. Advances in Intelligent Systems and Computing, vol. 350, pp. 63–72. Springer (2015). doi:10.1007/978-3-319-15796-2_7. http://dx.doi.org/10.1007/978-3-319-15796-2_7
13. Hryniów, K., Markowski, K.A.: Optimisation of digraphs creation for parallel algorithm for finding a complete set of solutions of characteristic polynomial. In: Proceedings of 20th International Conference on Methods and Models in Automation and Robotics, MMAR 2015, pp. 1139–1144. Miedzyzdroje, 24–27 Aug (Accepted 2015). doi:10.1109/MMAR.2015.7284039. http://dx.doi.org/10.1109/MMAR.2015.7284039
14. Kaczorek, T.: Positive 1D and 2D Systems. Springer, London (2001)
15. Kaczorek, T.: Polynomial and Rational Matrices. Springer, London (2007)
16. Kaczorek, T., Busłowicz, M.: Minimal realization problem for positive multivariable linear systems with delay. Int. J. Appl. Math. Comput. Sci. **14**(2), 181–187 (2004)
17. Kaczorek, T., Sajewski, L.: The Realization Problem for Positive and Fractional Systems. Springer, Berlin (2014). doi:10.1007/978-3-319-04834-5. http://dx.doi.org/10.1007/978-3-319-04834-5
18. Luenberger, D.G.: Introduction to Dynamic Systems: Theory, Models, and Applications. Positive Linear Systems. Wiley, New York (1979)
19. Markowski, K.A.: Deretmination of minimal positive realisation of one-dimensional continuous-time fractional linear system. In: Both-Rusu, R. (ed.) Proceedings of the International Symposium on Fractional Signal and Systems, pp. 7–12. U.T. Press (2015)
20. Markowski, K.A.: New digraphs-building algorithm for minimal realisations of one-dimensional fractional positive systems with delays. In: Proceedings of 19th International Conference on System Theory, Control and Computing, pp. 545–550 (2015)
21. Markowski, K.A., Hryniów, K.: Digraphs minimal positive stable realisations for fractional one-dimensional systems. In: Proceedings of 7th Conference on Non-integer Order Calculus and its Applications. 7th Conference on Non-integer Order Calculus and its Applications, pp. 105–118 (2015). doi:10.1007/978-3-319-23039-9_9. http://dx.doi.org/10.1007/978-3-319-23039-9_9
22. Markowski, K.A., Hryniów, K.: Solving minimal realisation problem of positive two-dimensional hybrid systems in the form of second fornasini-marchesini model. In: Proceedings of 2nd IEEE International Conference on Cybernetics, CD-ROM, pp. 172–177. Gdynia, Poland, 24–26 June 2015. doi:10.1109/CYBConf..7175927. http://dx.doi.org/10.1109/CYBConf.2015.7175927
23. Monje, C.A., Chen, Y., Vinagre, B.M., Xue, D., Feliu, V.: Fractional-Order systems and Control: Fundamentals and Applications. Springer (2010). doi:10.1007/978-1-84996-335-0. http://dx.doi.org/10.1007/978-1-84996-335-0

24. Nishimoto, K.: Fractional Calculus. Decartess Press, Koriama (1984)
25. Ortigueira, M.D.: Fractional Calculus for Scientists and Engineers. Academic Press, Springer, Netherlands (2011). doi:10.1007/978-94-007-0747-4. http://dx.doi.org/10.1007/978-94-007-0747-4
26. Podlubny, I.: Fractional Differential Equations. Academic Press, San Diego (1999)
27. Tepljakov, A., Petlenkov, E., Belikov, J.: FOMCON: a MATLAB toolbox for fractional-order system identification and control. Int. J. Microelectron. Comput. Sci. **2**(2), 51–62 (2011)

Research Study of the Micro Cogeneration System with Automatic Loading Unit

Adrian Chmielewski, Szymon Gontarz, Robert Gumiński, Jędrzej Mączak and Przemysław Szulim

Abstract This article presents the test bench research of a micro cogeneration system with the Stirling engine with an automatic loading system. The tests were conducted for nitrogen as a working gas. The work presents the influence of the load current on the change of the working gas pressure in the working chamber and on the changing voltage on the electric machine. A significant number of repetitions allowed for a statistical rendering of the research description with the use of such measures as: kurtosis, coefficient of variation, coefficient of asymmetry, and of the function of the probability density. The research allows for determining whether the conducted experiment is reproducible and how the chosen parameters influence the character of the system's work.

Keywords Automation · Stirling engine · Probability density function · Skewness · Kurtosis

A. Chmielewski (✉) · S. Gontarz · R. Gumiński · J. Mączak · P. Szulim
Faculty of Automotive and Construction Machinery Engineering, Institute of Vehicles, Warsaw University of Technology, 84 Narbutta St, Warsaw, Poland
e-mail: a.chmielewski@mechatronika.net.pl

S. Gontarz
e-mail: Szymon.Gontarz@simr.pw.edu.pl

R. Gumiński
e-mail: rgumin@simr.pw.edu.pl

J. Mączak
e-mail: jma@mechatronika.net.pl

P. Szulim
e-mail: p.szulim@mechatronika.net.pl

R. Szewczyk et al. (eds.), *Challenges in Automation, Robotics and Measurement Techniques*, Advances in Intelligent Systems and Computing 440, DOI 10.1007/978-3-319-29357-8_34

1 Introduction

The need to meet the requirements regarding protection of the natural environment [1] arouses growing interest in the small scale cogeneration appliances, the so-called micro cogeneration appliances [2], in which the amount of produced electric energy equals up to 50 kWe [1, 2]. Among these appliances [3] the micro cogeneration system with the Stirling engine [4, 5], discussed in this work, can be included. From the point of view of the user of such a system, information regarding its proper use is particularly vital, enabling the efficiency maximalization, and assuring the system's operation comfort (minimalization of vibrations and noise). Such an analysis allows for investigating the influence of the chosen work parameters, among others: pressure, voltage, and loading current of the electric machine, which substantially affect the system's character of work.

The analysis of reproducibility of the obtained current and voltage values, as well as gas pressure in the working chamber in statistical rendition, using the statistical measures (the mean value, standard deviation of kurtosis, skewness, and the coefficient of variation). The analyses were performed for different load values (7, 29, 50, 71, and 93 % of the maximal value of load applied during the experiment).

Each technical device is characterised by certain reproducibility of work, which is related to achieving the same or similar performance levels during the same load cycle. The concept of the random variable is used to describe the magnitudes changing randomly, i.e., the value of the variable is impossible to predict. Generally, it can be stated that the reality surrounding us is determined, and every phenomenon has its causes and nothing happens by chance. However, we are not always able to analyse all causes, or we analyse them with certain assumed precision. This results from the fact that either we do not know all the factors influencing a given magnitude (the phenomenon's course), or there is a significant quantity of them which causes that factors of lesser significance are omitted in the analysis. As a consequence, the real course of a phenomenon can divert from predictions. On the basis of the analysis of the technical object use, it can be stated whether it is characterised by predictability of work obtaining the same—located within the narrow range of variation work parameters, among others: rotational velocity at the maximal power, efficiency [1, 6, 7].

This work focuses on the analysis and determining the reproducibility of the micro cogeneration system. This analysis is particularly important for real cogeneration systems which can be used in prosumer households and places where there is a stream of a waste high-temperature heat.

The aim of the statistical analysis was to determine the reproducibility of the real object in the technical aspects. Such an approach allows for elimination of wrong conclusions resulting from uncertainty which came from random character of research results, for example occurring of the outliers, or changes in value of the measured physical magnitude, caused by the uncontrolled change in the conditions of the object's work.

The statistical analysis in combination with automatic loading unit could give the possibility to test cogeneration devices.

2 Testing Workbench with Automatic Measuring System

The laboratory workbench, where the tests were conducted, consisted of the single-action Alpha type Stirling engine (Fig. 1), an electric DC engine (electric engine rated power—500 W) loading the system powered with a belt transmission with the i = 1:4 ratio, and a measuring system by the National Instruments to register the chosen parameters. The measuring system comprised a gas pressure converter placed in the cold cylinder, a magneto-inductive sensor measuring velocity, the gauging thermocouples of the K-type located in: the compression space T_1, expansion space T_4, and on the regenerator from the side of the cooler T_2, as well as from the heater side T_3. The system was loaded to 550 W with an automatic loading system, working as an adjustable current source.

The working gas, nitrogen in this case, was delivered to the buffer space pbuf and to the working space pch from the pressure container (Fig. 2) until the preset pressure value of p = 0.6 MPa was achieved. During the tests, temperatures in the compression space T_1 (the preset value 301 K), expansion space T_4 (the preset value 910 K), temperatures on the regenerator from the cold side T_2 and the hot side T_3 were simultaneously registered. Rotational velocity of the Stirling engine (which was converted to the velocity of the electric machine with the known ratio between the electric machine and the Stirling engine), the current applied to the loading system, voltage on the electric machine, and pressure in the cold cylinder (compression space), were also registered. Figure 2 shows the diagram of a layout of the test bench.

During the first test, the maximal current load was determined, which could be applied to the loading system. It allowed for definition of extremes, as well as of the step with which the system will be loaded. The tests were conducted for 15 different values of load. Based on the above, the tests of work reproducibility of a micro cogeneration system were carried out in the next step. During the tests, the preset number of 30 repetitions of measuring sessions was performed (a session means

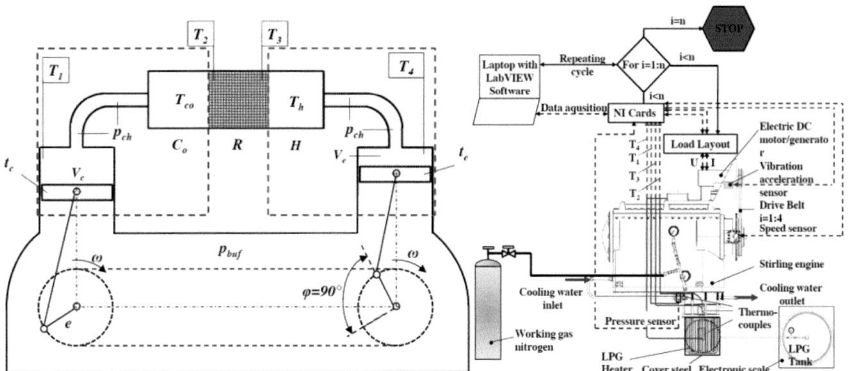

Fig. 1 Schematic representation of the Alpha type Stirling engine

tests for a specific load value). Among successive repetitions there were 10-s breaks in order to restore the system's idling. After performing the preset number of cycles, the obtained measuring data was registered on a hard drive disc of a PC (Fig. 2).

The time of signal registration for every load amounted to 2 s. As a consequence of the conducted test, 30 time flows for each load value were obtained.

In the purpose of statistical comparison of individual repetitions, apart from presenting and comparing the functions of probability density (30 for each load), the use of statistical measures was necessary. The system for applying load was designed based on the programmable current source (Fig. 3a). This is a typical application in which one assembly was multiplied 16-fold in order to increase the device's current load capacity. A simplified scheme of a single module is shown in Fig. 3b. A MOS-FET transistor with high load capacity was used to adjust current from the source, which in this case was a DC motor. Such a layout requires that voltage symbol on the clamps should be permanent, which was assured. It turned out that the system did not have the tendency to excitation. The measuring resistor was selected so that a single cell could apply a maximal load at the level of 10 A. The combined maximal load capacity amounted to 160 A. The maximal power which could be received by the source was not determined. Depending on the cooling conditions, powers of 500 W could be successfully diminished on the device. It should be emphasised here that the advantage of the device designed on the basis of the current source consists in automatic tuning to changing working

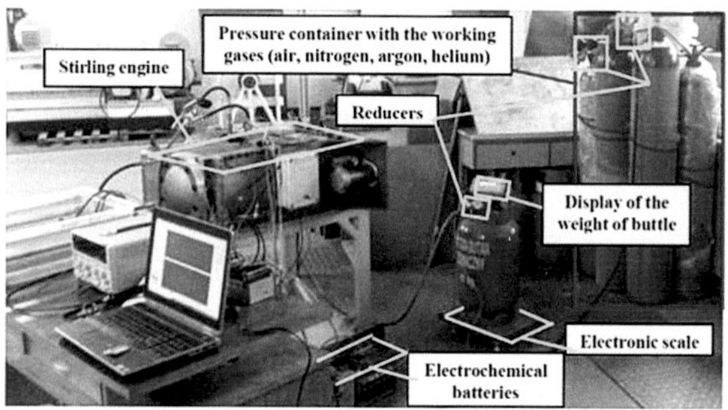

Fig. 2 Picture of the test bench

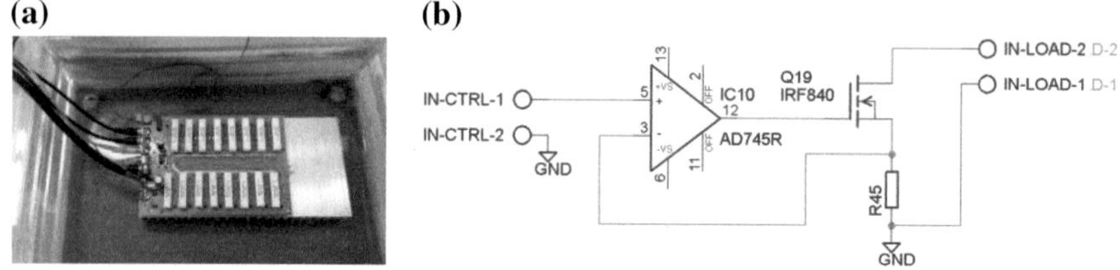

Fig. 3 System designed to apply load: **a** picture, **b** scheme of a single module

conditions of the loaded source. In the case of the electric DC motor, the moment generated on this motor has nearly linear dependence on the current directed through it. Because due to the nature of load work, the constant value of the current was maintained, and also the load moment of the tested Stirling engine was kept at the nearly constant level, even though the rotational velocity was very changeable. The reference voltage for operational amplifiers was set either manually (using a potentiometer) or from a PC by means of an analogue output card by National Instruments. It enabled the automatization of the engine loading process, and thus the automatization of the whole measuring process.

The experimental data obtained in the course of the tests was analysed using statistical measures for the registered parameters, among others: the current and voltage on the electric machine, as well as pressure in the compression space.

3 Test Bench Research

3.1 Theoretical Basis of the Analyses

The analyses regarded the period of normal use [7] of the object, i.e. the micro cogeneration system with the Stirling engine. Performing analyses of the selected sets of data, attention was focused on several characteristic features of the given trials, among which the normal and central moments can be included [8]. In this article, the measures using central moments M_k of k order, of x_1, …, x_k trials of n number, which are the mean values of powers of the x_i value deviations from the arithmetic mean μ:

$$M_k = \frac{1}{n} \sum_{i=1}^{n} (x_i - \mu)^k \quad (1)$$

A crucial feature of a given trial is an asymmetry coefficient S. It is a third central moment (k = 3 in expression 1) divided by the standard deviation raised to the third power. The asymmetry coefficient S allows for the estimation of symmetry of the observed results; it can be written as follows:

$$S = \frac{M_3}{\sigma^3} \quad (2)$$

In the case when S = 0, we deal with the symmetric distribution, however, when S > 0, the right skew is to be observed, and when the right skew occurs, and when S < 0, the left skew can be observed.

Another measure describing a given trial is Kurtosis (K). It addresses the flattening of the distribution around the mean value. Kurtosis is a quotient of the fourth order moment and the third power of the standard deviation minus three. Kurtosis can be written as follows:

$$K = \frac{M_4}{\sigma^4} - 3 \quad (3)$$

From the analysis of relationship (3) it ensues that if K = 0, the flattening similar to that in the normal distribution (Gaussian) occurs, in the case when K > 0—the distribution is more concentrated than normal, however, when K < 0—the distribution is less concentrated than normal.

During the data analysis, the coefficient of variation (CV) was also used, which is a quotient of the standard deviation σ and the value of the mean.

For better illustration of the analysed data in statistical rendering, the experimental functions of probability density were presented [8]. The higher the value of this function, the greater the concentration of the measuring data and smaller the dispersion (standard deviation).

4 Results of Test Bench Research

The obtained experimental data (among others: current, voltage and pressure) was analysed using statistical measures described in Sect. 3.1. Presentation of the results of the conducted analyses was limited to five loads of the micro cogeneration system (7, 29, 50, 73, and 93 % of maximal load).

Further in this article, the time flows of the discussed magnitudes (Fig. 4) are shown, as well as their statistical measures. Figure 4a shows an exemplary flow of the current signal for the loads: 7, 29, 50, 73, and 93 %. This corresponded to the current value 7 %–0.7 A, 29 %–1.75 A, 50 %–2.9 A, 71 %–3.92 A, 93 %–5 A.

Figure 4b shows an exemplary flow of the voltage signal on the electric machine for the load from 7 to 93 % of the maximal load. With the growing load of the system, the voltage value on the electric machine was decreasing from 12 V for 7 % load to 7,5 V for 93 % load. The drop in voltage during increasing the system's load was also accompanied by a decreasing rotational velocity.

On the basis of the tests, the following was determined: the conversion coefficient of voltage change into rotational velocity changes k_v = 100 revs/V, with the relation between the change of voltage and rotational velocity defined by the relationship:

$$\Delta n = \Delta u k_v \quad (4)$$

where:

Δn change in the rotational velocity of the Electric machine (revs/min),
Δu change in voltage on the electric machine (V),
k_v conversion coefficient of voltage change into rotational velocity change (revs/V).

Figure 4c illustrates the signal flow of the working gas pressure (nitrogen) measured in the compression space with the load from 7 to 93 % of the maximal load. The character of changes in pressure in the compression space was approaching sinusoidal.

With the growing system load from 7 to 93 %, a decrease in rotational velocity occurs, therefore extension of the period of the work cycle. For 7 % load, the period of the work cycle amounted to 0.097 s (f = 10.31 Hz), and for 93 % load, the period of the work cycle amounted to 0.11 s (f = 8.75 Hz).

Figure 5a shows the diagrams of standard deviation for pressure, and specific numerical values, which are presented in Table 1. Analysis of the Fig. 5a diagram, and of the numerical values from Table 1 allows for an observation that with the growing load, the standard deviation σ_p from the μ_p mean value increases, the coefficient of variation CV_p increases too.

The increasing values of the standard deviation for the load increase influence the diminishing character of the negative value of mean kurtosis μ_{Kp} (Fig. 5b and

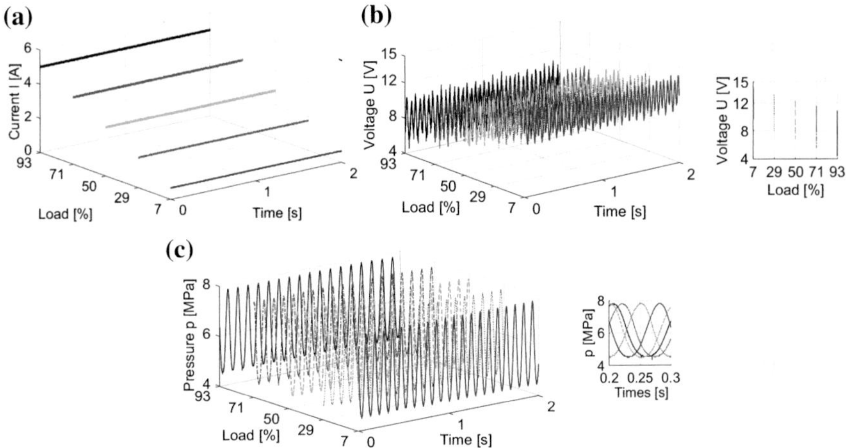

Fig. 4 Signal flow: **a** current, **b** voltage, **c** pressure changes in the compression space

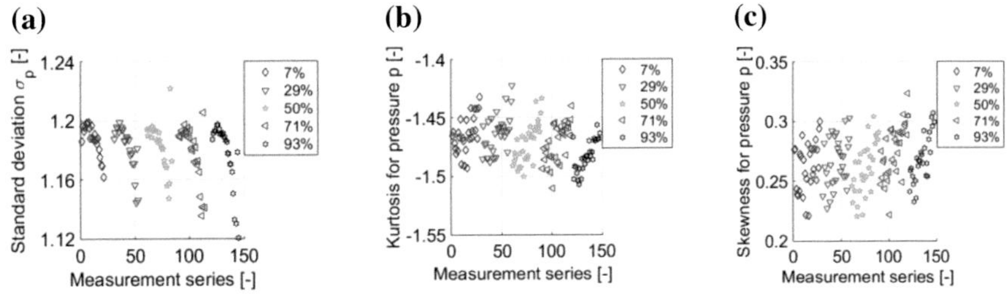

Fig. 5 Diagrams: **a** of standard deviation, **b** kurtosis and **c** skewness for pressure

Table 1). It is worth mentioning that the mean value of kurtosis is negative, which means that on the basis of analysis of the relationship (3), the distribution is less concentrated than normal (greater dispersion of results). It should also be emphasised that the kurtosis dispersion for pressure is decreasing, it decreases with the load growth (from 7 to 93 %), which is reflected in the decreasing coefficient of variation CV_{Kp} (Table 1).

Analysing the skewness value for pressure (Fig. 5c and Table 1) enables the observation that the right skew occurs, with the increasing, alongside the load, mean value μ_{SKEp}. (Table 1) with a nearly constant standard deviation σ_{SKEp}, and the coefficient of variation CV_{SKEp}.

Figure 6 shows the diagrams of the standard deviation (Fig. 6a), kurtosis (Fig. 6b), and skewness (Fig. 6c) for the increasing from 7 to 93 % maximal value of the load current. With the growing load current, the mean value of the standard deviation for current μ_i (Fig. 6a), and the coefficient of variation CV_i, whereas the spread around the mean value itself is constant for the loads above 7 %, $\sigma_i = 0.002$ (Table 2). It should be emphasised that the automatic loading system provides a very small spread of load current values, which highlights how precise the character of load control is (Fig. 3). The mean value of kurtosis for the current is negative, the distribution less concentrated than normal (Fig. 6b, Table 2), for the load greater than 7 %, the mean value of kurtosis increases. In the whole load range, the mean value of skewness μ_{SKEi} increases, too. The spread around the mean value σ_{SKEi} and CV_{SKEi} (for 7 % load—the left skew occurs, in the case of loads greater than 7 %—the right skew occurs—Fig. 5c and Table 2).

Table 1 Test results for pressure

Load (%)	μ_p	σ_p	CV_p	μ_{Kp}	σ_{Kp}	CV_{Kp}	μ_{SKEp}	σ_{SKEp}	CV_{SKEp}
7	1.187	0.011	0.008	−1.464	0.016	−0.011	0.262	0.021	0.079
29	1.182	0.015	0.013	−1.464	0.017	−0.011	0.266	0.021	0.078
50	1.183	0.016	0.013	−1.472	0.018	−0.012	0.261	0.023	0.088
71	1.179	0.019	0.017	−1.472	0.017	−0.011	0.275	0.022	0.080
93	1.174	0.025	0.022	−1.483	0.014	−0.009	0.270	0.021	0.078

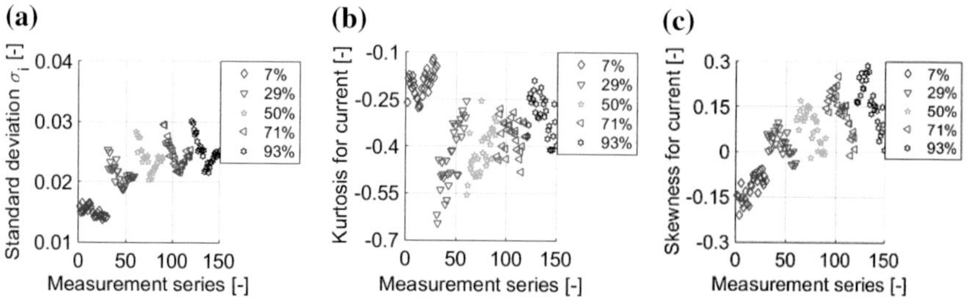

Fig. 6 Diagrams: of standard deviation **a**, kurtosis **b** and skewness **c** for the current

Figure 7 shows diagrams of the standard deviation (Fig. 7a), kurtosis (Fig. 7b) and skewness (Fig. 7c) growing from 7 to 93 % of the maximal value of load current. With the increasing load, the mean value for the standard deviation μ_u increases, and so does the dispersion σ_u (Table 3). The kurtosis mean value for the voltage, which is negative, grows too.

The mean value of skewness μ_{SKEu} is negative, declining for subsequent system loads, the standard deviation for the skewness σ_{SKEu} and the coefficient of variation CV_{SKEu} grow, too (Table 3).

Figure 8a shows the results of the load current influence on the value of the function of probability density. The test results and analyses indicate that with the growing system's load (Fig. 8a) the value of the function of probability density decreases, which means that the probability of occurrence of the value in proximity of the modal value decreases with the increase in load (as a consequence, the growing values of current that occur have a greater dispersion of the standard deviation of kurtosis and skewness).

Figure 8b presents the diagram of the function of probability density for the voltage. The flow of the voltage signal (Fig. 4b) is worth paying attention to, which is similar to sinusoidal. Such a character of the flow causes that the distribution of the probability density function (Fig. 8b) is multimodal. It is also worth emphasizing that the mean value occurs among the modal values (Table 3). Analogically to the case of the current, the increase in load results in greater dispersion of voltage around the mean μ_u, the value of the probability density function decreases (Fig. 8b).

Table 2 Test results for the load current

Load (%)	μ_i	σ_i	CV_i	μ_{Ki}	σ_{Ki}	CV_{Ki}	μ_{SKEi}	σ_{SKEi}	CV_{SKEi}
7	0.015	0.001	0.054	–0.202	0.042	–0.209	–0.123	0.042	–0.341
29	0.021	0.002	0.079	–0.433	0.106	–0.244	0.024	0.039	1.604
50	0.024	0.002	0.085	–0.422	0.071	–0.168	0.083	0.058	0.697
71	0.024	0.002	0.087	–0.366	0.052	–0.141	0.136	0.064	0.470
93	0.025	0.002	0.092	–0.299	0.064	–0.216	0.168	0.075	0.446

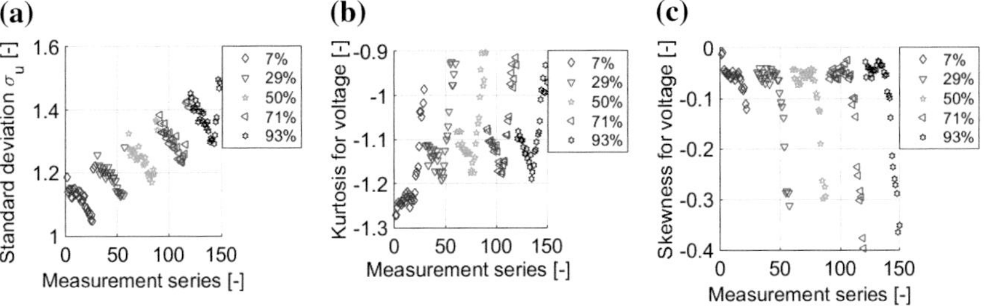

Fig. 7 Diagrams: **a** of standard deviation, **b** kurtosis and **c** skewness for voltage

Table 3 Test results for voltage

Load (%)	μ_u	σ_u	CV_u	μ_{Ku}	σ_{Ku}	CV_{Ku}	μ_{SKEu}	σ_{SKEu}	CV_{SKEu}
7	1.123	0.043	0.038	−1.193	0.085	−0.071	−0.059	0.025	−0.430
29	1.186	0.040	0.033	−1.097	0.084	−0.076	−0.102	0.086	−0.846
50	1.259	0.051	0.041	−1.077	0.083	−0.077	−0.101	0.083	−0.821
71	1.310	0.054	0.041	−1.081	0.081	−0.075	−0.118	0.112	−0.946
93	1.378	0.056	0.041	−1.088	0.069	−0.063	−0.107	0.105	−0.986

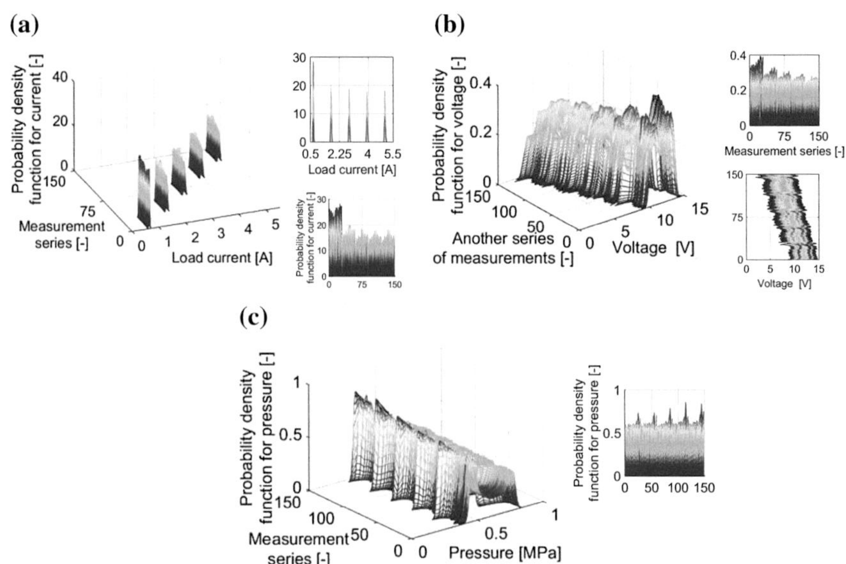

Fig. 8 Diagrams of the function of probability density for: current (**a**), voltage (**b**), pressure (**c**)

Figure 8c shows the diagrams of the function of probability density for pressure measured in the compression space. Similarly as in the case of voltage (Fig. 8b), also the distribution of the function of probability density for pressure is multi-modal, it is a consequence of the fact that the pressure flow is similar to sinusoidal. For 7 % load, the value of the function of probability density (Fig. 8c) amounted to 0.605. The growing value of the function of probability density, amounting to 0.841 for 93 % load (Fig. 8c), informs about a greater reproducibility of the system's work. Practically, it results from the fact that the system works with the same degree of compression, and a reproducible character of changes in the occurring phenomenon during the work cycle [3]. This can also be interpreted as an increase in the working area, and a change of the shape of the indicator diagram closed within the coordinates pressure—specific volume (p-v).

5 Summary and Conclusions

The use of the described measures enabled the interpretation of test reproducibility for the conducted measuring series. In the purpose of illustrating the probability of occurrence of specified values, among others: current, voltage, pressure (measured in the compression space), in the predetermined range of variation the function of probability density was used. In the case of periodic signals, similar to sinusoidal, the distributions were of multimodal character (negative and positive value concentration). Also, the fact should be highlighted, that with the growing system's load, the values of the function of probability density for pressure, whereas for the remaining analysed parameters, among others: current and voltage, these values were smaller (a greater deviation from the mean value occurred, as well as a wider variation range). On the basis of the conducted tests and analyses, it can be concluded that obtained results are reproducible with different system's load values (coefficient of variation for the standard deviation from the mean value does not exceed in any case 0.01, it is relatively small).

With the growing load, the period of the work cycle extends, which is accompanied by a drop in the rotational velocity of the crankshaft (also a decrease in voltage on the electric machine). A detailed description of the work cycle was discussed in [3, 9], with particular attention paid to pumping gas through heat exchangers [9]. Also, it is worth noticing that the functions of probability density achieve greater values for increasing load currents, too. This indicates a greater repeatability and predictability of the work of such a system.

It is worth emphasising that the automatic loading system presented in this work, designed on the basis of a programmable current source, can be used in numerous industry branches, first of all to automate and test the device in laboratory conditions before making a given device available for the end client. The purpose of using such a system can be for example: long-time testing of the device, and as a consequence, determining the warranty period for the given product on the basis of tests.

References

1. Chmielewski, A., Gumiński, R., Radkowski, S., Szulim, P.: Aspekty wsparcia i rozwoju mikrokogeneracji rozproszonej na terenie Polski (Aspects of support and development of distributed microcogeneration in Poland), Rynek Energii **114**(5), 94–101 (2014) (In Polish)
2. Directive 2012/27/EU of the European Parliment and of the Council of 25 October 2012 on energy efficiency, amending Directives 2009/125/EC and 2010/30/EU and repealing Directives 2004/8/EC and 2006/32/EC
3. Chmielewski, A., Gumiński, R., Radkowski, S., Szulim, P.: Experimental research and application possibilities of microcogeneration system with Stirling engine. J. Power Technol. **95** (Polish Energy Mix), 14–22 (2015)
4. Chmielewski, A., Lubikowski, K., Radkowski, S.: Simulation of energy storage work and analysis of cooperation between micro combined heat and power (mCHP) systems and energy storage. Rynek Energii **117**(2), 126–133 (2015)

5. Chmielewski, A., Gumiński, R., Radkowski, S.: Chosen properties of a dynamic model of crankshaft assembly with three degrees of freedom. In: 20th International Conference on Methods and Models in Automation and Robotics (MMAR), IEEE, pp. 1038–1043 (2015), ISBN: 978-1-4799-8700
6. García, D., González, M.A., Prieto, J.I., Herrero, S., López, S., Mesonero, I., Villasante, C.: Characterization of the power and efficiency of Stirling engine subsystems. Appl. Energy **121**, 51–63 (2014)
7. Radkowski, S., Szulim, P.: Analysis of vibration of rotors in unmanned aircraft. In: 19th International Conference on Methods and Models in Automation and Robotics (MMAR), IEEE, pp. 748–753 (2014)
8. Krysicki, W., Bartos, J., Dyczka, W., Królikowska, K., Wasilewski, M.: Rachunek Prawdopodobieństwa i statystyka matematyczna w zadaniach, PWN, Warszawa (2003) (In Polish)
9. Xiao, G., Chen, C., Shi, B., Cen, K., Ni, M.: Experimental study on heat transfer of oscillating flow of a tubular Stirling engine heater. Int. J. Heat Mass Transf. **71**, 1–7 (2014)

Research on a Micro Cogeneration System with an Automatic Load-Applying Entity

Adrian Chmielewski, Szymon Gontarz, Robert Gumiński, Jędrzej Mączak and Przemysław Szulim

Abstract The article concerns the testbed research of a micro cogeneration system based on a Stirling engine equipped with an automatic load-applying system. The article presents the influence that the load current has on acceleration of vibration of a micro cogeneration system's body. The research was conducted while using nitrogen as the working gas. Significant number of tests offered the possibility of providing the description of the results in statistical terms while using such measures as kurtosis, coefficient of variation, the asymmetry coefficient as well as the probability density function. The research offers the possibility of concluding whether the completed experiment is repeatable as well as for determining the impact that selected load changes have on acceleration of the micro cogeneration system's vibration.

Keywords Automation · Stirling engine · Vibration acceleration · Standard deviation · Measurement

A. Chmielewski (✉) · S. Gontarz · R. Gumiński · J. Mączak · P. Szulim
Faculty of Automotive and Construction Machinery Engineering,
Institute of Vehicles, Warsaw University of Technology,
84 Narbutta St, Warsaw, Poland
e-mail: a.chmielewski@mechatronika.net.pl

S. Gontarz
e-mail: Szymon.Gontarz@simr.pw.edu.pl

R. Gumiński
e-mail: rgumin@simr.pw.edu.pl

J. Mączak
e-mail: jma@mechatronika.net.pl

P. Szulim
e-mail: p.szulim@mechatronika.net.pl

R. Szewczyk et al. (eds.), *Challenges in Automation, Robotics and Measurement Techniques*, Advances in Intelligent Systems and Computing 440, DOI 10.1007/978-3-319-29357-8_35

1 Introduction

The paper presents the research related of a micro cogeneration system based on a Stirling engine. This paper presents the influence that loads (with a value of 7, 29, 50, 73 and 93 % of the maximum load) have on acceleration of vibration of a Stirling engine's body in three mutually perpendicular directions.

The tests and the analysis of their results enabled determination of the influence exerted by the analyzed parameters which accompany the system's operation [1–21].

The paper analyzes the repeatability of the obtained values of the body's acceleration in three mutually perpendicular directions while using statistical measures (the average value, the standard deviation of kurtosis, skewness and coefficient of variation as well as the probability density function).

2 Presentation of the Testbed with an Automatic Measuring System

The laboratory testbed used for conducting the tests was described in detail in [1, 16]. The experimental data obtained during the research was subjected to analysis while using the statistical measures for the recorded parameters of acceleration of the body's vibration (Fig. 1) in the following directions: x (parallel to the axis of the cylinders), y (parallel to the axis of the shaft) and z (perpendicular to the plane formed by the axis of the shaft).

The system used for applying the loads was constructed while relying on a programmable source of electrical current which is described and presented in detail in [1].

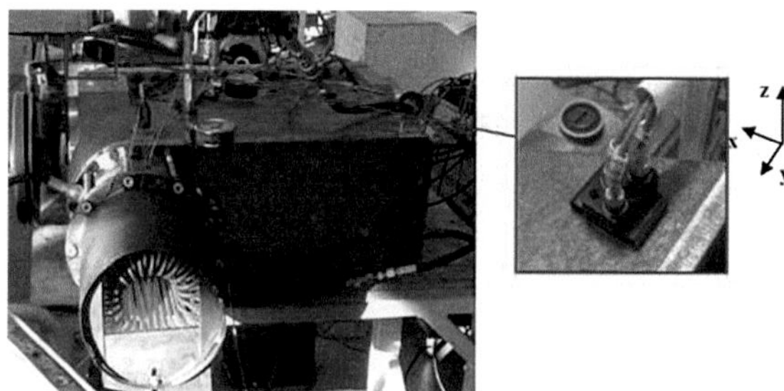

Fig. 1 Picture of the testbed with a tri-axle vibration acceleration sensor (for capturing vibration of the body)

3 Testbed Research

3.1 Results of the Testbed Research

The obtained data was subjected to analysis while using the statistical measures (described in detail in [1] for the recorded parameters) of vibration acceleration in three mutually perpendicular directions. The presentation of the results of the analyses has been restricted to five load values applied to the micro cogeneration system (5, 29, 50, 73 and 93 % of the maximum load).

The further part of the article presents the time runs for the load values ranging from 7 to 93 % of the maximum load for the following items: vibration acceleration in respective directions: *x* (Fig. 2a), *y* (Fig. 2b), *z* (Fig. 2c) as well as their statistical measures.

As expected, the highest acceleration values occur in the direction x, which is associated with the motion of the piston-crank system. The highest value of the acceleration occurs for the load of 50 %, and in the case of positive acceleration it is $a_{xmax29\ \%} = 33.08\ m/s^2$ (Fig. 2a) while in the case of negative acceleration it is lower and it has the value of $a_{xmin50\ \%} = -30.8\ m/s^2$. For the direction x the values of vibration acceleration decrease for loads higher than 50 %, and for the load of 93 % they do not exceed 26.2 m/s^2.

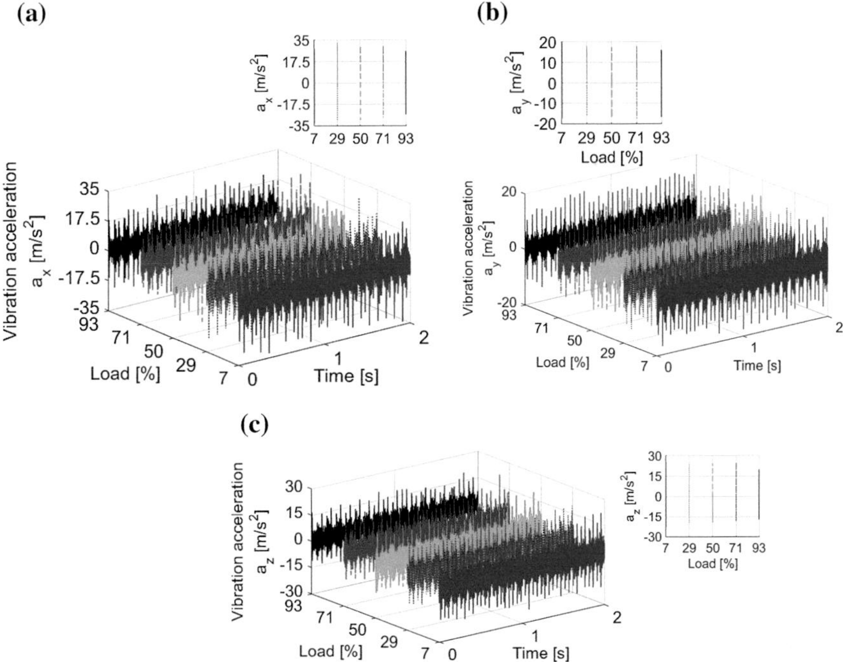

Fig. 2 Vibration acceleration signal curves for directions a_x (**a**), a_y (**b**) and a_z (**c**)

Figure 2b presents the change of vibration acceleration a_y in the direction y for loads ranging from 7 to 93 % of the maximum load. The highest acceleration value occurred for the load of 7 %, with the maximum value being $a_{ymax7\ \%} = 19.8\ m/s^2$ for positive acceleration (Fig. 2b). In the case of negative acceleration, the value was higher and amounted $a_{ymin7\ \%} = -17.67\ m/s^2$.

Values of vibration acceleration for the direction y decreased as the load increased and they did not exceed 16.15 m/s^2 for 93 % of the maximum load.

Figure 2c presents the changes of vibration acceleration for loads ranging from 7 to 93 % of the maximum load. The highest value of vibration acceleration was recorded for the positive load of 7 % and amounted $a_{zmax7\ \%} = 25.47\ m/s^2$ (Fig. 2c), while the lowest value was recorded for the load of 93 % ($a_{zmin93\ \%} = 19.85\ m/s^2$).

The values of vibration acceleration for this direction decrease as the load increases and in the case of the load of 93 % they do not exceed 20 m/s^2.

Figure 3a presents the graphs showing the standard deviation for vibration acceleration a_x, while the values of statistical measures are presented in Table 1. The analysis of graph Fig. 3a and the analysis of the figures found in Table 1 demonstrate that the standard deviation σ_{ax} from the mean value μ_{ax} decreases as the load increases. The decreasing standard deviation values for the growing load affect the nature of the growing and positive value of mean kurtosis μ_{Kax} (Fig. 3b and Table 1). The distribution is more concentrated than normal (with smaller scatter of the results) It should be also stressed that the scatter of kurtosis for the

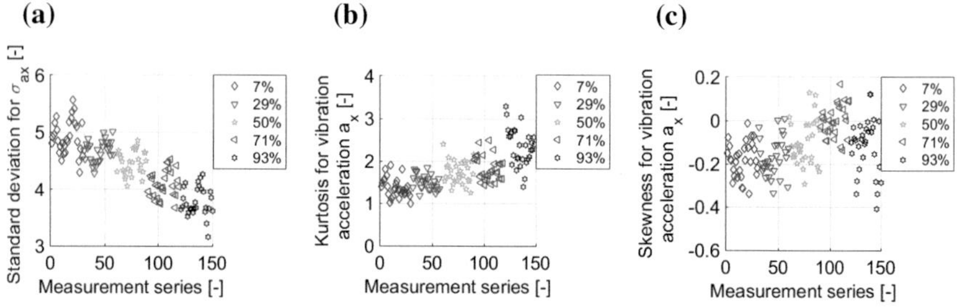

Fig. 3 Diagrams showing respective measurement values: **a** standard deviation, **b** kurtosis, and **c** skewness for acceleration in the direction x

Table 1 Test results for vibration acceleration a_x

Load (%)	μ_{ax}	σ_{ax}	CV_{ax}	μ_{Kax}	σ_{Kax}	CV_{Kax}	μ_{SKEax}	σ_{SKEax}	CV_{SKEax}
7	4.913	0.301	0.061	1.364	0.229	0.168	−0.183	0.082	−0.447
29	4.673	0.188	0.040	1.501	0.179	0.119	−0.169	0.089	−0.528
50	4.466	0.210	0.046	1.735	0.270	0.155	−0.101	0.105	−1.045
71	4.086	0.239	0.059	1.826	0.293	0.161	−0.001	0.074	−242.61
93	3.778	0.252	0.067	2.335	0.442	0.189	−0.119	0.130	−1.094

acceleration in the direction x is decreasing, and it decreases as the load increases (from 7 to 93 %).

The mean value of skewness μ_{SKEax}, the scatter around the mean value σ_{SKEax} as well as the coefficient of variation CV_{SKEax} (Fig. 3c and Table 1) grow along with the increase of the system's load.

Figure 4a presents the graphs showing standard deviation for acceleration in the direction y along with the detailed numerical values which are shown in summary Table 2. Analysis of graph Fig. 4a and of the values found in Table 2 show that as the load increases above 7 %, the standard deviation σ_{ay} from the mean value of μ_{ay} also increases. It is also the coefficient of variation CV_{ay} that increases. Growth of the load causes increase of the standard deviation, leading to growth of the positive mean value of kurtosis μ_{Kp} (Fig. 4b and Table 2). It is worth adding that the mean value of kurtosis is positive, which means that the distribution is more concentrated than normally (smaller scatter of the results). It should be also stressed that the scatter of kurtosis for the acceleration a is decreasing and it decreases along with the growth of the load (from 7 to 93 %), which is indicated by the decreasing value of CV_{Kay} coefficient of variation (Table 2).

While analyzing the values of skewness for the acceleration in the direction y (Fig. 4c and Table 1.) one may notice that there occurs right-hand side asymmetry whose mean value μ_{SKEay} grows along with the growth of the load (Table 1.) while the standard deviation σ_{SKEp} and the coefficient of variation CV_{SKEay} are nearly constant

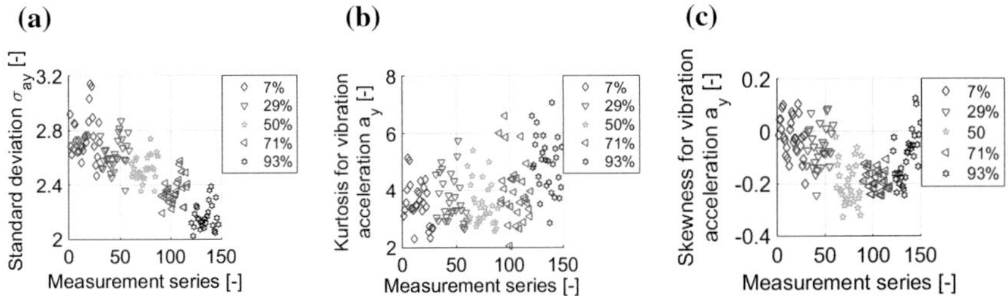

Fig. 4 Diagrams: **a** standard deviation, **b** kurtosis, and **c** skewness for acceleration in the direction a_y

Table 2 Test results for vibration acceleration a_y

Load (%)	μ_p	σ_p	CV_p	μ_{Kp}	σ_{Kp}	CV_{Kp}	μ_{SKEp}	σ_{SKEp}	CV_{SKEp}
7	1.187	0.011	0.008	−1.464	0.016	−0.011	0.262	0.021	0.079
29	1.182	0.015	0.013	−1.464	0.017	−0.011	0.266	0.021	0.078
50	1.183	0.016	0.013	−1.472	0.018	−0.012	0.261	0.023	0.088
71	1.179	0.019	0.017	−1.472	0.017	−0.011	0.275	0.022	0.080
93	1.174	0.025	0.022	−1.483	0.014	−0.009	0.270	0.021	0.078

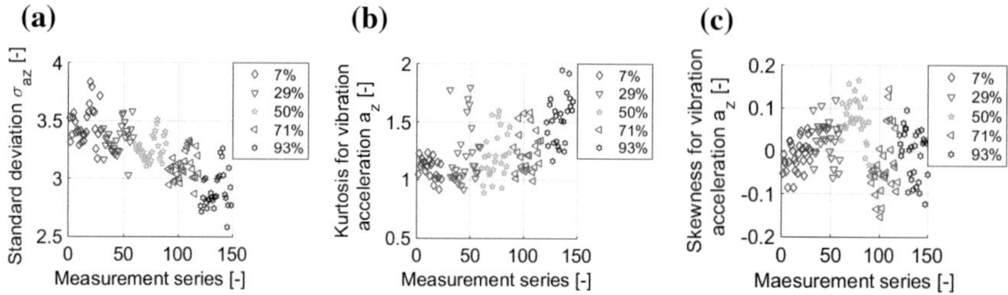

Fig. 5 Graphs: **a** standard deviation, **b** kurtosis, and **c** skewness for the acceleration in the direction a_z

Figure 5 presents the graphs showing standard deviation (Fig. 5a), kurtosis (Fig. 5b) and skewness (Fig. 5c) for the value a_z of acceleration in the direction z as the acceleration increases from 7 to 93 % of its maximum value. As the load increases, the mean value of standard deviation μ_{az} (Table 3) decreases. It is also the mean value of kurtosis (Fig. 5b) that increases and it is positive.

The mean value of skewness μ_{SKEu} is positive for the loads of 29 and 50 %, and it is negative for the remaining values of the load, while the standard deviation of skewness, σ_{SKEu}, increases for all load values (Table 3).

Figure 6 shows the influence that the load current has on the value of the probability density function for accelerations in three mutually perpendicular directions.

The results of the examination and the analyses demonstrate that as the load applied to the system grows, so does the value of the probability density function for accelerations in all directions (Fig. 6a–c), which means that the probability of occurrence of a value close the modal value increases along with the growth of the load.

The highest value of the probability density function exists for the acceleration in the direction y for a load equal to 93 % of the maximum load.

Table 3 Test results for vibration acceleration a_z

Load (%)	μ_z	σ_z	CV_z	μ_{Kz}	σ_{Kz}	CV_{Kz}	μ_{SKEz}	σ_{SKEz}	CV_{SKEz}
7	3.464	0.156	0.045	1.102	0.086	0.078	−0.012	0.036	−3.012
29	3.339	0.126	0.038	1.213	0.260	0.215	0.029	0.045	1.555
50	3.260	0.106	0.033	1.208	0.200	0.166	0.077	0.047	0.608
71	3.103	0.121	0.039	1.241	0.177	0.143	−0.032	0.075	−2.376
93	2.896	0.160	0.055	1.524	0.205	0.134	−0.006	0.058	−9.027

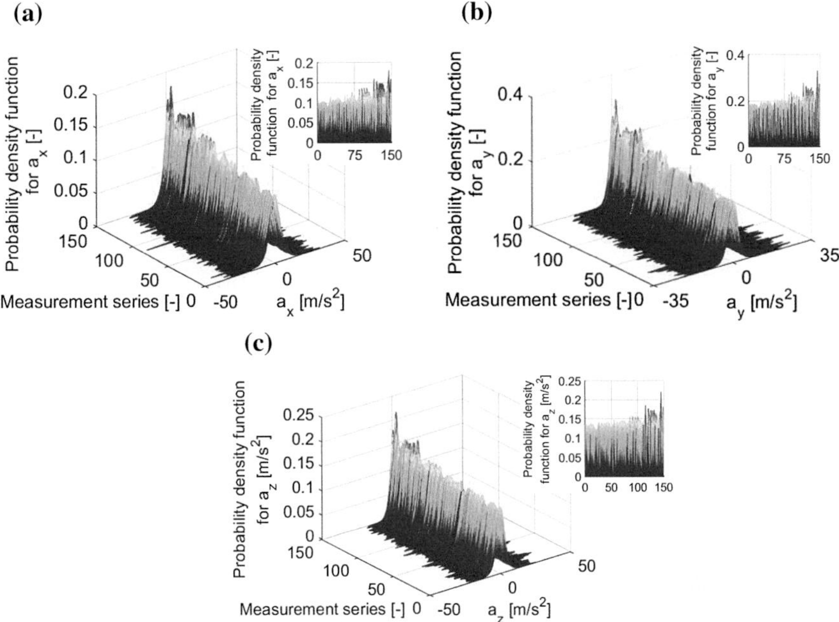

Fig. 6 Graphs showing the probability density function for acceleration of vibration in respective directions: **a** in direction *x*, **b** in direction *y*, **c** in direction *z*

4 Conclusions

Use of statistical measures enabled interpretation of recurrence of results of measurements for a conducted series of measurements. The probability density function has been used to illustrate the probability of occurrence of specific values of acceleration of the engine body's vibration in the tree mutually perpendicular directions. What should also be stressed is that as the load on the system increased, so did the values of the probability density function for acceleration of the engine body's vibration in all the directions (there occurred smaller standard deviation from the mean value and broader variability range). Based on the conducted research and analyses, one can state that the obtained results are recurrent.

It is extremely important from a cognitive point of view to note that as the system's load increased from 7 to 93 % of the maximum load, the vibration acceleration in all directions had lower values, which means that vibration acceleration decreased as the load on the micro cogeneration system increased. The reason was the longer work cycle which was accompanied by decrease of crankshaft's rotational speed. It is also worth noting that as the load increased from 7 to 93 %, the density probability functions achieved higher values, which is indicative of bigger recurrence and predictability of operation of such a system.

In no direction had the vibration acceleration exceeded the value of 3.4 g, nonetheless it is worth noting that the highest values of vibration existed in the direction which is parallel to the axis of the cylinders (direction x), with vibration coming from the movement of the piston-crank system. It should be added that the highest values of vibration acceleration occurred when the pistons were in their extreme positions while the working liquid (gas) was forced between the heat exchangers (the process is described in detail in [3, 6, 13, 14]).

The conducted research offers grounds for concluding that there exists a possibility of diagnosing the level of load of a micro cogeneration system by looking at the vibration acceleration trends. Decrease of vibration acceleration is indicative of more load being applied to a system, which indicates a greater repeatability and predictability of the work of such a system (the growing value of the probability density function for vibration acceleration). The research shows that in the case of bigger workloads (which correspond to higher efficiency), the micro cogeneration systems demonstrate lower values of vibration acceleration, which is very important for users of such systems.

References

1. Chmielewski, A., Gontarz, S., Gumiński, R., Mączak, J., Szulim, P.: Research study of the micro cogeneration system with automatic loading unit. Advances in Intelligent Systems and Computing. Springer (2016) (In print)
2. Chmielewski, A., Gumiński, R., Radkowski, S.: Chosen properties of a dynamic model of a crankshaft assembly with three degrees of freedom. In: 20th International Conference On Methods and Models in Automation and Robotics (MMAR), IEEE, pp. 1038–1043 (2015), ISBN: 978-1-4799-8700
3. Chmielewski, A., Gumiński, R., Radkowski, S., Szulim, P.: Experimental research and application possibilities of a micro-cogeneration system with a Stirling engine. J. Power Technol. **95**(Polish Energy Mix), 14–22 (2015)
4. Szabłowski, Ł., Milewski, J., Badyda, K.: Cooperation of energy sources in distributed generation. Rynek Energii. **6**(115), 120–131 (2014)
5. Li, T., DaWei, T., Li, Z., Du, J., Zhou, T., Jia, Y.: Development and test of a Stirling engine driven by waste gases for the micro-CHP system. Appl. Therm. Eng. **33–34**, 119–123 (2012)
6. Bert, J., Chrenko, D., Sophy, T., Moyne, L.L., Sirot, F.: Simulation, experimental validation and kinematic optimization of a Stirling engine using air and helium. Energy **78**, 701–712 (2014)
7. Chmielewski, A., Lubikowski, K., Radkowski, S.: Simulation of energy storage work and analysis of cooperation between micro combined heat and power (mCHP) systems and energy storage. Rynek Energii **117**(2), 126–133 (2015)
8. Cinar, C., Karabulut, H.: Manufacturing and testing of a gamma type Stirling engine. Renew. Energy **30**, 57–66 (2005)
9. Karabulut, H.: Huseyin, Yucesu S., Cınar C., Aksoy F., An experimental study on the development of a β–type Stirling engine for low and moderate temperature heat sources. Appl. Energy **86**, 68–73 (2009)
10. Renzi, M., Brandoni, C.: Study and application of a regenerative Stirling cogeneration device based on biomass combustion. Appl. Therm. Eng. **67**, 341–351 (2014)

11. Rogdakis, E.D., Antonakos, G.D., Koronaki, I.P.: Thermodynamic analysis and experimental investigation of a Solo V161 Stirling cogeneration unit. Energy **45**, 503–511 (2012)
12. Thombare, D.G., Verma, S.K.: Technological development in the Stirling cycle engines. Renew. Sustain. Energy Rev. **12**, 1–38 (2008)
13. Xiao, G., Chen, C., Shi, B., Cen, K., Ni, M.: Experimental study on heat transfer of oscillating flow of a tubular Stirling engine heater. Int. J. Heat Mass Transf. **71**, 1–7 (2014)
14. Batmaz, I., Ustun, S.: Design and manufacturing of a V–type Stirling engine with double heaters. Appl. Energy **85**, 1041–1049 (2008)
15. Chmielewski, A, et al.: Geometrical model of a cogeneration system based on a 1 MW gas engine. Combust. Engines **162**(3), 570–577 (2015). ISSN 2300-9896
16. Chmielewski, A., Gumiński, R., Mączak, J., Szulim, P.: Badania układu mikrokogeneracyjnego z silnikiem Stirlinga. Część II. (Research on the micro cogeneration system with a Stirling engine. Part II). Rynek Energii **120**(5), 53–60 (2015) (In Polish)
17. Gontarz, S., Szulim, P., Seńko, J., Dybała, J.: Use of magnetic monitoring of vehicles for proactive strategy development. Transp. Res. Part C **52**, 102–115 (2015)
18. Chmielewski, A., Gontarz, S., Gumiński, R., Maczak, J., Szulim, P.: Analiza wpływu parametrów eksploatacyjnych na drgania układu mikrokogeneracyjnego. Przeglad Elektrotechniczny (2016) (In Print)
19. Chmielewski, A., Gumiński, R., Maciąg, P., Mączak, J.: The use of Fuzzy Logic in the control of an inverted pendulum. Springer Proceedings in Mathematics and Statistics (2016) (In Print)
20. Szulim, P., Gontarz, S.: Using the surrounding magnetic field in the diagnosis of BLDC motors. J. Electr. Eng.—Elektrotechnicky Casopis (2016) (In Print)
21. Chmielewski, A., Gumiński, R., Mączak, J., Szulim, P.: Model-based research on the micro cogeneration system with Stirling engine. J. Power Technol. (2016) (In Print)

Part II
Robotics

Performance of Feedback Linearization Based Control of Bicycle Robot in Consideration of Model Inaccuracy

Joanna Zietkiewicz, Adam Owczarkowski and Dariusz Horla

Abstract This paper considers the influence of model inaccuracy on control performance when feedback linearization is used. For this purpose we use plant of bicycle robot. The problem is analysed in two ways: by simulations with artificially changed parameters and by comparison of simulated data with the results obtained from the real object. The collected data show that, even if the model differs from the real object, control method provides good results. This indicates that feedback linearization, method strongly relying on model, can be successfully used for some real plants.

Keywords Feedback linearization · Nonlinear control · Bicycle robot · Reaction wheel · LQR

1 Introduction

Nonlinear control is difficult to apply and Jacobian, in point, linearization is not sufficient in many cases. This problem can be solved by Feedback Linearization (FBL) method, that provides an exact and linear model. The disadvantage of the FBL is that the cancellation of nonlinearity requires the knowledge of the exact model. Therefore the question arises, whether the method is useful for a real object, for which the accuracy of a model is always limited.

For the purpose of our investigations we use a bicycle robot with inertial wheel, for which the control aim is to stabilise it in upright position, regardless of the initial

J. Zietkiewicz (✉) · A. Owczarkowski · D. Horla
Institute of Control and Information Engineering, Poznan University of Technology, Poznan, Poland
e-mail: Joanna.Zietkiewicz@put.poznan.pl

A. Owczarkowski
e-mail: Adam.J.Owczarkowski@doctorate.put.poznan.pl

D. Horla
e-mail: Dariusz.Horla@put.poznan.pl

R. Szewczyk et al. (eds.), *Challenges in Automation, Robotics and Measurement Techniques*, Advances in Intelligent Systems and Computing 440, DOI 10.1007/978-3-319-29357-8_36

position and disturbances. The model of this robot does not have a defined output, therefore the input-state feedback linearization is appropriate. However, although the model fulfills criteria for this kind of FBL (given in [2, 5]), we have decided to use linearization of 3 from 4 state variables. This allows us to obtain a less involved, but sufficient for control, linear model. The control policy relies on linear quadratic regulator. Other examples of control of this plant can be found in [1, 4, 6].

The simulations and experiment presented in the paper show that feedback linearization, in combination with linear quadratic control, provides a better control quality than when Jacobian linearization is used, even when inaccuracies of the model are taken into consideration. The results show that the method can be used for real plants. They can also be a good starting point for further investigation of robust control like e.g. presented in [9].

2 Experimental Setup

Figure 1 shows the real robot. It has two degrees of freedom (angle from the vertical position and angle of the reaction wheel) and one actuator (electric motor). The high torque low speed DC electric motor accelerates the inertia wheel causes reaction momentum which stabilizes the construction. There are a couple of sensors to measure the state of the machine two hundred times per second. The crucial thing is the main measurement unit—IMU (Inertial Measurement Unit). It consists of two sensors: a gyroscope and an accelerometer. The result is estimated by Kalman filter. The encoder is used to obtain the angle and velocity of the motor. Every calculation is made by high performance STM32 32-bit microcontroller. It calculates the whole control law. After each computational operation the information goes to the fast direct torque dc motor control unit.

The main goal of the robot is to keep itself in the vertical position. This robot is named as a bicycle, but it is actually a simpler version of the real bicycle [3]. It

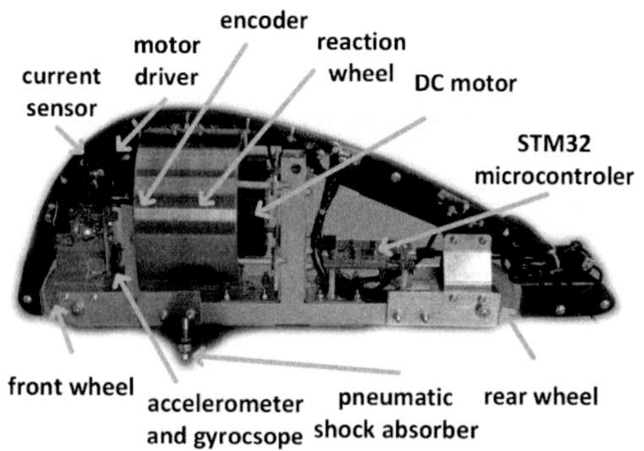

Fig. 1 The picture of the real robot

does not change the fact that the control of this machine is not simple. The control of this machine is not simple. It is underactuated—it has more degrees of freedom than actuators. It is also unstable like every inverted pendulum.

3 Mathematical Model

The continuous-time 2DOF model of the plant is presented by nonlinear difference equations (1)–(4) (time index is omitted for brevity)

$$\dot{x}_1 = x_2, \tag{1}$$

$$\dot{x}_2 = \frac{g\,h_r\,m_r\,\sin(x_1)}{I_{rg}} - \frac{b_r x_2}{I_{rg}} - \frac{b_I x_4}{I_{rg}} + \frac{k_m u}{I_{rg}}, \tag{2}$$

$$\dot{x}_3 = x_4, \tag{3}$$

$$\dot{x}_4 = \frac{k_m u}{I_I + I_{mr}} - \frac{b_I x_4}{I_I + I_{mr}}, \tag{4}$$

where (see Fig. 2 for the kinematic scheme of the object): $\underline{x}$—state vector, x_1—angle of the robot from the vertical, x_2—angular velocity of the robot, x_3—rotation angle of the reaction wheel, x_4—angular velocity of the reaction wheel, u—current of the motor. Parameters of the model are as follows: m_r—weight of the robot; moments of inertia: I_I—of the reaction wheel, I_{mr}—of the rotor, I_{rg}—of the robot related to the ground; h_r—distance from the ground to the center of mass (COM) of the robot, g—gravity constant, k_m—motor constant, friction coefficients: b_r—in the robot rotation, b_I—of the reaction wheel.

The model (1)–(4) can be expressed in simplified form

$$\begin{aligned}
\dot{x}_1 &= x_2, \\
\dot{x}_2 &= \psi_1\psi_4 u + \psi_1\psi_5 x_4 + \psi_2 x_2 + \psi_3 \sin x_1, \\
\dot{x}_3 &= x_4, \\
\dot{x}_4 &= \psi_4 u + \psi_5 x_4,
\end{aligned} \tag{5}$$

where $\psi_1 = \frac{I_I+I_{mr}}{I_{rg}}$, $\psi_2 = \frac{-b_r}{I_{rg}}$, $\psi_3 = \frac{g\,h_r\,m_r}{I_{rg}}$, $\psi_4 = \frac{k_m}{I_I+I_{mr}}$, $\psi_5 = \frac{-b_I}{I_I+I_{mr}}$.

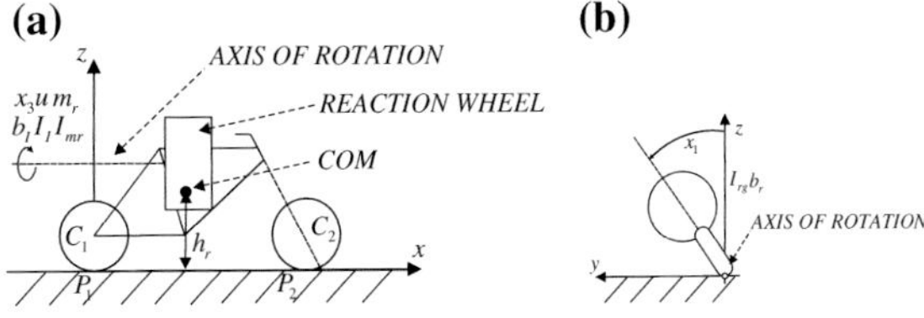

Fig. 2 The bicycle robot—kinematic scheme. **a** Main view. **b** Side view

The main goal of the control system is to keep the robot in the vertical position minimizing the cost function as much as possible. The mathematical model is expressed by four state variables: x_1, x_2, x_3 and x_4. The typical approach is to find the control law which can lead the whole state to zero value as fast as possible. However, after many experiments, it has turned out that the reaction wheel angle (the x_3) can be neglected and the robot still works properly. In this paper the control law is based on the whole state without the x_3.

Every parameter of the robot was identified by means of the previous experiments or appropriate measurements. Unfortunately, the model is inaccurate. It is a huge problem to find the control law for the real machine which is based on uncertain model.

4 Jacobian Linearization of the Model

The system (1)–(4) can be linearized by Jacobian matrix to the form: $\dot{\underline{x}}(t) = \boldsymbol{A}_l \underline{x}(t) + \boldsymbol{B}_l \underline{u}(t)$, where:

$$\boldsymbol{A}_l = \begin{bmatrix} 0 & 1 & 0 & 0 \\ \frac{gh_r m_r}{I_{rg}} & -\frac{b_r}{I_{rg}} & 0 & -\frac{b_l}{I_{rg}} \\ 0 & 0 & 0 & 1 \\ 0 & 0 & 0 & -\frac{b_l}{I_l + I_{mr}} \end{bmatrix}, \tag{6}$$

$$\boldsymbol{B}_l = \begin{bmatrix} 0 \\ \frac{k_m}{I_{rg}} \\ 0 \\ \frac{k_m}{I_l + I_{mr}} \end{bmatrix}. \tag{7}$$

The linealization point is equal to: $\underline{x}_0 = [0\,0\,0\,0]^T$.

5 Feedback Linearization

Feedback linearization (FBL) is a method that provides the exact linear model, mapping nonlinear dynamics, unlike a model obtained by Jacobian linearization. As the linear control theory is well known and easier to apply than nonlinear control, linear models are more convenient to use. FBL method is well explained in [2], the theoretical background can be found in [5]. Here we provide terms and definitions vital for the application performed for this paper.

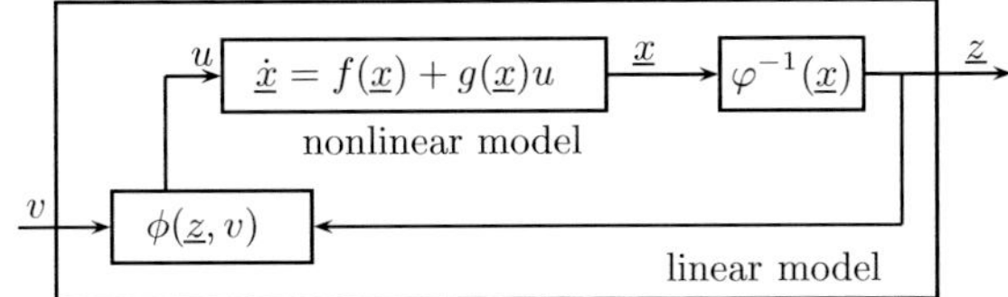

Fig. 3 Feedback linearization scheme

5.1 Definitions and FBL Procedure

FBL can be applied for nonlinear model, affine in input, in the form

$$\dot{\underline{x}} = f(\underline{x}) + g(\underline{x})u, \tag{8}$$

where $\underline{x} \in \mathscr{R}^n$, $u \in \mathscr{R}$ and where $f(\underline{x}) : \mathscr{R}^n \to \mathscr{R}^n, g(\underline{x}) : \mathscr{R}^n \to \mathscr{R}^n$ are smooth functions. This type of linearization illustrates the Fig. 3 and it is called input-state feedback linearization. The calculated linear model has the same order as the nonlinear model, n.

If, in addition, there exists a defined output,

$$y = h(\underline{x}), \tag{9}$$

where $h(\underline{x}) : \mathscr{R}^n \to \mathscr{R}$, FBL is designed in that way, that the output is one of the state variables of the linear model. In this case some additional conditions have to be fulfilled: the function $h(\underline{x})$ is smooth and the model (8) and (9) is minimumphase. This linearization is called input-output linearization and the order of the obtained linear model can be lower than the order of the nonlinear model.

In this paper we use a combination of the two above approaches. For the model (5) it is not necessary to linearize the whole state (4 variables), linearization of 3 variables is sufficient. Therefore, we have been searching for the artificial output function, for which the system is minimumphase and there is a possibility to linearize at least 3 variables of the state $\underline{x}$. The number of those variables defines relative degree.

Relative degree, r, $r \leq n$, is a number of derivatives of the output y, which has to be calculated, in order to obtain explicit function of the input u.

FBL procedure of calculating linear model consists of the following steps:

- use the output as the first variable, $z_1 = y$;
- derivative the output, r times, every time assigning the new variable,

$$z_2 = \dot{z}_1 = \frac{dh}{dt} = \frac{dh}{d\underline{x}}\left(f(\underline{x}) + g(\underline{x})u\right), \tag{10}$$

$$z_3 = \dot{z}_2 = \frac{d}{dt}\left[\frac{dh}{d\underline{x}}f(\underline{x})\right] = \frac{d}{d\underline{x}}\left[\frac{dh}{d\underline{x}}f(\underline{x})\right](f(\underline{x}) + g(\underline{x})u), \tag{11}$$

$$\vdots$$

$$z_r = \phi^{-1}(\underline{x}, u), \tag{12}$$

(because it was assumed that z_2 in the above equations is not the direct function of u, and therefore $\frac{dh}{d\underline{x}}g(\underline{x})u = 0$, in the Eq. (11) it is used $\frac{d}{dt}\left[\frac{dh}{d\underline{x}}f(\underline{x})\right]$ instead of $\frac{d}{dt}\left[\frac{dh}{d\underline{x}}(f(\underline{x}) + g(\underline{x})u)\right]$);

- assign the new input, $v = \phi^{-1}(\underline{x}, u)$; the new linear model has the canonical form

$$y^{(r)} = v; \tag{13}$$

- if $r < n$ find new additional $n - r$ variables fulfilling

$$\frac{dz_k(\underline{x})}{d\underline{x}}g(\underline{x}) = 0, \tag{14}$$

for $k = 1, \ldots, n - r$. This condition assures stability of the $n - r$ variables unobserved from the linear model (13) and, if it is fulfilled, means that the original nonlinear model (8) is minimumphase for the output y.

The dependence of $\underline{z}$ on $\underline{x}$, obtained from (10) and (11), can be expressed by $\underline{z} = \varphi^{-1}(\underline{x})$. Here $v = \phi^{-1}(\underline{x}, u)$ is the inverse of the $u = \phi(\underline{z}, v)$ given in the Fig. 3. The Fig. 3 presents the general idea of FBL and is also proper for the case, when only the part of the state space model is linearized.

5.2 FBL of the Bicycle Robot Model

For the model (5) we apply input-output FBL, except that the output is artificial, chosen function, for which $r = 3$. This allows us to control 3 variables, named: x_1, x_2, x_3. The artificial output chosen for the model is the following: $y = -\psi_1 x_4 - \psi_2 x_1 + x_2$. This is the first variable of the new state $\underline{z}$. Using the procedure of calculating FBL model we can obtain $\underline{z} = \varphi^{-1}(\underline{x})$ assigning:

$$\begin{aligned} z_1 &= \psi_1 x_4 - \psi_2 x_1 + x_2, \\ z_2 &= \psi_3 \sin x_1, \\ z_3 &= \psi_3 x_2 \cos x_1. \end{aligned} \tag{15}$$

The above mapping provides linear model in the form

$$\begin{aligned} \dot{z}_1 &= z_2, \\ \dot{z}_2 &= z_3, \\ \dot{z}_3 &= v, \end{aligned} \tag{16}$$

where $v = \phi^{-1}(\underline{x}, u)$:

$$v = \psi_3(\cos x_1(\psi_1\psi_5 x_4 + \psi_2 x_2 + \psi_3 \sin x_1 + \psi_1\psi_4 u) - x_2^2 \sin x_1). \tag{17}$$

To complement the new state $\underline{z}$ to order $n = 4$ we use an additional variable, which has to fulfill the condition (14). One of solutions is the following

$$z_4 = x_3, \tag{18}$$

and the dynamics of the variable z_4 is then expressed by the equation

$$\dot{z}_4 = \frac{z_3}{\psi_1\sqrt{\psi_3^2 - z_2^2}} - \frac{z_1}{\psi_1} \arcsin\frac{z_2}{\psi_3}. \tag{19}$$

It should be noted that additional difference equation (19) complementing the dynamics presented by (16) is no longer linear. However, to regulate the system we can use the model with 3 variables (16). To calculate linear model by FBL we use the nonlinear model with its structure, dynamics, and values of parameters. Therefore, the method is reliable on the exact knowledge of the nonlinear model. On the other hand, the accuracy of the model is always limited.

6 Discrete-Time Model of the Plant

In order to apply the controller on the hardware, the following multivariable model of the linearized plant must be introduced:

$$\underline{x}_{t+1} = \boldsymbol{A}\underline{x}_t + \boldsymbol{B}\underline{u}_t, \tag{20}$$

where the appropriate matrices have known sizes and are the result of step-invariant discretization of the continuous-time model of the (16), with chosen sample period T_S. The constrained control vector $\underline{u} \in \mathscr{R}^m$, and the state vector $\underline{x} \in \mathscr{R}^n$ are in discrete-time domain (denoted henceforth by subscript t).

7 Linear Quadratic Control

Control strategy considered in the paper is based on minimization of the performance index

$$J_x = \int_0^\infty \left(\underline{x}^T \boldsymbol{Q}_x \underline{x} + r_x u^2\right) dt. \tag{21}$$

The weights in this index are chosen adequately to physical meaning of the variables and their values are the following $\boldsymbol{Q}_x = diag\{10, 0.01, 0.1, 0.01\}$, $r_x = 1$.

With the use of the linear model (e.g. by Jacobian linearization) minimization of the (21) can be performed by the linear quadratic control. The method delivers the gain vector, k_x^T, used afterwards in the control law $u = -\underline{k}_x^T \underline{x}$.

In the case of linear model obtained through feedback linearization, (16) variables $\underline{z}$ and v have no longer direct physical meaning. Therefore the performance index requires weights beyond the diagonal of the weighting matrix $\boldsymbol{Q}_z$ and also in the vector $\underline{n}_z$. The index takes form

$$J_z = \int_0^\infty \left(\underline{z}^T \boldsymbol{Q}_z \underline{z} + r_z v^2 + \underline{z}^T \underline{n}_z v\right) dt, \tag{22}$$

with the control law $v = -\underline{k}_z^T \underline{z}$. The idea of the designing of J_z is that it should be possibly close to J_x. It is done by linear approximations of the variables $\underline{x}$ and u in functions of $\underline{z}$ and v made in the origin; approximations of the nonlinear functions obtained from (15) and (17). As the linear model (16) has only 3 state variables, the variable z_4 (x_3) is omitted in (22). This method of calculating weights, when FBL is applied, is described in [7, 8]. After appropriate calculations, the weights in function (22) are the following

$$\boldsymbol{Q}_z = \frac{1}{(\psi_1\psi_3\psi_4)^2} \begin{bmatrix} \psi_3^2(q_4\psi_4^2 + r_x\psi_5^2) & \psi_3(q_4\psi_2 q_4^2 + r_x\psi_5(\psi_2\psi_5 - \psi_3)) & -\psi_3(q_4\psi_4^2 + r_x\psi_5(\psi_2 + \psi_5)) \\ \psi_3(q_4\psi_2\psi_4^2 + r_x\psi_5(\psi_2\psi_5 - \psi_3)) & \psi_4^2(q_1\psi_1^2 + q_4\psi_2^2) + r_x(\psi_2\psi_5 - \psi_3)^2 & -q_4\psi_2\psi_4^2 - r_x(\psi_2 + \psi_5)(\psi_2\psi_5 - \psi_3) \\ -\psi_3(q_4\psi_4^2 + r_x\psi_5(\psi_2 + \psi_5)) & -q_4\psi_2\psi_4^2 - r_x(\psi_2 + \psi_5)(\psi_2\psi_5 - \psi_3) & \psi_4^2(q_2\psi_1^2 + q_4) + r_x(\psi_2 + \psi_5)^2 \end{bmatrix}, \tag{23}$$

$$r_z = \frac{r_x}{(\psi_1\psi_3\psi_4)^2}, \tag{24}$$

$$\underline{n}_z = \frac{1}{(\psi_1\psi_3\psi_4)^2}\begin{bmatrix} r_x\psi_3\psi_5 \\ r_x(\psi_2\psi_5 - \psi_3) \\ -r_x(\psi_2 + \psi_5) \end{bmatrix}, \tag{25}$$

where $q_k = Q(k, k)$, k = 1, 2, 4.

For the experiment, where the model takes discrete-time form, another performance index is used:

$$J_z = \sum_0^\infty \left(\underline{z}_t^T Q_z \underline{z}_t + r_z v_t^2 + \underline{z}_t^T \underline{n}_z v_t\right), \tag{26}$$

with the same weights as in (22).

8 Simulations and Experiment

In every figure the time is expressed in seconds. Performance index, J, used to evaluate the quality of the control, is similar to the index (21), but without the variable x_3, that is: $J = \int_0^{tmax}(q_1x_1^2 + q_2x_2^2 + q_4x_4^2 + r_xu^2)dt$, where $tmax$ is the simulation/experiment horizon. When FBL is used, every signal converges to 0 aside from signal x_3. Variable x_3 can be stabilized on arbitrary value.

In the first part of simulations, we compared control algorithm in two cases: with linear model obtained by Jacobian linearization and by feedback linearization. In the Fig. 4 time simulations are presented, starting from the initial point: $\underline{x}_0 = [\,0.065\ 0\ 0\ 0\,]'$. The Table 1 contains performance indices; the index J is always lower for FBL, especially for higher initial position of x_1.

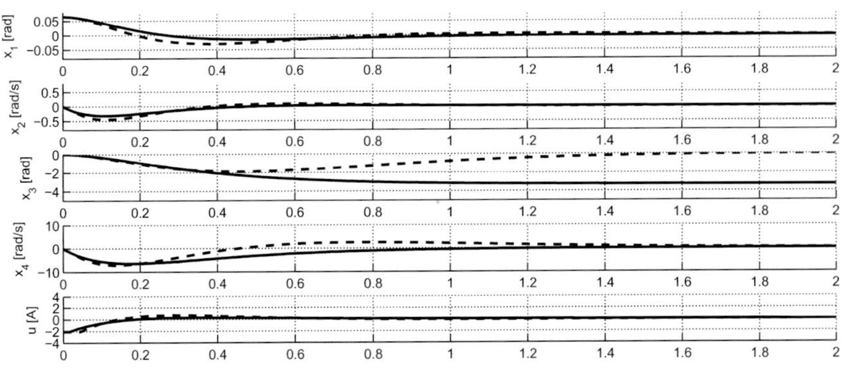

Fig. 4 Linear quadratic control with Jacobian linearization (*dashed line*) and feedback linearization (*solid line*)—a comparison

Table 1 Performance index, J, for several initial values of x_1 and two different linearization methods

	0.01	0.02	0.03	0.04	0.05	0.06	0.07	0.08	0.09	0.10
Jacobian lin.	0.0135	0.0541	0.1217	0.2163	0.3368	0.4851	0.6688	0.901	1.1953	1.5747
FBL	0.0095	0.0382	0.0859	0.1526	0.2385	0.3430	0.4688	0.6208	0.8100	1.0446

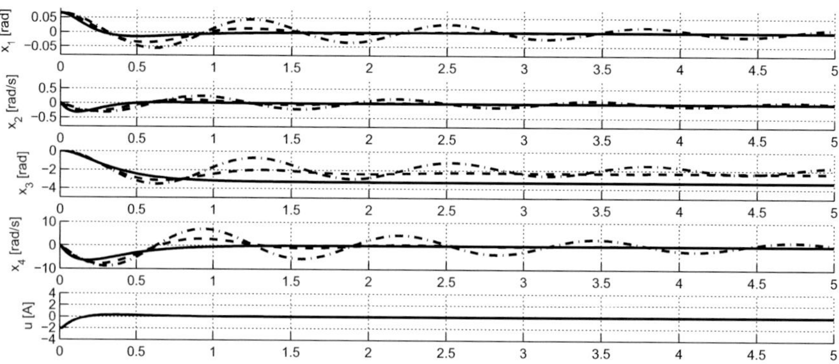

Fig. 5 Performance of control system with feedback linearization and model inaccuracy: 0 % (*solid line*) 20 % (*dashed line*) and 30 % (*dash-dot line*)

Table 2 Performance indices for several different parameters perturbation, when the initial value $x_1(0) = 0.65$ and FBL was used

	$\int x_1^2$	$\int x_2^2$	$\int x_3^2$	$\int x_4^2$	$\int u^2$	J
0 %	0.0005	0.0190	49.1011	14.2585	0.2555	0.4034
10 %	0.0007	0.0210	35.2981	15.8641	0.2555	0.4210
20 %	0.0010	0.0280	26.4128	22.0369	0.2555	0.4864
30 %	0.0032	0.0807	21.8595	65.8639	0.2555	0.9472

The Fig. 5 and the Table 2 show the behaviour of control system with FBL when the model parameters was perturbed. For this purpose $\psi_k^* = \psi_k \pm p\psi_k$ was used for calculations and control algorithm, and ψ_k for simulated object; p is the percentage perturbation. We simulated 32 combinations of adding/subtracting for every selected p. Presented results are those obtained for the worst combination. The Fig. 5 and the Table 2 indicate that, implementation of $p = 10\,\%$ resulted in increasing J only of 4.4 %; for $p = 20\,\%$ the performance was still reasonably well; for $p = 30\,\%$ the performance index J was increased of 135 % but the control system still remained stable. The Table 2 also contains the values of integrated squares of individual signals.

The Fig. 6 illustrates the comparison between the simulation results of the algorithm with FBL and the performance of the real plant. Inaccuracies of the model can concern not only parameters values but also the specificity of dynamics equations in the model.

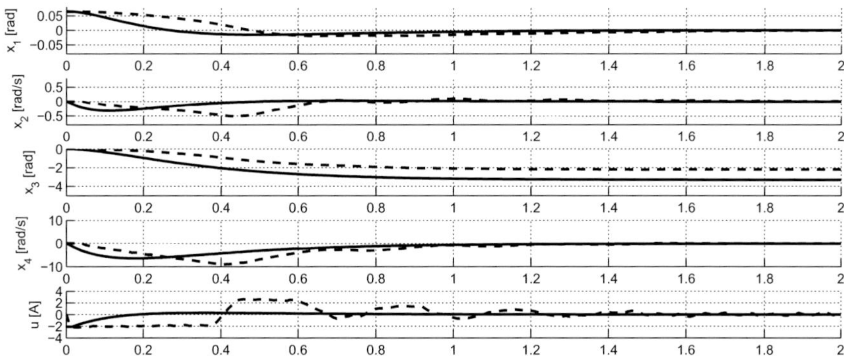

Fig. 6 Results of linear quadratic control with feedback linearization: experiment result on the real machine (*dashed line*) and simulation (*solid line*)

9 Conclusions

Feedback linearization designed for the bicycle robot has delivered an exact linear model and, with the linear quadratic control, has guaranteed a better control performance than with the Jacobian linearization. Simulations and experiment have showed that, despite the differences between the model and the real plant, the control algorithm has assured, to some extent, stable and good results. It demonstrates that the method can be applied for real plants and it can be also a good base for further research of control methods.

References

1. Afkhami, S., Yazdanpanah, M.J., Maralani, P.J.: Stabilization of inertia wheel pendulum using output feedback back-stepping. In: Proceedings of 2003 IEEE Conference on Control Applications, CCA 2003, vol. 2, pp. 977–982, 23–25 June 2003
2. Isidori, A.: Nonlinear Control Systems. Springer, London (1995)
3. Meijaard, J.P., Papadopoulos, J.M., Ruina, A., Schwab, A.L.: Linearized dynamics equations for the balance and steer of a bicycle: a benchmark and review. Proc. R. Soc. A Math. Phys. Eng. Sci. **463**, 1955–1982 (2084)
4. Qaiser, N., Iqbal, N., Qaiser, N.: Stabilization of inertia wheel pendulum using multiple sliding surfaces control technique. In: Multitopic Conference, 2006. INMIC '06, pp. 461–466. IEEE, 23–24 Dec 2006
5. Slotine, J.E., Li, W.: Applied Nonlinear Control. Prentice Hall, New Jersey (1991)
6. Spong, M.W., Corke, P., Lozano, R.: Nonlinear control of the reaction wheel pendulum. Automatica **37**, 184–1851 (2001)
7. Zietkiewicz, J.: Input and state constrained nonlinear predictive control. application to a levitation system. In: Proceedings of International Conference on Informatics in Control, Automation and Robotics, vol. 1, pp. 274–279 (2014)

8. Zietkiewicz, J.: Non-minimum phase properties and feedback linearization control of nonlinear chemical reaction. Proceedings of International Conference on Methods and Models in Automation and Robotics, pp. 489–494 (2015)
9. Zietkiewicz, J., Horla, D., Owczarkowski, A.: Robust actuator fault-tolerant LQR control of unmanned bicycle robot. A Feedback Linearization Approach, (submitted to Automation 2016)

Robust Actuator Fault-Tolerant LQR Control of Unmanned Bicycle Robot. A Feedback Linearization Approach

Joanna Zietkiewicz, Dariusz Horla and Adam Owczarkowski

Abstract Robust control must be able to cope with various system behavior subject to mismodeling and able to assure some performance level. In the paper, we propose to use actuator fault-tolerant control law to stabilize a bicycle robot model with inertial wheel to take into account the unmodeled uncertainty introduced by using linearized model in a LQR fashion with feedback linearization. Our proposal is illustrated by signal plots and performance indexes' values obtained from a set of experiments and is a natural extension of the results presented in the past.

Keywords Feedback linearization · LQ control · Robustness · Actuator failure

1 Introduction

Part of robust control papers present quadratic cost for uncertain systems measures, related to classical LQR, e.g. see [2] or [9, 12]. Since implementing control methods is inherently connected to either actuator or sensor failures or, simply, subject to modeling uncertainty or imperfect knowledge about the plant, it is vital to choose the algorithm that can both guarantee stability and certain level of performance. In the paper, the control method from [4, 14] is presented in feedback linearization framework in order to stabilize bicycle robot model (nonlinear) in an unstable equilibrium point, based on its feedback linearization. This approach differs from the one presented in [6], where convex optimization framework to linearized model was

J. Zietkiewicz (✉) · D. Horla · A. Owczarkowski
Institute of Control and Information Engineering, Poznań University of Technology, Poznań, Poland
e-mail: Joanna.Zietkiewicz@put.poznan.pl

D. Horla
e-mail: Dariusz.Horla@put.poznan.pl

A. Owczarkowski
e-mail: Adam.Owczarkowski@doctorate.put.poznan.pl

R. Szewczyk et al. (eds.), *Challenges in Automation, Robotics and Measurement Techniques*, Advances in Intelligent Systems and Computing 440, DOI 10.1007/978-3-319-29357-8_37

used or [7] where robust LQR control was used for linearized model again, mainly in presenting a new model of the plant, after feedback linearization omitting one, stable, mode of the model, and reducing its degree.

The robustness against actuator failure is used here to make the controller able to cope with uncertainty originating either from linearization or imperfect knowledge about the model of the robot, and can be easily used, as it has been shown by experiments with a different approach to linearization in [7] in real-time control, as it only requires off-line computation of feedback linearization compensator and state-feedback vector for LQR control with reformulated weighing matrices. Simulation results in the form of performance indices values are shown on the basis of a set of the performed computer simulations with a nonlinear bicycle robot model.

An introduction to the problem of inertial wheel stabilization can be found in [1, 8], and plain LQR control results can be found in [11, 13], but the approach presented in this paper enable one to take the uncertainty of the linearized model of the robot into account.

In the paper [15], concerning the same problem of stabilisation using feedback linearization but with no robustness considerations, basic assumptions, derivations of equations and complete description of the problem is included.

2 Model of the Robot, FBL Linearization

2.1 Continuous-Time 2DOF Model of the Robot

A complete mathematical model of the bicycle robot is defined by non-linear differential equations (1)–(4)

$$\dot{x}_1(t) = x_2(t), \tag{1}$$

$$\dot{x}_2(t) = \frac{g\, h_r\, m_r\, \sin\big(x_1(t)\big)}{I_{rg}} - \frac{b_r\, x_2(t)}{I_{rg}} + - \frac{b_I\, x_4(t)}{I_{rg}} + \frac{k_m\, u(t)}{I_{rg}}, \tag{2}$$

$$\dot{x}_3(t) = x_4(t), \tag{3}$$

$$\dot{x}_4(t) = \frac{k_m\, u(t)}{I_I + I_{mr}} - \frac{b_I\, x_4(t)}{I_I + I_{mr}}, \tag{4}$$

where (see Fig. 1 for the kinematic scheme of the object): $\underline{x}$—state vector, x_1—angle of the robot from the vertical, x_2—angular velocity of the robot, x_3—rotation angle of the reaction wheel, x_4—angular velocity of the reaction wheel, u—control signal. The remaining symbols can be found, e.g. in [4, 7].

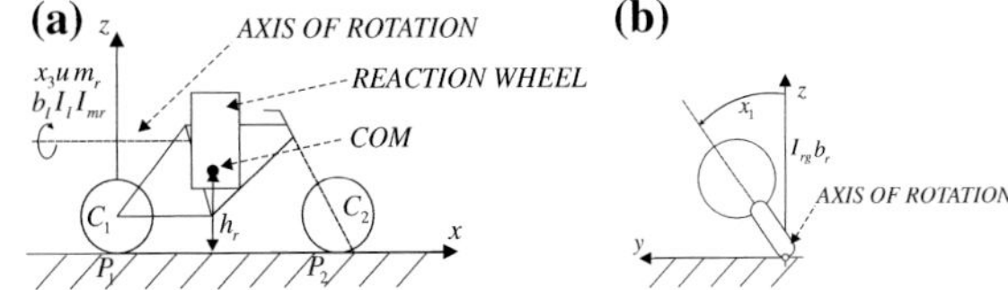

Fig. 1 The bicycle robot—kinematic scheme. **a** Main view. **b** Side view

The paper [6] presented the approach of applying linearization to nonlinear model of the plant based on Jacobian matrix and its discrete-time counterpart, whereas the paper [15] adopts the approach with FBL without robustness analysis. In this paper, we adopt LQR control law to FBL-linearized model of the plant that can be used both in continuous- and discrete-time controllers, comparing the results of four different approaches to LQR control.

2.2 Feedback Linearization of the Plant

Below, the feedback linearization of the model is presented, resulting in its new description, forcing the designer to redefine standard LQR control tuning parameters with respect to obtained linearization. This design part concerning new weighing matrices has been presented in Sect. 5 describing control algorithms.

Feedback linearization (abbr. FBL) enables one to obtain exact linear model, mapping nonlinear dynamics of the system [5, 10]. It is assumed here that it is applied to the system in the form

$$\dot{\underline{x}}(t) = \underline{f}(\underline{x}(t)) + \underline{g}u(t), \tag{5}$$

with $\underline{x} \in \mathscr{R}^n$, $u \in \mathscr{R}$ and smooth functions $\underline{f}(\underline{x}) : \mathscr{R}^n \rightarrow \mathscr{R}^n$, $\underline{g} \in \mathscr{R}^n$.

One can perform input-state linearization or input-output linearization of the system. According to [15] it is possible linearize 3 selected state variables, what results in defining artificial output first

$$y(t) = -\psi_1 x_4(t) - \psi_2 x_1(t) + x_2(t), \tag{6}$$

and it is the first variable of the new state $\underline{z}$. Using the procedure of calculating FBL model we can assign:

$$z_1(t) = \psi_1 x_4(t) - \psi_2 x_1(t) + x_2(t), \tag{7}$$
$$z_2(t) = \psi_3 \sin x_1(t), \tag{8}$$
$$z_3(t) = \psi_3 x_2(t) \cos x_1(t), \tag{9}$$

what results in linear model of the form

$$\dot{z}_1(t) = z_2(t), \tag{10}$$

$$\dot{z}_2(t) = z_3(t), \tag{11}$$

$$\begin{aligned} \dot{z}_3(t) = v(t) = \psi_3(\cos x_1(t)(\psi_1\psi_5 x_4(t) + \psi_2 x_2(t) + \psi_3 \sin x_1(t) \\ + \psi_1\psi_4 u(t)) - x_2^2(t) \sin x_1(t)), \end{aligned} \tag{12}$$

where ψ_i $(i = 1, \ldots, 4)$ have been derived in [15].

2.3 Discrete-Time Model of the Plant

In order to apply the LQR controller in discrete-time, the following multivariable model of the linearized plant must be introduced:

$$\underline{\zeta}_{t+1} = \boldsymbol{A}\underline{\zeta}_t + \underline{b}\mu_t, \tag{13}$$

$$\eta_t = \underline{c}^T \underline{\zeta}_t, \tag{14}$$

where the appropriate matrix $\boldsymbol{A}$ and vectors $\underline{b}$, $\underline{c}$ have known sizes and are the result of step-invariant discretization of the continuous-time model either as Jacobian-linearized model of (1)–(4) (see, e.g. [4, 7]) or as feedback-linearized model of the plant (see [15]), with chosen sample period T_S (the matrix and vector values are different for the two listed-above approaches). In the first case $\underline{\zeta} = \underline{x} \in \mathrm{R}^n$ is a full state vector of the model and $\mu = u$ is the constrained control signal. In the case of FBL linearization $\underline{\zeta} = \underline{z}$ and $\mu = v$. Artificial output refers to new variables resulting from FBL transformation, but because output signals are not used in both approaches, the definition of η is unnecessary here.

3 Actuator Failure Models

Initially, in our LQR approach to the problem, assuming that we have a linear representation of the plant derived by discretization of its nonlinear model, the control performance index is defined as in standard LQR control

$$J = \sum_{t=0}^{\infty} \left(\underline{x}_t^T \boldsymbol{Q} \underline{x}_t + R u_t^2 \right), \tag{15}$$

with weighing matrix $\boldsymbol{Q} \geq 0$, and $R \geq 0$. In the most general case, one can assume actuator failure model as in [14], i.e.,

$$u_t^k = (1 - \rho_t^k)\text{sat}(v_t;\, \alpha) \qquad (k = 1, 2, \ldots, g_F), \tag{16}$$

where ρ_t^k is an unknown constant from the span that will be defined in the further part of the text, index k denotes the k-th failure model, and g_F is the total number of failure models. The symbol u_t^k refers to the constrained control signal, assuming that an actuator failure takes place (in the other case, $u_t^k = v_t$). For any actuator failure model, including the situation for constraints imposed on the control signal, the constant ρ_t^k lies in $\rho_{-,t}^k \le \rho_t^k \le \rho_{+,t}^k$, and function sat defines the method of applying constraints (e.g., cut-off constraint).

Having taken a single model of failure into account, (16) can be transformed [3, 14] to

$$u_t^F = \varrho v_t \qquad (0 \le \varrho_- \le \varrho \le \varrho_+), \tag{17}$$

with $\varrho_- \le 1$ and $\varrho_+ \ge 1$. This enables one assume that uncertainty introduced by linearization (either FBL or linearization of the plant by Jacobian matrix), can be treated by the algorithm and that constraints imposed in control systems may both mimic actuator failure-like behavior of the model.

With reference to (16), $\varrho_- = \varrho_+$ means that there are no active constraints imposed on the control signal or no failures have taken place, and $u_t^F = v_t$. The case $\varrho_- > 0$ corresponds to partial failure, and $\varrho_- = 0$ to the outage case. It will be henceforth assumed that $u_t = u_t^F$.

4 LQR Control of the Model and Its Modifications

At first, a standard LQR control law applied to Jacobian-linearized discretized model of the plant

$$v_t = \underline{k}^T \underline{x}_t \tag{18}$$

is called reliable, i.e., and assuring that a specified value of the performance index (15) is not exceeded for the considered model of the plant, if it is connected with some matrix $\boldsymbol{P}$, the system (13) and (14), and if $\boldsymbol{P}$ satisfies the inequality [3]

$$\left(\boldsymbol{A} + \varrho \underline{bk}^T\right)^T \boldsymbol{P} \left(\boldsymbol{A} + \varrho \underline{bk}^T\right) - \boldsymbol{P} + R\varrho^2 \underline{kk}^T + \boldsymbol{Q} \le 0. \tag{19}$$

The closed-loop model of the system $\underline{x}_{t+1} = \left(\boldsymbol{A} + \varrho \underline{bk}^T\right) \underline{x}_t$ is then stable, and the performance index in an infinite horizon satisfies

$$J = \sum_{t=0}^{\infty} \underline{x}_t^T \left(\boldsymbol{Q} + R\varrho^2 \underline{kk}^T\right) \underline{x}_t \le \underline{x}_0^T \boldsymbol{P} \underline{x}_0. \tag{20}$$

If no robustness against actuator failure is taken into consideration, the optimal state-feedback vector $\underline{k}$ for the control law (18) is derived as a solution of the set of equations:

$$\underline{k}^T = -\left(\underline{b}^T \boldsymbol{P} \underline{b} + R\right)^{-1} \underline{b}^T \boldsymbol{P} \boldsymbol{A}, \tag{21}$$

$$\boldsymbol{P} = \boldsymbol{Q} + \boldsymbol{A}^T \boldsymbol{P} \boldsymbol{A} - \boldsymbol{A}^T \boldsymbol{P} \underline{b} \left(\underline{b}^T \boldsymbol{P} \underline{b} + R\right)^{-1} \underline{b}^T \boldsymbol{P} \boldsymbol{A}, \tag{22}$$

an optimal value $J_{\underline{k}}$ of the performance index (15), based on deriving $\underline{k}$ according to (21) and (22) being at the same time the upper boundary of (20), is $J_{\underline{k}} = \underline{x}_0^T \boldsymbol{P} \underline{x}_0$.

Secondly, substituting matrices $\boldsymbol{A}$, $\boldsymbol{Q}$, $\underline{b}$, state and state-feedback vectors and R one can use the same formulas to FBL-linearized system.

Thirdly, the LQI control law is also taken into account, but here only for FBL-linearized representation, where

$$v_t = \underline{k}_{\text{FBL}}^T \underline{z}_t + \underline{k}_I^T \underline{x}_{I,t}, \tag{23}$$

where $\underline{z}_t$ is a state vector of FBL representation, $\underline{x}_{I,t}$ is an appropriate integral of the state vector from the model, and $\underline{k}_I$ for the described problem is a zero vector with third element with value $a \neq 0$ (compensation of gyroscopic drift in x_3).

5 Optimal State-Feedback for FBL Control

It is to be stressed that computation of FBL for nonlinear plant results in reformulation of classical LQR-type control law. The form of the control law remains unchanged, but matrices defining performance indices need to be reformulated. The initial control performance index for control law for Jacobian-linearized discretized model is as follows

$$J = \sum_{t=0}^{\infty} \left(\underline{x}_t^T \boldsymbol{Q} \underline{x}_t + R u_t^2 + 2\underline{x}_t^T \boldsymbol{N} u_t\right)$$

and matrices $\boldsymbol{Q}$, $\boldsymbol{R}$, $\boldsymbol{N}$ are used to formulate the new performance index for diagonal $\boldsymbol{Q}$ only

$$J_{\text{FBL}} = \sum_{t=0}^{\infty} \left(\underline{z}_t^T \boldsymbol{Q}_z \underline{z}_t + R_z v_t^2 + 2\underline{z}_t^T \boldsymbol{N}_z v_t\right),$$

where [15] ($q_k = \boldsymbol{Q}(k,k)$, $k = 1, 2, 4$):

$$\boldsymbol{Q}_z = \frac{1}{(\psi_1\psi_3\psi_4)^2} \begin{bmatrix} \psi_3^2(q_4\psi_4^2 + r_x\psi_5^2) & \psi_3(q_4\psi_2 q_4^2 + r_x\psi_5(\psi_2\psi_5 - \psi_3)) & -\psi_3(q_4\psi_4^2 + r_x\psi_5(\psi_2 + \psi_5)) \\ \psi_3(q_4\psi_2\psi_4^2 + r_x\psi_5(\psi_2\psi_5 - \psi_3)) & \psi_4^2(q_1\psi_1^2 + q_4\psi_2^2) + r_x(\psi_2\psi_5 - \psi_3)^2 & -q_4\psi_2\psi_4^2 - r_x(\psi_2 + \psi_5)(\psi_2\psi_5 - \psi_3) \\ -\psi_3(q_4\psi_4^2 + r_x\psi_5(\psi_2 + \psi_5)) & -q_4\psi_2\psi_4^2 - r_x(\psi_2 + \psi_5)(\psi_2\psi_5 - \psi_3) & \psi_4^2(q_2\psi_1^2 + q_4) + r_x(\psi_2 + \psi_5)^2 \end{bmatrix} \tag{24}$$

$$R_z = \frac{R}{(\psi_1\psi_3\psi_4)^2}, \tag{25}$$

$$N_z = \frac{1}{(\psi_1\psi_3\psi_4)^2}\begin{bmatrix} r_x\psi_3\psi_5 \\ r_x(\psi_2\psi_5 - \psi_3) \\ -r_x(\psi_2 + \psi_5) \end{bmatrix}. \tag{26}$$

It can be noted that the last term of the expression under sum operator is not present in (15), and similarly is not taken into account into iterative algorithm. However, as it has resulted from simulation, its current value is neglectful, usually at least 3 orders less than the other two during transients.

Based on the algorithm from [14] (used successfully in [7] during experiment with a real robot), not cited in full here, one can compute in iterative manner for selected degree $0 \leq \delta < 1$ of robustness needed new solution to Riccati equation and new state-feedback vector

$$\underline{k}^T = -\gamma^{-1}\left(1-(X^{-1}-R_0)\left((1-\gamma_0^2)+\gamma_0^2R_0^{-1}X^{-1}\right)^{-1}\gamma_0^2R_0^{-1}\right)X^{-1}\underline{b}^T\boldsymbol{PA},$$

where $X = \underline{b}^T\boldsymbol{P}\underline{b} + R$ and the remaining symbols have been defined in [4].

6 Actuator Failure in Standard LQR Notation, Cut-Off Constraints and Modeling Errors

When amplitude constraints of control signal become active, computed control signal does not refer to constrained control signal, what can be treated as a special case of actuator failure, similarly as in the case of mismodeling or imperfect knowledge about the model. In such a situation, it is assumed that $\gamma = \alpha$ for and the constrained control signal becomes [3]

$$u_t^F = \text{sat}\left(\underline{k}^T\underline{x}_t; \alpha\right), \tag{27}$$

where sat is a function that defines constraints imposed on the control signal in the range of $\pm\alpha$.

7 Simulation Comparison of Control Performance

7.1 Experiment Conditions

The considered controllers work in discrete-time with their output fed to zero-order hold and based on discretized model of, Jacobian- or FBL-, linearized plant (13) and (14). These include: FBL + LQR controller with robustness against actuator failure, plain FBL + LQR controller, standard LQR controller and FBL + LQI controller with robustness against actuator failure.

During the simulations the following parameters have been fixed: $T_S = 0.02$ s, $\underline{Q} = \text{diag}\{1, 1, 0.01, 0.4\}, R = 10, 0 \leq \delta \leq 0.8, \rho_+ = 1 + \delta, \rho_- = 1 - \delta, \alpha = 2.1$ A, $R_0 = 0.1$ (iterative algorithm), $\underline{k}_I = [0, 0, 0.01, 0]^T$, $\underline{x}_0 = [-0.06\,\text{rad}, 0\,\text{rad} \cdot \text{s}^{-1}, 0\,\text{rad}, 0\,\text{rad} \cdot \text{s}^{-1}]^T$.

7.2 Performance Indices

In order to verify the behavior of the control system, the following performance indexes have been introduced:

$$J_{x_i} = \sum_{\tau=0}^{1000} x_{i,\tau}^2 \quad (i = 1, \ldots, 4), \tag{28}$$

$$J_u = \sum_{\tau=0}^{1000} u_\tau^2, \tag{29}$$

where τ is a number of sample of signals from sampled-data control system read every 0.01 s. The simulation time has been equal in all cases and set to 10 s. The first index is related to energy carried by $x_{i,t}$. In the case of x_1 that has to be stabilized at 0 (upright, unstable equilibrium point), similarly in the case of x_2 measuring angular velocity of the robot, x_3—rotation angle of the reaction wheel (related to energy consumption in steady-state) that also represents gyroscopic drift and its velocity x_4 small values of corresponding indexes represent smooth response to initial conditions.

In all approaches presented, the performance indexes refer to real state of the full nonlinear model, and control performance index to the control signal fed to the original nonlinear model.

7.3 Conclusion

In the Fig. 2, performance indexes plots have been presented for the considered four methods. As it can be seen optimal LQ control with Jacobian linearization approach results in the greatest energy consumption (the plot has been depicted as horizontal line for comparative purposes only, but its value is not a function of δ. As a result of the energy consumption, x_1 is closest to zero during simulations and x_3 varies slowly during simulations.

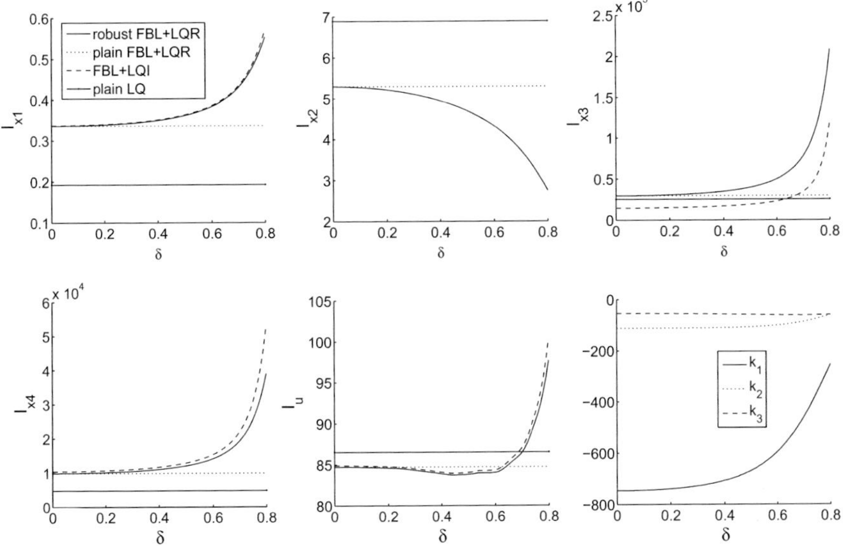

Fig. 2 Performance indices plots versus δ

Since energy usage is important, by introduction of FBL + LQR/LQI approach one can observe its local minimum for a range of δs, well below value for plain FBL + LQR algorithm. This causes, however, other performance to increase, but increase for $\delta < 0.3$ is not big. Based on the experiments presented in [7], one can expect that this minimum will be greater in real-world experiment and accompanied by minima in remaining performance indices. Introduction of LQI into control law assures one that uncontrollable stable non-oscillatory mode from FBL approach can be brought back to zero, what causes all the performance to increase, apart from the index connected to gyroscopic drift.

The values of 3-element state-feedback in the case of FBL have been shown in the right plot in the last row. By introduction of robustness, and increasing uncertainty parameter δ the controller becomes more cautious, decreasing the norm of the state-feedback vector, what is an intuitive result.

By observing plots presented in Fig. 3 one can see that plain LQR approach causes control signal to saturate, causing multiple phenomena, as e.g. oscillations in x_2. Robustness decreases speed of transients and allows smaller energy consumption. As it can be seen, using LQI approach causes gyroscopic drift to be compensated, and differences (as far as transients are concerned) are hard to be found between the plots, just on the basis of plots.

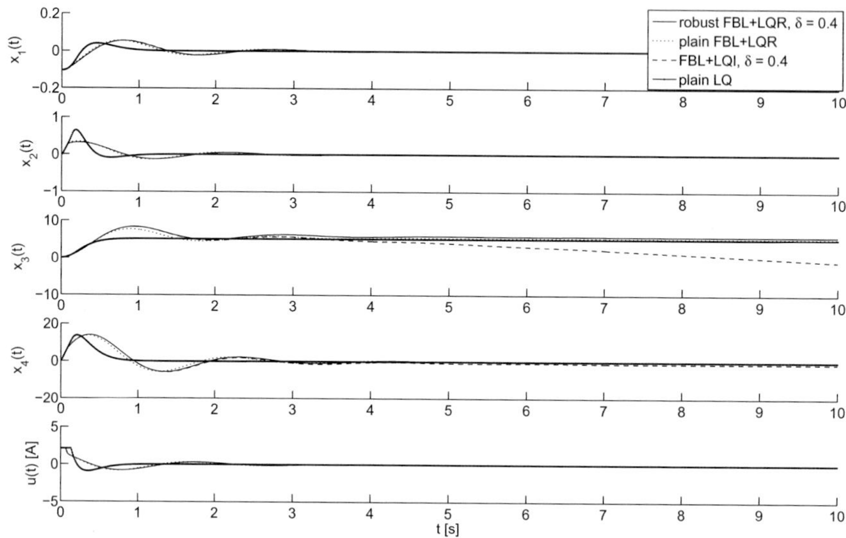

Fig. 3 Selected simulation results of four control methods

8 Summary

As it has been shown, by introduction robustness into control algorithm one can decrease energy consumption with minor changes in other performance indexes for small values of δ. Introduction of FBL enables one to use nonlinear control techniques to operate with control methods designed for linear models. It is to be borne in mind, however, that performance of the approach strongly relies on FBL procedure and values of weighing matrices. It is expected, on the basis of results of [7] that performance increase with a real-world robot, as experiments are planned, will be even better.

References

1. Åström, K.J., Block, D.J., Spong, M.W.: The Reaction Wheel Pendulum. Morgan & Claypool (2007)
2. Chang, S.S.L., Peng, T.K.C.: Adaptive guaranteed cost control of systems with uncertain parameters. IEEE Trans. Autom. Control **17**, 474–483 (1972)
3. Horla, D., Królikowski, A.: Discrete-time LQG control with actuator failure. In: Proceedings of the 8th International Conference on Informatics in Control, Automation and Robotics, CD-ROM. Noordwijkerhout (2011)
4. Horla, D., Owczarkowski, A.: Robust LQR with actuator failure control strategies for 4DOF model of unmanned bicycle robot stabilised by inertial wheel. In: Proceedings of the 15th International Conference IESM, CD-ROM. Seville (2015)
5. Isidori, A.: Nonlinear Control Systems. Springer, London (1995)

6. Owczarkowski, A., Horla, D.: A comparison of control strategies for 4DoF model of unmanned bicycle robot stabilised by inertial wheel. In: Szewczyk, R., et al. (eds.) Progress in Automation, Robotics and Measuring Techniques, Advances in Intelligent Systems and Computing, vol. 351, pp. 211–221
7. Owczarkowski, A., Horla, D.: Robust LQR and LQI control with actuator failure of 2DOF unmanned bicycle robot stabilized by inertial wheel. Int. J. Appl. Math. Comput. Sci. (accepted for publication)
8. Owczarkowski, A., Lis, M., Kozierski, P.: Tracking control of an inertial wheel pendulum by LQR regulation. In: Proceedings of the 19th International Conference on Methods and Models in Automation and Robotics, CD-ROM. Poland (2014)
9. Petersen, I.R., McFarlane, D.C.: Optimizing the Guaranteed Cost in the Control of Uncertain Systems. Brikhäuser, Boston (1992)
10. Slotine, J.E., Li, W.: Applied Nonlinear Control. Prentice Hall, New Jersey (1991)
11. Smerpitak, K., Ukakimparn, P., Trisuwananwat, T., Trakoonkootaworn, S.: An unmanned bicycle versus linear quadratic optimal controls. In: Proceedings of the 12th International Conference on Control, Automation and Systems, pp. 1337–1341. Korea (2012)
12. Xie, L., Soh, Y.C.: Guaranteed cost control of uncertain discrete-time systems. Control Theory Adv. Technol. **10**, 1235–1251 (1995)
13. Yang, J.H., Lee, S.Y., Kim, S.Y., Lee, Y.S., Kwon, O.K.: Linear controller design for circular motion of unmanned bicycle. In: Proceedings of the 11th International Conference on Control, Automation and Systems, pp. 893–897. Korea (2011)
14. Yang, Y., Yang, G.H., Soh, Y.C.: Reliable control of discrete-time systems with actuator failures. IEEE Proc. Control Theory Appl. **147**(4), 428–432 (2000)
15. Zietkiewicz, J., Owczarkowski, A., Horla, D.: Performance of feedback linearization based control of bicycle robot in consideration of model inaccuracy (accepted paper). Automation (2016)

Automated Drawing Recognition and Reproduction with a Multisensory Robotic Manipulation System

Anna Wujek and Tomasz Winiarski

Abstract Article presents a multisensory robotic system, that is reproducing contour drawings. Initially the system detects a sheet of paper with a reference drawing, determines contours of the drawing, and then draws contour image on a blank sheet of paper. The reproduction conserves features of the original drawing—shapes, location and scale. The system was created with usage of an embodied agent theory. In this article two main parts of a designed system are presented—vision module (virtual receptor) and control subsystem. System is verified on a modified industrial manipulator acting as a service robot with an eye in hand camera and a force/torque sensor mounted in the wrist.

Keywords Service robot · Controller design · Sensor fusion

1 Introduction

Service robots are useful in large variety of applications: technical, medical, domestic, even entertaining. In contrast to industrial robots, they are designed to work in partially unstructured and dynamically changing environment, also to interact with people [1]. Therefore, to successfully control service robots, it is crucial to provide data from various sensors [2]. People use mostly sight and touch to move and manipulate different types of objects, hence service robots are usually equipped with cameras and force/torque sensors. Image analysis, contact detection and force/torque measurement are vital to successfully move robot and manipulate objects effectively and safely.

A. Wujek (✉) · T. Winiarski
Warsaw University of Technology, Warsaw, Poland
e-mail: aniawujek@gmail.com
URL: http://robotics.ia.pw.edu.pl

T. Winiarski
e-mail: tmwiniarski@gmail.com

R. Szewczyk et al. (eds.), *Challenges in Automation, Robotics and Measurement Techniques*, Advances in Intelligent Systems and Computing 440, DOI 10.1007/978-3-319-29357-8_38

In this paper, we present a multisensorical robotic system, that is reproducing drawings. Basing on image analysis of a reference drawing, its contour representation is generated and drawn. Agent-based and component approaches have been used to design this system.

The paper is organised as follows. In Sect. 2, various drawing applications of robots and related work are discussed. Then, in Sect. 3, an experimental setup of our system and its general structure are described. Virtual receptor processing images is presented in Sect. 4, control subsystem and virtual effector controlling robotic arm are presented in Sect. 5. Section 6 presents the experimental verification of the proposed approach to reproduce drawings performed on a real robot. It contains description of the hardware and the software used during experiments, as well as the results of these experiments. Finally, Sect. 7 concludes paper and presents future work.

2 State of the Art

In recent years, a number of drawing applications for robots have been developed. Most of them are mainly entertaining and artistic projects, presenting computer vision and trajectory planning algorithms [3]. Many systems consist of robotic arm for manipulation and camera for image acquisition, however the other approaches are to reproduce an artist hand motion registered with force sensor [4]. Usually robot is prepared to draw specific objects, like human face [5]. Some authors focused on a robot behaviour to imitate human artist [6]. Many papers present robots painting with flexible brush for best artistic results, e.g [7], which requires precise control of pressure and slope of a brush. To perform more complex drawing, e.g. on a non-planar surface, robots are equipped with force sensors for contact detection [8]. Typically, there are two main problems, that authors of drawing applications for robots are facing: image analysis in order to determine contours of an object seen with a camera, and trajectory generation, that depends on surface shape, drawing tool and type of drawing.

Image processing and computer vision allows robot to behave autonomously and react to different situations without human interaction. Among other sensors, camera is the one, that provides a lot of useful data, that can be used for many purposes: determining position and orientation of a robot [9], location of people or objects around robot, objects or collision detection [10], quality control [11] etc. The key is to process images in an efficient and optimised way, so robot can react as fast as possible to avoid collisions and damages [12].

Concept of an embodied agent, that is an agent with a body interacting with real environment [1], is a way to describe a robotic system decomposed into a set of subsystems. This decomposition of the robotic system allows to distinguish: real and virtual receptors, real and virtual effectors, and control subsystem [13]. All of these components communicate with each other through communication buffers. In particular, in service robots real receptors include different types of 2D and 3D camera devices, providing data to be stored and processed with virtual receptors [14].

A robot hardware with low level joint controllers and encoders are robot's real effectors. Manipulator virtual effectors, general for various tasks but specific for the robot, execute the control loop in the task space of a robot, interpolate trajectory, etc. On top of the whole system, there is a control subsystem, specialized for specific tasks.

3 Experimental Setup and Ambodied Agent Structure

Experimental setup to reproduce drawings is presented in Fig. 1b. It consists of: a robotic arm with sensors (eye in hand camera and force/torque sensor mounted in the manipulator wrist), the tool able to hold a pen, and a horizontal platform with a paper sheet on which the robot can draw. Camera provides images of a drawing to be reproduced and images needed to locate a paper sheet. Force sensor is necessary to determine contact with surface during drawing, because the vision module produce only the rough approximation of the distance from the robot end–effector to the paper sheet [15].

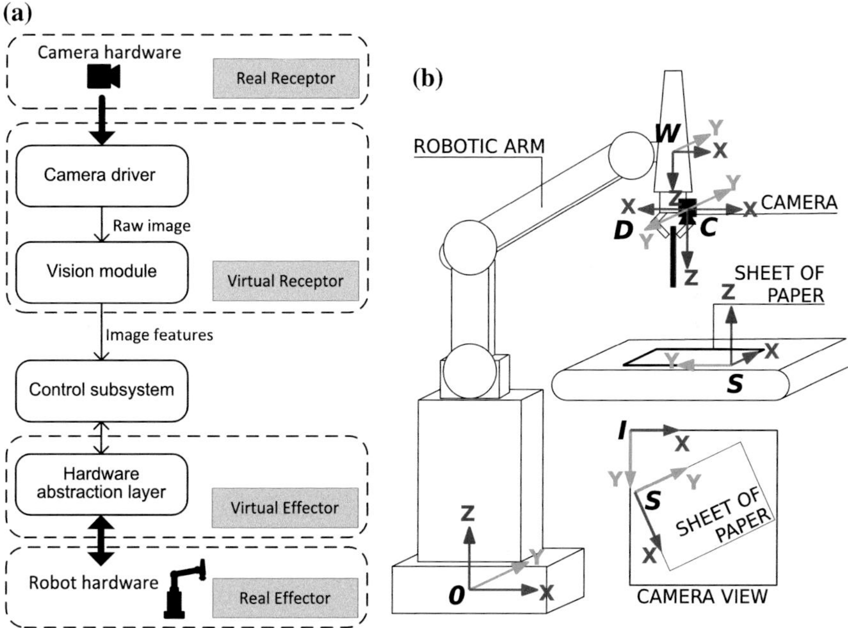

Fig. 1 **a** General system structure. **b** The experimental setup and the coordinate systems: *0*—base frame of a robot, *W*—frame of the sixth joint of a robot (wrist), *C*—camera frame, *D*—camera optical frame, *S*—frame of a sheet of paper, and *I*—frame of camera image

The overall structure of a developed embodied agent is depicted in Fig. 1a. Camera device, providing images for further analysis, is the real receptor connected directly to the virtual receptor, that consists of two elements: camera driver and vision module, processing images (Sect. 4). Robot hardware states the real effector, in this case performing a drawing, connected to the virtual effector (robot hardware abstraction layer). The control subsystem, specialised for a task of drawing, communicates with vision module and basing on that data prepares commands for the virtual effector.

4 Virtual Receptor

Image processing in drawing reproduction system consists of several phases (Fig. 2). The acquisition of image is performed with camera attached to robotic arm, controlled with camera driver. The obtained image is noised and distorted, which is a typical problem with optical devices. Its quality improvement is achieved with Gaussian blur algorithm to eliminate salt and pepper noise, and remap algorithm based on previously computed camera calibration parameters (focal lengths, principal points offsets and distortion coefficients) to eliminate distortion. The remap algorithm starts with computation of the joint undistortion and rectification transformation map_x and map_y, and then each pixel is remapped, according to (1)

$$\forall_{p' \in I} \; p'(i,j) = p(map_x(i,j), map_y(i,j)) \tag{1}$$

where $p'(i,j)$ is a pixel in the row i and the column j of output image I, $p(i,j)$ is a pixel of input image.

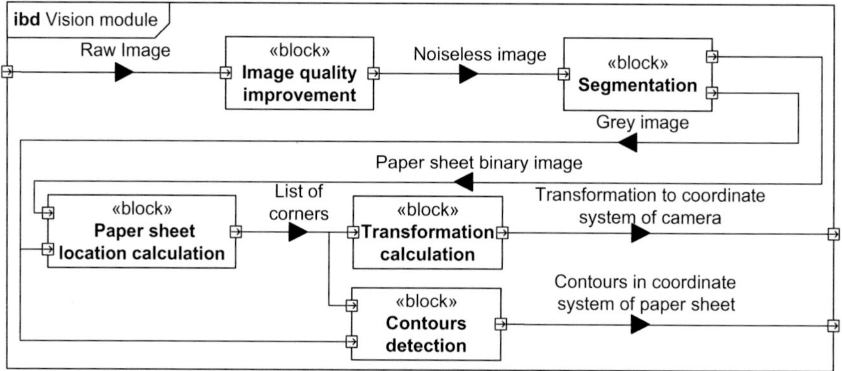

Fig. 2 Structure of a vision module, presenting sequence of image processing and data flow

The next step of image processing is a segmentation, that is classifying pixels of image into background, that will not be further processed, and objects to be processed. This information will be used to determine the location of a sheet of paper and to determine the features of a drawing. Segmentation consists of selection based on the colour of a pixel and morphology transformations, which smooth the image and improve its quality. Two types of morphology transformations are used: opening and closing. The result of this step is binary image, where white pixels show the area of paper sheet. Then, the position of paper sheet on image is determined. Paper sheet is rectangular, hence position can be specified by coordinates of four corners. At first, the edges in the binary image are found with Canny edge detector [16]. The result is binary image, where white pixels indicate edges. Then, system detects long straight lines in edges with modified probabilistic Hough transform [17]—one for each edge of a paper sheet. Each line is represented by a 4-element vector (x_1, y_1, x_2, y_2), where (x_1, y_1) and (x_2, y_2) are ending points. The intersections of these lines indicates the coordinates of corners. After small correction based on input image with function using Harris algorithm [18], that finds the most prominent corners in the image, coordinates of corners represented by 4-element vector of points $\{(x_1, y_1), \ldots, (x_4, y_4)\}$ are provided to the next phase of image processing.

The following step is the detection of contours. At first, rectangular mask with shape of paper sheet is put on input image so all pixels not belonging to area of paper sheet become black. Then, thresholding is performed to extract lines of drawing from the area of paper sheet. Due to the fact, that lines are thicker than one pixel, morphological skeleton [19] (topologically equivalent thinned version of the original contours) of lines is computed.

After morphology operation to smooth lines, the image is ready for contour detection with algorithm [20]. This algorithm produces a list of contours determined in image I:

$$^{I}l_{c'} = \{(p_{11}, p_{12}, \ldots), (p_{21}, p_{22}, \ldots), \ldots\} \tag{2}$$

where each contour consists of a list of its points and $p_{ij} = (x_{ij}, y_{ij})$. The distance between any two adjacent points of contour is 1 pixel,

$$dist(^{I}l_{c'}[j][k], {}^{I}l_{c'}[j][k+1]) = 1\ pixel \tag{3}$$

where j is j-th contour a k is its k-th point. Moreover, each contour is a closed curve, so even the contours of thin lines will be doubled. To avoid drawing the contour twice and to store all pixels of contours, new list of corrected contours $^{I}l_c$ in frame of image I is created with Algorithm 1, performed for each contour c, where $dist_{min}$, $dist_{max}$ are arbitrary determined minimal and maximal distances between two subsequent points in one contour (if $d < dist_{min}$, they are considered to be the same point, if $d > dist_{max}$, they are assigned to two different contours).

The next step of image processing is to represent contours in the frame of paper sheet S (origin is in one of the corners, X and Y axes overlap edges of paper sheet, Z axis is perpendicular to the surface of paper sheet). A 3×3 projective homography

H between frame of image I and frame S is found based on coordinates of paper sheet corners $x = \{(x_1, y_1), \ldots, (x_4, y_4)\}$ in frame I and corresponding model corners $x' = \{(0,0), (s_w, 0), (0, s_h), (s_w, s_h)\}$ in frame S, according to (4), where s_w and s_h are dimensions of paper sheet in m.

$$x' = Hx \tag{4}$$

Later, each point of list ${}^{I}l_c$ is transformed with projective homography H, according to (5).

$$\forall_{i,j} : {}^{I}l_c[i][j] \in {}^{I}l_c \; {}^{S}l_c[i][j] = H^{I}l_c[i][j] \tag{5}$$

After that, coordinate Z is added to each point to get list ${}^{S}l_c$, where each point is $(x_{ij}, y_{ij}, 0)$ (it is assumed that paper sheet with a drawing is a planar surface, and each point of a drawing is located on a surface indicated by X and Y axes).

Algorithm 1 Eliminate double contours

procedure ELIMINATECONTOURS(${}^{S}l_{c'}$)
  $m \leftarrow 0$ ▷ Index of contour in c
  for all i : ${}^{I}l_{c'}[i] \in {}^{I}l_{c'}$ **do** ▷ For each contour in list of contours
    $n \leftarrow 0$ ▷ Index of point in contour $c[m]$
    ${}^{I}l_c[m][n] \leftarrow {}^{I}l_{c'}[i][0]$ ▷ Get first point of a contour
    for all j : ${}^{I}l_{c'}[i][j] \in {}^{I}l_{c'}[i]$ **do** ▷ For each point in contour
      $d \leftarrow dist({}^{I}l_{c'}[i][j], {}^{I}l_c[m][n])$ ▷ Calculate distance between current point in c' and previous point in c
      if $dist_{min} < d < dist_{max}$ **then**
        $n \leftarrow n + 1$
        ${}^{I}l_c[m][n] \leftarrow {}^{I}l_{c'}[i][j]$ ▷ Append current point to ${}^{I}l_c$
      else if $d > dist_{max}$ **then**
        $n \leftarrow 0$; $m \leftarrow m + 1$
        ${}^{I}l_c[m][n] \leftarrow {}^{I}l_{c'}[i][j]$ ▷ Create contour in ${}^{I}l_c$ with current point
      end if
    end for
    $n \leftarrow 0$; $m \leftarrow m + 1$
  end for
end procedure

Last step is calculation of homogeneous transformation matrix ${}^{D}_{S}T$ relating frames D and S. The model of a paper sheet is created based on previously provided s_w and s_h and coordinates of corners in input image. This model consists of two vectors: $\{(x_1, y_1), \ldots, (x_4, y_4)\}$, that are coordinates of corners of paper sheet in image and $\{(0,0,0), (s_w, 0, 0), (0, s_h, 0), (s_w, s_h, 0)\}$, that are corresponding coordinates of model paper sheet in coordinate system of paper sheet. Based on that, using Perspective'n'Point algorithm [21], homogeneous transformation matrix ${}^{D}_{S}T$ is computed. After image processing, the resultant data is sent to control subsystem. The virtual receptor output buffer consists of the list of contours to be drawn ${}^{S}l_c = \{(p_{11}, p_{12}, \ldots), (p_{21}, p_{22}, \ldots), \ldots\}$, where $p_{ij} = (x_{ij}, y_{ij}, 0)$, related to the frame of paper sheet S, and transformation matrix ${}^{D}_{S}T$.

5 Control Subsystem and Virtual Effector

Control subsystem prepares commands to generate and execute trajectory in the virtual effector. Based on the data received from virtual receptor, current robot position and placement of a camera, coordinates representing contours to be drawn in robot base coordinate system are calculated, according to (6) and (7),

$$ {}^{0}_{S}T = {}^{0}_{W}T \, {}^{W}_{C}T \, {}^{C}_{D}T \, {}^{D}_{S}T \tag{6} $$

$$ \forall_{i,j} : {}^{S}l_c[i][j] \in {}^{S}l_c \quad {}^{0}l_c[i][j] = {}^{0}_{S}T \, {}^{S}l_c[i][j] \tag{7} $$

where ${}^{0}l_c[i][j] = p(x_{ij}, y_{ij}, z_{ij})$ is the point in base coordinate system, ${}^{S}l_c$ is the list of points provided by virtual receptor (vision module). The resultant list of contours is drawn with the Algorithm 2:

Algorithm 2 Draw contours

procedure DRAWCONTOURS(${}^{0}l_c$)
    moveToJointPosition(startingPosition) ▷ Move the robot arm to the starting position using joint interpolated motion.
    for all i : ${}^{0}l_{c[i]} \in {}^{0}l_c$ **do**
        moveToCartesianPose(${}^{0}l_{c[i][0]}$) ▷ Move over first point of a drawing using Cartesian interpolated motion.
        forceController() ▷ Move the robot end–effector downward until contact with a paper sheet occurs using force-controlled motion.
        $k \leftarrow 0$
        for all j : ${}^{0}l_c[i][j] \in {}^{0}l_c[i]$ **do** ▷ Plan the trajectory
            $t = time({}^{0}l_c[i][j], {}^{0}l_c[i][j+1])$ ▷ Time needed to get to next point
            $trajectory[k] \leftarrow CartesianTrajectoryPoint(t, {}^{B}l_c[i][j])$ ▷ Append structure with time and coordinates to vector of trajectory points.
            $k \leftarrow k + 1$
        end for
        moveAlongCartesianTrajectory(trajectory) ▷ Execute trajectory planned in previous step using Cartesian interpolated motion.
        moveRelToCartesianPose() ▷ Move the robot end–effector upward.
    end for
end procedure

6 Verification

Task of drawing requires robotic arm with a pen attached to its end-effector (Sect. 3). The modified 6DOF IRb-6 manipulator (Fig. 3a) was used to perform the tests. The robot was equipped with modern electronics [22] and software components replacing the industrial ones. The gripper had two coupled, parallel fingers (Fig. 3b). GIGE camera was attached to manipulator end-effector. The system was calibrated according to procedure [23]. For the purposes of this project, a compliant tool to hold a pen was designed (Fig. 3c).

The camera attached to the robot wrist (Fig. 3b) was a Point Grey Blackfly BFLY-PGE-14S2C-CS with 1.4MPix resolution. The camera was equipped with LG Security LC-M13VM2812IRD lens with F1.4 shutter and focal length 2.8–12 mm. The camera parameters (aperture, focal length and focus) were adjusted with moveable rings. Additional acquisition parameters (e.g. brightness) were set with camera driver.

In general, our robot controller is implemented with usage of Robot Operating System (ROS) [24] (set of software libraries and tools that help building robot applications) and OROCOS [25] (set of modular, configurable components for real-time robot applications). To control robots, IRPOS (IRb-6 robot virtual effector API) was used. Image processing was performed with DisCODe [26] (*Distributed Component Oriented Data Processing*), a framework for fast sensor data processing. A large number of drawings have been reproduced to test the designed system. All of them were correctly analysed and reproduced with only small imperfections (if input drawing was properly prepared, and requirements regarding lighting and paper position have been met). Test drawings included random shapes (e.g. lines, rectangles, triangles, circles, dots) and drawings of contours of particular things (e.g. castle, text). Different colours, thicknesses and distances between shapes has been tested as well as complex shaped consisting of many crossing lines. Experiments indicated, that any contour drawing, even the very complex one, can be correctly reproduced. The example of drawing prepared by human, contours found with vision module and final drawing are shown in Fig. 4.

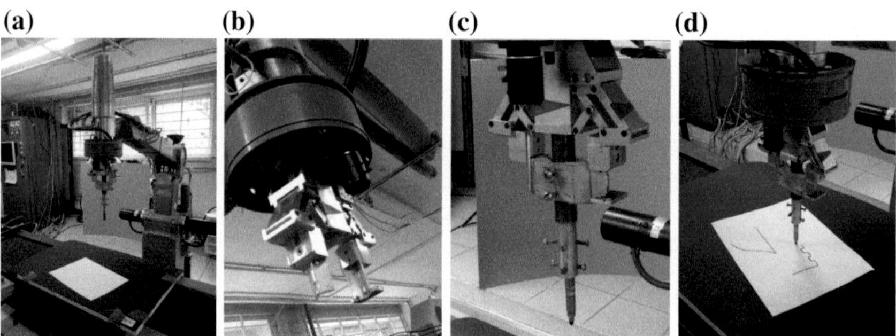

Fig. 3 The real robot performing the task of drawing (http://youtu.be/BvF8Cou4Qpc). **a** IRp-6 robot. **b** End-effector and camera. **c** Compliant tool. **d** Drawing

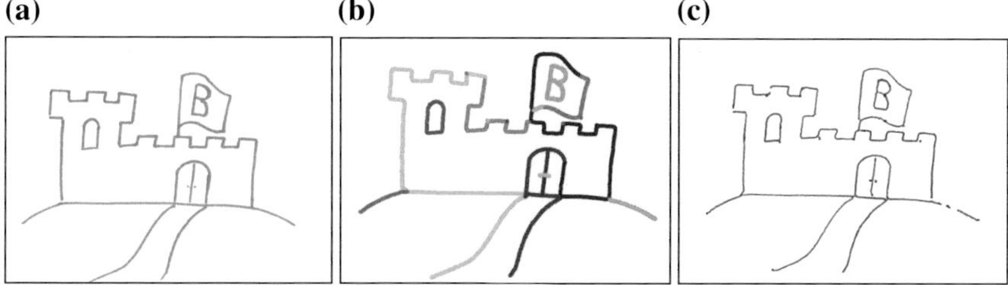

Fig. 4 Example of original drawing and reproduction. **a** Original drawing (by human). **b** Contours found with vision application. **c** Reproduced drawing (by robot)

7 Conclusions

Robotic arm with appropriate control system can successfully reproduce contour drawing. The performed tests indicated, that for our experimental station optimal thickness of lines on input drawing is 1–5 mm (if it is less the line is ignored, if it is more the reproduced line is deformed), and the minimal distance between endings of two separated lines is 4 mm (if it is less the neighbouring lines are connected in the result drawing). During the experiments some problems have been found: dark dots in the places, where pen touches paper for the first time (the problem is partially solved by the compliant pen holder), undesirable breaks in lines and small transition of the whole result drawing. All these problems do not significantly affect the system. Nevertheless, they can be fixed during system development.

There are many interesting modifications and improvements, that can be applied to the system: recognition of filled areas besides contours, drawing recognition and reproduction with several different colours, drawing on inclined or non-planar surface. Thanks to an embodied agent based decomposition of a system into two main modules (vision and control) different robots (real and virtual effectors) can be used for task of drawing with the same image processing subsystem (virtual receptor).

Acknowledgments This project was funded by the National Science Centre according to the decision number DEC-2012/05/D/ST6/03097.

References

1. Zieliński, C., Winiarski, T.: Motion generation in the MRROC++ robot programming framework. Int. J. Robot. Res. **29**(4), 386–413 (2010)
2. Winiarski, T., Banachowicz, K., Seredyński, D.: Multi-sensory feedback control in door approaching and opening. In: Filev, D., Jabłkowski, J., Kacprzyk, J., Krawczak, M., Popchev, I., Rutkowski, L., Sgurev, V., Sotirova, E., Szynkarczyk, P., Zadrozny, S. (eds.) Intelligent Systems'2014. Advances in Intelligent Systems and Computing, vol. 323, pp. 57–70. Springer International Publishing (2015)

3. Jean-Pierre, G., Said, Z.: The artist robot: a robot drawing like a human artist. In: 2012 IEEE International Conference on Industrial Technology (ICIT), pp. 486–491 (March 2012)
4. Zieliński, C., Winiarski, T.: General specification of multi-robot control system structures. Bull. Pol. Acad. Sci.—Tech. Sci. **58**(1), 15–28 (2010)
5. Lin, C.Y., Chuang, L.W., Mac, T.T.: Human portrait generation system for robot arm drawing. In: IEEE/ASME International Conference on Advanced Intelligent Mechatronics. AIM 2009, pp. 1757–1762 (July 2009)
6. Tresset, P., Leymarie, F.F.: Portrait drawing by paul the robot. Comput. Graph. **37**(5), 348–363 (2013)
7. Junyou, Y., Guilin, Q., Le, M., Dianchun, B., Xu, H.: Behavior-based control of brush drawing robot. In: 2011 International Conference on Transportation, Mechanical, and Electrical Engineering (TMEE), pp. 1148–1151 (Dec 2011)
8. Jain, S., Gupta, P., Kumar, V., Sharma, K.: A force-controlled portrait drawing robot. In: 2015 IEEE International Conference on Industrial Technology (ICIT), pp. 3160–3165 (March 2015)
9. Chenavier, F., Crowley, J.: Position estimation for a mobile robot using vision and odometry. In: 1992 IEEE International Conference on Robotics and Automation. Proceedings., vol. 3, pp. 2588–2593 (May 1992)
10. Yagi, Y., Kawato, S., Tsuji, S.: Real-time omnidirectional image sensor (copis) for vision-guided navigation. IEEE Trans. Robot. Autom. **10**(1), 11–22 (1994)
11. Brosnan, T., Sun, D.W.: Improving quality inspection of food products by computer vision—a review. J. Food Eng. **61**(1), 3–16 (2004)
12. Sharp, C., Shakernia, O., Sastry, S.: A vision system for landing an unmanned aerial vehicle. In: IEEE International Conference on Robotics and Automation. Proceedings 2001 ICRA, vol. 2, pp. 1720–1727 (2001)
13. Zieliński, C., Kornuta, T., Winiarski, T.: A systematic method of designing control systems for service and field robots. In: 19-th IEEE International Conference on Methods and Models in Automation and Robotics, MMAR'2014, pp. 1–14. IEEE
14. Kasprzak, W., Kornuta, T., Zieliński, C.: A virtual receptor in a robot control framework. In: Recent Advances in Automation, Robotics and Measuring Techniques. Advances in Intelligent Systems and Computing (AISC). Springer (2014)
15. Staniak, M., Winiarski, T., Zieliński, C.: Parallel visual-force control. In: Proceedings of the IEEE/RSJ International Conference on Intelligent Robots and Systems, IROS '08 (2008)
16. Canny, J.: A computational approach to edge detection. IEEE Trans. Pattern Anal. Mach. Intell. (6), 679–698 (1986)
17. Kiryati, N., Eldar, Y., Bruckstein, A.M.: A probabilistic hough transform. Pattern Recognit. **24**(4), 303–316 (1991)
18. Mikolajczyk, K., Schmid, C.: An affine invariant interest point detector. In: Computer Vision—ECCV 2002, pp. 128–142. Springer (2002)
19. Udrea, R.M., Vizireanu, N.: Iterative generalization of morphological skeleton. J. Electron. Imaging **16**(1), 010501–010501 (2007)
20. Suzuki, S., et al.: Topological structural analysis of digitized binary images by border following. Comput. Vis., Graph., Image Process. **30**(1), 32–46 (1985)
21. Quan, L., Lan, Z.: Linear n-point camera pose determination. IEEE Trans. Pattern Anal. Mach. Intell. **21**(8), 774–780 (1999)
22. Walcki, M., Banachowicz, K., Winiarski, T.: Research oriented motor controllers for robotic applications. In: Kozłowski, K. (ed.) Robot Motion and Control 2011 (LNCiS) Lecture Notes in Control & Information Sciences, vol. 422, pp. 193–203. Springer Verlag London Limited (2012)
23. Winiarski, T., Banachowicz, K.: Automated generation of component system for the calibration of the service robot kinematic parameters. In: 20th IEEE International Conference on Methods and Models in Automation and Robotics, MMAR'2015. IEEE (2015)
24. Quigley, M., Conley, K., Gerkey, B., Faust, J., Foote, T., Leibs, J., Wheeler, R., Ng, A.Y.: ROS: an open-source robot operating system. In: ICRA Workshop on Open Source Software, vol. 3 (2009)

25. Bruyninckx, H., Soetens, P., Koninckx, B.: The real-time motion control core of the orocos project. In: IEEE International Conference on Robotics and Automation. Proceedings. ICRA '03, vol. 2, pp. 2766–2771 (Sept 2003)
26. Stefańczyk, M., Kornuta, T.: Handling of asynchronous data flow in robot perception subsystems. In: Simulation, Modeling, and Programming for Autonomous Robots. Lecture Notes in Computer Science, vol. 8810, pp. 509–520. Springer (2014)

Visual Marker Based Shape Recognition System for Continuum Manipulators

Jan Fraś, Sebastain Tabaka and Jan Czarnowski

Abstract Soft robotics is young and popular research area. Continuum robots do not have traditional joints and their movement is generated by smooth deformation of their body. They are made of flexible materials which results in very complex shapes they can take. Due to complex kinematics and soft materials they are composed of, soft manipulator sensing is very demanding issue. Despite a number of custom sensors that have been proposed, data gathered by them is not sufficient for effective shape reconstruction. In this paper authors propose a system that enables efficient shape reconstruction for soft pneumatic manipulator. System is based on custom physical model and vision system composed of video camera and two-dimensional optical markers.

1 Introduction

Soft continuum robotics is a young robotics research area. It focuses on design and control of flexible manipulators that have no traditional prismatic or rotational joints. Such manipulators are manufactured using innovative soft materials that are safe in contact with external objects and enables the robot to achieve shapes that are

Part of the work described in this paper is funded by the Seventh Framework Programme of the European Commission in the framework of EU project STIFF-FLOP.

J. Fraś (✉) · S. Tabaka · J. Czarnowski
Przemysłowy Instytut Automatyki i Pomiarów, PIAP, al.Jerozolimskie 202, 02-486 Warsaw, Poland
e-mail: jfras@piap.pl

S. Tabaka
e-mail: stabaka@piap.pl

J. Czarnowski
e-mail: czarnowski.jan@gmail.com

R. Szewczyk et al. (eds.), *Challenges in Automation, Robotics and Measurement Techniques*, Advances in Intelligent Systems and Computing 440, DOI 10.1007/978-3-319-29357-8_39

unreachable for conventional manipulators. These properties make soft manipulators very promising from challenging tasks point of view especially for minimal invasive surgery or soft objects manipulation [1, 2]. From the control point of view the main difference between soft and conventional robotics is their continuum kinematics. Continuum manipulator movement is generated by smooth deformation of its body. Since there are no rigid parts connected by well defined joints, the shape reconstruction is not trivial and thus, control of the manipulator shape is difficult as well. Soft manipulators can be driven by tendons or fluids, but in both cases actuation results not only in rotation or translation of some manipulator part, but much more complex deformation of the manipulator structure. In traditional case manipulator any deformation is immediately recorded by encoders housed in joints and due to that can be easily encompassed. What makes the shape detection task really complex is that every external force applied to the manipulator causes it to deform and the deformation is distributed along it, so it can not be easily compensated. Moreover there are no reliable sensors available for proper shape detection. In traditional robotics sufficient sensors are for example rotational encoders, but that is no such case. Many sensors were proposed but since soft robot can deform to very complex shapes the data they provide individually is not satisfying for effective manipulator control [3–5].

In this paper we present a specific soft manipulator design, custom physical model of it and an external single-camera positioning system that coupled with the model provides a complete robot shape reconstruction. Due to low hardware requirements (single camera) the system can be applied to minimal invasive surgery.

2 Manipulator Design

The manipulator used in the experiment is composed of two identical modules, three degrees of freedom each (Fig. 1). The module is able to bend in any direction and elongate.

Single module consists of three symmetrically deployed pneumatic actuators and an empty central channel for pressure pipes and possible actuator or sensors cables. Module body is made of EcoFlex 0050 silicone and manufactured by molding in

Fig. 1 Single module design. On the *left* module; on the *right* module cross-section, reinforced actuation chambers and central channel visible

steps. Each actuation chamber is composed of two symmetrical cylinders reinforced by single nylon thread applied in a tight helix around each cylinder (Fig. 1). Such reinforcement limits radial cylinder expansion when pressurized and allows its elongation in the same time. Pressurization of a single actuation chamber causes it to elongate, which results in module bending. Cylindrical shape was chosen for the actuator because spherical cross-section is the only shape that does not change its geometry while internal pressure applied. Cylinders are coupled in order to gain the activation capabilities by increasing the chamber cross-section area. For more information see [6].

3 Manipulator Modeling

In order to determine the shape of the manipulator, a proper model had to be designed and implemented. Since the manipulator arm is designed to operate in tight spaces, there is high probability of contact with other bodies. This fact makes calculating the influence of external forces acting on the arm a must. This significant factor is not considered by the popular Constant Curvature model of continuous robots [7, 8], what renders it unusable in this case. For the model described in this paper, following assumptions had been made: the segment is made of homogeneous material of known stiffness, with three pressure chambers hollowed out; the dimensions of cross-section of the pressure chambers are constant (provided by the braiding); the pressure in chambers is constant at any point. The other influence of the braiding has been neglected. Since the arm consists of two separately actuated segments (modules), equations for a single segment are presented.

The segment is controlled by changing the pressure in the chambers. Forces resulting from the inner pressure are parallel with module Z-axis (perpendicular to the cross section) in each cross-section. Therefore the resulting moment causes pure bending of the module. Because the chamber diameter and pressure is constant throughout the whole module length, the resulting bending moment is also constant. Elongation at any point along the module's axis can described using the Hooke's Law. The force causing module stretching is calculated by adding forces resulting from each chamber pressure.

The external forces influence has to also be represented in bending, torsion and elongation calculation. For simplification the external force is assumed to act on the module tip. External force causes additional moment to appear in all module's cross sections. The force and moment at specific cross section depend on its position. Knowing the relative orientation of tip frame in the frame of the cross section and combining it with the force and moment resulting from pressures, one can calculate the overall force and moment acting on that cross section. Those values can be then used to determine the elongation and curvature of the module at point P. Shape of the module can be determined by integrating the elongation, twist and bend of every point along the module axis. For more information see [5].

4 External Absolute Positioning System

There are commercial systems available for absolute pose measurement. Depending on the desired application systems have different properties and provide different kinds of data. One of the systems is the Aurora Electromagnetic Tracing Systemi [9] that provides sub-millimetric and sub-degree tracking for a specific marker provided with the system. The main problem with the system is that it generates the electromagnetic field in order to measure the marker pose and thus is very sensitive to presence of metallic objects in the measurement idea, which makes it useless in many cases. Other option is to measure the manipulator shape using the surface detecting systems such as Microsoft Kinect [10]. In this case the measurement device has significant dimensions and can not be applied to many scenarios as well (for example minimal invasive surgery). Due to the limitations of the current systems the proposed solution is based on ordinary vision camera and visual two dimensional markers. Size is not an issue for present cameras, and even very small camera provides high image quality. Moreover optical camera does not affect the measured area in any way (no physical contact or any field generation). Moreover cameras are already present in medical operational field, so medical application of such solution would not require any additional equipment or certification.

The system is based on two-dimensional graphical markers. These markers are built from white and black squares, which can be grouped to form larger monochrome areas in which marker's identification number (ID) is coded (Fig. 2). To calculate the marker's position and orientation and detect its ID the Chilitags [11] computer programming library is used.

The system is equipped with video camera (GigE, 1936×1458, 30 fps) and camera lens (manual focus, focal length 6 mm). Markers are attached around each segment's base and a tip on rings (Fig. 3). First segment's tip is connected to the second segment's base.

The marker detection library detects 2D markers on the image acquired from the camera and calculates their position and orientation in a camera's coordinate frame. Both camera's and markers' position were measured and are well-known. Having all these data and transformations it is possible to obtain the soft-manipulator's global position and orientation. Translation and rotation of a single marker is always connected with the soft-manipulator's displacement.

4.1 *Marker Detection Precision*

With a static lighting conditions and correct camera calibration, the quality of single marker detection depends mainly on a marker's size, distance from the camera

Fig. 2 An example of two-dimensional marker

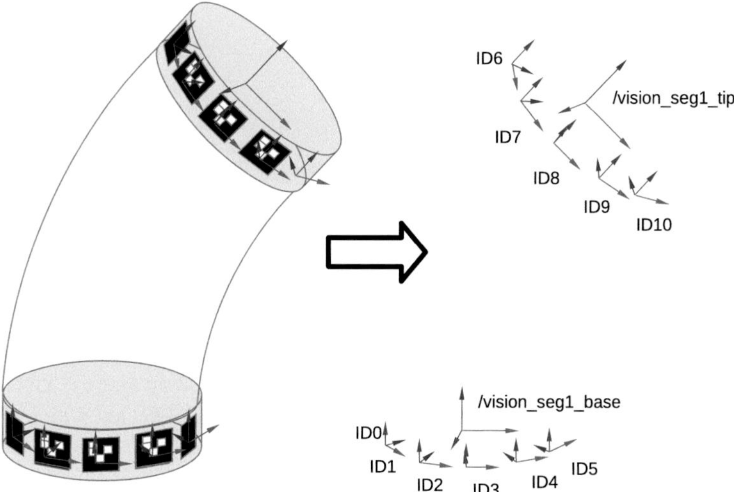

Fig. 3 Vision system principle

and its orientation. Bad marker detection results in incorrect soft-manipulator's final position calculation.

Markers on the soft-manipulator are placed on a ring, so the camera does not see all of them at the same angle. Observations of the soft-manipulator shape recognition system behavior showed that marker which was perpendicular to the camera's optical axis gave more unstable position due to the problems with a right marker's perspective detection. Therefore, small detection errors on the image plane had a great effect on the marker's rotation. Sometimes there were also problems with markers which were under the high angle to the camera optical axis.

4.1.1 Marker Detection Validation

To select the best marker's size for the soft-manipulator which would be detected without much noise caused by a distance from the camera and marker orientation research was made. Three sizes of a marker were checked: 0.5, 1 and 2 cm. Using larger markers makes no sense, because they would be too large for the manipulator. Each marker has also been checked on a three different distances from the camera: 15, 30 and 45 cm. The camera was always in a static position, only markers were moved and rotated around its vertical axis.

For each setup 50 samples were gathered and analyzed. Data analysis was based on calculating the average of detected marker's orientation (Eq. 1) and their standard deviation (Eq. 2).

$$\mu = \frac{1}{N}\sum_{i=1}^{N} x_i \tag{1}$$

$$S_n = \sqrt{\frac{1}{N}\sum_{i=1}^{N}(x_i - \mu)^2} \tag{2}$$

After changing distance of the markers to the camera, there was a need to refocus camera to get the most possible sharp marker's edges. After this operation camera stayed calibrated during the rest of the experiments.

According to the experiment, the smallest marker (0.5 cm) gives good results only for the shortest distance from the camera (15 cm). However, if an angle is lower than 20°, detection problems occur (Fig. 5). Marker sometimes is detected with a flipped orientation (Fig. 4) and its position is very noisy.

For a 30 cm distance from the camera marker was nearly unrecognizable. It was only detected when the angle was lower than 60°, but detection quality was very poor (Fig. 5). For a further distance (45 cm) marker was completely not detected.

Fig. 4 The smallest marker's orange axis (Z) is flipped over in a wrong direction due to detection problems

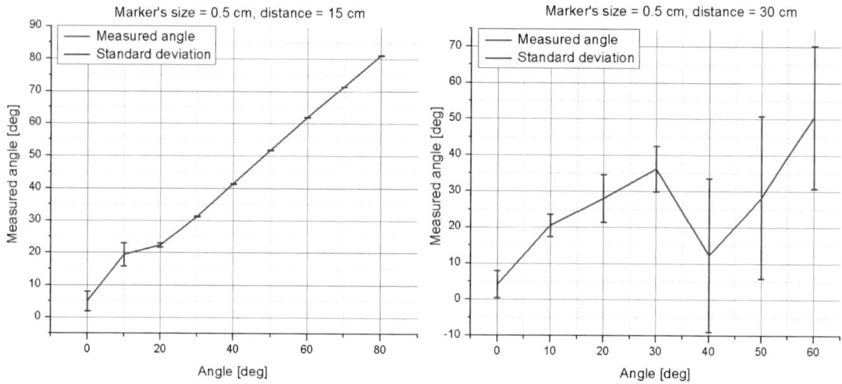

Fig. 5 Marker's size 0.5 cm detection quality

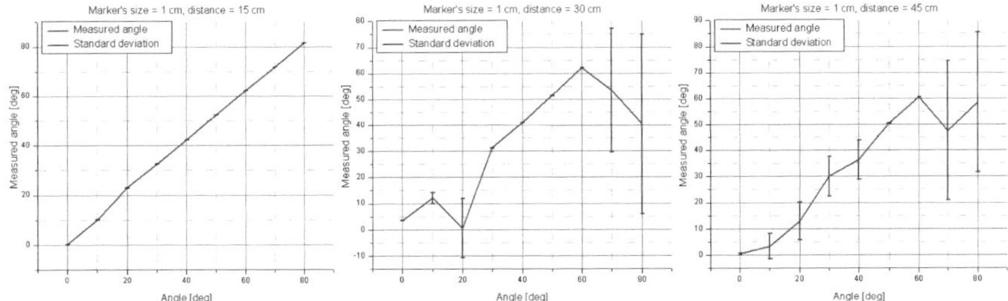

Fig. 6 Marker's size 1 cm detection quality

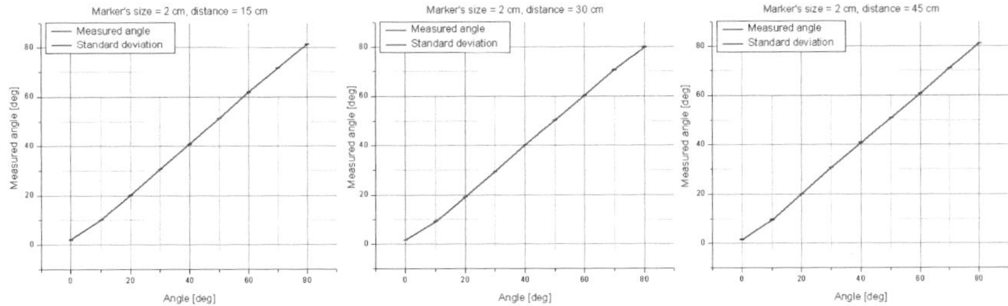

Fig. 7 Marker's size 2 cm detection quality

Larger marker's size (1 cm) gives much better results for a distance from the camera equals 15 cm (Fig. 6). Marker was perfectly detected even when the angle was high. Unfortunately, that marker's size was not enough for further distances from the camera (30 and 45 cm) (Fig. 6). There was a lot of noise in a detected marker's orientation. Best results were achieved for a marker's size 2 cm. For an every distance detection quality was very satisfying (Fig. 7).

5 Soft Robotics System Setup

As explained in the introduction, the shape reconstruction quality has great impact on the manipulator controlling possibilities. In this paper we propose an efficient system that provides complete shape reconstruction for the manipulator controlling task. System overview is presented in Fig. 8.

The system consist of a pressure supply, vision marker detection component and shape reconstruction software. Pressure supply provides actuation chambers with pressure and gives pressure measurement for shape reconstruction. The shape reconstruction block requires the pressures and forces acting on the manipulator. Pressure values are measured and provided directly by the valves powering the actuators, but force measurement is not easy task. Some custom sensors for soft robotics has been proposed, but all of them suffers form issues that makes them useless in real use

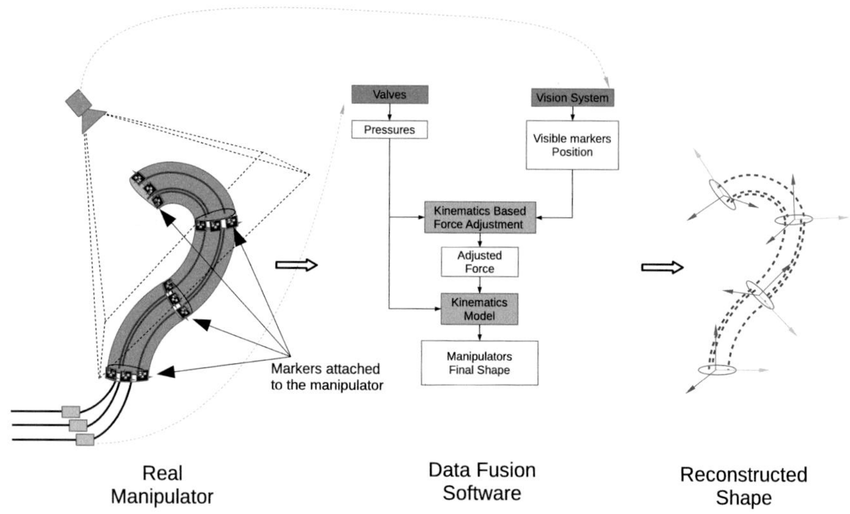

Fig. 8 Shape recognition system overview

cases. Some of them are able to measure force with good precision, but are able to measure the force only in finite number of discrete points. Therefore any force applied in between the sensors can not be measured and its effect on the manipulator is unnoticed by the model. Other solution are sensors that are capable to measure torque. There are commercial devices available, but the manipulator construction requires circular sensor shape and some room in the sensor axis for wires. There was a custom promising sensor proposed [4], but it is still in the development stage, and current performance is not sufficient for reliable shape reconstruction. Due to those issues a counterfactual solution has been proposed. The solution is based on the kinematics described above and the absolute pose measurement provided by the vision system. The measured pressure values (that are assumed to be of good quality) are supplied to the model, that produces the initial shape approximation. Error of the calculated poses of predefined points along the manipulator that markers are attached to can be calculated in relation to the measurement. Then the position error gradient is obtained in the external disturbances space. The disturbance space can be arbitrary defined. The simplest option is to assume that a three-dimensional force can act in any of the defined points. Using the gradient descent algorithm the forces that minimize the position error are obtained. The forces and the pressures are then supplied to the kinematics algorithm and the overall shape is calculated.

The solution provides force approximation that guarantee proper manipulator shape reconstruction for some range of forces. The reconstructed shape minimizes error of the position of known points and provides smooth approximation in between those points.

5.1 Experimental Setup

The proposed algorithm has been tested in three test cases. In the first case only pressure has been applied to the lower manipulator segment. In second one the force only was acting on the manipulator. The force was applied to the upper part of the tip module. At the last case the pressure and the force was acting on the manipulator simultaneously. All the cases are presented in Figs. 9, 10 and 11. In the first case the real manipulator is bended significantly more than it would result from internal pressures. Since manipulator is made of silicone the gravity is able to generate an bending as shown. The algorithm successfully approximates the gravitation force, and resulting shape is similar to the real manipulator configuration. In the second case no pressure is provided. Thus the simulation expects its body to be straight. The forces deduced from vision system bends the simulated manipulator in similar way the real ones as well. Simulation without forces is presented in green, yellow color represents the simulated manipulator shape with forces taken into account. The approximated forces are represented by red arrows. The algorithm was launched for two possible points that simulated forces could act on: tip of the first and the tip of the second module.

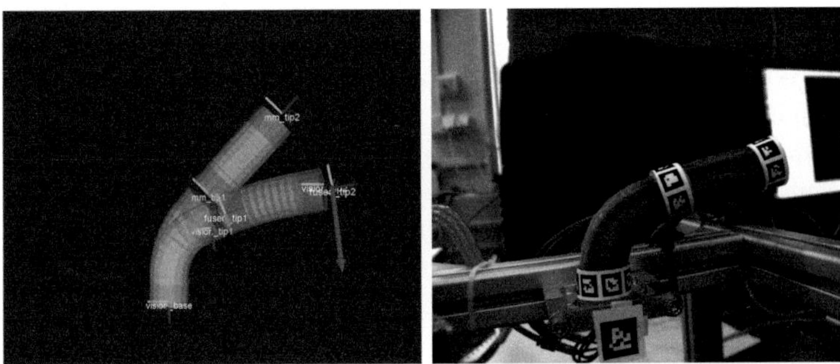

Fig. 9 Actuated lower module, no external force acting on the manipulator. Gravity influence visible

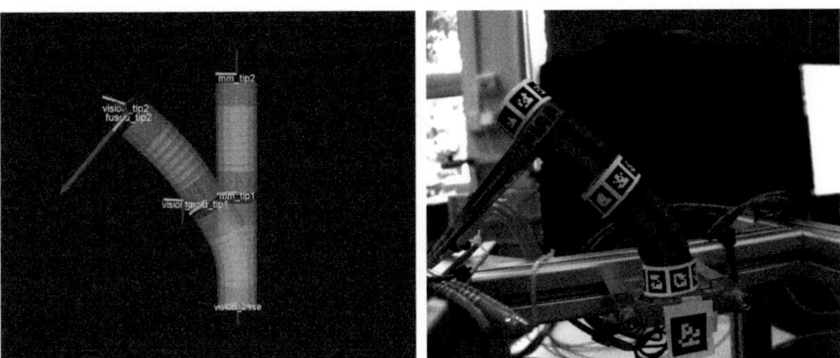

Fig. 10 No pressure supplied to the manipulator. External force acting on the upper module

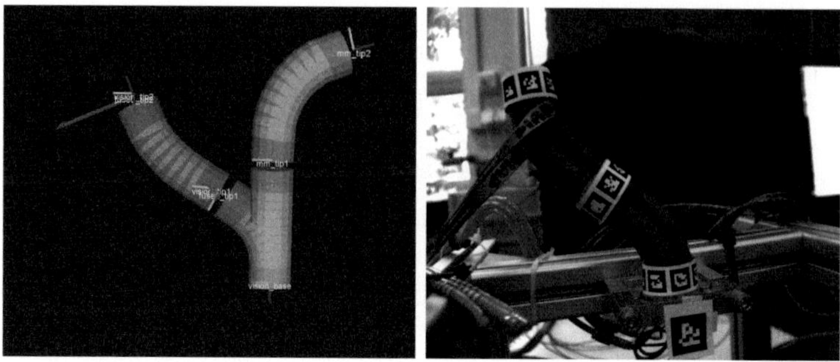

Fig. 11 Upper module actuated. Force acting on the upper module

6 Conclusion and Future Work

Since the force values are only approximated and the number of points those position are measured is finite, the accuracy of the reconstructed shape is not in 100 % correct. The force can be applied in any point on the manipulator surface, but algorithm assumes that there are only a few such points. There are other assumptions made that simplifies the real module, and those are the error source too. Moreover the marker reconstruction is not perfect so even if the model was perfect, the position of points that it would fit to are not perfectly measured too.

The proposed system potential usability has been shown, but additional work is required. In the future authors plan to perform quantitative tests with support of some commercial absolute measurement system (Aurora, Kinect, etc.) to describe the algorithm operational space. Different marker position and number configuration will be tested.

References

1. Degani, A., Choset, H., Wolf, A., Zenati, M.: Highly articulated robotic probe for minimally invasive surgery. In: International Conference on Robotics and Automation (2006)
2. Cianchetti, M., Ranzani, T., Gerboni, G., Nanayakkara, T., Althoefer, K., Dasgupta, P., Menciassi, A.: Soft robotics technologies to address shortcomings in today's minimally invasive surgery: the STIFF-FLOP approach. Soft Robot. **1**(2), 122–131 (2014)
3. Searle, T.C., Althoefer, K., Seneviratne, L., Liu, H.: An optical curvature sensor for flexible manipulators. In: International Conference on Robotics and Automation (2013)
4. Noh, Y., Sareh, S., Back, J., Wurdemann, H., Ranzani, T., Secco, E., Faragasso, A., Liu,H., Althoefer, K.: A three-axial body force sensor for flexible manipulators. In: IEEE International Conference on Robotics and Automation (2014)
5. Fraś, J., Czarnowski, J., Maciaś, M., Główka, J.: Static Modeling of Multisection Soft Continuum Manipulator for Stiff-Flop Project. Springer (2014)
6. Fras, J., Czarnowski, J., Macias, M., Glowka, J., Cianchetti, M., Menciassi, A.: New STIFF-FLOP module construction idea for improved actuation and sensing. In: IEEE International Conference on Robotics and Automation, Seatle (2015)

7. Webster, R.J., Jones, B.A.: Design and kinematic modeling of constant curvature continuum robots: a review. Int. J. Robot. Res. **29**, 1661–1683 (2010)
8. Rolf, M., Steil, J.J.: Constant curvature continuum kinematics as fast approximate model for the bionic handling assistant. In: IEEE/RSJ International Conference on Intelligent Robots and Systems (2012)
9. Aurora, http://www.ndigital.com/medical/products/aurora/
10. Microsoft Kinect, https://dev.windows.com/en-us/kinect
11. École Polytechnique Fédérale De Lausanne, http://chili.epfl.ch/softwarei

Survey of Turning Systems Used in Lower Extremity Exoskeletons

Dymitr Osiński and Danuta Jasińska-Choromańska

Abstract The paper presents a survey of kinematics of exoskeletons clasping the lower limbs, which enable realization of motion not only within the sagittal plane, but in coronal and transverse planes as well. The emphasis is put on the ways of realizing the turn. The survey is supported by kinematic schemes of the described designs. Appropriate solutions have been compared, indicating their advantages and disadvantages. The survey will be helpful while choosing the way of realizing the turn in the 'Veni-Prometheus' System for Verticalization and Aiding Motion.

Keywords Orthotic robot · Exoskeleton · Lower extremity · Kinematics survey · Turning system · Turning module

1 Introduction

1.1 Exoskeletons and Orthotic Robots

Exoskeleton is an external skeleton. Human, contrary to some animals, has no natural exoskeleton; nevertheless numerous designs of exoskeletons for humans have been elaborated. Such devices can both enhance capabilities of the human body, as well as restore a lost physical fitness. First designs that can be deemed successful and mature have been developed quite recently, at the turn of the 21st century. Three basic applications of the exoskeletons can be distinguished: military, civil and medical. Of course, these categories are not mutually exclusive; selected

D. Osiński (✉) · D. Jasińska-Choromańska
Faculty of Mechatronics, Warsaw University of Technology, Warsaw, Poland
e-mail: d.osinski@mchtr.pw.edu.pl

D. Jasińska-Choromańska
e-mail: danuta@mchtr.pw.edu.pl

R. Szewczyk et al. (eds.), *Challenges in Automation, Robotics and Measurement Techniques*, Advances in Intelligent Systems and Computing 440, DOI 10.1007/978-3-319-29357-8_40

designs of power-assisted suits may be applied, playing few various roles. The purpose of military exoskeletons is to ensure an advantage for the soldiers over the enemy, enabling them to move faster, to get tired slower and multiplying their strength. Bionic suits of a general civil purpose may offer similar capabilities, however they are rather designed to assist workers while carrying heavy loads or to provide support for them while working. External skeletons from the first two categories are designed mainly for healthy and fully able individuals. As opposed to them, medical exoskeletons can serve the sick. Medical exoskeletons are listed among orthotic robots. In a support mode, medical exoskeletons clasping the lower limbs can restore lost motor functions, enabling their users to move using both legs and to keep a straight posture. A rehabilitation mode may ensure a partial unloading of the limb or moving it according to a predefined profile, supporting the healing process in this way.

1.2 The 'Veni-Prometheus' System for Verticalization and Aiding Motion

A 'Veni-Prometheus' System for Verticalization and Aiding Motion has been designed and built at the Institute of Micromechanics and Photonics, Faculty of Mechatronics, Warsaw University of Technology (Fig. 1). This orthotic robot is designed for the handicapped suffering from paresis or paralysis of the lower limbs. The device clasps the lower limbs of the user, allowing the following motions to be performed within the sagittal plane: extension and flexion of the hip and knee joints as well as dorsal and plantar flexion of the ankle joints. Hip and knee articulations are driven, whereas articulations of the ankle joints are passive. Actuators (3) are electric DC motors coupled with reducers driving the articulations through flexible connector transmissions. The supply system—a battery of lithium accumulators, and the control system are placed within the backpack (1) of the device.

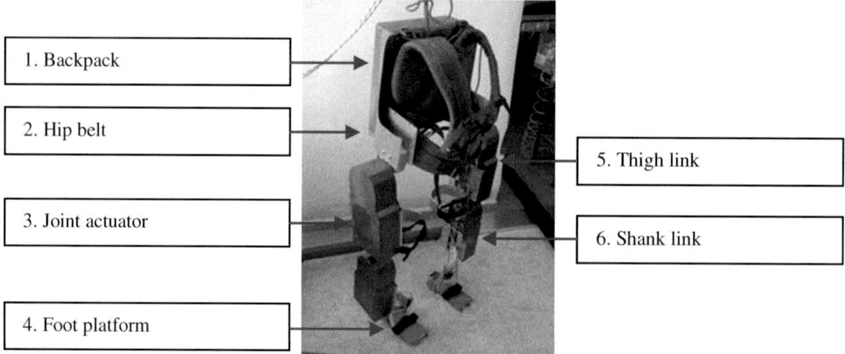

Fig. 1 The 'Veni-Prometheus' System for Verticalization and Aiding Motion

The backpack and the links of the device are connected by means of the hip belt (2). The orthotic robot is fastened to the user by means of Velcro fasteners and shoulder-straps of the backpack. At present, the system can realize five functions: walking, sitting down, raising, going up/down the stairs. Further works are carried out in order to enhance capabilities of the device, including a capability of realizing a turn [1, 2].

2 Survey of Turning Mechanisms Used in Exoskeletons

We have carried out a survey of chosen existing designs and solutions employed in active exoskeletons clasping the lower limbs, which are equipped with systems that enable turning by means of application of joints, which make it possible to move the members not only within the sagittal plane, but within the coronal and the transverse planes as well. In order to improve the readability of the text, it has been supplemented with basic kinematic schemes of the exoskeletons, created on the basis of the available information. The schemes do not reconstruct exact dimensions, however approximate proportions have been kept, thus making their comparison easier. The schemes were created using standard symbols of kinematic pairs.

2.1 *Kinematics of BLEEX Exoskeleton*

The Berkeley Lower Extremity Exoskeleton (Fig. 2) is designed for soldiers, rescuemen and firemen. The project had started in 2000 and had been financed by

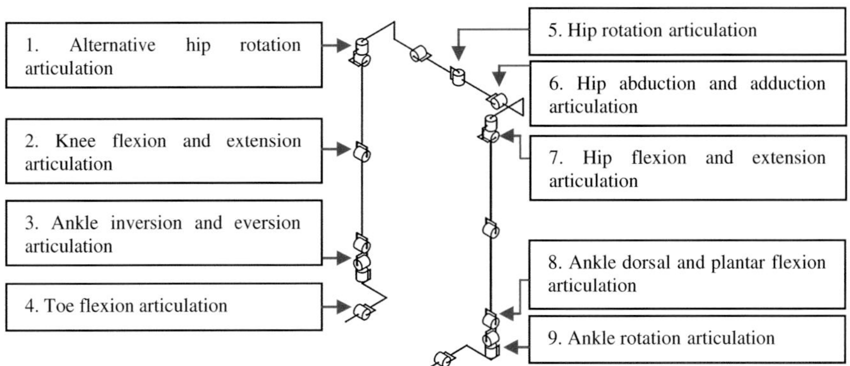

Fig. 2 Kinematic scheme of the BLEEX Exoskeleton

DARPA. The exoskeleton was premiered in 2004. It consists of two driven lower limbs, a backpack for carrying loads, which is its integral part, and a power supply unit. It is designed for a healthy user having full motor ability. It enables standing, walking (also on a tilted ground), running, squatting, tilting the trunk and turning. The drive of the device with a mass of 45 kg is a hybrid unit—combustion engine providing hydraulic drive for the actuators and electric supply for the control computer. Average power consumed while operating the exoskeleton is of 1143 W used for driving the actuators and 200 W used for supplying the computers.

The exoskeleton employs linear hydraulic actuators for driving the flexion/extension articulations of the hip joint (7), the knee joint (2) and the ankle joint (8), as well as abduction/adduction articulations of the hip joint (6). The other featured degrees of freedom, i.e. rotation of the hip joint (1), (5), inversion/eversion (3) and internal/external rotation (9) of the ankle joint as well as toe flexion (4) are not driven. However, they are equipped with springs, which try to set the joints in a neutral position. Choice of the articulations to be driven was based on a clinical analysis of gait—it was decided to make active those articulations, which consume the highest amount of energy while walking. To drive all the articulations would complicate the mechanical structure and the control algorithms. Besides, it would increase the energy consumption. Articulations of hip, knee and ankle joints were chosen to be active. The joints enable motion within the sagittal plane, and besides abduction/adduction of the hip joint, used for displacing the gravity center to the sides while taking steps.

Design of the hip belt enables rotation of the limbs in two ways: the main way uses the single articulation (5) located in the middle of the belt. An alternative method employs two separate articulations (1) located directly over the flexion/extension articulations of the hip joint (7). The abduction/adduction articulation (6) of the hip joint is located at the back part of the hip belt. The knee enables flexion and extension of the limb. The foot platform along with the ankle joint of the exoskeleton enable also flexion of the metatarsus bones, apart from providing three degrees of freedom related to capabilities of motion of the human ankle joint. Rotation axes of the members within the sagittal plane as well as abduction and adduction axes of the hip joint overlap the rotation axes of human limbs. Other rotation axes do not precisely overlap the rotation axes of human joints. While turning, hip rotation and ankle joint rotation are used the most [3, 4].

2.2 *Kinematics of Ekso Exoskeleton*

The Ekso exoskeleton (Fig. 3) was developed by Ekso Bionics Company. The first Ekso exoskeleton was completed in 2012. It is designed for individuals affected by injuries of the spinal cord, cerebral stroke, paresis of the lower limbs, hemiparesis or a traumatic brain damage. The task of the device is to teach the patient gait patterns again. Available functions are: standing, sitting and rising, walking. The exoskeleton supports also appropriate way of putting one's feet on the ground.

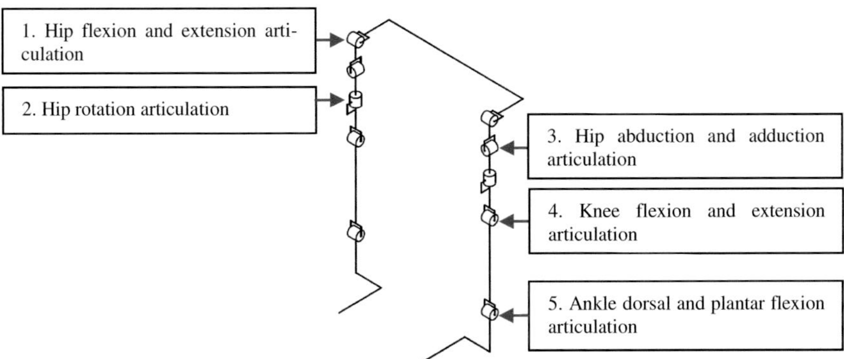

Fig. 3 Kinematic scheme of the Ekso Exoskeleton

The design is intended for users with height of 150–190 cm, mass up to 100 kg and hip width below 460 mm. The mass of the device is of ca. 23 kg. In order to use the exoskeleton it is necessary for the patient to also use crutches. The power supply of the exoskeleton are batteries placed in the backpack. Motors are located in the hip belt and on the hip links. Initiation of performing steps may be started both by means of a button at the crutches or a walker, or a wired remote control panel (operated by a therapist) or by tilting the user of the exoskeleton.

The exoskeleton is equipped with drives of the hip joints (1) and the knee joints (4) within the sagittal plane. Nevertheless, movements within other planes are possible as well. However, these degrees of freedom are passive (with no drives). If the user's use of his legs is strong enough, these degrees of freedom may be released, enabling him to perform relevant movements by himself, otherwise the degrees of freedom are blocked. The hip rotation is ensured by a possibility of turning the thigh connector (2). Additionally, the exoskeleton is equipped with an articulation located below the driven flexion/extension articulation of the hip joint, which makes it possible to spread the legs while seated (3). This movement is used while putting on or off the exoskeleton, and the articulation is immoveable while walking. In the ankle joint, the dorsal/plantar flexion axis overlaps the axis of the user's joint (5). The ankle joints make it possible to adjust the neutral position and the rigidity while performing a motion within the sagittal plane. The active articulations are driven by brushless DC motors coupled with a mechanism converting rotary motion into linear motion, based on ball screw-nut gear. A flexible connector wrapped on a cam is used to transmit motion from the nut to the articulation. Shape of the cam has been so designed that the arm of force ensures appropriate driving torque of the articulation over the whole range of its motion. The flexion/extension drive of the hip joint makes it possible to develop a maximal torque of 82 Nm and maximal angular speed of 14.9 rad/s. The flexion/extension drive of the knee joint makes it possible to develop a maximal torque of 83 Nm and maximal angular speed of 15.6 rad/s.

A patient devoid of the use of his legs is able to perform a turn, however an assistance of a therapist is necessary in this case. The therapist can hold the exoskeleton using grips located at the back and help the user by turning his body when one foot is erected. However, while performing such motion some slides of the foot sole occur, as the turn takes place about it [5, 6].

2.3 Kinematics of IHMC Mobility Assist Exoskeleton

The Mobility Assist Exoskeleton (Fig. 4) created at the Florida Institute for Human and Machine Cognition offers enhancement of strength of healthy users and a possibility of rehabilitation of the sick. In the second case, the user must lean on crutches. To drive the active articulations, series elastic actuators are used. They consist of a brushless DC motor coupled with a harmonic drive by means of a flexible connector and springs. The developed torque is of 40 Nm (yet a maximal instantaneous value can reach 80 Nm), the maximal angular speed is of 6.8 rad/s. The actuators are supplied from an external source of power, connected with the exoskeleton by means of a power cable. The sensory layer comprises sensors of member positions (optical encoders), torque sensors at the articulations and buttons on the feet responding to a contact with ground. The design provides a possibility of being adjusted to dimensions of the user so that the articulations could be located at appropriate positions. Three modes of operation are available: without assistance, performance enhancement and gait rehabilitation.

The device clasping the lower limbs is equipped with two driven degrees of freedom of the hip joint, i.e. extension/flexion (3) and abduction/adduction (2) and a driven knee joint (4). Besides, it comprises passive articulations—dorsal/plantar flexion of the ankle joint (5) as well as hip rotation (1). The rotation center of the second articulation overlaps the rotation axis of the joint owing to application of curved roller bearings. Both articulations are restored to the primary position by

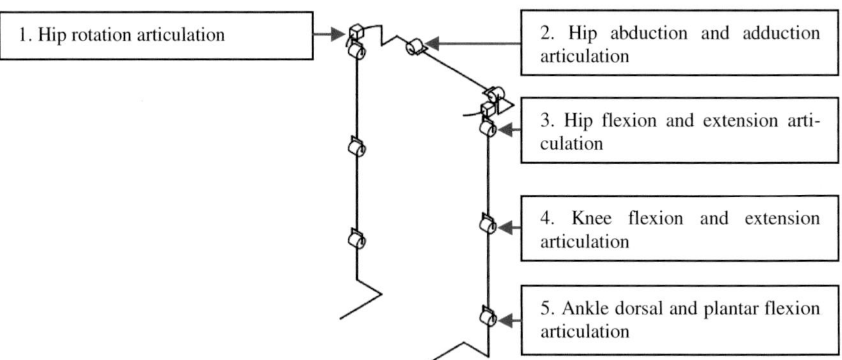

Fig. 4 Kinematic scheme of the IHMC Mobility Assist Exoskeleton

means of springs. The abduction/adduction articulation of the hip joint is located at the back part of the hip belt, at the same level as the flexion/extension articulation of that joint. The articulation enabling hip rotation is located at the side part of the hip belt, behind the flexion/extension articulation of that joint.

While turning by using hip rotation, due to application of curved roller bearings in the exoskeleton, the rotation center approximately overlaps the rotation axis inside the joint, in spite of the fact that the articulation is located outside the outline of the user's leg [7].

2.4 Kinematics of Mindwalker Exoskeleton

The Mindwalker (Fig. 5) is an outcome of cooperation between French, Italian, Belgian, Dutch and Icelandic universities. It is a medical exoskeleton designed for the handicapped. The user can take steps without any support by crutches or a walker, what leaves his hands free. The exoskeleton employs a non-invasive brain-computer interface based on neural networks, encephalography and electromyography. The exoskeleton with a mass of 28 kg is electrically powered. The control computer and its supply are located in the backpack, whereas the power supply for the actuators is drawn outside the exoskeleton, from an external source. Brushless DC motors are equipped with mechanisms converting rotary motion into linear motion. A kinematic string features a serial compliancy in a form of spiral flat springs made of titanium. The maximal torque developed is of 100 Nm.

The exoskeleton is equipped with drives of the hip joints enabling flexion and extension (3) as well as abduction and adduction (2), and knee joints enabling flexion and extension (4). The other degrees of freedom, i.e. hip rotation (1) and dorsal and plantar flexion of the ankle joint (5) are not driven, yet equipped with stiffeners restoring their neutral position. A driven articulation located at the back of the hip belt enables abduction/adduction of the hip joint. By means of a coupling

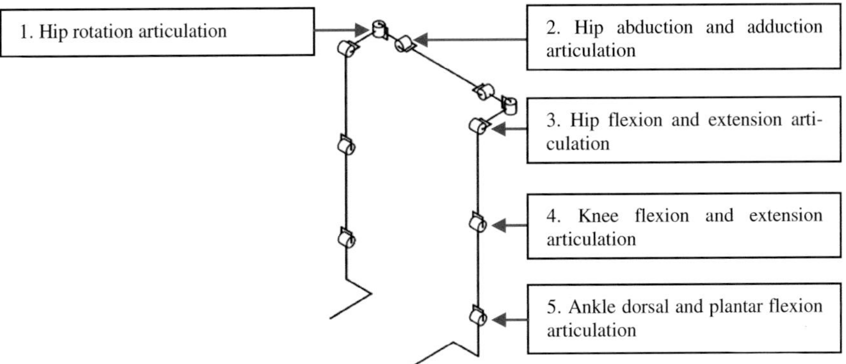

Fig. 5 Kinematic scheme of the Mindwalker Exoskeleton

link integrated with the hip belt, the articulation is connected with articulation of internal/external rotation of the hip joint, which is not driven. The articulation, aligned by means of springs, is located at the back external side of user's hip. It is connected with the flexion/extension articulation of the hip joint by means of a coupling link. Next, the articulations connected by means of a thigh connector with the flexion/extension articulation of the knee joint. Successively, it is connected by means of a shin connector with the dorsal/plantar flexion articulation of the ankle joint. These three articulations are driven. The exoskeleton enables inversion/eversion of the ankle joint, however it is devoid of a separate articulation—instead of that, compliancy of the foot platform is used for enabling a motion within a small angular range.

The flexion/extension and abduction/adduction axes of the hip articulations are located at the same level and overlap rotation axes of the user's joint. The axis of hip rotation of the exoskeleton articulation is shifted with respect to the rotation axis of the user's hip joint. Due to a small angular range of this motion and the permissible shifts between user's body members and the connecting members of the exoskeleton, such solution does not result in big loads of human joints while performing a turn. An interesting feature of the exoskeleton is the possibility of abduction and adduction in the hip articulation owing to application of an articulation of the ankle joint, which enables inversion/eversion within smaller angular range. Due to such configuration an edge mating is possible between the sole and the ground, when a leg movement to the side is performed [8–10].

2.5 *Kinematics of Rex Exoskeleton*

The Rex exoskeleton (Fig. 6) manufactured by the Rex Bionics Company is designated the first self-supported device that facilitates gait, which does not require to use crutches nor a walker, and thus leaves user's hands free. The device is offered in two versions. A rehabilitation version Rex Rehab is designed for exercises supervised by a therapist, and Rex Personal version may be operated by the user independently. Potential users are patients with injuries of the spinal cord, muscular dystrophy (muscle atrophy), multiple sclerosis, post-polio syndrome, backbone damage. Height of the user should be in the range of 146–193 cm, his mass must not exceed 100 kg and width in hips 380 mm. The mass of the personal version of the exoskeleton is of 38 kg (48 kg in the case of the rehabilitation version). The exoskeleton is manually controlled by means of an interface with a joystick. The batteries are placed in the hip belt, and ensure operation of the device for ca. 2 h. Functions of the device are the following: standing, sitting, walking on a flat as well as tilted surface, striding, turning and going up/down the stairs. Owing to the fact that the design is stable, the drives do not have to be powered while standing.

The exoskeleton is driven by electric motors coupled with mechanisms converting rotary motion into linear one, based on screw-nut gear. The driven degrees of freedom are applied at the following joints: the hip joint with two degrees of

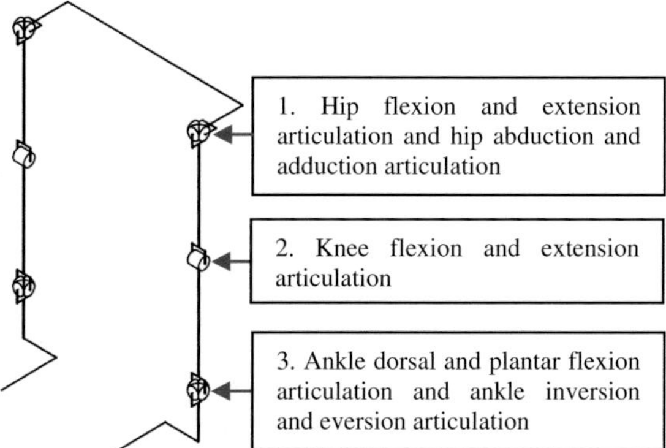

Fig. 6 Kinematic scheme of the Rex Exoskeleton

freedom—flexion/extension and abduction/adduction (1), knee joint (2) and ankle joints enabling dorsal/plantar flexion and inversion/eversion (3). Three drives are located within the thigh connector (they perform motion in the hip and knee joint), and two are placed on the shin connector (they perform motion in the ankle joint). Flexion/extension rotation axes of articulations at the hip, knee and ankle joint overlap axes of the respective joints. Abduction/adduction axes of articulations at the hip joint as well as inversion/eversion axes of the ankle joint are shifted with respect to axes of the respective joints. We presume that designers of the exoskeleton deemed such shifts acceptable because of small ranges of motion.

Despite the lack of any capability of motion within the transverse plane (i.e. an internal/external hip rotation or abduction/adduction of the ankle joint) this exoskeleton enables turning, which is realized by taking small steps, where both feet are in contact with the ground. One foot is flexed dorsally and the other is flexed plantarly. Due to this fact, the weight of the user is supported on the heel of one foot and the toes of the other foot. An opposite motion of the feet while passing over the ground results in a turn of the user. Since rotation is not possible, the result is a momentary parallel orientation of the feet while performing the whole maneuver, what implies that the feet slip on the ground. In order to realize a turn about 90° it is necessary to take two steps and it takes ca. 10 s. Due to taking small steps, the radius of turning is small as well [11, 12].

3 Discussion

Structure of most of the exoskeletons is a kinematic chain of articulations with one degree of freedom only, which are interconnected in such a way as to reconstruct the motion of human joints as accurately as possible. Rotation axes of the articulations do not always overlap rotation axes of human joints, and kinematic strings of particular designs are different. However, solutions employed in all the designs can be indicated. These are among other things a tendency to drive only a limited number of articulations or a design of the hip belt clasping the lower back part of the user's back. Driving only selected articulations surely diminishes complication of the design and its mass, and thus reduces the energy consumption and increases the reliability, at the cost of limiting the motor capabilities. One can also distinguish a trend to apply articulations of one degree of freedom that are connected in series instead of using single kinematic pairs featuring more degrees of freedom. It can be noted that the capability of abducting/adducting the hip joint was usually realized by applying an articulation in a back part of the hip belt, which rotation axis overlaps the rotation axis of the user's joint. Other articulations, except for those that operate within the saggital plane, usually feature the rotation axis shifted with respect to the rotation axis of the joint.

Techniques of turning applied in the considered exoskeletons are different. The Rex exoskeleton employs an opposite motion of the legs to perform turning, when the soles of the foot platforms slip on the ground. In order to help disabled person using Ekso exoskeleton make a turn, a support of a therapist is needed, as he makes the device turn using grips located at the back. Techniques of turning employed in the Mindwalker exoskeletons and designs of IHMC are not know precisely. However, on the basis of their kinematics and video recordings presenting operation of these exoskeletons, it is presumed that, first of all, a capability of hip rotation is used to perform a turn. It seems that the BLEEX exoskeleton enables a motion that resembles the natural one the most, however it must be kept in mind that this is a design intended to assist soldiers, created for individuals with legs fully able.

4 Conclusion

The survey of mechanical structures of the exoskeletons focused on devices featuring degrees of freedom enabling movements of their members not only within the sagittal plane, made it possible to indicate some features common for these structures, as well as the existing differences between them. It should be noted that the analysis has included only a few from among few tents of exoskeletons that exist at present, which clasp the lower limbs. The reason for that is the fact that only chosen designs are equipped in mechanisms that enable turning. The discussed findings will be used while designing a turning mechanism for the 'Veni-Prometheus' System for Verticalization and Aiding Motion.

References

1. Jasińska-Choromańska, D., et al.: Mechatronic system for verticalization and aiding the motion of the disabled. Bull. Pol. Acad. Sci. Tech. Sci. **61**(2), 419–431 (2013)
2. Osiński, D., Zaczyk, M., Jasińska-Choromańska, D.: Conception of turning module for orthotic robot. Advanced Mechatronics Solutions. Adv. Intell. Syst. Comput. **393**, 147–152 (2016)

3. Kazerooni, H., Herr, H.: The berkeley lower Extremity Exoskeleton. J. Dyn. Syst. Meas. Contr. **128**(1), 14–25 (2006)
4. Zoss, A., Kazerooni, H., Chu, A.: On the mechanical design of the berkeley lower Extremity Exoskeleton (BLEEX). In: IEEE/RSJ International Conference on Intelligent Robots and Systems, pp. 3465–3472 (2005)
5. Ekso Bionics. http://www.eksobionics.com/ekso, 12.09.2015
6. Swift, T.: Control and Trajectory Generation of a Wearable Mobility Exoskeleton for Spinal Cord Injury Patients. PHD Thesis. University of California, Berkeley (2011)
7. Kwa, H., et al.: Development of the IHMC mobility Assist Exoskeleton. In: IEEE International Conference on Robotics and Automation, IEEE ICRA, pp. 2556–2562 (2009)
8. Gancet, J., et al.: MINDWALKER: going one step further with assistive lower limbs exoskeleton for SCI condition subjects modeling, design, and optimization of mindwalker series elastic joint. In: 4th IEEE RAS & EMBS International Conference on Biomedical Robotics and Biomechatronics, pp. 1794–1800 (2012)
9. Wang, L., Meijneke, C., van der Kooij, H.: Modeling, design, and optimization of mindwalker series elastic joint. In: IEEE International Conference on Rehabilitation Robotics, pp. 1–8 (2013)
10. Wang, L., et al.: Actively controlled lateral gait assistance in a lower limb exoskeleton. In: IEEE/RSJ International Conference on Intelligent Robots and Systems, pp. 965–970 (2013)
11. Irving, R, Little, R.: Mobility aid. Europe. Patent. EP 2231096. Publ. 2013
12. Rex Bionics. http://www.rexbionics.com/. Accessed 12 Sept 2014

Exposure Control Algorithm for Maker Detection in Robotic Application

Mateusz Maciaś, Jakub Główka, Artur Kaczmarczyk, Paweł Solecki and Sebastian Tabaka

Abstract Fiducial markers are used in many applications like augmented reality or robotics. In this paper we try to improve marker detection in robotics application by implementing algorithm for controlling camera exposure. Method of measuring quality of marker is introduced. Exposure control algorithm is implemented and tested. Test results are analyzed.

Keywords Markers · Automatic exposure control · Augmented reality

1 Introduction

Fiducial markers have been used in robotics for quite some time. That technique can be used in place of more complex image recognition algorithms that allow general feature detection and tracking [7].

Most of research done in this area is concentrated on image processing algorithms that detect, decode and track markers, on marker types, id generation etc. In this paper we are concentrating on camera settings. Modern cameras have integrated automated exposure control algorithms, but those algorithm are generic, aimed for general purpose camera usage [2, 6].

M. Maciaś (✉) · J. Główka · A. Kaczmarczyk · P. Solecki · S. Tabaka
Przemysłowy Instytut Automatyki i Pomiarów PIAP, Al. Jerozolimskie 202, 02-486 Warszawa, Poland
e-mail: mmacias@piap.pl
URL: http://www.piap.pl

J. Główka
e-mail: jglowka@piap.pl

A. Kaczmarczyk
e-mail: akaczmarczyk@piap.pl

S. Tabaka
e-mail: stabaka@piap.pl

R. Szewczyk et al. (eds.), *Challenges in Automation, Robotics and Measurement Techniques*, Advances in Intelligent Systems and Computing 440, DOI 10.1007/978-3-319-29357-8_41

We are interested in shifting dynamic range of camera, so that all useful information gets into the picture. For real world situation expected light values can range, from $10^{-3}\frac{cd}{m^2}$ (starlight) to $10^{5}\frac{cd}{m^2}$ (direct sun) [2] that translates to 90 dB. Typical camera can have dynamic sensitivity from 55 to 100 dB (using multi-slope response feature) [3], but this is often reduced to 32 dB that corresponds to use of 8 bits for storing single pixel.

In terms of camera settings two parameters are interesting, gain and exposure time. Since raising gain increases noise, we will set it to minimal and use only exposure time. In theory we could change iris swing, but on most industrial cameras this can be done only manually and adding this function would increase camera cost.

Since we have specific needs such as proper marker detection, image quality doesn't always mean visual quality. One of solution used in similar cases is trying to use only fragment of frame as input for exposure control algorithm [5, 9] and in our case this fragment is detected marker.

2 Markers

There are many marker detection algorithms and implementations. In this paper most popular square markers are used. Those are easy to create and there are variety of algorithm capable of detecting them published on open licences. Comparison of some of them and description of marker detection techniques can be found in [4, 11]. Example marker is shown on Fig. 1.

For this work we used two marker detection libraries: Aruco [4] and Alvar [1]. Experiments results were gathered using Alvar. The first step of the Alvar algorithm is to find the boundaries of the potential markers using adaptive thresholding method [10], then the edges of all potential markers are marked using edge filters. Finally, system is checking whether potential markers have exactly four straight lines and four corners.

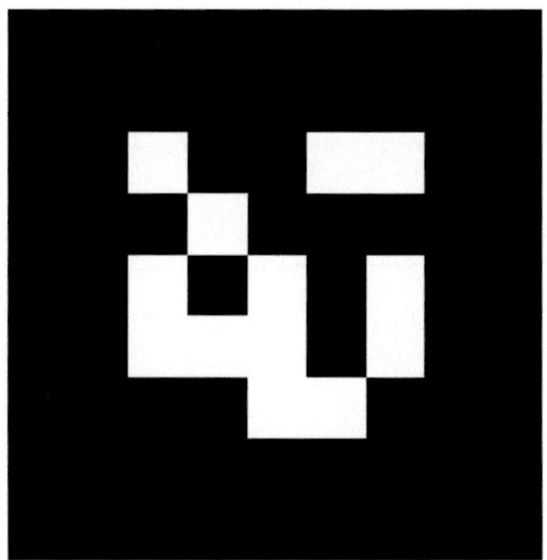

Fig. 1 Marker with id 94

3 Marker Detection Quality

For any exposure control algorithm some kind of error value is required. In simplest form this can be value calculated as difference of threshold value and mean pixel illumination for entire frame [8]. Threshold value set as a parameter of that approach is often mid tone, (i.e. 128 for 8 bits). However that method is more relevant for visual quality of the picture. For our algorithm histogram based error calculation method is proposed.

Using information from marker detection algorithm (eg. location of marker corners in original frame), and video frame it is possible to calculate histogram not for entire frame, but for marker area. Marker histogram provides information about its contrast that should be high in order to obtain proper detection.

Calculated histogram is smoothed using simple filter with window, to allow operation when marker is far and is visible as only few pixels. The histogram is a sum of two normal distributions, since marker area should have only white and black pixels. So after filtration sum of two Gaussian functions are fitted in form:

$$g(x) = K1 \cdot e^{(-\frac{(x-M1)^2}{2 \cdot S1^2})} + K2 \cdot e^{(-\frac{(x-M2)^2}{2 \cdot S2^2})} \tag{1}$$

where:

- K1 and K2 are heights of function peaks,
- M1 and M2 are position of centres of peaks,
- S1 and S2 are standard deviations.

Proposed error value could be calculated as

$$err = (M2 - M1)/HIST\_SIZE \tag{2}$$

where $HIST\_SIZE$ is size of histogram, and in this case is $HIST\_SIZE = 255$. This normalizes error to one, since images we are using are 8 bits per pixel. Obtained error value is correlated with contrast. During contrast enhancement, the borders of marker are clearer and marker detection quality is increased.

In order to verify efficacy of proposed approach experiment was performed in setup like on Fig. 2. The static camera located in in first room with the artificial lights was looking at single marker in second room without additional lightening. Marker was detected with varying exposure values, and error was calculated and shown on Fig. 6. This curve has single maximum, so can be used as control input (Figs. 3, 4 and 5).

Fig. 2 Picture with marker

Fig. 3 Histogram for marker area from Fig. 2

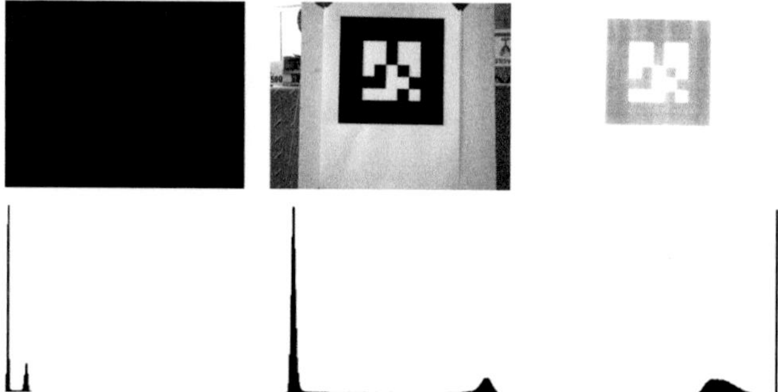

Fig. 4 Image shows 3 cases of marker (*top row*) and their histograms (*bottom row*), taken with various exposition. Image on the *left* is too dark, image on the *right* too bright

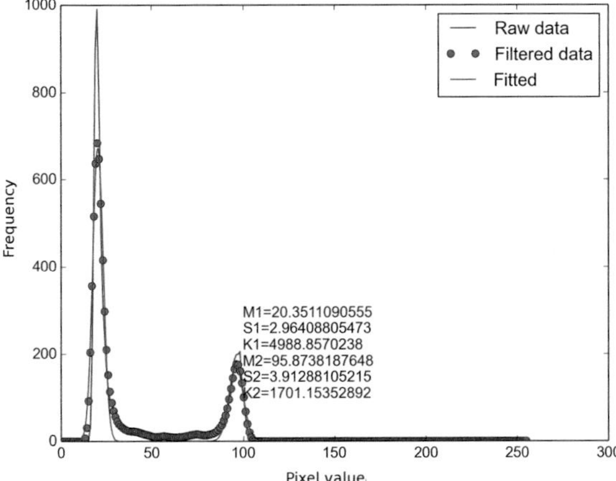

Fig. 5 Histogram with filtration and fitted curve. Numerical values are Gaussian functions coefficients

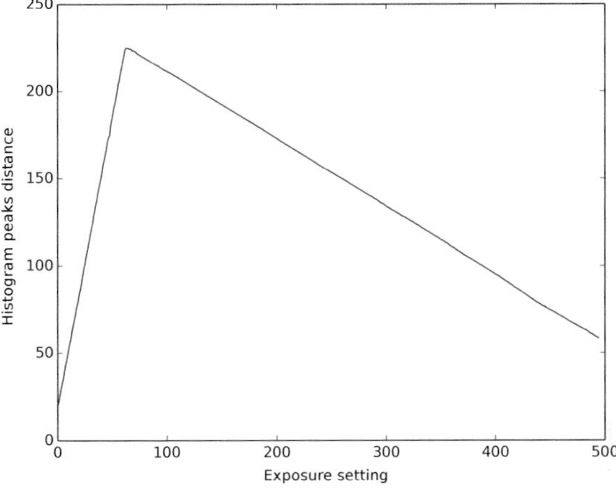

Fig. 6 Distance between histogram peaks, for various exposures

4 Algorithm

Basic algorithm is shown on Fig. 7. Raw video data from camera is streamed to marker detection component, that is one of the libraries mentioned earlier. This components detects markers, and provide information on detected markers to error calculation components. This component is responsible for determining how well markers are being seen with current exposure value. Exposure calculation has two variants. Were no markers have been detected, and when there is at least one marker. When marker is detected calculated error is used as an input to standard PID algorithm, that is controlling camera exposure. When no markers are detected, exposure is changed in fixed steps, to cover wider dynamic range and possibly find markers.

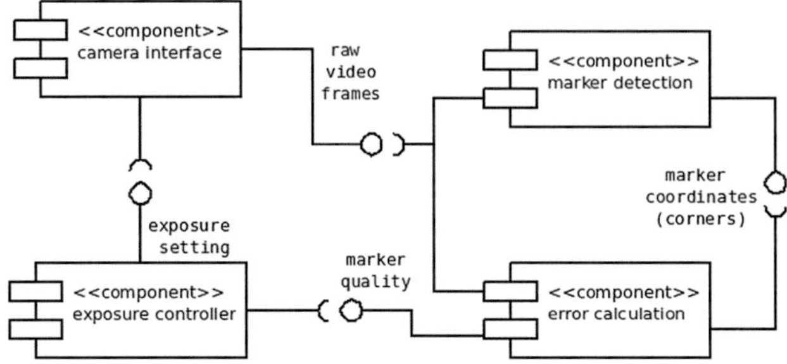

Fig. 7 Algorithm components

5 Experimental Setup

For experiments 2.8M pixels monochromatic Manta G-282B camera was used. All images were acquired uncompressed. Camera parameters:

- Sensor size 1/1.8,
- Sensor type CCD,
- Gain from 0 to 32 dB,
- Exposure from 41 μs to 60 s,
- Dynamic range 61.12 dB.

Camera is shown at left side of Fig. 8, and is looking at marker that is being moved forward and backwards into dark corridor. Room with camera has good artificial lightning conditions. Test was started when marker was located 4 m in front of camera, then it was slowly moved backwards. Runs were performed with using both camera automatic exposure control algorithm (build into camera), and algorithm proposed here. Two sample images are shown on Fig. 9 (both taken with marker at distance 5.5 m from camera). Image on left shows regular automatic exposure control, and entire image could be considered good, but marker itself is dark. During entire test with automatic algorithm exposure has not been changed much. That is

Fig. 8 Experimental setup. *Left* picture shows Manta G-282B camera, *right* shows marker used

Fig. 9 Images taken during experiments. *Left* one shows picture taken with standard exposure control algorithm, *right* one with proposed one

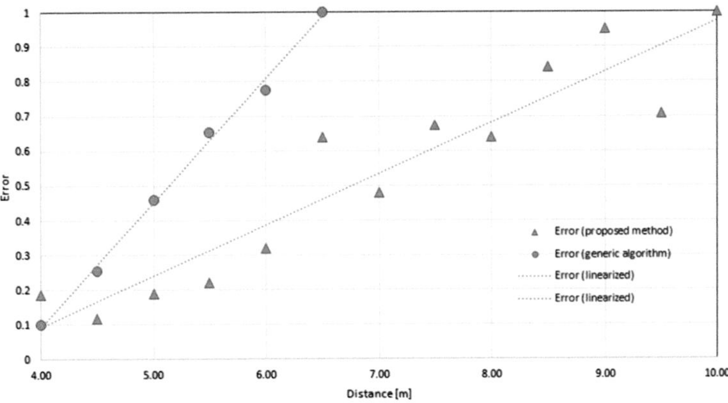

Fig. 10 Plot showing error with two tests (for generic algorithm and generic algorithm build into camera)

because marker is small in comparison to entire scene and thus has little influence on the exposure control. Image on right was taken with and proposed algorithm, scene is too bright (over exposed), but marker black and white parts still keep good contrast.

Algorithms are compared on Fig. 10. The error on the y axis is inversely proportional to contrast of marker area. Blue points are locations where marker was still detected in our method, orange points represented detected locations of generic method. Plot show error growing with distance, but with proposed algorithm we were able to track marker up to 9.5 m. With regular AGC this was possible only to 6 m. What is worth noting is that with good lightning conditions and same camera settings marker can be seen at 14 m.

6 Summary

We have presented a novel custom algorithm to control exposure of camera used in vision system for detection of planar bitonal markers. By using specific error function we gain ability to track marker in changing light conditions. In the presented experiment we have compared new method with automatic gain and exposure algorithm built-in in the camera. The results revealed the advantages of our algorithm in changing scenes and illumination which could be translated to typical situations in mobile robotic applications. However there is still room for further research.

7 Further Work

Issue of multiple markers has to be studied further. Currently if multiple markers are seen, algorithm calculates new exposure value from the one that is the nearest to camera. This is not always optimal behaviour:

- In next frame marker can leave observation area (eg. due to robot movement)
- After adapting exposure to this marker, all other markers can leave camera dynamic range
- Marker can be observed from angle, and no being optimal for position calculation algorithm

Additional work is also required in implementation. Current implementation uses generic least square fit for fitting sum of two Gaussian functions, which is not optimal in means of complexity and can sometimes give wrong results by fitting to local minima instead of global.

References

1. Alvar—a library for virtual and augmented reality. http://virtual.vtt.fi/virtual/proj2/multimedia/alvar/index.html
2. Battiato, S., Messina, G., Castorina, A.: Exposure correction for imaging devices: an overview, pp. 323–349
3. Darmont, A.: Methods to extend the dynamic range of snapshot active pixel sensors. In: Society of Photo-Optical Instrumentation Engineers (SPIE) Conference Series, Society of Photo-Optical Instrumentation Engineers (SPIE) Conference Series, vol. 6816, p. 3
4. Garrido-Jurado, S., Muñoz-Salinas, R., Madrid-Cuevas, F.J., Marí-n-Jiménez, M.J.: Automatic generation and detection of highly reliable fiducial markers under occlusion, vol. 47(6), pp. 2280–2292
5. Liang, J., Qin, Y., Hong, Z.: An auto-exposure algorithm for detecting high contrast lighting conditions. In: 7th International Conference on ASIC. ASICON '07, pp. 725–728 (2007)
6. Muehlebach, M.: Camera auto exposure control for VSLAM applications
7. Nguyen, H., Ciocarlie, M., Hsiao, K., Kemp, C.C.: ROS commander (ROSCo): behavior creation for home robots. In: IEEE International Conference on Robotics and Automation

8. Nourani-vatani, N., Roberts, J.: Automatic camera exposure control. In: Proceedings of the Australasian Conference on Robotics and Automation 2007, pp. 1–6 (2007)
9. Pan, S., An, X.: Content-based auto exposure control for on-board CMOS camera. In: 11th International IEEE Conference on Intelligent Transportation Systems. ITSC 2008, pp. 772–777
10. Pintaric, T.: An adaptive thresholding algorithm for the augmented reality toolkit. In: Proceedings of the 2nd IEEE International Augmented Reality Toolkit Workshop (ART) (2003)
11. Siltanen, S.: Theory and applications of marker-based augmented reality. VTT Sci. **3** (2012)

Evaluating Map-Based RGB-D SLAM on an Autonomous Walking Robot

Dominik Belter, Michał Nowicki and Piotr Skrzypczyński

Abstract This paper demonstrates an application of a Simultaneous Localization and Mapping algorithm to localize a six-legged robot using data from a compact RGB-D sensor. The algorithm employs a new concept of combining fast Visual Odometry to track the sensor motion, and a map of 3-D point features and robot poses, which is then optimized. The focus of the paper is on evaluating the presented approach on a real walking robot under supervision of a motion registration system that provides ground truth trajectories. We evaluate the accuracy of the estimated robot trajectories applying the well-established methodologies of Relative Pose Error and Absolute Trajectory Error, and investigate the causes of accuracy degradation when the RGB-D camera is carried by a walking robot. Moreover, we demonstrate that the accuracy of robot poses is sufficient for dense environment mapping in 3-D.

Keywords RGB-D SLAM · Factor graph · OctoMap · Walking robot

1 Introduction

Autonomous walking robots need accurate self-localization with respect to six degrees of freedom in order to plan and execute their complicated motion patterns in previously unknown environments. Such robots start to be considered for Urban Search And Rescue (USAR) missions, due to their high mobility, which includes the ability to overcome rough terrain and man-made obstacles. In an USAR mission the robot has to plan its motion, create a map of unknown environment, and self-localize

D. Belter (✉) · M. Nowicki · P. Skrzypczyński
Institute of Control and Information Engineering,
Poznań University of Technology, ul. Piotrowo 3A, 60-965 Poznań, Poland
e-mail: dominik.belter@put.poznan.pl

M. Nowicki
e-mail: michal.nowicki@put.poznan.pl

P. Skrzypczyński
e-mail: piotr.skrzypczynski@put.poznan.pl

R. Szewczyk et al. (eds.), *Challenges in Automation, Robotics and Measurement Techniques*, Advances in Intelligent Systems and Computing 440, DOI 10.1007/978-3-319-29357-8_42

with respect to this map. These functions need to be tightly integrated in order to achieve full autonomy and to ensure reliable motion execution [6]. Our earlier experiments demonstrated that it is necessary to achieve self-localization accuracy of about the size of the robot's foot to ensure robust registration of the range data into an elevation map that enables foothold planning [1]. Moreover, an autonomous walking robot requires a dense representation of the terrain for planning. While some state-of-the-art 3-D self-localization systems build voxel-based maps, they usually rely on massive parallel processing to handle the huge amount of range data [22]. Such approach is not suitable for most legged robots, due to the power consumption of a high-end GPU. Therefore, an autonomous legged robot needs a Simultaneous Localization and Mapping (SLAM) algorithm that is accurate, runs in real-time without hardware acceleration, and is coupled with a dense environment mapping.

In this paper we investigate a RGB-D SLAM system based on factor graph optimization implemented on an autonomous six-legged robot. This system uses a compact and affordable structured light RGB-D sensor. This class of sensors has been positively evaluated in our previous works for self-localization of USAR robots [5], and for legged robots in particular [13]. We consider a new architecture of the RGB-D SLAM system, which differs from our recent developments [3, 5] and other similar solutions [8, 9] by employing a map of 3-D point features that are included in the factor graph optimization process. The flexible structure of the factor graph enables real-time optimization of the map in a background process delegated to a separate thread [4]. While there are other RGB-D SLAM systems that use a map of point features [12, 15], they directly match the current perception to features in a large map, whereas we use a fast Visual Odometry (VO) algorithm to obtain a first guess of the camera pose, which improves matching robustness.

The evaluation focuses on the trajectory accuracy achieved in experiments on a real walking robot. We demonstrate that the proposed approach to RGB-D SLAM yields robot pose estimates that are accurate enough to build a dense, 3-D map of the environment employing a voxel-based mapping algorithm. Moreover, we identify the practical problems related to using a RGB-D camera mounted on a legged robot, and show that tighter coupling between the SLAM system and the control strategy of the robot is beneficial to the accuracy of the estimated trajectories.

2 Map-Based RGB-D SLAM

The Poznan University of Technology (PUT) RGB-D SLAM system is based on the factor graph optimization approach and a persistent, scalable map of 3-D point features. This system follows the non-linear optimization approach to state estimation, which is currently considered to be the state-of-the-art method in SLAM. The software[1] architecture of PUT RGB-D SLAM is presented in Fig. 1a. The front-end

---

[1]Source code is available at https://github.com/LRMPUT/PUTSLAM/tree/release.

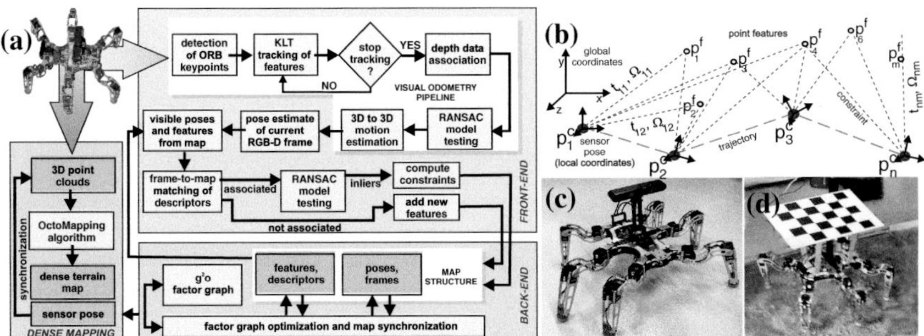

Fig. 1 Block scheme of the PUT RGB-D SLAM with dense mapping system (**a**), structure of the feature-based map (**b**), and the Messor II robot with the Asus Xtion (**c**) and the chessboard marker (**d**)

and the back-end work asynchronously, and get synchronized only on specific events. Owing to this architecture the system efficiently uses a multi-core CPU without additional hardware acceleration.

2.1 Back-End

In PUT RGB-D SLAM the back-end holds a map (Fig. 1b), which contains 3-D point features augmented by their local visual descriptors [16]. The map contains also the sensor poses related to the keyframes, which were used to extract the map features. The back-end is based on the g^{2}o general graph optimization library [11]. The factor graph for optimization in the back-end is constructed from the map data. The observed features and the sensor poses are related by measurement constraints in $\mathbb{R}^3$. The initial locations of the sensor poses $\mathbf{p}^c_i$ ($i = 1 \ldots n$) are computed using the VO pipeline. The sensor poses are represented by Cartesian positions (x, y, z) and quaternions (q_w, q_x, q_y, q_z) parameterizing the sensor orientation. The factor graph has two types of vertices: $\mathbf{p}^f$ representing the point features, and $\mathbf{p}^c$ representing the sensor poses. The initial positions of features $\mathbf{p}^f_j$ ($j = 1 \ldots m$) in the sensor coordinate system are determined from the RGB-D measurements. Sensor poses are expressed in the global frame of the map, while the features are represented relatively to the poses from which they have been detected. The edge $\mathbf{t}_{ij} \in \mathbb{R}^3$ represents measurement constraints between the ith pose and the jth feature, and the edge $\mathbf{T}_{ik} \in$ SE(3) is a rigid transformation constraint resulting from the estimated motion between poses i and k. The pose-to-pose constraints are introduced to the factor graph whenever the number of re-observed features in the map is below a threshold. The accuracy of each constraint is represented by its information matrix $\mathbf{\Omega}$, which can be obtained by inverting the covariance matrix of the measurement, or can be set to identity, if we assume anisotropic spatial uncertainty in the measurements. The sequence of the sensor (robot) poses $\mathbf{p}_1^c, \ldots, \mathbf{p}_n^c$ and feature positions $\mathbf{p}_1^f, \ldots, \mathbf{p}_m^f$ that satisfies the set of constraints is computed as:

$$\underset{\mathbf{p}}{\operatorname{argmin}} F = \sum_{i=1}^{n} \sum_{j=1}^{m} \mathbf{e}_{f(i,j)}^T \boldsymbol{\Omega}_{ij} \mathbf{e}_{f(i,j)} + \sum_{i,k=1,i \neq k}^{n} \mathbf{e}_{c(i,k)}^T \boldsymbol{\Omega}_{ik} \mathbf{e}_{c(i,k)} \tag{1}$$

where $\mathbf{e}_{f(i,j)} = e(\mathbf{p}_i^c, \mathbf{p}_j^f, \mathbf{t}_{ij})$ and $\mathbf{e}_{c(i,k)} = e(\mathbf{p}_i^c, \mathbf{p}_k^c, \mathbf{T}_{ik})$ are error functions for the feature-to-pose and pose-to-pose constraints, respectively. These functions are computed for the estimated pose of the vertex and measured pose of the vertex which comes out from the measurement $\mathbf{t} \in \mathbb{R}^3$ for feature-to-pose or $\mathbf{T} \in SE(3)$ for pose-to-pose constraints. We use the implementation of Preconditioned Conjugate Gradient method from the g^{2}o library [11] to solve (1). We buffer new sensor poses, features, and measurements in a smaller temporary graph during the graph optimization process. Thus, we can synchronize the optimized factor graph with the map of features when the back-end finishes the on-going graph optimization session.

2.2 Front-End

The core part of the front-end is a fast VO pipeline that yields a sensor displacement guess. The VO estimates the SE(3) transformation between two poses employing sparse optical flow tracking with the Lucas-Kanade algorithm [19]. We use ORB features [14] to initialize the keypoints for tracking, as due to the computing efficiency considerations local descriptors of the features extracted in the VO pipeline are then re-used for matching between the features from a new keyframe and the map. Correct matching features in two keyframes are estimated from the initial set of keypoints associated by optical flow tracking by using the preemptive RANSAC framework. The RANSAC samples three pairs of keypoints from the initial set, and using the Umeyama algorithm [21] estimates the candidate transformation, which is then verified using the remaining pairs of keypoints. The implementation of the tracking-based VO algorithm is described in more detail in our earlier work [3].

Once the current sensor pose is computed by VO, the front-end attempts to match features detected in the current keyframe to features in the map. To robustly determine feature-to-map correspondences the guided matching approach is applied—the feature matches are considered only in a small neighborhood of the predicted features in the image plane, and then verified by typical matching based on the ORB descriptors. With a set of possible matches, the inliers are determined again by RANSAC. Thanks to the constraints established between the current keyframe and the features already included in the map (from previous, possibly distant keyframes) our SLAM algorithm handles local metric loops implicitly, without the need to identify already visited places by their appearance. However, closing large loops by this method is problematic, as the amount of drift in the predicted sensor pose may be too large for the guided matching approach.

2.3 Dense Environment Mapping

The PUT RGB-D SLAM system creates a sparse map of 3-D features, which is an efficient environment representation for self-localization. However, for motion planning the autonomous walking robot needs a dense 3-D, or at least 2.5-D environment representation [6]. Therefore, we coupled our SLAM system with a 3-D occupancy grid mapping algorithm. Having in the SLAM map the optimized robot/sensor poses from which the measurements have been made, we project point clouds associated to the keyframes into a global coordinate system. Although the point cloud or surfel-based representation of the RGB-D data can provide a good visualization of the scene, they are highly redundant and inefficient whenever the map has to be queried by the motion planning algorithm for occupancy information at particular coordinates. Thus, we convert the point clouds to the efficient octree-based OctoMap occupancy map representation proposed in [10]. In OctoMap the voxels are stored in a tree structure that results in a compact memory representation and allows the planning algorithms to query the occupancy map at various resolutions. Moreover, OctoMap is a probabilistic framework, which efficiently handles noisy measurements and uncertain sensor poses. We use the algorithm and open source software from [10] added to our system to create OctoMaps from registered point clouds. In the experiments presented in this paper we have used 10 mm voxel grid resolution.

2.4 Implementation on the Walking Robot

The Messor II robot [2] uses the Asus Xtion Pro Live RGB-D sensor based on the PrimeSense structured light camera—the same that is employed in the more popular Kinect sensor, which is however bigger, heavier, and requires an additional power source. The compact and lightweight Asus Xtion, which is powered by USB only is much more suitable for a walking robot.

On the walking robot the Asus Xtion is mounted on a mast fabricated using the 3-D printing technology (Fig. 1c). The sensor is located at the elevation of 40 cm above the ground, assuming the neutral posture of the robot. Since the RGB-D sensor should also measure the ground area in front of the robot for terrain mapping [6], it is slightly tilted down. This configuration allows the sensor to perceive the shape of the terrain at some distance from the robot, and to observe objects located farther away. Such objects are important for SLAM, as they provide more salient features than the terrain surface. Although the robot has an Inertial Measurements Unit (IMU) that can estimate the attitude of the robot's trunk with respect to the gravity vector, in the present implementation of our SLAM system these measurements are not used.

Currently the Messor II robot has only an embedded computer (PandaBoard) for its control software, thus the SLAM system runs on a laptop PC. Although for the experiments presented in this paper the robot was tethered to the laptop and an external power source, it is possible to use a WiFi connection and on-board batteries for full autonomy.

3 Evaluation Procedure

There are publicly available data sets, such as the TUM RGB-D benchmark [20], which facilitate benchmarking of RGB-D SLAM algorithms and enable comparison to state-of-the-art approaches. The PUT RGB-D SLAM was evaluated on the TUM RGB-D benchmark, demonstrating on some sequences the trajectory reconstruction accuracy better than the best results published so far in the literature [4]. However, the TUM RGB-D benchmark and other similar data sets, obtained either using a handheld Kinect sensor or a sensor mounted on a wheeled robot, do not allow us to evaluate robustness of our SLAM system to problems specific to walking robots: legged locomotion is discrete, which often causes uncontrolled oscillations of the attitude and vibratory motion of the platform, which makes it hard to keep the line of sight of any on-board sensor horizontal. Moreover, the RGB-D sensor on a walking robot is usually mounted at quite low height (a tall mast cannot be used due to the robot stability issues), which limits the effective field of view.

Therefore, we decided to test the PUT RGB-D SLAM on a real six-legged robot.[2] Some results from tests on the same walking robot were already published for the earlier, pose-based variant of our RGB-D SLAM [5], however, these results were only qualitative due to the lack of ground truth trajectories. We demonstrate the trajectory reconstruction accuracy achieved on the Messor II robot with quantitative results, applying the Absolute Trajectory Error (ATE) and Relative Pose Error (RPE) metrics proposed in [20]. The ATE error compares the distance between the estimated and ground truth trajectories, and is computed from the Root Mean Square Error (RMSE) for all nodes of the ground truth and the recovered trajectory. The RPE error corresponds to the local drift of the trajectory. To compute RPE the relative transformation between the neighboring point of the ground truth trajectory $\mathbf{T}_i^{\mathrm{GT}}$ and the recovered trajectory $\mathbf{T}_i$ is computed, and the relative error for each point of the trajectory is given by:

$$\mathbf{E}_i = \left(\mathbf{T}_i^{\mathrm{GT}}\right)^{-1} \mathbf{T}_i. \tag{2}$$

Taking the translational or rotational part of $\mathbf{E}_i$ we obtain the translational or rotational RPE, respectively.

The ground truth trajectories were obtained using an overhead system of cameras [17]. Although this multi-camera vision system consists of five high-resolution Basler acA1600 cameras equipped with low-distortion, aspherical 3.5 mm lenses, for the presented experiments we have used only the central camera, which was possible because of the small area of the terrain mockup traversed by the walking robot (Fig. 2a). Employing only one camera simplified the registration procedure, as mutual calibration of the overhead cameras was not necessary [18]. The central camera was calibrated using a chessboard pattern according to the rational lens model [7]. We synchronized the acquisition of RGB-D frames by the Asus Xtion on the robot with the frame rate of the overhead camera (they both work at 15 Hz, due to

[2]Data set is publicly available at http://lrm.put.poznan.pl/putslam/.

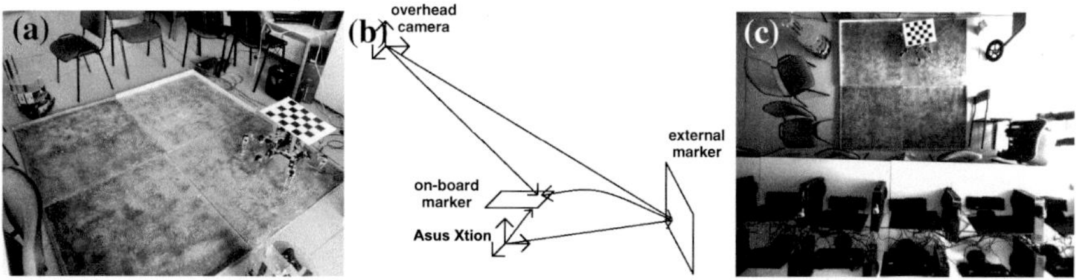

Fig. 2 Evaluation experiment: experimental set up with a flat terrain mockup and the Messor II robot (**a**), system calibration concept (**b**), and an exemplary view from the overhead camera of the PUT Ground Truth system (**c**)

limitations of the camera) to avoid errors due to trajectory interpolation, which arise in other RGB-D data sets [20].

The Messor II robot used in the experiments was equipped with the Asus Xtion sensor and a rigidly attached chessboard marker facing up (Fig. 1d). The rigid transformation between the Asus Xtion coordinate system, and the coordinates of the chessboard marker was estimated applying the calibration algorithm from [18], which uses images of an external calibration marker observed by the on-board sensor and images registered by an external camera observing both the external marker, and the marker attached to the robot (Fig. 2b). During the experiments overhead camera images (Fig. 2c) were registered synchronously with the Asus Xtion frames. The chessboard pattern of known size was extracted from these images and its pose with respect to the overhead camera was determined. Then, this pose was transformed to the pose of the sensor using the transformations obtained by calibration. All artificial markers were used only by the PUT Ground Truth system, and the operation of PUT RGB-D SLAM didn't depend on them.

4 Results

All experiments with the Messor II robot have been conducted on a flat terrain mockup of the size 2×2 m. Some common objects (furniture, lab equipment) have been located outside the mockup to make the environment richer in salient features, which are needed by our SLAM to operate reliably. All test trajectories resembled rectangles of the lap size similar to the width of the mockup, but executed at different translational speeds of the robot, making also more or less sharp turns at the corners of the trajectory. The layout of objects around the mockup area was different for each experiment.

In the first experiment the Messor II robot used the default tripod gait. This is the fastest statically stable gait for a six-legged machine, thus the robot moved at 95 % of its maximal speed. As the robot was teleoperated using a joystick, the step length was variable and the effective velocity of the robot's trunk with the sensor

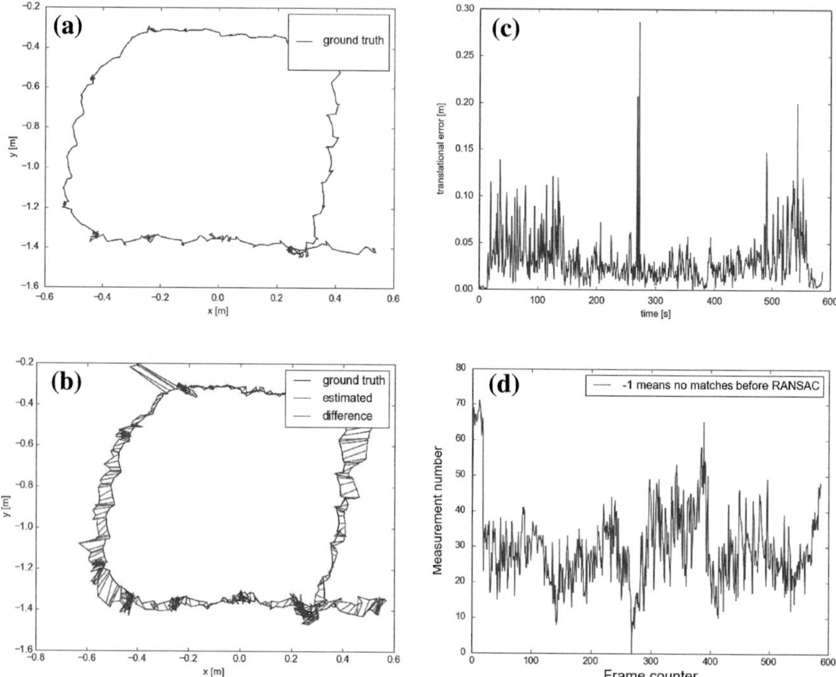

Fig. 3 Results for the robot walking at 95 % of its maximal speed (`messor2_1`): ground truth trajectory (**a**), estimated trajectory with ATE (**b**), transitional RPE (**c**), and the number of matched map features (**d**)

varied as well, but the average velocity was about 0.15 m/s. The resulting ground truth trajectory is shown in Fig. 3a. The estimated trajectory and the ATE values are visualized in Fig. 3b. The ATE RMSE in this experiment was about 9.5 cm, which is a value bigger than expected for reliable dense mapping and motion planning. Plot of the translational RPE (Fig. 3c) reveals that at some frames (e.g. frame 280, pointed by the arrow) the relative error peaked to more than 0.25 m. The moments of increased relative error coincide with sudden drops of the number of measurements between the new frame and the map (i.e. the number of established feature-to-pose constraints, Fig. 3d), which usually means that the VO-based guess of the camera pose was very poor. This leads to sensor poses that are weakly connected to the map features, and to an increased error in the optimized trajectory, as seen at the ATE plot (Fig. 3a, pointed by the arrow).

Careful inspection of the recorded RGB frames revealed that the moments of much degraded performance directly coincide with the frames containing an excessive amount of motion blur. An example of such corrupted RGB-D data from this experiment is shown in Fig. 4. While frames (a) and (e) are sharp enough, the frame (c) between them is blurred to the extent that makes it impossible to track the keypoints with the Lucas-Kanade algorithm. The motion blur is apparently caused by

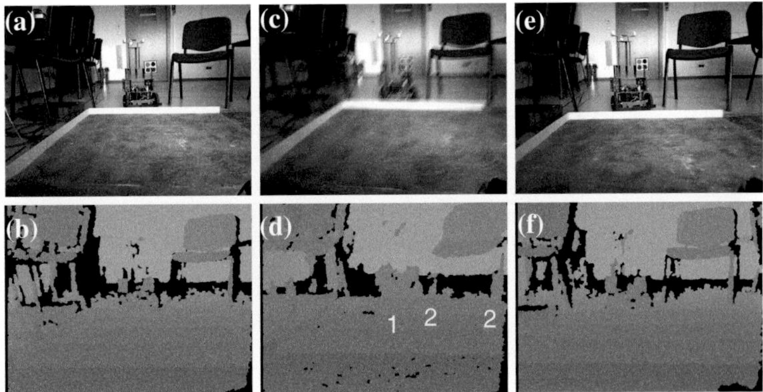

Fig. 4 RGB (**a**, **c**, **e**) and depth (**b**, **d**, **f**) frames from the Asus Xtion mounted on the Messor II robot acquired during the experiment at 95 % of maximal speed

quite large amplitude vibrations of the robot's trunk observed when it is supported by three legs in the swing phase of the fast tripod gait. What is interesting, the motion blur degrades also the depth images, as seen in Fig. 4d—some details seen on the neighboring depth frames, like the wheeled robot (denoted '1') or legs of the chair (denoted '2') are completely blurred or missing entirely. This effect makes extraction of the 3-D point features from the blurred frames extremely unreliable.

If the speed of the robot is reduced to 45 % of the maximal value (average velocity of about 0.09 m/s) the trajectory is recovered with slightly better accuracy, as shown in Fig. 5. The robot still takes some frames when it is supported by three legs only, but due to the reduced speed, the trunk vibrations have smaller amplitude. As the result, the number of measurements between the current keyframe and the map never drops to zero (Fig. 5d).

Better accuracy and a much smoother recovered trajectory can be obtained if we couple the acquisition of RGB-D frames for SLAM with the gait of the robot. This can be fully implemented only if we let the motion execution algorithm to control the RGB-D sensor, which was impossible in the presented experiments. However, when we have changed the default tripod gait to a slower crawl gait, which moves only one leg of the robot at once, we were able to avoid excessive motion blur. In this case the average velocity of the robot was only 0.05 m/s, but we were able to achieve the ATE RMSE of about 5 cm, which seems to be enough for terrain mapping (almost the size of the robot's foot). Results of this experiment are depicted in Fig. 6. Table 1 summarizes the quantitative results for all the three experiments presented in this paper.

In order to demonstrate that the self-localization accuracy achieved by PUT RGB-D SLAM on the real walking robot is sufficient for dense environment mapping we produced an voxel-based occupancy grid from the points measured in the third experiment (`messor2_3` trajectory). In Fig. 7a and b we show screenshots from a dynamic visualization of the point cloud registered with the recovered trajectory.

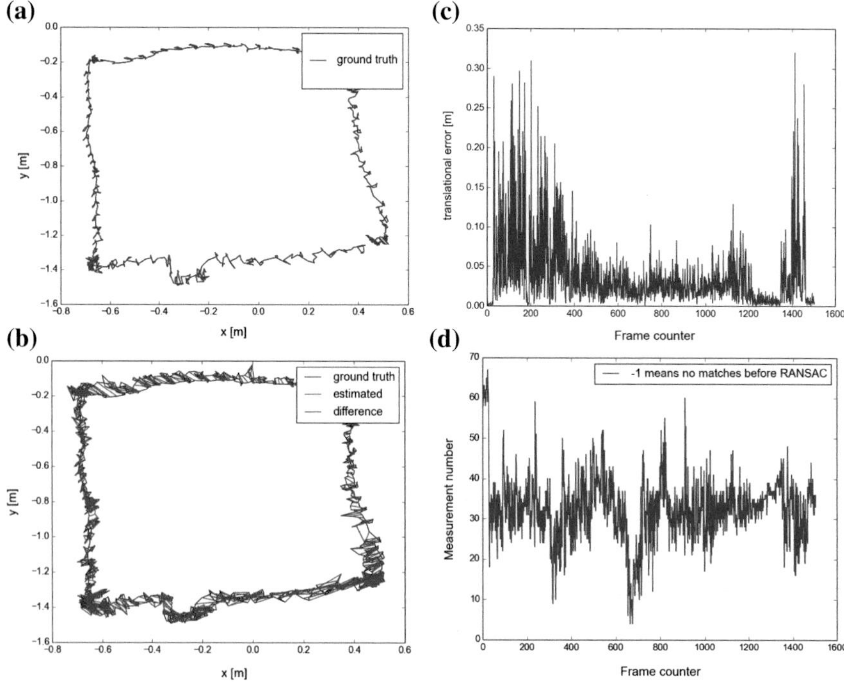

Fig. 5 Results for the robot walking at 45 % of its maximal speed (`messor2_2`): ground truth trajectory (**a**), estimated trajectory with ATE (**b**), transitional RPE (**c**), and the number of matched map features (**d**)

This visualization can run on-line along with PUT RGB-D SLAM, and is presented in a movie clip available at http://lrm.cie.put.poznan.pl/aut2016.mp4. The registered point clouds are transformed into the more compact OctoMap representation presented in Fig. 7c and d using the OctoMap viewer [10].

5 Conclusions

A new approach to the RGB-D SLAM problem, which employs the factor graph optimization and builds a persistent map of 3-D features has been presented and evaluated for the use on a real autonomous walking robot. The SLAM algorithm uses visual odometry to compute the sensor motion guess, instead of direct tracking of the sensor against the map. The VO pipeline implemented using fast optical flow tracker results in a very accurate RGB-D SLAM running at the RGB-D camera frame rate without GPU acceleration, which makes it suitable for on-board implementation in various robots that require accurate self-localization in 3-D, such as autonomous walking robots in USAR missions.

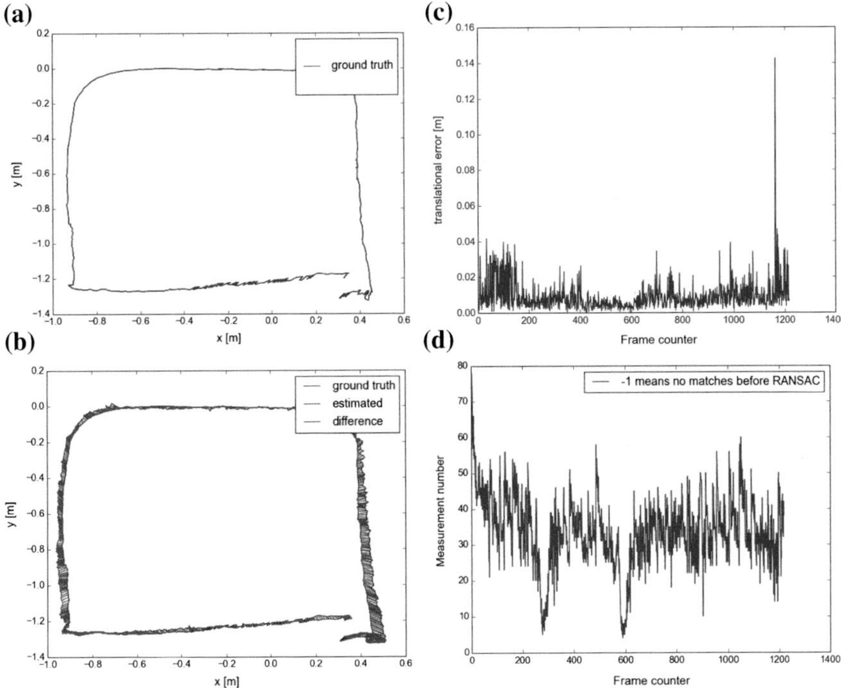

Fig. 6 Results for the robot acquiring RGB-D frames when it was in the support phase of crawl gait (`messor2_3`): ground truth trajectory (**a**), estimated trajectory with ATE (**b**), transitional RPE (**c**), and the number of matched map features (**d**)

Table 1 ATE and RPE for the PUT RGB-D SLAM measured on three sequences

Messor II experiment RGB-D data sequence	ATE RMSE (m)	ATE std dev. (m)	Translat. RPE RMSE (m)	Rotat. RPE RMSE (deg)
`messor2_1`	0.095	0.052	0.054	4.96
`messor2_2`	0.069	0.033	0.041	5.66
`messor2_3`	0.053	0.025	0.012	5.42

Our experiments revealed that the most pronounced problems specific to the multi-legged platform are the unstable motion and excessive vibrations, which corrupt the RGB-D data acquired in-motion. One solution to these problems, that was to some extent implemented in our experiments is to couple the frame acquisition process with the gait control of the robot. Another one, which we plan to research in the near future is to compensate the motion of the platform by using data from the on-board IMU. Obviously, both approaches may be then combined for even better results.

Fig. 7 Visualization of RGB-colored point clouds registered with the estimated robot trajectory `messor2_3` (**a**, **b**), and the RGB-colored voxel-based map created from these data with the OctoMapping algorithm (**c**, **d**)

We have also demonstrated experimentally, that the accuracy of PUT RGB-D SLAM is good enough for dense mapping of the environment from a real walking robot. In further research we will demonstrate that the obtained occupancy map can be used for efficient motion planning with the algorithms that have been already tested on simpler elevation grid maps [6].

Acknowledgments This work was financed by the Polish National Science Centre under decision DEC-2013/09/B/ST7/01583.

References

1. Belter, D., Skrzypczyński, P.: Precise self-localization of a walking robot on rough terrain using parallel tracking and mapping. Ind. Robot: Int. J. **40**(3), 229–237 (2013)
2. Belter, D., Walas, K.: A compact walking robot—flexible research and development platform. In: Szewczyk, R., et al. (eds.) Recent Advances in Automation, Robotics and Measuring Techniques, AISC 267, pp. 343–352. Springer (2014)
3. Belter, D., Nowicki, M., Skrzypczyński, P.: On the performance of pose-based RGB-D visual navigation systems. In: Cremers, D., et al. (eds.) Computer Vision—ACCV 2014, LNCS 9004, pp. 407–423. Springer (2015)

4. Belter, D., Nowicki, M., Skrzypczyński, P.: Accurate map-based RGB-D SLAM for mobile robots. In: Reis, L., et al. (eds.) Robot 2015: Advances in Robotics, AISC. Springer (2015)
5. Belter, D., Nowicki, M., Skrzypczyński, P., Walas, K., Wietrzykowski, J.: Lightweight RGB-D SLAM system for search and rescue robots. In: Szewczyk, R., et al. (eds.) Progress in Automation, Robotics and Measuring Techniques, AISC 351, pp. 11–21. Springer (2015)
6. Belter, D., Łabęcki, P., Skrzypczyński, P.: Adaptive motion planning for autonomous rough terrain traversal with a walking robot. J. Field Robot., Wiley Early View (2015)
7. Claus, D., Fitzgibbon, A.: A rational function lens distortion model for general cameras.In: Proceedings of the IEEE International Conference on Computer Vision and Pattern Recognition (2005)
8. Endres, F., Hess, J., Sturm, J., Cremers, D., Burgard, W.: 3-D mapping with an RGB-D camera. IEEE Trans. Robot. **30**(1), 177–187 (2014)
9. Henry, P., Krainin, M., Herbst, E., Ren, X., Fox, D.: RGB-D mapping: using kinect-style depth cameras for dense 3D modeling of indoor environments. Int. J. Robot. Res. **31**(5), 647–663 (2012)
10. Hornung, A., Wurm, K., Bennewitz, M., Stachniss, C., Burgard, W.: OctoMap: an efficient probabilistic 3D mapping framework based on octrees. Auton. Robots **34**(3), 189–206 (2013)
11. Kümmerle, R., Grisetti, G., Strasdat, H., Konolige, K., Burgard, W.: g2o: a general framework for graph optimization. In: IEEE International Conference on Robotics and Automation, Shanghai, pp. 3607–3613 (2011)
12. Maier, R., Sturm, J., Cremers, D.: Submap-based bundle adjustment for 3D reconstruction from RGB-D data. In: Pattern Recognition, LNCS 8753, pp. 54–65. Springer (2014)
13. Nowicki, M., Skrzypczyński, P.: Experimental verification of a walking robot self-localization system with the Kinect sensor. J. Autom., Mobile Robot. Intell. Syst. **7**(4), 42–51 (2013)
14. Rublee, E., Rabaud, V., Konolige, K., Bradski, G.: ORB: an efficient alternative to SIFT or SURF. In: IEEE International Conference on Computer Vision, pp. 2564–2571 (2011)
15. Scherer, S., Zell, A.: Efficient onbard RGBD-SLAM for autonomous MAVs. In: Proceedings of the IEEE/RSJ International Conference on Intelligent Robots and Systems, Tokyo, pp. 1062–1068 (2013)
16. Schmidt, A., Kraft, M., Fularz, M., Domagala, Z.: Comparative assessment of point feature detectors and descriptors in the context of robot navigation. J. Autom., Mobile Robot. Intell. Syst. **7**(1), 11–20 (2013)
17. Schmidt, A., Kraft, M., Fularz, M., Domagala, Z.: The registration system for the evaluation of indoor visual SLAM and odometry algorithms. J. Autom., Mobile Robot. Intell. Syst. **7**(2), 46–51 (2013)
18. Schmidt, A., Kasinski, A., Kraft, M., Fularz, M., Domagala, Z.: Calibration of the multi-camera registration system for visual navigation benchmarking. Int. J. Adv. Robot. Syst. **11**(83) (2014)
19. Shi, J., Tomasi, C.: Good features to track. In: IEEE Conference on Computer Vision and Pattern Recognition, Seattle, pp. 593–600 (1994)
20. Sturm, J., Engelhard, N., Endres, F., Burgard, W., Cremers, D.: A benchmark for the evaluation of RGB-D SLAM systems. In: IEEE/RSJ International Conference on Intelligent Robots and Systems, Vilamoura, pp. 573–580 (2012)
21. Umeyama, S.: Least-squares estimation of transformation parameters between two point patterns. IEEE Trans. Pattern Anal. Mach. Intell. **13**(4), 376–380 (1991)
22. Whelan, T., Johannsson, H., Kaess, M., Leonard, J., McDonald, J.: Robust real-time visual odometry for dense RGB-D mapping. In: IEEE International Conference on Robotics and Automation, Karlsruhe, pp. 5704–5711 (2013)

Reasoning with Four-Valued Logic in Multi-robotic Search-and-Rescue Problem

Łukasz Białek, Jacek Szklarski, Marta M. Borkowska and Michał Gnatowski

Abstract The aim of this paper is to present application of non-standard logic in the process of reasoning regarding the state of environment in a multi-robotic system. We have investigated the possibility that the usage of four-valued logic can serve as an alternative to the traditional probabilistic approach when handling the uncertainties. As a proof of concept, we have considered a typical search-and-rescue scenario in which a group of n robots performs a search for k victims localized in random positions in an initially unknown building. The building is represented in the form of a graph while the uncertainties are modeled by parameters denoting probabilities of victim observation and random blocking of a robot transition between neighboring nodes (which precludes the possibility for the robot to know anything about the actual graph structure with full confidence). The simulation results show that the unique features of applied reasoning language give significant improvement over typical approaches based on the two-valued logic. The results have a preliminary character, in the sense that up to now, only numerical experiments have been performed—applying the idea in a real multi-robotic system is currently underway.

Keywords Search-and-rescue problem · Reasonong engine · Four-valued logic · 4QL · Simulation

Ł. Białek (✉)
Faculty of Mathematics, Informatics and Mechanics, University of Warsaw,
Banacha 2, 02-097 Warsaw, Poland
e-mail: lukasz.bialek@mimuw.edu.pl
URL: http://www.mimuw.edu.pl

J. Szklarski · M.M. Borkowska · M. Gnatowski
Institute of Fundamental Technological Research, Polish Academy
of Sciences (IPPT PAN), Pawińskiego 5B, 02-106 Warsaw, Poland
URL: http://www.ippt.pan.pl

M. Gnatowski
e-mail: michal.gnatowski@gmail.com

R. Szewczyk et al. (eds.), *Challenges in Automation, Robotics and Measurement Techniques*, Advances in Intelligent Systems and Computing 440, DOI 10.1007/978-3-319-29357-8_43

1 Introduction

Throughout years, methods and techniques used during rescue missions have improved significantly. Human rescuers are often equipped with, or even replaced by, machines. This greatly increases safety and makes it possible to perform rescue operations in severe conditions. With the advancement of technology, the problem of searching and rescuing victims became a real issue in the field of robotics [1]. The aim of this paper is to propose a theoretical model of a multi-robot system with reasoning engine based on a new, multi-valued reasoning scheme. Mathematically, the problem is equivalent to a variation of a pursuit-evasion game known from graph theory. According to our knowledge, up till now, problems of this kind have not been tackled with the use of such non-classical logical approach.

The presented results have been obtained with the use of a relatively new language, called 4QL (*four valued query language*). It has been introduced in 2011 [2, 3] and since then has been actively developed [4, 5]. The unique features of the language are four truth values (in addition to the classical *true* and *false*, also *inconsistent* and *unknown* were introduced), and the possibility of negation not only in the body of the rule but also in its head—which naturally introduces inconsistency into the language. Moreover, the 4QL language is based on OWA (*Open World Assumption*) which means, that all the facts not present in the knowledge base are assumed to be *unknown* and not *false*. We show that using this kind of logic naturally applies when an uncertain, dynamic environment is concerned.

Of course, in the field of non-deterministic search-and-rescue problems, there are many known solutions based, e.g., on probabilistic or fuzzy reasoning. However, one of the advantages of our approach is that distribution of probabilities in the graph does not necessary have to be known, which is often the case in any probabilistic reasoning. In a dynamical, unknown environment that serves as a basis for testing the proposed solution, providing reliable probability functions would be very difficult or close to impossible. As a substitute of a quantitative approach, we prefer to use symbolic reasoning, which acts on logic facts and does not need any additional characteristics. Moreover, it is worth noticing that the four-value logic provides a native support of inconsistencies. This is also the case, for example, when *Intuitionistic Fuzzy Sets* [6] are taken into account, but for our application the sets cannot be easily applied because of their quantitative characteristic. Moreover, in the classical approaches, there is yet another problem connected with the treatment of unknown knowledge and resolving inconsistencies, whereas 4QL provides a natural solution to this. Finally, reasoning with 4QL uses polynomial time (see [7] for proof) while e.g. probabilistic reasoning often requires an exponential size of distribution of probabilities (when all conditional probabilities are concerned). This makes it suitable to apply 4QL in problems, where timing is of critical importance (as in rescue missions).

The initial state of the research presented here has been published in conference proceedings [8], with a very positive feedback. Community response encouraged us to continue our work in the field and verify applicability of 4QL in more complex test scenarios. The most significant improvement is the introduction of dynamic graph modification by applying "blockades" on graph links (which can appear and disappear randomly). Previously, any considered graph was static, and this extension allowed us to fully use the unique features of 4QL. The change included modifications of 4QL modules, which are described in further parts of this paper. We also present a novel way of using 4QL's *unknown* truth value in rescue reasoning. Finally, search strategies have been altered to match new blockades with a set of new strategies implemented including a naive, random strategy implemented for comparison.

The paper is organized as follows: in Sect. 2 we define the problem, then in Sect. 3, a detailed description regarding application of four-valued logic is presented. Section 4 briefly discusses the simulation framework, which is followed by results of numerical experiments.

2 Problem Definition

The motivation for the simulation model described in this document comes from a basic search-and-rescue problem related to a building search. We consider a building which suffered some kind of damage, and some of its parts (unknown) became inaccessible. In the model, the building is represented as a graph, where rooms and other spaces are represented as nodes, and doors and small passages as links. Passages and doors may become blocked at any moment of a simulation (with probabilities as parameters). In order to "save survivors", who can be located in nodes, we introduce a team of rescue robots into the building, which traverse rooms and corridors and react to unexpected situations. Similar problems have been considered in the past (e.g. [9, 10]). Our approach is based on a paradigm *"first simulate then act"*, which means, that in order to achieve the best result we must first consider a large amount of possible scenarios, and then evaluate strategies by simulating robot movements. After evaluating a large number of scenarios, the best strategy is chosen and will be applied in real life.

Figure 1 presents one of several building specifications used during the numerical experiments. Let us imagine an exemplary situation: the rescue robots have focused on searching through the big hall, 3.2.1, where the probability of finding survivors is the highest. After a successful search, the robots notice that both exits from the hall are blocked (either by collapsed passages or a simple error in robot perception). It is easy to see, that if the robots were using standard, two-valued logic, not much would be left to do. Robots can try to examine doors one after another to try to reduce the possibility of getting a sensor error, however, even if two results are obtained, one has to be chosen (either *true* or *false*). Moreover, even if it is decided that, e.g., *true* should be chosen when a sensor error is detected, the decision may not be optimal. On the other hand, sometimes there are passages not placed on official plans of buildings (or passages are created by a disaster itself). For example, there might be a second corridor from hall 3.2.1 to corridor 1.4 and robots, which have only positive (connection exists) and negative (no connection) facts, will have a very hard time finding it. In the following, we consider how this examples can be treated with 4QL.

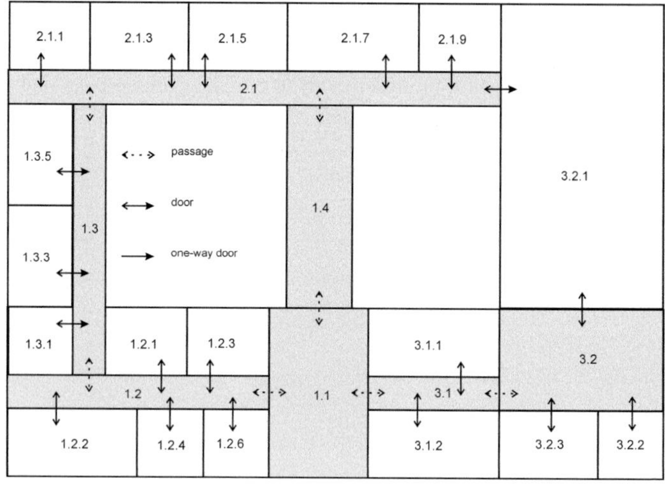

Fig. 1 A schematic representation of a single floor of a building to be searched. In the numerical model, the building is represented as a graph where nodes correspond to spaces and links to passages

3 Reasoning Engine

3.1 4QL *Characteristics*

The 4QL language is a rule-based query language [8]. Its remarkable feature is that it allows one to use rules with negation in conclusions and premises of rules. It is based on a simple and intuitive semantics and provides uniform tools for lightweight versions of well-known forms of commonsense (non-monotonic) reasoning. In addition, 4QL is tractable w.r.t. data complexity and allows one to express all queries computable in deterministic polynomial time.

4QL supports a modular and layered architecture. As the underpinning principle, as stated above, openness of the world is assumed (OWA), which may lead to lack of information. Negative facts and conclusions of rules may lead to inconsistencies. To reduce the *unknown*/*inconsistent* zones, *modules* and *external literals* provide means for:

- application-specific disambiguation of inconsistent information;
- a use of Local Closed World Assumption, LCWA (thus also Closed World Assumption (CWA), whenever needed);
- implementing various forms of common sense and defeasible reasoning.

A brief comparison of 4QL and SQL/PROLOG is provided in Table 1. One of the main differences of 4QL and SQL comparing to PROLOG is that the former are database languages while the latter one is a programming language offering full expressiveness. A similar characteristics to PROLOG is shared by many other business rules engines (for their overview see, e.g., [11]).

Table 1 Comparison of main features of 4QL and SQL/PROLOG (see [8])

Feature	4QL	SQL	PROLOG
OWA versus CWA	OWA, LCWA, CWA	CWA	CWA
PTime, LogSpace complexity	Non-recursive 4QL	Standard SQL	Not guaranteed
PTime, PSpace complexity	Full 4QL	Recursive SQL	Not guaranteed
Incomplete information	Full support	Partial support	No support
Inconsistent information	Full support	No support	No support
Negative conclusions	Allowed	Not allowed	Not allowed

3.2 Underlying Logic

Full theoretical description of four-valued logic and 4QL reasoning mechanisms is available in [7]. However, for a better understanding the the mechanisms used, we present here some basics of the underlying logic of the language itself. To begin with, let us present truth tables for 4QL:

∧	f	u	i	t
f	f	f	f	f
u	f	u	u	u
i	f	u	i	i
t	f	u	i	t

∨	f	u	i	t
f	f	u	i	t
u	u	u	i	t
i	i	i	i	t
t	t	t	t	t

→	f	u	i	t
f	t	t	t	t
u	t	t	t	t
i	f	f	t	f
t	f	f	t	t

	¬
f	t
u	u
i	i
t	f

To achieve a layered structure of a program, 4QL introduces modules as a way of modeling different sources of information. A module can query another module for values of literals, but queried module has to be in lower layer. It is done by using, so called, *external facts* [12]. The construction of 4QL language assumes that a module can have, in addition to domains, relations and facts, one or more rules defined in section `rules`. Each rule consists of one or more *conditions* and exactly one *conclusion*. Conclusion and conditions are separated with an ':-' sequence. Conditions can be separated with a comma which is equal to logical ∧, or with |, which is equal to logical ∨. During the reasoning phase, for each rule, its conditions are matched with facts that are present in the knowledge base. Afterwards, logical operators are applied (using rules for ∧, ∨ and ¬) and based on the logical result of the operation the rule's conclusion is either reasoned or not. Below, we present the main idea with the use of a short example.

Let us consider rule "*rescueVictim :–foundVictim.*" (saying "if a victim is found, then rescue it"). This rule has only one condition (*foundVictim*), but in general it can be an advanced logical expression. In the beginning, the knowledge base contains no information about *foundVictim* so it has an *unknown* value. According to the table for "→" any logic value can be assigned to *rescueVictim* to satisfy the rule. However, during the development of the 4QL language also common sense was important (why should we rescue a victim when it is unknown if there are any?). Therefore, reasoning based on *unknown* facts is not possible (although, it is possible using *in* notation, see [7]). Apart from maintaining intuitive behaviour, this will keep the rule satisfied. The same common sense problems occur, when trying to reason from *false* facts—therefore it is also not possible.

At some point in time, *foundVictim* may become *true*. According to the truth table, either a *true* or *inconsistent* conclusion can be derived. The design of 4QL specifies, that *true* conditions create *true* conclusions. The other options was introduced to allow *inconsistent* conclusions to be supported by *true* conditions (when there are two rules with *true* conditions, one supporting a conclusion and the other supporting its negation).

Finally, the rule's condition may become *inconsistent*, which will cause, that *rescueVictim* will not be supported by the rule any more. To regain that, the *rescueVictim* has to become *inconsistent* too.

The reasoning process is continued until all rules are satisfied and all positive and inconsistent facts are supported by some rule.

3.3 Reasoning Modules

The core of the developed simulator is constructed from two components—the knowledge base and the reasoning engine. The elements are directly taken from INTER4QL application [4] (4QL interpreter available on [5]). The knowledge base contains modules, which are responsible for managing experiment states, building graphs, connections, and assuring that the experiment will end.

Currently, the knowledge base contains the following modules, initially introduced in [8]:

- '`staticMap`': contains all the static information about the experiment
- '`dynamicMap`': receives all the dynamic information from the outside world
- '`dynamicMapClosure`': combines the facts from all of the modules into data for the PLANNER component
- '`disaster`': stores all current data about the disaster environment
- '`robots`': contains information about the current state of robots
- '`victims`': contains all the information about the found victims

During the experiment, sensors from (virtual) robots send data to the PLANNER module, which cooperates with the database. In the beginning of the experiment, all initial data are put into the '`staticMap`' module, which is later used as the constant database about the initial building structure. There is no inconsistency assumed there.

Throughout the whole experiment, robots use their sensors to discover the environment. All their findings are passed to the '`dynamicMap`' module. Initially, most of the facts in the module are *unknown* (except those, which were resolved from '`staticMap`'). It is commonly known that a robot's sensor may send erroneous data, for example because of some interferences on the communication pipeline between the robot and the database or there could just simply be a problem with the robot's perception. Therefore, inconsistencies are expected in this module.

An the same time, all the other data about the experiment (victims found, robot localization) are stored in their modules (names of modules are self-explanatory). Although inconsistencies may occur for that data, in practical simulations no such thing ever happens.

Let us consider the following two modules:

```
module dynamicMapClosure:
  ...
  rules:
    connected(X,Y,Z):-
      dynamicMap.connected(X,Y,Z)
        in {true, unknown}.
    -connected(X,Y,oneWay):-
      dynamicMap.connected(X,Y,oneWay),
      staticMap.linkName(X,Y,LinkID),
      disaster.threshold(T),
      disaster.probability(LinkID,Z),
      math.gtR(Z,T).
    // twoWay connections are analogous
    ...
    blocked(X):-
      dynamicMap.blocked(X)
        in {true, incons}.
end.

module disaster:
  ...
  relations:
    // passage attractiveness, start = 1/2
    attractive(id,noOfSuccesses,noOfTries).
    // probability of unblocked passage
    probability(id,probabilityOfSuccess).
    // when reached, cuts off graph branch
    threshold(real).
  ...
end.
```

These modules are particularly important and we will look at them more closely. The first one, '`dynamicMapClosure`', was historically devoted to the local closing of the world data kept in the knowledge base. It has evolved into a module, which extended its initial responsibilities by combining data achieved from '`staticMap`', '`dynamicMap`' and '`disaster`' modules into data which can be read by the PLANNER component. There are two major facts reasoned here which are crucial for the simulation process—*connected* and *blocked*. The first of them, after reasoning regarding the part of each step of the experiment, keeps information about the non-blocked connections in the graph of the building. The entire work connected with checking if the current patch in graph is available is moved over to INTER4QL engine, and it deals with all inconsistencies which may appear in other modules.

Let us look at lines 5 and 6 of the listing above. The code will allow non-existing connections to be added into the connection graph (because the *unknown* keyword is present in the value set). However, connections which were reported to be inconsistent will be ignored and the robots will not try to use them. This place may be freely changed to allow various connections to appear in the graph. Nothing forbids, e.g., adding inconsistent connections by adding the *incons* keyword to the logic set. It can be clearly seen, that with little effort we get lots of possibilities of modules extension. The same can be applied to lines 16–17, which can remove some nodes from the graph in the reasoning stage of generating the instructions (in the example above, it is reasoned that blocked nodes are those which have been confirmed to be blocked or whose block state is inconsistent).

As it can be seen, in lines 7–12 it is checked whether the current connection is not permanently removed from the graph by exceeding the threshold. When this happens, a negative fact is added to the database, which transforms the fact into an *inconsistent* one. This information could then be used for further computation, but for the PLANNER it is sufficient.

The second module, '`disaster`', is used to keep data related to the passages and their current states. We use a simple *probability sampling* technique for approximating the passage's current state. This allows for rechecking previously blocked nodes which is a great improvement over the previous version of the simulator described in [8] (the strategies used earlier assumed that blockades are permanent, i.e., a static graph assumption).

3.4 Exemplary Scenario Analysis

Let us now return to the exemplary situation presented in the problem specification. During the initialization of the experiment, facts about connections in the graph are obtained from an external data source and placed into the '`staticMap`' module. In addition to the positive facts about connections, the application generates all possible connection facts for other pairs of places and stores them with logic values equals to *unknown*. To reduce the amount of facts, which may be enormous and significantly reduce application performance, domains are used. This allows for omitting unknown facts which mix different types of data (like room identifiers and victim identifiers which obviously cannot be connected). On that stage, the database contains fact '`connected("3.2.1", "1.4") : unknown`' which will be used for finding the hidden connection.

When all the facts are present in the '`staticMap`' module, data is moved to '`dynamicMap`'. Although fact '`connected("3.2.1", "1.4") : unknown`' itself will not be reasoned (the rule "copies" only positive connections), but it will appear during the generation of *unknown* facts (*unknown* facts are generated for each relation in each module based on domain elements). Elements are added to the domains when a new fact is discovered, so "3.2.1" and "1.4" will be present in the room identifier's domain (there is a connection for both of them).

Finally, '`dynamicMapClosure`' has to decide which connections will be included in the connection graph. The code snippet shown earlier assumes that all the unknown connections are transformed to existing ones, so eventually the robots will try to go from hall "3.2.1" to corridor "1.4". And this is what we would expect. Note, that if 2-valued logic were used, this would not be possible.

Let us now discuss reactions for inconsistent facts. As it was stated in the example, robots can check a passage several times to minimize the possibility of a sensor error (which, however, takes resources, i.e., time). All the data are passed directly to '`dynamicMap`', where it can become *true*, *false* or *inconsistent* (initially all the passages are unknown—when a robot tries to use a passage, it's state is moved to the correct module). Then, '`dynamicMapClosure`' has to decide if the given node is blocked or not. This can be realized by an arbitrary decision that changes *inconsistent* into *true* or *false*. Using 4QL, advanced selection criteria can be declared, which may make the choice depending on advanced criteria (i.e. if the node has a neighbor which is not visited, then select *true*; otherwise select *false*).

4 MultiLight System

The effectiveness of the proposed approach was tested in a simulation environment, called "MultiLight", it's overall schema is presented in Fig. 2. The simulator is constructed using three modules: GUI, OBSERVER and PLANNER. This section will briefly discuss the mechanisms and technologies used in the system.

4.1 GUI *Module*

Since a 4QL-based interface had been applied to the PLANNER module, it was paramount to check, whether the obtained search strategies were rational from a common sense perspective. This goal was achieved in a twofold manner. First, by visually observing the movement of robots inside the building, properties were evaluated and a particular search strategy was chosen. Then, the data on each experiment was

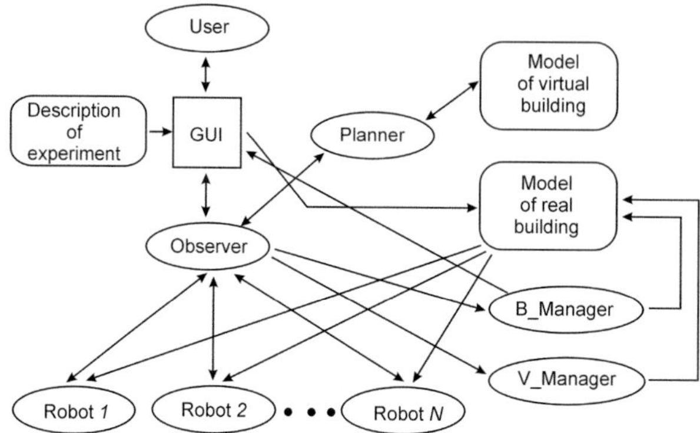

Fig. 2 Block chart of the MultiLight simulator

stored in the log-files. Those files can by analyzed later in order to judge the outcome of experiment.

4.2 OBSERVER *Module*

The OBSERVER module simulates the environment and is responsible for managing random blockades and victims. At the beginning of the experiment, it gets data from the GUI module, filters it and sends a part of the information to the PLANNER module. The filtered data contains information on the building structure, number and starting position of robots but doesn't contain information on victims and blockades.

After starting an experiment, OBSERVER receives commands from PLANNER, which is dedicated to robots and keeps control of all of them. Cyclically, it sends out information on the current position and mode of each robot to PLANNER and GUI modules. PLANNER uses this information for commanding robots and eventually re-planning their routes. The GUI module needs the information from OBSERVER to present the robots' current positions and modes.

PLANNER in its commands indicates a sequence of rooms which are yet to be checked and decides how to get to those rooms. This sequence is a list of graph nodes, which are: rooms, corridors and doors. If PLANNER decides, that all of the rooms have been checked, it sends a signal to the OBSERVER module, indicating that the experiment came to an end.

Robots may be in one of three modes: *waiting*, *running* and *searching*. At the beginning of the experiment, a robot is in *waiting* mode. After receiving a command from PLANNER, it switches into *running* mode, goes to the indicated room, switches to *searching* mode and starts searching the room. Switching into *waiting* mode is possible in one of three cases: after finishing a task, when a robot finds a blockade or when PLANNER cancels the current task.

4.3 PLANNER *Module*

While the previous two modules were responsible for user interaction and inter-module communication, the PLANNER is used for managing robots and generates instruction for them.

The construction of the module contains two main parts: the 4QL knowledge base with a reasoning engine and a set of graph-traversing algorithms for generating final instructions. The first part has already been discussed, let us focus on instruction generation.

In the beginning, when experiment initialization is performed, the OBSERVER module sends all the necessary data to the PLANNER, which transfers them to the '`staticMap`' module. Data in '`dynamicMap`', '`robots`' and '`disaster`' modules are automatically filled with default values. The reasoning process is then started. After completion of the process, the knowledge base contains full information on connections in the building. The information is used to build an internal graph and run one of the graph-traversing algorithms. Similarly, when OBSERVER sends a new cyclic message with the status of robots, all the facts are automatically placed into the '`dynamicMap`' module (and possibly in the '`robots`' and '`victims`' modules as well) and the reasoning part is started. Then the graph is built and the next stage is started.

In order to generate instructions for robots, some graph traversing has to be done. 4QL language itself is not designed for defining such actions and therefore some other, external algorithms were implemented. Algorithms combined with the advanced reasoning engine used for creating input for them, from now on, will be called *strategies*.

Implemented graph-search strategies can be found below:

1. `DFS`: Standard DFS [13, pp. 603–610] algorithm
2. `BFS`: Standard BFS [13, pp. 594–601] algorithm
3. `DAGM`: Dijkstra Algorithm [13, pp. 658–662] with Greedy Matching Method
4. `DAHM`: Dijkstra Algorithm with Hungarian Matching Method [14]
5. `ADAGM`: Abortable Dijkstra Algorithm with Greedy Matching Method
6. `ADAHM`: Abortable Dijkstra Algorithm with Hungarian Matching Method
7. `RND`: A Random Strategy

The first two strategies were used, as a simple and effective ways to determine way to rooms. Strategies 3–6 have been altered to support randomly disappearing blockades, by repeatedly trying to go through a blocked passage (persistence in doing this is controlled by a simulation parameter). A random strategy means that robots try to move to any of the neighboring nodes with equal probability. It is important to benchmark against a random strategy in order to evaluate how much gain would more reasonable strategies give (comparing against random strategies is a standard procedure in pursuit-evasion games).

5 Experiments

As was presented in previous sections, several major improvements have been introduced when compared to [8]. First of all, introducing possibility of blockades to disappear (false-positive and false-negative blockade perception) made test environment dynamic, which allowed for creating much more interesting test scenarios. Moreover, we have introduced two additional building structures, which may occur in real-life rescue missions. This section will present new test buildings and achieved results.

Figure 1 from Sect. 2 (problem definition) depicts original building (*B0*) developed during initial phase of our implementation. It contains two cycles connected with corridors in the middle. This building, due to introduced dynamic blockades, can be transformed to one-cycle or no-cycle building. It can also be then restored to original state (blockades disappear) which was particularly interesting to us.

Figures 3 and 4 show newly added buildings, which were used in latest test sessions. Due to it's tree-like structure (no cycles), building *B1* is perfect for testing strategies against retrying previously blocked paths. Building *B2* also does not contain cycles, but it does not contain any corridors as well. Therefore blockades are much more invasive into building's structure by removing many possible paths at once.

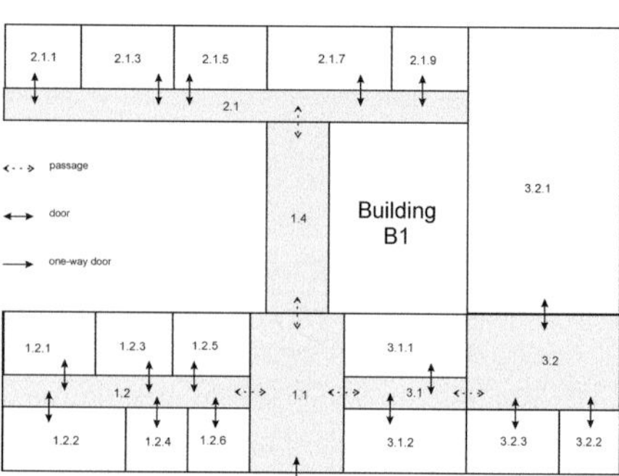

Fig. 3 Building *B1*—tree structure

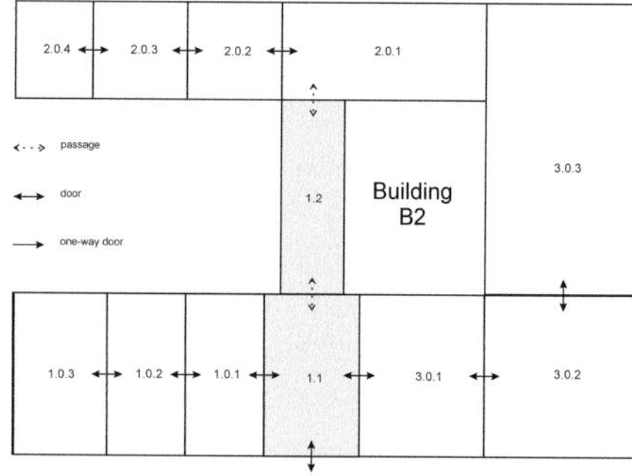

Fig. 4 Building *B2*—enfilade structure

The proposed strategies were evaluated by performing simulations on networks with varied topologies and for a wide range of initial parameters, i.e., number of robots, their initial location, number and placement of victims and parameters characterizing probabilities of false positives (P_{pos}) and false negatives (P_{neg}) regarding the perception of the blockades. In order to make a fair judgment, all the strategies were tested on the same set of initial conditions (i.e., positions of robots and blockades).

Figures 5 and 6 correspond to the results achieved for building *B0*. During those tests all strategies were compared to newly implemented random strategy (RND), which was prepared to see how much better our solution is when compared to totally random one. Prepared plots depict average number of steps required to finish the searching process (dashed lines) and room coverage for each strategy (solid lines). The room coverage is simply the number of visited rooms (by one or more robots) divided by the total number of rooms. Presented values are calculated as an average of results for 100 different initial conditions for the same graph.

The most obvious conclusion is that all the strategies outperform the random one. As for coverage, the winner is DAGM, however for $P_{pos} = 0.1, P_{neg} = 0.1$ ADAGM has a bit smaller coverage, while being notably faster. DFS and BFS in this case have unacceptable coverage rates.

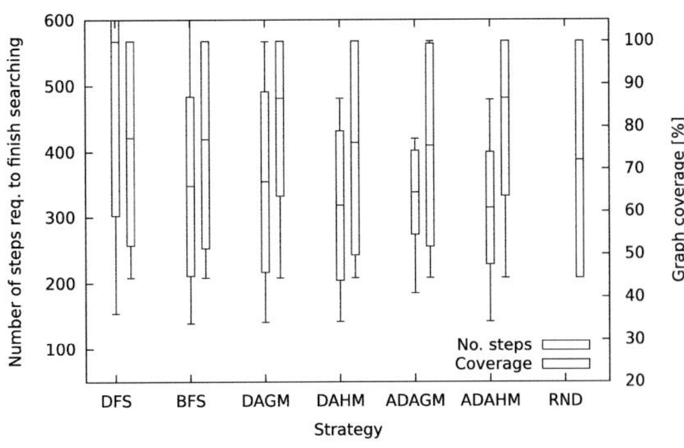

Fig. 5 Measurement of efficiency of the proposed strategies: averaged number of steps required to finish the searching process and average graph coverage. The values of candlesticks correspond to: minimum, mean-σ, mean, mean+σ and maximum. Here $N_{robots} = 4$, $P_{pos} = 0, P_{neg} = 0.1$

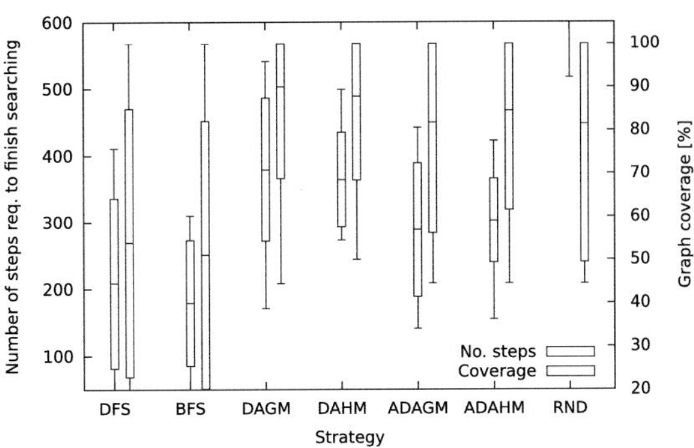

Fig. 6 As in Fig. 5, but here $P_{pos} = P_{neg} = 0.1$

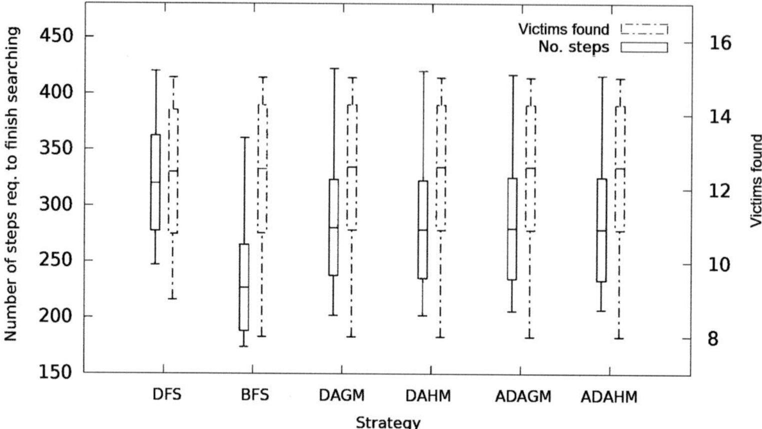

Fig. 7 Simulation results for building *B0* (4 robots, 15 victims and up to 2 blockades)

During real search mission, graph coverage is not the most important thing—number of rescued victims is. Therefore our next tests were focusing on how fast each strategy is and how many victims were rescued (strategy "RND" is omitted, no new conclusions were found). We have obtained a great amount of graphs, but here we will present only the most interesting ones for 4 robots, 15 victims and up to 2 blockades in the building ($P_{pos} = 0.1, P_{neg} = 0.1$).

Figure 7 compares 6 strategies during test scenario in building *B0* (please note a change in line styles—solid lines represent number of steps and dashed lines represent number of victims rescued). Surprisingly, when graph coverage is not considered, BFS strategy is the best—it rescues statistically as many victims as the others, but uses much less steps to do that.

For *B1* however (Fig. 8), ADHM takes the lead. We will not present here graph for *B2*—it is sufficient to say, that BFS proved to be the best for this kind of building.

Having all the data, we have noticed an interesting dependency between type of the building and strategy used (each type of building has it's own best strategy). This led us to making a guess: *"After coming across an unexpected blockade it may be worth stopping all the robots and regenerate instructions for them (possibly using different strategy)"*. In order to prove it, we have developed 36 new strategies *XY* ($X, Y \in \{1..6\}$) which basically meant "start searching with strategy *X*; when a blockade is encountered stop all robots, change strategy from *X* to *Y* and regenerate instructions for all robots using new strategy"'.

For further tests we have chosen building *B0*. The same tests were run as before, but this time including new 36 strategies. Tests were altered in such a way, that blockades could only occur on a cycle, so that structure of building was changed. This means that as the beginning tests were performed using clear graph with 2 cycles (for which BFS earlier was the best) and then it changed to graph without cycles (where ADAHM and DAHM proved earlier to be the leaders). Our expectations

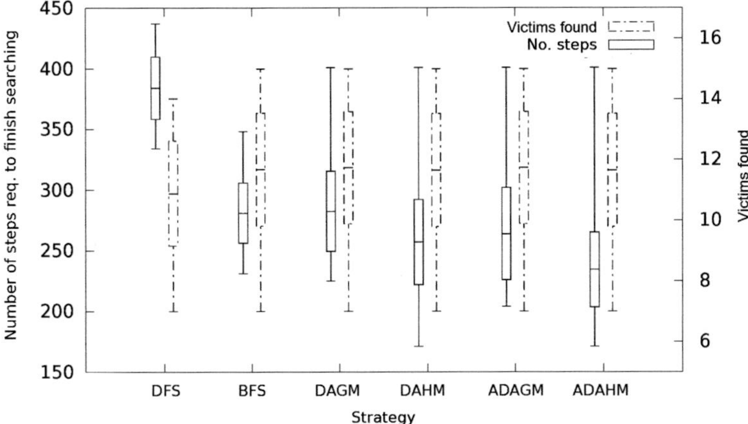

Fig. 8 Simulation results for building *B1* (test conditions the same as in Fig. 7)

Fig. 9 Improvement for number of steps between standard BFS strategy (2) and changeable one (25) for building *B0* (4 robots, 15 victims, 2 blockades removing the cycles)

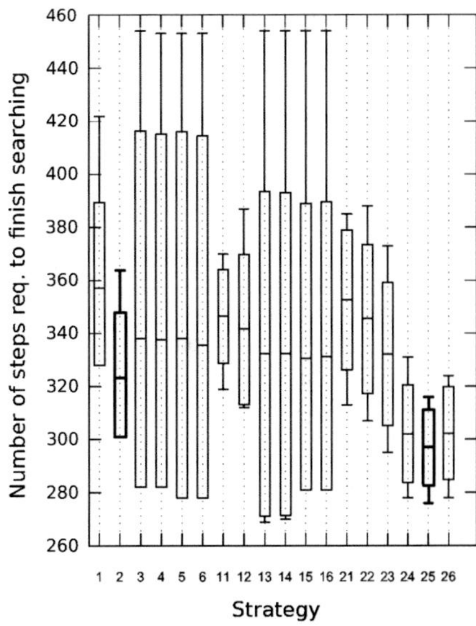

were confirmed after tests. For such scenario BFS is the best from all strategies 1 to 6 (strategy 2), but strategies BFS→DAHM/ADAGM/ADAHM (numbers 24/25/26 on the graph) are even better (see Fig. 9). What is more important, number of rescued victims is not influenced by strategy change. Unfortunately whole graph with all 42 strategies is too big to place here. It is sufficient here to say, that all other strategies did not make any improvement when compared to standard, non-changeable strategy.

6 Conclusions

The presented results show that a standard graph-traverse algorithm can deal with a dynamic environment in an exceptionally good way, provided that it is supported by a reasoning engine which can handle inconsistencies and incomplete knowledge. The problem considered by us is widely discussed in the literature, however the solutions proposed by other authors are based on a classical logic which cannot handle any unexpected observations in a natural way. Our approach uses a new, actively developed, query language. The development of the planning solution is supported by improvements of INTER4QL itself, which allows for the best adjustments to the problem.

This introductory paper is devoted to describing the general idea and demonstrating results of numerical experiments. Naturally, future work will be focused on implementing this solution in a real multi-robot system. Currently, we are working on incorporating the link between the OBSERVER module and a group of 4 real mobile robots.

Acknowledgments This work was supported by the Polish National Science Centre grant 012/05/B/ST6/03094.

References

1. Murphy, R., Tadokoro, S., Nardi, D., Jacoff, A., Fiorini, P., Choset, H., Erkmen, A.: Search and Rescue Robotics in Springer Handbook of Robotics, pp. 1151–1173 (2008)
2. Małuszyński, J., Szałas, A.: Living with inconsistency and taming nonmonotonicity. In: Datalog Reloaded, LNCS, vol. 6702, pp. 384–398. Springer (2011)
3. Małuszyński, J., Szałas, A.: Partiality and inconsistency in agents' belief bases. In: Proceedings of the KES-AMSTA, Frontiers of Artificial Intelligence and Applications, vol. 252, pp. 3–17. IOS Press (2013)
4. Białek, Ł.: The 4QL Interpreter. http://4ql.org/wp-content/uploads/2013/10/inter4ql.pdf (2013)
5. 4QL. http://4ql.org
6. Atanassov, K.T.: Intuitionistic fuzzy sets. Fuzzy Sets Syst. **20**, 87–96 (1986)
7. Małuszyski, J., Szałas, A.: Logical foundations and complexity of 4QL, a query language with unrestricted negation. J. Appl. Non-Class. Log. **21**(2), 211–232 (2011)
8. Białek, Ł., Borkowski, A., Borkowska, M., Gnatowski, M., Dunin-Kęplicz, B., Szałas, A., Szklarski, J.: Coordinating multiple rescue robots. Krajowa Konferencja Robotyki (Polish National Robotics Conference) (2014)
9. Murphy, R.R.: Disaster Robotics. MIT Press (2014)
10. Tadokoro, S.: Rescue Robotics; DDT Project on Robots and Systems for Urban Search and Rescue. Springer (2010)
11. Lämmel, U.: Business rules make business more flexible. In: Proceedings of the 3rd Conference on Baltic Business and Socio-Economic Development BBSED, Tallinn, Estonia 17–20. June (2007)

12. Szałas, A.: How an agent might think. Logic J. IGPL **21**(3), 515–535 (2013)
13. Cormen, T.H., Leiserson, C.E., Rivest, R.L., Stein, C.: Introduction to Algorithms, 3rd edn. MIT Press and McGraw-Hill (2009)
14. Kuhn, H.W.: The Hungarian method for the assignment problem. Naval Res. Logistics **2**, 83–97 (1955). doi:10.1002/nav.3800020109

Modern Sensors in Mobile Robot Teleoperation

Tomasz Pachana, Radosław Trzcionkowski and Wojciech Turek

Abstract Fast development of sensing devices able to collect data about human actions creates new possibilities for building robot teleooperation interfaces. The devices for capturing gestures, movements or even thoughts can replace joysticks and buttons and simplify expressing decisions. In this paper a series of experiments is presented which involve testing four different interfaces in the task of controlling a mobile robot with manipulator. Results of the experiments, which involved ten volunteers, show that use of modern sensing devices can improve efficiency and intuitiveness of teleoperation.

1 Introduction

The development of robotics inevitably directs towards giving machines more and more autonomy in all aspects of their operation. Fully autonomous decision making is definitely the future of robotics. However, before autonomous robots are able to perform responsible tasks which require expert knowledge, humans support in the decision making process is required.

Most of the robots used in solving real-life problems, which operate in unknown environments and perform complex tasks, require constant supervision of a human operator. The class of remotely controlled robots find applications in many situations where presence of a human would be dangerous or simply impossible. This includes inspection of contaminated buildings, searching places on fire or location with a danger of explosion [1]. Robots can also penetrate locations where humans could not enter, like ventilation shafts or a rubble during rescue actions.

The process of remote controlling robot's actions is called teleoperation. The schema of major elements and their interactions in a teleoperated mobile robotic system is shown in Fig. 1. The operator of the system is provided with information about the state of the robot and the surrounding environment. The information is

T. Pachana · R. Trzcionkowski · W. Turek (✉)
AGH University of Science and Technology, Krakow, Poland
e-mail: wojciech.turek@agh.edu.pl

R. Szewczyk et al. (eds.), *Challenges in Automation, Robotics and Measurement Techniques*, Advances in Intelligent Systems and Computing 440, DOI 10.1007/978-3-319-29357-8_44

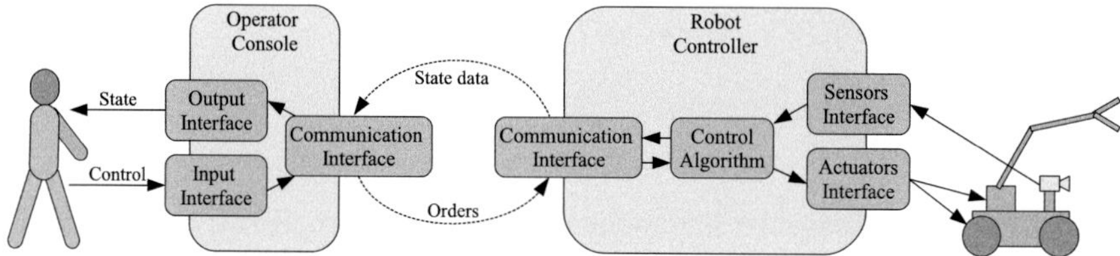

Fig. 1 Schema of interactions between the elements of the teleoperated robotic system

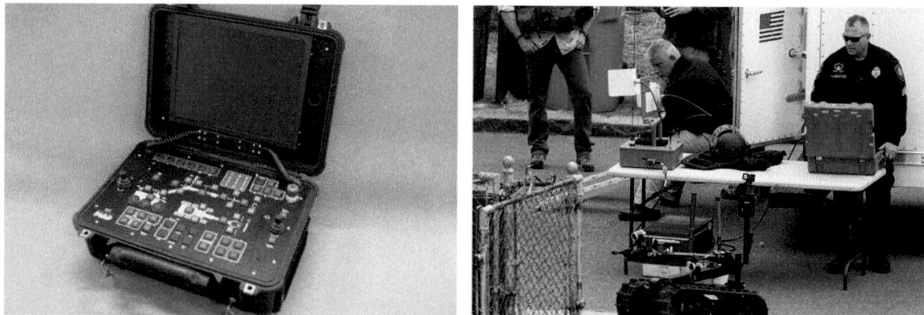

Fig. 2 Control console of PIAP Ibis robot and an example of complex way of mobile robot teleoperations. *Sources* http://antyterroryzm.com, http://dailymail.co.uk

extracted from data collected by robot's sensors. It is presented to the operator by some type of output interface. The operator has to decide what action to perform next and express his decisions using available input interface. Typically this kind of interface has a form of a dashboard with a set of buttons and joysticks, as show in Fig. 2.

In this paper we focus on input interfaces of the operator's console. It seems that the ability of expressing decisions is crucial for effective utilization of robot's capabilities. Adding new types of actuators or increasing the number of manipulator degrees-of-freedom will not bring expected gain in effectiveness if the new features are controlled by additional sets of buttons and joysticks.

We believe that use of modern input devices, which can acquire data concerning humans actions, can lead to more effective and more intuitive methods for expressing decisions in mobile robot teleoperation. In order to verify this hypothesis we have performed a set of experiments involving a group of volunteers, who compared various control interfaces.

2 Existing Approaches to Mobile Robot Teleoperation

There are several features a good teleoperation interface should provide. The most important is the ability to use all available features of the robot, which seems obvious. This can be achieved by providing a separate input device for each degree of freedom of a robot. The interface should also allow to perform tasks in efficient and safe way, which often requires managing several degrees of freedom simultaneously. Achieving this will require the teleoperation method to be *intuitive*. According to [2] intuitiveness, it is subjective and specific for particular personal issue so it should be treated as such during any kinds of reviews and tests. In this section, we would like to introduce widely applied approaches of mobile robot teleoperations and consider if they meet the conditions presented above.

The most commonly applied teleoperation controllers are in great majority based on joysticks and keyboards equipped with a big amount of specific-function-dedicated buttons. Such a device usually takes shape of a big case, often equipped additionally with screen presenting video from mounted on the robot camera to make operation process possible to be carried remotely. Example of such device can be seen in Fig. 2. It assures lack of any kind of functional limitations due to it's complexity but also because of the same reason it cannot be too intuitive for most users. Working with such a device definitely requires long technical training and even though it may still be quite difficult to master simultaneous use of several features.

Authors of [3] present a rescue mobile robot with many sensors of different kinds. The robot is semi-autonomous, offering a remote-controlled mode when needed. A simple game-pad is used as an input device of the teleoperation interface, showing the possibility of using such devices instead of dedicated controllers. Still the device offers only a set of buttons and two joysticks.

More complex solution was presented in [4]. Its core concept was to control mobile robot equipped with six degrees of freedom (DOF) gripper using well known joystick and a video camera which provided images processed by a tracking algorithm. The system utilized reversed-kinematics algorithm to allow the operator for changing gripper's position not by controlling every servomotor completely separately but by defining desired tool position. Huge advantage of such approach is bringing control of more DOFs into a single controlling device. The authors claim that the system introduced by them may be successfully applied for controlling not only manipulating systems but also e.g. flying or riding vehicles.

As an example of modern, experimental system, idea proposed by [5] can be taken. Researchers attempted to construct human's arm mapping system which can recognize movements of the arm and use them to control computer programs and electronic devices. Using this strategy for operating robot's gripper can definitely be the most intuitive from of all described above because of direct projection of arm's movements to those performed by the gripper. It can be freely combined with any kind of robot's position changing system using second hand. At that moment it handles 4 DOF but the system is still under development. Such approach seems to be the best starting point for the further research.

The limitations of the existing teleoperation interfaces is one of the reasons for adding autonomous behaviours to the control software of teleoperated robots. A good survey of this type of solutions can be found in [1]. Different approaches use various levels of autonomy, supporting the operator in simple tasks, correcting errors or reducing the number of robot's DOFs to control.

To sum-up, we can notice that there are many kinds of possible ways of mobile robots teleoperations. However, it is important to indicate that the majority of production-ready solutions are based on joysticks combined with complex keyboards. It seems that this approach usually can be inefficient and hard to learn. Fast development of modern sensors in recent years encouraged us to test other options and see if it can be possible to utilize them for mobile robots teleoperations.

3 Modern Sensors for Human Actions Acquisition

Recent years brought fast development of sensing devices able to collect data about human actions very efficiently and to process it in order to extract valuable information. Applications of such information can already be found in modern video games, TV controllers, medical devices and probably in near future they will find much more fields.

The devices can use audio or video sensors (cameras), accelerometers but also more sophisticated sensors like time-of-flight cameras, thermographic cameras or even electroencephalographs. These data can be used for detecting gestures, body positions, movements, facial expressions, voice commands, emotions and thoughts. This section presents a brief review of modern senses interpretation devices which might become useful in robots teleoperations.

Most widely spread devices in this class are motion game controllers, such as Nintendo Wii Remote or PlayStation Move, presented in Fig. 3. Every game consoles producer represents different way of mapping human body actions to digital signal, however core concept of all of them is based on collecting data from accelerometers and support it with analyzing video signal from a camera. The sensors provide information about location in space, orientation and velocity of movement.

This type of devices have to be constantly held in hand. Because of that specific way of performing the control it may be difficult to do that for the longer period of time.

Similar information about hand movement can be collected from the Myo Armband sensor (Fig. 3). This device leaves both hands free and provides additional information about muscle tension in the arm, which can be mapped into particular poses of fingers. Our tests show that the signal provided by this device is very noisy.

Another approach, which origins in video game industry, is to utilize a specialized cameras to track movements. This approach is implemented by the Microsoft Kinect sensor (Fig. 4) which uses a standard camera and infrared sensor which provides depth information. The approach seems very promising as it leaves both hands free, however its accuracy is definitely not sufficient for precise control.

Similar approach has been implemented in the LeapMotion sensor, which tracks the position of hands. High accuracy combined with very frequent measurements lead to ability of recognizing really gentle and exact movements of all fingers. Possibility of tracking both hands simultaneously gives LeapMotion big advantage above other hand motion controllers.

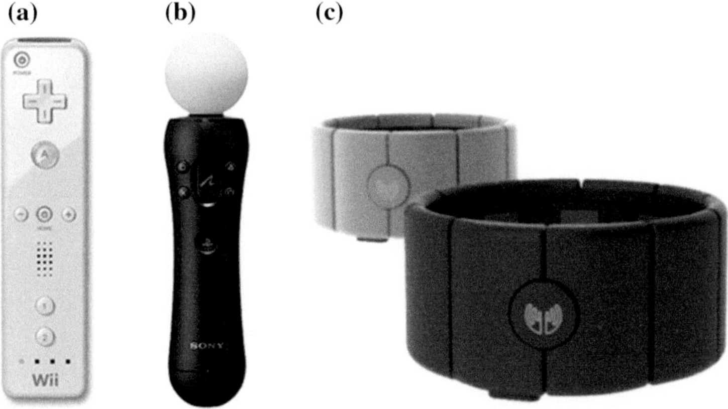

Fig. 3 Hand motion sensors: Wii Remote (**a**), PlayStation Move (**b**) and Myo Armband (**c**) *Sources* https://www.nintendo.com, https://www.playstation.com, https://www.myo.com

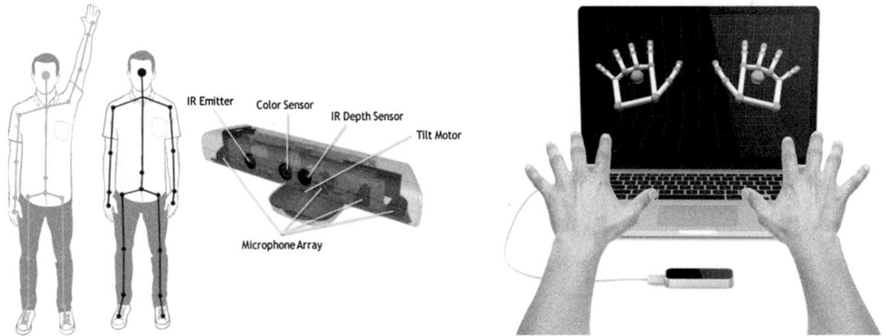

Fig. 4 Pose trackers based on video and distance sensor: Microsoft Kinect and Leap Motion. *Sources* https://www.microsoft.com, https://www.leapmotion.com

Promising results can also be observed with haptic devices, which capture movements of an object touched by a human relatively to a fixed base. The object is physically connected to the base using several joints, which can measure position and apply force when needed. Some research on the applications of such devices in teleoperation has been already performed [6].

The most experimental approach is being applied by electroencephalography (EEG) devices, which provide direct brain-computer interface. There are already devices available which implement this technology, like e.g. Emotiv EPOC, and some attempts of applying in robotics were performed [7]. The authors report low precision in measurements together with significant noise, so it seems to be rather the technology of the future.

The development of the modern devices for collecting data about human actions encourages research on their applicability in robots teleoperation. In the following sections we are reporting our experiences with LeapMotion compared to common joystick and keyboard set.

4 Created Mobile Platform and Teleoperation Stands

The mobile platform used in the experiments is based on the CAPO robot [8]. The robot is a four-wheel mobile platform with independently propelled wheels. It is controlled by an on-board computer running Linux operating system. The robot can communicate using WiFi interface, which was the basis for sending commands from the control stands. In order to increase motion agility we added third degree of freedom to the motion control—the robot has been equipped with Mecanum wheels making it possible to move aside. The robot has also been equipped with a 6-DOF mechanical gripper by Lynxmotion which allows object manipulation. Core robot's components as well as the whole construction is shown in Fig. 5.

Based on the results of reviews presented in previous sections we have selected three input devices to be compared: joystick and keyboard as the representatives of common teleoperation controllers and LeapMotion as an example of modern gestural sensor with high precision. We have composed four test stands using them. Each of those consists of two separate devices to simultaneously control both robot's position and gripper's operations. Every stand is suitable for both right- and left-handed people. The first device from each presented pair controls robot's motion, the second controls the manipulator.

The very first stand is a mix of well know for everyone machine's control device—joystick and modern movement mapping sensor—LeapMotion. Joystick is a commonly applied device in the control of RC models, planes, simulator games etc. Because of that we can be sure that the operator will not be limited in any way by physical specification of it. LeapMotion has been chosen as the gripper's controller because of the natural way of usage—it gives the feeling of grabbing the object with the operator's own hand. Thanks to that he/she can highly concentrate on the task

Fig. 5 CAPO robot internals, version with Mecanum wheels and Lynxmotion Gripper

Fig. 6 One of the tested control stands: keyboard for controlling robot's movements and LeapMotion for controlling the gripper

instead of the way of making the control. Performing the operation in this case can be done with simple putting palm above it and imitation of real gripper movements.

In the next stand, which is presented in Fig. 6, joystick was replaced with keyboard. It has been chosen for it's equality to the previous controller popularity. LeapMotion was selected again for controlling the gripper to be able to compare comfort of using joystick and keyboard for changing robot's position.

The third stand represent the most common approach which is used with manipulation robots: joystick and keyboard. The last stand is the most experimental one—LeapMotion controls robots movements while the keyboard controls the gripper. The position of a hand above the LeapMotion sensor is converted into the velocity of the robot. The gripper keeps changing its position only as long as particular button stays pressed.

5 Experiments and Results

The purpose of the experiments was to determine if modern sensors can be more effective and more intuitive in mobile robot teleoperation than typical interfaces based on joysticks and buttons. In order to verify the hypotheses three types of measures were proposed:

- quantitative measure of time required to fulfil a particular task,
- qualitative measure defined as the average rank of a particular control stand assigned by the operator.
- integrated measure of intuitiveness.

The intuitiveness of the particular control stand was calculated as:

$$I = \frac{n * t_{max}}{\sum_{i=1}^{n}(t_i * r_i)} \tag{1}$$

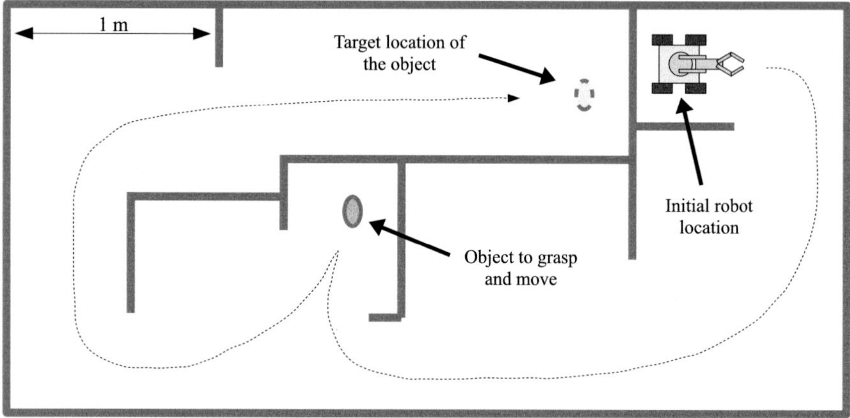

Fig. 7 Plan if the environment used in the experiments

where n is the number of testers, t_{max} is the maximum time of task execution and t_i and r_i are the time and the rank of the ith tester.

The selected task is depicted in Fig. 7. The robot was to be moved from its initial location to the position suitable for grasping an object. Then the object was to be moved to its final destination.

The path was around 10 m long and it included several turns and narrow passages. The location of the object required precise positioning of the robot before grasping was possible.

Each person performed the task four times, using all four control stands. The maximum time to fulfil each task was set to 10 min. After all four tasks were finished each person rated the stands by defining order of subjective intuitiveness.

We selected two groups of people to perform the task. First group of 4 persons consisted of people working in the lab very often, who know the specifics of the robot ant particular control device. The second group of 6 persons included people unfamiliar with the robot, laboratory and the control stands—the experiment was their first contact with the hardware.

The results of the experiments are summarized in Table 1.

The results lead to several interesting conclusions. First of all there is no direct relation between the rank assigned to a stand by the operator and the time needed to fulfill the task. Even average values show that Stand 1 received better ranks, leading to far worse execution times than Stands 2 and 3. Stands 2 and 3 lead to similar execution times, receiving significantly different rates.

All the measures show that Stand 2 provides the best method for controlling the robot in this particular task. The Leap-Motion sensor made it possible to control the 6-DOF gripper fast and accurately. Surprisingly this stand is significantly better than Stand 1, which means that it is easier to control robot's motion with buttons than with joystick.

Table 1 Results of the experiments with four different control stands

	Stand 1		Stand 2		Stand 3		Stand 4	
	Motion: joystick gripper: leapmotion		Motion: keyboard gripper: leapmotion		Motion: keyboard gripper: joystick		Motion: leapmotion gripper: keyboard	
	Time (s)	Rank	Time (s)	Rank	Time (s)	Rank	Time (s)	Rank
Tester 1	194	3	154	1	212	2	250	4
Tester 2	267	3	181	2	187	1	484	4
Tester 3	191	1	207	2	150	3	253	4
Tester 4	380	3	235	2	336	1	600	4
Tester 5	180	2	107	1	148	3	283	4
Tester 6	471	3	268	2	251	1	439	4
Tester 7	161	2	148	1	199	3	386	4
Tester 8	199	3	141	1	189	2	600	4
Tester 9	600	1	600	3	361	4	600	2
Tester 10	203	1	233	2	247	3	600	4
Average	**284.6**	**2.2**	**227.4**	**1.7**	**228**	**2.3**	**449.5**	**3.8**
Intuitiveness	**0.387**		**0.522**		**0.457**		**0.143**	

The analysis of collected opinions of the testers lead to several other conclusions. First of all almost all testers pointed that using Leap-Motion tires hand, which limits maximum time of precise control. This issue can probably be solved by using properly designed armrest.

Moreover, different testers complained about too large or too little sensitivity of joystick and Leap Motion. Different ranges of peoples' movements can probably be automatically compensated by signal filtering and controllers adaptation.

6 Conclusions

Fast development of sensing devices able to collect data about human actions creates new possibilities for building robot teleoperation interfaces. The simple experiments presented in this paper show that use of modern sensing devices can improve both the efficiency and the intuitiveness of the teleoperation. Use of the LeapMotion sensor instead of a joystick for controlling the 6-DOF arm with gripper resulted in 14 % gain in intuitiveness.

Variety of modern data acquisition devices encourages further research in this area. These can include tests involving other devices and development of new methods of communicating with robots which would go beyond direct control of particular actuators.

Acknowledgments The research leading to these results has received funding from the Polish National Science Centre under the grant no. 2011/01/D/ST6/06146 and from the AGH-UST statutory fund no. 11.11.230.124.

References

1. Liu, Y., Nejat, G.: Robotic urban search and rescue: a survey from the control perspective. J. Intell. Robot. Syst. **72**(2), 147–165 (2013)
2. Nielsen Norman Group: Usability articles. http://www.nngroup.com/articles/ (2015). Accessed 2015–09-28
3. Mano, H., Kon, K., Sato, N., Ito, M., Mizumoto, H., Goto, K., Chatterjee, R., Matsuno, F.: Treaded control system for rescue robots in indoor environment. In: IEEE International Conference on Robotics and Biomimetics. ROBIO 2008, pp. 1836–1843. IEEE (2009)
4. Hu, A.-P., McMurray, G., Matthews, J.M., Marshall, M.: Systems and methods for operating robots using visual servoing (2012). US Patent App. 13/584,594
5. Fernandes, V.B.P.: Wearable intertace for teleoperation of robotarms—WITRA. https://challenge.toradex.com/projects/10157-wearable-intertace-for-teleoperation-of-robot-arms-witra (2015). Accessed 2015–09-28
6. Surez-Ruiz, F., Galiana, I., Tenzer, Y., Jentoft, L., Howe, R., Ferre, M.: Grasp mapping between a 3-finger haptic device and a robotic hand. In: Auvray, M., Duriez, C. (eds.) Haptics: Neuroscience. Devices, Modeling, and Applications, Lecture Notes in Computer Science, vol. 8618, pp. 275–283. Springer, Berlin (2014)
7. Paszkiel, S., Błachowicz, A.: The use of BCI for controlling the mobile robot. Pomiary, Automatyka, Robotyka **16**(2), 270–274 (2012)
8. Szominski, S., Gadek, K., Konarski, M., Blaszczyk, B., Anielski, P., Turek, W.: Development of a cyber-physical system for mobile robot control using erlang. In: 2013 Federated Conference on Computer Science and Information Systems (FedCSIS), pp. 1441–1448, Sept 2013

Robot System Design Procedure Based on a Formal Specification

Cezary Zieliński and Maksym Figat

Abstract This paper presents a robot control system design procedure based on a formal specification. The specification describes both the structure and the activities of a robot system in a formal manner. Using this specification the code of the robot system controller can be produced. The paper shows how. Moreover it has been shown that the complexity of the robot control system is concentrated in two of its elements: transition functions and the structure of the FSM. The reduction of the complexity of transition functions simultaneously increases the number of states of the FSM and vice versa, so it is up to the designer of the system to choose the adequate complexity level of this pair.

Keywords Robot system specification · Controller code generation

1 Introduction

The diversity of robot hardware and its applications is vast. New system configurations and requirements necessitate either new robot control system design or modifications of the already existing ones. Writing and maintaining control software is a tedious process, thus any tools facilitating this process are more than welcome. This paper focuses on a robot control system development procedure based on a formal specification of both the structure of the created system and its activities. Conversion from the specification to the resulting controller code is also delved into. Specification of the software that is to be created is of paramount importance to the discussion of what the designer wants to create. For that purpose a specification language has to be defined, so that the parties taking part in such a discussion have

C. Zieliński (✉) · M. Figat
Warsaw University of Technology, Warsaw, Poland
e-mail: C.Zielinski@ia.pw.edu.pl
URL: http://robotics.ia.pw.edu.pl

M. Figat
e-mail: M.Figat@ia.pw.edu.pl

R. Szewczyk et al. (eds.), *Challenges in Automation, Robotics and Measurement Techniques*, Advances in Intelligent Systems and Computing 440, DOI 10.1007/978-3-319-29357-8_45

a common understanding of the discussed topic. Just this makes the subject interesting, but it becomes even more so if the resulting specification can be readily converted into code of the controller. To devise automatic means of code generation first design and architectural patterns [1] must be produced. This paper focuses on this topic.

2 State of the Art

Designing robot systems is challenging, because integration of knowledge from different domains is required. As a result there is a growing demand for DSLs (Domain Specific Languages) in robotics. DSL is a specialised language, which is focused on a certain domain [2]. Usually a complex program can not be written by using just one DSL. A GPL (General Purpose Language) supplemented with several different DSLs is necessary. DSLs are valuable, because they are focused on the domain that the program pertains to, thus they facilitate the process of program development. They also improve productivity of a developer and communication between domain experts. The survey of DSLs [3] classifies known robot DSL approaches using Precision Placement Test (PPT). This paper shows that the majority of DSLs concentrates on coordination, motion control and robot structures, while almost none deal with perception, reasoning, planning, manipulation and grasping or control architectures.

Usually robot system development is divided into system specification, implementation and testing. The process of robot system development was formalized and structured in the BRICS project [4]. In [5] a complete process of robot-control architecture development based on a model-driven engineering approach was presented. Many different tools were created in order to integrate the process of creation of robotic systems. An appropriate system specification can be easily transformed into its implementation. One of the well known tools, which enables code generation, is a Genom [6]. It is meant for designing real-time architectures, providing an encapsulation of software functions into independent components, integration of heterogeneous functions with different real-time constraints, integration of homogeneous functions into a control architecture with typical behaviours (starting, ending and error handling) and standard interfaces. Another tool enabling code generation is Robotic Modeling Language (RobotML) [7]. It eases the whole process of robot system design, simulation and deployment to several platforms. Further, MontiArcAutomaton framework [8] provides integrated, platform independent structure and behaviour modelling language with an extensible code generation. Specification of a robot system heavily relies on the knowledge about the task that is to be realised by that system. The paper [9] focuses on the knowledge organised as an ontology created for robotic tasks. The robot task is expressed using a language resembling natural languages [10]. An attempt of code generation from symbolic descriptions of the object was introduced in [11]. In robotics, where different subdomains require code generation, a conceptual model for code generator composition is required [12]. Another framework integrating robotic subdomains, such as kinematics, dynamics, decision making algorithms, interfaces to hardware, is Operational Software Components for Advanced Robotics (OSCAR) [13]. It requires significant manual labour, thus another method for rapid assembly of software was introduced: Feature Oriented Programming (FOP) [14]. It models and specifies programs in term of features, i.e. capabilities of the software, what is significant for the end user. An attempt of applying FOP to robot controller software was presented in [15], where the product variants were generated from a declarative specification.

Robotics is decomposed into many different subdomains. The imbalance of development of certain subdomains is well visible. Still not much research on robotic system specification is conducted. Existing tools, which tread on this topic are primarily focused on code generation instead of defining a method of problem specification in a formal manner. Such a tool is necessary for a systematic discussion of design variants. Moreover, the specification method should be easy to transform into implementation code. Thus this paper focuses on robot control system design procedure based on a formal specification.

3 An Embodied Agent

Each robotic system can be represented as a set of agents a_j, where $j = 1, \dots, n_a$, is a designator of an agent and n_a is the cardinality of the set of agents. Agents having physical bodies are named embodied agents [16–19]. An embodied agent is composed of five types of subsystems (Fig. 1):

- Real effectors $E_{j,h}$—are the devices affecting the environment; they are designated as $h = 1, \dots, n_E$,

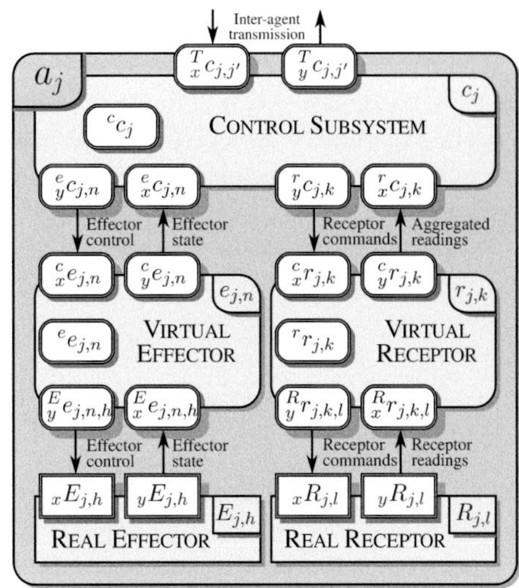

Fig. 1: Internal structure of an agent a_j

- Real receptors $R_{j,l}$—are the devices gathering information about the environment; they are designated as $l = 1, \dots, n_R$,
- Virtual effectors $e_{j,n}$—present the real effectors to the control subsystem in an adequate form; each virtual effector has its own identifier $n = 1, \dots, n_e$,
- Virtual receptors $r_{j,k}$—acquire sensor data from the real receptors and transform it into an aggregated form required by the control subsystem; each virtual receptor has its own identifier $k = 1, \dots, n_r$,
- Control subsystem c_j—realizes the task of the agent, coordinates the associated subsystems and communicates with the other agents $a_{j'}, j \neq j'$.

An embodied agent contains only a single control subsystem, while the number of other subsystems can be zero or more. The general structure of an embodied agent is presented in Fig. 1. The control subsystem as well as the virtual effectors and receptors use three types of buffers. Buffers to transmit and receive data to/from other components, and an inner buffer called the internal memory.

Herein a systematic denotation method is used to name both the subsystems and their buffers. In the assumed notation a one-letter symbol located in the centre (i.e E, R, e, r, c) designates a subsystem. To reference its buffers or to single out the state of this component at a certain instant of time extra indices are placed around this central symbol. The left superscript designates the subsystem to which the buffer is connected. The right superscript designates the time instant at which the state is being considered. The left subscript tells us whether this is an input (x) or an output (y) buffer and if the left subscript is missing it refers to the internal memory of the subsystem. The right subscript contains elements separated by comas. They represent respectively: agent name, subsystem name and buffer element name. Buffer elements can also be referenced by placing their names in square brackets to the right of the main symbol. Example: ${}^{c}_{x}e^{i}_{j,1}$ denotes the contents of the first ($n = 1$) virtual effector input buffer of the agent a_j transferred from the control subsystem at an instant i. In the same manner functions are denoted. The central symbol for any function is f, the left superscript designates the owner of the function and the subsystem that this function produces the result of its computations for, the right superscript: τ, σ refer to the terminal and initial conditions respectively (each one of them being a predicate). When the right superscript is missing then the function is a transition function. The list of right subscripts designates a particular function.

A transition function of a subsystem $s \in \{c, e, r\}$ of the agent a_j is defined as:

$$\left[{}^{s}s^{i+1}_{j,\zeta},\ {}_{y}s^{i+1}_{j,\zeta}\right] := {}^{s}f_{j,\zeta,\omega}({}^{s}s^{i}_{j,\zeta},\ {}_{x}s^{i}_{j,\zeta}) \tag{1}$$

where ω is the designator of a particular transition function. The symbol ζ disappears for the control subsystem and stands for n in the case of a virtual effector and k in the case of a virtual receptor (because there is only one control subsystem for an agent and possibly many virtual effectors and many virtual receptors, which must be distinguished). The calculation of a transition function must be associated with the input of its arguments and output of its results, thus an elementary action ${}^{s}\mathcal{A}_{j,\zeta,\omega}$ consists in:

- calculation of the transition function (1) for a given subsystem. This function takes as input the current data from the internal memory ${}^{s}s^{i}_{j,\zeta}$ and the input buffers ${}_{x}s^{i}_{j,\zeta}$, and calculates the values which are subsequently inserted into the output buffers ${}_{y}s^{i+1}_{j,\zeta}$ and the internal memory ${}^{s}s^{i+1}_{j,\zeta}$,
- ${}_{y}s^{i+1}_{j,\zeta} \rightarrow {}_{x}s_{j^{\bullet}}$, i.e. output of the results to the associated subsystems, where $j^{\bullet} \in \{j, j'\}$, denotes all subsystems connected to s_j through appropriate buffers,
- ${}_{y}s_{j^{\bullet}} \rightarrow {}_{x}s^{i}_{j,\zeta}$, i.e. input of the current state from the connected subsystems.

It should be noted that there are as many transition functions as there are elementary actions defined for a particular subsystem. A behaviour of a subsystem consists in an iterative execution of an elementary action. Each iteration has the same duration, defining the sampling period of the considered subsystem. The iterations are terminated when a certain predicate ${}^{s}f^{\tau}_{j,\zeta,\xi}$, called the terminal condition is fulfilled (ξ is the designator of a particular terminal condition).

A behaviour ${}^{s}\mathcal{B}_{j,\zeta,\gamma}$ is defined by an elementary action ${}^{s}\mathcal{A}_{j,\zeta,\omega}$ and a terminal condition ${}^{s}f^{\tau}_{j,\zeta,\xi}$, where γ and ω are the designators of a specific behaviour and elementary action respectively.

$$ {}^{s}\mathcal{B}_{j,\zeta,\gamma} \triangleq {}^{s}\mathcal{B}_{j,\zeta,\gamma}\left({}^{s}\mathcal{A}_{j,\zeta,\omega}, {}^{s}f^{\tau}_{j,\zeta,\xi}\right), \tag{2} $$

As any subsystem can execute many behaviours, there must be a method of switching between them. Once a certain behaviour is terminated, the choice of the next behaviour is based on a predicate named an initial condition ${}^{s}f^{\sigma}_{j,\zeta,\alpha}$, where $\alpha = 1, \ldots, {}^{s}n_{\sigma}$ and ${}^{s}n_{\sigma}$ is the overall number of initial conditions of a subsystem. The activities of each subsystem are described by a finite state machine (FSM). With each of its states a behaviour is associated. When the terminal condition ${}^{s}f^{\tau}_{j,\zeta,\xi}$ is satisfied the behaviour ${}^{s}\mathcal{B}_{j,\zeta,\gamma}$ terminates iterating the elementary action ${}^{s}\mathcal{A}_{j,\zeta,\omega}$ and the FSM switches from the state ${}^{s}S_{j,\zeta,u}$ to ${}^{s}S_{j,\zeta,\beta}$, if the initial condition ${}^{s}f^{\sigma}_{j,\zeta,\alpha}$ is fulfilled (u and β are the FSM state designators).

There are two types of behaviours:

- ${}^{s}_{*}\mathcal{B}_{j,\zeta,\gamma}$—which repeats its defining elementary action zero or more times, i.e. its terminal condition testing before calculating the transition function.
- ${}^{s}_{+}\mathcal{B}_{j,\zeta,\gamma}$—which repeats its defining elementary action one or more times, i.e. its terminal condition testing after calculating the transition function.

An FSM is represented by a state transition table, a graph or a transition vector. The columns of the transition table are labeled by initial conditions whereas the rows by the FSM states. Each table item contains the next FSM state. In the graph approach, states are represented by nodes and the directed arcs between nodes are labeled by initial conditions. If the number of initial conditions ${}^{s}n_{\sigma}$ is large the transition tables tend to be huge, as their columns are labeled by all possible variations with repetitions of their values (0 or 1). On the other hand graphs are good for illustration, but directly not suitable for future implementation. Thus an intermediate form has been devised—a transition vector. A transition vector ${}^{s}L_{j,\zeta}$ is defined as a vector containing lists of pairs composed of the next FSM state and its initial condition.

$$ {}^{s}L_{j,\zeta} = [\,{}^{s}L_{j,\zeta,1}, \dots , {}^{s}L_{j,\zeta,n_S}], \tag{3} $$

where n_S is the number of states of the FSM of a subsystem s, and ${}^{s}L_{j,\zeta,u}$, $u = 1, \dots , n_S$, is a list of pairs:

$$ {}^{s}L_{j,\zeta,u} = [({}^{s}f^{\sigma}_{j,\zeta,\alpha}, {}^{s}S_{j,\zeta,\beta}), \dots] \tag{4} $$

Each list ${}^{s}L_{j,\zeta,u}$ contains pairs composed of the next FSM state ${}^{s}S_{j,\zeta,\beta}$ and the initial condition ${}^{s}f^{\sigma}_{j,\zeta,\alpha}$. If the FSM is in the state ${}^{s}S_{j,\zeta,u}$, and the terminal condition of the associated behaviour is satisfied, then if the initial condition ${}^{s}f^{\sigma}_{j,\zeta,\alpha}$ is satisfied, the FSM switches to a successor state ${}^{s}S_{j,\zeta,\beta}$. The transition vector ${}^{s}L_{j,\zeta}$ as defined by (3) and (4) represents the graph of an FSM, but without an association of its states and behaviours. If with each state a behaviour is associated a Moore type automaton will result. The mentioned association is produced by a behaviour vector (equivalent of an output table of a Moore type automaton):

$$ {}^{s}\mathcal{O}_{j,\zeta} = [\,{}^{s}\mathcal{B}_{j,\zeta,u}, \dots , {}^{s}\mathcal{B}_{j,\zeta,\gamma}] \tag{5} $$

where the number of components of this vector is n_S. The ordering of components in the transition vector and behaviour vector is the same, i.e. the respective components of those vectors pertain to the same state, thus a one–to–one relationship occurs.

4 Procedure of Developing Robot System Specification

The main purpose of this specification method is to describe the robot system in a formal manner, so that based on this specification we can generate the code of the robotic system controller. System design is always an iterative process. Below an iterative method of specifying a robot systems is presented. The design process progresses both by stepwise refinement and backtracking, where an increased detail of specification possibly requires modifications of the previous stages in the project. In the following the stages of the procedure are enumerated, but it is up to the designer to choose its starting point and the level of the required detail. In other words, all of the enumerated stages must be realised, but their order is up to the designer's preference.

The primary input for the design procedure is the task or class of tasks that the robotic system has to execute. The task has to be expressed at an ontological level adequate for future implementation of the system. Thus either the description should be at an ontological level involving the description of hardware used or at the level of concepts appearing in the environment of the system. In the latter case a transformation from the actions performed by environmental objects to the actions of the robot hardware must be provided [20, 21].

The design procedure consists of the following stages:

- Taking into account the class of tasks that is to be realised choose the necessary effectors and receptors—those will be the elements of the real effectors and real receptors,
- Again, taking into account the considered class of tasks and the chosen hardware subdivide the system into a number of agents and assign to each one of them the real effectors and receptors (in the process of division take into account the transmission delays of inter-agent transmission and the necessary computational power of the hardware on which the agent's code will be executed),
- Define the virtual receptors and effectors (i.e. views of receptors and effectors that will be presented to the control subsystem),
- Define the contents of the communication buffers and the internal memory of each subsystem (it should be noted that the output buffer of one subsystem has to be exactly the same as the input buffer of the one it is connected to),
- Define the sampling rate of each subsystem,
- Determine the behaviours of each subsystem,
- Define the transition function and the terminal condition defining each behaviour,
- Build an FSM switching the behaviours for each subsystem—at this stage initial conditions should be defined too.

It is obvious that an FSM can have fewer states if the associated behaviours are more complex or the FSM will have more states if the behaviours are simplified and thus more numerous. This is one of many design decisions that have to be taken. The number of agents is also dictated by their complexity—again complexity versus numerousness. In the design process, those decisions can be, and often are, changed, thus an iterative procedure with backtracking is utilised. The definition of the transition function and the terminal as well as the initial conditions is based on their arguments, which belong to the communication buffers and the internal memory, thus often the definition of the structure of the buffers and the definitions of those functions are created in an iterative process.

The above procedure is highly dependent on the task to be realised. However some hints can be provided as to how to deal with certain aspects. An FSM can be specified as the transition vector (3). The number of states of an FSM is equal to the number of behaviours of the considered subsystem. The directed arcs protruding from each state are labeled by initial conditions. The set of those initial conditions is usually a small subset of all initial conditions. Moreover those initial conditions have to fulfill the following conditions to form a well defined set.

- For any state ${}^{s}S_{j,\zeta,u}$ any two initial conditions ${}^{s}f^{\sigma}_{j,\zeta,\alpha_1}$ and ${}^{s}f^{\sigma}_{j,\zeta,\alpha_2}$ appearing on the list of successor pairs ${}^{s}L_{j,\zeta,u}$, where $\alpha_1 \neq \alpha_2$, $\alpha_1, \alpha_2 = 1, \ldots, {}^{s}n_{\sigma}$:

$$ {}^{s}f^{\sigma}_{j,\zeta,\alpha_1} \wedge {}^{s}f^{\sigma}_{j,\zeta,\alpha_2} = \text{FALSE} \tag{6} $$

This ensures that the designed FSM is deterministic, i.e. there is no ambiguity as to the choice of the next state.

- For each state ${}^{s}S_{j,\zeta,u}$ the logical sum of all initial conditions appearing on the list of successor pairs ${}^{s}L_{j,\zeta,u}$:

$$ {}^{s}f^{\sigma}_{j,\zeta,\alpha_1} \vee {}^{s}f^{\sigma}_{j,\zeta,\alpha_2} \vee \cdots \vee {}^{s}f^{\sigma}_{j,\zeta,\alpha_{n_\alpha}} = \text{TRUE} \tag{7} $$

  where n_α is the number of pairs in the ${}^{s}L_{j,\zeta,u}$ list. This ensures that there is always a transition from the current state of the FSM, i.e. at least one of the initial conditions is TRUE. Both conditions (6) and (7) together assure that always there is one and only one such transition. Formula (7) also assures that there is at least one directed arc emerging from each node of the graph representing the considered state ${}^{s}S_{j,\zeta,u}$ (then the initial condition labeling this single directed arc is labeled by TRUE).

The sampling period of virtual effectors and virtual receptors (single iteration period of the behaviours of the considered subsystem) depends on the hardware employed. The choice of hardware in the case of virtual receptors should take into account both the data acquisition time and the data aggregation time. In the case of virtual effectors both the delay introduced by the hardware driver and the control data transformation time should be taken into account. Those times should be computed based on the allowable reaction time of the system. In general the faster the sampling rate the better—this improves both reactivity and stability of the system. However one should note that an increase in sampling rate usually requires more expensive equipment. Based on the sampling periods of the virtual effectors and receptors the sampling period of the control subsystem is established. Synchronization of the operation of the subsystems was discussed in [22].

5 Transformation of the Specification into Code

The advantage of a specification expressed in terms defined in the previous sections is that those terms can be directly transformed into code using patterns. Herein the following assumptions are made.

- As the transition functions ${}^{s}f_{j,\zeta,\omega}$, terminal ${}^{s}f^{\tau}_{j,\zeta,\xi}$ and initial ${}^{s}f^{\sigma}_{j,\zeta,\alpha}$ conditions are defined in terms of mathematical concepts and operators their transformation into code is straightforward.
- The name of a function and its designator (e.g. an address or a pointer to it) will not be distinguished here, both for the purpose of brevity and because of contextual obviousness.
- Vectors ${}^{s}L_{j,\zeta}$ and ${}^{s}\mathcal{O}_{j,\zeta}$ are indexed by states ${}^{s}S_{j,\zeta,\beta}$ instead of ordinal numbers, for clarity.
- Behaviours ${}^{s}\mathcal{B}_{j,\zeta,\gamma}$ are represented as triplets $(\mathtt{bt}, {}^{s}f_{j,\zeta,\omega}, {}^{s}f^{\tau}_{j,\zeta,\xi})$, where $\mathtt{bt}$ stands for behaviour type, i.e. ${}^{s}_{*}\mathcal{B}$ or ${}^{s}_{+}\mathcal{B}$; $\mathtt{bt} \in \{*, +\}$. Instead of an elementary action

${}^{s}\mathcal{A}_{j,\zeta,\omega}$ a corresponding transition function ${}^{s}f_{j,\zeta,\omega}$ can be used directly in this data structure, because there is only one pattern of a function executing an elementary action and with just one parameter (${}^{s}f_{j,\zeta,\omega}$) at that.

- Three functions are defined: `first`, `second`, `third` extracting the first, second and third element of an n-tuple respectively.
- All the functions have access to all subsystem buffers, i.e. ${}^{s}s^{i}_{j,\zeta}$ and ${}_{x}s^{i}_{j,\zeta}$.

The code executing any elementary action ${}^{s}\mathcal{A}_{j,\zeta,\omega}$ is defined in (listing 1.1). It calls the specific transition function (updating internal memory ${}^{s}s^{i+1}_{j,\zeta}$ and output buffers ${}_{y}s^{i+1}_{j,\zeta}$), propagates data to the other subsystems and receives the data from the connected subsystems. This code pattern relies on (1).

Listing 1.1: Definition of an Elemntary Action pattern

```
void ${}^{s}\mathcal{A}_{j,\zeta}({}^{s}f_{j,\zeta,\omega})${
/*Calculate the new output values*/
  [${}^{s}s^{i+1}_{j,\zeta}$, ${}_{y}s^{i+1}_{j,\zeta}$] := ${}^{s}f_{j,\zeta,\omega}({}^{s}s^{i}_{j,\zeta}, {}_{x}s^{i}_{j,\zeta})$;
/*Send the results to the associated subsystems*/
    ${}_{y}s^{i+1}_{j,\zeta} \rightarrow {}_{x}s_{j\cdot}$;
    i := i+1;
/*Get the data from the associated subsystems*/
    ${}_{y}s_{j\cdot} \rightarrow {}_{x}s^{i}_{j,\zeta}$;
}
```

There are two types of behaviours presented in (listings 1.2–1.3). Each one of them iteratively calls the elementary action and checks the terminal condition. The one in (listing 1.2) executes the elementary action at least once, whereas the one in (listing 1.3) might not be executed at all. The number of iterations depends on the current value of the terminal condition.

Listing 1.2: Behaviour ${}^{s}_{+}\mathcal{B}_{u}$ pattern

```
void ${}^{s}_{+}\mathcal{B}_{j,\zeta}({}^{s}f_{j,\zeta,\omega}, {}^{s}f^{\tau}_{j,\zeta,\xi})${
   do{
   /*Execute an elementary action*/
      ${}^{s}\mathcal{A}_{j,\zeta}({}^{s}f_{j,\zeta,\omega})$;
   }
   until(${}^{s}f^{\tau}_{j,\zeta,\xi}$);
}
```

Listing 1.3: Behaviour ${}^{s}_{*}\mathcal{B}_{u}$ pattern

```
void ${}^{s}_{*}\mathcal{B}_{j,\zeta}({}^{s}f_{j,\zeta,\omega}, {}^{s}f^{\tau}_{j,\zeta,\xi})${
   while(!${}^{s}f^{\tau}_{j,\zeta,\xi}$){
   /*Execute an elementary action*/
      ${}^{s}\mathcal{A}_{j,\zeta}({}^{s}f_{j,\zeta,\omega})$;
   }
```

```
}
```

For each agent j (a_j) and subsystem s an FSM function is defined in accordance with the pattern outlined in (listing 1.4). At system initiation this function is called with a parameter reflecting the required initial state of the FSM, e.g. ${}^sS_{j,\zeta,1}$. The state machine switches control between FSM states based on the current value of the cs (current state) variable (equivalent to the designator of a node of the graph of the FSM).

Listing 1.4: FSM pattern

```
void FSM(${}^sL_{j,\zeta}$, ${}^s\mathcal{O}_{j,\zeta}$,cs){
   while(true){
      cs = node(${}^sL_{j,\zeta}$, ${}^s\mathcal{O}_{j,\zeta}$, cs)
   }
}
```

The node function is the workhorse of the subsystem. First it tests whether the ${}^s_+\mathcal{B}$ or ${}^s_*\mathcal{B}$ type of behaviour should be invoked. Once the decision is made the right behaviour type is used with the adequate transition function and terminal condition as arguments. Upon termination of this behaviour the next FSM state must be chosen. This is done on the basis of the initial condition being the first element of the pairs on the list ${}^sL_{j,\zeta,\beta}$.

Listing 1.5: Node pattern

```
state node(${}^sL_{j,\zeta}$, ${}^s\mathcal{O}_{j,\zeta}$, ${}^sS_{j,\zeta,\beta}$) {
      /*Execute the correct type of behaviour*/
      if (first(${}^s\mathcal{O}_{j,\zeta}[{}^sS_{j,\zeta,\beta}]) = +$)
         then ${}^s_+\mathcal{B}_{j,\zeta}$(second(${}^s\mathcal{O}_{j,\zeta}[{}^sS_{j,\zeta,\beta}]$), third(${}^s\mathcal{O}_{j,\zeta}[{}^sS_{j,\zeta,\beta}]$))
         else ${}^s_*\mathcal{B}_{j,\zeta}$(second(${}^s\mathcal{O}_{j,\zeta}[{}^sS_{j,\zeta,\beta}]$), third(${}^s\mathcal{O}_{j,\zeta}[{}^sS_{j,\zeta,\beta}]$));
   /*Initiate the list elemnt pointer*/
      le = ${}^sL_{j,\zeta}[{}^sS_{j,\zeta,\beta}]$;
      while(le != NIL) {
         /*Test the initial condition to search for
            the next state*/
         if (first(${}^sL_{j,\zeta,\beta}$)) then
            return second(${}^sL_{j,\zeta,\beta}$) /*next state*/
         else le = next (${}^sL_{j,\zeta,\beta}$[le]);
      }
}
```

where next extracts the pointer to the next element. Hence the whole of the specification can be transformed into code of a high-level procedural or object-oriented programming language.

6 Conclusions and Future Work

This paper shows that the complexity of robot control systems is concentrated in two of its elements: transition functions and the structure of the FSM. In reality, for the same task the reduction of the complexity of one of those elements simultaneously increases the complexity of the other. If the definitions of transition functions are kept simple, the number of them increases and thus the number of behaviours increases, increasing the number of FSM states. On the other hand if the designer prefers complex multi-tiered definitions of transition functions the number of behaviours decreases and thus the number of states of the FSM. Besides that the complexity of the structure of the control system is low—it relies on simple patterns shown in listings 1.1–1.5. Both the definitions of transition functions and the data structures describing the FSM graph rely on mathematical concepts that can be easily transformed into a programming language code. All this encourages us now to pursue the path heading towards automatic code generation of robot control systems.

The correctness of the presented pattern of developing specification was verified on many systems, e.g. recently on a mobile robot REX [23] and on a system with variable multi-agent structure, where not only its structure varied, but also the supervisory responsibilities switched between agents [24].

Acknowledgments This project was funded by the National Science Centre grant number DEC-2012/05/D/ST6/03097.

References

1. Kaisler, S.: Software Paradigms. Wiley Interscience (2005)
2. Fowler, M.: Domain Specific Languages, 1st edn. Addison-Wesley Professional (2010)
3. Nordmann, A., Hochgeschwender, N., Wrede, S.: A survey on domain-specific languages in robotics. In: Brugali, D., Broenink, J., Kroeger, T., MacDonald, B. (eds.) Simulation, Modeling, and Programming for Autonomous Robots. Lecture Notes in Computer Science, vol. 8810, pp. 195–206. Springer International Publishing (2014)
4. Bischoff, R., Guhl, T., Prassler, E., Nowak, W., Kraetzschmar, G., Bruyninckx, H., Soetens, P., Haegele, M., Pott, A., Breedveld, P., Broenink, J., Brugali, D., Tomatis, N.: Brics—best practice in robotics. In: Robotics (ISR), 2010 41st International Symposium on and 2010 6th German Conference on Robotics (ROBOTIK), pp. 1–8 (2010)
5. Trojanek, P.: Model-driven engineering approach to design and implementation of robot control system. In: 2nd International Workshop on Domain-Specific Languages and models for ROBotic systems (DSLRob'11), arXiv preprint arXiv:1302.5085 (Sept 2011)
6. Fleury, S., Herrb, M., Chatila, R.: Genom: a tool for the specification and the implementation of operating modules in a distributed robot architecture. In: Proceedings of the 1997 IEEE/RSJ International Conference on Intelligent Robots and Systems. IROS '97, vol. 2, pp. 842–849 (1997)
7. Dhouib, S., Kchir, S., Stinckwich, S., Ziadi, T., Ziane, M.: Robotml, a domain-specific language to design, simulate and deploy robotic applications. In: Simulation, Modeling, and Programming for Autonomous Robots, pp. 149–160. Springer (2012)

8. Ringert, J.O., Rumpe, B., Wortmann, A.: Montiarcautomaton: modeling architecture and behavior of robotic systems. In: In Workshops and Tutorials Proceedings of the 2013 IEEE International Conference on Robotics and Automation (ICRA), 6–10 May 2013
9. Stenmark, M., Malec, J.: Knowledge-based instruction of manipulation tasks for industrial robotics. Robot. Comput.-Integr. Manuf. **33**, 56–67 (2015). Special Issue on Knowledge Driven Robotics and Manufacturing
10. Stenmark, M., Nugues, P.: Natural language programming of industrial robots. In: 2013 44th International Symposium on Robotics (ISR), pp. 1–5 (2013)
11. Stenmark, M., Malec, J., Stolt, A.: From high-level task description to executable robot code. In: Proceedings of the 7th IEEE International Conference Intelligent Systems IS'2014, pp. 189–202 (2014)
12. Ringert, J.O., Roth, A., Rumpe, B., Wortmann, A.: Code generator composition for model-driven engineering of robotics component and connector systems. In: Proceedings of the1st International Workshop on Model-Driven Robot Software Engineering (MORSE 2014) (2014)
13. Kapoor, C., Tesar, D.: A reusable operational software architecture for advanced robotics. In: 12th CSIM-IFToMM Symposium on Theory and Practice of Robots and Manips (1998)
14. Batory, D.: A tutorial on feature oriented programming and product-lines. In: Proceedings of the 25th International Conference on Software Engineering. ICSE '03, pp. 753–754. IEEE Computer Society (2003)
15. Jung, E., Kapoor, C., Batory, D.: Automatic code generation for actuator interfacing from a declarative specification. In: 2005 IEEE/RSJ International Conference on Intelligent Robots and Systems (IROS 2005), pp. 2839–2844 (2005)
16. Brooks, R.A.: Intelligence without reason. Artif. Intell.: Crit. Concepts **3**, 107–163 (1991)
17. Zieliński, C., Winiarski, T.: General specification of multi-robot control system structures. Bull. Polish Acad. Sci.—Tech. Sci. **58**(1), 15–28 (2010)
18. Kornuta, T., Zieliński, C.: Robot control system design exemplified by multi-camera visual servoing. J. Intell. Robot. Syst. **77**(3–4), 499–524 (2015)
19. Zieliński, C., Kornuta, T., Winiarski, T.: A systematic method of designing control systems for service and field robots. In: 19-th IEEE International Conference on Methods and Models in Automation and Robotics, MMAR'2014, pp. 1–14. IEEE
20. Zieliński, C., Kornuta, T.: An object-based robot ontology. In: Advances in Intelligent Systems and Computing (AISC), vol. 323, pp. 3–14. Springer (2015)
21. Kornuta, T., Winiarski, T., Zieliński, C.: Specification of abstract robot skills in terms of control system behaviours. In: Szewczyk, R., Zieliński, C., Kaliczyńska, M. (eds.) Progress in Automation, Robotics and Measuring Techniques. Vol. 2 Robotics. Advances in Intelligent Systems and Computing (AISC), vol. 351, pp. 139–152. Springer (2015)
22. Trojanek, P., Kornuta, T., Zieliński, C.: Design of asynchronously stimulated robot behaviours. In: Kozłowski, K. (ed.) 9th Workshop on Robot Motion and Control (RoMoCo), vol. 129–134 (2013)
23. Janiak, M., Zieliński, C.: Control system architecture for the investigation of motion control algorithms on an example of the mobile platform rex. Bull. Polish Acad. Sci.—Tech. Sci. **63**(3), 667–678 (2015)
24. Szlenk, M., Zieliński, C., Figat, M., Kornuta, T.: Reconfigurable agent architecture for robots utilising cloud computing. In: Szewczyk, R., Zieliński, C., Kaliczyńska, M. (eds.) Progress in Automation, Robotics and Measuring Techniques. Vol. 2 Robotics. Advances in Intelligent Systems and Computing (AISC), vol. 351, pp. 253–264. Springer (2015)

Performance of Coaxial Propulsion in Design of Multi-rotor UAVs

Adam Bondyra, Stanisław Gardecki, Przemysław Gąsior and Wojciech Giernacki

Abstract There are many different types of propulsion systems developed for multi rotor *UAV*s. One of the most interesting designs is so called X8 quadrocopter, which extends original quadrotor concept to 8 motors, arranged in 4 coaxial pairs. The advantage of this solution is increased lift of platform, with reasonable volume of platform kept. However, this design suffers from the loss of efficiency due to coaxial propellers' configuration, because the lower propeller loses thrust working in prop wash of upper propeller. This paper presents the experimental verification of performance of such propulsion system in practical terms of designing multi rotor platforms, comparing to design with 8 isolated propulsion units. In addition, its advantages versus classic quadrotor concept is shown. The series of experiments with different motors and sizes of propellers were conducted to estimate efficiency of coaxial propulsion regarding useful thrust generated by each configuration.

Keywords Coaxial propellers · X8 quadrotor · Multirotor · Octoquad

1 Introduction

The multirotor *UAV*s are widely used in many commercial and scientific applications. Depending from the purpose of given solution, different drone designs differ in size, lifting capabilities and maximum flight time [1–3]. One of the key issues in

A. Bondyra (✉) · S. Gardecki · P. Gąsior · W. Giernacki
Institute of Control and Information Engineering,
Poznań University of Technology, ul. Piotrowo 3A, 60-965 Poznań, Poland
e-mail: adam.bondyra@put.poznan.pl

S. Gardecki
e-mail: stanislaw.gardecki@put.poznan.pl

P. Gąsior
e-mail: przemyslaw.gasior@cie.put.poznan.pl

W. Giernacki
e-mail: wojciech.giernacki@put.poznan.pl

R. Szewczyk et al. (eds.), *Challenges in Automation, Robotics and Measurement Techniques*, Advances in Intelligent Systems and Computing 440, DOI 10.1007/978-3-319-29357-8_46

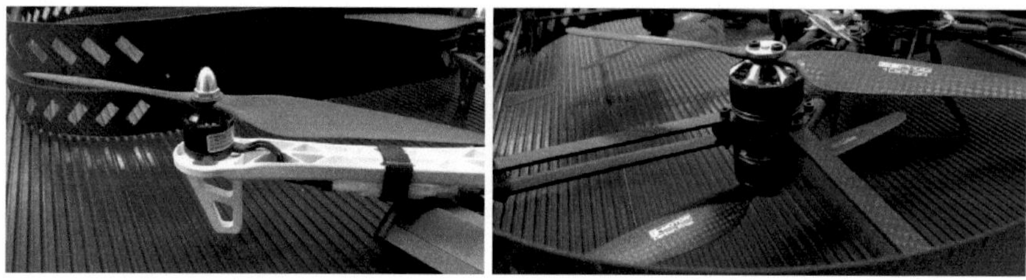

Fig. 1 Single propeller in *DJI Flamewheel 450* quadrotor and coaxial design in *Falcon V5* [9]

designing a micro multi-rotor vehicle is an adequate choice of propulsion system. The original quadrotor concept, introduced in 2000s [4], evolved into many various solutions. Nowadays, the tri-, quad-, hexa- and octocopters are available on the market, as well as many custom solutions with different propellers' configuration [5–7]. In our research concerning development of drones, we focus on few multirotor designs, with so called x8 quadrotor or octoquad among them. This configuration of multirotor extends original quadcopter concept by increasing the total thrust output of platform thanks to additional set of motors. On each side, there are two identical rotors installed, one above another. The propellers rotate in opposite directions, which equalizes the momentum of platform. The upper propeller works as a tractor, while the lower unit is a pusher. As a result, the total thrust of propulsion unit is increased with similar physical volume in comparison to single propeller [8].

However, considering that a lower propeller operates in a prop wash of upper unit, the total thrust performance of coaxial propulsion is lower comparing to two separated propellers. In this paper we present experimental verification of this performance in practical terms of multirotor design with different propellers' sizes and motors intended for various classes and sizes of multi rotor aerial platforms. A large data set of experimental results was obtained during the tests on a motor test bench. A series of over 60 tests was conducted where the thrust, rotational speed and motor's power were measured in relation to control signal. Later, the data was processed to show the efficiency loss and provide practical information about the performance of coaxial propulsion. The article is arranged in following order: related work and multi rotor designs are described in second chapter. Third chapter is a description of experimental setup on which our data was obtained. The series of experiments is shown in details as well. Next chapter presents the results and analysis of experimental data, when last two chapters contain evaluation of our octoquad designs and present plans for future research (Fig. 1).

2 Related Work

There are various designs of x8 quadcopter available on the market. These designs are usually focused on very high lifting capability. From some point, the logical way to keep reasonable size of platform and increase the total thrust output is to apply the x8 configuration. One of the most popular middle-sized octoquads is Dragan-Flyer X8 platform, equipped with coaxial pairs of 16 and 15 inch propellers. The platform weighs about 1.7 kg with payload capability of 0.8 kg and maximum flight time of 20 min without payload [10]. In work of Sharft et al., the ground effect for this specific *UAV* model was analyzed. In addition, the average loss of total thrust outcome of coaxial propulsion unit in comparison to design with 8 isolated propellers was estimated at about 14 %, with equal rotational speeds of both motors [11]. Another example of octoquad configuration in commercial platform is Hammer XB8 *UAV*, with excessive lifting capability of 15 kg in reasonable size of about 2 m in diameter. C. Simoes in his paper [8] focused on performance of coaxial propulsion in *UAV*s, describing efficiency of such propulsion design according to Glauert's theory with varying motors' speed or spacing and propellers' diameter and pitch. However, the analysis was made only for 12–14 inch propellers and one motor model, where current designs utilize wider range of sizes with different rotational speeds. In this paper, we present wider analysis of performance of coaxial propulsion in multi rotor *UAV*s, focusing on optimal choice of propulsion for practical applications.

3 Experimental Setup

In order to collect data required for analysis of propulsion units, the custom test bench was constructed, based on load cell for thrust measurements. The instrument described in [12] was modified and adapted to new propulsion units presented in following sections. The size of test bench was extended to fit bigger propellers and the laser tachometer, allowing measurement of motor's rotational speed, was introduced.

3.1 Test Bench

Constructed test bench is based on *L*-shaped steel frame with pivot in the connection point of both arms. The propulsion unit is mounted on top of the frame, while the load cell is located on opposite side of *L*-shape. With rotor producing thrust, the steel frame rotates and applies pressure on load cell. The *Control Board* module measures load cell's voltage using 12-bit ADC converter and handles the calibration process. Thrust measurement can be calibrated using remote PC software using set of precise weights. The load cell itself is supplied with ±11 V symmetrical voltage and provides measurement of up to 6 kg of force with resolution of 1000 divisions. The measurement system is equipped with safety circuit based on high-power *MOSFET* transistors allowing to cut off power in case of emergency. Supply voltage of propulsion system is measured using conventional multimeter with range of 0–40 V, resolution of 10 mV and accuracy of ±0.8 %+1 digit. To obtain the motor's current, we used digital clamp meter with measuring range of 40 A, resolution of 0.01 A and accuracy of ±2 %+5 digits. External laser tachometer with corresponding reflec-

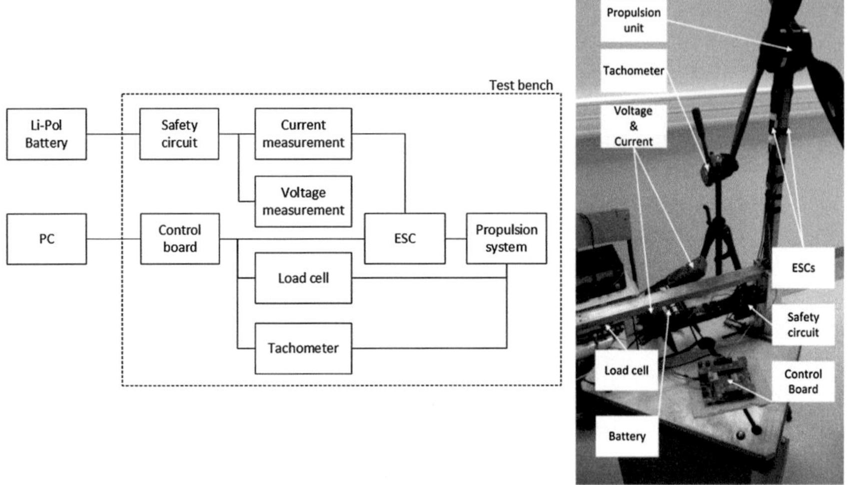

Fig. 2 Test bench block diagram and actual view

tive stickers on motors is used to measure rotational speed with resolution of 1 rpm and accuracy of ±0.05 % + one digit. This solution allows us to measure rotational speed up to 99 999 rpm. The tachometer's sampling time of 0.8 s is sufficient for this experiment. The concept and overview on test bench is shown in Fig. 2.

3.2 Analyzed Propulsion Units

Three different propulsion setups were tested, varying at size of propellers and motor models. Distance between propellers' disks in coaxial configuration was constant for each propulsion unit type. Each tested setup contains single or dual set of identical *BLDC* motors, propellers and *ESCs (Electronic Speed Controllers)*.

First Propulsion Setup

The first configuration of analyzed propulsion unit was based on *MN3110-15* brushless electric motor with a propeller with 10 inches of diameter and 3.3 inches of pitch. Single unit provides about 1.03 kg of thrust with nominal rotational speed of 11544 rpm when powered with 14.8 V lithium-polymer battery at maximum continuous power of 481 W.[1] This setup, considering propeller's size and motor power, is quite common in most small, market-available quadrotors.

[1] http://www.rctigermotor.com/html/2013/Navigator_0910/35.html.

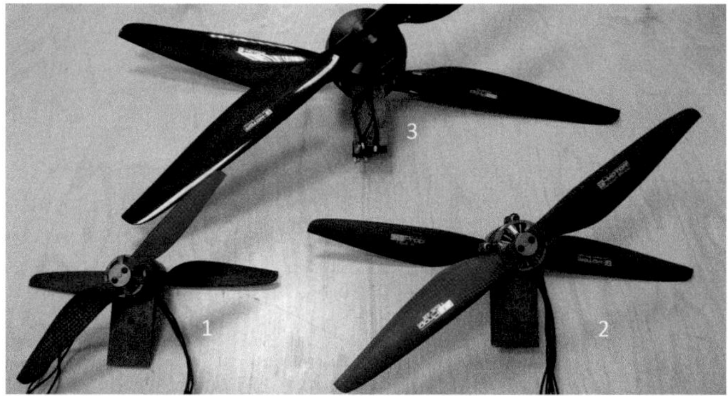

Fig. 3 Comparison of tested propulsion units *1* MN3110 + 10″prop., *2* MN4014 + 16″prop., *3* U8 Pro + 26″prop.

Second Propulsion Setup

Second configuration includes *MN4014* BLDC motor and 16 × 5.4 propeller. This setup is focused on maximizing thrust while maintaining small volume of platform. The solution is similar to the one applied in DraganFlyer X8 platform. This unit produces up to 3 kg of thrust with maximum power consumption of 900 W.[2]

Third Propulsion Setup

Last propulsion unit's configuration utilizes energy-efficient *U8-16 Pro* BLDC motor with 26 inch propeller. This solution, with much lower nominal rotational speed of about 2500 rpm and much bigger propeller's span, provides up to 2 kg of thrust with maximum power of 300 W (Fig. 3).

3.3 Experiment Assumptions

In overall, 6 series of experiments were performed, for three different propulsion unit setups. For every setup, the single and coaxial configuration were analyzed. To maintain high data reliability, ten identical sequences were performed in each phase and the results were averaged. Each sequence was conducted with constant supply voltage provided for both motor and measuring equipment. The calibration process was performed before each sequence and the data was stored in PC connected via USB with *Control Board*. Afterwards, all gathered data was post-processed and analyzed in *MATLAB R2014a* software. Obtained results are shown in the next section.

[2]http://www.rctigermotor.com/html/2013/Navigator_0910/40.html.

4 Experimental Results

One of the basic experimental assumptions was to maintain equal rotational speed of both motors in coaxial configuration, the same as in single propeller configuration. The Fig. 4 shows the record of rotational speeds for three different classes of motors and propellers analyzed during the experiment, with different range of effective operational speeds. The speeds of propellers does not differ more than 3 %, which rules out the factor of difference in rotational speed affecting performance of coaxial propulsion.

In order to show the performance of coaxial propulsion, the useful thrust in coaxial and single configuration was compared. By useful thrust, we understand the measured thrust generated by propellers minus the weight of complete propulsion unit. The comparison of useful thrust to power of propulsion unit is shown on Fig. 5. We conclude that pair of motors in coaxial arrangement produce less thrust comparing to single motor when operating at the same total power of propulsion unit. However, the loss in efficiency varies depending on propeller's size, but does not change significantly with increasing rotational speeds of propellers.

In order to estimate performance of coaxial design in practical application of multirotor propulsion system we calculated the efficiency of tested configurations. The

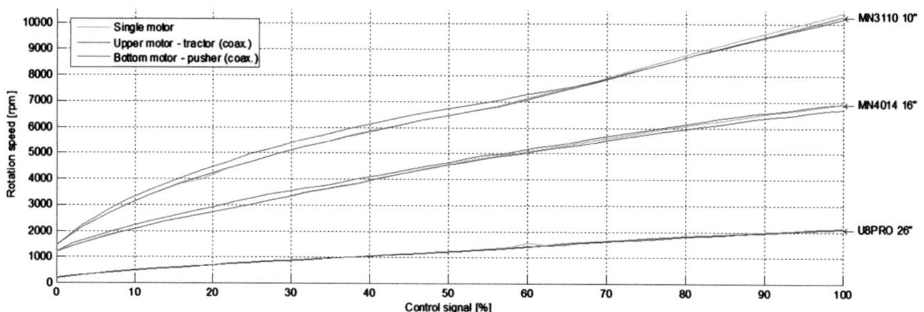

Fig. 4 Rotational speeds of propulsion units analyzed during experiment

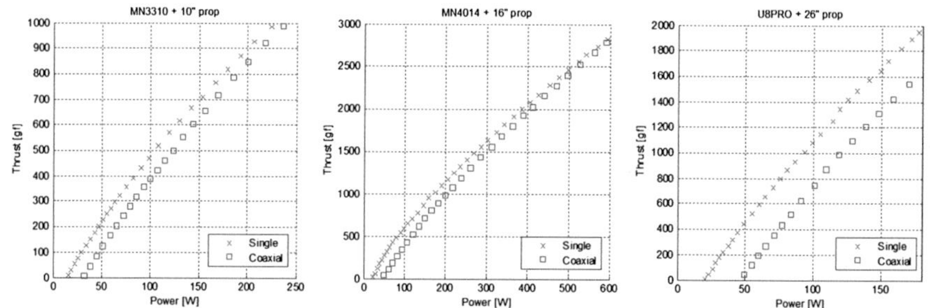

Fig. 5 Thrust produced by different propulsion units in single and coaxial configurations

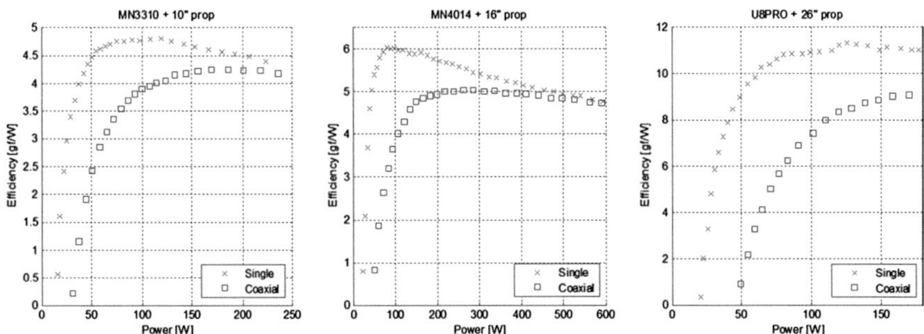

Fig. 6 Efficiency in single and coaxial setup

efficiency factor, given in grams of force per watts, shows how much thrust can be achieved with 1 W of electric power. The results are given on Fig. 6.

First of all, we conclude that bigger propellers provide significant gain in efficiency. However, the performance of coaxial propulsion decreases. We estimated average loss in efficiency for coaxial motor pair comparing to single unit. For the biggest tested propulsion setup, the coaxial propulsion has a efficiency at about 77 % of single unit. For 16 inch propeller and MN4014 motor and 10 inch propeller with MN3110 motor, these values estimate at about 85 %. Having in mind that 10 and 16 inch configuration are quite similar, with smaller difference in propellers' size and rotational speed, we assume that efficiency of coaxial propulsion decreases for bigger, low-speed propulsion. In most common sizes of *UAV* multirotor's propulsion (varying from 8 to 16 inches of propeller's diameter [1, 10]), the coaxial configuration of propellers requires about 17 % more power to generate the same thrust, considering useful lift including motors' weight.

However, one of the biggest advantages of octoquad configuration is ability to provide significantly higher thrust for the platform in smaller volume. The comparison of maximum thrust for both versions of propulsion system—classic quadrotor concept and octoquad configuration—is shown in Table 1. The maximum thrust capability of coaxial design is reduced by mass of additional motor and propeller.

Table 1 Maximum thrust in quadrotor and octoquad configuration

Configuration	Quadrotor (4 × 1)		Octoquad (4 × 2)		Thrust gain (%)
	Thrust (gf)	Power (kW)	Thrust (gf)	Power (kW)	
MN3110 + 10" prop.	3940	0.89	5539	1.35	136
MN4014 + 16" prop.	12904	2.57	18542	4.23	144
U8PRO + 26" prop.	9399	0.71	13389	1.28	142

4.1 Conclusions

- Introducing octoquad configuration allows to increase lift capabilities by about 40 %, considering weight of additional motor unit and slight rise in vehicle's volume, comparing to classic quadrotor design.
- Coaxial propulsion does not operate on maximum motor's power, probably because of the lower propeller operating in prop wash of upper unit.
- The loss in efficiency for coaxial propulsion is not that significant, because lower than double thrust gain comes with less power consumption. Considering grams per watt ratio, coaxial propulsion units needs about 17 to 29 % of more power to produce the same thrust. However, this values vary for different rotational speeds. In addition, different sizes of propellers have different loss in efficiency, with best results for smaller, high-speed propellers.

5 Evaluation

Based on experimental results, we conclude that octoquad configuration of *UAV*'s propulsion is worth considering, when maintaining small volume with significant thrust is essential for the design. Presented analysis of coaxial propulsion for quadrotor platforms found a practical application in our projects. The *MN3310* paired with 10 inch propeller setup was already applied in our *Falcon V5* quadrotor [9], providing 6.5 kg of total thrust in octoquad configuration. Another example of our application of coaxial propulsion is the *Dropter* platform [13], with maximum total thrust of 21 kg, developed for European Space Agency's *StarTiger* project.

6 Future Work

Experiments performed with third propulsion setup (*U8-16 Pro* BLDC motor with 26 inch propeller) introduce possibility to develop energy efficient multi rotor platform, focused on maximizing flight time. Comparing experimental data with first drafts of mechanical design gives over 1 h of estimated maximum operational time.

Performed experiments are planned to be enhanced towards testing different propellers' pitches and diameters in symmetrical and unsymmetrical configuration of coaxial propulsion.

In addition, dual propulsion systems creates opportunity to explore the subject of redundancy and fail-safe control algorithms for multi rotors. When one of the motors in coaxial pair fails, it is still possible to continue flight and maintain stability, with completely changed control laws and different state estimation, which appears as promising field of research.

References

1. Pounds, P., Mahony, R.: Design principles of large quadrotors for practical applications. Robot. Autom. ICRA'09 (2009)
2. Ranganathan, T., Thondiyath, A., Kumar, S.P.S.: Design and analysis of an underwater quadrotor - AQUAD. Underwater Technology (UT) (2015)
3. Sinha, P., Esden-Tempski, P., Forrette, C.A., Gibboney, J.K., Horn, G.M.: Versatile, modular, extensible vtol aerial platform with autonomous flight mode transitions. In: Aerospace Conference (2012)
4. Pounds, P., Mahony, R., Corke, P.: Modelling and control of a quad-rotor robot. In: Proceedings of Australasian Conference on Robotics and Automation (2006)
5. Onen, A.S., Senipek, M., Mutlu, T., Gungor, O., Uzunlar, I.O., Kurtulus, D.F., Tekinalp, O.: Modeling and controller design of a VTOL UAV. Unmanned Aircr. Syst. (ICUAS) (2015)
6. Magnussen, O., Hovland, G., Ottestad, M.: Multicopter UAV design optimization. Mechatron. Embed. Syst. Appl. (MESA) (2014)
7. Er, M.J., Yuan, S., Wang, N.: Modeling and controller design of a VTOL UAV. Control Autom. (ICCA) (2013)
8. Simoes, C.: Optimizing a coaxial propulsion system to a quadcopter. Technical Report. https://fenix.tecnico.ulisboa.pt/downloadFile/563345090412782/Resumo.pdf
9. Bondyra, A., Gardecki, S., Gsior, P., Kasiński, A.: Falcon: a compact multirotor flying platform with high load capability. Adv. Intell. Syst. Comput. **351**, 35–44 (2015)
10. DraganFly Innovations Inc.: Draganflyer X8 helicopter tech specification. http://www.draganfly.com/uav-helicopter/draganflyer-x8/specifications/
11. Sharf, I., Nahon, M., Harmat, A., Khan, W., Michini, M., Speal, N., Trentini, M., Tsadok, T., Wang, T.: Ground effect experiments and model validation with draganflyer X8 rotorcraft. In: Proceedings of International Conference on Unmanned Aircraft Systems (ICUAS) (2014)
12. Gardecki, S., Kasiński, A.: Badanie i dobór elektronicznych jednostek napdowych wielowirnikowego robota latajcego. Pomiary Automatyka Kontrola, Oficyna wydawnicza PAK, Warszawa, z. 58, s. 80 83 (2012)
13. Lutz, T., Gu, C., Gardecki, S., Cordes, F., Hormigo, T., Bljes, A., Bolz, J., De Maagt, P.: Startiger dropter project: development and flight experiment of a skycrane-like terrestrial lander demonstrator. In: GNC 2014: 9th International ESA Conference on Guidance, Navigation & Control Systems

Concept Analysis and Development of an Innovative Remotely Controlled Portable Tyre Puncturing Device (R-TPD)

Agnieszka Spróńska, Adam Wołoszczuk, Paweł Sałek, Mateusz Maciaś and Jakub Główka

Abstract The aim of this paper is to present the concept development and demonstrator of a remotely controlled tyre puncturing device in the context of vehicle stopping and the EU 7th Framework Programme AEROCEPTOR project objectives. First, the general problem statement is presented, identifying major legal, ethical and technical issues, related to the project assumptions and the developed solution. Subsequently, the concept of a remotely controlled portable tyre puncturing device is introduced and described, from the early concept development up to current stage of a technology demonstrator at TRL6. Insight into construction of the device and its principle of operation, as well as brief summary of field tests results up-to-date are presented. Observations, conclusions, development trends and recommendations are described in the conclusions.

Keywords Tyre puncturing · Vehicle stop · Remote operation

A. Spróńska (✉) · A. Wołoszczuk · P. Sałek · M. Maciaś · J. Główka
Przemysłowy Instytut Automatyki i Pomiarów PIAP, Warsaw, Poland
e-mail: agnieszka.spronska@piap.pl

A. Wołoszczuk
e-mail: adam.woloszczuk@piap.pl

P. Sałek
e-mail: pawel.salek@piap.pl

M. Maciaś
e-mail: mateusz.macias@piap.pl

J. Główka
e-mail: jakub.glowka@piap.pl

R. Szewczyk et al. (eds.), *Challenges in Automation, Robotics and Measurement Techniques*, Advances in Intelligent Systems and Computing 440, DOI 10.1007/978-3-319-29357-8_47

1 Introduction

To protect citizens from crime, violence and terrorism, it is sometimes necessary to stop non-cooperative vehicles, i.e. those that are used for a criminal or illegal activity, such as trafficking of people, arms and dangerous, illegal substances or for conducting terrorist attacks (e.g. car bombs), as well as stolen ones or driven by an intoxicated/violating the traffic driver, who does not stop to a Law Enforcement Agency (LEA) officers' (such as Police or Border Guards) call.

Many of those situations result in a "hot pursuit" and as various studies show [1–9] the interception of the vehicles is often dangerous and threatens the lives and health of those who intercept, and it can also result in the loss of life of the offenders or third parties, such as hostages or people who just happen to stand or pass by.

Review of numerous regulations and guidances of LEAs and Ministries of Interior [10–19] shows that all of these authorities are fully aware of the complexity of this issue and that they actively strive to find the balance between the function of crime control and apprehending criminals and the risks that it poses to society. The competing values of crime combat and safety constitute a recurrent theme in all pursuit regulations. The technical and tactical part of the examined regulations is much less detailed when compared to parts concerning safety and operations management. However, it's possible to identify that the four main tactics and special devices being commonly mentioned in police pursuit policies: stationary roadblocks, rolling road block, controlled contact (such as PIT manoeuvre) and tyre deflation devices.

Review of the relevant technology market shows that the research over new technologies assisting vehicle stopping tactics is constantly ongoing, proposing solutions based either on the concepts of affecting the mechanical parts of the vehicle (such as wheel/axis entanglement devices, e.g. SQUID, VLAD, X-NET, PVAB or Pit-BUL [20–24]) or its electronics, where the emerging research is related to application of the electromagnetic pulse (EMP), which is an intense burst of electromagnetic energy caused by an abrupt, rapid acceleration of charged particles, usually electrons and may cause damage to a sensitive hardware such as on-board computer of the vehicle.

Two of the currently ongoing EU 7th Framework Programme projects are dealing with the research on the latter—SAVELEC (Safe control of non-cooperative vehicles through electromagnetic means) and AEROCEPTOR (UAV based innovative means for land and sea non-cooperative vehicles stop).

While SAVELEC is strictly focused only on the EMP technology for car-stopping application [25], the AEROCEPTOR project is aimed at development of a new and innovative concept of operation to remotely and safely control, slow and stop non-cooperative vehicles in both land and sea scenarios, by the means of Remotely Piloted Aerial System (RPAS) equipped with the set of relevant payloads. The scope of the project is taking into account both the developments in EMP technology as well as application of other vehicle stopping technologies in an innovative way, i.e. by their deployment from the air [26].

The main R&D activities in the project are related to development of an innovative RPAS control system and research on RPAS technology modifications to comply with AEROCEPTOR concept, research on new vehicle stopping technologies and adaptation of the existing payloads performing ground/maritime vehicle control to their airborne version needed in AEROCEPTOR as well as rulemaking and regulatory assessment activities, to study the needed regulatory framework allowing RPAS to operate (certification and integration into aerospace) complying with the necessary levels of safety and security and with the European regulations.

In addition to the technical development the project objectives encompass another goal, which is providing a comprehensive and holistic security study on the topic, ensuring that the new developed means will be safe and will meet all the ethical and Social Science and Humanities (SSH) standards.

This paper describes a part of the research and development works conducted within the AEROCEPTOR project over the vehicle-stopping devices in land scenarios that result in technology demonstrator of TRL 6.

2 Problem Statement

As mentioned in the previous section, a significant part of the AEROCEPTOR project has been devoted to the research over the existing means for vehicle stopping and the technical developments necessary for their adaptation to be deployed from the airborne platforms. In that context and for the purpose of this paper, which does not intend to deal with the problematic of the RPAS technology development or application, it shall be considered that the platforms can either be manned or unmanned, taking into account any airborne platforms that may be in disposition of the LEA forces.

A thorough study over the existing vehicle stopping devices has been conducted in the course of the project, in order to determine the selection of those, which may be of the highest relevance to the project objectives and create the most promising results, also with consideration of the legal aspects and LEAs operational practice.

It needs to be underlined, that despite the project's scope encompasses both the land and maritime scenarios, from the point of view of the research results presented in this paper, only the ground vehicles stopping technologies are relevant and described in more detail, although some of the findings may relate to both environments.

State of the art review shows that the technical means used for stopping non-cooperative vehicles can be grouped up in two main sets: direct and indirect means, where direct means are those directly affecting moving vehicle and indirect are those, which application is directed towards influence on the vehicle driver—such as loudspeakers (sound emitters, distributing specific tones and volume, turned on the driver can make him temporary deaf), high power lights (affecting the sense of sight), or malodorants (specific chemical substances, which spread extreme stench, causing temporary incapacity).

Since a great focus in AEROCEPTOR project has been put on the compliance with the ethical and social aspects of the developed technologies and one of the project objectives is to ensure safety of all the persons involved in the action, including the offenders, all the indirect means were excluded from the application considerations at the early stage, as they impose the highest risk of injuring the vehicle driver and passengers as well as other persons present within or nearby the scene.

Further study over the existing technologies revealed that the direct means can be divided into three main subgroups:

- **Entanglement devices (either wheel or axles)**—where a tangled wheel or axle is being blocked by increasing force of tightening bonds, which finally make the element unable to spin, which results in car stopping.
- **Tyre puncture devices**—where the air is drained from the tyres, so the car is constantly loosing speed and comfort of driving. Special construction of these devices prevents the tyre from exploding, as the air is released slowly.
- **Emerging technologies**—which are currently on a very low level of maturity, such as EMP technology or foam polymer/polymer-based paste sprayed onto the ground to tangle car tyres and several other systems.

The only category that is already legitimate in LEAs operational practice and finds reference in the reviewed law regulations (as described in the previous section) are the tyre puncture devices. Also, the lecture of the available reports and statistics (as mentioned previously as well as [27, 28]) confirmed that the tyre deflators are one of the most efficient method of car stopping techniques used by LEAs.

Two relevant tyre puncture devices were recognized as relevant for AEROCEPTOR project purposes, i.e.: spike strips and caltrops.

Spike strips (traffic spikes, tyre shredders, one-way traffic treadles, stingers, stop sticks) are devices composed of a collection of 35–75 mm (1.5–3 in.) long metal barbs, teeth or spikes pointing upward. The barbs are designed to puncture and flatten tyres when a vehicle is driven over them, they can be either hollow or solid. Hollow barbs are designed to become embedded in the tyres and allow air to escape at a steady pace in an attempt to reduce the risk of the driver crashing into oncoming traffic or the surroundings. Spike strips make halting the vehicles by police very effective. After the drop down of pressure in tyres, vehicle becomes low driveable and further drive is very difficult. There are also prototypes of the remotely-controlled spike strips advertised on the market [29], disguised as a small suitcase and once placed on the roadside—unlike traditional spike strips—does not require an officer to stand nearby to deploy or retract the device. However, though commonly used by LEA forces and proved efficient, in the context of the air deployment spike strips pose a lot of problems, since they must be placed with preservation of the certain orientation in reference to geometric features of the environment and on the specific spot along the road.

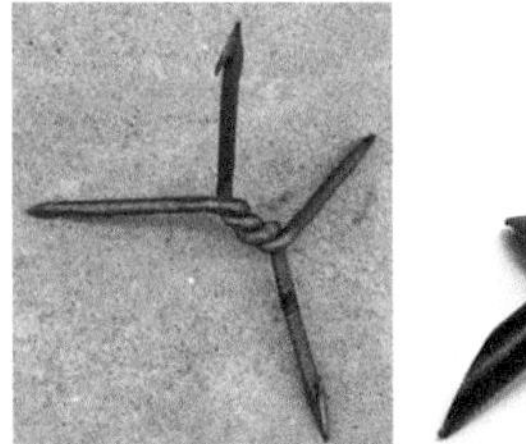

Fig. 1 **a** Simple caltrop (http://en.wikipedia.org/wiki/File:Caltrop_from_Vietnam_1968.jpg), **b** modern caltrop—the hollow spikes allow air to escape at a steady pace (http://en.wikipedia.org/wiki/File:Caltrop.jpg)

Therefore, as the most adequate alternative of the tyre puncturing device for the given application the caltrop has been identified. Caltrop (also known as caltrap, galtrop, galthrap, galtrap, calthrop) is a device made up of two or more sharp nails or spines arranged in such manner that one of them always points upward from a stable base (Fig. 1). Caltrops are used against wheeled vehicles with pneumatic tyres.

Caltrops have a very significant advantage, since they are small and easily deployable from ground level, as they can be simply thrown or dropped onto a surface in a random order, without the need of precise positioning. Therefore, there are very suitable to be deployed from the air.

However, the caltrops as recognised an antipersonnel non-lethal weapons are prohibited for production, possession and usage in some countries. As much as they are easily deployable, they are difficult to collect and can uncontrollably damage vehicles and property of uninvolved third parties or even injure people. As such, they were non-applicable in AEROCEPTOR project, as one of the project main goals is to comply with the safety and highest ethical standards. Consequently, there was a need for an effective device or tool able to stop the non-cooperative vehicles that is both easy deployable from the air and has safety features allowing to minimise collateral damage to people, animals or third party vehicles. Therefore, a new concept of a Remotely Controlled Portable Tyre Puncturing Device (R-TPD) was developed, leading to a technology demonstrator, which is presented in the next sections of this paper.

3 Concept and Evolution

The idea behind R-TPD was to deliver a remote tyre puncturing device that would allow stopping a non-cooperative car with, unlike traditional methods such as barricades or spike strips, minimum exposure to traffic. The way to achieving this goal was to provide R-TPD with a functionality of remote opening and closing on demand. In closed state, the device would be harmless to any vehicle running over it, while in open mode it would puncture the tyre of the vehicle. The time of opening and closing execution should be kept at a possible minimum. Remote operation is considered a primary and fundamental functionality of the device, provided that in closed state the device is harmless both to traffic and the user and the spike in open state is always directed upwards.

Secondarily, the device should be robust enough to withstand the pressure of a car running over it as well as shocks and inertia resulting from tossing it on the ground. Finally, the device shall be deployed from an airborne platform, which involves drops from a significant height of approximately 20 m (line of trees for safe air navigation). As the device is intended to be used in groups, it should also be compact and lightweight, keeping the user's and/or the flying platform's encumbrance at a minimum.

To obtain an adequate performance, the device should be equipped with fundamental components comprising of: a housing, a spike, a drive for the spike, radio and controller PCBs as well as a battery. Implementation of the current solution was preceded by elaboration on several concepts.

The original idea, was the 'penknife' construction sheltering the single spike, embedded on rotary axis. The spike would be driven by miniature DC motor. The sides of the device were wedge-shaped to eliminate the possibility of the device's random positioning in horizontal manner after deployment. However, due to the narrow base surface the device appeared unstable. The other downside of this solution was the necessity of 'knowing' the vertical orientation of the device in order to position the spike in proper direction in the open state. This would impose the installation of a sensor (MEMS or visual sensor as shown on the Fig. 2 and associated circuit board in an already small volume of a small-sized casing.

The next idea addressed the issue of the device's stability by introducing a cross-shaped housing (Fig. 3). This would lead to an increase of the device's size, but it shall also improve its stability. Taking into consideration the sizes of electronic circuits at hand, the size of the cross-shaped device would even surpass the initial assumptions. The other option was to reduce the size of electronics by implementing a dedicated integrated PCB.

As an alternative to cross-shaped housing, the new concept of spade-shaped housing was elaborated. This concept allowed to install all necessary PCBs and components inside the housing as well as to reinforce the housing with supporting posts, in order to distribute the external compressive force resulting from a car running over it.

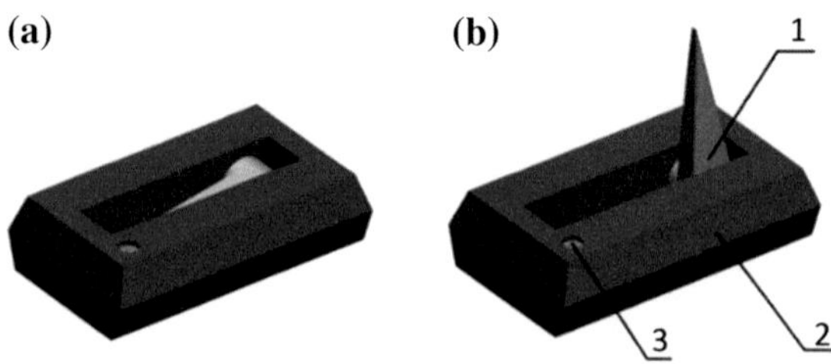

Fig. 2 Early concept of R-TPD in closed state (**a**) and open state (**b**). *1* Spike, *2* main body, *3* light sensor

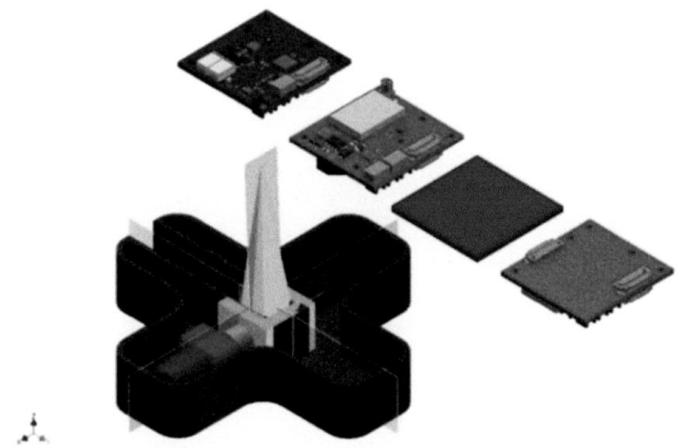

Fig. 3 Cross-shaped R-TPD with a set of pre-designed PCBs

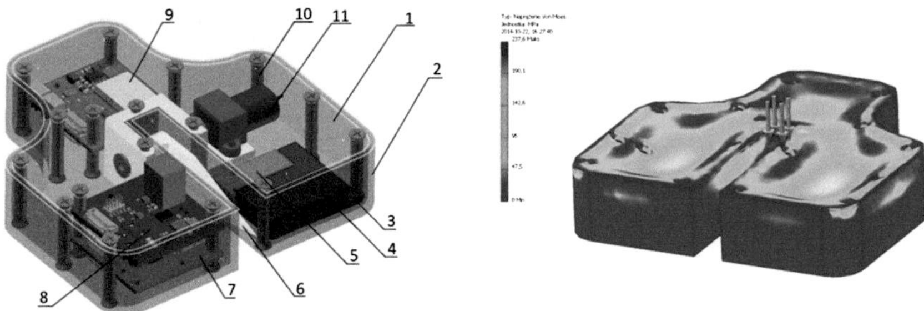

Fig. 4 Spade-shaped R-TPD design and FEM analysis. *1* Cover, *2* main body, *3* inductor, *4* battery, *5* IMU, *6* spike, *7* battery balancer PCB, *8* motor control PCB, *9* radio PCB, *10* support post, *11* DC motor

It is important to note, that in this concept the electronic modules used, were pre-designed by PIAP for universal mobile robotics applications.

Initial FEM stress analysis of this spade-shaped housing (Fig. 4), confirmed that it would withstand an external force resulting from static load from a car. Analysis also revealed the stress concentration in proximity of the support posts. The spade-shaped device is shown on Fig. 5.

Dynamic run-over tests were performed on physical prototype, with the housing made of Ultem (printed with 3D printer) and, subsequently, made of POM-C. It is necessary to note, that the housing, in order not to suppress the radio signal, should be entirely made of dielectric material. The dynamic run-over tests proved, that both Ultem and POM-C housings are capable of withstanding the forces imposed by the car running over them with the approximate speed of 30 km/h. However, shell-like construction of the housing made it vulnerable to shocks (for example resulting from dropping the device from height of approximately 20 m and thus potentially dangerous for traffic in case of shattering of the housing (exposed supporting posts could potentially damage the tyre).

Fig. 5 Spade-shaped R-TPD prototype

Tests have shown that the device was successful in puncturing the tyre, not causing it to explode and enabling the smooth deflation. However, in several cases, the puncturing spike was stuck in the tyre which may cause a significant threat to traffic—the device rotated together with a tyre and finally centrifugal force caused the device to detach and move freely in random direction. This occurrence was unacceptable in terms of safety for traffic and surroundings. Drop tests from 11 m height on solid surface confirmed that the shell-like housing is vulnerable to shocks and strikes, which result in cracks in the device structure.

Similarly to cross-shaped device, this concept also imposed the necessity of implementing orientation sensing system, in order to ensure the proper spike direction in open state.

The most important conclusions from dynamic tests and analysis included: the necessity of detaching the spike thrust in the tyre, solid and homogenous housing sheathing the spike with venting channels and elaboration of orientation-independent spike that will be always directed upwards in open state, regardless of which surface (bottom or top) the device is laying on. Implementation of these conclusions resulted in the current version of R-TPD.

4 Demonstrator

Taking into account conclusions and observations from previous version of the device, the project team decided to demonstrate the solution that would address spotted gaps. The concept of the device comprising two solid covers connected via rotary joint was elaborated. The rotary joint shaft acts also as the spike support. The spike was designed to rotate freely around the shaft, and the base of the shaft was shaped to form a separable rotary connection with the shaft. This enables the spike to detach from the shaft, once thrust into tyre. The spike features vent channels to prevent the tyre from exploding. The device is opened by the DC motor driving the solid covers in relation to each other—when the covers are open the spike is exposed, when the covers are closed the spike is safely contained. The spike itself is not driven by the motor—the vertical position of the spike is achieved by push of the bottom edge of the spike by the rotating solid cover. Regardless of the part of the cover on which the device is laying, it will always open with the spike directed upwards in open state—thus the necessity of orientation sensing system has been eliminated. A significant development of electrical modules has been made, compared to previous concept—all functions have been integrated into single module. Small sized dedicated PCB, integrating radio, motor controller and battery controller was implemented, allowing to keep the device small-sized and compact. Two pairs of neodymium magnets prevent the covers from undesired dislocation.

The system context of using the R-TPD in field, assumes the deployment of multiple R-TPDs, opened remotely from operator's console. The devices are intended to be transported in stacks, held in holdfast box. The device is equipped with two pairs of brazen contacts which are used to charge the battery and enable charging multiple R-TPDs stacked in the holdfast box. When attached to airborne platform, the holdfast box will be equipped with release mechanism, allowing the R-TPDs a free fall on the ground.

The overview of the R-TPD is presented of Fig. 6, the design of the R-TPD is shown on Fig. 7 and the opening sequence of R-TPD is shown on Fig. 8—closing is performed analogically in a reverse way.

At the current stage, the R-TPD has been shock-tested when falling from various heights, with maximum at approximately 13 m. It has proved its performance after each drop. R-TPD demonstrator used for proof-testing is shown on Fig. 9.

Electronic control unit designed for the demonstrator device consists of the following components: radio module, microprocessor, motor driver, current monitor, supply/charging block and Li-Ion battery. Block diagram is shown in Fig. 10.

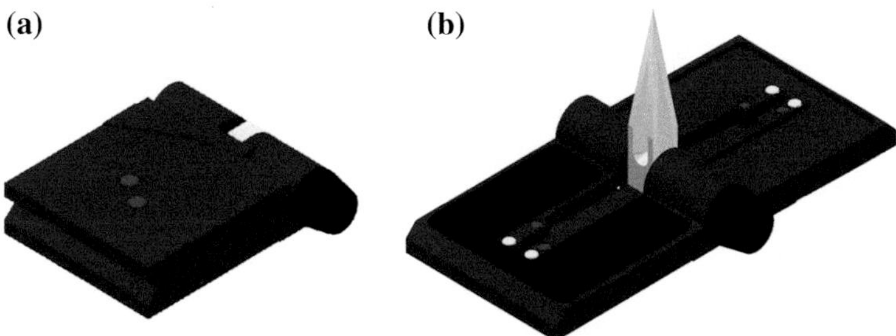

Fig. 6 R-TPD demonstrator in closed state (**a**) and open state (**b**)

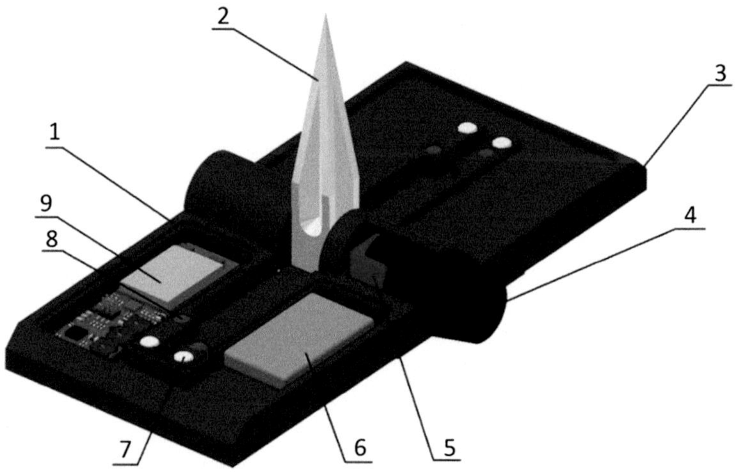

Fig. 7 R-TPD demonstrator design. *1* Main body, *2* spike, *3* rotating cover, *4* motor cover, *5* DC motor, *6* battery, *7* neodymium magnet, *8* contact, *9* control and communication module

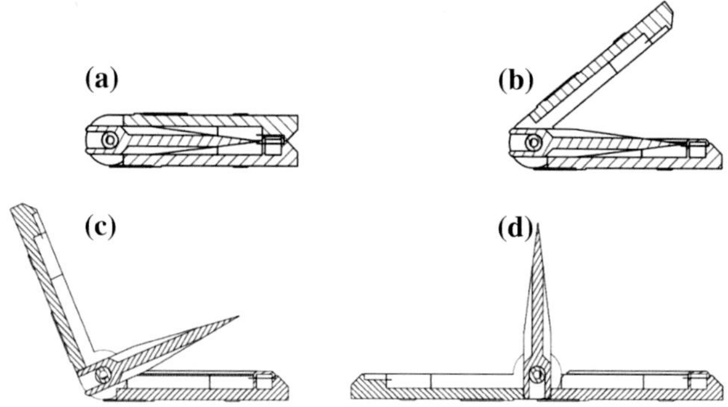

Fig. 8 R-TPD demonstrator principle of operation: **a** closed state (motor idle), **b** opening phase (motor on), **c** opening phase and spike latching (motor on), **d** open state (motor set back to idle)

Typical usage of the R-TPD implies the design focused on two basic configurations:

- STORAGE—when the device is not used, and
- MISSION—when the device is expected to react to operator's requests with a minimal latency.

In the first situation the main objective is to reduce power consumption to avoid frequent re-charging of the battery. However, there is no strict requirement on latency. The second situation takes place when the operator places the device (or a set of them) on the flying platform. Starting from there, latency becomes a critical

Fig. 9 R-TPD physical demonstrator

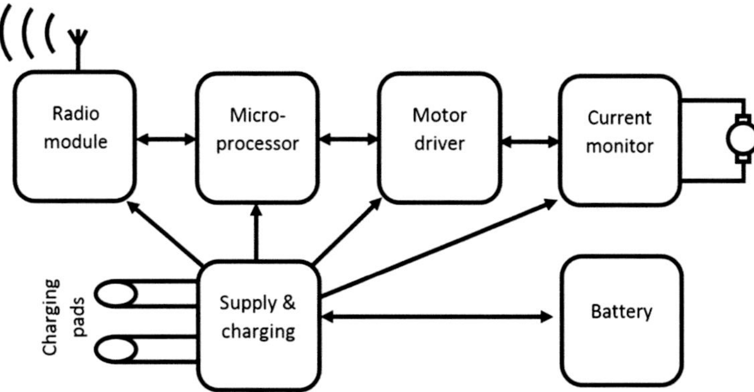

Fig. 10 Block diagram of electronic control unit

parameter. However, since mission duration is rather short, it is not necessary to reduce the power consumption.

During STORAGE configuration the radio module stays in a shut-down mode, while the microprocessor wakes up from the sleep mode periodically and performs cyclic procedures: battery level and charging status check. Every 60 s the radio module is switched to active mode and R-TPD sends it's state as a broadcast radio message (STATE message), in compliance with the IEEE 802.15.4 standard. If no request has been received during 100 ms, the radio is turned off and the device stays in STORAGE state.

Transition from STORAGE to MISSION state is controlled by the operator. If such request is registered by the operator's console unit, the console will respond to the next STATE message with a REQSTATE_MISSION request. After receiving the REQSTATE_MISSION message the R-TPD changes it's state to MISSION. In this state the radio module and the microprocessor are always on, what allows the device to provide the minimum latency for user requests. Transition from MISSION to STORAGE state can result from one of the following conditions:

- REQSTATE_STORAGE message has been received from the console,
- no messages have been received during last 180 s.

The second condition allows to shut down the console unit without sending the REQSTATE_STORAGE message. Particularly, this can happen when battery level of the console is low or communication between the console and the R-TPD was lost and there is no technical way for requesting the change of state. To maintain the regular operation of the device in MISSION state, the console re-sends a REQSTATE_MISSION message every 10 s.

Since radio module remains in MISSION state, the operator's console can send REQOPEN request at any moment. After receiving REQOPEN message, the R-TPD starts the motor and measures it's current. The motor stall is detected using a threshold for current value, which is set to 200 mA. Reaching the threshold causes disabling of the motor driver and R-TPD is now in the OPEN mode. At any moment, the console can send a REQCLOSE request. This will cause the R-TPD to start the motor with an opposite polarisation. After detecting the increased value of current the device is in the CLOSED mode.

The first tests revealed a need for an additional state of device. If a non-intentional request of the STORAGE state occurred, the operator would have to wait 60 s before changing to MISSION state again, since the radio module in STORAGE state is disabled for this time interval and the device is not able to receive messages. This problem has been solved by introducing additional state. After receiving the REQSTATE_STORAGE message the device state is set to MISSION_END. The radio module is still enabled and the device can accept a new REQSTATE_MISSION request. If no message is received within 60 s, the state is changed to STORAGE.

As mentioned before, the R-TPDs will be used in sets. Sets are defined on the console and each set consists of multiple devices identified by their unique 64-bit MAC number. If the console receives a STATE message from any device of the set, the whole set becomes available for the user and is shown on the HMI of the console. All devices belonging to one set are controlled by one button. It is also possible to use many sets in the same mission, i.e. to open them sequentially: if first attempt fails, then open second set, etc.

5 Conclusions

This paper describes the concept analysis and development of a new device able to stop a car through cautious puncturing of tyres with preservation of the safety of other public space users by enabling remote control of the puncturing element.

Applied system engineering approach is described and demonstrated in a form of the prototype which performance was measured and results are discussed with regard to the identified requirements. Important part of presented research is the concept of operation and intended use of R-TPDs as a set of devices housed in holdfast box.

Further development of the platform can be conducted in two directions. One is the engineering effort in order to miniaturise the solution in terms of size and weight. Both parameters are crucial in application of the device in airborne operations. The second development path is focused on concept of use and tests related to effectiveness of application of large sets of the device in specific operational cases. That second approach will cover such issues as optimisation of R-TPD's deployment in terms of time, set size and number of sets, localisation on the road and other in typical applications and scenarios.

Following the conducted research the identified functionalities that should be analysed and optimised in further design are related to the remote control part of the R-TPD and safety features that it should provide. The challenge spotted during project concept analysis that hasn't been solved within the proposed device solution is the insensitivity of the vehicles equipped with tyre auto healing mechanisms, RSC type tyres or other than pneumatic types of tyres. However, the proposed idea of an radio-activated small device can be adopted for such cases as well. The proposed folding design could enclose the other type of wheel affecting mechanisms such as self-attaching and tyre-deforming element. That concepts should be further explored in relation to the state-of-the-art in wheel solutions in automotive industry.

All efforts described in this paper, experience gained during presented research as well as data collected are expected to conclude in development of operable prototype of a safe device able to stop non-cooperating vehicles. The aim of the future work is to prove the concept in a real case scenario within further research and development activities.

Acknowledgments The research leading to these results has received funding from the European Community's Seventh Framework Programme (FP7/2007-2013) under grant agreement no. 285144.

References

1. Alpert, G.: Pursuit driving: planning policies and action from agency, officer, and public information. Police Forum **7**(1997), 1–12 (1997)
2. Alpert, G.P., Dunham, R.G.: Policing hot pursuits: the discovery of aleatory elements. J. Criminal Law Criminol. **80**, 521–539 (1989)
3. Department of the California Highway Patrol Pursuit Study, Operational Planning Section (1983)
4. Lum, C., Fachner, G.: Police Pursuits in an Age of Innovation and Reform. The IACP Police Pursuit Database. www.theiacp.org/Portals/0/pdfs/Publications/Police%20Pursuit.pdf

5. Rivara, F.P., Mack, C.D.: Motor vehicle crash deaths related to police pursuits in the United States. Inj. Prev. **10**, 93–95 (2004)
6. Frank, T.: High-speed police chases have killed thousands of innocent bystanders. www.usatoday.com/story/news/2015/07/30/police-pursuits-fatal-injuries/30187827/
7. Rayner, G.: Number of police chase deaths soars. www.dailymail.co.uk/news/article-59483/Number-police-chase-deaths-soars.html
8. Edwards, R.: 66 per cent increase in people killed in police chases. www.telegraph.co.uk/news/5843287/66-per-cent-increase-in-people-killed-in-police-chases.html
9. Roper, M.: Nine crashes a day caused by police. www.mirror.co.uk/news/uk-news/nine-crashes-a-day-caused-by-police-789745
10. California Law Enforcement Vehicle Pursuit Guidelines, California Commission on Peace Officer Standards and Training; Printed 1995/Revision 2007, POST2007TPS-038
11. Code of Practice on the Management of Police Pursuits, Secretary of State for the Home Department (2011)
12. High Speed Police Pursuit in Perth. A report to the Police Department of Western Australia, Ross Homel 1990, Australia
13. Zarządzenie Nr 1355 Komendanta Głównego Policji z dnia 20 grudnia 2007 r. w sprawie metod i form organizowania i prowadzenia przez Policję pościgów i zorganizowanych działań pościgowych (Ordinance No. 1355/2007 of the Chief Constable on forms and methods and performance of police hot pursuits and organised pursuit actions), Dziennik Urzędowy Komendy Głównej Policji nr 1/2008
14. Wytyczne Nr 1 Komendanta Głównego Policji z dnia 20 kwietnia 2005 r. w sprawie ochrony bezpieczeństwa osobistego policjantów oraz osób postronnych podczas zatrzymywania osób poruszających się pojazdami (Instruction No. 1/2005 of the Chief Constable on the personal safety of police officers and third persons during apprehension of persons travelling with vehicles), Dziennik Urzędowy Komendy Głównej Policji nr 10/2005
15. Zarządzenie Nr 40 Komendanta Głównego Straży Granicznej z dnia 13 października 2011 r. w sprawie zasad i form organizowania i prowadzenia pościgów oraz działań pościgowych przez Straż Graniczną (Ordinance No. 40/2011 of the Commander in Chief on forms and methods and performance of border guards hot pursuits and organised pursuit actions), Dziennik Urzędowy Komendy Głównej Straży Granicznej nr 11/2011
16. Zarządzenie Nr 14 Komendanta Głównego Straży Granicznej z dnia 10 lutego 2015 r. zmieniające zarządzenie w sprawie zasad i form organizowania i prowadzenia pościgów oraz działań pościgowych przez Straż Graniczną (Ordinance No. 14/2015 of the Commander in Chief modifying ordinance on forms and methods and performance of border guards hot pursuits and organised pursuit actions), Dziennik Urzędowy Komendy Głównej Straży Granicznej 2015 poz. 15
17. Ontario Regulation 266/10 "Suspect Apprehension Pursuits" made under the POLICE SERVICES ACT, 28 (2010)
18. The Management of Police Pursuits Guidance, Association of Chief Police Officer of England, Wales and Northern Ireland (UK) (2009)
19. Grygutis, A., Sobiech, P., Wilisowski, A.: Prowadzenie działań pościgowych przez Policję (Execution of pursuit actions by Police), Wydawnictwo Szkoły Policji w Katowicach (2009)
20. Safe Quick Undercarriage Immobilization Device SQUID. www.dhs.gov/squid-long-and-sticky-arms-law, www.squid-technologies.com/auto.html, www.darkgovernment.com/news/new-dhs-vehicle-stopper-uses-tentacles/
21. Vehicle Lightweight Arresting Device: VLAD. http://jnlwp.defense.gov/current/VLAD.html, http://www.pica.army.mil/pmccs/combatmunitions/nonlethalsys/nonlethalcapset/counterper/4nlc_vlad.htm, www.liberatedmanuals.com/TM-5-4240-536-10.pdf
22. Qinetiq webpage. http://www.qinetiq.com/what/products/Pages/vehicle-arresting-systems.aspx, http://eng.1september.ru/articlef.php?ID=200701002
23. Portable Vehicle Arresting Barrier—PVAB. http://www.gd-ots.com/PVAB.html
24. Pit-Ballistic Undercarriage Lanyard: Pit-BUL. http://www.wired.com/autopia/2013/03/remote-spike-strip/

25. SAVELEC project webpage. http://savelec-project.eu/
26. AEROCEPTOR project website. http://www.aeroceptor.eu/
27. Drum, D.T.: Police Pursuits: Chasing Disaster, American Public University System. https://www.apus.edu/content/dam/online-library/masters-thees/drum2011.pdf (2011)
28. Managing Police Pursuits, Findings from IACP's Police Pursuit Database. Executive brief 2004. International Association of Chiefs of Police, http://www.theiacp.org/portals/0/pdfs/ManagingPolicePursuitsExecBrief.pdf
29. NightHawk. http://blutube.policeone.com/videos/6232520-nighthawk-remotely-deployed-spike-strip-demo/

Fast Self-collision Detection Method for Walking Robots

Tomasz Augustyn and Dominik Belter

Abstract The paper presents fast collision detection method for walking robots. In the paper we present the module which uses triangle mesh of the robot to detect collisions between parts of the robot. To this end, the triangle to triangle intersection test is applied. To speed up the computation the bounding box test is carried out at the beginning. We show the properties and performance of the collision detection module. Then, we propose the method which uses Gaussian mixture to determine self-collision model. The method is significantly faster than the method which uses triangle meshes but less precise. The collision detection methods can be applied during motion planning as well as during execution of the planned motion to detect infeasible configurations of the robot.

Keywords Walking robots · Self-collisions detection · Gaussian mixture

1 Introduction

To safely execute a task the robot should plan its motion in advance. During execution of the planned motion a wheeled or tracked robot might collide with obstacles only. In addition to this, the parts of a walking robot might also collide between each other during walking. Thus, the robot should plan the sequence of body configurations and take into account not only collisions between the robot and obstacles in the environment but also self-collisions.

Recently, we presented the motion planning method for six-legged walking robots [1, 4]. To check if the planned path is secure we check not only self-collisions but also stability and the kinematic workspace of the robot. These modules play

T. Augustyn · D. Belter (✉)
Institute of Control and Information Engineering,
Poznan University of Technology, ul. Piotrowo 3A, 60-965 Poznan, Poland
e-mail: dominik.belter@put.poznan.pl

T. Augustyn
e-mail: tomasz.augustyn@student.put.poznan.pl

R. Szewczyk et al. (eds.), *Challenges in Automation, Robotics and Measurement Techniques*, Advances in Intelligent Systems and Computing 440, DOI 10.1007/978-3-319-29357-8_48

important role in the sampling-based motion planning framework. To determine the next step of the robot the planner randomly select random configuration from the given workspace. By using modules which check the feasibility of the desired motion we can verify if the planned step is possible to execute. Moreover, by using self-collision detection module the robot can stop the executed motion if the risk of self-collision is detected. As a result, the collision detection software allows to avoid damages of the robot.

Self-collision detection for walking robots is a difficult task due to the complex mechanical design of the robot and high number of degrees of freedom. The Messor robot (Fig. 1) used in this research has 24 degrees of freedom [2]. Each leg has three degrees of freedom. The legs of the robot share common workspace. Thus, not only parts of the same leg might collide but also these parts might collide with the robots' trunk and with neighboring legs. The controller of the robot should be equipped with the module to detect such situations. For the Messor robot the problem is 18 dimensional. The input of the collision detection module expects 18 reference values for the servomotors and returns information which parts of the robot collide.

In this paper we present the method for a priori collision checking. We show how to detect collisions between parts of the robot at the stage of motion planning. Despite the collision free motion planning the robot might still collide due to either inaccuracies in trajectory execution or unexpected events in dynamic environment. The collision detection during motion execution is beyond the scope of this paper. Collision detection at this stage can be handled by robots' sensors and dedicated control strategies [7, 10, 20].

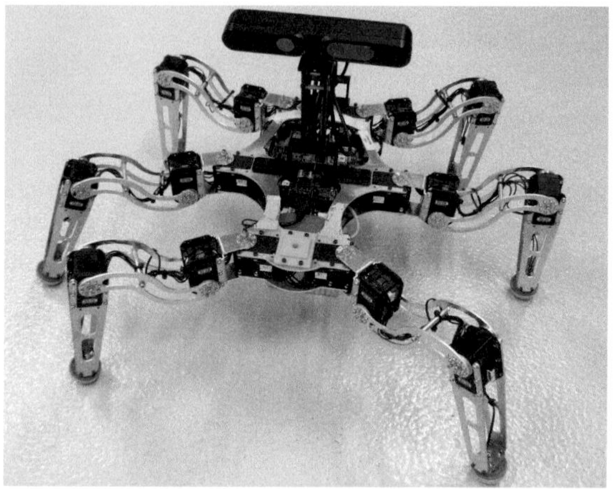

Fig. 1 Messor 2 robot used in this research

2 Related Work

Most collision detection methods approximate the shape of the object by simple geometrical shapes [8, 21]. The most popular shapes are spheres, boxes or polytopes. The collision checking is the fastest for the spheres and axis-aligned bounding boxes (ABB). In these two cases collision checking is very inaccurate if the shape of the object is long or the object changes its orientation in space. In this case more precise models are used e.g. object-oriented bounding boxes (OBB) and discrete orientation polytopes (k-DOPs) (cf. Fig. 3b, c). OBB and k-DOPs better represent bounding volumes of the objects but require more computational effort. The difference between ABB and OBB is presented in Fig. 3b, c.

The bounding volumes methods are inappropriate if the shape of the object is complex. In this case hierarchical bounding volumes (HBV) are applied [8, 16]. Hierarchical methods use bounding boxes and k-DOPs to approximate smaller parts of the objects. Then, the bounding primitives are grouped into bigger parts. The whole object is represented by the hierarchy of smaller parts. At the top of the hierarchy single OBB or k-DOPs is used. The shape of the robot can be better approximated using HBV. However, the collision detection time is not constant. The required computational effort increases when the objects start to collide.

The simplest representation using axis-aligned bounding boxes can be also precise and fast. In this case the size of each box in the representation of the object should be small. Thus, the complex objects like mobile robots or robotic arms consist of thousands of bounding boxes. Despite the fact that the collision detection for a single box is very fast, the same task for the whole robot is very slow. The problem can be solved by implementation of the collision checking procedures in the Graphics Processing Unit (GPU) [11]. The parallel implementation enables on-line self-collision detection for the robot and collision checking between the robot and environment [12]. This approach also enables efficient motion planning in dynamic environment [13].

The most precise collision detection is provided by methods which use triangle mesh surfaces to model the objects. These methods are also easy to apply for the real robot. The mesh-based model of the robot can be easily obtained from the Computer-Aided Design (CAD) software model of the mechanical design. To detect collision the triangle-to-triangle intersection test has to be performed. Up to now, many efficient algorithms which solve this problem have been proposed [22, 23]. The efficient implementations of these methods are also available [9]. In practice, these algorithms can perform continuous collision queries in few milliseconds on models composed of thousands of triangles [17]. We show how to obtain similar properties using learning technique.

3 Paper Contribution

The paper presents universal collision detection module for walking robots. The method can be also used for other types of robots e.g. robot manipulators, robotic hands. In this work we show how to use CAD model of the robot to detect collisions between robots' mechanical components. Then, we show the performance of the system in the typical tasks. Finally, we propose an approximation-based method for collision detection which allows to significantly improve the speed of the collision checking procedure.

4 Collision Detection

The collision detection model uses triangle mesh surfaces to detect collisions. We use the 3D Studio (3ds) files exported from CAD software used to design the robot. Each kinematic link is exported as a separate part. The current version of the collision detection module can model two types of robots: six-legged Messor and four-legged StralETH robot. The mesh models for the robots are presented in Fig. 2. Both models are generated using robots' configuration files (xml) and four mesh objects: robots' trunk and parts of the legs: tibia, femur and coxa.

4.1 CAD Model-Based Collision Detection

In this work we don't detect collisions between the robot and the environment. Detection between parts of the robot and the ground is rather slow because the ground model changes during walking on an uneven terrain. In this case the application of GPU is advantageous [12, 19]. On the other hand, the foothold selection module is in most cases sufficient to prevent collision with the ground. Thus, the collision

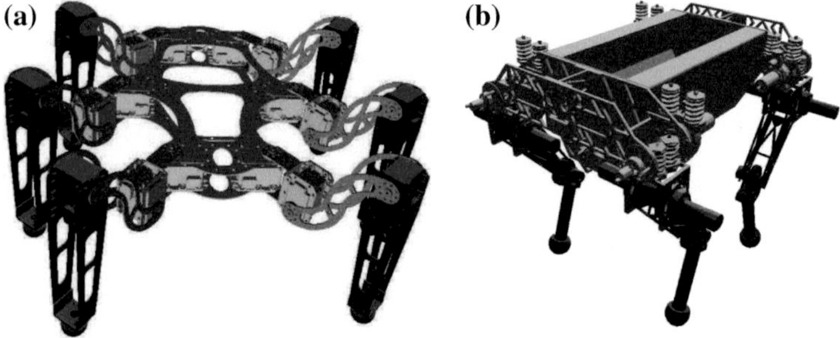

Fig. 2 Mesh models used for collision detection: Messor 2 (**a**) and StarlETH (**b**) robots [14]

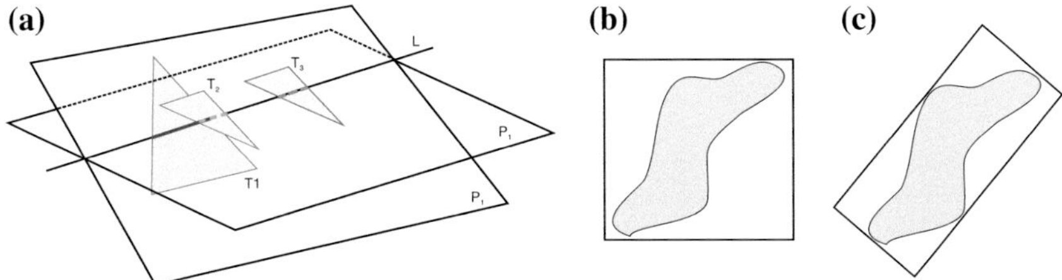

Fig. 3 Concept of the triangle to triangle intersection test [18] (**a**) and the difference between ABB (**b**) and OBB (**c**)

detection module is used only to prevent collisions between parts of the robot. The core of the collision detection module is based on the Coldet library [9]. The collision model of the robot is constructed from numerous triangles. When two bodies collide, there is a line which is common for two triangles. In order to detect a collision the module conducts triangle-triangle intersection test for oriented bounding box (OBB) algorithm [18]. It is fast and reliable when it is used to detect collision between robots' parts (the model is fixed, only its configuration may vary).

The concept of the triangle to triangle intersection test is presented in Fig. 3a. The triangle T_1 is located on the plane P_1. Triangles T_2 and T_3 are located on the plane P_2. If triangles are not co-planar the intersection of planes P_1 and P_2 is a line L. Then one can find intersection intervals on line L for each triangle (red line for triangle T_1 and green dotted line for triangles T_2 and T_3 in Fig. 3). If intersection intervals along line L overlap the triangles collide (triangle T_1 and T_2 in Fig. 3). If intersection intervals along line L don't overlap the triangles also don't collide (triangle T_1 and T_3 in Fig. 3) [18].

The collision detection software also imports the kinematic model of the robot. Thus, the user can provide the configuration of the robot (reference values of servomotors) and check the collision between parts. What is more, visual inspection of robots' configuration is possible using OpenGL visualization. The software returns information about parts which collide for the given configuration. In graphical mode the parts which collide are red-colored. The user interface of the application is provided in Fig. 4d.[1]

The collision detection module for the walking robots is mainly used by controller and motion planner. Thus, collision checking for robots' parts should be fast. We performed the series of tests which checked response time of the collision detection module. As the response time depends on the number of colliding parts we defined seven configurations of the robot. The defined configurations of the robot are presented in Fig. 4d. The first configuration is collision free. The number of colliding parts increases gradually from 0 to 19 for the following configurations. For the last configuration each part of the robot collides at least once with other part.

[1] The collision detection software is available at https://github.com/LRMPUT/CollisionDetection.

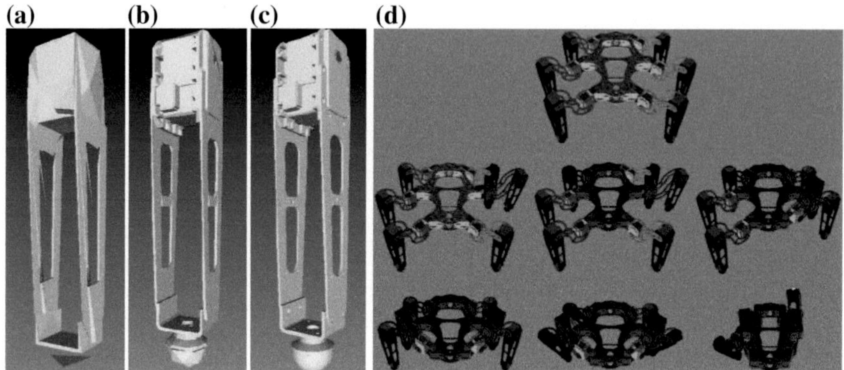

Fig. 4 Triangle mesh models of the last link of the robots' leg: model$_{2\,\%}$ (**a**), model$_{40\,\%}$ (**b**), model$_{100\,\%}$ (**c**) and the defined set of robots' configurations (**d**)

The time of the collision checking depends also on the complexity of the triangle mesh model. Thus, we defined four models of the robot. The first model consists of 50883 vertices (model$_{100\,\%}$). Then we reduced the complexity of the model by limiting the number of vertices. The number of vertices in second model (model$_{70\,\%}$) is reduced by 30 % to 35012 vertices. Similarly, we created models containing 40 % and 2 % of the vertices from initial model (model$_{40\,\%}$ and model$_{2\,\%}$). The complexity of models is presented in Table 1. The triangle mesh models of the last link of the robots' leg for model$_{2\,\%}$, model$_{40\,\%}$ and model$_{100\,\%}$ are presented in Fig. 4. Despite the fact that the last model (model$_{2\,\%}$) consists of only 2 % of the vertices from the initial model, the shape of the part is still preserved (cf. Fig. 4a).

The results of the experiment are presented in Table 2. In all experiments we used the computer with Intel Core i7-2630QM CPU 2.00 GHz microprocessor. As

Table 1 Complexity of the models used to verify the response time of the collision detection module

Model	Model$_{100\,\%}$	Model$_{70\,\%}$	Model$_{40\,\%}$	Model$_{2\,\%}$
Number of vertices	50883	35012	19688	908
Number of faces	100627	70434	43523	2008

Table 2 Collision detection time (ms) for various complexity models and different number of colliding parts

Number of collisions/model	0	3	7	9	13	16	19
Model$_{100\,\%}$	3.95	4.13	4.50	4.58	5.04	6.37	7.18
Model$_{70\,\%}$	2.46	2.94	3.13	3.18	3.62	4.94	5.28
Model$_{40\,\%}$	1.92	2.13	2.27	2.33	2.75	3.44	3.97
Model$_{2\,\%}$	1.35	1.41	1.75	1.74	1.95	2.08	2.22

expected, the collision detection time increases with the number of colliding parts. For the full model of the robot the maximal detection time is 7.18 ms. In practice the collision test is much faster. When the configuration of the robot is collision-free the test takes 3.95 ms. Thus, the self-collision detection test can be used with frequency higher than 100 Hz. We use the collision detection module on board of the robot. Before the reference values are sent to the servomotors the controller checks if the reference position is collision-free. The robot can stop the motion if the parts collide and prevent the robot from being broken down.

The execution time of collision checking procedure can be reduced by using simplified model of the robot. Collision checking for the model$_{2\,\%}$ which consists of 2 % of the triangles from the initial model is three times faster. The checking procedure takes 1.35 ms if configuration is collision-free and 2.22 ms when 19 parts of the robot collide.

The results show that the presented collision detection methodology is fast and precise and can by used by the on-board controller of the robot. On the other hand, the simplification of the model does not bring significant improvement in the computational overhead. The computation time is still above 1 ms. Similarly, when parts of the robot don't collide the computation still takes more than 1 ms. This situation is caused by the mechanical structure of the robot. The parts are close to each model. The fast method which uses bounding boxes always detects collision thus it's necessary to perform computationally expensive triangle-to-triangle intersection test for all meshes.

4.2 Approximation of Collision Detection Model

To speed up collision checking we propose the application of the Gaussian Mixture (GM) regression method. The GM is used to find the relation between configuration of the robot $\mathbf{q}$ and the collision value f_c:

$$f_c = f(\mathbf{q}). \tag{1}$$

Collision value f_c is a binary value which encodes collision and lack of collisions. When f_c is used the information on which part collides is lost. To find the relation between f_c and configuration of the robot $\mathbf{q}$ a least-squares Gaussian-mixture approximation is used:

$$f_c = \sum_{j=0}^{m} c_j \cdot \phi_j(\mathrm{q}) = \sum_{j=0}^{m} c_j \cdot \exp\left(\sum_{i=1}^{N} \lambda_{i,j}(q_i - a_{i,j})^2 \right). \tag{2}$$

where m represents the complexity of the model. The m value is directly connected to the number of multi-dimensional Gaussians used for regression. The polynomial (2) allows to detect collision between part of the robot with given configuration $\mathbf{q}$.

To determine the c_j values, the Least-Squares Fitting method [6] is applied on the set of $\mathbf{p}_r$ points obtained in the data acquisition stage. The method uses the Gram matrix $\mathbf{G}$:

$$\mathbf{G} = \mathbf{V}^T \cdot \mathbf{V}, \ \ \mathbf{c} = \mathbf{G}^{-1} \cdot \mathbf{V}^T \cdot \mathbf{r}, \tag{3}$$

where

$$\mathbf{c} = [c_1, c_2, \ldots, c_m]^T \tag{4}$$

is the vector of c_j coefficients of the length m, and

$$\mathbf{V} = \begin{bmatrix} \phi_1(\mathbf{q}_{r_1}) & \phi_2(\mathbf{q}_{r_1}) & \ldots & \phi_j(\mathbf{q}_{r_1}) \\ \ldots & \ldots & \ldots & \ldots \\ \phi_1(\mathbf{q}_{r_k}) & \phi_2(\mathbf{q}_{r_k}) & \ldots & \phi_j(\mathbf{q}_{r_k}) \end{bmatrix} \tag{5}$$

is a Vandermonde matrix, while $\mathbf{r}$ is a vector of collision values f_c corresponding to the respective $\mathbf{q}_r$ learning points.

The centroid $\mathbf{a}_i$ for each n-dimensional Gaussian function should be found, as well as the width λ_i of each kernel. In practice, an identical width for all Gaussian kernels should be avoided [5]. Their widths should depend on the positions of the centroids, which in turn depend on the data distribution in the input space [5]. In the formula (2), the width of the q-th Gaussian kernel changes in all dimensions.

To find the parameters of Gaussian mixture the Particle Swarm Optimization (PSO) is used [15]. One particle stores exactly the whole information about the position (mean values) and width λ_i (which is related to the variance) of each Gaussian in the approximation polynomial. During the algorithm operation the PSO method moves n-dimensional Gaussians through the search space and changes their variance to find the best-fitted polynomial base. The Least-Squares Fitting method finds appropriate weights to compose the final approximation function. In all experiments presented in this paper the population size is 300. The maximal number of iterations of the PSO method is 40.

The configuration space of the robot is 18-dimensional. Considering the fact that approximation in high-dimensional space is challenging and problematic for Gaussian mixture we use two separate types of functions (1). First function is used to detect collision for a single leg. The dimension of the input vector is three ($dim(q) = 3$). The output is a binary information about collision. The function should return 1 if parts of the leg collide and 0 if the configuration of the leg is collision-free. We determine separate function for each leg because the leg collides with the robots' trunk for different reference configuration (the kinematic model of each leg is the same but mounting point on the robots' trunk is different).

We also take into account collisions between legs. In this case we approximate a six-dimensional function (1). The arguments of the function are configurations of two neighboring legs. The functions return 1 if legs collide and 0 if legs don't collide. We determine four separate functions. The functions are determined for all neighboring legs on the left and the right side of the robot. The collision functions for front and rear legs are not determined because they are too far and they can not collide.

To learn single leg collision model we use 40000 samples. The samples are obtained from the mesh-based collision model. The number of Gaussian functions in the mixture is 150. Despite the fact that we learn binary output the mixture returns real values. We discretize the output to obtain binary values. The threshold is set to 0.5. All values below threshold are considered as 0 (lack of collisions) and all values above threshold are considered as 1 (leg collide). In spite of the fact that the Gaussian mixture is designed to approximate continuous functions it also performs well for functions with the binary output. The number of correctly classified collisions for the testing dataset is 94.93 %. The average error is 0.15.

The number of correctly classified collisions for neighboring legs is 87.23 %. The average error is 0.25. In neighboring legs collision model we use 220 Gaussian functions. By using Gaussian mixture-based collision model we can detect collisions between components of the robot. The collision checking procedure, which uses Gaussian mixture, takes 25 μs. It means that the approximated collision model is 40 times faster than the module which uses triangle meshes.

We can combine GM-based and triangle mesh-based collision models to obtain fast and precise collision checking procedure. The GM model can be used instead of OBB test to perform initial test. If fast GM method detects collision we can run precise triangle mesh-based method to detect which parts of the robot collide. In this case the execution time changes from 25 μs to few milliseconds. However, in most cases the legs of the robot don't collide and the proposed procedure is very fast. To increase accuracy of classification we are going to apply deep learning method instead of Gaussian mixture.

5 Conclusions and Future Work

In the paper we present the self-collision detection method for walking robots.[2] The precise collision model is based on the triangle mesh extracted from CAD model of the robot. The model allows to check collisions between parts at the planning stage before the reference configuration is sent to the servomotors. The method can be applied in the motion planning module to check if the planned sequence of robots configurations is feasible. The collision detection method is fast and can be also used on the controller of the robot.

In the paper we also show that the Gaussian mixture can be used to approximate the collision model in order to significantly reduce the execution time of the collision checking procedure. When Gaussian mixture is used the information about colliding parts is lost. We obtain information that the parts collide but the identifiers of these parts are unknown. However, the binary information about collisions is sufficient in

[2]Video is available on http://lrm.cie.put.poznan.pl/coldetAut2016.wmv.

many applications. For example the sampling-based methods used for motion planning mostly require binary information about collisions for the given position or configuration of the robot [3].

In future we are going to combine collisions and workspace of the robot in a single Gaussian mixture model. Then, we are going to use gradient-based optimization to maximize kinematic margin of each leg and the robot during walking on rough terrain. This approach is possible for Gaussian mixture model because the model is differentiable. Such approach is not possible for the triangle mesh-based model.

References

1. Belter, D., Łabecki, P., Skrzypczyński, P.: Adaptive motion planning for autonomous rough terrain traversal with a walking robot. J. Field Robot. (in print)
2. Belter, D., Walas, K.: A compact walking robot—flexible research and development platform, recent advances in automation. In: Szewczyk, R., Zieliński, C., Kaliczyńska, M. (eds.) Robotics and Measuring Techniques, vol. 267, pp. 343–352 (2014)
3. Belter, D., Skrzypczyński, P.: Posture optimization strategy for a statically stable robot traversing rough terrain. In: IEEE/RSJ 2012 International Conference on Intelligent Robots and Systems, pp. 2204–2209. Vilamoura, Portugal (2012)
4. Belter, D., Skrzypczyński, P.: Integrated Motion Planning For A Hexapod Robot Walking on Rough Terrain, 18th IFAC World Congress. Milan, Italy (2011)
5. Benoudjit, N., Archambeau, C., Lendasse, A., Lee, J., Verleysen, M.: Width optimization of the Gaussian kernels in radial basis function networks. In: Proceedings of European Symposium on Artificial Neural Networks, pp. 425–432. Bruges (2002)
6. Dahlquist, G., Björck, A.: Numerical Methods. Series in Automatic Computing. Prentice Hall, New Jersey (1974)
7. De Luca, A., Flacco, F.: Integrated control for pHRI: collision avoidance, detection, reaction and collaboration. In: IEEE RAS/EMBS International Conference on Biomedical Robotics and Biomechatronics, pp. 288–295 (2012)
8. Ericson, C.: Real-time Collision Detection. CRC Press (2004)
9. Geva, A.: ColDet 3D Collision Detection. http://sourceforge.net/projects/coldet/ (2015)
10. Haddadin, S., Albu-Schäffer, A., De Luca, A., Hirzinger, G.: Collision detection and reaction: a contribution to safe physical human-robot interaction. In: IEEE/RSJ International Conference on Intelligent Robots and Systems, pp. 3356–3363 (2008)
11. Hermann, A., Klemm, S., Xue, Z., Roennau, A., Dillmann, R.: GPU-based real-time collision detection for motion execution in mobile manipulation planning. In: IEEE International Conference on Advanced Robotics, pp. 1–7. Montevideo (2013)
12. Hermann, A., Bauer, J., Klemm, S., Dillmann, R.: Mobile manipulation planning optimized for GPGPU voxel collision detection in high resolution live 3D-maps. In: 41st International Symposium on Robotics, ISR/Robotik 2014, pp. 1–8. Munich, Germany (2014)
13. Hermann, A., Drews, F., Bauer, J., Klemm, S., Roennau, A., Dillmann, R.: Unified GPU voxel collision detection for mobile manipulation planning. In: IEEE/RSJ International Conference on Intelligent Robots and Systems, pp. 4154–4160. Chicago (2014)
14. Hutter, M., Gehring, C., Bloesch, M., Hoepflinger, M., Remy, C.D., Siegwart, R.: StarlETH: a compliant quadrupedal robot for fast, efficient, and versatile locomotion. In: Proceedings of the International Conference on Climbing and Walking Robots (CLAWAR), pp. 483–490 (2012)
15. Kennedy, J., Eberhart, R.C.: Particle swarm optimization. In: Proceedings of IEEE International Conference on Neural Networks, pp. 1942–1948. Piscataway (1995)

16. Klosowski, J.T., Held, M., Mitchell, J.S.B., Sowizral, H., Zikan, K.: Efficient collision detection using bounding volume hierarchies of k-DOPs. IEEE Trans. Vis. Comput. Graph. **4**(1), 21–36 (1998)
17. Min, T., Kim, Y.J., Manocha, D.: C2A: controlled conservative advancement for continuous collision detection of polygonal models. In: IEEE International Conference on Robotics and Automation, pp. 849–854 (2009)
18. Möller, T.: A fast triangle-triangle intersection test. J. Graph. Tools **2**, 25–30 (1997)
19. Pan, J., Manocha, D.: GPU-based parallel collision detection for fast motion planning. Int. J. Robot. Res. **31**(2), 187–200 (2012)
20. Plagemann, C., Fox, D., Burgard, W.: Efficient failure detection on mobile robots using particle filters with Gaussian process proposals. In: Proceedings of the 20th International Joint Conference on Artificial Intelligence, pp. 2185–2190 (2007)
21. Schwarzer, F., Saha, M., Latombe, J.C.: Exact Collision Checking of Robot Paths. Springer (2004)
22. Shen, H., Heng, P.A., Tang, Z.: A fast triangle-triangle overlap test using signed distances. J. Graph. Tools **8**(1), 3–15 (2003)
23. Tropp, O., Tal, A., Shimshoni, I.: A fast triangle to triangle intersection test for collision detection. Comput. Animation Virtual Worlds **17**(5), 527–535 (2006)

Incremental Version Space Merging Approach to 3D Object Model Acquisition for Robot Vision

Jan Figat and Włodzimierz Kasprzak

Abstract A *concept learning* algorithm is developed, which uses the visual information generated by a *virtual receptor* in a robotic system (e.g. symbolic image segments) to create learning examples. Its goal is to detect similarities in the training data and to create an appropriate object model. The *version-space*, intended to describe the possible concept hypotheses, is generated by a novel IVSM-ID algorithm, the *incremental version space merging with imperfect data*, that deals with partly imperfect and noisy training data—a common problem in computer vision systems. The generated model takes the form of a graph of constraints with fuzzy predicates. The approach is verified by learning concepts of elementary surface and solid primitives on base of segmented RGB-D images, taken for various light conditions and for different exposure times.

Keywords Inductive learning · Version spaces · 3D objects · Model acquisition · Robot perception

1 Introduction

Automatic model acquisition, that uses machine learning techniques, is an important technology issue in autonomous robot vision. It is expected, that reliable visual scene recognition will require a mutually supporting integration of image-driven and model-based analysis [1]. So far, the dominating approach in image-related object recognition, is to cluster low-level image primitives like 2-D or 3-D point features [2, 3], or to aggregate image regions or surface patches [4]. Recently, increased

J. Figat (✉) · W. Kasprzak
Institute of Control and Computation Engineering, Warsaw University of Technology, Nowowiejska 15/19, 00-665 Warsaw, Poland
e-mail: J.Figat@ia.pw.edu.pl
URL: http://robotics.ia.pw.edu.pl

W. Kasprzak
e-mail: W.Kasprzak@elka.pw.edu.pl

R. Szewczyk et al. (eds.), *Challenges in Automation, Robotics and Measurement Techniques*, Advances in Intelligent Systems and Computing 440, DOI 10.1007/978-3-319-29357-8_49

impact of the model-based analysis paradigm is observed [5], as robot vision systems reach more abstract, symbolic data representation levels (in terms of concepts representing 3D objects and scenes). Hence, efficient model acquisition techniques are needed in robot vision.

For a long time period, a large gap existed between available machine learning algorithms [6, 7], designed for high-level symbolic knowledge, and low-level image training data available in robot vision. Recently, the growing use of RGB-D sensors in machine perception, encourages the design of new learning methods that may close such observed technology gap. The representation of 3D objects seems to be the interface level between the two domains—machine learning [8] and robot control [9].

On the pother hand, *Concept learning* is an important aspect of automatic model acquisition, considered in the machine intelligence field [6]. It is a well-known problem of inductive machine learning. In order to create a desired concept, which is initially unknown, a background knowledge (typically expressed in some concept description language) and appropriate training data must be given. Training data must contain positive and negative examples of a concept to be learned. A popular form of concept description is the *version-space*, defined by Mitchell [6]. *Version-space* is a subset of hypotheses consistent with the training examples, where consistent means that the examples are correctly classified. The most popular method for *version-space* creation is the *Candidate Elimination Algorithm* and its various modification, such as analytically-generalized positive examples [10] or analytically-generalized negative examples [11]. In this approach, imperfect and noisy data can not be taken into account.

A generalized version of the space learning algorithm for noisy and uncertain data was introduced by Hong [12]. In this algorithm, each training instance can be thought of as partially positive and partially negative, once a certainty factor is attached to each training instance to represent its uncertainty. Hong's algorithm allows the use of the learning examples with partial adherence to the positive training set, but does not allows the use of uncertain data within learning examples. Accordingly, the use of imperfect data, which often occurs in robotics, will be very difficult for mentioned algorithm.

A popular approach to *version-space* learning is based on the *Incremental Version Space Merging (IVSM)* algorithm proposed by Hirsh [13–15]. *IVSM* computes the intersection of two *version-spaces*, while using only *boundary set* representations (S and G). This computation is done in a two-step process. First, the set of minimal generalizations of pairs from S_I and S_n are assigned to the set $S_{I \cap n}$. Secondly, the set of maximal generalizations of pairs from G_I and G_n are assigned to the set $G_{I \cap n}$. The second step removes the overlaid specific elements from the set $G_{I \cap n}$ and the overlaid general elements from the set $S_{I \cap n}$.

In this paper, a concept learning algorithm is developed, based on visual information, which generates models of elementary 3D surfaces and volumes (solids). A new object concept is created together with its symbolic model expressed in terms of a graph of constraints with fuzzy predicates. To create the *version-space*, which is intended to describe the different possible concepts, a modification of the IVSM algorithm is proposed, called IVSM-ID, that deals with partly imperfect and noisy training data—a common problem in computer vision systems.

A *virtual receptor* is applied [16], to generate contour and region hypotheses in images from the training set. Homogeneous regions and line segments are detected, such as straight lines and elliptic arcs. Examples of objects, that can be defined using such primitives, are cuboid, cylinder, cone and generalized cone.

2 The Application Problem

The problem of automatic creation of a 3D solid model can be decomposed into following subproblems:

1. Acquisition of training images—images of given solid type taken for many particular solids and from different views;
2. Image segmentation—detecting typical image structures and relationships, occurring between them;
3. Counting the frequency of relationships in the training data;
4. Assigning a label to every learning example (positive or negative)—based on the frequency of relationship detection;
5. Version-space creation—usage the proposed algorithm called *IVSM-ID* (see Sect. 3);
6. Learning of attribute evaluation functions—an *attribute* is a vector of values being unique to every instance of a given concept. This problem is not within scope of this paper. A function approximation can be learned using artificial neural networks, for example.

One of the assumptions in this model learning methodology is that the set of possible relationships is well known. This is part of the *common-sense knowledge*, a typical approach in the AI domain. It is also worth to mention that the relationship can be symmetric or not.

In the images of given object, a particular relationship appears with various a priori probability or membership function value (in fuzzy set theory). An example of such vague relationship can be the parallelism between two straight lines visible in the image. Lines which are parallel in a real environment are most often not exactly parallel in the image. These lines can be seen to be inclined at an angle. Considering this angle's distance from 180°, the relationship "quality" can be defined on a scale from 0 to 1, tailored into bins, while the final pdf can be estimated from the frequency of such quality bins.

Before moving to the main contribution of this paper—the version space learning algorithm—let us draw some remarks to the training data preparation steps.

Ad 1. An RGB-D sensor, such as the *Kinect* device, is used to acquire RGB images with corresponding 3-D point clouds.

Ad 2. Typical image segments and their relationships are detected in the RGB-D image. For example, the specified segments for generalized cone are as follows: a Polyline, a circle or ellipse part, a texture class or a region. Examples of relationships for these segments are as follows: “shared point”—a bilateral relationship, “inclusion”—an unidirectional relationship.

Ad 4. Learning examples, which were classified below the frequency of detected relationship threshold, will be labeled as *negative learning examples*. While those examples, which were classified above this threshold, will be classified as *positive learning examples*. The input to *version-space learning* should consists of both *positive learning examples* and *negative learning examples*.

Version-space is learned while using the proposed modification of Hirsh’s algorithm *IVSM*, the IVSM-ID. This novel algorithm is described in the Sect. 3.2.

3 Incremental Version-Space Learning

3.1 Basic Approach—IVSM

In the *version space* approach to concept learning both positive and negative examples are needed. A positive learning example affects the **S-set** of the version-space boundary sets, while a negative learning example affects the **G-set**. These learning examples should be characterized in an appropriate way, such that the taught concepts can be differentiated from each other.

The *Incremental version-space merging* (IVSM) algorithm uses previously created version-spaces. It is acceptable that version-spaces were taught for each of examples. As an input to this algorithm two version-spaces are given. First is the previously learned VS_n, and the second is the new one VS_I (not used previously—containing new information).

The basic IVSM algorithm proceeds as follows:

1. Creation of the version-space for new piece of information (VS_I),
2. Intersection of newly created version-space (VS_I) with the version-space generated from past information (VS_n) (see Eq. 1):

$$VS_{n+1} = VS_n \cap VS_I \tag{1}$$

3. Returning to the first step, if the new piece of information is given.

For the first creation of the *version-space* intersection, as the *version-space* for past data, the initial *version-space* contains all concept descriptions in the language [6], as it was in the initial step of the Mitchell’s *Candidate Elimination Algorithm*. The IVSM finds boundary sets $S_{I \cap n}$ i $G_{I \cap n}$, as the result of the intersection of version-spaces VS_I and VS_n, as shown on Fig. 1a.

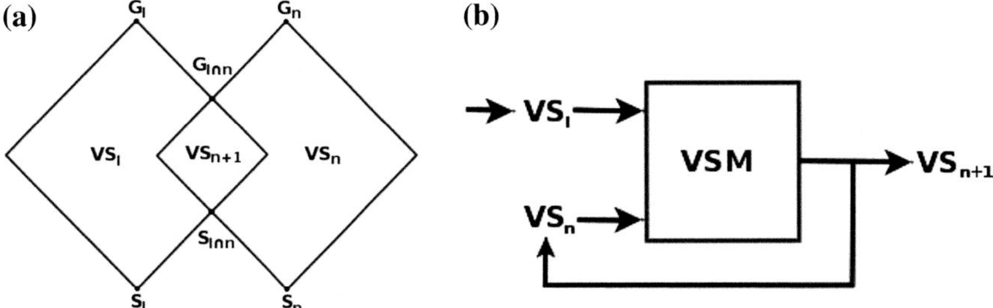

Fig. 1 Incremental version-space merging (IVSM). **a** Intersection of the version-spaces. **b** IVSM diagram

3.2 The Novel IVSM-ID Algorithm

The *Incremental Version Space Merging* algorithm has been modified to take into account the incomplete compatibility of a concept with all the positive observations, by using fuzzy measures. As it is known, the relationships detected in the image are very often not ideal, especially the properties: parallel and perpendicular. Actually, these relationships are satisfied with some probability.

It is possible, for both the basic algorithm *IVSM* and the new proposed one, called *IVSM-ID*, to create version-spaces for all of the learning examples before running the real concept learning algorithm. The analysis of this set and the observation how frequently a particular predicate appears in such a set, leads to the estimation of their prior fuzzy membership functions in version spaces. The subsequent modification of the IVSM learning takes into account these fuzzy measures. The proposed *IVSM-ID* algorithm proceeds as follows:

1. Estimate fuzzy membership functions for given predicates in the training set.
2. FOR *sample* $\in$ *training_set* DO:
   (a) Creation of the version-space for new sample (piece of information (VS_I) for the fuzzy data,
   (b) Intersection of newly created version-space (VS_I) with the version-space generated from past information (VS_n),
   (c) Continue from the first step, if a new sample is given.

The *IVSM* uses the Candidate Elimination Algorithm, while the *IVSM-ID* uses its modification, in case of fuzzy data, which were not supported in the original algorithm.

Ad 2a. The version-space can be formed for deterministic or non-deterministic data. In later case, real numbers from the set $[0, 1]$ can characterize the learning examples. In case of non-deterministic data, the boundary sets will be obtained as follows:

a. For the positive learning example:

   - a S-set is a most specific one, so it contains the positive example—Size of a S-set is equal to one,
   - the G-set is the most general one, so it contains all possible values of variables (the most general)—Size of the G-set is equal to one,

b. For the negative learning example:

   - a S-set is empty,
   - the G-set contains all of the possible variable values excluding the one appearing in the negative learning example—Size of the G-set can be much bigger than one (the size of G-set depends on the set of permissible values). Instead of values 0 or 1, the set $[0,x) \cup (x,1]$ will be used in the G-set, where $x \in (0,1)$.

Ad 2b. A new important issue is the definition of set intersection for fuzzy data. During the tests, it appeared that it is sufficient to use the two functions, $\min(A, B)$ and $\max(A, B)$, defined in fuzzy logic, in order to determine the intersection $A \cap B$ for the fuzzy data. The decision of function selection is taken on the basis of the S-sets. The G-set is specified, while the S-set is generalized, if possible.

4 Concept Learning for the Cone

The result of image segmentation consists of a list of symbolic segment descriptions, preferably boundaries, homogeneous regions and key points. New image segments are compared with segments detected in previous images. When succeeded, this positive match increases a probability that the new segments may be linked to the same solid type as the previous ones.

If we are using a fixed camera, i.e. its position does not change during the acquisition of training samples, the new image segments will be located in the same image place as in previous images, but if the camera orientation or position are changed, the corresponding camera coordinate transformation must be estimated first before a comparison of detected segments is started. When creating a boundary set for a side view of a solid, it is important that camera looks at the solid in the same way. We say, that the same **modality** of an object must be observed in the training set. Thus, a version-space is created for a particular view type of a particular solid type.

Negative learning examples may come from the observation of a solid of the same type and the same modality as the positive examples. The set of segment attributes and relationships between segments, which appeared less frequently than expected (below some predefined threshold), are designated as negative learning examples.

4.1 The Training Images

The training images contain several objects of the same type, collected by a single camera in a fixed position. During image capture, the light conditions and camera's exposure time were changed. Segment detection is performed by a software package, called *virtual receptor*. The version space for a cone will correspond to the largest homogeneous area detected in the image, enclosed by an elliptic arc and straight line segments, see Fig. 2. The corresponding image segments are shown in Fig. 3.

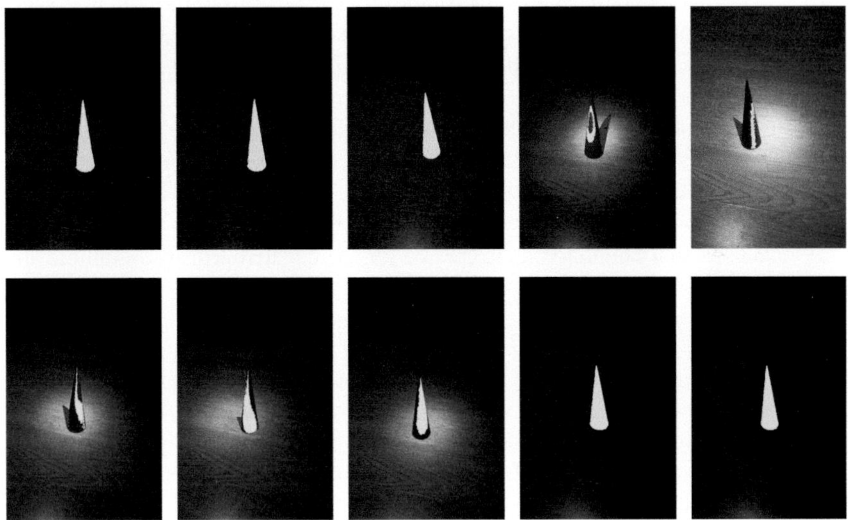

Fig. 2 Example of homogeneous image areas representing the cone's *side view*

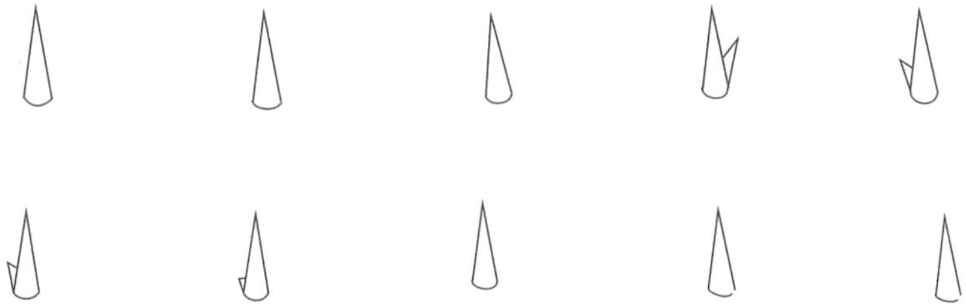

Fig. 3 Examples of segments assigned to a cone concept

4.2 Version-Spaces Created by IVSM-ID

This chapter contains exemplary results of the *Incremental Version Space Merging—with Imperfect Data (IVSM-ID)* algorithm. The input data to this algorithm consists of segment attributes and relationships, and their allocation to the class of positive or negative learning example. Actual relationships are characterized by a membership function value, expressing their level of relationship type satisfaction. Therefore these values belong to the unit interval $I([0, 1])$.

Examples presented in the Table 1 were used for the process of the boundary set creation. The membership values were established on the basis of the data frequency found by analysing more examples. The learning examples were obtained for images of the same view type of same type solids, acquired under different light conditions or exposure time.

Positive learning examples hold the predicates characterized by fuzzy variables with sufficiently similar and high values. Predicates which occur rarely were classified as negative learning examples.

Table 2 presents version spaces created for the learning examples, containing fuzzy data. The first boundary set is created for the negative learning example and the second one—for the positive one.

The version space created by the IVSM-ID algorithm for the data presented in Table 1 is given in Table 3. This result was created from version spaces corresponding to single learning examples. The fuzzy level of boundary set satisfaction is

Table 1 Learning examples

Example number	Number of ellipses	Number of polylines	Number of contact points	Homogeneous area is inside segments?	Homogeneous area occupancy	Learning example class
1	Zero	One	Zero	0	0	–
2	One	One	One	0.89	0.4	–
3	One	One	One	0.90	0.8	+
4	One	One	Two	0.99	0.89	+
5	One	One	One	0.89	0.1	–
6	One	One	Two	0.95	0.97	+
7	One	One	Two	0.92	0.9	+
8	One	One	One	0.99	0.95	+
9	One	One	Two	0.99	0.95	+
10	One	Two	Three	0.88	0.42	–
11	One	Two	Two	0.8	0.3	–
12	One	Two	Two	0.5	0.4	–
13	One	One	Two	0.99	0.95	+

Table 2 Example of version spaces created for data shown in Table 1

Instance Boundary Set for the 2-nd learning example:
G={<zero,any-polylines-no.,any-contact-points-no.,?,?>,
<two,any-polylines-no.,any-contact-points-no.,?,?>,
<any-ellipses-no.,two,any-contact-points-no.,?,?>,
<any-ellipses-no.,any-polylines-no.,zero,?,?>,
<any-ellipses-no.,any-polylines-no.,two,?,?>,
<any-ellipses-no.,any-polylines-no.,three,?,?>,
<any-ellipses-no.,any-polylines-no.,any-contact-points-no.,[0, 0.89) or (0.89, 1],?>,
<any-ellipses-no.,any-polylines-no.,any-contact-points-no.,?,[0, 0.4) or (0.4,1]>}
S={<@,@,@,@,@>}

Instance Boundary Set for the 3-rd learning example:
G={<any-ellipses-no.,any-polylines-no.,any-contact-points-no.,?,?>}
S={<one,one,one,0.90,0.8>}

Table 3 Output of the IVSM-ID algorithm for data shown in Table 1

	Instance Boundary Set
S:	(<one,one,one,0.90,0.8>,
	<one,one,two,0.92,0.89>)
G:	(<one,any-polylines-no.,any-contact-points-no.,?,?>,
	<two,any-polylines-no.,any-contact-points-no.,?,?>,
	<any-ellipses-no.,two,any-contact-points-no.,?,?>,
	<any-ellipses-no.,any-polylines-no.,one,?,?>,
	<any-ellipses-no.,any-polylines-no.,two,?,?>,
	<any-ellipses-no.,any-polylines-no.,three,?,?>,
	<zero,any-polylines-no.,any-contact-points-no.,?,?>,
	<any-ellipses-no.,any-polylines-no.,zero,?,?>,
	<any-ellipses-no.,any-polylines-no.,any-contact-points-no.,(0.89, 1],?>,
	<any-ellipses-no.,any-polylines-no.,any-contact-points-no.,?,(0.42, 1]>,
	<any-ellipses-no.,one,any-contact-points-no.,?,?>)

checked during the process of making the decision about the concept membership of tested example. For this purpose, both the S-set with its generalization and the G-set with its specialization are taken into consideration.

5 Summary

The paper presents a new concept learning algorithm from image data, based on the idea of *Incremental Version Space Merging* [10]. The main novelty of presented approach is the adaptation of an inductive learning methodology, defined for high-level symbolic data, to imperfect training data extracted from images. Hence, relationship instances are not evaluated simply by Boolean values, but fuzzy "quality" scores can be assigned to them. The proposed algorithm (called *Incremental Version Space Merging with Imperfect Data (IVSM-ID)*) deals with stochastic or fuzzy data measures.

The approach was verified by learning concepts of elementary solid primitives on base of segmented RGB-D images. The learning algorithm iteratively develops two version space sets—the specialization set S and the generalization set G—and decides about concept membership of every tested example.

In future work the presented methodology will be adopted for complex object modelling. One will detect the necessary primitives in the RGB-D image—basic solids with their attributes—localization, size, texture. The IVSM-ID is expected to learn different parts of a complex object.

Acknowledgments The authors gratefully acknowledge the support of the National Centre for Research and Development (Poland), grant no. PBS1/A3/8/2012.

References

1. Sagerer, G., Niemann, H.: Semantic Networks for Understanding Scenes. Springer (1997)
2. Pangercic, D., Haltakov, V., Beetz, M.: Fast and robust object detection in household environments using vocabulary trees with sift descriptors. In: IEEE/RSJ International Conference on Intelligent Robots and Systems (IROS). San Francisco, CA, USA (2011)
3. Collet, A., Martinez, M., Srinivasa, S.S.: The moped framework: object recognition and pose estimation for manipulation. Int. J. Robot. Res. **30**(10), 1284–1306 (2011)
4. Richtsfeld, A., Mörwald, T., Prankl, J., Zillich, M., Vincze, M.: Learning of perceptual grouping for object segmentation on RGB-D data. J. Vis. Commun. Image Represent. **25**(1), 64–73 (2014)
5. Heisele, B., Kim, G., Meyer, A.J.: Object recognition with 3D models. In: Proceedings of the British Machine Vision Conference, pp. 29.1–29.11 (2009). doi:10.5244/C.23.29
6. Mitchell, T.M.: Version spaces: a candidate elimination approach to rule learning. In: Proceedings of the 5th International Joint Conference on Artificial Intelligence—vol. 1. IJCAI'77, San Francisco, CA, USA, Morgan Kaufmann Publishers Inc., pp. 305–310 (1977)
7. Kodratoff, Y., Michalski, R.S.: Machine Learning: An Artificial Intelligence Approach, vol. 3. Morgan Kaufmann (2014)
8. Jordan, M., Mitchell, T.: Machine learning: trends, perspectives, and prospects. Science **349**(6245), 255–260 (2015)
9. Zieliński, C., Winiarski, T.: General specification of multi-robot control system structures. Bull. Pol. Acad. Sci. Tech. Sci. **58**(1), 15–28 (2010)
10. Hirsh, H.: Combining empirical and analytical learning with version spaces. In: Proceedings of the Sixth International Workshop on Machine Learning, pp. 29–33 (1989)
11. Carpineto, C.: Analytical negative generalization and empirical negative generalization are not cumulative: a case study. In: Proceedings of the European Working Session on Learning on Machine Learning. EWSL-91, New York, USA. Springer, New York, Inc., pp. 81–88 (1991)
12. Hong, T.P., Tsang, S.S.: A generalized version space learning algorithm for noisy and uncertain data. IEEE Trans. Knowl. Data Eng. **9**(2), 336–340 (1997)
13. Hirsh, H.: Incremental version-space merging. In: Incremental Version-Space Merging: A General Framework for Concept Learning. The Kluwer International Series in Engineering and Computer Science, vol. 104, pp. 9–14. Springer, US (1990)

14. Haym Hirsh, N.M., Pitt, L.: Version spaces without boundary sets. In: Proceedings of the 14th National Conference on Artificial Intelligence and 9th Innovative Applications of Artificial Intelligence Conference (AAAI-97/IAAI-97), pp. 491–496 (1997)
15. Hirsh, H., Mishra, N., Pitt, L.: Version spaces and the consistency problem. Artif. Intell. **156**(2), 115–138 (2004)
16. Kasprzak, W., Kornuta, T., Zieliński, C.: A virtual receptor in a robot control framework. In Szewczyk, R., Zieliński, C., Kaliczyńska, M. (eds.) Recent Advances in Automation, Robotics and Measuring Techniques. Advances in Intelligent Systems and Computing, vol. 267, pp. 399–408. Springer International Publishing (2014)

Simulation-Based Design of Mobile Ad Hoc Network for Tracking and Monitoring

Piotr Bazydło, Ewa Niewiadomska-Szynkiewicz, Kamil Czerwiński and Patryk Rękawek

Abstract This paper addresses issues of the application of computer simulation to design a wireless network for tracking and monitoring purposes. The network is formed by mobile robots carrying sensors. To achieve a common goal it has to automatically adopt to changes in the environment and ensure the permanent connectivity with the central decision unit. The focus is on high coverage of given region of interest and motion trajectory planning avoiding obstacles. Two calculation schemes for computing target positions of robots, i.e., off-line (centralized) and on-line (distributed) are investigated. They employ different optimization algorithms to calculate the positions of robots. Simulators of two types of ad hoc networks for detecting, tracking and monitoring of a heavy gas cloud were implemented based on V-REP and MATLAB tools. The results of presented case studies illustrate the performance of simulation-based approach to ad hoc networks design. Advantages and disadvantages of different computing schemes and selected optimization methods are described and discussed.

Keywords MANET · Robot-assisted deployment and monitoring · Mobile robotics · Global optimization · Multiple-start optimization

---

P. Bazydło (✉)
Industrial Research Institute for Automation and Measurements PIAP, Warsaw, Poland
e-mail: pbazydlo@piap.pl

E. Niewiadomska-Szynkiewicz · P. Rękawek
Institute of Control and Computation Engineering, Warsaw University of Technology, Warsaw, Poland
e-mail: ens@ia.pw.edu.pl

P. Rękawek
e-mail: patryk.rekawek@gmail.com

K. Czerwiński
Emerson Process Management Power and Water Solutions, Warsaw, Poland
e-mail: kamil.czerwinski@emerson.com

R. Szewczyk et al. (eds.), *Challenges in Automation, Robotics and Measurement Techniques*, Advances in Intelligent Systems and Computing 440, DOI 10.1007/978-3-319-29357-8_50

1 Introduction

Ad hoc network is a wireless decentralised structure network comprised of devices, which autonomously set up a network without any external network infrastructure necessary to transmit data. In general, ad hoc networks can be classified by their application. Wireless Sensor Network (WSN) and Mobile Ad hoc Network (MANET) are two common types of an ad hoc network. WSN is a dynamically configurable and distributed systems composed of numerous smart sensing devices (network nodes) deployed densely over a significant area [1–3]. Every node is equipped with one or multiple sensors and radio transceiver (in order to exchange data between nodes). WSNs are usually stationary or quasi-stationary. MANET [4, 5] comprises self-configuring mobile wireless communication devices which combine the roles of terminals and routers. Each device moves in a workspace and direct communication between each pair of nodes is usually not possible. The lack of fixed network infrastructure components both in WSN and MANET allows creating unique topologies and enables the dynamic adjustment of individual nodes to the changing environment in order to execute assigned tasks. This feature is crucial in case of emergency and monitoring systems. However, for the networks to operate in an ad hoc mode in practice, several basic issues must be solved. The most important ones are: network redundancy, inadequate nodes deployment leading to the communication problems and important data loss. The important characteristic of MANET is its mobility. The mobility modes are necessary to calculate the motion trajectories and simulate MANETs applications. Numerous mobility models have been developed and described in literature [5–7]. It is worth noting that design, development and evaluation of MANET is a non-trivial task, especially as it is envisioned to be deployed in a large scale. The complexity and the ad hoc operating mode limit the applicability of purely analytic analysis. Therefore, investigation of MANETs is achievable by resorting either to software simulators or to testbed networks.

The main contribution of this paper was to develop a simulator of MANET formed by mobile robots equipped with radio transceivers and sensors for detection, tracking and monitoring of heavy gas clouds. Various approaches to robot-assisted sensors deployment are described in literature [8–12]. The most important tasks while development mobile sensing network are: optimal coverage of a region of interest, connectivity with the central dispatcher and motion trajectory planning. We present and evaluate through simulation two MANET systems exploring a workspace with obstacles and tracking of a heavy gas cloud. The focus is on application of global optimization methods and RRT (Rapidly Exploring Random Tree) algorithm [13] for computing motion trajectories for all robots. The second objective of this paper is to demonstrate the usefulness of a software environment composed of the V-REP simulation platform and MATLAB in developing new MANET applications.

2 Problem Formulation

The purpose of the research was to develop a simulator of MANET formed by N mobile robots (R_n, $n = 1, \ldots, N$), each one equipped with radio transceiver and sensors for exploring entire workspace W for a gas cloud, and next tracking and monitoring the detected cloud. All measurements were transmitted to a central unit of the system—a base station. Numerous obstacles could be located in W.

The problem to solve was how to guide mobile robots to explore the workspace avoiding obstacles, detect the gas cloud and cover this cloud. With regard to calculation scheme and deployment process organization we can distinguish pre-defined deployment and self-organizing deployment. In pre-defined deployment all spatial positions of sensors are a priori (off-line) determined by the base station and the mobile robots are forced to move avoiding obstacles in advisable direction to reach the target positions, and a monitoring system is created. In the self-organizing (on-line) deployment all robots move in a workspace to establish a network topology that guarantee sensing and communication coverage. They are not assigned to fix spatial positions. In our research we have developed a family of algorithms for robot-assisted deployment. We have implemented two variants of our MANET operation utilizing two computation schemes, i.e., off-line (centralized) and on-line (distributed). In case of centralized deployment target positions of robots and motion trajectories are off-line calculated by the central supervisor (base station). This option requires full knowledge of the workspace. In case of unknown environment the on-line deployment with calculations performed locally by robots is usually the only possible option to create the sensing network.

3 MANET for Tracking and Monitoring

3.1 Off-Line Deployment

The target positions of all sensing robots are calculated off-line. A base station is responsible for all calculations and the nodes deployment. A full knowledge about the whole workspace and devices used to create the sensing system are available. A map of a workspace W is provided by a camera carried by a drone. The aim is to calculate the optimal deployment of robots carrying sensors in the area of interest (gas cloud) based on the global data about W The base station computes target positions of all robots and calculates collision-free motion trajectories to these points in W. The temporary loss of connectivity between robots is acceptable—robots do not exchange any information between each other until they reach desired positions. Finally all robots form a sensing network inside the cloud. The topology of this network is determined by the base station. The network follows the cloud in case of its displacement.

Hence, the following operations have to be performed to create MANET for tracking and monitoring:

1. Acquiring current map of a workspace.
2. Map processing to extract knowledge about the location of a gas cloud and all obstacles in the workspace (image processing operations).
3. Computing of optimal positions for all robots.
4. Displacement of robots to the target positions.
5. Recalculation of robot positions and shifting the network after the cloud displacement.

The image processing operations are not the subject of this paper, hence we will not focus on them. We simply assume that the whole workspace is divided into $L \cdot M$ small cells. This grid is mapped into two matrices A and B with binary values: $a_{ij} \in \{0, 1\}$, $b_{ij} \in \{0, 1\}$, $i = 1, \ldots L$, $j = 1, \ldots, M$. A maps the contaminated area and B maps the area covered by robots. The position of a central point of a cell $cellA_{ij}$ of A is described by the geographical coordinates $\left[x_i^A, y_j^A\right]$, while the position of $cellB_{ij}$ of B is described by the coordinates $\left[x_i^B, y_j^B\right]$, where i denotes a number of row and j a number of column of respectively, matrices A and B. In case of the matrix A, $a_{ij} = 0$ denotes the area free from contamination, and $a_{ij} = 1$ the contaminated area, while in case of the matrix B, $b_{ij} = 0$ denotes the cell not covered by any sensing device and $b_{ij} = 1$ the cell covered by at least one sensor. The main objective of the deployment is to maximize sensors coverage of the contaminated area. Hence, the goal is to determine the positions of robots that minimize the sensing scene uncovered by sensors. The following optimization problem is formulated and solved by the base station:

$$\min_{x_n, y_n \in R, n=1,\ldots N} \sum_{i=1}^{L} \sum_{j=1}^{M} |a_{ij} - b_{ij}| + Pf, \tag{1}$$

where

$$b_{ij} = \begin{cases} 1 & if \quad \exists_{n=1,\ldots N} \sqrt{(x_i^B - x_n)^2 + (y_j^B - y_n)^2} \leq r_n^s \\ 0 & otherwise \end{cases} \tag{2}$$

where $[x_n, y_n]$ denotes the reference point of the nth robot (e.g. the location of the nth detector), r_n^s is a sensing range of the nth robot. It is assumed that sensing coverage of the nth detector is a disc of a radius r_n^s. Equation (2) determines whether $cellB_{ij}$ is covered by at least one robot. Pf denotes the penalty for positioning robots within obstacles. The more similar matrices A and B, the better solution is obtained. The concept of the method is depicted in Fig. 1.

It is obvious that all mobile robots have to avoid collisions with obstacles. To guarantee collision free movement the safety margin is introduced around each obstacle (see Fig. 2). The value of the objective function is increased with high

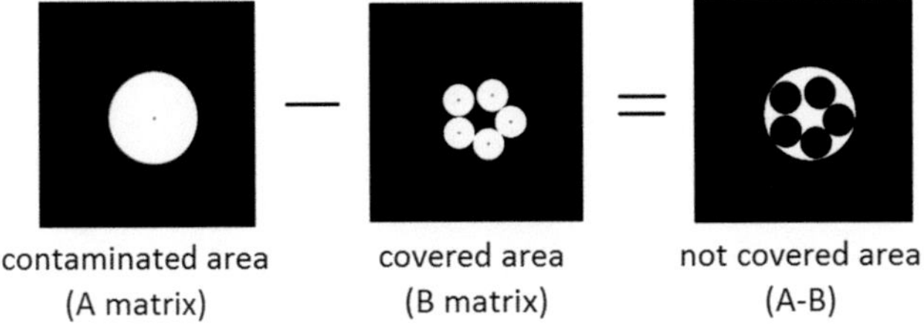

Fig. 1 Graphical interpretation of the of-line deployment strategy

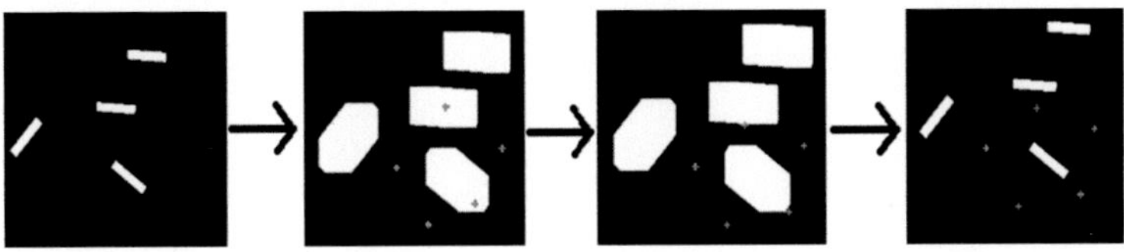

Fig. 2 Influence of the penalty function on the solution of the problem (1), (2)

value of the penalty function *Pf* in case of determined position inside the forbidden regions. The example solutions of the problem (1), (2) are presented in Fig. 2.

3.2 On-Line Deployment

Let us assume that a map of a workspace is unknown. In such application scenario the on-line deployment with self-organizing robots can be applied. The problem to be solved is how to determine the optimal positions of all robots carrying sensors to cover the monitoring scene and imply connectivity among all these devices and the base station. It is assumed that robots carrying sensors move in space and identify on-line all obstacles. Moreover, each robot broadcasts its reference point location, and known reference points of its neighbors. Every time step each robot calculates its new position in a workspace based on the information gathered from other robots. In general, the deployment process is divided into two phases (1) workspace exploration (searching for the gas cloud), (2) coverage of the detected cloud. Hence, in the first phase all robots explore the environment in order to find the area to monitor. After detection a contaminated area monitoring and tracking tasks are performed.

Phase 1: In the exploration phase the leader of the robot team randomly samples its surrounding and moves to the region with higher degree of contamination. The leader is equipped with 3D laser or other specialized sensors [14], and is responsible for a gas cloud detection. Other robots follow the leader. Moving group of robots form a commonly used triangular grid topology, i.e., three neighboring robots self-organize into and equilateral triangle formation (cf. Fig. 3a). This triangular grid

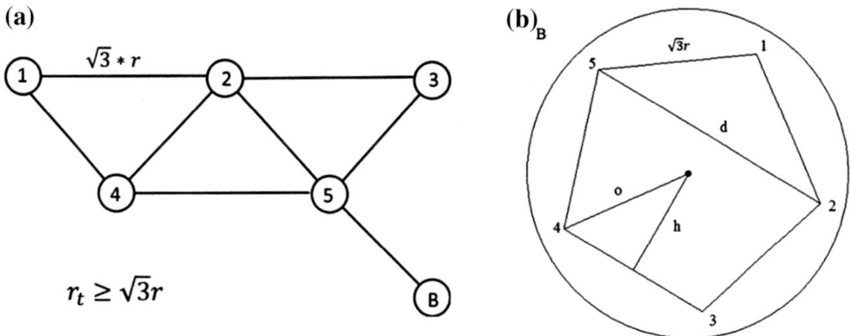

Fig. 3 Triangular grid topology formed by five robots: **a** exploration phase, **b** monitoring phase; *B* a base station

topology is maintained until the gas cloud is detected. The geometric properties of this type of topology are described in [11]. From this analysis it can be derived that maximum distance *dist* between each pair of neighboring nodes that implies both sensing coverage and permanent connectivity with the base station can be calculated as follows: $dist = \sqrt{3}r^s$ for $r^t \geq \sqrt{3}r^s$ or $dist = r^t$ otherwise. It is assumed that both sensing and transmission coverage of the detector and radio transceiver are discs of radiuses r^s and r^t. The numerical optimization is employed to calculate new positions of robots. The optimization problem with the objective function ensuring the maintenance of a triangular grid formation is formulated. The goal of each robot is to solve this problem based on the gathered knowledge about the sensing system. The solution is a vector of new geographical coordinates of a given robot.

Phase 2: After a gas cloud detection the network system is switched into the monitoring and tracking phase. The robots explore the cloud and calculate the optimal positions to guarantee high coverage of the region of interest. The goal is to surround the detected gas cloud. The robots are deployed on the border of the cloud and create as big as possible polygon, Fig. 3b. The larger surface of the polygon, the higher coverage of the gas cloud.

The numerical optimization is employed to calculate the positions of robots. After reaching destination points all robots start to monitor the area and transmit measurements to the base station.

In summary, the following operations have to be performed to create MANET for tracking and monitoring:

1. Creation of the triangular grid formation of robots carrying sensors.
2. Exploration of the workspace in search of a gas cloud:
   - the direction of the leader shifting is randomly selected,
   - all other robots follow the leader (a new position of each robot is calculated based on the local data collected by a robot),
   - the ad hoc network switches into monitoring and tracking after detecting the contaminated area.

3. Creation of a polygon-shaped topology for monitoring. The goal—covering an area of highest degree of contamination.
4. Recalculation of robot positions and shifting the network after the cloud displacement.

3.3 Evaluation of Deployment Strategies

The following criteria can be used to evaluate a performance of the robot-assisted monitoring system. Several metrics have been provided in literature. The widely used are defined below.

1. Contaminated area coverage is a key metric for evaluating the examined sensing system.

$$M_{cov} = \frac{A_{covered}}{A_{total}} * 100\,\% \tag{3}$$

where $A_{covered}$ denotes an area covered by the network and A_{total} is a sensing scene.
2. Total path length of all robots (from the starting point to the target position).

$$M_{len} = \sum_{i=1}^{N} L_i \tag{4}$$

where L_i denotes the length distance traveled by the ith robot.
3. Quality of radio transmission from the leader of the robots team to the base station

$$M_{eff} = \frac{P_{sent}}{P_{received}} * 100\,\% \tag{5}$$

where P_{sent} is a number of packets sent by the leader of the robot team, while $P_{received}$ is a number of packet received by the base station.

4 MANET Simulators and Performance Evaluation

4.1 Implementation

The main objective of our research was to create simulators of mobile networks for contaminated area monitoring, and compare both described above deployment strategies through simulation experiments. We implemented two simulators of MANETs formed by sensing robots equipped with radio transceivers, GPS system gas and detectors. The implementation was done in the Virtual Robot Experimentation Platform V-REP (http://www.coppeliarobotics.com/) and the MATLAB environment. V-REP is an advanced tool, designed for simulation and implementation of algorithms from the field of robotics [15]. It provides tools for 3D visualization of the created environment. It cooperates with different devices and software platforms, e.g. Robot Operating System (ROS) and MATLAB. Robotics, Vision & Control Toolbox [16] provides mechanisms for linking MATLAB and V-REP, and data exchange between both platforms. We used modules provided by the V-REP system to simulate the working scenes, wireless transmission, gas detection and movement of robots. The positions of all robots in the working scene were calculated using MATLAB. Two optimization solvers, i.e., particle swarm optimization (PSO) and genetic algorithm (GA) were used to this purpose, and compared [17, 18]. The optimization algorithm was repetitively triggered and executed by each robot, just after reaching the temporary destination point. The architecture of the mobile sensing network simulator is depicted in Fig. 4. In general, it is composed of five main components:

1. working scene simulator,
2. mobile robots simulators,
3. gas detectors and wireless transmission simulators,
4. optimization solver module,
5. image processing module.

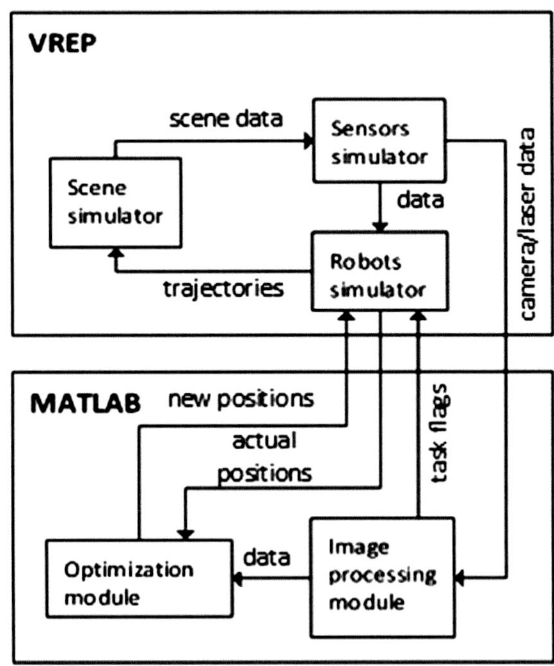

Fig. 4 The architecture of sensing network simulator

4.2 *Case Study Results*

Both off-line (pre-defined) and on-line (self-organized) deployment strategies were evaluated through simulations, and compared. In this paper, we present the simulation results of establishing MANET for gas cloud monitoring. Our MANET is formed by five robots carrying sensors, i.e., R1, R2, R3, R4, R5 and one robot that plays a role of a leader—R6. The objective of the sensing system was to detect the contaminated area and perform comprehensive and permanent monitoring. The Rapidly-exploring Random Tree (RRT) algorithm implemented in the V-REP platform was used to robots motion trajectories calculation.

4.2.1 Off-Line Deployment

The first series of tests was performed for off-line deployment assuming pre-defined positions of sensing robots. The workspace *W* was a square space with four obstacles. Two of the obstacles were located inside the contaminated area. The result of example simulation of the monitoring system formation process when target positions of all robots were calculated off-line by the base station is presented in Fig. 5.

GA and PSO optimization algorithms were used to calculate the target positions of robots. The results—values of metrics defined in Sect. 3.3 calculated for five executions of simulation—obtained for both solvers are collected in Tables 1 and 2.

It can be seen that the performance metrics calculated for network topologies determined by the GA solver are worse than the metrics obtained for network

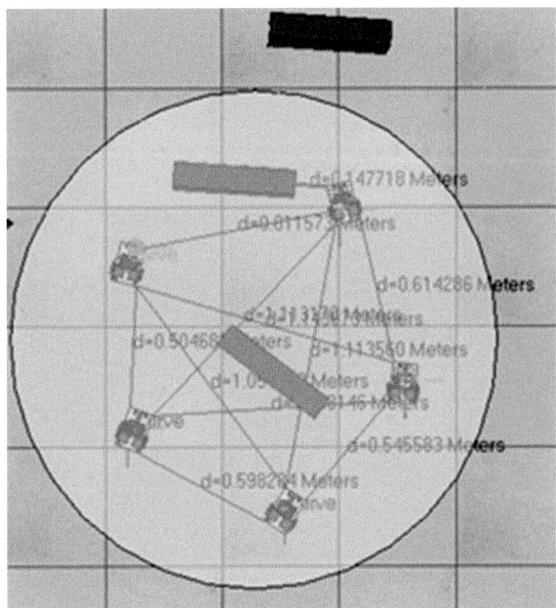

Fig. 5 Off-line deployment of sensing robots

Table 1 Performance metrics for off-line deployment (GA algorithm)

Experiment	Total time of deployment (s)	Total path length (m)	Coverage (%)
Test 1	50	9.3	58.586
Test 2	90	14.2	61.330
Test 3	110	12.4	60.430
Test 4	65	10.5	53.122
Test 5	83	11.7	61.761

Table 2 Performance metrics for off-line deployment (PSO algorithm)

Experiment	Total time of deployment (s)	Total path length (m)	Coverage (%)
Test 1	60	10.50	61.8830
Test 2	100	14.40	61.8801
Test 3	90	12.45	61.8764
Test 4	160	16.80	61.8830
Test 5	110	13.35	61.8830

topologies determined by the PSO solver. The results obtained using PSO were close to optimal for the considered team of robots (the maximum possible coverage in this case study was about 62 %). Moreover, the similar results were obtained in all experiments. every execution. The disadvantage of PSO was a long time of deployment. The total time of deployment was longer than in case of the application of the GA algorithm.

4.3 On-Line Deployment

The next series of tests was performed for on-line deployment of sensing robots. In this variant the target positions and motion trajectories were calculated directly by each robot. The aim was to detect the gas cloud as fast as possible and cover the cloud. The workspace W was a square space with four obstacles see Fig. 6.

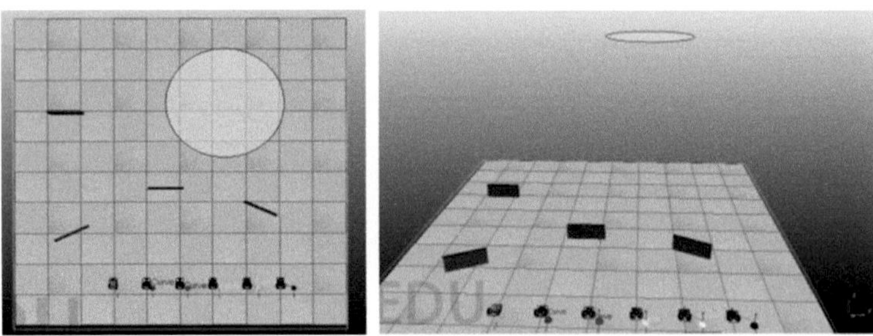

Fig. 6 The plan of the workspace W

It is obvious that to speed up the contaminated area detection the mobile robots should cover (every time step) as much of the workspace while maintaining the triangular grid formation. In our simulation study we assumed homogenous devices with the same sensing r^s and transmission ranges r^t. Moreover, we assumed that the transmission range exceeded the sensing range, the following inequality was satisfied $r^t \geq \sqrt{3}r^s$. Finally, in the exploration phase the following optimization problem was formulated and solved by the nth sensing robot (n = 1,…,5):

$$\begin{aligned} \min_{x \in R^{10}} & \left|dist_{12} - \sqrt{3}r^s\right| + \left|dist_{14} - \sqrt{3}r^s\right| + \left|dist_{23} - \sqrt{3}r^s\right| + \left|dist_{25} - \sqrt{3}r^s\right| \\ & + \left|dist_{24} - \sqrt{3}r^s\right| + \left|dist_{35} - \sqrt{3}r^s\right| + \left|dist_{45} - \sqrt{3}r^s\right|, \end{aligned} \tag{6}$$

$$dist_{nk} = \sqrt{(x_n - x_k)^2 + (y_n - y_k)^2}, \quad n \neq k, \quad n,\ k = 1, \ldots, 5, \tag{7}$$

under the constraints ensuring the network connectivity:

$$\begin{aligned} & dist_{12} \leq r^t; dist_{14} \leq r^t; dist_{23} \leq r^t; dist_{25} \leq r^t; \\ & dist_{24} \leq r^t; dist_{35} \leq r^t; dist_{45} \leq r^t; dist_{56} \leq r^t. \end{aligned}$$

In the above formulas x_n and y_n denote the geographical coordinates of the nth robot and $dist_{nk}$ denotes the Euclidean distance between robots n and k.

After detecting the cloud all robots were switched in the monitoring phase. The pentagon with the maximum surface area calculated for the sensing range r^s was created. The following optimization problem was formulated and solved by the nth robot (n = 1,…,5), and each robot was placed in the calculated vertices of the pentagon.

$$\begin{aligned} \min_{x \in R^{12}} & \left|dist_{12} - \sqrt{3}r^s\right| + \left|dist_{23} - \sqrt{3}r^s\right| + \left|dist_{34} - \sqrt{3}r^s\right| + \left|dist_{45} - \sqrt{3}r^s\right| \\ & + \left|dist_{51} - \sqrt{3}r^s\right| + \left|dist_{56} - \sqrt{3}r^s\right| + |dist_{m1} - o_1| + |dist_{m2} - o_2| \\ & + |dist_{m3} - o_3| + |dist_{m4} - o_4| + + |dist_{m5} - o_5| + |dist_{25} - d| \\ & + |dist_{24} - d| + |dist_{14} - d| + |dist_{13} - d| + |dist_{35} - d|, \end{aligned} \tag{8}$$

$$dist_{mn} = \sqrt{(x_m - x_n)^2 + (y_m - y_n)^2}, \tag{9}$$

$$o_i = \sqrt{\left(\frac{1}{2}\sqrt{3}r^s\right)^2 + \left(\frac{1}{2}\sqrt{3}r^s * cot(36°)\right)^2}, \quad d = \frac{\sqrt{5}+1}{2}\sqrt{3}r^s \tag{10}$$

Table 3 Performance of distributed variant

Experiment	Total time of deployment (s)	Total path length (m)	Transmission quality (%)
Test 1	300	44.40	86
Test 2	340	40.65	90
Test 3	840	112.00	84
Test 4	240	39.30	75
Test 5	440	57.50	73

under the following constraints ensuring connectivity:

$$\begin{aligned} &dist_{12} \leq r^t; dist_{23} \leq r^t; dist_{34} \leq r^t; dist_{45} \leq r^t; dist_{56} \leq r^t; \\ &dist_{16} \leq r^t; dist_{21} \leq r^t; dist_{32} \leq r^t; dist_{43} \leq r^t; dist_{54} \leq r^t. \end{aligned} \tag{11}$$

In the above equations, x_m and y_m denote coordinates of the assumed (based on available measurements) epicenter of the gas cloud, o_i is a target distance between the pentagon center and ith vertex, d a diagonal of the pentagon. The (8)–(11) optimization problem was solved by each robot. The active-set algorithm with multiple-starting points was applied to calculate the optimal positions of all robots.

The results of five simulation experiments are collected in Table 3. The snapshots of exploration and monitoring phases simulation are presented in Fig. 7.

It can be seen that the total time of deployment and total path length significantly increased in case of on-line deployment. It was due to a lack of a global knowledge about the working space W. Robots had to explore W to detect the gas cloud and adopt to the application scenario. The coverage was about 61 %.

In case of on-line deployment the quality of data transmission to the base station is one of the most important metrics. The transmission losses were about 25–10 % during the exploration phase, that was on the acceptable level. No losses were observed in the monitoring phase. Some problems were observed during obstacles omitting. The destination point and motion trajectory were calculated by different algorithms. Hence, a wrong path could be selected by the robot (see Fig. 8). In some scenarios it can cause a loss of connectivity. However, in general, the on-line deployment strategy gave the satisfactory solutions.

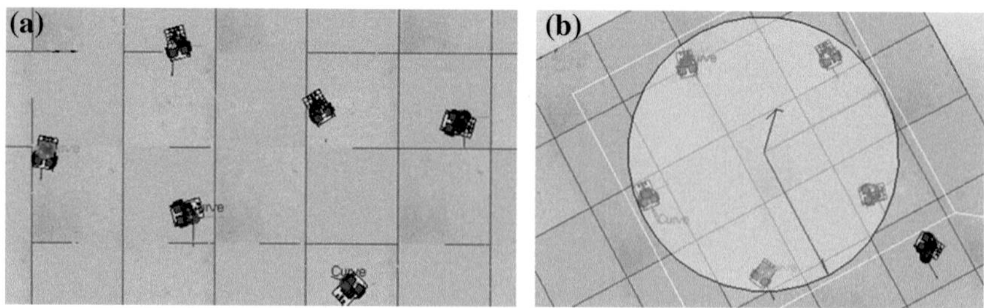

Fig. 7 MANET simulation **a** exploration phase **b** monitoring phase

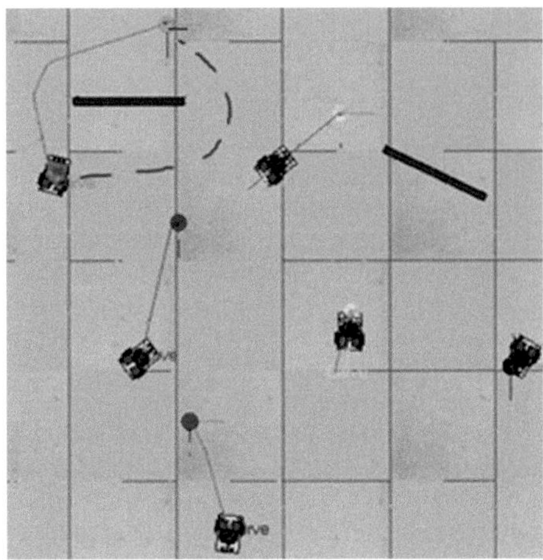

Fig. 8 A loss of connectivity in the exploration phase (selection of the wrong path)

5 Summary

The paper summarizes the results of research concerned with the simulation-based design and development of a mobile ad hoc network for detecting and monitoring a gas cloud. The advantages of using multiple mobile robots for wireless sensors deployment are addressed. Two deployment strategies were validated and compared through simulation: off-line, pre-defined deployment and on-line, self-organizing deployment. Various optimization methods were used to calculate the optimal positions of robots. The observation is that the accurate information about the workspace and the sensing area and adequate optimization strategy is essential for a high quality deployment. The off-line approach with PSO algorithm gave high quality deployment. On the other hand an on-line deployment utilizing self-organizing capabilities of MANET is more resistant and flexible. The important goal of this paper was to stress the role of the computer simulation in the development of new applications of modern ad hoc networks. Simulation can successfully support the development of the mobile monitoring system with the required characteristics.

References

1. Akyildiz, I., Vuran, M.: Wireless Sensor Networks. Wiley, UK (2010)
2. Marks, M., Niewiadomska-Szynkiewicz, E., Kołodziej, J.: High performance wireless sensor network localization system. Int. J. Ad Hoc Ubiquitous Comput. **17**(32), 122–133 (2014)
3. Niewiadomska-Szynkiewicz, E.: Energy aware communication protocols for wireless sensor networks. LNCS, vol. 7776, pp. 135–149. Springer (2013)
4. Aggelou, G.: Mobile Ad Hoc Networks. McGraw-Hill, USA (2005)
5. Roy, R.: Handbook of Mobile Ad Hoc Networks for Mobility Models. Springer (2010)

6. Choset, H., Lynch, K.M., Hutchinson, S., Kantor, G., Burgard, W., Kavraki, L., Thrun, S.: Principles of Robot Motion: Theory, Algorithms, and Implementations. MIT Press (2005)
7. Kasprzak, W., Szynkiewicz, W., Zlatanov, D., Zielińska, T.: A hierarchical csp search for path planning of cooperating self-reconfigurable mobile fixtures. Eng. Appl. Artif. Intell. **34**, 85–98 (2014)
8. Niewiadomska-Szynkiewicz, E., Sikora A.: Simulation-based evaluation of robot-assisted wireless sensors positioning. LNCS, vol. 351, pp. 181–190. Springer (2015)
9. Fletcher, G., Li, X., Nayak, A., Stojmenovic, I.: Back-tracking based sensor deployment by a robot team. In: IEEE SECON, pp. 1–9 (2010)
10. Tuna, G., Gungor, V., Gulez, K.: An autonomous wireless sensor network deployment system using mobile robots for human existence detection in case of disasters. Ad Hoc Netw. **13**, 54–68 (2014)
11. Pompili, D., Melodia, T., Akyildiz, I.F.: Deployment analysis in underwater acoustic wireless sensor networks. In: Proceedings of WUWNet'06, Los Angeles (2006)
12. Ramadan, R., El-Rewini, H., Abdelghany, K.: Optimal and approximate approaches for deployment of heterogeneous sensing devices. EURASIP J. Wirel. Commun. Netw. **1** (2007)
13. Amato, N.M., Wu, Y.: Randomized roadmap method for path and manipulation planning. In: Proceedings of IEEE International Conference on Robotics and Automation, vol. I, pp. 113–120
14. Batog, P., Wolczowski, A.: Chemical scanner for mobile robot navigation. J. Autom. Mobile Robot. Intell. Syst. **9**, 5–11 (2015)
15. Rohmer, E., Singh, S.P.N., Freese, M.: V-REP: a versatile and scalable robot simulation framework. In: Proceedings of the International Conference on Intelligent Robots and Systems (IROS) (2013)
16. Corke, P.I.: Robotics, Vision & Control. Springer (2011)
17. Kennedy, J., Eberhart, R.: Particle swarm optimization. In: Proceedings of IEEE International Conference on Neural Networks, vol. IV, pp. 1942–1948 (1995)
18. Mitchell, M.: An Introduction to Genetic Algorithms: MIT Press (1996)

Control of an Electro-Hydraulic Manipulator by Vision System Using Central Point of a Marker Estimated via Kalman Filter

Piotr Owczarek, Jarosław Gośliński, Dominik Rybarczyk and Arkadiusz Kubacki

Abstract In the paper, the vision system for control a electro-hydraulic manipulator is presented. The authors have proposed a Kalman Filter (KF) to estimate a central position of markers. Two different methods of initial estimation of markers position are used. First one is based on central point of marker's mass. Second refers to a circle fitting method of binary object. These two methods theoretically give the same position of a central point of given marker, however, in case of image distortion results will have different errors. This can occur when operator is holding marker in a hand, and cover some parts of it. Therefore it is important to develop new robust method for marker tracking. Authors proposed Kalman Filter to estimate central point of a marker by making a fusion of information provided by two initial estimation methods. The conducted research proved that KF reduces total manipulator's control error even in situation where 30 % of marker area is invisible.

Keywords Vision system · Camera · Geometric fit · Pratt · Kasa · Taubin · Kalman filter · Marker detection · Electro-hydraulic manipulator

---

Please note that the LNCS Editorial assumes that all authors have used the western naming convention, with given names preceding surnames. This determines the structure of the names in the running heads and the author index.

P. Owczarek (✉) · D. Rybarczyk · A. Kubacki
Institute of Mechanical Technology, Poznań University of Technology,
60-965 Poznań, Poland
e-mail: piotr.owczarek@put.poznan.pl
URL: http://www.put.poznan.pl

D. Rybarczyk
e-mail: dominik.rybarczyk@put.poznan.pl

A. Kubacki
e-mail: arkadiusz.j.kubacki@doctorate.put.poznan.pl

J. Gośliński
Institute of Control and Information Engineering, Poznań University of Technology,
60-965 Poznań, Poland
e-mail: jaroslaw.a.goslinski@doctorate.put.poznan.pl
URL: http://www.put.poznan.pl

R. Szewczyk et al. (eds.), *Challenges in Automation, Robotics and Measurement Techniques*, Advances in Intelligent Systems and Computing 440, DOI 10.1007/978-3-319-29357-8_51

1 Introduction

Nowadays vision systems are fest-growing technology which is getting more and more usefull in everyday of life (mobile phones—camera with face recognition algorithm, smart TV—geasture control as well as industry e.g. quality control [1], products classification [2], monitoring systems [3], items sorting in cooperation with robots [4, 5]. Many of them need a special algorithms which will be robust for disturbances. These algorithms are divided into two groups. First refers to features detection [6, 7]. Second group is focused on marker detection. Also three groups of markers are known. The passive markers [8], active [9, 10] and coded [11]. Passive markers are made of color paper or special material which allows to reflect the light, it is usually used in 3D measurement vision system [12]. The reflective marker is illuminated by infrared light, which should be robust for Sun rays. In many cases, when some parts of markers (either passive or active) are invisible, the recognition is difficult, or in worst scenerio impossible. Therefore most popular markers have circle-like shape [13], it is claimed that this is an optimal configuration [14]. However, when the user accidentally covers small parts of it, the coupled vision system does not guarantee that markers will be well captured. That can yield problems with detection and, what is more, tracking of markers central points. Thus it is important to search for a new implementable algorithm to recognize and filter markers features. The presented study is a continuation of a research given in [15]. The authors proved that in the problem of marker's position tracking, the Kalman Filter turned out to be more accurate than complementary filter. Therefore here, only the previously undertaken methodology will be considered and applied in control of a electro-hydraulic manipulator.

2 The Testbed

The tests were performed on a electro-hydraulic manipulator equipped in proportional valves, an industrial controller and the vision system.

Images were captured by a fast camera 3iCube IC1500CU (Fig. 1) [16] with special software called *Adaptive Vision Studio*, which uses SEE multicore instructions [17]. This camera was connected to the PC computer by USB 3.0. The focal length of the camera lens is 3.5 mm. The max resolution of camera is equal to 2592×1944

Fig. 1 3iCube CMOS camera [16]

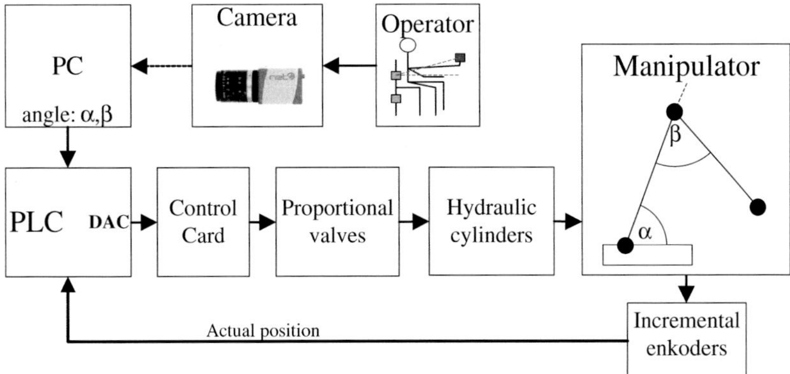

Fig. 2 Block diagram of research stand

yielding 15 fps, but in current research the camera's resolution was reduced to 640×480 px and image capture was performed with frequency up to 50 Hz. Calculations were performed on the personal computer. Next, the data with coordinates were sent to PLC (Programmable Logic Controller—B&R company) via TCP/IP protocol. The position of hydraulic cylinder was controlled by proportional valves connected with dedicated control card. Input signal in the card was in a range of −10..+10 VDC. The positions of angles of manipulator's arms were measured by incremental encoders with resolution of 3600 impulses per revolution. The block diagram of the stand is shown in Fig. 2.

The main algorithms of image processing and estimation of a central position of the markers is computed in two phases. The first phase is a basic processing of image:

- image capture from the camera (RGB image),
- Gaussian smoothing,
- HSV filtration,
- image thresholding to region,
- region erode,
- region dilate,
- split region into blobs.

The second phase refers to marker's features extraction. In this phase, a few steps can be highlighted. Firstly it is necessary to calculate the size of region:

$$S = \sum_{i=1}^{n} \sum_{j=1}^{m} p(i,j) \tag{1}$$

where: p—is a position of the point on an image, $n \times m$—is a size of region, $p(i,j) = 1$—when object exist, $p(i,j) = 0$—when object does not exist, Next, the center of mass of the region (x_m, y_m) must be found. It can be simply calculated with the following:

$$x_m = \frac{\sum_{i=1}^{n}\sum_{j=1}^{m} k}{S}, k = \begin{cases} \text{i - when object} \\ \text{0 - in other case} \end{cases} \quad (2)$$

$$y_m = \frac{\sum_{i=1}^{n}\sum_{j=1}^{m} k}{S}, k = \begin{cases} \text{j - when object} \\ \text{0 - in other case} \end{cases} \quad (3)$$

In the literature several fitting algorithms can be found. All of them find best circle and based on that, the central points of the markers are established. The most popular are proposed by Kasa [18], Pratt [19] and Taubin [20]. The main task of those algorithms is to minimize the cost of function:

$$\mathcal{F}(a, b, R) = \sum d_i^2 \quad (4)$$

$$d_i = r_i - R, \quad r_i = \sqrt{(x_i - a)^2 + (y_i - b)^2}, \quad (5)$$

where: a, b—center of circle, R—radius of circle.

The performance of these algorithms was verified by Al-Sharadhad in [21]., but Chernov proved, that algorithms proposed by the Pratt and Taubin are more stable and accurate than the Kasa [22]. Therefore in this article the Authors will focus only on Pratt's algorithm.

3 Implementation of Kalman Filter

The Kalman Filter is well-known recursive algorithm, which was originally used for linear problems [23]. During the last fifty years, the KF evaluated to deal with nonlinear models, however here the classic type of the KF is used. The overall framework of the KF assumes two-stages process. In the first one, namely prediction, the model data is exploited. During that stage some forecasts on state vector behavior are made. In next stage, the corrections are being imposed. The KF operate on process model which can be written in a state-space representation:

$$\begin{aligned} \underline{x}_k &= A \cdot \underline{x}_{k-1} + B \cdot \underline{u}_k \\ \underline{y}_k &= C \cdot \underline{x}_k \end{aligned} \quad (6)$$

where $\underline{x}_k$ denotes state vector, A, B and C refer to process, input and output matrices, respectively. In the fusion process proposed by the authors, the state vector contains positions (x,y) and velocities (v_x,v_y) of a central of the marker.

$$\underline{x}_k = \begin{bmatrix} x \\ v_x \\ y \\ v_y \end{bmatrix}, \quad (7)$$

Matrices A and B are time-variant and thus they were extended with subscripts. The main process is incorporated in A_k and B_k. It simply reflects the dynamics of the central point and it is given by:

$$\mathbf{A}_k = \begin{bmatrix} \mathbf{A}_k^z & 0_{2\times2} \\ 0_{2\times2} & \mathbf{A}_k^z \end{bmatrix}, \mathbf{B}_k = \begin{bmatrix} \mathbf{B}_k^z & 0_{2\times1} \\ 0_{2\times1} & \mathbf{B}_k^z \end{bmatrix}, \quad (8)$$

where: $\mathbf{A}_k^z = \begin{bmatrix} 1 & dt \\ -\frac{1}{dt} & 0 \end{bmatrix}, \mathbf{B}_k^z = \begin{bmatrix} 0 \\ \frac{1}{dt} \end{bmatrix},$

The output matrix C_k directly transfers the state vector to the output vector:

$$C = I_{4\times4}, \quad (9)$$

The Kalman Filter equations are divided into two stages: The first is state prediction, which yields the a priori state vector $\hat{\underline{x}}_k^-$, next it is necessary to compute a priori covariance matrix P_k^-. One will notice, that by applying the first estimation of central point positions in $\underline{u}_k$ the a priori estimation of velocities will be calculated. The KF is prepared in that way, so it can estimate the velocity of point coordinates (in X and Y axis) based on previous (k-1) a posteriori positions and the input. Also the positions are upgraded by integrating the a posteriori velocities.

In the second stage of Kalman Filter, namely correction, three main tasks are being executed. The first one is the Kalman gain K_k calculation, then the prior state vector $\hat{\underline{x}}_k$ is corrected with measurement innovation. Finally the covariance matrix P_k is updated, obtaining the a posteriori covariance P_k.

The Kalman Filter is tunned via the Q_k and R_k matrices. The Q_k refers to process noise, while the R_k stand for measurement noise.

4 The Experiment

The main aim of the research was to verify a new method of markers position estimation in case of simulated disturbances like covering or shading. During the experiment the saved video was used with dynamic disturbance like covering some parts of marker.

Similar situation can occur when a person holds the marker in a hand and makes some movements. The incomplete information of marker's position can be observed. The offline study with saved video is useful in comparison of obtained results via different methods. The recorded track for the same for all experiments. In order to capture an image in a wide area, the camera was placed in a distance of 1.3 m against

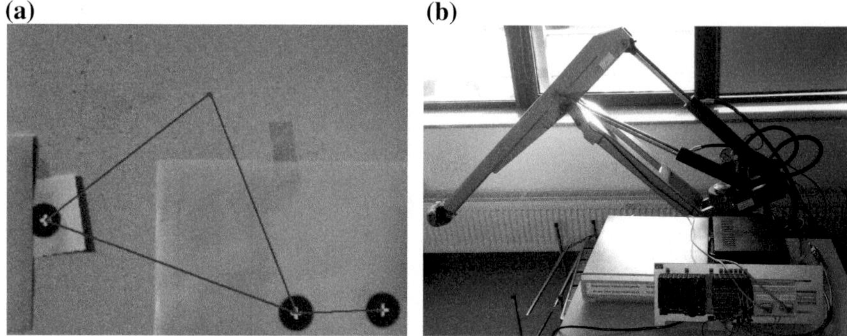

Fig. 3 Control of electro-hydraulic manipulator with vision system. **a** Image after processing. **b** Electro-hydraulic manipulator

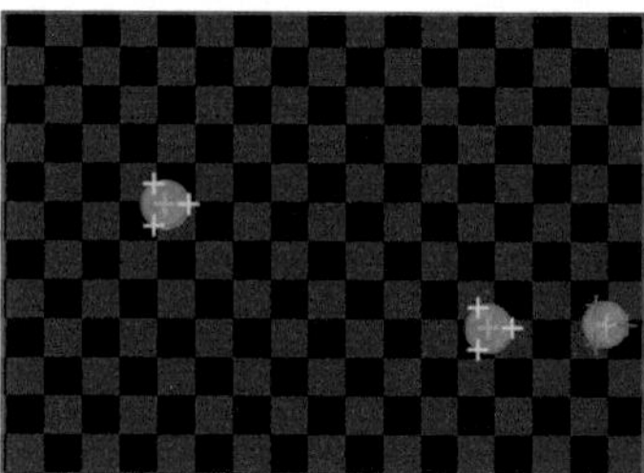

Fig. 4 Centers of the markers

the marker. In the Fig. 3 the processed image and manipulator's trajectory were shown. As it was presented in the Fig. 3a, three red markers were used to control an electro-hydraulic manipulator. It simply increases efficient, while only one color must be filtrated. In the bottom right hand corner two markers are visible. Those markers are used to scale pixels from the image to manipulator's metric units. The last marker is directly used to control the tool center point (TCP) of the manipulator.

The image in processing application was shown in Fig. 4. A background of image was deleted to extract marker's features. Three points placed around markers were used to circle fitting algorithm with results in the fitted circles. In the center of markers a results of a center of mass algorithm was presented. In Fig. 5 the output charts from image processing and output of control of the electro-hydraulic manipulator were shown. As it can be noted, in Fig. 3b a fluctuating pixel in X axis in time domain in occurance of disturbances has appeared. In Fig. 5b an angle computed from inverse kinematic was presented. The last, Fig. 6 is the output trajectory from electro-hydraulic manipulator.

In Fig. 7 absolute positions were compared, these remains the most interesting result. In Fig. 7a the position computed from vision system were presented. First positions were taken from pixels, next inverse kinematic was computed. To check results, forwards kinematic was computed. The fluctuating noise on chart is visible,

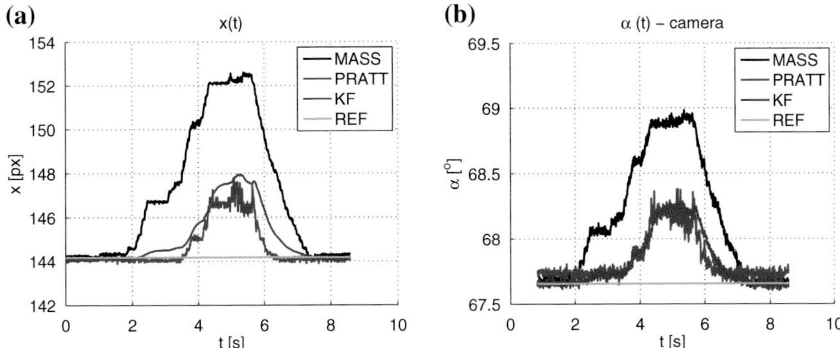

Fig. 5 Control of electro-hydraulic manipulator: input image from vision system. **a** Pixels in axis X. **b** Angle from vision system

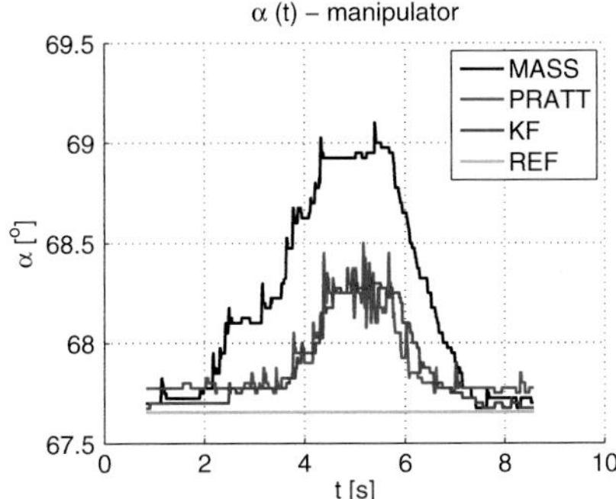

Fig. 6 Control of electro-hydraulic manipulator: angle

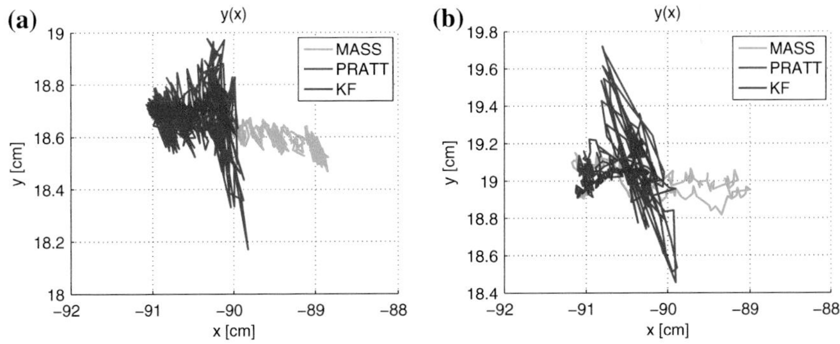

Fig. 7 Position of the TCP of the manipulator. **a** Position from camera. **b** Position from manipulator

Table 1 Absolute error of position manipulator

$e[cm]$	CF	CM	KF
Camera	1.9725	2.4711	1.3243
Manipulator	2.4827	2.5288	1.3441

because the reference signal with scale was taken from two markers which were disrupted by position measurement errors.

The output trajectory from the manipulator was shown in Fig. 7b. The line is more stable, because electro-hydraulic drives are in general inertial objects and cause natural signal filtration.

The overall results were presented in Table 1. As it can be seen, the best results are provide by Kalman filter marked by blue color. A comparison of total amplitude of oscillation in both axes i.e. absolute error was shown in Table 1.

The total error of positioning of the TCP in electro-hydraulic manipulator was reduced by more than 40 %.

5 Conclusions

In the article three different methods i.e. circle fitting (proposed by Pratt), central point of mass, and the Kalman filter applied in estimation of a central position of the marker were presented. The first part of the article recalls state of the art. In the second section the testbed and the methods for markers central point estimation were described. To undertake the experimental studies, the research stand was built. The images were captured from camera with frequency up to 50 Hz and the resolution of 640×480. In the third section the mathematical equations which were used to implementation of the KF were described. In the fourth section the results of the experiment were shown. First part consist of image processing results and electro-hydraulic manipulator trajectory, therefore the charts with angle and the positions of manipulator tip were compared. The main aim of the research was to implement and compare three estimation methods i.e. circle fitting, center of mass and the Kalman filter. All of these methods returns central points of markers. The KF uses two inputs of estimated positions in X and Y axis from other proposed methods. The results prove that KF method is suitable in case of estimation of markers positions and yields better results than CF or CM. However, when input parameters are over the range, the estimation results cannot be valid and the total error can increase. Because errors can appear caused by incompleteness of the markers positions information, all filters are important. The KF method improves the estimation provided by CF or CM method, also by using the KF the output signal has smooth characteristic. What is more, the solution given in the paper ensures good results even though when some image parts are incomplete or require prior reconstruction. The total error in the worst scenario, (30 % of marker is invisible) was reduced by more than 40 %.

Acknowledgments The work is partially sponsored by the project "Scholarship support for PH.D. students specializing in majors strategic for Wielkopolska's development", Sub-measure 8.2.2 Human Capital Operational Programme, co-financed by European Union under the European Social Fund.

The work described in this paper was funded from 02/23/DS-PB/1208 project (Nowe techniki w urzadzeniach mechatronicznych).

References

1. Nandi, C.S., Tudu, B., Koley, C.: An automated machine vision based system for fruit sorting and grading. In: 2012 Sixth International Conference on Sensing Technology (ICST), pp. 195–200 (2012)
2. Akbar, H., Prabuwono, A.S.: The design and development of automated visual inspection system for press part sorting. In: International Conference on Computer Science and Information Technology. ICCSIT 08, pp. 683–686 (2008)
3. Govardhan, P., Pati, U.C.: NIR image based pedestrian detection in night vision with cascade classification and validation. In: International Conference on Advanced Communication Control and Computing Technologies (ICACCCT), pp. 1435–1438 (2014)
4. Bodhale, D., Afzulpurkar, N., Thanh, N.T.: Path planning for a mobile robot in a dynamic environment. In: IEEE International Conference on Robotics and Biomimetics (ROBIO), Bangkok,pp. 2115–2120 (2009)
5. Hong, S.M., Jang, W.S., Son, J.K., Kim, K.S.: Evaluation of two robot vision control algorithms developed based on N-R and EKF methods for the rigid-body placement. In: International Conference on Advanced Intelligent Mechatronics, Wollongong, Australia, July 9–12, pp. 938–943 (2013)
6. Schomerus, V., Rosebrock, D., Wahl, F.M.: Camera-based lane border detection in arbitrarily structured environments. In: 2014 IEEE Intelligent Vehicles Symposium Proceedings, pp. 56–63 (2014)
7. Chugo, D., Hirose, K., Nakashima, K., Yokota, S., Kobayashi, H., Hashimoto, H.: Camera-based navigation for service robots using pictographs on the crossing point. In: IECON 2012—38th Annual Conference on IEEE Industrial Electronics Society, pp. 4154–4159 (2012)
8. Sampe, I.E., Amar Vijai, N., Tati Latifah, R.M., Apriantono, T.: A study on the effects of lightning and marker color variation to marker detection and tracking accuracy in gait analysis system. In: 2009 International Conference on Instrumentation, Communications, Information Technology, and Biomedical Engineering (ICICI-BME), pp. 1–5 (2009)
9. Kim, D., Choi, J., Park, M.: Detection of multi-active markers and pose for formation control. In: 2010 International Conference on Control Automation and Systems (ICCAS), pp. 943–946 (2010)
10. Fukuzawa, M., Hama, H., Nakamori, N., Yamada, M.: High-speed distance measurement between moving vehicles with NIR-LED markers. In: 11th International Conference on Computer and Information Technology. ICCIT 2008, pp. 516–520 (2008)
11. Gherghina, A., Olteanu, A., Tapus, N.: A marker-based augmented reality system for mobile devices. In: Roedunet International Conference (RoEduNet), 2013 11th, pp. 1–6 (2013)
12. PONTOS—Dynamic 3D Analysis. http://www.gom.com/metrology-systems/system-overview/pontos.html
13. Mochizuki, Y., Imiya, A., Torii, A.: Circle-marker detection method for omnidirectional images and its application to robot positioning. In: IEEE 11th International Conference on Computer Vision. ICCV 2007, pp. 1–8 (2007)
14. Khler, J., Pagani, A., Stricker, D.: Detection and Identification Techniques for Markers Used in Computer Vision, in Modeling and Engineering. Bodega Bay, CA, USA, o.A. (2011)

15. Owczarek, P., Goski, J.: An estimation of central points of circle markers in a vision system by using Kalman filter and complementary filter. In: Proceedings of The 20th International Conference on Methods and Models in Automation and Robotics, between 24th and 27th Aug 2015
16. NET New Electronic Technology GMBH. http://net-gmbh.com/
17. Adaptive Vision Studio. https://www.adaptive-vision.com/en/home/'
18. Corral, C.A., Lindquist, C.S.: On implementing Kasas circle fit procedure. IEEE Trans. Instrum. Meas. **47**(3), 789–795 (1998)
19. Pratt, V.: Direct least-squares fitting of algebraic surfaces. Comput. Graph. **21**, 145–152 (1987)
20. Taubin, G.: Estimation of planar curves, surfaces, and nonplanar space curves defined by implicit equations with applications to edge and range image segmentation. IEEE Trans. Pattern Anal. Mach. Intell. **13**(11), 1115–1138 (1991)
21. Ali, N.C.: Al-Sharadqah. Error analysis for circle fitting algorithms. Electron. J. Stat. **3** (2009)
22. Chernov, N., Lesort, C.: Least squares fitting of circles. J. Math. Imaging Vis. **23**(3), 239–252 (2005)
23. Kalman, R.E.: A new approach to linear filtering and prediction problems. Trans. ASME J. Basic Eng., Ser. D **82**, 35–45 (1960)

Perception Subsystem for Object Recognition and Pose Estimation in RGB-D Images

Tomasz Kornuta and Michał Laszkowski

Abstract RGB-D sensors have become key components of all kind of robotic systems. In this paper we present a perception subsystem for object recognition and pose estimation in RGB-D images. The system is able to recognize many objects at once, disregarding whether they belong to one or many classes. Next to the detailed description of the principle of the system operation we present several off-line and on-line experiments validating the system, including verification in the task of picking up recognized objects with IRp-6 manipulator.

Keywords RGB-D image · SIFT · Object recognition · Pose estimation · Object picking

1 Introduction

Recent advent of RGB-D sensors, providing color images supplemented with depth maps, accelerated progress in diverse areas of robotics. The RGB-D based solutions are currently used as key components of many applications, starting from social robotics (e.g. in human-robot interaction by recognition of human posture [1] or hand gestures [2]), mobile robots (e.g. for detection of obstacles during navigation [3] or localization and mapping [4]), walking machines (e.g. for terrain classification [5]) up to industrial robots (e.g. for robust estimation of object poses in bin picking task [6]).

T. Kornuta (✉)
IBM Research, Almaden, 650 Harry Rd, San Jose, CA 95120, USA
e-mail: tkornuta@gmail.com

M. Laszkowski
Institute of Control and Computation Engineering, Warsaw University of Technology,
Nowowiejska 15/19, 00-665 Warsaw, Poland
e-mail: mlaszkow@gmail.com

R. Szewczyk et al. (eds.), *Challenges in Automation, Robotics and Measurement Techniques*, Advances in Intelligent Systems and Computing 440, DOI 10.1007/978-3-319-29357-8_52

But most of all, RGB-D sensors significantly facilitated the object recognition [7], thus impacted the broad range of service robotics applications, enabling execution of such non-trivial tasks as two-handed folding of towels [8] and clothes [9] or cutting fruits and vegetables [10].

During last years many diverse methods of recognition of objects in RGB-D images and point clouds were developed. On the one hand there is a variety of methods depending on extraction of a object class independent features (e.g. FPFH [11]), whereas on the other there are several propositions of utilization of features specific for given class of objects [12]. Next, there are also approaches that instead of hard-coded routines use machine learning and neural networks for learning of RGB-D image patches and relations between them [13].

In this paper we present a perception subsystem able to recognize household objects possessing rich textures, enabling their subsequent grasping and manipulation. Thus aside of object detection the emphasis was put on robust pose estimation. The idea was to use features extracted from RGB images and use depth map for their transformation into sparse cloud of features, used subsequently for recognition.

The paper is structured as follows. In the next section we present the general principle of operation of the proposed RGB-D perception subsystem, followed by detailed description of key components. Next we validate the system on experiments involving both off-line (on test dataset) and on-line (working with images acquired from Kinect in real-time). Finally, we present a robotic system, consisting of the developed RGB-D perception subsystem working in conjunction with controller of an IRp-6 manipulator, able to successfully and robust grasp and pick up diverse objects lying in random positions.

2 RGB-D Perception Subsystem

Dataflow diagram of the developed object recognition system is presented in Fig. 1. The input to the system consists of an RGB image with the associated depth map. The general principle of operation relies on finding correspondences between features of the object model(s) and features extracted from the scene, followed by generation of object hypotheses by grouping of correspondences and hypothesis verification by rejection of weak hypotheses that are in conflict of the stronger ones. The returned verified hypotheses contain, beside others, estimated object poses, their projections (clouds in the sensor reference frame) and confidences. The details of the system are presented in the following subsections.

The recognition pipeline assumes the presence of a priori created three-dimensional models of objects, stored in the form of dense colour point clouds (Fig. 2a–c) supplemented with sparse clouds of features (Fig. 2d). Those sparse clouds consist of SIFT (Scale Invariant Feature Transform) [15] features, extracted from the colour images and enhanced by Cartesian coordinates.

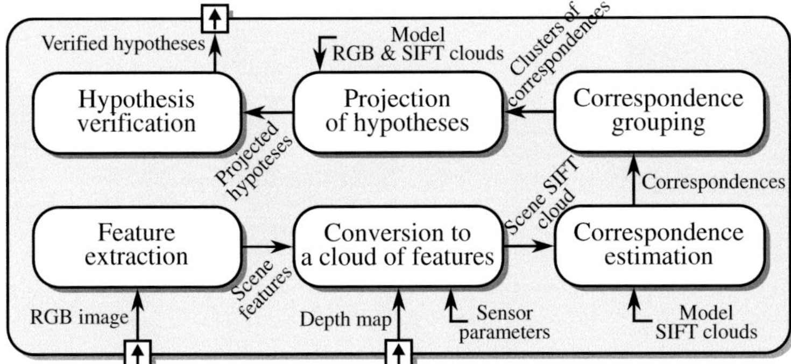

Fig. 1 Dataflow diagram of the developed perception subsystem

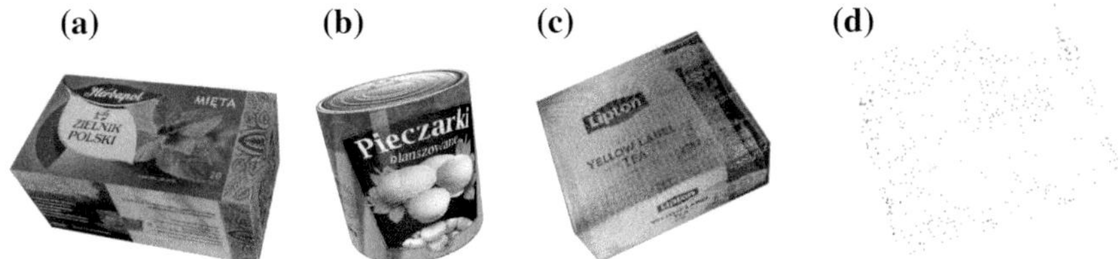

Fig. 2 Selected models of objects from 3D Models Dataset [14]: **a** Herbapol mint tea. **b** Foodcan 2. **c** Lipton yellow label tea. **d** Lipton yellow label tea—sparse cloud of SIFT features

2.1 Feature Extraction

In the first step of the recognition pipeline SIFT features are extracted from the acquired RGB image. Subsequently the data from depth map associated with RGB image is used for computation of Cartesian coordinates of every of SIFT features, resulting in a sparse cloud of features representing the scene.

2.2 Correspondence Estimation

In the next step the descriptors of features extracted from the scene are matched with descriptors of features of all loaded object models. This problem is typically solved by the k-Nearest Neighbours algorithm. As the comparison is made in a high-dimensional space (SIFT descriptor is a vector os 128 elements), we decided to use FLANN (Fast Library for Approximate Nearest Neighbors) [16], which is an efficient implementation of k-NN algorithm. Exemplary results of correspondence estimation is presented in Fig. 3a. The red dots indicate scene features, green dots indicate model features, whereas green lines represent found reciprocal correspondences.

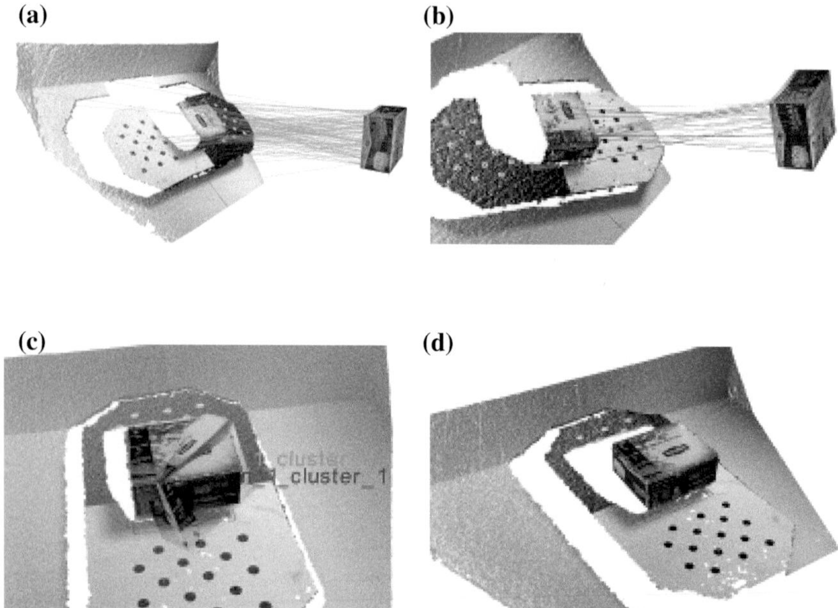

Fig. 3 Verification of the perception subsystem on the task of recognition of a single object (Lipton yellow label tea) on a scene from test scenes dataset [14]. **a** Estimated correspondences. **b** Clustered correspondences (2 clusters). **c** Projected hipotheses (2 hypotheses). **d** Hypothesis remaining after verification

2.3 Correspondence Grouping

Considering that there might be several instances of object belonging to the same class (thus fitting to the same model) present in the image, as well as we want to recognized objects of many different classes at once, it is required to cluster somehow the correspondences constituting different hypotheses. For that purpose we applied the Geometric Consistency Grouping [17] algorithm, which clusters correspondences on the basis of relations between points belonging to the model and scene. In opposition to correspondence estimation, in grouping only Cartesian coordinates of features are taken into consideration. The idea is that object hypothesis clusters correspondences with similar transformations between the model points and scene points—if there is a correspondence that fit to this hypothesis (i.e. when the projection of the model point with hypothesis transformation matches the scene point of the considered correspondence) then it is added to the cluster. An exemplary result of correspondence grouping is presented in Fig. 3b. Please notice two colours of correspondences, which indicate the presence of two different clusters, hence two object hypotheses.

2.4 Hypothesis Projection

Hypothesis projection is the transformation of the model point clouds (both dense colour point cloud and sparse feature cloud) into the scene reference frame. For each projection we use the transformation associated with given hypothesis found in the correspondence grouping step. Exemplary projections are presented in Fig. 3c.

2.5 Hypothesis Verification

Finally, having several hypotheses it is required to reject the ones that are weak and/or are in conflict with other hypotheses. In the presented system we decided to incorporate the Greedy Verification [18] algorithm. This method counts the number of features belonging to given projection of model that fit into the scene points—disregarding whether there was underlying (i.e. earlier found) correspondence for given pair or not. Hypothesis is considered as valid if number of inliers is greater from a outliers (the required ratio in fact may by parametrized).

3 Experimental Verification of the Perception Subsystem

3.1 Off-Line Verification

The first phase of verification consisted of three types of experiments: validation of object detection on scenes with a single object of interest present, simultaneous detection of many objects present on scene and validation of estimation of pose of detected objects. In those experiments we used models from 3D Model Dataset, scenes were taken from Test Scenes Dataset, whereas for the pose estimation tests we used the Pose Estimation Test Dataset—all belonging to the WUT Visual Perception Dataset [14]. Exemplary results for a single object present in the scene are presented in Fig. 3. Please note that correspondence grouping resulted in two clusters, hence two hypotheses. However, the weaker hypothesis that was in conflict was rejected in the hypothesis verification step.

3.2 On-line Verification with Kinect Sensor

Having system working off-line we performed several experiments on-line, with Kinect sensor streaming consecutive RGB-D images. In Fig. 4 we present results of subsequent steps of recognition pipeline in the experiment with two objects of different classes being present in the scene. Analogically to Fig. 3, at the end the system was able to recognize both objects (and successfully rejected false hypotheses).

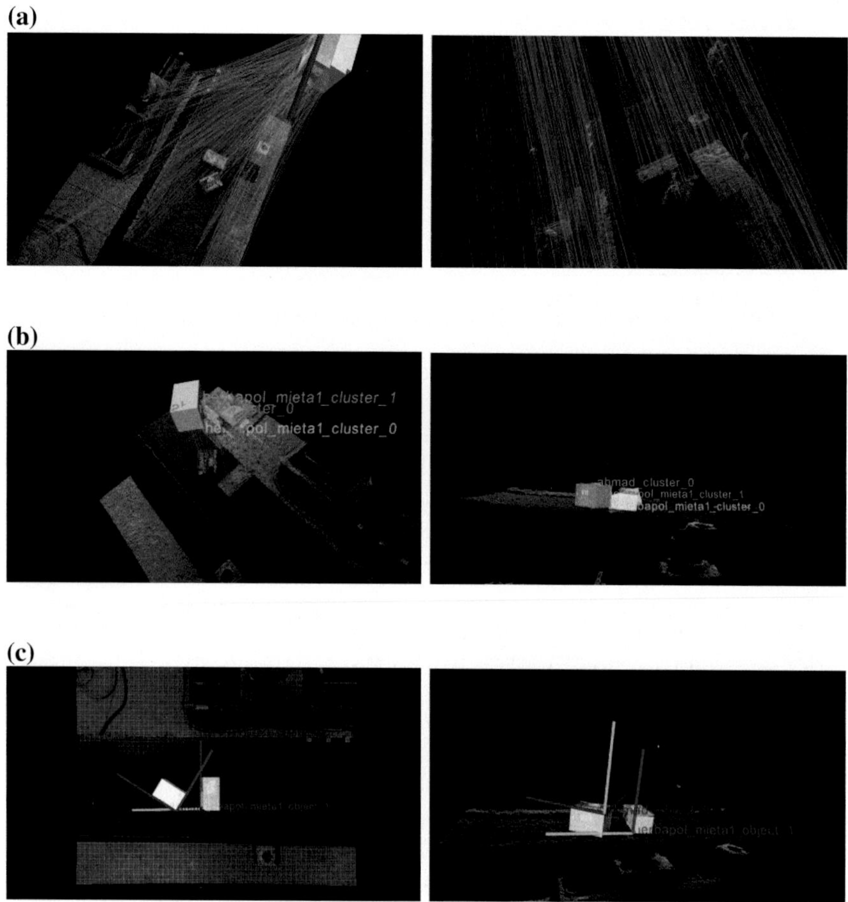

Fig. 4 On-line verification of the perception subsystem on images acquired from Kinect sensor with many different objects of interests present in the scene and many models of objects (*left* column—*top* view, *right* column—*side* view). **a** Estimated scene-models reciprocal correspondences. **b** Found hypotheses (note 2 hypotheses for Mieta tea). **c** Hypotheses remaining after verification (with object reference frames)

3.3 On-line Verification in the Object Picking Task

Finally we verified the system in the object picking task. As this task was executed just for the purpose of verification of the perception subsystem and its integration with the robot controller, we assumed that we do not required any sophisticated grasp planning, as presented in e.g. [19].

3.3.1 Hardware Configuration

In the experiments we decided to use the IRp-6 6DOF manipulator, controlled by the modern hardware components that recently replaced the original, industrial ones [20]. The manipulator ends-effector is equipped with a two-finger gripper and supplemented with a Point Grey Blackfly BFLY-PGE-14S2C-CS camera, attached to the robot wrist (not used in our experiments). Additionally, there was a Kinect sensor mounted above the manipulator workspace. The calibration of the system was performed according to procedure presented in [21].

3.3.2 Control Subsystem

In Fig. 5 we presented the structure of the developed control system. It was implemented on the basis of Robot Operating System (ROS) [22], with OROCOS [23] used for implementation of the low-level robot controller (IRPOS) and DisCODe (Distributed Component Oriented Data Processing) [24] used for implementation of the presented perception subsystem. Each of those two subsystems work as a separate ROS node.

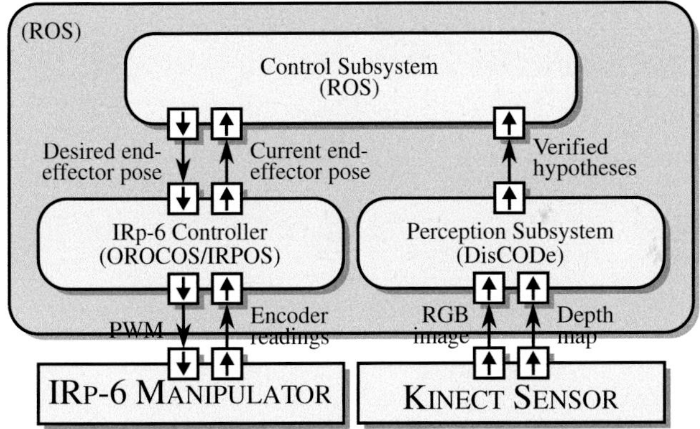

Fig. 5 Structure of the control system developed for the purpose of object picking task

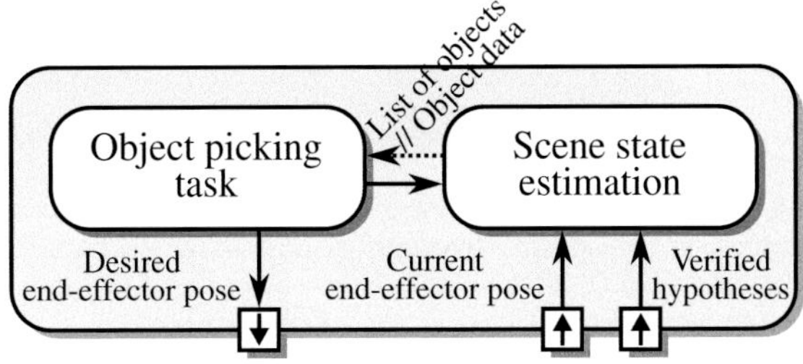

Fig. 6 Structure of the control subsystem

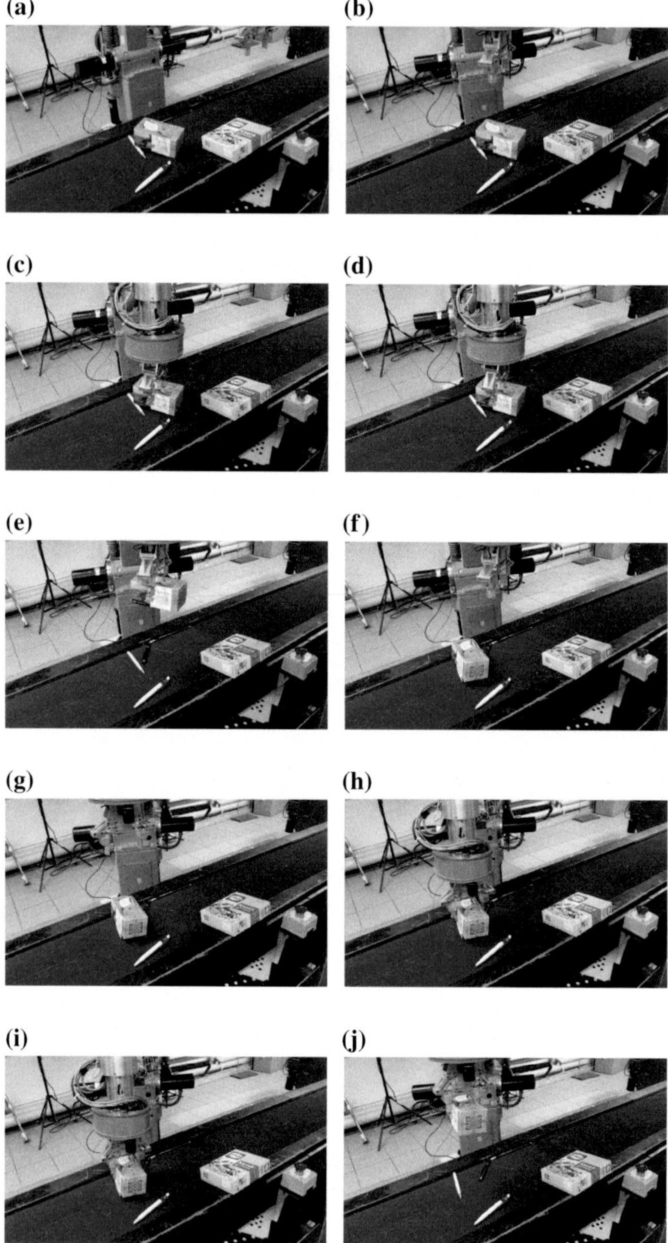

Fig. 7 Consecutive steps of the experiment with picking an exemplary object. The whole experiment can be seen at https://youtu.be/z-FB5itNz9M. **a** Start pose. **b** Pregrasp pose reached. **c** Grasp pose reached. **d** Object grasped. **e** Object picked up. **f** Object dropped. **g** Pregrasp pose reached. **h** Grasp pose reached. **i** Object grasped. **j** Object picked up

The high-level control subsystem, governing the execution of the task, was implemented as three ROS nodes (Fig. 6). The **Scene state estimation** node is responsible for aggregation of information regarding object present in the scene. In particular, it merges the consecutive hypotheses returned by perception subsystem. It offers two ROS services: returning list of names and confidence scores of objects present in the scene and returning data of object with given name. The **Object picking task** node is responsible for realization of the picking task according to the following algorithm. If there are any objects in the retrieved object list, it selects the one with highest confidence, retrieves its pose and model (mesh, dimensions etc.). Next it generates several grasp poses and selects the one with approach from the top. For this grasp pose it generates pregrasp pose and orders the manipulator to move to it. Next it orders the manipulator to reach the grasp pose, close the gripper (this depends on the object dimensions) and move back to pregrasp pose. Finally it commands opening of the gripper, which results in dropping the object in a random position. After this the system starts the picking operation once again. Consecutive steps of an experiment with picking up of an exemplary object are presented in Fig. 7.

In the **Object picking task** node we used IRPOS, a high-level API facilitating programming of diverse tasks executed by IRp-6 robot (e.g. executing motion expressed as end-effector Cartesian trajectory).

4 Summary

The article described the perception subsystem able to recognize and estimate poses of diverse textured objects in RGB-D images. The presented both off-line and on-line experiments proved the correct operation of the system.

The developed subsystem was integrated with controller of a IRP-6 manipulator and the resulting solution managed to successfully recognize and pick up different objects appearing in the scene. Currently there is ongoing work on integration of the perception subsystem with controller of the dual-handed Velma service robot and more sophisticated grasp planning algorithms presented in [19].

The major limitation of the system results of from used SIFT features, extracted straight from colour image. Thus we plan to further expand the system by adding pipeline depending on features extracted from depth map/point cloud.

Acknowledgments This project was funded by the National Science Centre according to the decision number DEC-2012/05/D/ST6/03097. Tomasz Kornuta is supported by the IBM Research, Almaden through the IBM PostDoc/LTS Programme.

References

1. Dziergwa, M., Kaczmarek, P., Kędzierski, J.: RGB-D sensors in social robotics. J. Autom. Mobile Robot. Intell. Syst. **9**(1), 18–27 (2015)
2. Ramey, A., González-Pacheco, V., Salichs, M.A.: Integration of a low-cost RGB-D sensor in a social robot for gesture recognition. In: Proceedings of the 6th International Conference on Human-robot Interaction, pp. 229–230. ACM (2011)
3. Stefańczyk, M., Banachowicz, K., Walęcki, M., Winiarski, T.: 3D camera and lidar utilization for mobile robot navigation. J. Autom. Mobile Robot. Intell. Syst. **7**(4), 27–33 (2013)
4. Belter, D., Nowicki, M., Skrzypczyński, P., Walas, K., Wietrzykowski, J.: Lightweight RGB-D slam system for search and rescue robots. In: Progress in Automation, Robotics and Measuring Techniques, pp. 11–21. Springer (2015)
5. Walas, K.: Terrain classification and negotiation with a walking robot. J. Intell. Robot. Syst. **78**(3–4), 401–423 (2015)
6. Großmann, B., Siam, M., Krüger, V.: Comparative evaluation of 3D pose estimation of industrial objects in RGB pointclouds. In: Computer Vision Systems, pp. 329–342. Springer (2015)
7. Ren, X., Fox, D., Konolige, K.: Change their perception: RGB-D for 3-D modeling and recognition. IEEE Robot. Autom. Mag. **20**(4), 49–59 (2013)
8. Maitin-Shepard, J., Cusumano-Towner, M., Lei, J., Abbeel, P.: Cloth grasp point detection based on multiple-view geometric cues with application to robotic towel folding. In: 2010 IEEE International Conference on Robotics and Automation (ICRA), pp. 2308–2315 (2010)
9. Doumanoglou, A., Kargakos, A., Kim, T.K., Malassiotis, S.: Autonomous active recognition and unfolding of clothes using random decision forests and probabilistic planning. In: 2014 IEEE International Conference on Robotics and Automation (ICRA), pp. 987–993. IEEE (2014)
10. Lenz, I., Knepper, R., Saxena, A.: DeepMPC: learning deep latent features for model predictive control. In: Proceedings of Robotics: Science and Systems, Rome, Italy (July 2015)
11. Rusu, R., Blodow, N., Beetz, M.: Fast point feature histograms (FPFH) for 3D registration. In: IEEE International Conference on Robotics and Automation. ICRA'09, pp. 3212–3217. IEEE (2009)
12. Harasymowicz-Boggio, B., Chechliński, Ł., Siemiątkowska, B.: Nature-inspired, parallel object recognition. In: Progress in Automation, Robotics and Measuring Techniques, pp. 53–62. Springer (2015)
13. Bo, L., Ren, X., Fox, D.: Unsupervised feature learning for RGB-D based object recognition. In: Experimental Robotics, pp. 387–402. Springer (2013)
14. Stefańczyk, M., Laszkowski, M., Kornuta, T.: WUT visual perception dataset—a dataset for registration and recognition of objects. In: Szewczyk, R., Zieliński, C., Kaliczyńska, M. (eds.) Recent Advances in Automation, Robotics and Measuring Techniques. Advances in Intelligent Systems and Computing (AISC). Springer (2016) (Under review)
15. Lowe, D.G.: Distinctive image features from scale-invariant keypoints. Int. J. Comput. Vis. **60**(2), 91–110 (2004)
16. Muja, M., Lowe, D.G.: Fast approximate nearest neighbors with automatic algorithm configuration. In: VISAPP (1), pp. 331–340 (2009)
17. Chen, H., Bhanu, B.: 3D free-form object recognition in range images using local surface patches. Pattern Recognit. Lett. **28**(10), 1252–1262 (2007)
18. Aldoma, A., Tombari, F., Rusu, R.B., Vincze, M.: OUR-CVFH-Oriented, Unique and Repeatable Clustered Viewpoint Feature Histogram for Object Recognition and 6DOF Pose Estimation. Springer (2012)
19. Seredynski, D., Winiarski, T., Banachowicz, K., Zielinski, C.: Grasp planning taking into account the external wrenches acting on the grasped object. In: 2015 10th International Workshop on Robot Motion and Control (RoMoCo), pp. 40–45. IEEE (2015)

20. Walcki, M., Banachowicz, K., Winiarski, T.: Research oriented motor controllers for robotic applications. In: Kozłowski, K. (ed.) Robot Motion and Control 2011 (LNCiS) Lecture Notes in Control and Information Sciences, vol. 422, pp. 193–203. Springer Verlag London Limited (2012)
21. Winiarski, T., Banachowicz, K.: Automated generation of component system for the calibration of the service robot kinematic parameters. In: 20th IEEE International Conference on Methods and Models in Automation and Robotics, MMAR'2015, pp. 1098–1103. IEEE (2015)
22. Quigley, M., Gerkey, B., Conley, K., Faust, J., Foote, T., Leibs, J., Berger, E., Wheeler, R., Ng, A.: ROS: an open-source robot operating system. In: Proceedings of the Open-Source Software Workshop at the International Conference on Robotics and Automation (ICRA) (2009)
23. Bruyninckx, H.: The real-time motion control core of the OROCOS project. In: Proceedings of the IEEE International Conference on Robotics and Automation, pp. 2766–2771. IEEE (Sept 2003)
24. Stefańczyk, M., Kornuta, T.: Handling of asynchronous data flow in robot perception subsystems. In: Simulation, Modeling, and Programming for Autonomous Robots. Lecture Notes in Computer Science, vol. 8810, pp. 509–520. Springer (2014)

The Comparison of Keypoint Detectors and Descriptors for Registration of RGB-D Data

Aleksander Kostusiak

Abstract In many computer vision tasks it is almost impossible to process all data from camera in a reasonable amount of time. Because of that, only a subset of this data is typically used, which is defined by interest points. In this paper we present a comparison of some popular keypoint detector/descriptor pairs in the context of frame-to-frame trajectory estimation (visual odometry) on RGB-D data. We focus on recently introduced keypoint detection and description algorithms. In the test application the unavoidable trajectory drift is not compensated, e.g. by a graph-based back-end. Thus, one can directly see how different approaches to determine the location of keypoints influence the quality of the recovered trajectory. The presented system uses a publicly available RGB-D dataset and applies well-established metrics to assess the quality of the trajectories.

Keywords RGB-D SLAM · Point detectors · Point descriptors · Evaluation

1 Introduction

As for now one of the most difficult and popular task in robotics is to estimate the trajectory of the robot. In order to achieve that, many different techniques has arrived, like simple wheel odometry, more complicated visual odometry [1] and finally full SLAM systems [2], that give us not only the trajectory, but also a map of the environment. On the other hand one could use GPS to achieve high precision, but here are some limitation to this: if it is online we lose precision (the signal is

A. Kostusiak (✉)
Poznań University of Technology, Institute of Control and Information Engineering, Poznan, Poland
e-mail: aleksander.m.kostusiak@doctorate.put.poznan.pl

R. Szewczyk et al. (eds.), *Challenges in Automation, Robotics and Measurement Techniques*, Advances in Intelligent Systems and Computing 440, DOI 10.1007/978-3-319-29357-8_53

disturbed and we will not be able to contact reference stations in time), we have to perform it outdoor, in the moment when there are at least 3 satellites visible. Because many mobile robots should perform as well in indoor situations as in the outdoor ones, at any given time, we can't rely on GPS systems.

The best way to calculate the trajectory would be to use a camera system, and match every point seen in the last frame to the content of the current frame. This is of course computationally very expensive and instead of doing so, only some salient points are used, denoted usually as keypoints.

The advent of Microsoft Kinect device significantly increased the interest in the use of RGB-D data, as the price was set very low as for this kind of device. The affordable price and relatively good quality of the depth data from the Kinect-type devices resulted in new applications of the existing computer vision algorithms for navigation, and arrival of new ones. Many of the existing and new SLAM or visual odometry techniques applied to RGB-D data are based on the use of keypoints. Thus, it is important to know which keypoint detector and descriptor performs the best with this class of data. Since each of the authors of a navigation algorithm compares his work to the others on a different dataset (most likely owned by him) in such a manner, that it looks better than the rest in comparison, there is the need to perform independent experiments.

2 Related Work

Up to now there are many works focused around keypoint detection and description dated back almost to the half of the last century. It is no surprise, that along with new algorithms, new evaluation appeared as well, naming just a few: work of Mikolajczyk and Schmid [3], Mikolajczyk et al. [4], Schaeffer and Cameron [5]. But those works do not compare interest point detectors and descriptors in the contexts of trajectory estimation. The works of Filipe and Alexandre [6, 7] provide us with comparison of 3D keypoints by using the repeatability measure but not in the context of robot navigation. Schmidt et al. [8] compare several recently introduced keypoint detector and descriptor pairs in the context of visual navigation, but taking into account only the photometric (RGB) data, and using their proprietary performance metrics. In the work of Manoj et al. [9] 3D feature detectors and descriptors (from depth data) are compared via absolute trajectory error and relative pose error metrics, which we use as well in our comparison. However, in [9] the trajectory is calculated by frame-to-model tracking what results in a suppressed drift of trajectory, comparing it to the frame-to-frame matching approach.

Also some works on RGB-D SLAM attempt to compare results achieved when applying different detectors/descriptors in the SLAM algorithm implementation. For instance Belter et al. [10] evaluate performance of a pose-based RGB-D SLAM system applying FAST, SURF and ORB algorithms. This comparison gives a valuable insight in the role of the keypoint detector performance in RGB-D SLAM, but does not take into account more recent algorithms, such like BRISK or KAZE/AKAZE.

As stated in [11], the uncertainty of RGB-D keypoint position is dominant in feature-based RGB-D SLAM.

3 System Architecture and Algorithms

We use visual odometry to estimate frame to frame trajectory obtained with the use of SURF [12], KAZE [13], AKAZE [14], BRISK [15] and ORB [16] algorithms from OpenCV library. In order to show the best results, that one could expect from trajectory estimate we compare visual odometry trajectories against RGB-D SLAMv2 [17] system, which also uses three different detectors and descriptors: SURF, SIFT [18] and ORB.

3.1 Architecture

In order to compare detectors and descriptors in terms of quality of trajectory that can be estimated with the points detected by them, a simple frame-to-frame odometry system has been implemented. In this simple visual odometry system, we firstly detect keypoints on the RGB image for two subsequent images, and then we use depth data to obtain their positions in the real world, then we calculate local transformation with the use of Kabsch algorithm [19, 20]. After that we update the global transformation. The flowchart in Fig. 1 shows the main building blocks of the program. Default parameters of the detectors have been changed, in order to detect more keypoints in the RGB images, however, these parameters are not changed anymore during the experiment and are the same for all benchmark sequences.

In the initialisation process, the first RGB and D frames are loaded, the keypoints are extracted, described, and associated with corresponding depth information. Then, in the loop, the previously newest frame becomes the old one, a new frame is loaded from the dataset and new keypoints are extracted, described, and associated with the corresponding depth information—if the depth information is missing the interest point is discarded. Afterwards, the keypoints from old and new frames are matched (the cross-matching check is performed). The next step is to run RANSAC (3 paired sets of points are randomised) algorithm twice to eliminate mismatched keypoints.

Then rotation and translation is computed with the use of all remaining matched points. At the end of the loop the global position of the robot is updated and some of the benchmark data saved. If there are no more frames in the sequence the program saves last statistics and prepares for closure.

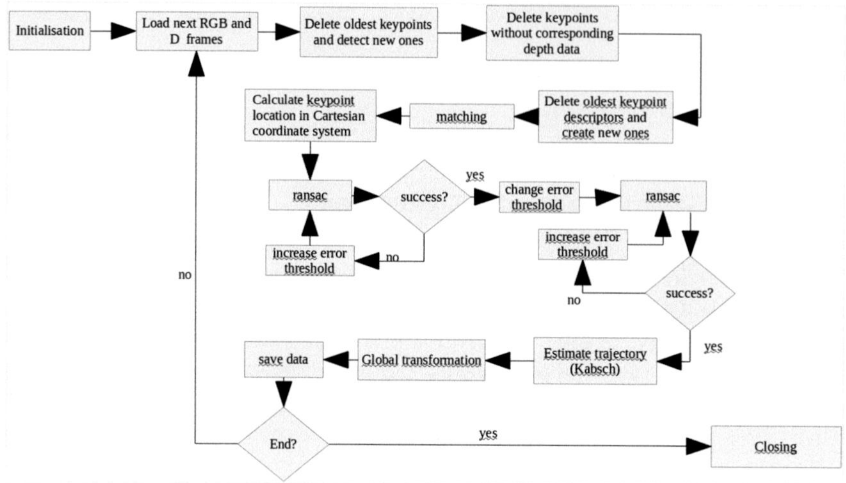

Fig. 1 Flowchart of the visual odometry program

3.2 *Kabsch Algorithm*

This algorithm allows to determine the translation and rotation between two sets of paired points. It consists of 3 main steps: calculating centroids, rotation [21] and translation [22] calculation.

3.2.1 Centroids

Firstly we have to calculate the centroid—the average point of all points in the set.

$$p_c = \frac{1}{N} \sum_{i=1}^{N} p_i \tag{1}$$

where $\boldsymbol{p_c}$ is the average point of all points in the set, $\boldsymbol{p_i}$ is the ith point in the set, and N is the total number of points in the set.

Now we have to translate all points from the set P, so that the average point aligns the origin of the coordinate system:

$$p_i^{'} = p_i - p_c \tag{2}$$

Now we have centred set P′.

3.2.2 Rotation

To calculate rotation we perform the following steps:

1. Create 3 × N **A** and **B** matrices containing previously matched and centred points from the sets a and b.
2. Calculate covariance matrix $\mathbf{C} = \mathbf{AB}^{\mathrm{T}}$.
3. Compute SVD of the covariance matrix $\mathbf{C} = \mathbf{VSW}^{\mathrm{T}}$.
4. Compute $d = \mathrm{sign}(\det(\mathbf{C}))$.
5. Compute optimal rotation matrix:

$$R = \boldsymbol{W}\begin{bmatrix} 1 & 0 & 0 \\ 0 & 1 & 0 \\ 0 & 0 & d \end{bmatrix} V^T \tag{3}$$

3.2.3 Translation

Having computed the centroid for both of sets and rotation **R** from set A to B we can compute translation as [22]:

$$t = -R\,\boldsymbol{Centroid_A} + \boldsymbol{Centroid_B} \tag{4}$$

where **Centroid** A and **Centroid** B are the average points of set A and B, **R** is rotation from a to b and **t** stands for translation. It is important to emphasize that the presented algorithm finds the best estimation of rotation and translation from point set A to B but not for the robot/camera. The motion of our device is just opposite. We can transpose rotation matrix and change the sign of translation or can use as set A points coming from the newest image and for set B the old ones.

3.3 *KAZE Detector/Descriptor*

This algorithm, presented in the [13], is similar to the well-known SIFT algorithm [18]. In order to build the scale space it uses additive operator splitting AOS scheme and variable conductance diffusion to reduce the noise but preserve sharpness. The resolution is not changed. To find the interest point, the algorithm looks for the maxima in scale and spatial locations of responses of scale-normalized determinant of the Hessian at multiple scale levels.

In order to obtain rotation invariant descriptor the main orientation is calculated in a similar way to SURF—the first order derivatives L_x and L_y are weighted with a Gaussian centred on the interest point. Then these responses are represented as points in vector space and summed within sliding circle fragment. The longest vector stands for orientation.

Descriptor is a modified version of M-SURF—for a detected feature first order derivatives are computed over a rectangular grid, which is divided into 4 × 4 overlapping subregions. Each sample in the rectangular grid is rotated accordingly to the dominant orientation as well as derivatives. After weighting, descriptors from each region are added to the descriptor sub-vector, which are then weighted again and added to create the final description.

3.4 AKAZE Detector/Descriptor

This algorithm is supposed to perform better than KAZE and is created by the same authors [14]. Here the scale space is built in almost the same manner as in SIFT, but with the use of Fast Explicit Diffusion algorithm instead of Gaussian kernel.

To detect an interest point the determinant of Hessian is calculated (5) for each image in the nonlinear scale space and weighted accordingly. The second order derivatives are computed with the help of Scharr filter. The local maxima are picked as salient point candidates and compared with other candidate points within a $\sigma x \sigma$ window from sublevel i + 1 and i − 1.

$$L^i_{Hessian} = \sigma_i^2 / 2^{o^i} \left(L^i_{xx} L^i_{yy} - L^i_{xy} L^i_{xy} \right) \tag{5}$$

where i is the number of filtered image, L is the image luminance, o is octave number and s sublevel. Authors introduce the M—LDB descriptor which starts by estimating orientation by using a histogram method, same as in KAZE. After that the pattern is rotated accordingly and finally the descriptor is generated by performing binary tests of average areas and the mean of the horizontal and vertical derivatives in the areas.

3.5 SURF Detector/Descriptor

This algorithm performs the same steps as SIFT, but in a different manner [12].

Firstly, to speed up whole process, the original image is transformed into integral image representation. In order to build up scale space, the original image does not need to be subsampled, instead the Hessian determinant is approximated by convolving integral image with different sizes of approximated Gaussian kernels. The last step to detect an interest point is to perform non-maxima suppression in a 3 × 3 × 3 neighborhood. Then the maxima of the determinant of the Hessian matrix are interpolated in the scale and image space.

The Haar wavelet responses are calculated in x and y directions within a sampled, circular neighbourhood of radius corresponding to the scale at which the keypoint was detected. Then these responses are represented as points in vector space and then summed within a sliding circle fragment. The longest vector stands for orientation.

Next is to construct square region centred around the interest point. The region is split into smaller 4 × 4 regions, in which the Haar wavelet responses are calculated for 5 × 5 regularly spaced points, in x and y directions, interpolated accordingly to orientation and weighted with a Gaussian. Then in each subregion the responses of Haar wavelet are summed up:

$$v = \left(\sum d_x,\ \sum d_y,\ \sum |d_x|,\ \sum |d_y|\right). \tag{6}$$

The final descriptor consist of vectors originating from all of the sub regions.

3.6 ORB Detector/Descriptor

The main idea behind the detector part of this algorithm [16] is to effectively use FAST [23] with a circular radius of 9 combined with the Harris corner measure to preserve N best keypoints (the threshold is firstly set to detect at least N points) and scale pyramid of the image. To achieve scale invariance the authors of [16] employ scale pyramid of the image to produce FAST features. The intensity centroid is used as orientation measure in order to be rotation invariant.

Descriptor is formed as a set of binary tests according to a pattern which is a result of machine learning process, rotated by the corresponding keypoint dominant orientation. Test for two paired points looks like this:

$$\tau(p; x, y) := \begin{cases} 1: p(x) < p(y) \\ 0: p(x) \geq p(y) \end{cases} \tag{7}$$

where p(x) is the intensity of p at the point x.

3.7 BRISK Detector/Descriptor

Like other algorithms, BRISK [15] is creating scale space—here by down sampling original image. In each level of this scale space, the FAST detector [23] is applied to find features. 9 of 16 pixels have to be consequently sufficiently brighter or darker that the central pixel, to consider it as interest point. Then the non-maxima suppression is applied, firstly the candidate point must have better FAST score than 8 surrounding pixels, and secondly must have better score than pixels in the patches above and below (due to different discretization the patch is interpolated). Next, for each detected maximum subpixel scale refinement is performed in order to obtain the final keypoint. Finally the image coordinates of interest point are re-interpolated.

Given the refined keypoint location in the image and its floating point scale, BRISK descriptor firstly smoothes the image with the Gaussian and scales the pattern properly. Then it uses its pattern to perform binary tests between long distance pairs to determine the rotation. Finally the pattern is rotated and a series of brightness test, now based on short distance pairs, is performed.

4 Experimental Results

To perform the evaluation of keypoint detectors and descriptors we use TUM RGB-D dataset [24, 25]. It has many sequences recorded both traditionally and in "rosbag" format. The following sequences are used in the evaluation: freiburg1_room, freiburg1_desk, freiburg3_long_office_household. To compare estimated trajectories with ground-truth we use ATE and RPE error evaluation tools provided in [25]. We employ the same measures of the estimated trajectory error as in [9]—ATE (absolute trajectory error) and RPE (relative pose error) and we use same dataset but not exactly the same sequences. Moreover, the RGBDSLAMv2 system [17] is used to compare visual odometry results (containing drift) to the ones that optimize trajectory and reduce drift. The RGBDSLAMv2 could be run only with every sixth frame, due to high memory demands. It also changes detector parameters during runtime.

For the freiburg_room sequence Figs. 2 and 3 show ATE and RPE trajectory errors, for the other sequences we present only the summary of quantitative results in tables, which contain ATE and RPE errors, times of detection, description, matching, RANSAC and Kabsch algorithm running time, average numbers of inliers and outliers.

4.1 Freiburg_Room

In this sequence we can see, that permanently set up BRISK performs poorly—maybe it could work better with varying parameters. The other algorithms seem to perform head to head here (Tables 1 and 2).

The results of RGBDSLAMv2 ORB version are unsatisfying- they are not that better than frame to frame odometry, where no graph optimisation is performed.

ORB occurred to be the fastest algorithm in this sequence for odometry usage. AKAZE was just behind it and slightly slower was SURF and then BRISK. KAZE was the slowest one here.

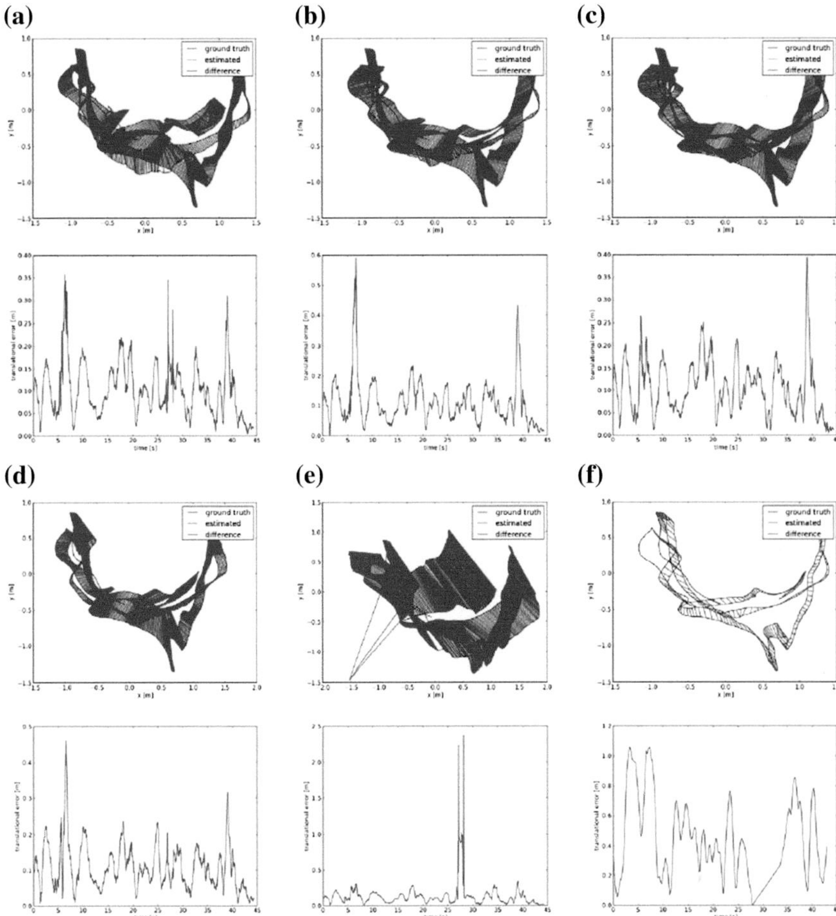

Fig. 2 In the *top* and *third row* freiburg1_room absolute trajectory errors of estimated trajectory versus ground truth. In the *second* and *bottom row* freiburg1_room relative pose errors of estimated trajectory versus ground truth. *First two rows* from the *left*: From *left*: **a** AKAZE, **b** KAZE, **c** SURF. *Second two rows* from the *left* **d** ORB, **e** BRISK, **f** RGB-D SLAM v2 (SURF)

If it goes for inliers to outliers ratio—all algorithms had ratio of around 3, while BRISK had the ratio of 2.

4.2 *Freiburg_Desk*

In this sequence BRISK occurred to be worst again, if it goes for ATE errors, the RPE ones were not much different from the others. SURF and ORB occurs here to

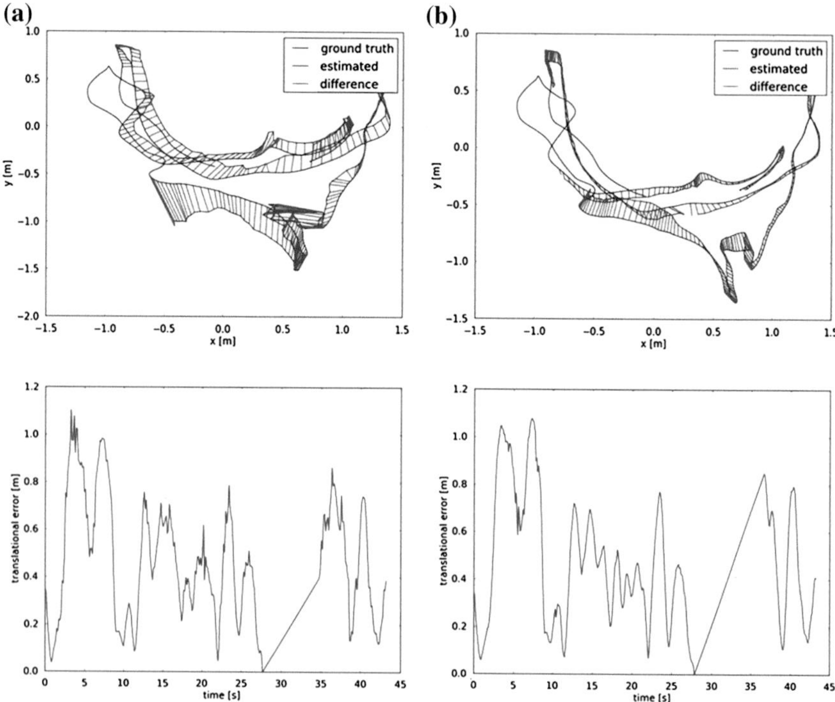

Fig. 3 In the *top row* freiburg1_room absolute trajectory errors of estimated trajectory versus ground truth. In the *bottom row* freiburg1_room relative pose errors of estimated trajectory versus ground truth. From *left*: **a** RGBDSLAMv2 (ORB), **b** RGBDSLAMv2 (SIFT)

Table 1 Results of juxtaposition of estimated trajectory and groundtruth freiburg1_room

	ATE rmse (m)	RPE rmse (m)	RPE rmse (°)
AKAZE	0.334	0.119	4.100
KAZE	0.373	0.137	3.986
SURF	0.377	0.123	3.939
ORB	0.379	0.127	3.873
BRISK	0.725	0.206	9.841
RGB-D SLAM (ORB)	0.285	0.533	53.06
RGB-D SLAM (SIFT)	0.095	0.537	54.61
RGB-D SLAM (SURF)	0.124	0.540	53.91

work the best, but AKAZE is just behind them. The results of RGB-D SLAMv2 are approximately 3 times better than the lowest ATE error of frame to frame odometry (Tables 3 and 4).

All algorithms of feature detector and description have the inliers to outliers ratio equal at least 2.5, when BRISK has this ratio around 2.

Table 2 Statistics of detectors descriptors pairs freiburg1_room

	Average time (s)					Average number of inliers	Average number of outliers
	Detection	Description	Matching	RANSAC	Kabsch		
AKAZE	0.107	0.090	0.052	0.607	4.528e−05	232.77	86.08
KAZE	0.476	0.338	0.009	0.342	2.692e−05	171.36	67.62
SURF	0.109	0.185	0.020	0.630	4.370e−05	203.12	75.66
ORB	0.009	0.011	0.083	0.653	3.799e−05	255.50	110.99
BRISK	0.006	0.004	0.066	0.750	3.419e−05	157.65	77.31

Table 3 Results of juxtaposition estimated trajectory and groundtruth freiburg1_desk

	ATE rmse (m)	RPE rmse (m)	RPE rmse (°)
AKAZE	0.096	0.074	2.487
KAZE	0.101	0.069	2.379
SURF	0.085	0.070	2.201
ORB	0.085	0.080	3.246
BRISK	0.202	0.099	3.089
RGB-D SLAM (ORB)	0.063	0.719	39.551
RGB-D SLAM (SIFT)	0.042	0.704	41.756
RGB-D SLAM (SURF)	0.034	0.705	41.773

Table 4 Statistics of detectors descriptors pairs freiburg1_desk

	Average time (s)					Average number of inliers	Average number of outliers
	Detection	Description	Matching	RANSAC	Kabsch		
AKAZE	0.111	0.103	0.094	1.103	5.950e−05	314.02	121.37
KAZE	0.477	0.353	0.016	0.562	3.363e−05	230.50	92.72
SURF	0.114	0.218	0.033	0.886	5.092e−05	261.80	100.77
ORB	0.010	0.011	0.085	0.844	3.744e−05	260.38	108.90
BRISK	0.008	0.005	0.115	1.428	4.152e−05	217.73	105.71

The fastest algorithm here is ORB, but just a little slower SURF, KAZE and AKAZE (high time spent in RANSAC tests means that error threshold was raised many times). BRISK was the slowest one.

4.3 Freiburg3_Long_Office_Household

Here the best algorithms, in case of ATE errors, were SURF and AKAZE. The others were a little worse. RPE errors are here bigger than in previous sequences—

Table 5 Results of juxtaposition of estimated trajectory and groundtruth freiburg3_long_office_household

	ATE rmse (m)	RPE rmse (m)	RPE rmse (°)
AKAZE	0.861	0.192	10.832
KAZE	1.067	0.181	10.787
SURF	0.892	0.183	10.618
ORB	1.040	0.207	11.016
BRISK	1.190	0.188	10.714
RGB-D SLAM (ORB)	0.404	0.342	20.846
RGB-D SLAM (SIFT)	0.056	0.341	20.517
RGB-D SLAM (SURF)	0.089	0.338	20.417

Table 6 Statistics of detectors descriptors pairs freiburg3_long_office_household

	Average time (s)					Average number of inliers	Average number of outliers
	Detection	Description	Matching	RANSAC	Kabsch		
AKAZE	0.110	0.096	0.066	1.389	5.340e−05	296.60	122.46
KAZE	0.458	0.334	0.011	0.738	3.131e−05	222.61	97.44
SURF	0.116	0.205	0.024	1.290	4.703e−05	240.06	104.94
ORB	0.010	0.012	0.110	1.246	4.704e−05	341.66	150.93
BRISK	0.007	0.004	0.098	1.535	4.401e−05	231.33	113.04

maybe because of its length and drift natural for frame-to-frame estimation (Tables 5 and 6).

ORB was the fastest algorithm again, then comparably SURF, BRISK and AKAZE. The slowest was KAZE (about 50 % slower than ORB).

Here one can see advantage of using SLAM techniques, ATE errors for SURF and SIFT versions are one order of magnitude smaller. The inlier to outlier ratio remains the same for BRISK = 2, for the other algorithms it is now varying form 2.25 for ORB to around 2.5 for AKAZE.

5 Conclusions

Taking into account all of experimental results and statistics we can say that the best algorithms for trajectory estimation are SURF, AKAZE and just a little worse ORB. Each has its own assets and drawbacks. SURF and AKAZE allow to achieve the same quality of trajectory estimation in almost the same time. But taking into account that SURF is patented, the AKAZE algorithm is a good, free alternative.

The fastest detector/descriptor algorithm here is ORB. It allows for detection and description in one order of magnitude faster than other algorithms, but it seems that the greedy for best results RANSAC procedure slowed it down. If other method was used, the whole process should be faster but most probably the drift (measured by ATE) would be bigger. The aim in these experiments was to achieve the best frame-to-frame trajectory with the use of certain keypoint detector and descriptor, without further optimisation of the pose graph. Finding a good compromise between the running speed and trajectory estimation accuracy, e.g. by applying some population-based search algorithm to optimize the parameters, could be an interesting topic of further research. Taking into account graph optimisation which the RGB-D SLAM v2 system performs, it seems that ORB does not find the best points for optimisation process.

References

1. Scaramuzza, D., Fraundorfer, F.: Visual odometry: part I the first 30 years and fundamentals. IEEE Robot. Autom. Mag. **18**(4), 80–92 (2011)
2. Skrzypczynski, P.: Simultaneous localization and mapping: a feature-based probabilistic approach. Int. J. Appl. Math. Comput. Sci. **19**(4), 575–588 (2009)
3. Mikolajczyk, K., Schmid, C.: A performance evaluation of local descriptors. IEEE Trans. Pattern Anal. Mach. Intell. (PAMI) **2**, 1115–1125 (2005)
4. Mikolajczyk, K., Tuytelaars, T., Schmid, C., Zisserman, A., Matas, J., Schaffalitzky, F., Kadir, T., Gool, L.: A comparison of affine region detectors. Int. J. Compu. Vis. (IJCV) **65**(1), 43–72 (2005)
5. Schaeffer C.: A Comparison of Keypoint Descriptors in the Context of Pedestrian Detection: FREAK vs. SURF vs. BRISK, Stanford University CS Department (2013)
6. Filipe, S., Alexandre, L.A.: A comparative evaluation of 3D keypoint detectors. In: 9th Conference on Telecommunications, Conftele 2013, pp. 145–148 (2013)
7. Filipe, S., Alexandre, L.A.: A comparative evaluation of 3D keypoint detectors in a RGB-D object dataset. In: Proceedings of the 9th International Conference on Computer Vision Theory and Applications (2014)
8. Schmidt, A., Kraft, M., Fularz, M., Domagala, Z.: Comparative assessment of point feature detectors and descriptors in the context of robot navigation. J. Autom. Mobile Robot. Intell. Syst. **7**(1), 11–20 (2013)
9. Manoj, P.S., Bingbing, L., Weisi, L., Qayyum, U.: Sparse depth odometry: 3D keypoint based pose estimation from dense depth data. In: Conference: IEEE International Conference on Robotics and Automation (ICRA), pp. 4216–4223 (2015)
10. Belter, D., Nowicki, M., Skrzypczynski, P.: On the performance of pose-based RGB-D visual navigation systems. In: Cremers, D. et al. (eds.) Computer Vision—ACCV 2014, LNCS 9004, pp. 407–423. Springer (2015)
11. Belter, D., Skrzypczynski, P.: The importance of measurement uncertainty modeling in the feature-based RGB-D SLAM. In: Proceedings of the 10th International Workshop on Robot Motion and Control, Poznan, pp. 308–313 (2015)
12. Bay, H., Ess, A., Tuytelaars, T., Van Gool, L.: Speeded-up robust features (SURF). Comput. Vis. Image Underst. (CVIU) **110**(3), 346–359 (2008)
13. Alcantarilla, P.F., Bartoli, A., Davison, A.J.: KAZE features. In: European Conference on Computer Vision (ECCV), Fiorenze, Italy (2012)
14. Alcantarilla, P.F., Nuevo, J., Bartoli, A.: Fast explicit diffusion for accelerated features in nonlinear scale spaces. In: British Machine Vision Conference (BMVC), Bristol, UK (2013)
15. Leutenegger, S., Chli, M., Siegwart, R.: BRISK: binary robust invariant scalable keypoints. Int. Conf. Comput. Vis. **2011**, 2548–2555 (2011)

16. Rublee, E., Rabaud, V., Konolige, K., Bradski, G.: ORB: an efficient alternative to SIFT or SURF. In: ICCV '11 Proceedings of the 2011 International Conference on Computer Vision, pp. 2564–2571 (2011)
17. Endres, F., Hess, J., Sturm, J., Cremers, D., Burgard, W.: 3D mapping with an RGB-D camera. In: IEEE Transactions on Robotics, pp. 177–187 (2014)
18. Lowe, D.G.: Distinctive image features from scale-invariant keypoints. Int. J. Comput. Vis. **60** (2), 91–110 (2004)
19. Kabsch, W.: A solution for the best rotation to relate two sets of vectors. Acta Crystallographica **32**, 922 (1976)
20. Kabsch, W.: A discussion of the solution for the best rotation to relate two sets of vectors. Acta Crystallographica **A34**, 827–828 (1978)
21. Kavraki, L.E.: Molecular Distance Measures, OpenStax-CNX (2003). cnx.org/contents/1d5f91b1-dc0b-44ff-8b4d-8809313588f2@23/Molecular_Distance_Measures
22. Nghia, H.: Finding optimal rotation and translation between corresponding 3D points (2011). http://nghiaho.com/?page_id=671
23. Rosten, E., Drummond, T.W.: Machine learning for high-speed corner detection. In: Leonardis, A., Bischof, H., Pinz, A. (eds.) ECCV 2006, Part I. LNCS, vol. 3951, pp. 430–443. Springer (2006)
24. Sturm, J., Engelhard, N., Endres, F., Burgard, W., Cremers, D.: A benchmark for the evaluation of RGB-D SLAM systems. In: Proceedings of the International Conference on Intelligent Robot Systems (2012)
25. http://vision.in.tum.de/data/datasets/rgbd-dataset/tools#evaluation

Nao Robot Navigation System Structure Development in an Agent-Based Architecture of the RAPP Platform

Wojciech Dudek, Wojciech Szynkiewicz and Tomasz Winiarski

Abstract This paper focuses on the development of a navigation system structure for the Nao humanoid robot in an agent-oriented distributed architecture. The proposed navigation system is a part of the RAPP framework, a cloud based robotics platform. The RAPP framework is an open-source software platform to support the creation and delivery of robotic applications, which are expected to increase the versatility and utility of robots. All navigation tasks are defined and divided into separate components. The robot navigation system consists of a relative localisation based on Extended Kalman Filter (EKF) using both IMU and odometry measurements, visual QR-code based global localization, path planning, and motion control components. A proper allocation of navigation components, in the four-agent structure of the RAPP platform, is the main goal of this work. Navigation system components are implemented using Robot Operating System and Nao robot programming framework—NAOqi. Experimental results for the Nao robot are presented to show the validity of the proposed approach.

Keywords Humanoid navigation · Nao robot · Agent system

1 Introduction

Mobile robots as companions of human beings everyday life are very complex systems. They must be equipped with some extra perceptual abilities and human-like manual skills to be able to perform diverse tasks in the environment that is easily accessible by people [1]. Robot servants can be specialized in a specific task (e.g. vacuum cleaners, robotic lawn mowers, ironing robots) or can be versatile and satisfy as many owner's requests as possible. It should be noted that the latter concept requires much more powerful computation capabilities then the former one.

W. Dudek (✉) · W. Szynkiewicz · T. Winiarski
Warsaw University of Technology, Warsaw, Poland
e-mail: wojciechdudek.mail@gmail.com
URL: http://robotics.ia.pw.edu.pl

R. Szewczyk et al. (eds.), *Challenges in Automation, Robotics and Measurement Techniques*, Advances in Intelligent Systems and Computing 440, DOI 10.1007/978-3-319-29357-8_54

Additionally, one cannot predict all possible tasks that will be demanded of a robot companion. To answer these issues a concept of cloud infrastructure appears—a cloud possessing abilities to redistribute computations on the one hand, and store knowledge on the other. Such an infrastructure, besides the ability to act as a data storage, should also contain robot skills, and distribute them among robots. Developing such a robot system, that will resolve above mentioned issues using the cloud infrastructure, is the goal of the RAPP (*Robotic Applications for Delivering Smart User Empowering Applications*) project [2]. The RAPP project will provide an open-source software platform to support the creation and delivery of Robotic Applications (RApps), which are expected to increase the versatility and utility of robots. The RAPP platform will provide computational capabilities and storing place for diverse RApps for different robots, in particular, for the affordable humanoid robot Nao used in this project. The general structure of the RAPP control system is presented in [3].

Humanoid robot autonomous navigation in domestic environments remains a challenging task. This paper describes the design, implementation and verification of the RAPP platform based on the structure of a navigation system of the Nao robot, in a social robot environment. A proper allocation of the navigation components, in the RAPP platform structure, is the main goal of this work. Our navigation system, alike the approach in [4], is divided into a local odometry based and global visual localizations. However, we additionally employed an EKF component to estimate the Nao pose using odometry and IMU (Inertia Measurement Unit) data. For the global localization purposes, the approach based on QR-code visual landmarks is used [5]. There are several known approaches to the Nao robot navigation using vision systems [6, 7]. An image-based visual servoing approach to humanoid robot guidance through corridors is proposed in [7]. In [8] a visual SLAM for Nao orientation estimation is presented. There are also several approaches that require Nao to be equipped with a laser range finder and use a SLAM algorithm [9]. The work [10] describes an interface between the Nao and the cloud service and demonstrates a use case in which several Nao robots download and execute abstract plans from RoboEarth cloud platform. Our robot navigation system is implemented using Robot Operating System middleware [11] and Nao robot programming framework (NAOqi) and satisfies all RAPP platform requirements.

2 Agent Oriented Distributed Robot Control System

Nowadays, most robot control systems are based on robot programming frameworks [12]. Frequently, these tools define just the middleware, i.e. provide inter-module communication structure, letting the designer to define the architectural structure. Most of those systems have a fixed structure [13, 14]—the modules of the system are not created or destroyed. RAPP platform approach enables exchange of its components and provides capabilities of the cloud.

An agent should be considered as anything that is aware of its environment and having an internal imperative to realize a certain task acts upon that environment.

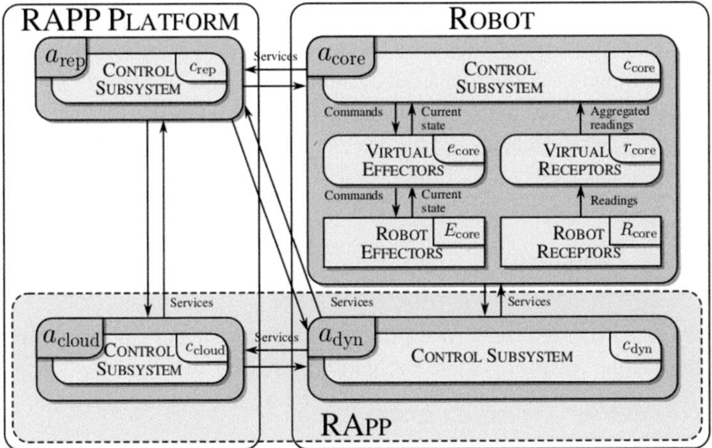

Fig. 1 General 4-agent system structure [3]

Taking into consideration the RAPP platform, the repository agent a_{rep} provides the RApp store service, enabling robots to download RApps and use certain computational services. The core agent a_{core} is a robot specific subsystem that handles its effectors and receptors. It behaves as a driver, enabling the RApps to use robot hardware layer. Virtual receptors aggregate variety of data from sensors and supply the system with an information about the surrounding environment. Virtual effectors execute high-level instructions using devices and return the robot current state. The control subsystem executes the task instructions and the manages virtual receptors and effectors. The core agent, communicating with the repository agent a_{rep}, is able to download RApps. RApps are either composed of a single agent or two agents. The second case, where the RApp contains two agents, emerges when additional computational capabilities are needed. The repository agent a_{rep} can provide simple services to the dynamic agent a_{dyn} and the core agent a_{core} directly. If a more complex service or set of services is necessary to perform a task, a cloud agent a_{cloud} is invoked. In consequence, the RAPP application is composed of two agents. The first one is created as a dynamic agent a_{dyn}, performing its work on the robot. The second one—the cloud agent a_{cloud}–provides extra computational services in the cloud. The system establishes communication between the dynamic agent a_{dyn} and the cloud agent a_{cloud}, so a structure composed of two agent is created. In consequence of the above, four agent system is established (Fig. 1).

3 Navigation System Structure and Behaviour

Autonomous navigation is an essential capability of any service or personal robot that performs high level tasks involving mobility. There is a variety of navigation systems. Usually they are based on a fixed structure. We propose a new approach to robot navigation using an agent oriented distributed robot control system [3]. Specification of our navigation modules is based on System Modelling Language (SysML) [15]. It is an extension of popular UML, thus SysML is well defined and additionally supports specification, analysis, verification and validation processes for a wide range of devices and systems.

3.1 System Functions

Taking into consideration the RAPP project requirements, the Nao robot properties and limitations, the following basic navigation functions were established:

1. MoveTo[*callable*]—move to a specified point,
2. MoveVel[*callable*]—move with a specified velocity,
3. MoveStop[*callable*]—stop Nao walking,
4. MoveJoint[*callable*]—move a specified joint to the desired angle,
5. GetPosition[*callable*]—return current Nao position and orientation,
6. Global-path-planning[*callable*]—plan the global path from the start position to final position,
7. MoveAlongPath[*callable*]—move along a specified global path,
8. EKF-localization[*continuous*]—robot position estimation using EKF,
9. Obstacle-detection[*continuous*]—detect an obstacle and stop the robot motion,
10. QR-localization[*callable*]—estimate the robot position using QR-code markers,
11. LookAtPoint[*callable*]—look at a specified point with respect to 3D coordinate frame,
12. TakePredefinedPosture[*callable*]—a specified, predefined posture,
13. Rest[*callable*]—a safe posture and remove motors stiffness.

The above functions form the navigation API. Most of them can be launched with arguments [*callable*], but two of them work continuously [continuous]. The system is designed to execute a single callable function at a time. Those functions imply the behaviours of the agent's subsystems.

3.2 System Components

Components of the navigation system have been chosen and designed to provide the navigation functionalities (Table 1).

The role of the components is as follows:

1. *camera_server*—responsible for image delivery,
2. *obstacle_detector*—responsible for obstacle detection,
3. *state_server*—delivers the current state of the Nao robot,
4. *execution_server*—handles basic motion tasks,

Table 1 Component use by specific navigation functions 1..13, as they are listed in Sect. 3.1

	1	2	3	4	5	6	7	8	9	10	11	12	13
camera_server										+			
obstacle_detector	+	+	+				+		+				
state_server							+	+					
execution_server	+	+	+	+			+		+		+	+	+
robot_localization					+		+	+		+	+		
global_planner						+							
map_server						+							
QR_code_detection										+			
estimation_server					+		+	+		+	+		
move_server	+	+	+	+			+		+		+	+	+

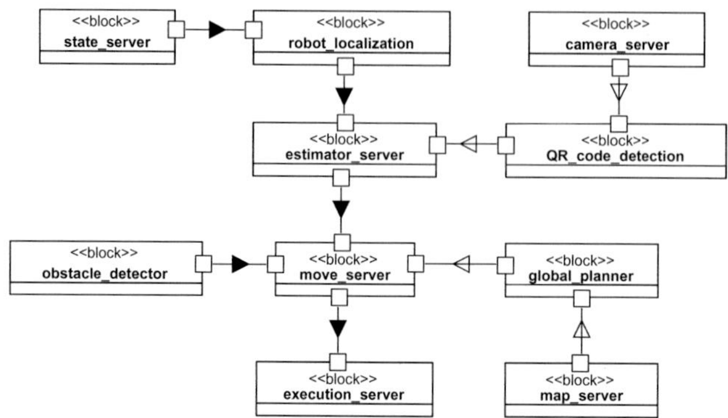

Fig. 2 Interactions between the navigation system components

5. *map_server*—2D environment map publisher,
6. *robot_localization*—incremental localization handler, based on Extended Kalman Filter [16],
7. *QR_code_detection*—provides the homogeneous transformation matrix relating the robot coordinate frame to the QR-code coordinate frame,
8. *estimation_server*—localization requests handler, driver for *robot_localization* module,
9. *global_planner*—global path planner,
10. *move_server*—motion services handler

Basing on the navigation functions (Sect. 3.1), the component roles described above, and the use of those components by the specific system functions (as shown in Table 1), we determine the communication links between these components (Fig. 2). Establishing these links is necessary for proper distribution of the components in the RAPP platform (Sect. 3.3). Figure 2 shows which components interact with delay-sensitive data or their algorithm is sensitive to Wi-Fi connection latency spikes (black arrow), and which communication links can be realized using Wi-Fi (white arrow).

3.3 Distribution of Components in the RAPP Platform

The RAPP platform structure (Fig. 1) is composed of four instances of the control subsystem—one in each agent, virtual effectors and virtual receptors. All virtual effector and receptor components were located in the core agent a_{core}. The component distribution starts with these virtual subsystems. Then, the rest of navigation components is distributed among the control subsystems of the four agents. We performed several tests with different structure configurations to determine the criteria of proper component distribution. These criteria are formulated in three questions and are presented in the flow diagram (Fig. 3). This diagram is designed to guide

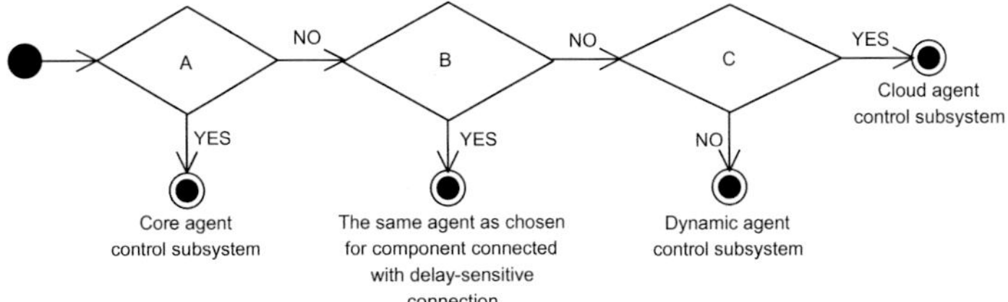

Fig. 3 Allocation procedure of the control system components in the RAPP platform. (*A*) Does the component perform robot core functionality or is associated with a specific robot hardware/software? (*B*) Does the component interact with delay-sensitive data during intercomponental communication with its predecessor? (*C*) Is the algorithm computationally intensive?

developers how to allocate components in the four-agent RAPP platform structure. It should be noted that the criterion B depends on the intercomponental communication parameters, thus the proper order of components distribution is required, starting from the components interacting with the components associated to the virtual receptors and effectors. Then, components connected with these components, and so on.

The presented allocation procedure was verified during validation of our navigation system (Sect. 3.4). It is obvious that the procedure can fail, when navigation functions are decomposed into components in an improper way. For example, let us assume that the first component, allocated in the cloud agent, is connected to the second component by a sensitive link. Next, the second component following the distribution procedure (Fig. 3) is allocated in the core agent. According to the system structure (Fig. 1), there is no direct connection between a cloud agent and components directly associated with a specific robot—a core agent.

3.4 Distribution of Navigation Components of the Nao Robot

The procedure of components allocation starts with the virtual effectors and virtual receptors of the core agent. There are two virtual receptors, each of them is composed of one component. The first one, that delivers camera images to the control subsystem, is composed of the *camera_server* component and the second one, that delivers information about detected obstacles, is composed of the *obstacle_detector* component. The virtual effector is composed of three components: *state_server*, *execution_server* and *robot_localization*. Next, following the procedure, *move_server* (motion services handling) and *estimator_server* (localization services handling should be located in the core agent control subsystem. However, *global_planner*, requires high computational power, but it is not sensitive to WiFi communication delays and does not provide robot core functionality. Therefore, such a component should be placed in the cloud agent control subsystem. In Fig. 4 the distribution of navigation system components in the RAPP infrastructure is presented. Navigation

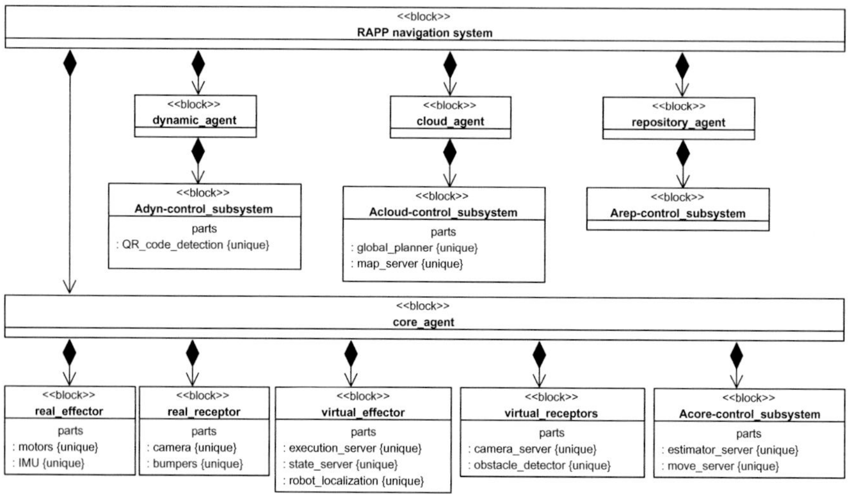

Fig. 4 Navigation system in the RAPP platform structure

modules provide basic robot services, so most of the components are allocated in the core agent. The *QR_code_detection* component delivering common functionality, that is not directly associated with a specific robot, is allocated to the dynamic agent. It enables every robot, connected to the RAPP platform, to download this module and use it to localize itself in the map.

4 Navigation System Structure Implementation and Verification

Implementation of the navigation system components is based on the Robot Operating System (ROS) middleware [11]. We use some ROS packages to perform navigation and localization services, i.e. the *robot_localization* [16] as the EKF implementation, and the *global_planner* [17] as the global path planning module. Virtual effector and receptor components communicate with the Nao robot using NAOqi framework. In Fig. 5 virtual effector interfaces structure is presented. In the *move_server* component several system functions are implemented, e.g. *moveVel*, *moveAlongPath*, *moveJoint*, and *takePredefinedPosture*. Moreover, *estimation_server* handles localization requests.

The navigation system was verified on an example of a hazard detection task [3]. To detect the opened doors and the turned on lights hazards, the dynamic agent was created. The dynamic agent invokes several navigation functions in the following sequence: 1. Stand up 2. Localize the Nao via QR-code 3. Move the robot to the doors—specified point in the global coordinate frame 4. Localize the Nao via

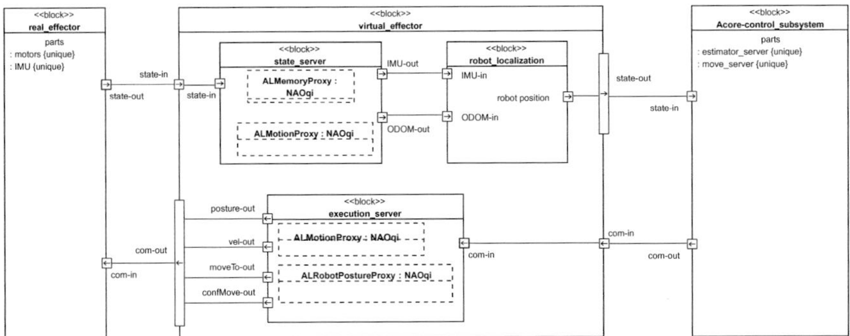

Fig. 5 Virtual effector interfaces

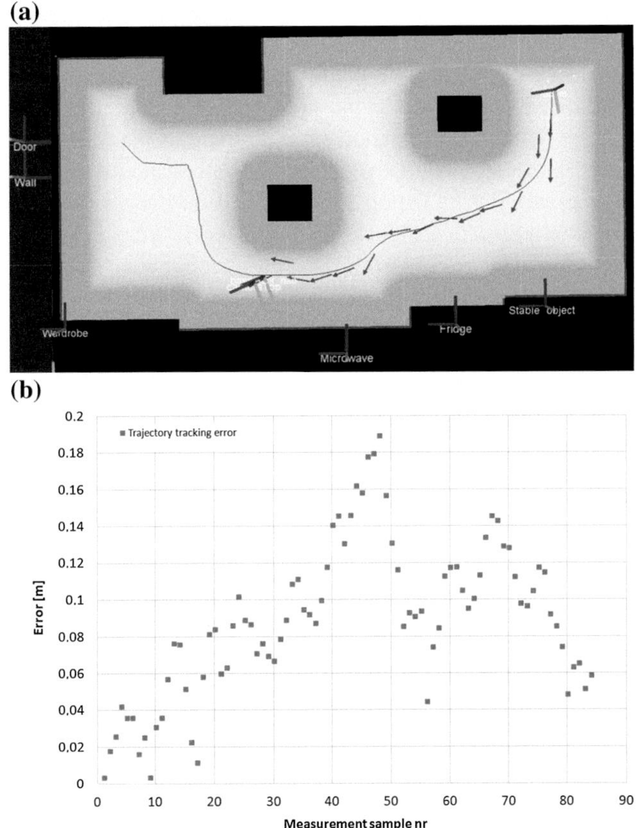

Fig. 6 Path tracking performance in the hazard detection task. **a** Visualisation of the path tracking service. **b** Path tracking error

QR-code 5. Sit down 6. Look at the lamp—specified point in the global coordinate system The whole task was completed successfully. Performed experiments show that the Nao robot localizes itself with good accuracy, the path is collision-free and the EKF-based pose estimation enables the robot to precisely track the desired path. As shown in Fig. 6, illustrating the representative experiment execution, the Nao follows the desired path with good enough accuracy. The maximum path tracking error is 19 cm, and it is a satisfactory result for the Nao robot.

5 Conclusions

In this paper we presented a structure of an autonomous robot navigation system in an agent-based distributed architecture. Based on the performed experiments we proposed a procedure that enables appropriate allocation of the control subsystem components in the RAPP cloud robotics platform. Experiments demonstrated that the proper components distribution model must be carefully chosen. This paper provides an insight into partitioning of navigation system functions into those that should be assigned to core agent, to dynamic agent, and those that should be allocated to the cloud agent. The capabilities of the distributed navigation system were presented on the example of hazard detection task. In the future work, we plan to improve visual localization technique and provide an efficient obstacle avoidance algorithm to avoid collisions with non-stationary objects, including people moving around.

Acknowledgments This work is funded by the FP7 Collaborative Project RAPP (Grant Agreement No. 610947), funded by the European Commission.

References

1. Winiarski, T., Banachowicz, K., Seredyński, D.: Multi-sensory feedback control in door approaching and opening. In: Intelligent Systems'2014. Advances in Intelligent Systems and Computing, vol. 323, pp. 57–70. Springer International Publishing (2015)
2. Psomopoulos, F., Tsardoulias, E., Giokas, A., Zieliński, C., Prunet, V., Trochidis, I., Daney, D., Serrano, M., Courtes, L., Arampatzis, S., Mitkas, P.: Rapp system architecture. In: IROS 2014—Assistance and Service Robotics in a Human Environment, Workshop in conjunction with IEEE/RSJ International Conference on Intelligent Robots and Systems, Chicago, Illinois, Sept 14, pp. 14–18 (2014)
3. Zieliński, C., Szynkiewicz, W., Figat, M., Szlenk, M., Kornuta, T., Kasprzak, W., Stefańczyk, M., Zielińska, T., Figat, J.: Reconfigurable control architecture for exploratory robots. In: Kozłowski, K. (ed.) 10th International Workshop on Robot Motion and Control (RoMoCo), pp. 130–135. IEEE (2015)
4. Wei, C., Xu, J., Wang, C., Wiggers, P., Hindriks, K.: An approach to navigation for the humanoid robot nao in domestic environments. In: Towards Autonomous Robotic Systems, vol. 8069, pp. 298–310. Springer, Berlin (2014)

5. Figat, J., Kasprzak, W.: NAO-mark vs QR-code recognition by NAO robot vision. In: Szewczyk, R., Zieliński, C., Kaliczyńska, M. (eds.) Progress in Automation, Robotics and Measuring Techniques. Vol. 2 Robotics. Advances in Intelligent Systems and Computing (AISC), vol. 351, pp. 55–64. Springer (2015)
6. Gouda, W., Gomaa, W., Ogawa, T.: Vision based slam for humanoid robots: a survey. In: 2013 Japan-Egypt International Conference on Electronics, Communications and Computers (JEC-ECC), pp. 170–175. IEEE (2013)
7. Faragasso, A., Oriolo, G., Paolillo, A., Vendittelli, M.: Vision-based corridor navigation for humanoid robots. In: 2013 IEEE International Conference Robotics and Automation (ICRA), pp. 3190–3195 (2013)
8. Wirbel, E., Bonnabel, S., Fortelle, A.D.L., Moutarde, F.: Humanoid robot navigation: getting localization information from vision. J. Intell. Syst. **23**(2), 113–132 (2014)
9. Wen, S., Othman, K.K.M., Rad, A., Zhang, Y., Zhao, Y.: Indoor SLAM using laser and camera with closed-loop controller for NAO humanoid robot. Abstr. Appl. Anal. **2014**, 8 p (2014)
10. Johannssen, F.: Nao in the cloud. Knowledge sharing for robots via cloud services. In: Workshop on Software Development and Integration in Robotics ICRA 2013 (2013)
11. Foundatioin, O.S.R.: Robot Operating System. http://ros.org/. [Online; Accessed 21-May-2015]
12. Zieliński, C., Winiarski, T.: Motion generation in the MRROC++ robot programming framework. Int. J. Robot. Res. **29**(4), 386–413 (2010)
13. Zieliński, C., Kornuta, T., Trojanek, P., Winiarski, T., Walcki, M.: Specification of a multi-agent robot-based reconfigurable fixture control system. Robot Motion and Control 2011 (Lecture Notes in Control and Information Sciences), vol. 422, pp. 171–182 (2012)
14. Zieliński, C., Winiarski, T.: General specification of multi-robot control system structures. Bull. Polish Acad. Sci.—Tech. Sci. **58**(1), 15–28 (2010)
15. SysML.org: SysML documentation. http://sysml.org/ (2003) [Online; Accessed 21-May-2015]
16. Moore, T., Stouch, D.: A Generalized Extended Kalman Filter Implementation for the Robot Operating System. http://wiki.ros.org/robot_localization?action=AttachFile&do=view&target=robot_localization_ias13_revised.pdf [Online; Accessed 17-Sept-2015]
17. Marder-Eppstein, E., Berger, E., Foote, T., Gerkey, B., Konolige, K.: The office marathon: robust navigation in an indoor office environment. In: 2010 IEEE International Conference on Robotics and Automation (ICRA), pp. 300–307 (2010)

WUT Visual Perception Dataset: A Dataset for Registration and Recognition of Objects

Maciej Stefańczyk, Michał Laszkowski and Tomasz Kornuta

Abstract Modern robots are typically equipped with many sensors with different modalities, e.g. RGB cameras, Time-of-Flight cameras or RGB-D sensors. Thus development of universal, modality-independent algorithms require appropriate datasets and benchmarks. In this paper we present WUT Visual Perception Dataset, consisting of five datasets, captured with different sensors with the goal of development, comparison and evaluation of algorithms for automatic object model registration and object recognition.

Keywords Dataset · Object recognition · Object detection · Registration · RGB-D · Kinect · Stereopair · RGB Camera

1 Introduction

Visual perception is one of the most important robot senses, enabling service robots to act in unstructured and dynamically changing environments. It is a key issue in many state-of-the-art service robotics applications, like fetching drinks [1], folding clothes [2] or cutting food [3]. Along with the progress in the field of visual sensors and advent of RGB-D sensors [4, 5] we observe the flourishing of novel algorithms and solutions [6]. This in turn requires proper tools for their comparison and benchmarking, with datasets playing an essential role.

M. Stefańczyk · M. Laszkowski · T. Kornuta (✉)
Institute of Control and Computation Eng., Warsaw University of Technology, Nowowiejska 15/19, 00-665 Warsaw, Poland
e-mail: tkornuta@gmail.com

M. Stefańczyk
e-mail: stefanczyk.maciek@gmail.com

M. Laszkowski
e-mail: mlaszkow@gmail.com

T. Kornuta
IBM Research, Almaden, 650 Harry Rd, San Jose, CA 95120, USA

R. Szewczyk et al. (eds.), *Challenges in Automation, Robotics and Measurement Techniques*, Advances in Intelligent Systems and Computing 440, DOI 10.1007/978-3-319-29357-8_55

In this paper we present WUT Visual Perception Dataset[1]—a multi-modal dataset created with the purpose of development, testing and comparison of different visual perception algorithms for registration of object models and their recognition (i.e. detection and pose estimation). The paper is structured as follows. After a brief review of robotic-perception related benchmarks and datasets we present the hardware and software used during the process of data acquisition. Next we present each of the five particular datasets constituting the WUT Visual Perception Dataset and conclude with exemplary applications using the dataset.

2 Benchmarking the Machine Perception

Computer vision and robotic communities have a long history of benchmarks and challenges that stimulated progress on those fields. For example, the first edition of the USC-SIPI image database was released in 1977 and is still maintained to this day.[2] It contains several images that are today considered as classics (Lena, Baboon, Airplane etc.) and appeared in hundreds of publications and books devoted to computer vision, image processing and pattern recognition.

Another prominent example is the PASCAL Visual Object Classes (VOC),[3] along with the associated PASCAL VOC challenges on object class recognition (classification), that were organized between years 2005–2012 [7].

Yet another example is the ImageNET dataset, distributed for the purpose of ImageNET challenge [8], organized annually since 2010.[4] The challenge consists of two competitions: classification of objects (on dataset of fully labelled data for 200 categories), and classification with object localization (1000 categories).

There are also dozens of other datasets created for different, more specific purposes. Examples include the family of Middlebury Stereo Datasets,[5] developed since 2001 [9] and devoted to evaluation and comparison of two-frame stereo correspondence algorithms, or a simple, yet well recognized dataset for comparison of different keypoint detectors and feature extractors [10].

Along with the rising popularity of RGB-D sensors several datasets for RGB-D based perception appeared. One of the very first was the Washington RGB-D Object dataset,[6] containing images of 300 instances of household objects from 51 categories (250,000 RGB-D with masks in total) along with test scenes, useful for the purpose of object model generation, instance and category recognition [11].

[1] http://robotyka.ia.pw.edu.pl/datasets/.

[2] http://sipi.usc.edu/database/database.php.

[3] http://host.robots.ox.ac.uk/pascal/VOC/.

[4] http://image-net.org/challenges/LSVRC/2015/.

[5] http://vision.middlebury.edu/stereo/data/.

[6] http://rgbd-dataset.cs.washington.edu/.

Next, there is the Willow Garage Dataset, published for the purpose of Solutions in Perception Challenge,[7] a challenge launched by NIST in cooperation with Willow Garage in 2011. The dataset contains about 35 objects for training along with scenes for evaluation of algorithms for identification and determination of the positions (with ground truth containing object poses).

It is also worth to mention the Object Segmentation Database (OSD),[8] consisting of objects on a table top organized in several learn—and test-sets, created for the purpose of evaluation of RGB-D images segmentation [12].

Finally, as RGB-D sensors significantly facilitated SLAM (Simultaneous Localization and Mapping) [13], there are several publicly available datasets containing sequences of RGB-D images along with the ground truth (e.g. consecutive sensor poses). Most popular examples of such datasets are TUM RGB-D SLAM Dataset [14] and ICL-NUIM RGB-D Benchmark Dataset [15], the latter being a set of trajectories in artificially generated (rendered) scenes.

In our past research on robot perception we used several of those datasets, e.g. we used dataset from [10] for comparison of binary descriptors [16], we used OSD in our work on recognition of cuboid and cones in RGB-D images [17], ICL-NUIM RGB-D Benchmark Dataset for evaluation of our surfel-based V-SLAM system [18]. Finally, we used the Washington RGB-D Dataset from the very beginning of our research on RGB-D-based object recognition for the purpose of their grasping and manipulation.

However, as new robots and sensors with different modalities started to appear in our laboratory we realized that despite its diversity, the publicly available datasets do not fit anymore to our needs. It is related to the fact that we are currently working on unified pipeline for object recognition, being universal to three kind of visual sensors associated with our robots—monocular RGB cameras, binocular stereopairs and Kinect sensors (Fig. 1). So having learned the best practices and ideas from the previously utilized datasets we developed our own, multimodal dataset, presented in the following sections.

3 Data Acquisition

3.1 Turntable

As a turntable we have used a fiducial board with patterns similar to one used in Object Recognition Kitchen (ORK). Two dot patterns are used to determine the board position with respect to the camera frame regardless of occlusion (however it is required that at least one pattern should be fully visible in every view). During the view acquisition the object is placed at the centre of the board, and its

[7]http://www.acin.tuwien.ac.at/forschung/v4r/mitarbeiterprojekte/willow/.

[8]http://www.acin.tuwien.ac.at/forschung/v4r/mitarbeiterprojekte/osd/.

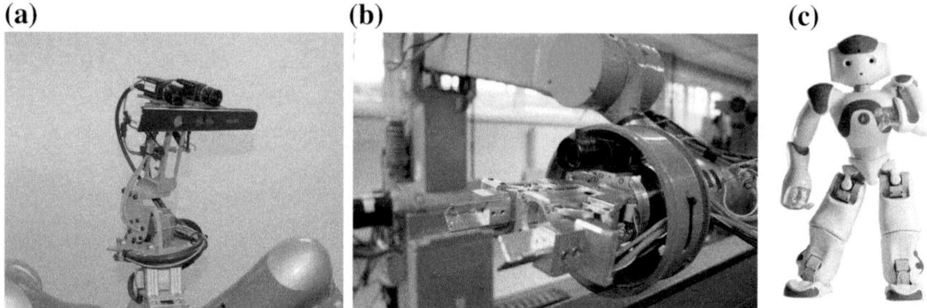

Fig. 1 WUT robots equipped with visual sensors with diverse modalities **a** Velma (high resolution RGB stereopair and Kinect sensor integrated with its head) **b** Modified IRp-6 manipulator (with RGB camera integrated with its end-effector) **c** NAO (equipped with two independent RGB cameras)

coordinate frame is set to the one shown in Fig. 2a. This enables relatively accurate view point pose estimation and makes all resulting point clouds easy to move to common, board-centred coordinate frame. Besides, this facilitates the process of object segmentation—assuming that the objects lies on the xy plane of this coordinate system, segmentation consists of selection of points lying inside axis-aligned bounding box with positive z. The axis of rotation of the fiducial board is aligned with z axis of the coordinate frame.

3.2 *Monocular Camera*

The first hardware setup is presented in Fig. 2b and consists of a modified IRp-6 manipulator (with 7DOF) supplemented with a RGB camera integrated with its gripper. The utilized camera is PointGrey BFLY-PGE-13S2C-CS with 72° horizontal field of view, acquiring images with resolution of 1288×964px.

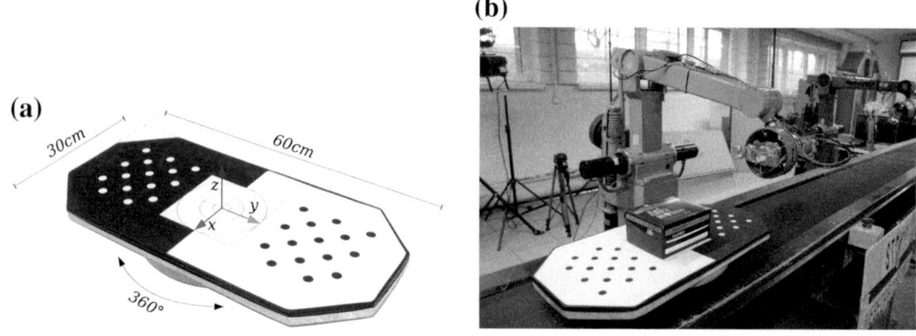

Fig. 2 **a** Turntable with two patterns **b** Modified IRp-6 manipulator with camera during acquisition

3.3 Structured Light

The second hardware setup consists of a single Microsoft Kinect sensor. The size of the board along with rather wide horizontal angle of sensor view (62°) and lower limit of depth acquisition of the sensor enables us to acquire RGB images along with depth maps for both detection of the board and generation of point cloud. The structured light technique gives dense depth information disregarding the lack of object texture, however, the low resolution of depth image results in quite small pointclouds constituting the object, containing typically only several thousand of points (Fig. 3) out of almost 300 thousand points returned in each frame.

3.4 Textured Stereovision

The third hardware setup is more complex and consists of a narrow-wide combination of three cameras with additional texture projector (Fig. 4a). The stereo setup is composed of two PointGrey BFLY-PGE-13S2C-CS cameras with long-focal lenses (narrow, 33° horizontal field of view) with 10 cm baseline. As a result objects of size 15 cm, registered from the distance of 70 cm, occupy almost the whole usable image area (about 200 thousand pixels per object).

In order to make the resulting depth map as dense as possible we used EFFI-Lase LED projector with pseudo-random cloud of dots pattern, which resulted in a solution that can be classified as textured stereo [19].

As the projected pattern is white and influences the colour information, for each object view we collect two pairs of images— one with projector and other with diffused light (Fig. 5a). The first pair is used for generation of dense pointcloud of the object (Fig. 5b), whereas the second is used to colourize this cloud and to extract keypoints with descriptors in further utilization in object recognition, model registration etc.

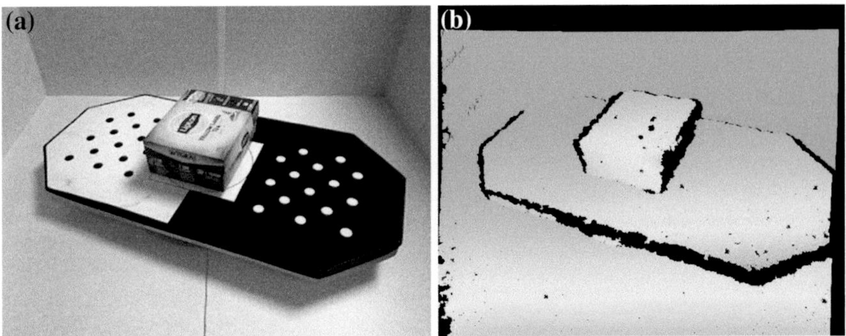

Fig. 3 Image registered with Microsoft Kinect: **a** colour image **b** depth map, black means no data

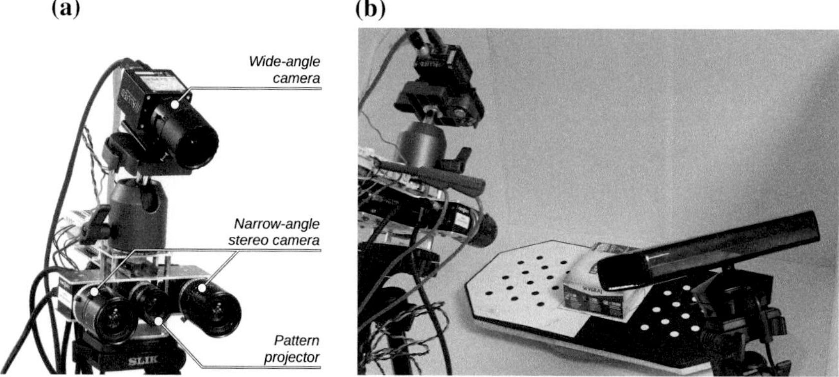

Fig. 4 **a** Multi-camera setup with additional texture projector **b** Exemplary scene during data acquisition

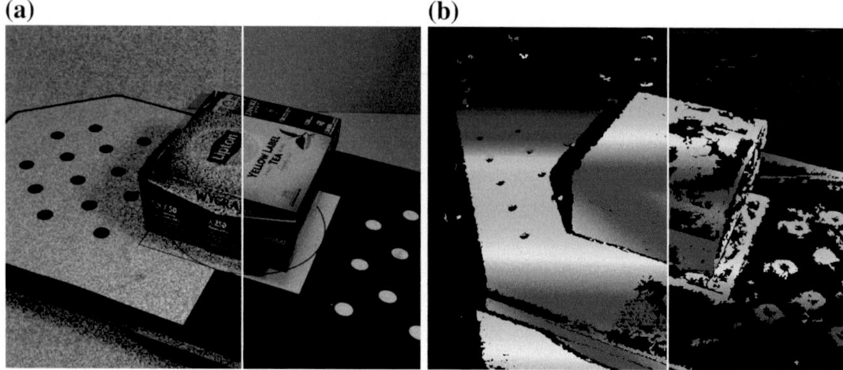

Fig. 5 **a** Image registered with projected pattern (*left*) and diffused light (*right*) **b** Depth reconstructed from projected pattern (*left*) and diffused light (*right*)

Because the narrow field of view of stereopair makes it impossible to capture the whole board, so we decided to incorporate another camera into the system. We used Prosilica GC1290C with 56° lens, mounted above and slightly tilted in relation to the stereo pair. Such a configuration enables us to capture the whole board and subsequently determine its pose.

Please note that despite the actual number of used cameras for the simplicity throughout the rest of this paper we will refer to this setup as to stereo.

4 Datasets

4.1 3D Model Dataset

The first dataset consists of two parts. First part contains 9 artificially generated models of objects of two types: cuboids and cylinders, stored as dense colour point clouds, supplemented with sparse clouds of features (currently SIFTs). Those models were generated for the purpose of testing object recognition methods (with high quality models) and may also be used as ground truth for algorithms of automatic registration of object models from different viewpoints. Exemplary models are presented in Fig. 6.

The second part contains high resolution images from which the models were generated: six images for each model of cuboidal objects and three images for cylindrical objects, along with the YML files containing description of their geometry and mappings of images to model faces.

4.2 Turntable Dataset

This dataset contains data acquired from the multi-sensor setup (as presented in Fig. 4b). There were 13 objects placed on the turntable and for each object we acquired 12 views from different viewpoints (each rotated by around 30°). For each view we acquired (Fig. 7): RGB image along with the associated depth map from Kinect sensor, RGB image from wide-angle camera, pair (left-right) images from stereopair (when the diffused light was turned on) and a pair of images from stereopair (when the texture projector was turned on). Additionally, the dataset offers data enabling calibration of internal parameters of wide, left and right cameras as well as calibration of Cartesian transformations between all cameras.

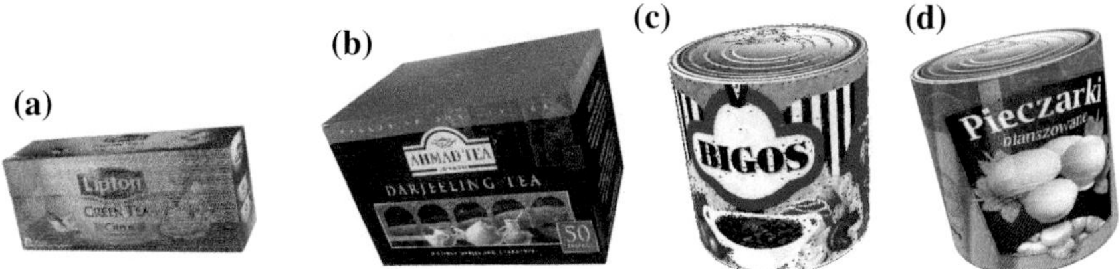

Fig. 6 Point clouds of exemplary objects: **a** Tea 1 **b** Tea 2 **c** Foodcan 1 **d** Foodcan 2

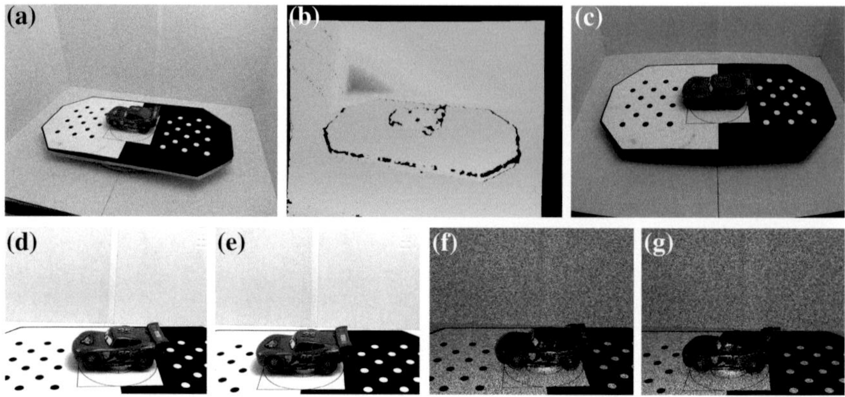

Fig. 7 Exemplary images of one of the objects captured: **a**, **b** RGB and depth map (rainbow) from Kinect **c** wide-angle camera **d**, **e** stereopair with diffused light **f**, **g** stereopair with projected texture

4.3 Approaching Sensor Dataset

This dataset contains images of six single objects placed on the turntable. However, in this case the turntable remained still, whereas the motion of the RGB camera was induced by the manipulator. For each of the objects two trajectories were performed (approach from the above Fig. 8a and from the side Fig. 8b), each consisting of 107 consecutive camera poses. Each trajectory, aside of images, contains a file containing camera poses. Dataset contains also images enabling calibration of the internal parameters of the used camera.

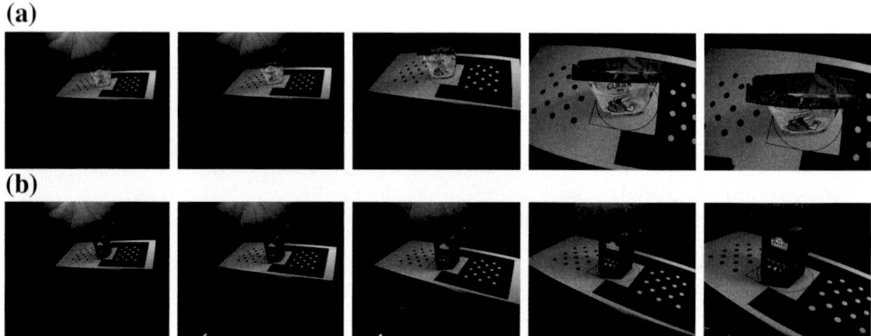

Fig. 8 Exemplary images the two types of trajectories. **a** Approach from the above of the object, **b** approach from the side of the object

Fig. 9 Exemplary images from different test scenes

4.4 *Pose Estimation Test Dataset*

One of the things we need to test is object localization. Ground-truth object position is thus required. This dataset consists of 9 objects, put on turntable in a way, that point $(0, 0, 0)$ of object coordinate frame (one of the box corners) was aligned with origin of coordinate frame of the board, and both had the same orientation. For each image object pose in camera frame was first detected from pattern localization and then hand tuned to match as close as possible. Those poses are stored in dataset along with corresponding images.

4.5 *Test Scenes Dataset*

Last dataset contains test scenes for the purpose of evaluation of object recognition in the case of presence of many different, occluding each other objects. This dataset is simple set of multiple view for different scenes. At the moment there are 7 scenes with multiple objects, both known (existing in previous datasets) and unknown (Fig. 9).

5 Summary

In this paper we presented the WUT Visual Perception Dataset—a dataset containing images acquired with different types of visual sensors. The dataset is publicly available at http://robotyka.ia.pw.edu.pl/datasets/. It is currently used for the purpose of development and evaluation of diverse algorithms, e.g. in the research on automatic reconstruction of object models from many views. In particular, it was used for verification of ideas during implementation of the generic semantic-based object recognition system described in [20] as well as for testing the RGB-D perception subsystem presented in [21].

In the future we plan to extend the dataset by adding data acquired with ToF cameras, which will enable comparison of diverse visual perception algorithms on dataset acquired by the three, currently most popular methods of depth acquisition.

Acknowledgments The authors acknowledge the financial support of the National Centre for Research and Development grant no. PBS1/A3/8/2012, Poland. Tomasz Kornuta is supported by the IBM Research, Almaden through IBM PostDoc/LTS Programme.

References

1. Bohren, J., Rusu, R.B., Jones, E.G., Marder-Eppstein, E., Pantofaru, C., Wise, M., Mosenlechner, L., Meeussen, W., Holzer, S.: Towards autonomous robotic butlers: lessons learned with the PR2. In: 2011 IEEE International Conference on Robotics and Automation (ICRA), pp. 5568–5575. IEEE (2011)
2. Maitin-Shepard, J., Cusumano-Towner, M., Lei, J., Abbeel, P.: Cloth grasp point detection based on multiple-view geometric cues with application to robotic towel folding. In: 2010 IEEE International Conference on Robotics and Automation (ICRA), pp. 2308–2315 (2010)
3. Lenz, I., Knepper, R., Saxena, A.: DeepMPC: Learning deep latent features for model predictive control. In: Proceedings of Robotics: Science and Systems, Rome, Italy (2015)
4. Stefańczyk, M., Kornuta, T.: Acquisition of RGB-D images: methods (in Polish). Pomiary Automatyka Robotyka PAR **18**(1), 82–90 (2014)
5. Kornuta, T., Stefańczyk, M.: Acquisition of RGB-D images: sensors (in Polish). Pomiary Automatyka Robotyka PAR **18**(2), 92–99 (2014)
6. Ren, X., Fox, D., Konolige, K.: Change their perception: RGB-D for 3-D modeling and recognition. Rob. Autom. Mag. IEEE **20**(4), 49–59 (2013)
7. Everingham, M., Eslami, S.M.A., Van Gool, L., Williams, C.K.I., Winn, J., Zisserman, A.: The pascal visual object classes challenge: a retrospective. Int. J. Comput. Vis. **111**(1), 98–136 (2015)
8. Russakovsky, O., Deng, J., Su, H., Krause, J., Satheesh, S., Ma, S., Huang, Z., Karpathy, A., Khosla, A., Bernstein, M., et al.: Imagenet large scale visual recognition challenge. Int. J. Comput. Vis. 1–42 (2014)
9. Scharstein, D., Szeliski, R.: A taxonomy and evaluation of dense two-frame stereo correspondence algorithms. Int. J. Comput. Vis. **47**(1–3), 7–42 (2002)
10. Mikolajczyk, K., Schmid, C.: A performance evaluation of local descriptors. IEEE Trans. Pattern Anal. Mach. Intell. **27**(10), 1615–1630 (2005)
11. Lai, K., Bo, L., Ren, X., Fox, D.: A large-scale hierarchical multi-view RGB-D object dataset. In: 2011 IEEE International Conference on Robotics and Automation (ICRA), pp. 1817–1824. IEEE (2011)
12. Richtsfeld, A., Mörwald, T., Prankl, J., Balzer, J., Zillich, M., Vincze, M.: Towards scene understanding–object segmentation using RGBD-images. In: Computer Vision Winter Workshop (2012)
13. Thrun, S., Leonard, J.J.: Simultaneous Localization and Mapping. In: The Handbook of Robotics, pp. 871–890. Springer (2008)
14. Sturm, J., Engelhard, N., Endres, F., Burgard, W., Cremers, D.: A benchmark for the evaluation of RGB-D SLAM systems. In: 2012 IEEE/RSJ International Conference on Intelligent Robots and Systems (IROS), pp. 573–580. IEEE (2012)
15. Handa, A., Whelan, T., McDonald, J., Davison, A.J.: A benchmark for RGB-D visual odometry, 3D reconstruction and SLAM. In: 2014 IEEE International Conference on Robotics and Automation (ICRA), pp. 1524–1531. IEEE (2014)
16. Figat, J., Kornuta, T., Kasprzak, W.: Performance evaluation of binary descriptors of local features. In: Chmielewski, L., Kozera, R., Shin, B.S., Wojciechowski, K. (eds.) Proceedings of the International Conference on Computer Vision and Graphics, vol. 8671 of Lecture Notes in Computer Science, pp. 187–194. Springer, Berlin (2014)

17. Kornuta, T., Stefańczyk, M., Kasprzak, W.: Basic 3D solid recognition in RGB-D images. In: Szewczyk, R., Zieliński, C., Kaliczyńska, M. (eds.) Recent Advances in Automation, Robotics and Measuring Techniques, vol. 267 of Advances in Intelligent Systems and Computing (AISC), pp. 421–430. Springer (2014)
18. Wilkowski, A., Kornuta, T., Stefańczyk, M., Kasprzak, W.: Efficient generation of 3D surfel maps using RGB-D sensors. Int. J. Appl. Math. Comput. Sci. (AMCS) (2015) (accepted)
19. Konolige, K.: Projected texture stereo. In: International Conference on Robotics and Automation (ICRA), pp. 148–155. IEEE (2010)
20. Stefańczyk, M., Pietruch, R.: Hypothesis generation in generic, model-based object recognition system. In: Szewczyk, R., Zieliński, C., Kaliczyńska, M. (eds.) Recent Advances in Automation, Robotics and Measuring Techniques. Advances in Intelligent Systems and Computing (AISC). Springer (2016) (Under review)
21. Kornuta, T., Laszkowski, M.: Perception subsystem for object recognition and pose estimation in RGB-D images. In: Szewczyk, R., Zieliński, C., Kaliczyńska, M. (eds.) Recent Advances in Automation, Robotics and Measuring Techniques. Advances in Intelligent Systems and Computing (AISC). Springer (2016)

Navigation for Satellites

Mariusz Jacewicz and Robert Głębocki

Abstract Navigation is a main challenge for the operation of an autonomous satellite. This paper deals with the problem of pose estimation for real-time vision based navigation system. Presented method is able to estimate the target pose with high accuracy and does not require any landmark on the object. The results of numerical simulation were presented. The delays introduced by guidance system into the control system of an active satellite were tested. Experimental results showed that the proposed method of navigation introduced a small delay in the control system. Created system is also robust to occlusions and light variations in space. The approach will be further evaluated and tested in the continued work.

Keywords Mobile robot · Autonomous navigation · Control systems

1 Problem Description

The problem of pose estimation of known object is one of the most research issues in vision navigation. A lot of modern tracking methods are applicable to pose estimation problem but not at every conditions [1]. This paper describes the results of analysis, implementation and testing of simulation intended for vision-based guidance, navigation and control applications such as rendezvous and docking of satellites. In this document markerless local features based tracking system has been studied. It was done under the randezvous of satellites on Earth orbit project conducted by Warsaw University of Technology. Navigation of satellites for proximity operations require very accurate measurements without human involvement. A servicing satellite is sending to capture target object and perform servicing

M. Jacewicz (✉) · R. Głębocki
Warsaw University of Technology, Warsaw, Poland
e-mail: mariuszadamjacewicz@gmail.com

R. Głębocki
e-mail: rgleb@meil.pw.edu.pl

R. Szewczyk et al. (eds.), *Challenges in Automation, Robotics and Measurement Techniques*, Advances in Intelligent Systems and Computing 440, DOI 10.1007/978-3-319-29357-8_56

tasks [2]. There must be a high level of autonomy of the system. Described system focus on final phases of randezvous. It was assumed an image of an object taken by a calibrated camera in each step of time is known, and was assumed a 3D representation of an object is known. It was proposed a solution for tracking 3D rigid objects that is based on local features and promises better computational performance [3]. Proposed method is more accurate than other methods presented in literature because is more stable and robust on the tracking failures. Projected system is also cheaper than other proposed earlier solutions.

Calculating the pose of an object with using of calibrated camera is computer vision problem dating to 1841 but an important element of many computer vision systems [4]. Orientation of a camera given a set of n 2D-to-3D point correspondences, is a well studied problem in photogrammetry. The problem of determining the absolute position and orientation of a camera from a set of 2D-to-3D point correspondences is known as the Perspective-n-Point (PnP) problem. The PnP problem is to get the position and orientation of a camera given its intrinsic parameters and a set of n correspondences between 3D points and their 2D projections. The minimal number of correspondences to solve PnP problem is three. The original Grunert's [4] formulation appears more or less in each P3P problem. Given a pair of 2D-to-3D correspondences gives a single constraint on the distance. The basic relation is based on the cosine law. Quan build a 4th degree polynomial to solve the cosine law equations for the P3P problem [5]. Next it was shown how to solve this problem when available are four or five point correspondences. He constructed polynomials from different triplets of points and rewrote them into a matrix form with polynomial coefficients in the matrix and monomials in a vector. By analyzing the kernel or the coefficient matrix and exploiting monomial dependencies, he solved the problem linearly for 4 and 5 points. Fischler found that P4P problem with non-coplanar points had many solutions. P4P problem with coplanar points had only one solution [6]. For P5P problem there were as many as two solutions. For more than 6 correspondences, it became Direct Linear Transformation (DLT) problem. First approaches to visual tracking were based on tracking of the outer contour of a target [7]. Contour-based trackers gained significant attention on the early vision based navigation systems. There were proposed algorithms which were based on fiducial markers or LEDs on the target object to make simpler the registration task [8, 9]. It this case it is assumed that fiducials are observable at all times [10]. The 3D positions of the fiducials in the world coordinate system are assumed to be known. When one or more blobs are missing, these approaches do not mention these cases and are not able to get right results of pose. This approach in case of satellites is impractical because many existing satellites have not these fiducial markers. The other popular approach is based on three-dimensional models. In this case pose computation is achieved by minimization the distance between 3D model edges and the corresponding edge features in the image. The most weakness of approach based on 3D model is reliance on detailed geometric model. The system must either be initialized by hand or require the camera to be close to a specified position when objects are made with cylindrical, spherical and complex shapes as it is often in case of satellites.

This makes the system easily broken. Moreover when the object is complex there are achieved low frame per second rates. To reject outliers algorithms such as Random Sample Consensus (RANSAC) are implemented to achieve robustness to illumination conditions in space [6]. Active sensors like Light Detection and Ranging (LIDAR) were used for automatic relative pose estimation. These sensors are expensive. Camera sensors are cheaper and can provide accurate capabilities to obtain relative pose. The use of the interface ring used to attach the satellite to launch vehicle has been proposed for capturing the satellite. This has a disadvantage because it is restricted only for proximity operations where the target satellite is visible from the interface side. Capturing should be performed autonomously because there are communication delays between on-orbit systems and ground. Feature matching computer vision approaches have been developed but they are very computational intensive and cannot be used during whole mission [11]. Critical sensitivity to illumination, relative orientation and occlusions of target had been observed. Learned database is also used on Orbital Express and the algorithms are based on edges in this case. Lepetit suggested for tracking objects in 3D by using corner features with a single camera. This approach was robust to camera displacements and partial occlusion. Drawback of this method was camera should be close enough to one of key frames and there is a real problem when then tracking must be initialized again for example after tracking failure. Limitation of described above methods lies in their long execution time. Moreover, the existing approaches are very computationally expensive. In many cases system is based on artificial features. This makes tracking far more difficult: finding and following feature points or edges can be difficult because there are too few of them on many typical objects. It is much more better to rely on naturally present features, such as edges, corners, or texture. In this work authors assumed that satellite has been detected. Proposed system is planned to deals with finally stages of satellite randezvous from far proximity operations, when the satellite is about 2 km from satellite to the contact of satellites. This paper addresses the design of a spaceborne monocular vision-based navigation system for on-orbit-servicing and formation-flying applications. Someone can divide PnP algorithms according to the method used to solve them to:

- Non-Iterative, which usually formulate the problem as system of linear or non-linear equations, which is solved in a sequence of operations using linear algebra. These methods deliver exact solution in the absence of rounding errors.
- Iterative which search in a parameter space and optimize some objective function, for example the Euclidean distance between reprojected 3D points and known measurements. Iterative approaches usually minimize an error function but may fall into local minimum and result in pose ambiguity. Among iterative approaches, Dementhon presented Pos with Iteration (POSIT) algorithm to solve PnP problem for more than four non-coplanar correspondences. Lu introduced iterative algorithm which minimized a 3D space error and was faster than other iterative ones. But it often fell into local minimum and resulted in pose ambiguity. In order to get accurate pose estimation results, iterative approaches are good choices.

Depending on the number of point correspondences between 2D and 3D space, someone can split camera pose algorithms into:

- minimal, which use the smallest possible set of point correspondences between 2D and 3D space to calculate camera pose
- non-minimal, which use more point correspondences to linearize the task or to return the more precise result.

Minimal algorithms are usually used to filter out incorrect correspondences, which are called outliers, for example using RANSAC paradigm [6]. Once correct correspondences are identified one can use non-minimal algorithms to improve the final solution. To find correct correspondences, it is important to have a fast algorithm which is using as small number of measurements as possible to calculate the camera pose. It is because such an algorithm is executed inside the RANSAC loop many times and the number of iterations is proportional to the size of the minimal set. When more than a minimal number of measurements it could be possible to better estimate is pose. Technically it is possible to extend this solution to more than 5 points, but this solution is not very practical since the number of possible triplets grows exponentially with the number of input points. It is known that three point correspondences are sufficient to recover the camera rotation and translation in the case of calibrated camera—that is why the name of problem is P3P—and there are up to four real solutions to the problem. The most convenient method calls Direct Linear Method evaluating the rotation and the translation is by simply disregarding the constraints existing on and estimating the parameters of matrix R and the parameters of t. The parameters can be estimated from at least 6 linearly independent measurements. Although simple this method is not practical since it produces a true rigid transformation only in the ideal cases when the image are free of noise. In the case when measurements are noisy this method does not give the correct solution. Furthermore not all information in the system is being exploited the fact that the transformation is rigid thus the solution obtained is not expected to be optimal. Several methods solve the problem analytically when few measurements are given and when the model points are in a specific configuration. Fischler give a closed form solution for three or four coplanar model points [6]. He evaluated the lengths of the segments connecting the camera focal point and the points on object and derived the extrinsic parameters. Assumption that correspondences are obtained is reasonable for problems such as hypothesis verification where the pose estimation of an object serves to confirm or reject an interpretation hypothesis. Likewise in problems of motion tracking it is often assumed that the matching is known when the object features can be applied. However the assumption that the matching is known is not acceptable in all pose estimation problems particularly when dealing with recognition. One of the methods suggested to solve the matching problem is to evaluate the pose and the matching simultaneous during the iterative process the pose of the object is estimated from a partial interpretation and this pose estimate is used to eliminate irrelevant interpretations at the next interpretation stage. When correspondences are known, pose vectors can be computed in each iterative loop by minimizing the object function. Camera pose can be calculated from various kinds of image measurements, for example from a set of 2D projections of 3D points or 3D lines, from a combination of points and lines, projections of known planar objects like chessboards, coplanar circles, intersections of parallel lines (vanishing points), edges of models and more. The estimation from rich objects, like lines or circles, might appear more accurate, but it is needed to solve non-trivial computer vision tasks such as detection, and to compensate the fact that image is affected by distortion from optics, leaving lines curved or circles deformed. The aim is to estimate the pose of a passive object using its known three-dimensional model and single two-dimensional images collected on-board the active spacecraft. Star trackers or infrared cameras can be used for increased dynamic range to accurately navigate with respect to a target space vehicle at low cost from zero to several tens of kilometers [11]. Furthermore the unknown correspondences between image and model features result in a large search space for ambiguity resolution, and thus in a significant computational load. A number of requirements for future monocular vision-based systems can be derived. Ideally, a pose estimator should rely on a minimum number of image features, be robust to ambiguous pose solutions, compensate for image noise and model errors, not rely on the uncooperative satellite dynamics, and offer multi-stage solutions of increasing accuracy from lost-in-space all the way to fine navigation. After the extension of the initial set of matches through projection of the model onto the image, the estimation problem becomes over-determined. The definition of a proper spacecraft model is a fundamental step of the pose estimation strategy. On one hand, the spacecraft model has to be as minimalist as possible to reduce the system complexity and the search space for matching. The estimate of the initial pose is certainly the most challenging task of the pose estimation process. Many authors assume coarse a priori knowledge of the relative position and orientation to aid the vision navigation system, whereas others refer to this task as the weakest and least robust in the estimation chain. Aside the client spacecraft 3D model, here no assumption on the relative translational and rotational motion is made. Using a single monocular image, and utilizing knowledge of the target spacecraft, estimation of the target's six relative rotation and translation parameters with respect to the camera are found.

2 Proposed Method

In this section the proposed method was described. The problem of pose estimation is formulated as follow: Given model of the object and a calibrated camera, find the pose of the object with respect to camera.

Presented method of solution is applicable to monocular camera systems. There are given photos of known object which is seen from different camera locations at every steps of times. The features used to describe the target object are in 3D, while the projection of a feature found in the image is in 2D. The pinhole camera model is used to done the projection of the 3D coordinates of the object features with respect to the camera frame to the 2D coordinates found in the image. This transformation from the 3D coordination to the 2D coordinate is also called a perspective projection. There are 2 sets of points: a 3D set representing the model denoted by X_M, and a 2D set detected from the image denoted by X_I. Assume it is known that each point X_I in the 2D set is the image local feature which is X_M in the 3D set. Unfortunately, the correspondence between the 2D features and 3D is usually not known. This leads to a closely related to the pose estimation problem known as the correspondence problem. It is the process of finding out which features in a set correspond to a feature in another set. The problem of pose estimation without correspondence is also known as the simultaneous pose and correspondence problem. If the position of the target is approximately known, the correspondence problem becomes simpler because one can project the geometric representation onto the image plane and associate each projected model feature to the closest image feature to obtain the correspondence. A feature representation of an object is more effective to the pose determination. Local features every entity is locally evaluated and represents a small part of the object. Examples for local features are corners maximum curvature segment points and corners. In this case at least 3 model points and three corresponding image points are required in order to determine the pose of an object. However in order to reduce the effects of measurement noise on the precision of the solution several studies including presented work use a larger number of points. The advantage in using point features in the pose estimation solution is the relative ease of extracting these features. Unknown are the six parameters that can describe relative pose of two objects: three coordinates x, y, z which describes the linear translation of object in relation to camera and three angles of rotation Φ, Θ, Ψ (roll, pitch, yaw) which describes the mutual angular orientation of two objects in space.

There were defined two main coordinate systems. Both are right hand side (clockwise) oriented Cartesian coordinated systems. The first one is camera coordinate system. The origin O_C of this coordinate system is located in optical center. The $O_C x_C$ axis of this coordinate system is oriented downward, the $O_C y_C$ axes is oriented on right and the $O_C z_C$ axis completes the right handed coordinate system (Fig. 1). The position of each pixel on photo is given in camera coordinate system. The second one is model coordinate system. The origin O_M of this coordinate system is located on the surface of object. $O_M x_M$ axis is oriented in front of the target, the $O_M y_M$ axis on right and the $O_M z_M$ axis completed the coordinated system. Origin of this coordinate system is translated from origin of camera coordinate system by vector $t = \begin{bmatrix} x \\ y \\ z \end{bmatrix}$ in Cartesian coordinates. The down scripts C by coordinates means that coordinates are referred in camera coordinate frame. Down subscripts M means that coordinates are in model coordinate frame. The image plane is parallel to the $O_C x_C$ and $O_C y_C$ axes of camera coordinate system at distance f from the optical center. Object is rotated and displaced with respect to camera coordinate system.

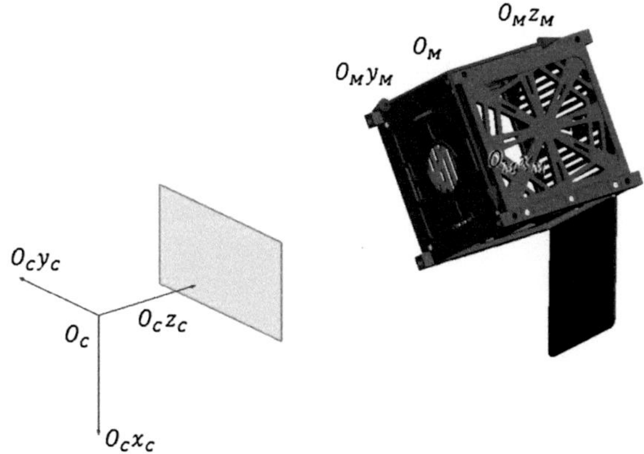

Fig. 1 Pose estimation problem [8]

The equation of mapping 3D points in camera coordinate frame to a image plane in homogenous coordinates is defined as follow:

$$\begin{bmatrix} u \\ v \\ 1 \end{bmatrix} = \begin{bmatrix} \alpha_u & s & u_0 & 0 \\ 0 & \alpha_v & v_0 & 0 \\ 0 & 0 & 1 & 0 \end{bmatrix} \begin{bmatrix} x_C \\ y_C \\ z_C \\ 1 \end{bmatrix} \tag{1}$$

α_u and α_v are the scale factors in the u- and v-coordinate directions. They are proportional to the focal length f of the camera: $\alpha_u = k_u f$ and $\alpha_v = k_v f$, where k_u and k_v are the number of pixels per unit distance in the u and v directions [12]. The camera internal parameters are known. This was designated in camera calibration process. In most 3D tracking methods, the internal parameters are assumed to be fixed and known, which means that the camera cannot zoom, because it is difficult to distinguish a change in focal length from a translation along the camera Z-axis. In visual system for servicing satellites it was assumed that there will be two cameras. First camera will be operated when the object is far. Second camera will be operated when the object will be near. For additional details about camera models the interested reader is referred to the photogrammetric literature.

Then small letters u and v describes the coordinates in image plane. Coordinates in image plane and 3D space were described in homogenous coordinates.

The transformation of two object consists of two components: a component that describes the rotation of the model coordinate frame relative to the camera coordinates and a component that describes the relative translation between the two coordinate frames. The coordinates of points of object are referred in model coordinate frame so before mapping on image plane they must be transferred to camera coordinate system:

$$X_C = \begin{bmatrix} R & t \\ 0 & 1 \end{bmatrix} X_M \tag{2}$$

Unknown are $\boldsymbol{R}$ and $\boldsymbol{t}$, left side and last vector on right are known. $[\boldsymbol{R}\ \boldsymbol{t}]$ is the 3×4 external parameters matrix, and corresponds to the Euclidean transformation from a world coordinate system to the camera coordinate system. $\boldsymbol{R}$ represents a 3×3 rotation matrix, and $\boldsymbol{t}$ a translation. The translator component is usually described by a three dimensional vector $\boldsymbol{t}$ which gives the translation in the X, Y and Z axis directions. However the rotational component can be described in several methods which influence the choice of computational method to be used for the estimation process. The angular orientation of object was parametrized by using of Euler angles. These three angles form three free parameters that describe any rotation transformation. A rotation matrix $\boldsymbol{R}$ can always be written as the product of three matrices representing rotations around X, Y, and Z axes. There is singularity when the coordinate frames are rotated mutual by pitch angle equal $\pi/2$. So in expanded form the Eq. (2) is defined as follow:

$$\begin{bmatrix} x_C \\ y_C \\ z_C \\ 1 \end{bmatrix} = \begin{bmatrix} \cos\Theta\cos\Psi & \sin\Phi\sin\Theta\cos\Psi - \cos\Phi\sin\Psi & \cos\Phi\sin\Theta\cos\Psi + \sin\Phi\sin\Psi & x \\ \cos\Theta\sin\Psi & \sin\Phi\sin\Theta\sin\Psi + \cos\Phi\cos\Psi & \cos\Phi\sin\Theta\sin\Psi - \sin\Phi\cos\Psi & y \\ -\sin\Theta & \sin\Phi\cos\Theta & \cos\Phi\cos\Theta & z \\ 0 & 0 & 0 & 1 \end{bmatrix} \begin{bmatrix} x_M \\ y_M \\ z_M \\ 1 \end{bmatrix} \tag{3}$$

The proposed method work in a such manner as described downwards. At first, a photo of real object is taken. Next there are detected local features on this photo. Local features of object is usually associated with a change of image properties simultaneously, although it is not necessarily localized exactly on this change. To handle as wide as possible a range of viewing conditions, feature point extraction should be insensitive to scale, viewpoint, and illumination changes. The local features of the object are extracted by using of Scale Invariant Feature Transform (SIFT) detector and descriptor proposed by Lowe [13]. Algorithm extracts features and is for object recognition based on local 3D extrema in the scale-space pyramid build with difference-of-Gaussian filters. First the location and scale of the key-points are determined precisely by interpolating the pyramid of Difference-of-Gaussians used for the detection. The input image is successively smoothed with a Gaussian kernel and sampled. The difference of Gaussian representation is obtained by subtracting two successive smoothed images. The Gaussian kernel and its derivatives are the only possible smoothing kernels for scale space analysis. To achieve image rotation invariance, an orientation is also assigned to the keypoint. It is taken to be the one corresponding to a peak in the histogram of the gradient orientations within a region around the keypoint. All dig levels are constructed by combining smoothing and subsampling. The local 3D extrema in the pyramid representation determine the localization and the scale of the interests points. This method is quite stable under viewpoint changes, and achieves an accuracy of a few degrees. An image is transformed into a group a local features. On the exit of this algorithm there is known the two dimensional vector of coordinates of each feature and the second vector which contains the radius of each feature and the angle of orientation in radians. It represents the left side on equation of mapping (1).

To further explore the methods of solving the pose estimation problem, one must be able to model the target object. It is usually described by a set of features, and a feature is any descriptor of a shape. Next, similar to the previous step, the local features of the object from model of target are extracted. During an offline training stage, a database of interest object points was build. Their positions on the object surface are known. Images in which the object has been manually registered were used for this purpose. At runtime, SIFT features are extracted from the current frame, matched against the database, resulting in a set of 2D-3D correspondences.

The next task is the pose estimation of the object. It was assumed earlier a set of point correspondences with model points and image points. The object pose can then be estimated from such correspondences. The unknowns are the translation vector t and rotational vector R. They have been found iteratively by using POSIT algorithm. This algorithm need focal length, which is assumed to be same for x and y, 4 or more non-coplanar 2D-3D correspondences. This algorithm estimates pose uses a scaled orthographic projection, which resembles the real perspective projection at convergence. Such approximation leads to a linear equation system. This gives the rotation and translation directly, and there is no the need of a starting pose. A scale value is introduced for each correspondence, which is iteratively updated. What is known is the distribution of the feature points on the object and the images of these points by perspective projection. If someone could build SOP images of the object feature points from a perspective image someone could apply the POS algorithm to these Scale Orthographic Projection (SOP) images and we would obtain an exact object pose. Computing exact SOP images requires knowing the exact pose of the object. However someone can apply POS to the actual image points we then obtain an approximate depth for each feature point and we position the feature points on the lines of sight at these depths. Then we can compute an SOP image. At the next step we apply POS to the SOP image to obtain an improved SOP image. Repeating these steps we converge toward an accurate SOP image and an accurate pose. More details about POSIT algorithm can be found in [14].

3 Experiments

This section described experiments which were made. Experiments were tested in MATLAB software.

It was used a set of photos of real object. It were presented the tracking results of the real object. Experiments were conducted on various image sequences and show advantages of the chosen approach. The main goal of experiments was to obtain the measurements of position and orientation of object and check how accurate is algorithm. Others goals were to confirm the robustness to varying occluding conditions. Experiments were tested as follow. It was used a plane with example image, which was mounted on mounting stand, which can be precisely translated and rotated due to camera coordinate frame. The object was placed on calibration grid so measured parameters were compared with proposed methods. Typical object was shown on image below. It was used low cost camera. The source of light was fixed. It was provided uniform illumination for all the positions of tested object. There were measured of position and one rotational degree of freedom on ground. Next the result of ground truth measurement were compared with calculated results. It was expected that ground truth measurements should be very close to calculated results. Expected translation error should be below a few millimeters. Similar, rotational error should be under 4°.

These internal parameters can be estimated during an offline camera calibration stage. Camera was calibrated the using Caltech calibration method. The motion pattern is visible in the figure. The ground truth error varies around 1 mm in space.

The six plots presents results for a first chosen example. The camera has a distance of approximately 220 mm to the object. At the beginning the object is not moving. Next the object is moving manually till second 37. The object was constrained and not moved along x and y axes. On the x axes of first three plots there is given time in seconds and on y axes the measured translation in millimeters. On the next plots on x axes there are given, similar as upper figures, time and on y orientation in degrees. Green line shows ground truth and blue line shows the measurements (Fig. 2).

Ground truth (green line) should be very close to measured results for all degrees of freedom. On the first plot (upper left) there was presented the linear translation of object along x axis of the camera coordinate system. Vision based measurement indices that there was an translation about 6 mm. It may be caused by camera mounting. On the second plot there was presented linear translation along y axis. In reality there was no translational motion along y but from vision system measurements one can see that maximum difference for y axis is about 9 mm. Possible cause of this errors is nature of presents method. There is possible to try reduce the

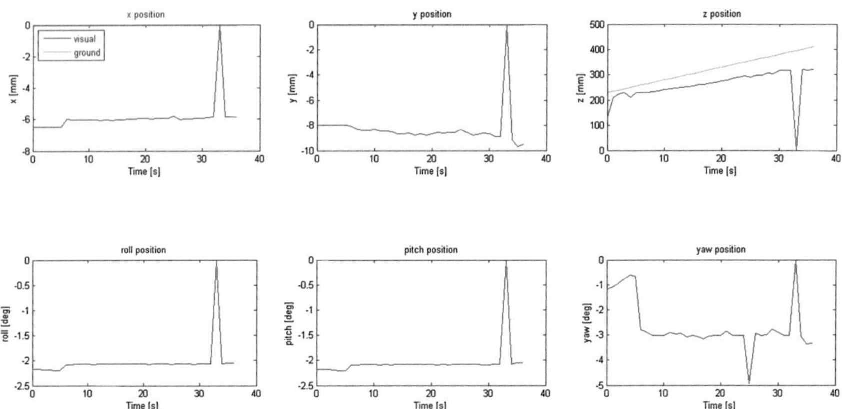

Fig. 2 Tests results of the first experiment. Position and orientation of the object

errors if better correspondence generation algorithm will be obtained. On third plot error for z axis is about 85 mm, which is much more than for x and y axis. This is measured in direction perpendicular for image plane and it is distance from camera coordinate system and the object. It was expected that this error should be smaller. Next three plots presents rotations around three axes of object coordinate system. In ground truth measurements there were no rotation about x and y axes. Four plot presents that there was an rotation about 3°. On the fifth plot there is small error 3 mm between both ground truth rotation and vision based measurement. In the case of rotation around z axis there was quite small error. After 5 s error is three times bigger than at the beginning. The peak at 34 s is caused by software because one frame was lost during simulation.

There results were as expected. It worth nothing that rotational errors were with negative sign. Translational errors were under ten millimeters for x and y axes and bigger for z axis.

Next, there was conducted second experiment. Object was moved but in other manner as in first experiment. The object was moved mainly along z axis, from position far to closer from camera (Fig. 3).

Similar as in the first case there are presented six plots. First three presents linear translations along axes of camera coordinate system and three presents angular orientation of object. First plot presents linear translation along x axis. There is significant error between both measurements. At the end of simulation difference is about 100 mm. This is caused by mounting of camera. There were small motions caused by imperfectly camera to ground. On the second plot both lines green and blue are very close because object was constrained on y axis and could not pitch and yaw. On the next three plots there are presented measurements for angular orientation of object. For roll motion there is error about 2°. Very similar results were obtained for pitch. For both vision based measurements there is a peak in 6 s. The sixth plot presents yaw. In this case error is 5 mm.

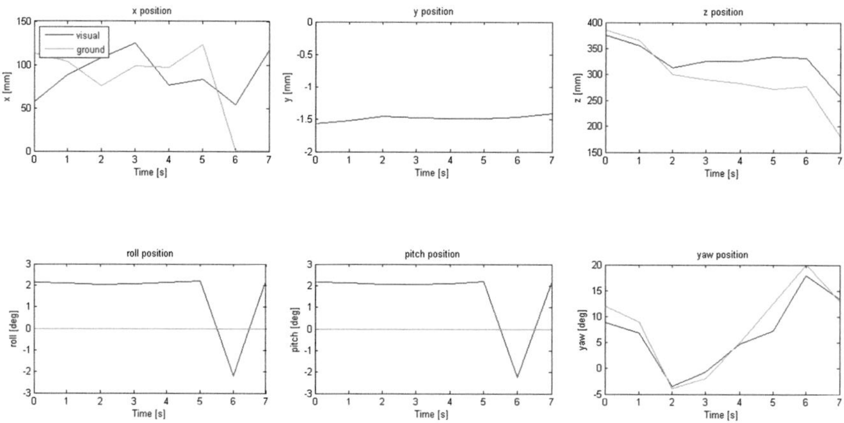

Fig. 3 Tests results for second case

The experiments take about 13 ms per frame on modern CPU. It was shown that introduced method is able to run in real time. The camera bed should be more precisely screwed because it introduced serious problems. It is planned to still work on proposed method.

4 Conclusions

This paper describes the results of analysis and testing of a real-time pose estimation system intended for vision based navigation for satellites. Presented method is more accuracy than earlier proposed methods. The promising result of numerical simulation have been obtained. The linear translation error was amount of few millirs. The rotational error was in the worst case 4°. Approach will be further evaluated and tested in the continued work. There is planned to still work on all system.

Acknowledgments The following research was done within project (agreement No. PBS2/B3/17/2013). Sponsored by National Centre for Research and Development.

References

1. Adamski, M.: Bezzałogowe Statki Powietrzne, cz. II, Konstrukcja, wyposażenie i eksploatacja. WSOSP-Dęblin (2015)
2. Głębocki, R., Kopyt, A., Kicman, P.: Navigation module for mobile robot. doi:10.1007/978-3-319-15847-1_9 (2015)
3. Vogt, R., Głębocki, R., Adamski, M.: Sterowanie lotem w ujęciu systemowym. WSOSP-Dęblin (2015)
4. Grunert, J.: Das pothenotische Problem in erweiterter Gestalt nebst über seine Anwendungen in Geodäsie. Grunerts Archiv für Mathematik und Physik (1841)
5. Quan, L., Lan, Z.-D.: Linear N4-point pose determination. In: 6th International Conference on Computer Vision, Bombay (1998)
6. Fischler, M.: Random sample consensus: consensus: a paradigm for model fitting with applications to image analysis and automated cartography. Comm. ACM **24**(6), 381–395 (1988)
7. Rosenhahn, B.: Foundations about 2D-3D pose estimation problem
8. Cropp, A.: Estimating pose of known target satellite. Electr. Lett. **36**(15), 1331–1332 (2000)
9. Simulation of the Docking Phase for the SMART-OLEV Satellite Servicing Mission. In: 9th International Symposium on Artificial Intelligence, Robotics and Automation in Space (iSAIRAS)
10. Jasiobedzki1, P.: Pose determination and tracking for autonomous satellite capture. In: Proceeding of the 6th International Symposium on Artificial Intelligence and Robotics and Automation in Space, Quebec (2001)
11. Jörgensen, J., Harr, J,: Prisma: an autonomous formation flying mission. In: ESA Small Satellite Systems and Services Symposium, Sardinia (2006)
12. Vincent, P.: 3D model-based tracking for space autonomous randezvouz
13. Mikolajczyk, K.: Scale and affine invariant interest point detectors. Int. J. Comput. Vis. **60**(1), 63–86 (2004)
14. Petersen, T.: A comparison of 2D-3D pose estimation methods. Aalborg University, Ballerup (2008)

Detection and Recognition of Compound 3D Models by Hypothesis Generation

Artur Wilkowski and Maciej Stefańczyk

Abstract In the paper there is proposed an integrated object detection and recognition system, based on object description given in semantic form [5]. The objects are described in a generic way in terms of parts and relations between them. The Bayesian inference system is utilized, so each object detection and recognition score has probabilistic interpretation. There are designed basic 3D models founded on the inference framework. Object instances are then detected and recognized in real-world Kinect RGBD images.

Keywords 3-D objects recognition · Point cloud · RGBD image analysis · Constraint satisfaction

1 Introduction

Scene analysis is one of the most fundamental areas of robot perception. Detection and pose estimation of the 3D object has been widely discussed in literature. Successful solutions include e.g. utilization of linear segments [6] or feature points [7]. In case when the depth data is available also Point-Cloud based algorithms can be employed [1, 4]. An important feature of such algorithms is that they relay mostly on well defined (e.g. CAD) object models.

More generic solution are proposed that take into account variability of object shapes within object class. They utilize a composition paradigm that refers to most

A. Wilkowski (✉)
Industrial Research Institute for Automation and Measurements,
Al. Jerozolimskie 202, 02-486 Warsaw, Poland
email: awilkowski@piap.pl
URL: http://www.antiterrorism.eu/

M. Stefańczyk
Institute of Control and Computation Engineering,
Nowowiejska 15/19, 06-665 Warsaw, Poland
email: mstefanc@elka.pw.edu.pl
URL: http://robotyka.ia.pw.edu.pl/

R. Szewczyk et al. (eds.), *Challenges in Automation, Robotics and Measurement Techniques*, Advances in Intelligent Systems and Computing 440, DOI 10.1007/978-3-319-29357-8_57

objects in a real world (e.g. bike is composed of two wheels, the frame, pedals etc.). One such solution is given in [8]. Here, objects are described by a set of parts and relations between them, and have desirable generalization property [2]. A typical drawback of such approaches is that object parts are loosely connected (e.g. there can be only verified if specific object parts form cluster in the image), so in general they are unable to estimate proper object pose and parameters of the object as a whole.

To fill this gap we propose an integrated, complex object detection and recognition system, based on object description given in semantic form [5]. The proposed framework is built on the following principles: Firstly, the objects are described in a generic way in terms of the parts that constitute objects and relations between them. Secondly, no assumption is made regarding the nature of particular parts (they can be e.g. surfaces, edges, textures). Thirdly, the Bayesian inference system is utilized, so each object detection and recognition score has probabilistic interpretation.

The core of the system structure is described in [11]. In this paper we concentrate on the demonstration of practical application of the proposed framework and inference system to the problem of 3D object detection and recognition in RGBD images.

Article is structured as follows. Section 2 gives an overview of model structure, and the inference scheme. In the next Sect. 3 there are introduces two generic 3D models that are to be recognized by the system. Experimental results and their discussion is given in Sect. 4. In the last section there are given conclusions and directions for future work (Sect. 5).

2 Inference System Description

Model Structure Model M is built from parts P and constraints C defined between them. Each part must have assigned class ($class(P)$) and unique role in model ($role(P)$). Thus, model is a tuple consisting of two sets:

$$M = \left\{P = \left\{p_1 \dots p_n\right\}, C = \left\{c_1 \dots c_m\right\}\right\}$$

Constraints Constraints between the parts are defined by telling which predicate is to be used on which subset of parts. Predicates in our system returns extended information, i.e. from the $[0 \dots 1]$ range, instead of classical $\{True, False\}$, where 1 means full constraint satisfaction and 0 total inconsistency of given set of parts with examined predicate. It enables us to treat constraint satisfaction result as an intermediate score in further processing steps, giving finally overall score for current model hypothesis.

$$pred(p_1, \dots, p_k) \in [0, 1]$$

Hypotheses Each assignment of existing instances to model parts, made in a way that for each part of the model exactly one instance of proper class is selected, is called hypothesis $H(M)$. It can be defined as:

$$H(M) : \forall i \in \{1..n\} \exists j \in \{1..k\} : p_i \leftarrow in_j \wedge class(p_i) = class(in_j)$$

where n is the number of parts in the model and k is the number of already recognized instances (e.g. segments). Overall hypothesis score is calculated by taking the product of all constraints for given assignment.

$$score\,(H\,(M)) = \prod_m eval(c_m, H)$$

Inference During inference there are generated object hypotheses basing on the object parts detected in the data source (RGBD image for instance). Then object hypotheses are evaluated according to the model and constraints. The set of the best hypotheses is returned. More detailed description of the process is given in [11].

3 Concepts 3D

RGBD Image Segmentation In order to extract 3D concepts from RGBD image, image segmentation procedure is first applied. In our approach a method proposed in [9] is utilized. From four-tier complex system proposed in the publication we utilize only two lowest-level procedures due to their ability to perform noise-robust segmentation of 3D point cloud. The two steps are given as follows

- *Signal level: pixel clustering*. The output of this step are planar patches. Clustering is performed at three different pyramid levels from coarse to fine. Then a parametrized Minimum Description Length criterion is applied to see whether a given set of points is better represented by a single patch at corase level of several patches at finer level.
- *Primitive level: parametric surface model creation*. The output of this step are planar and parametric surfaces (parametrized as B-splines). Neighbouring planar patches (output of the previous step) are combined into parametric surfaces. Also at this step the MDL criterion is utilized to find a compromise between the number of surface parameters and actual surface-data matching cost.

Although the actual surface parameters are not subsequently used in our system their application at the segmentation stage ensures output that is robust to noise with good generalization properties. The segmentation module produces also information concerning neighbor relations of segments which are useful to generate some Concept attributes.

For subsequent processing there are selected only segments contain a fixed minimum number of points.

3D Concepts Description In order to carry out recognition of model objects, the following hierarchy of basic concepts was defined:

- 'Surface'—represents a generic surface consisting of neighboring points sharing similar geometric properties
  - *Cylinder*—represents a cylindrical surface
  - *Plane*—represents a planar surface

Extraction of 3D Concept Instances During processing an effort is made to assign scene 3D points to *Cylinder* and *Plane* concepts. As a general rule one segment (result of the segmentation procedure) is mapped to a single concept instance. Planar segments are mapped onto *Plane* concept. For non-planar segments a Random Sample Consensus Method (RANSAC) [3] is used to verify if the segment can represent *Cylinder* concept.

The RANSAC procedure assumes that we have available a set of points with associated normal vectors. In each iteration of the procedure two points (with normals) are selected to propose a hypothesis concerning parameters of the cylindrical surface. Then, there is evaluated a proportion of points in the segment consistent with the hypothesis and if it is the best hypothesis found so far it is memorized. The segment is decided to be a *Cylinder* instance if the number of inliers associated with the best hypothesis exceeds certain threshold $t_{cylinder}$. Obviously, besides the classification decision, the RANSAC procedure provides also the resulting cylinder parameters than can be directly used as attributes of the *Cylinder* concept.

For the sake of completness, note that two points with two normals are enough to compute cylinder hypothesis. The two normals define a plane parallel to cylinder base, and so the direction of the cylinder main axis. In addition the normals projected on this plane define two lines. Their intersection specify a point on the cylinder main axis.

Attributes The *Surface* concept uses attributes $\{id, normal, neighbors\}$. Below there is given the meaning of the attributes:

- *Concept::id*—id number associated with the surface. Among other things it acts as a link to underlying points associated with the given surface
- *Concept::normal*—a normal vector describing orientation of the surface. The exact interpretation depends on the particular surface type
- *Concept::neighbors*—a list of neighboring surfaces, the surfaces that touch given surface

The *Cylinder* concept inherits attributes $\{id, normal, neighbors\}$ from *Surface* concept and introduces a new attribute *radius*. The meaning of *Cylinder* attributes are as follows:

- *Cylinder::id*—similar to *Surface::id*
- *Cylinder::normal*—a vector aligned with the main axis of the cylinder
- *Cylinder::neighbors*—similar to *Surface::neighbors*
- *Cylinder::radius*—the radius of the geometrical cylinder

The 'Plane' concept inherits attributes $\{id, normal, neighbors\}$ from *Surface* concept and introduces new attributes $\{centroid, extent\}$. The meaning of *Plane* attributes are as follows:

- *Plane::id*—similar to *Surface::id*
- *Plane::normal*—a normal vector to the planar segment
- *Plane::neighbors*—similar to *Surface::neighbors*
- *Plane::centroid*—geometric centroid of all points belonging to the planar segment
- *Plane::extent*—2-dimensional extent of the planar segment (see: computation details below), if segment is rectangular this corresponds to lengths of rectangle sides

Computation of Attributes Most concept attributes are straightforwardly derived from the segmentation procedure (*id*, *neighbors*) or from particular object matching method (e.g. *Cylinder::normal* and *Cylinder::radius*). In order to compute a normal vector and an extent of a planar segment there is adopted the following procedure.

First the Principal Component Analysis (PCA) method is applied to the point cloud segment associated with the Concept. In result we obtain three eigenvectors $\mathbf{v}_1$, $\mathbf{v}_2$, $\mathbf{v}_3$ of the point covariance matrix together with associated eigenvalues λ_1, λ_2, λ_3, arranged so that $\lambda_1 > \lambda_2 > \lambda_3$. Then the vector $\mathbf{v}_3$ corresponding to the smallest eigenvalue λ_3 is selected as the plane normal vector. The direction of the normal vector computed is selected in a way that it points in the direction of the sensor origin (this will be important for convexity relations between planes).

Next, the assumption is made that the planar segment is perfectly flat and rectangle-shaped. Therefore, points constituting the plane can be represented as a product of two independent variables of a uniform distribution. The variances in each direction are then exactly λ_1 and λ_2. Having in mind that the standard deviation of $U(0, 1)$ distribution is $\sqrt{\frac{1}{12}}$, and the standard deviation of $U(0, x)$ is $\sqrt{\frac{1}{12}} \cdot x$ for any interval length x, we may express lengths of each side of the rectangle by the means of the computed variances, namely

$$\begin{cases} u_1 = \sqrt{\frac{1}{12}}\sqrt{\lambda_1} \approx 3.5\sqrt{\lambda_1} \\ u_2 = \sqrt{\frac{1}{12}}\sqrt{\lambda_2} \approx 3.5\sqrt{\lambda_2} \end{cases}$$

Therefore the 2-dimensional extent of the planar segment can be approximated by pair $e = (u_1, u_2) = (3.5\sqrt{\lambda_1}, 3.5\sqrt{\lambda_2})$. This approximation is exact when the planar segment is rectangle-shaped, for oval segments it will tend to slightly underestimate the extent of the segment. However, experiments shown that this approach is sufficient to successfully compare shape sizes of different concepts.

Constraints The following contraints were defined in order to perform part matching.

- *Touch*—applies to all *Surface*-derived concepts s_1, s_2, returns: 1 if $s_1.id \in s_2.neighbors$ or $s_2.id \in s_1.neighbors$ and 0 otherwise
- *Perpendicular*—applies to all *Surface*-derived concepts s_1, s_2, returns: $1 - ((s_1.normal)^T \cdot s_2.normal)^2$
- *Parallel*—applies to all *Surface*-derived concepts s_1, s_2, returns: $((s_1.normal)^T \cdot s_2.normal)^2$
- *Sizematch*—applies to *Plane* and *Cylinder* concepts, the value of the predicate depends on the actual type of the operands
  - if p_1 and p_2 are both *Plane* concepts, there is computed the largest (but not greater than 1) proportion of extents

$$Sizematch(p_1, p_2) = \max_{i,j=1,2}\left(\min\left(\frac{p_2.extent.u_i}{p_1.extent.u_j}, \frac{p_1.extent.u_j}{p_2.extent.u_i}\right)\right)$$

  - if p is *Plane* concept and c is *Cylinder* concept, there is computed the largest (but not greater than 1) proportion of planar segment extent and the cylinder diameter

$$Sizematch(p, c) = \max_{i=1,2}\left(\min\left(\frac{p.extent.u_i}{2 \cdot c.radius}, \frac{2 \cdot c.radius}{p.extent.u_i}\right)\right)$$

  - if both c_1 and c_2 are cylinder concepts, the results is the proportion of their diameters (or radii) not larger than 1

$$Sizematch(c, c) = \min\left(\frac{c_1.radius}{c_2.radius}, \frac{c_2.radius}{c_1.radius}\right)$$

- *Convex*—applies to *Plane* plane concepts p_1 and p_2, returns: 1.0 if $(p_2.centroid - p_1.centroid)^T \cdot p_1.normal < 0$ and $(p_1.centroid - p_2.centroid)^T \cdot p_2.normal < 0$, otherwise it returns 0. This predicate checks actually if the line connecting centroids of the two planar segment lies 'on the same side' of planar segments as their normal vectors (remind - the normal vectors points in the observer's direction).

Models Two simple models *tube-with-a-lid* and a *box* were designed in order to evaluate proposed Concept classes and Constraint predicates.

Tube-with-a-lid model represents a round container covered by a lid. It is composed of two concept instances *Cylinder* and *Plane*. These concepts are related by the following constraints: *Touch*, *Parallel* and *Sizematch*. A graph representing this model is given in Fig. 1.

Box model represents a rectangular box. As of now only the object modalities with all three sides visible are detected and recognized. The model is composed of a sin-

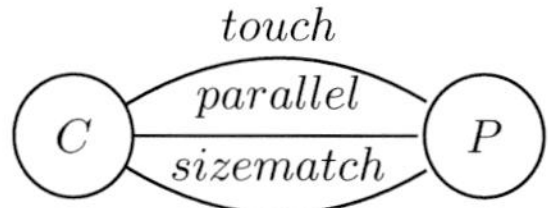

Fig. 1 Tube model. Nodes *C* and *P* denote *Cylinder* and *Plane* concepts correspondingly. Edges are labelled by constraints

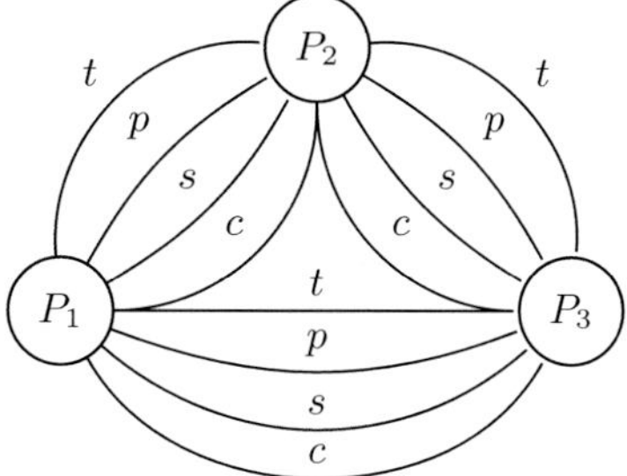

Fig. 2 Box model. Nodes P_i denotes different *Plane* concept instances. Edges are labelled by constraints *t-Touch, p-Perpendicular, s-Sizematch, c-Convex*

gle concept *Plane* available in three instances. Every pair of instances is related by the following constraints: *Touch*, *Perpendicular*, *Sizematch*, *Convex*. A graph representing this model is given in Fig. 2.

4 Results

4.1 Dataset

The Datasets Images for testing computer vision algorithms consists of (13 objects, 12 views per objects) Image resolution was 1280 × 960 [10]. Objects positioned on a turntable against white background are observed from fixed viewpoint. For each object, the turntable was rotated 30° each. For the purpose of following experiments there were selected two sequences of images 'inka_1_kinect' (short: 'Inka'), containing images of a round coffee container and 'lipton_2_kinect' (short: 'Lipton') containing images of a tea-box. Sample images are given in Fig. 3. The 'Inka' images are used to evaluate detection of a *Tube-with-a-lid model* and 'Lipton' images are used to evaluate detection of a *Box model.*

4.2 Detection and Recognition of 3D Objects

The results of detection of the *tube-with-a-lid* model on the image from 'Inka' sequence is given in Fig. 4. The correct object receives a very high score (close to 1)

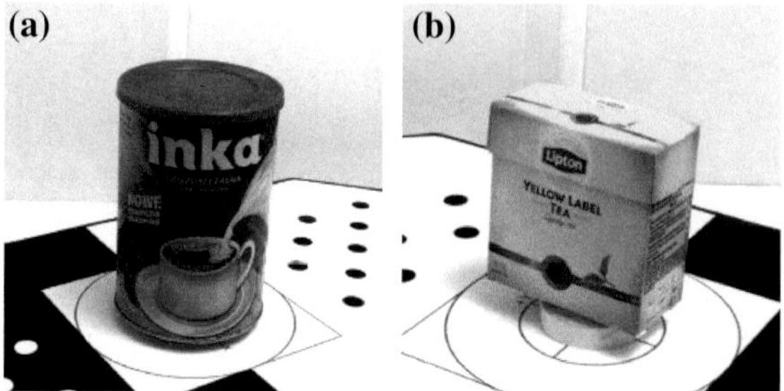

Fig. 3 Detected and recognized objects: **a** Inka coffee container, **b** Lipton tea-box

due to a very good alignment between the cylinder and the lid and good results of their size comparision. The second best match (Fig. 4c) receives much lower score (0.35). This example shows that the model is indifferent to whether the lid is found at the top or at the bottom of the cylinder. Here, this is the *Sizematch* predicate that enables to differentiate between the lid and the table the container is standing on.

The results of detection of the *box* model on the image from 'Lipton' sequence is given in Fig. 5. The correct object receives a relatively high score (close to 0.9). The second best match (Fig. 5c) receives insignificant score (0.09), which forms a clear decision boundary.

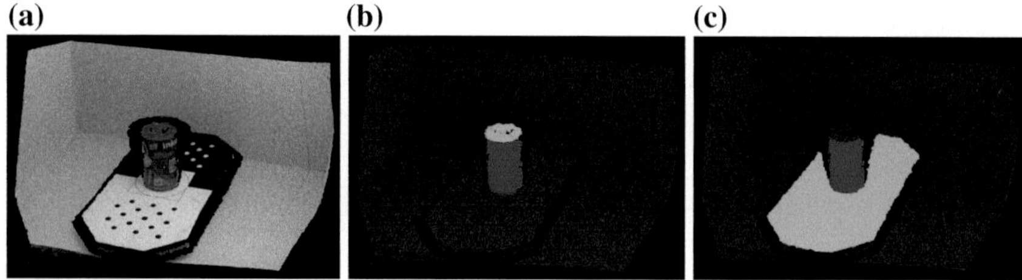

Fig. 4 Detection of a *tube-with-a-lid model*: **a** input point cloud, **b** best match (*score* = 0.99), **c** second best match (*score* = 0.35)

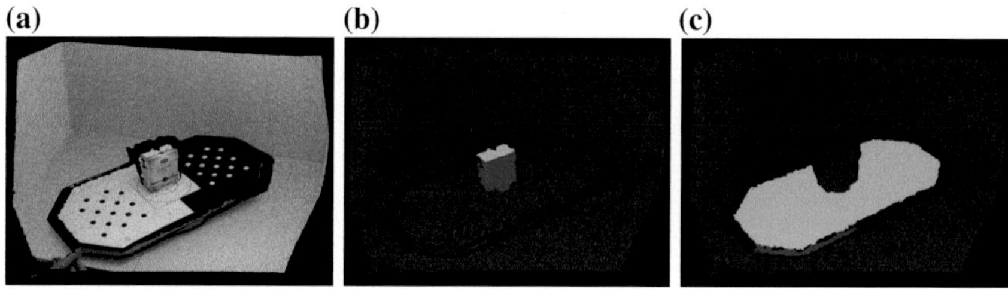

Fig. 5 Detection of a *box model*: **a** input point cloud, **b** best match (*score* = 0.89), **c** second best match (*score* = 0.09)

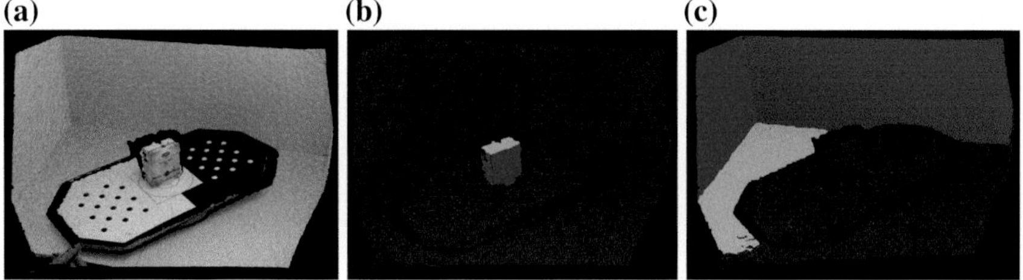

Fig. 6 Detection of a *box model* withouth *Convex* predicate: **a** input point cloud, **b** best match (*score* = 0.89), **c** second best match (*score* = 0.51)—the 'wall corner' structure receives score above 0.5

The next set of examples (Fig. 6) demonstrates the importance of the *Convex* predicate for box recognition. Without such predicate (as in Fig. 6) the 'wall-corner' structures tend to receive good scores (which can be very high if planes are aligned well and are of comparable sizes), while utilization of such predicate practically eradicate such constructs from the set of well-scored hypotheses.

5 Conclusions

In the paper a complex object detection and recognition system, based on object description given in semantic form was presented. The features of the system include genericity of object description, flexibility in design of primitive concepts and model objects and a presence of an effective Bayesian inference system. There was also proposed a specification of two Concepts (*Plane* and *Cylinder*) that were subsequently used to detect and recognize objects in RGBD images. Two different models 'coffee container' and 'tea-box' were utilized in recognition, however the proposed concept attributes and contraint predicates are generic and easily extensible, so they may serve for recognition of much more complex models.

Planned future work includes extension of the number of geometric primitives (e.g. by Cone, Torus) as well asother not easily parametrized, custom objects. Successive works will also encompass evaluation of the inference system on more complex objects build up of ten or more parts.

Acknowledgments The authors gratefully acknowledge the support of the National Centre for Research and Development (Poland), grant no. PBS1/A3/8/2012.

References

1. Aldoma, A., Tombari, F., Di Stefano, L., Vincze, M.: A global hypotheses verification method for 3d object recognition. In: Fitzgibbon, A., Lazebnik, S., Perona, P., Sato, Y., Schmid, C. (eds.) Computer Vision—ECCV 2012, Lecture Notes in Computer Science, vol. 7574, pp. 511–524. Springer, Berlin (2012). http://dx.doi.org/10.1007/978-3-642-33712-3_37

2. Felzenszwalb, P.F., Girshick, R.B., McAllester, D., Ramanan, D.: Object detection with discriminatively trained part-based models. IEEE Trans. Pattern Anal. Mach. Intell. **32**(9), 1627–1645 (2010)
3. Fischler, M.A., Bolles, R.C.: Random sample consensus: a paradigm for model fitting with applications to image analysis and automated cartography. Commun. ACM **24**(6), 381–395 (1981). http://doi.acm.org/10.1145/358669.358692
4. Hinterstoisser, S., Lepetit, V., Ilic, S., Holzer, S., Bradski, G., Konolige, K., Navab, N.: Model based training, detection and pose estimation of texture-less 3d objects in heavily cluttered scenes. In: Computer Vision-ACCV 2012, pp. 548–562. Springer (2013)
5. Kasprzak, W., Pietruch, R., Bojar, K., Wilkowski, A., Kornuta, T.: Integrating data- and model-driven analysis of rgb-d images. In: Filev, D., Jabłkowski, J., Kacprzyk, J., Krawczak, M., Popchev, I., Rutkowski, L., Sgurev, V., Sotirova, E., Szynkarczyk, P., Zadrozny, S. (eds.) Intelligent Systems'2014, Advances in Intelligent Systems and Computing, vol. 323, pp. 605–616. Springer International Publishing (2015)
6. Lowe, D.G.: Three-dimensional object recognition from single two-dimensional images. Artif. Intell. **31**(3), 355–395 (1987)
7. Lowe, D.G.: Object recognition from local scale-invariant features. In: The Proceedings of the Seventh IEEE International Conference on Computer Vision, 1999, vol. 2, pp. 1150–1157. IEEE (1999)
8. Opelt, A., Pinz, A., Fussenegger, M., Auer, P.: Generic object recognition with boosting. IEEE Trans. Pattern Anal. Mach. Intell. **28**(3), 416–431 (2006)
9. Richtsfeld, A., Mörwald, T., Prankl, J., Zillich, M., Vincze, M.: Learning of perceptual grouping for object segmentation on rgb-d data. J. Vis. Comun. Image Represent. **25**(1), 64–73 (2014). http://dx.doi.org/10.1016/j.jvcir.2013.04.006
10. Stefańczyk, M., Laszkowski, M., Kornuta, T.: WUT visual perception dataset: a dataset for registration and recognition of objects. In: Progress in Automation, Robotics and Measuring Techniques. Advances in Intelligent Systems and Computing (AISC). Springer (2016)
11. Stefańczyk, M., Pietruch, R.: Hypothesis generation in generic, model-based object recognition system. In: Proceedings of the International Conference Automation 2016 (2016) (submitted)

Autonomous Agricultural Robot—Conception of Inertial Navigation System

Marcin Jasiński, Jędrzej Mączak, Stanisław Radkowski, Sebastian Korczak, Roman Rogacki, Jarosław Mac and Jan Szczepaniak

Abstract The aim of the paper was to propose conception of autonomous robot for sowing and wide row planting. Autonomous work of the robot in range of traction and agronomic processes will be implemented on the basis of data from a many sensors (cameras, position and distance). Positive test results will allow for the use of the robot in organic crops requiring mechanical removal of weeds or in crops with application of selective liquid agrochemicals limited to the minimum. Until control systems can be perfected and development cost recouped, the growth in autonomous field production systems will come in fits to starts. So that very important is mentioned in this paper, inertial navigation system.

Keywords Agriculture robot · Care of plants · Autonomous work · Inertial navigation

1 Introduction

There are a few constructions of autonomous agricultural robots. The best known of them in series production are Vibro Crop Robotti produced by Kongskilde and ANATIS produced by Carre.

M. Jasiński (✉) · J. Mączak · S. Radkowski
Institute of Vehicles, Warsaw University of Technology, Warsaw, Poland
e-mail: jachuu@simr.pw.edu.pl

S. Korczak
Institute of Machine Design Fundamentals, Warsaw, Poland
e-mail: sebastian.korczak@simr.pw.edu.pl

R. Rogacki · J. Mac · J. Szczepaniak
Industrial Institute of Agricultural Engineering in Poznan, Poznan, Poland
e-mail: rogacki@pimr.poznan.pl

R. Szewczyk et al. (eds.), *Challenges in Automation, Robotics and Measurement Techniques*, Advances in Intelligent Systems and Computing 440, DOI 10.1007/978-3-319-29357-8_58

In 2013 Kongskilde presented the Vibro Crop Robotti a new autonomous agricultural platform for automated and semi-automated task performance in plant production. The Kongskilde Robotti can be equipped with implements for precision seeding, ridging discs and mechanical row crop cleaning units [1] (Fig. 1).

In 2013 Carre presented the agricultural robot ANATIS that assists farmers in their daily work by automatically hoeing between and within crop rows grown in beds, the autonomous vehicle also monitors and collects information on the number and types of weeds, crop density, light levels, soil moisture content, and even the temperature of the air and soil. ANATIS is powered by quiet-running electric motors [2] (Fig. 2).

Syndicate of Industrial Institute of Agricultural Engineering in Poznan, with the Institute of Vehicles of Warsaw University of Technology and PROMAR company from Poznan was started a design of autonomous farm robot for sowing and cultivation of wide row planting.

The aim of the project is to develop the structure and operation procedures of an autonomous robot for sowing and wide row planting and conduct laboratory and exploitation tests on an experimental model. Autonomous work of the robot (traction and agronomic processes) will be implemented on the basis of data from different kinds of sensors (cameras, position, distance and others). Positive test results will allow utilisation of the robot in organic crops requiring mechanical removal of weeds or in crops with application of selective liquid agrochemicals limited to the minimum. The use of the vision system, supported by the maps and known coordinates of the sown seeds, will allow for their care on an early stage of plant development. The applicability of the robot to onerous work in organic farming may encourage farmers to discontinue the use of herbicides in crops include sugar beet, corn, etc.

Fig. 1 The autonomous agricultural platform Kongskilde Vibro Crop Robotti [1]

Fig. 2 The agricultural robot ANATIS [2]

In recent years most of the initial work done on control architectures of mobile robots was carried out in the aerospace and artificial intelligence research laboratories to accomplish military missions and space explorations. Unlike industrial robots, where the environment is controlled and structured, the working environment of robots is relatively unstructured, unpredictable and dynamic. An intelligent, robust and fault tolerant control architecture [3] is essential to ensure safe and desired operation of the robot.

A behaviour based (BB) [4] control approach provides an autonomous mobile robot the intelligence to handle complex world problems using simple behaviours. Complex behaviours of a robot emerge from simple behaviours. In this context behaviour is being defined as a response to a stimulus [5]. BB control structure can be either reactive or deliberative in nature. Reactive behaviours are part of reactive control architectures where the behaviour responds to stimuli and develops control actions. Deliberative behaviours on the other hand are pre-defined control steps which are executed to accomplish a given task. Associating these behaviours to actual actions of an agricultural robot is crucial to understand the capabilities of a robot. The importance of decomposition of agricultural tasks into robotic behaviours was illustrated by Blackmore [6]. For the robot to tackle unknown environments and attain assigned goals both reactive and deliberative behaviours are important [7] and thus a robust fault tolerant intelligence is achievable with a combination of reactive and deliberative behaviours.

The implements mounted on the machine to perform a particular operation (spraying, weeding, etc.) are conventionally constructed and safeguarded for the use of a manned tractor. The implement manufacturer did not foresee that an automatic machine without a driver would operate and control the implement.

According to this, achieving the objectives of the project will require the achievement of the following specific objectives of the practice [8, 9]:

- development of the robot platform,
- selection of precision drills and development of methods of positioning sown seeds,
- development of tools to allow the processing of weeding both the surface and the surface spacing between crops in a row,
- development of tools to selectively spray the surface and the surface of the soil,
- development of the control system and control algorithms autonomous robot field in terms of traction and agronomic processes.

The autonomous farm robot should be able to work in following working conditions:

- terrain: empowered field, field roads, mud, sand, grassy ground, rocky ground or other hardened,
- work in the open 24 h/day,
- work in areas with varying degrees of lighting and visibility,
- temperature: 5–40 °C,
- weather: average rainfall, moderate wind, fog,
- typical obstacles in the open area.

Projected robot will enable complex care of field crops including: red beet, sugar beet, sweet corn, cabbage, lettuce, forest nurseries, orchard, production of vegetables and ornamental plants. Additionally it should enable the mechanical destruction of weeds and, if necessary, precise application of crop protection formulations and fertilizers.

Robot constructed by us will be have a smart weeders, equipped in LPS system, which uses digital image analysis for steering working tools, for mechanical weed control.

2 The Construction of the Robot

The concept of the autonomous agricultural robot is based on the four wheel drive platform powered and steered by the hydraulic system unit based on a Diesel engine. The robot will have a modular design (Fig. 3):

- The carrier will be four wheeled vehicle, able to work remotely and autonomous. The drive wheels provide hydraulic motors and planetary gears. This way of transfer of power will get the high ground clearance of the machine (free space for tools) and independent suspension—important when working in difficult terrain. In addition, the speed differentiation of individual wheels and their appropriate setting will result in high manoeuvrability.

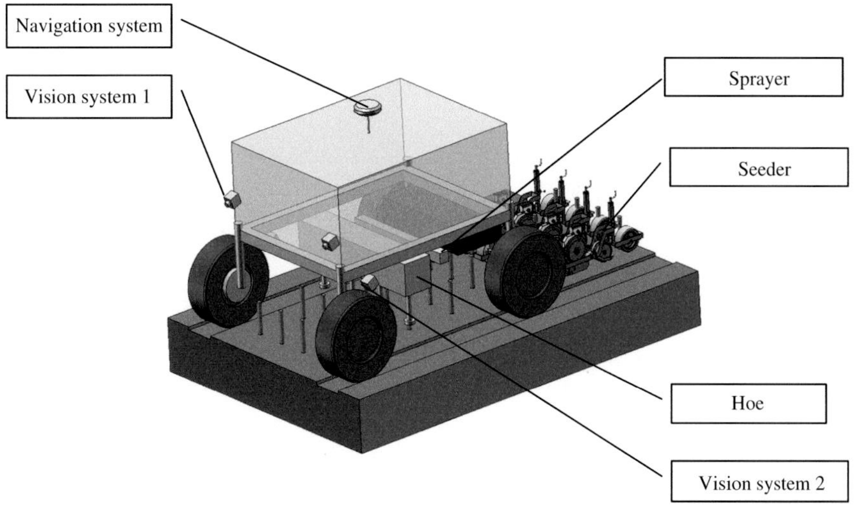

Fig. 3 Model of the autonomous robot [10]

- For precision seeding seeders will be applied with the seed position recording system in the field.
- During the sowing or weeding sprayer will be used to distribute liquid fertilizer or other liquid preparations for plant protection. The application will be executed surfactants (for plants or soil), or to the soil (of the depth below the surface of soil) in a continuous or selective based on the optical identification and location of objects.
- The main working unit will be the active hoe for precise control of weeds in both the inter-rows and row crops.

3 Conception of Inertial Navigation System

Main sensor system is based on a specialized GPS receiver providing position information with an accuracy of less than 100 mm. This system will be used to: control speed of the robot, guidance and maintenance robot on the designated path, precision seeding—the exact information on where sowing the seeds will be used to build maps of seeds, which will be used as supporting information for precision weeding, and to control the position of and operation of key components. For tests a typical GPS (10 Hz) will be used.

The front camera view will be used to increase positioning accuracy of the robot. It will allow corrections of the robot path regarding the rows of plants. The vision system will also be used for detection of non-moving objects. Simultaneously second vision unit will also be used for acquiring camera images immediately before active hoes and sprayer. Additionally information from the acceleration sensors and encoders built-in wheels will be used in navigation purposes. To determine the angular acceleration the IMU (Inertial Measurement Unit) will be required. This will enable a:

- trajectory correction of the robot,
- precise work of active hoe,
- position adjustment and precise dosing of liquid fertilizer plant health products.

The exact position of the robot is obtained from the fusion of signals from precision GPS (in the test version standard GPS receiver Ublox NEO7) and integrated system of inertial and magnetic sensors VN-100 [11]. VN-100 is a complete AHRS (attitude heading reference system) system integrating measurements from three axial sensors: acceleration, angular velocity and Earth magnetic field. All of the sensors have temperature compensated sensitivity and common values. Moreover, all the skewness of axes were calibrated. VN-100 device provides accurate information from all the sensors and estimates of the spatial orientation angles, DCM (direct cosine matrix) transformation matrix, and estimated values of linear accelerations and angular velocities in the absolute coordinates (NED—north/east/down) independent from the angular orientation of the sensor. It is planned that two AHRS systems will be placed on one machine: first related to the vehicle and the second with a connection to the tools (e.g. seeder). AHRS integrated with the vehicle will provide the detailed momentary angular orientation and acceleration of the body while the second set allows better estimation of the momentary position of the tool thus allowing precise localisation of the seeds. Detailed information about their positions, stored in the internal database, will be used in further fieldwork related to the care of plants.

In conclusion it is planned to determine the precise orientation of two coordinate systems, one associated with the vehicle, the second associated with the working equipment. Information about the absolute position of the robot will be acquired from the GPS system. This information will supplement the measurements from two separate AHRS systems. Fusion of the information will be based on the Kalman filter [12], and specifically on its extended version (Unscented Kalman Filter UKF [12]). One of the expected problems can be significant distortion of the magnetic field caused by the presence of the ferromagnetic structure of the robot body. Such distortions can significantly interfere with the AHRS systems and include additional errors to the estimated values. In case of detection of such problems first step towards its solution will be to use additional magnetic field sensors placed at an appropriate distance from the structure. The experience of the authors show that such solutions are often sufficient. As an additional sensor Honeywell HMC5983 sensor [13] could be used exhibiting good thermal stability (analysed in papers [14, 15]). In the case of the ineffectiveness of this solution it is planned to skip the measurement of magnetic sensors AHRS algorithm VN-100 (this mode is available). You'll lose the ability to estimate the Yaw angle informing about the course the robot directly from AHRS. This information will be able to be estimated from the following GPS measurements. Another expected problem may be caused by the vibration of the machine. VN-100 system is based on the sensors created with MEMS technology. These sensors are sensitive to the vibrations. In extreme situations vibration may lead to significant distortion of the measurements and estimation errors. This problem was analysed in [16, 17]. The way to reduce the impact of vibration on the sensor is the use of a suitable sensor system vibration damping suspension. It appears that such a solution at relatively low dynamics of the robot will be sufficient. In terms of improving the estimation of the position of the robot seems to be an interesting opportunity to take measurements of the angle of rotation of the wheels and the data fusion algorithm. However, due to a large hilly area and the possibility of uncontrolled slip the opportunity to take that information should be examined at the stage of prototype testing. The inclusion of information from encoders and combining it with measurements of AHRS system and a simple GPS system may have to provide enough accurate information about the position of the robot, that there would be no need to equip the robot in expensive precise GPS system.

During the preliminary phase of the project Authors are planning to test possibility of usage of several low cost sensors for moving objects detection:

360° Laser Scanner Development Kit (RPLIDAR), RoboPeak [18]. RPLIDAR is a low cost 360° 2D laser scanner (LIDAR) solution developed by RoboPeak (Fig. 4). The system can perform 360° scan within 6 m range. The produced 2D point cloud data can be used in mapping, localization and object/environment modelling.

- RPLIDAR's scanning frequency reached 5.5 Hz when sampling 360 points each round. And it can be configured up to 10 Hz maximum.
- RPLIDAR is basically a laser triangulation measurement system. It can work excellent in all kinds of indoor environment and outdoor environment without sunlight.

HC-SR04 (Cytron Technologies) ultrasonic distance sensor with range of 5 m. The HC-SR04 ultrasonic sensor uses sonar to determine distance to an object like bats or dolphins do. It offers excellent range accuracy and stable readings in an easy-to-use package. Its operation is not affected by sunlight or black material.

Fig. 4 360° Laser Scanner Development Kit (RPLIDAR), RoboPeak [16]

IR Camera [19]. The Wii Remote includes a 128 × 96 monochrome camera with built-in image processing. The camera looks through an infrared pass filter in the remote's plastic casing. The camera's built-in image processing is capable of tracking up to 4 moving objects, and these data are the only data available to the host. Raw pixel data is not available to the host, so the camera cannot be used to take a conventional picture. The built-in processor uses 8× subpixel analysis to provide 1024 × 768 resolution for the tracked points. Wii Remote has a ranging 0.3 m and effective field of view of about 33° horizontally and 23° vertically.

4 Mobile Platform

Low cost autonomous mobile platform for testing the inertial navigation system is under constructions now. This mobile platform is based on four driving modules connected with two main chassis plates (Fig. 5). It will be equipped with:

- two lithium polymer batteries,
- dual high-power DC motor controller,
- myRIO controller (RT system part and FPGA board for controlling peripherals).

Driving module is based on a high torque standard size analogue servo motor with 1:1 gear train (Fig. 6). Construction parts, gears and wheels are prototyped from PLA material (Fig. 7) using additive methods (RepRap type 3D printer). The main driving motor consists of a high-power 6 V brushed DC motor combined with a 75:1 planetary gearbox. An integrated quadrature encoder on the motor shaft provides 48 counts per revolution. A wheel with 120 mm diameter will have rotation velocity of 130 rpm.

Fig. 5 Full model render of the navigation testing platform

Fig. 6 Steering module—3D model render

Fig. 7 Steering module—picture (*upside down*)

5 Summary

Currently, several autonomous vehicles development programs are underway targeting various agricultural production sectors. Until control systems can be perfected and development cost recouped, the growth in autonomous field production systems will come in fits to starts. So that very important is mentioned in this paper, projects initial assumptions.

Mobile platform for testing the inertial navigation system is under construction now.

In the next stages of the project design tasks are planned: selecting the target concept robot based on numerical analysis and developing the concept of the control system and autonomous robot control algorithms.

Work financed from NCBiR years 2015-2018 funds.

References

1. http://www.kongskilde.com/ro/en/News/Year%202013/09-09-2013%20-%20New%20automated%20agricultural%20platform%20-%20Kongskilde%20Vibro%20Crop%20Robotti. Accessed 16 Nov 2015
2. http://www.trp.uk.com/carre-farm-machinery/carre-meadow-maintenance/anatis.html. Accessed 16 Nov 2015
3. Radkowski, S., Jasiński, M., Gumiński, R., Szulim, P.: Use of fault tolerant control systems in agriculture machinery. Mechanization in Agriculture. Year LXI, Issue 7/2015, pp. 3–6
4. Shearer, S.A., Pitla, S.K., Luck, J.D.: Trends in the automation of agricultural field machinery
5. http://www.clubofbologna.org/ew/documents/KNR_Sherear.pdf. Accessed 21 June 2015
6. Arkin, R.C.: Behaviour-based robotics (1986)
7. Blackmore, B.S., Fountas, S., Vougioukas, S., Tang, L., Sorensen, C.G., Jorgensen, R.: A method to define agricultural robot behaviours. In: Mechatronics and Robotics Conference (MECHROB), pp. 1197–1200 (2004)
8. Jasiński, M., Mączak, J., Radkowski, S., Rogacki, R., Mac, J., Szczepaniak, J., Pawłowski, T: Conception of autonomous robot for sowing and wide row planting. Mechanization in Agriculture. Year LXI, Issue 2/2015, pp. 21–24
9. Jasiński, M., Mączak, J., Radkowski, S., Rogacki, R., Mac, J., Szczepaniak, J., Pawłowski, T.: Autonomous agricultural robot: initial assumptions of project. Zeszyty Naukowe IP **2**(102), 49–56 (2015)
10. Rogacki, R., Szczepaniak, J., Mac, J., Grzechowiak, R.: Functional and structural assumptions of autonomous farm robot platform. In: Proceedings of XIV International Technical Systems Degradation Conference, Liptovský Mikuláš, Slovak Republic, April 8–11, pp. 179–180 (2015)
11. VN-100 Rugged—VectorNav Technologies. http://www.vectornav.com/products/vn100-rugged. Accessed 29 Oct 2015
12. Kalman, R.E.: A new approach to linear filtering and prediction problems. J. Basic Eng. **82**, 35–45 (1960)
13. Zhao, L., Xia, L., Liu, F., Cheng, Y.: Application of UKF for MEMS IMUs and fluxgate sensors based attitude and heading reference system of carriers. In: 2nd IEEE Conference on Industrial Electronics and Applications 2007. ICIEA 2007, pp. 2278–2283 (2007). 'HMC5983_3_Axis_Compass_IC.pdf' [Online]. https://aerospace.honeywell.com/~/media/Images/Plymouth%20Website%20PDFs/Magnetic%20Sensors/Data%20Sheets/HMC5983_3_Ax6is_Compass_IC.ashx. Accessed 18 Nov 2015
14. Gontarz, S., Mączak, J., Szulim, P.: Online monitoring of steel constructions using passive methods. In: Tse, P.W., Mathew, J., Wong, K., Lam, R., Ko, C.N. (eds.) Engineering Asset Management; Systems, Professional Practices and Certification, pp. 625–635. Springer International Publishing (2015)
15. Szulim, P., Gontarz, S.: Calibration of magnetic field sensors used for diagnosis of steel construction. J. Electr. Eng. **66**(6) (2015) (in print)

16. Radkowski, S., Szulim, P.: Analysis of vibration of rotors in unmanned aircraft. In: Awrejcewicz, J., Szewczyk, R., Trojnacki, M., Kaliczynska, M. (eds.) Mechatronics: ideas for industrial applications, vol. 317, pp. 363–371. Springer, Berlin (2015)
17. Radkowski, S., Szulim, P.: Analysis of vibration of rotors in unmanned aircraft. In: 19th International Conference on Methods and Models in Automation and Robotics. MMAR 2014, pp. 748–753 (2014)
18. http://www.robopeak.com. Accessed 29 Oct 2015
19. http://wiibrew.org/wiki/Wiimote#IR_Camera. Accessed 29 Oct 2015

Fast Grasp Learning for Novel Objects

Dawid Seredyński and Wojciech Szynkiewicz

Abstract This paper presents a method for fast learning of dexterous grasps for unknown objects. We use two probabilistic models of each grasp type learned from a single demonstrated grasp example to generate many grasp candidates for new objects with unknown shapes. These models encode probability density functions representing relationship between fingers and object local features, and whole hand configuration that is particular to a grasp example, respectively. Both, in the training and in the grasp generation stage we use an incomplete 3D point cloud from a depth sensor. The results of simulation experiments performed with the BarrettHand gripper and several objects of different shapes indicate that the proposed learning approach is applicable in realistic scenarios.

Keywords Grasp learning · Probabilistic models · Kernel density estimation

1 Introduction

Automatic robotic grasping of unknown objects is a complex process and still remains an open research problem. Grasp synthesis for novel objects deals with the problem of computing candidate grasps without an object shape model given a priori. In this paper, we present a method for fast learning of grasp models from a single demonstrated grasp example. Two separate models are learned, the contact model to describe the relationship between fingers and object local features, and the hand configuration model to represent whole hand configuration during approach to grasp. These models are represented by probability distribution functions and used to generate a great number of possible grasp candidates for new objects given an object partial point cloud obtained from a 3D sensor. In this work, we follow an original approach presented in [1] to address grasp synthesis in two steps.

D. Seredyński (✉) · W. Szynkiewicz
Warsaw University of Technology, Warsaw, Poland
e-mail: dawid.seredynski@gmail.com
URL: http://robotics.ia.pw.edu.pl

R. Szewczyk et al. (eds.), *Challenges in Automation, Robotics and Measurement Techniques*, Advances in Intelligent Systems and Computing 440, DOI 10.1007/978-3-319-29357-8_59

However, we propose some important modifications and improvements to the original method of Kopicki et al. [1]. In our approach, the probability density functions are calculated in the neighborhood of the contact points, thus the error caused by the dependence between position and orientation is less significant. We also use different kernel function for modeling the orientation in the $SO(3)$ space, to achieve simpler sampling method. Finally, we modified the specification of some equation for clarity. We verify the performance of the proposed method to plan feasible grasps for the BarrettHand and a set of test objects with simple and more complex shapes.

2 Related Work

There is a quite large body of work dealing with grasp planning for robots [2–4] and grasp contact point estimation for various tasks [5]. Grasp synthesis methods for novel objects do not assume availability of object models or other form of grasp experience. They focus on identifying some local or global features of the object in sensory data for generating and selecting grasp candidates. Most of the learning-based approaches use data-driven model to learn mappings between object shape and feasible hand configurations [1, 6–9].

Pelossof et al. [9] use a support vector machine (SVM) to learn the grasp quality manifold for a specific hand and simple object shapes. The manifold represents the mapping from grasp parameters and object shape to the grasp quality and new optimal grasps are found through interpolation on the manifold.

Saxena et al. [10] propose a system that infers a point at where to grasp an object directly as a function of its image. They apply a simple logistic regressor trained on large amounts of synthetic, labeled training data to predict good grasping points in a monocular image.

In [6] authors propose a method that allows a robot to learn to formulate grasp plans from visual data obtained from a 3D sensor. The proposed method relies on the identification of prototypical parts by which objects are often grasped. When a novel object appears, the robot tries to fit the prototypical parts to a 3D image provided by a depth sensor that partially captures the object.

The approaches presented in [7] and [8] maintain a database of grasp examples and combine learning by trial and error on real world data with a part-based representation of the object. Herzog et al. [7] store a set of local templates of the parts of the object that have been in contact with the object during the human demonstration. If a local part of an object is similar to a template in the database, then the associated grasp hypothesis is performed. In [8] an object part is represented as a set of points weighted according to an isotropic 3-D Gaussian probability distribution with a given standard deviation. When a new object is shown, the goal of the method is to find the subpart that is most likely to afford the demonstrated action. In approach proposed in [1] the robot learns a mapping from simple local 3D shape features to a complete parametrisation of the robot hand pose and its fingers. It uses two separate models, i.e., the contact model to express the relationship between finger phalanges and object local features, and the hand configuration model to represent whole hand configuration during approach to grasp pose.

Approaches proposed in [11, 12] assume that a complete 3D shape model of the object can be created from sensory data. Hillenbrand and Roa [12] formulate the problem of transferring functional grasps between objects of the same category as pose alignment and shape warping. They assume that there is a source object given on which a set of functional grasps is defined. The transfer is achieved through warping the surface geometry of the source object onto the target object, and along with it the contact points of a grasp. Ben Amor et al. [11] propose an imitation learning approach for learning and generalizing grasping skills based on human demonstrations. They use the warping method to transfer grasps taught by a human hand to contact points for a robot hand on a new object.

3 Problem Formulation

Experiments presented in [1] showed the performance of the method for the DLR Hand-Hit II, the anthropomorphic 12-DOF gripper. The authors proposed the kernel density estimation method for estimating various probability density functions (PDFs). The whole procedure of learning and generating grasps for novel objects consists of four stages [1]:

- grasp training on a small set of example grasps: the Contact Model (M_i) for each link and the Hand Configuration Model (C) for whole gripper are generated,
- estimation of PDF of gripper links' poses for a novel object (the Query Density Q_i),
- sampling the Query Density Q_i and the Hand Configuration Model C: a set of grasps is generated,
- grasps optimization and selection

The learning stage requires a small number of example grasps (Fig. 1) and even one example of each grasp type is sufficient. In this section we introduce some of the concepts from [1] and we propose a number of modifications. The following symbols are used for clarity:

1. ${}^{\mathcal{A}}_{\mathcal{B}}p$—translation vector of the origin of a frame $\mathcal{B}$ expressed in the reference frame $\mathcal{A}$,
2. ${}^{\mathcal{A}}_{\mathcal{B}}q$—orientation of a frame $\mathcal{B}$ with respect to (w.r.t.) the reference frame $\mathcal{A}$.
3. ${}^{\mathcal{A}}P$—position of the point P expressed in the reference frame $\mathcal{A}$,

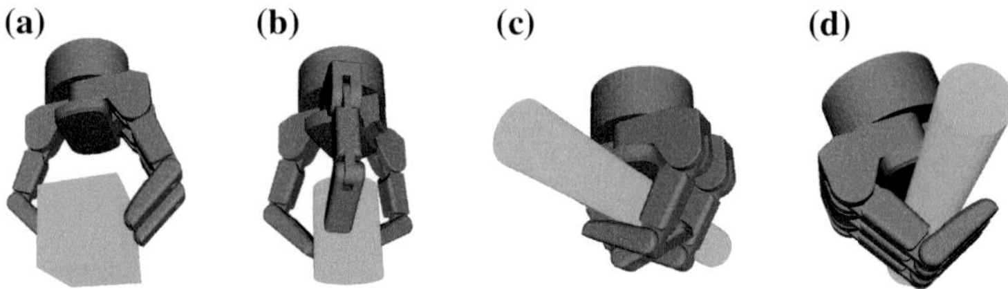

Fig. 1 The grasp examples used for learning: **a** pinch grasp, **b** spherical grasp, **c** power grasp, **d** hook grasp

3.1 Object Model

Object Model is a PDF of object's surface. It is used to estimate the shape and surface features of an object. Given a point cloud of an object with K_O points, local principal curvatures $r_j \in \mathbb{R}^2$ near the jth point and the reference frame $\mathcal{F}_j$ of the surface feature at the jth point, the PDF is estimated as:

$$O\left({}^{\mathcal{O}}_{\mathcal{F}}p, {}^{\mathcal{O}}_{\mathcal{F}}q, r\right) \simeq \sum_{j=1}^{K_O} w_j \mathcal{N}_3\left({}^{\mathcal{O}}_{\mathcal{F}}p | {}^{\mathcal{O}}_{\mathcal{F}}p_j, \sigma_p\right) \Theta\left({}^{\mathcal{O}}_{\mathcal{F}}q | {}^{\mathcal{O}}_{\mathcal{F}}q_j, \sigma_q\right) \mathcal{N}_2\left(r|r_j, \sigma_r\right), \quad (1)$$

where $\mathcal{N}_n$ is an n-variate isotropic Gaussian kernel, Θ is a kernel function for orientation in $SO(3)$, $w_j = 1/K_O$, $\mathcal{O}$ is the object frame, and σ_p, σ_q and σ_r are the kernel bandwidths. The Object Model is used in the grasp training stage to model the example object and in the grasp transfer stages to model a novel object.

3.2 Contact Model

Contact Model is a PDF that estimates the probability of finding the ith gripper link in a pose $\mathcal{L}_i$ w.r.t. the frame $\mathcal{F}_j$ of the jth local surface feature with the given principal curvatures r_j. The Contact Model is computed for each link i of the gripper for each example grasp. The PDF is estimated as:

$$M_i\left({}^{\mathcal{F}}_{\mathcal{L}}p, {}^{\mathcal{F}}_{\mathcal{L}}q, r\right) \simeq \frac{1}{Z} \sum_{j=1}^{K_O} w_{ij} \mathcal{N}_3\left({}^{\mathcal{F}}_{\mathcal{L}}p | {}^{\mathcal{F}}_{\mathcal{L}}p_{ij}, \sigma_p\right) \Theta\left({}^{\mathcal{F}}_{\mathcal{L}}q | {}^{\mathcal{F}}_{\mathcal{L}}q_{ij}, \sigma_q\right) \mathcal{N}_2\left(r|r_j, \sigma_j\right), \quad (2)$$

where Z is the normalizing constant and w_{ij} is the weighting function:

$$w_{ij} = \begin{cases} \exp\left(-\lambda d_{ij}^2\right), & \text{if } d_{ij} < \delta_i, \\ 0, & \text{otherwise,} \end{cases} \quad (3)$$

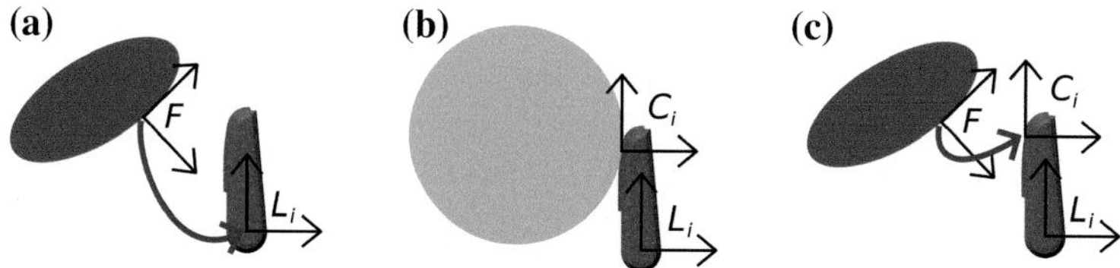

Fig. 2 The visualization of relative pose computation between a surface feature and a link of the gripper. In the original work [1] the pose F_i of the link i relative to the feature frame F is used in the PDFs of Contact Model and Query Density (**a**). We propose that an additional frame C_i, near the contact area, should be introduced during the grasp learning (**b**) for better approximation of the joint distribution of position and orientation (**c**)

where $\lambda \in \mathbb{R}^+$ is a parameter, d_{ij} is the distance from the jth point of the object's point cloud to the ith gripper link, and δ_i is the threshold distance. The weighting function w_{ij} causes that only the closest surface features to the gripper link are taken into account. The closer is the feature, the bigger is the weight. The transformation $({}^{\mathcal{F}}_{\mathcal{L}}p_{ij}, {}^{\mathcal{F}}_{\mathcal{L}}q_{ij})$ is shown in Fig. 2a. The conditional probability $M_i\left({}^{\mathcal{F}}_{\mathcal{L}}p, {}^{\mathcal{F}}_{\mathcal{L}}q|r\right)$ and marginal probability $M_i(r)$ can easily be computed [1].

3.3 Query Density

Given the Contact Model M_i for the ith link for a specified grasp example and the Object Model O of a novel object, the Query Density PDF Q_i can be estimated:

$$Q_i({}^{\mathcal{O}}_{\mathcal{L}}p, {}^{\mathcal{O}}_{\mathcal{L}}q) \simeq \sum_{j=1}^{K_{Q_i}} w_{ij} \mathcal{N}_3\left({}^{\mathcal{O}}_{\mathcal{L}}p|\, {}^{\mathcal{O}}_{\mathcal{L}}p_{Q,ij}\right) \Theta_3\left({}^{\mathcal{O}}_{\mathcal{L}}q|\, {}^{\mathcal{O}}_{\mathcal{L}}q_{Q,ij}\right), \tag{4}$$

where K_{Q_i} is the number of kernel centers, $\left({}^{\mathcal{O}}_{\mathcal{L}}p_{Q,ij}, {}^{\mathcal{O}}_{\mathcal{L}}q_{Q,ij}\right)$ is the jth kernel center and weights w_{ij} are normalized $\sum_j w_{ij} = 1$. The kernel centers are generated by sampling the object model $O\left({}^{\mathcal{O}}_{\mathcal{F}}p, {}^{\mathcal{O}}_{\mathcal{F}}q, r\right)$ and sampling the Contact Model M_i with conditional probability for r: $M_i\left({}^{\mathcal{F}}_{\mathcal{L}}p, {}^{\mathcal{F}}_{\mathcal{L}}q|r\right)$. For each pair of sampled poses $\left({}^{\mathcal{O}}_{\mathcal{F}}p, {}^{\mathcal{O}}_{\mathcal{F}}q\right)$ and $\left({}^{\mathcal{F}}_{\mathcal{L}}p, {}^{\mathcal{F}}_{\mathcal{L}}q\right)$ the kernel center pose $\left({}^{\mathcal{O}}_{\mathcal{L}}p, {}^{\mathcal{O}}_{\mathcal{L}}q\right)$ is calculated with the weight $w_{ij} = M_i(r)$. After the K_{Q_i} kernel centers are generated, the weights w_{ij} are normalized.

3.4 Modifications

We propose some modifications to the original method of [1]. First of all, the estimation of Query Density (Eq. 7) assumes the independence of the position and orientation, and calculates the joint distribution of the pose as the product of two distributions. Such an assumption can be treated as an approximation, but this is generally incorrect. The pose $\mathcal{L}_i$ of the link depends on the surface feature pose $\mathcal{F}_j$ (Sect. 3.3) sampled form the Object Model. In the most cases the distance between the surface feature and the origin of the link frame is large enough, that even small changes in orientation result in relatively large changes in the position of the link w.r.t. the contact point (Fig. 2a). As the sampled pose of the link is dependent on the surface features near the contact point, the translation and the orientation cannot be sampled independently for a given surface feature. We propose a better approximation, that takes into account the position of the contact point (Fig. 2b). During the grasp learning procedure, the mean position of the contact points near the ith link is calculated for each Contact Model:

$$ {}^{\mathcal{L}}_{C}p_i = \sum_{j=1}^{K_O} w_{ij} \, {}^{\mathcal{L}}P_{ij} / \sum_{j=1}^{K_O} w_{ij}, \tag{5} $$

where ${}^{\mathcal{L}}P_{ij}$ is the position of the jth closest point on the object's surface expressed in the frame $\mathcal{L}_i$, and w_{ij} is the weight of the jth contact point. The orientation of the frame C is the same as the orientation of the ith link frame ${}^{\mathcal{L}}_{C}q_i = {}^{\mathcal{L}}_{\mathcal{L}}q_i$. The proposed PDF for the Contact Model is estimated as

$$ M'_i \left({}^{\mathcal{F}}_{C}p, {}^{\mathcal{F}}_{C}q, r\right) \simeq \frac{1}{Z} \sum_{j=1}^{K_O} w_{ij} \mathcal{N}_3 \left({}^{\mathcal{F}}_{C}p \,|\, {}^{\mathcal{F}}_{C}p_{ij}, \sigma_p\right) \Theta \left({}^{\mathcal{F}}_{C}q \,|\, {}^{\mathcal{F}}_{C}q_{ij}, \sigma_q\right) \mathcal{N}_2 \left(r | r_j, \sigma_j\right) \tag{6} $$

The difference between M_i and M'_i is that M'_i estimates the probability of finding the ith contact point in the pose C_i w.r.t. the frame $\mathcal{F}_j$ of the jth local surface feature (Fig. 2c). The ith link pose $\mathcal{L}_i$ can be calculated by combining transformations $\left({}^{\mathcal{F}}_{C}p, {}^{\mathcal{F}}_{C}q\right)$ and $\left({}^{\mathcal{L}}_{C}p_i, {}^{\mathcal{L}}_{C}q_i\right)^{-1}$.

The Query Density should also be changed to take into account the contact point frame C_i:

$$ Q'_i({}^{\mathcal{O}}_{C}p, {}^{\mathcal{O}}_{C}q) \simeq \sum_{j=1}^{K_{Q_i}} w_{ij} \mathcal{N}_3 \left({}^{\mathcal{O}}_{C}p \,|\, {}^{\mathcal{O}}_{C}p_{Q,ij}\right) \Theta_3 \left({}^{\mathcal{O}}_{C}q \,|\, {}^{\mathcal{O}}_{C}q_{Q,ij}\right), \tag{7} $$

where $\left({}^{\mathcal{O}}_{C}p_{Q,ij}, {}^{\mathcal{O}}_{C}q_{Q,ij}\right)$ is the jth kernel center for the pose of the contact point. The kernel centers are generated by sampling the object model $O\left({}^{\mathcal{O}}_{\mathcal{F}}p, {}^{\mathcal{O}}_{\mathcal{F}}q, r\right)$ and sampling the Contact Model M'_i with conditional probability for r: $M'_i\left({}^{\mathcal{F}}_{C}p, {}^{\mathcal{F}}_{C}q|r\right)$. For each pair of sampled poses $\left({}^{\mathcal{O}}_{\mathcal{F}}p, {}^{\mathcal{O}}_{\mathcal{F}}q\right)$ and $\left({}^{\mathcal{F}}_{C}p, {}^{\mathcal{F}}_{C}q\right)$, the kernel center pose $\left({}^{\mathcal{O}}_{C}p, {}^{\mathcal{O}}_{C}q\right)$ is calculated with the weight $w_{ij} = M_i(r)$. The result of sampling the Q'_i distribution is the transformation $\left({}^{\mathcal{O}}_{C}p, {}^{\mathcal{O}}_{C}q\right)$ from the object frame $\mathcal{O}$ to the contact point frame C_i for the ith link. The transformation from the object frame $\mathcal{O}$ to the ith link frame $\mathcal{L}_i$ can be calculated by combining the sampled transformation $\left({}^{\mathcal{O}}_{C}p, {}^{\mathcal{O}}_{C}q\right)$ and the transformation $\left({}^{\mathcal{L}}_{C}p_i, {}^{\mathcal{L}}_{C}q_i\right)^{-1}$ between the contact point frame C_i and the link frame

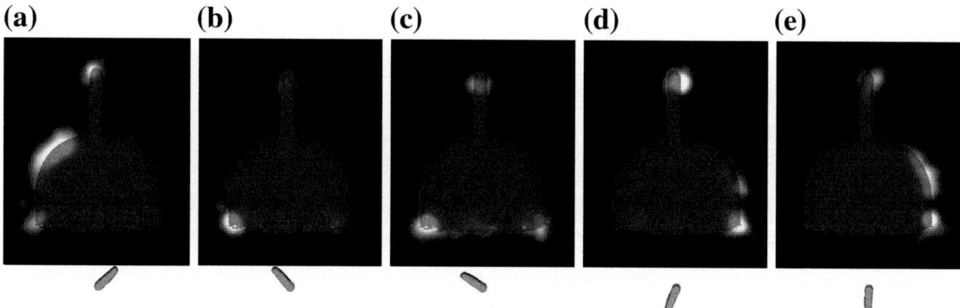

Fig. 3 The visualization of Query Density $Q'_i\left({}^{\mathcal{O}}_{C}p_i \mid {}^{\mathcal{O}}_{C}q_i\right)$ for the power grasp and for a kettle object (the *red* shape). *White* area denotes the points with the biggest value of the PDF. The orientation ${}^{\mathcal{O}}_{C}q_i = {}^{\mathcal{O}}_{\mathcal{L}}q_i$ is visualized below each function plot as a rotated *i*th link

$\mathcal{L}_i$. The visualization of the Query Density Q'_i calculated for the power grasp of the kettle object is shown in Fig. 3.

Kopicki et al. [1] proposed the von Mises–Fisher distribution as the kernel function in the $SO(3)$ space. Although there is analytical equation for the probability density function for such distribution, there is a problem with generating samples. The procedure of generating samples require rejection sampling method, which is slow for highly concentrated distributions. We propose a kind of Gaussian distribution for the axis-angle representation of the orientation, with the kernel function

$$\Theta(q|\mu_q, \sigma_q) = \mathcal{N}_1\left(A_{q\to a}(q^*\mu_q)|0, \sigma_q\right), \tag{8}$$

where q is the rotation unit quaternion, μ_q is the kernel mean quaternion, σ_q is the kernel bandwidth and $A_{q\to a}$ is the operator that extracts the angle of the axis-angle representation. This probability distribution can be easily sampled without the need of the rejection sampling method.

4 Experiments

All experiments were performed in the simulation environment. The experimental system was simulated with the DART (Dynamic Animation and Robotics Toolkit). The set of test objects included object models from the original work [1] (a kettle, a mug, containers) and some shape primitives such as a cuboid and cylinders.

The gripper used for tests was BarrettHand BH-280. It is a 4-DOF under actuated gripper with 8 joints coupled in pairs. Each of the three fingers has two flexion joints: the proximal joint and the coupled distal joint. Two fingers together have one additional DOF, i.e. the two coupled spread joints.

The grasp examples for the BarrettHand gripper used in the learning step are shown in Fig. 1. The Contact Models together with the Configuration Model were generated for each grasp example. They were used to generate the Query Densities for the Object Model of a novel object. Then, the Query Density was used to generate a set of grasps that were later optimized as described in [1]. The highest-ranked grasps were selected for the collision checking procedure. If the gripper in the initial configuration for the given grasp was in collision with the environment, the pose and configuration were randomly altered in some small range several times. Every altered configuration was checked against the environment collisions. Only the grasps with collision-free configuration were selected for the simulation (Table 1). Execution of each selected grasp was performed in dynamics simulation in two steps:

- perform the grasp: close the fingers on the object,
- lift-up: move the gripper against the gravity force.

The grasp was considered successful if the object could be lifted and held by the gripper above the specified height above the ground plane. The results of the grasping simulation are presented in Table 2. The Coulomb friction coefficient was set to relatively low value $\mu = 0.1$ during all experiments. This caused that only the most stable grasps were successful. The low friction force caused the change of some grasps types during the finger closure (Fig. 4) and (Table 3).

Table 1 Percentage of generated grasps qualified to the simulation verification

Object	Grasp model			
	Pinch	Spherical	Power	Hook
Container1	26.5	21.8	2.0	6.0
Container2	31.9	29.6	1.2	11.7
Cuboid	33.2	13.4	0.0	0.6
Cylinder (lying)	15.8	0.8	0.0	5.0
Cylinder (standing)	79.1	15.5	48.3	59.6
Jar	69.2	32.7	0.0	2.6
Kettle	16.8	34.8	25.0	28.4
Mug	49.4	46.1	0.0	1.4

Table 2 Percentage of successful simulated grasps

Object	Grasp model			
	Pinch	Spherical	Power	Hook
Container1	54.0	84.3	100.0	47.6
Container2	76.1	65.5	0.0	47.4
Cuboid	96.1	76.9	–	0.0
Cylinder (lying)	13.1	0.0	–	26.7
Cylinder (standing)	27.1	35.6	100.0	72.3
Jar	51.3	91.8	–	0.0
Kettle	41.1	44.2	100.0	84.5
Mug	53.4	83.8	–	0.0

Table 3 Examples of successful grasps generated for various objects and grasp models

Object	Grasp model			
	Pinch	Spherical	Power	Hook
Container1				
Container2			Unlikely	
Cuboid			Unlikely	Unlikely
Cylinder (lying)		Unlikely	Unlikely	

(continued)

Table 3 (continued)

Object	Grasp model			
	Pinch	Spherical	Power	Hook
Cylinder (standing)			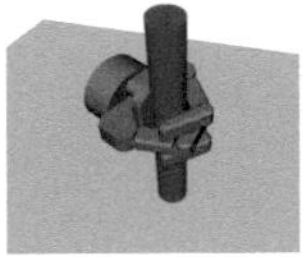	
Jar	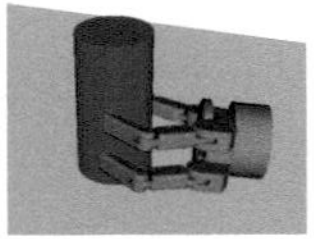	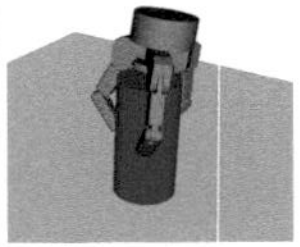	Unlikely	Unlikely
Kettle				
Mug	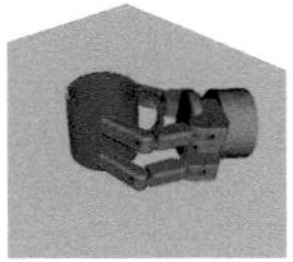	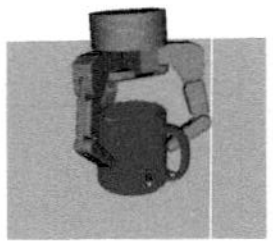	Unlikely	Unlikely

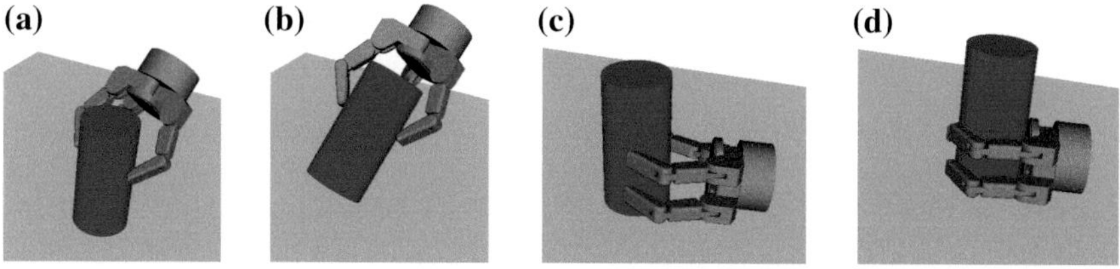

Fig. 4 The example cases when the grasp type changes during the manipulation due to low friction coefficient: **a**, **b** pinch grasp changes into spherical–like grasp and the orientation of the object changes, **c**, **d** pinch grasp changes into power–like grasp as the object slips towards the palm during the fingers closure

5 Conclusions

The method presented in this paper is a modification of [1]. We proposed a simple solution to the position and orientation dependency problem, a robust way to calculate and sample Gaussian distribution in the $SO(3)$ space, and a different specification of equations with designation of the reference frames for clarity. The application of the method was also presented. The experimental system and all models were implemented using DART toolkit for dynamics simulation and ROS for visualization and support of robot modeling tools, such as Unified Robot Description Format (URDF). The grasp learning and simulation results has been presented.

Acknowledgments This project was financially supported by the National Centre for Research and Development grant no. PBS1/A3/8/2012.

References

1. Kopicki, M., Detry, R., Adjigble, M., Stolkin, R., Leonardis, A., Wyatt, J.L.: One-shot learning and generation of dexterous grasps for novel objects. Int. J. Robot. Res. (2015) 0278364915594244
2. Bohg, J., Morales, A., Asfour, T., Kragic, D.: Data-driven grasp synthesis-a survey. IEEE Trans. Rob. **30**(2), 289–309 (2014)
3. Szynkiewicz, W.: Robot grasp synthesis under object pose uncertainty. J. Autom. Mobile Robot. Intell. Syst. **9**(1), 53–61 (2015)
4. Seredyński, D., Winiarski, T., Banachowicz, K., Zieliński, C.: Grasp planning taking into account the external wrenches acting on the grasped object. In: 10th International Workshop on Robot Motion and Control (RoMoCo), pp. 40–45 IEEE (2015)
5. Winiarski, T., Banachowicz, K., Seredyński, D.: Multi-sensory feedback control in door approaching and opening. In: Intelligent Systems'2014 of Advances in Intelligent Systems and Computing, vol. 323, pp. 57–70. Springer International Publishing (2015)
6. Detry, R., Ek, C., Madry, M., Kragic, D.: Learning a dictionary of prototypical grasp-predicting parts from grasping experience. In: 2013 IEEE International Conference on Robotics and Automation (ICRA), pp. 601–608 (2013)
7. Herzog, A., Pastor, P., Kalakrishnan, M., Righetti, L., Bohg, J., Asfour, T., Schaal, S.: Learning of grasp selection based on shape-templates. Auton. Robots **36**(1–2), 51–65 (2014)

8. Kroemer, O., Ugur, E., Oztop, E., Peters, J.: A kernel-based approach to direct action perception. In: 2012 IEEE International Conference on Robotics and Automation (ICRA), pp. 2605–2610 (2012)
9. Pelossof, R., Miller, A., Allen, P., Jebara, T.: An SVM learning approach to robotic grasping. In: Proceedings. ICRA '04. 2004 IEEE International Conference on Robotics and Automation, vol. 4, pp. 3512–3518 (2004)
10. Saxena, A., Driemeyer, J., Ng, A.Y.: Robotic grasping of novel objects using vision. Int. J. Robot. Res. **27**(2), 157–173 (2008)
11. Ben Amor, H., Kroemer, O., Hillenbrand, U., Neumann, G., Peters, J.: Generalization of human grasping for multi-fingered robot hands. In: 2012 IEEE/RSJ International Conference on Intelligent Robots and Systems (IROS), pp. 2043–2050 (2012)
12. Hillenbrand, U., Roa, M.: Transferring functional grasps through contact warping and local replanning. In: 2012 IEEE/RSJ International Conference on Intelligent Robots and Systems (IROS), pp. 2963–2970 (2012)

Prediction-Based Visual Servo Control

Michał Walęcki and Cezary Zieliński

Abstract In service robotics manipulator trajectories must be generated on the run, basing on the information gathered by sensors. This article discusses visual servoing applied to robot arm control, in a task of following a moving object with robot arm. The paper proposes a control system structure based on adaptive Kalman filter prediction algorithm and manipulator joint trajectory generator. Moreover, it shows how to build it using agent-based approach.

Keywords Robot object motion tracking · Robot visual servo control · Robot trajectory generation

1 Introduction

Service robots are being developed to assist human beings in their everyday activities. They are intended to operate in typical human environments, e.g.: homes, offices, hotels, hospitals. The control system of a service robot must be able to execute the diversity of tasks allotted to it by its user. As opposed to classic control approach applied to industrial robots, where robots follow preprogrammed trajectories, in the case of service robots their behaviors are based on reactions to events occurring in the environment. Those are hard to model or predict, so preprogrammed trajectories cannot be utilised in this case. The trajectories have to be generated on the run using information gathered by sensors. Manipulation is a primary human activity. It is based both on visual and tactile perception. As service robots must operate in the same environment and perform similar manipulation tasks to those that humans do, the robotics research community delves into visual servo control and force control of robots—sometimes combined, e.g. [1].

M. Walęcki (✉) · C. Zieliński
Warsaw University of Technology, Warsaw, Poland
e-mail: m.walecki@elka.pw.edu.pl
URL: http://robotics.ia.pw.edu.pl

R. Szewczyk et al. (eds.), *Challenges in Automation, Robotics and Measurement Techniques*, Advances in Intelligent Systems and Computing 440, DOI 10.1007/978-3-319-29357-8_60

In the original approach the visual tracking structures had no closed loop joint control and relied only on sensory feedback to drive the manipulator [2]. With the development of computer hardware and rising capabilities of image processing algorithms, object tracking based on motion model approach gained popularity. In this approach object motion prediction algorithm was separate from robot motion control system. Object tracking algorithms included, amongst others, various versions of Kalman filter [3, 4].

This article discusses visual servoing applied to control of a robot arm. It shows how to build a visual servo control system using agent-based approach [5], incorporating an extended Kalman filter for object tracking and prediction, and a joint trajectory generator for smooth robot arm motion. The object motion tracking algorithm is described using transition functions.

2 The Control Structure of a Visual Servo

The robot system that is considered here was initially specified as an embodied agent in [5]. An *embodied agent* is decomposed into five types of internal subsystems: real effectors E, real receptors R, virtual effectors e, virtual receptors r and a control subsystem c. The former two of the above subsystems form the agent's corporeal body, whereas the latter three form its control system. The subsystems communicate with each other through dedicated input and output communication buffers (the leading subscripts x and y distinguish them, while lack of such a subscript indicates internal memory).

The activities of virtual effectors e, virtual receptors r and the control subsystem c are represented by finite state machines (FSMs). With each node of the graph of the considered FSM a particular behaviour of the subsystem is associated. Each behaviour forms a pattern parametrized by two functions specific to it: its transition function and its terminal condition. Both of those take as input arguments the input buffers of the specific subsystem and its internal memory. The transition function produces the next contents of the output buffers and the internal memory. Each iteration of the subsystem behaviour consists of the computation of the transition function, the transmission of the results stored in the output buffers to the other subsystems and the input of the data produced by the other subsystems into the input buffers. The duration of a single such iteration determines the control cycle (sampling period) of the subsystem. The terminal condition is a predicate determining when the iteration of the activities of the behaviour should stop.

In the proposed embodied agent a_{rs} (where subscript rs stands for the robot system) (Fig. 1) a stand alone camera is represented by the real receptor $R_{\mathrm{rs,c}}$ (where subscript c stands for the camera). It is interfaced with the control subsystem by the virtual receptor $r_{\mathrm{rs,c}}$, which is responsible for image analysis. The real effector $E_{\mathrm{rs,m}}$ (where subscript m stands for the manipulator) represents a modified IRp6 manipulator with its control electronics, capable of position control in joint space.

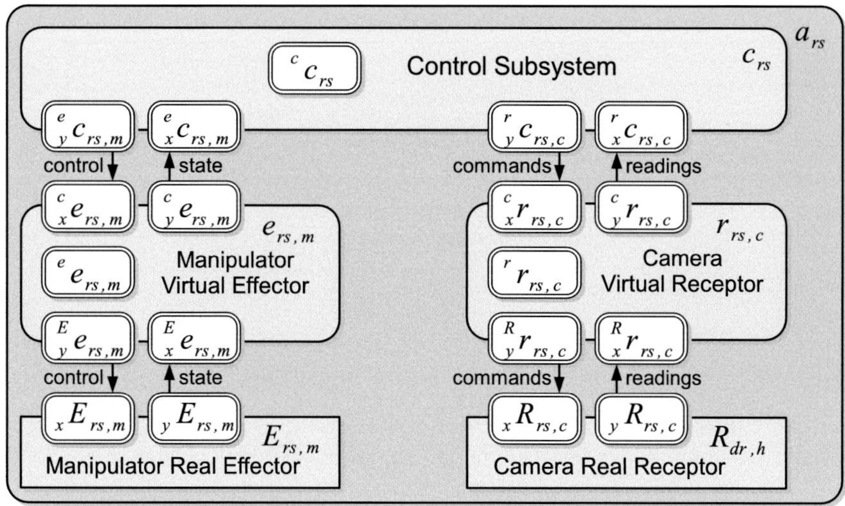

Fig. 1 Inner structure of an embodied agent representing the considered system

The virtual effector $e_{\mathrm{rs,m}}$ runs a joint-space trajectory generator, which provides the desired manipulator joint position in every control cycle.

3 Behaviors and Transition Functions

To facilitate subsequent implementation, a system modeling language [6] was used to specify the behaviors of the subsystems of the embodied agent a_{rs} [7].

3.1 *Control Subsystem*

The FSM, depicted in Fig. 2, governs the activities of the control subsystem c_{rs}. It invokes two behaviors. ${}^{c}\mathcal{B}_{\mathrm{rs},0}$ is an initial behavior, by which the control subsystem c_{rs} requests the tracked object pose measurement from the virtual receptor $r_{\mathrm{rs,c}}$, and then waits for the measurement. Occurrence of the fact that the object pose is acquired from the virtual receptor is the terminal condition of the ${}^{c}\mathcal{B}_{\mathrm{rs},0}$ behavior. It causes a switch of the FSM to ${}^{c}\mathcal{B}_{\mathrm{rs},1}$.

The goal of ${}^{c}\mathcal{B}_{\mathrm{rs},1}$ is to predict the pose of the tracked object on a time horizon of one observation. It calculates the goal joint position and velocity for the manipulator, so that its tool center point will follow the object. As prediction is based on estimation, a Kalman filter can be employed. It is a proper tool not only for visual tracking [8], but also for the prediction of the tracked object position [9]. The problem of motion tracking in human oriented environment is rarely Gaussian, due to

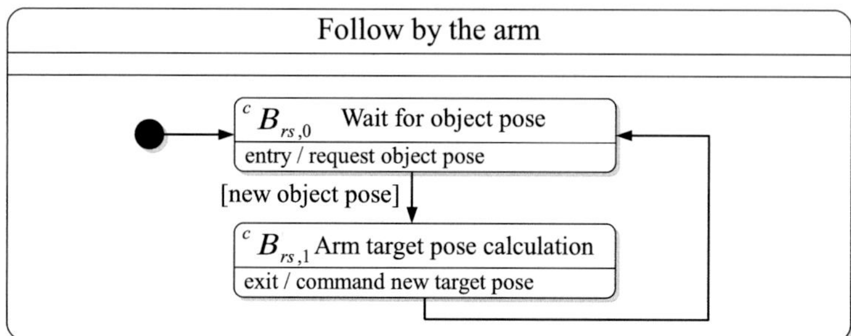

Fig. 2 FSM of the activities of the control subsystem c_{rs} while executing the task *follow by the arm*

the scene clutter and occlusion. Moreover, when tracking is applied to an object held by human, its motion dynamics are highly nonlinear, with step change of acceleration. That may suggest using an extended Kalman filter, however the lack of knowledge of the nonlinear motion model would lead to poor motion prediction.

Thus, a successful approach to the maneuvering target tracking problem is to assume that object trajectories can be piecewise approximated by a kinematics description [10]. The object motion, treated as a piecewise constant acceleration process (the acceleration remains constant for each process cycle of length T), is acted upon by a system noise ω which perturbs this constant acceleration trajectory [11]. This system noise stands for inaccuracies in constant acceleration assumption and any other random factors affecting the trajectory. The assumption that the noise is white results in a random walk model for the acceleration process. For each position coordinate in Cartesian space the object motion dynamics may be presented as:

$$\ddot{x}_{i+1} = \ddot{x}_i + T\omega_i \tag{1}$$

$$\dot{x}_{i+1} = \dot{x}_i + T\ddot{x}_i + \frac{T^2}{2}\omega_i \tag{2}$$

$$x_{i+1} = x_i + T\dot{x}_i + \frac{T^2}{2}\ddot{x}_i + \frac{T^3}{6}\omega_i \tag{3}$$

However, when highly manoeuvring targets are taken into account, the assumptions of a piecewise constant or slowly time evolving system noise variance would be violated and the algorithm would not be applicable. For this reason, an input term is added to the acceleration process, which models step changes of its variation [11, 12].

$$\ddot{x}_{i+1} = \ddot{x}_i + {}^{\ddot{x}}u_i + T\omega_i \tag{4}$$

The input term ${}^{\ddot{x}}u_i$ is estimated by an algorithm that uses a quadratic polynomial fitting to the last trajectory points [11].

This implies a process equation in the form:

$$\begin{bmatrix} x_{i+1} \\ \dot{x}_{i+1} \\ \ddot{x}_{i+1} \end{bmatrix} = \begin{bmatrix} 1 & T & \frac{T^2}{2} \\ 0 & 1 & T \\ 0 & 0 & 1 \end{bmatrix} \begin{bmatrix} x_i \\ \dot{x}_i \\ \ddot{x}_i \end{bmatrix} + \begin{bmatrix} 0 \\ 0 \\ {}^{\ddot{x}}u_i \end{bmatrix} + \begin{bmatrix} \frac{T^3}{6} \\ \frac{T^2}{2} \\ T \end{bmatrix} \omega_i \tag{5}$$

and a measurement equation:

$$y_i = \begin{bmatrix} 1 & 0 & 0 \end{bmatrix} \begin{bmatrix} x_i & \dot{x}_i & \ddot{x}_i \end{bmatrix}^T + \nu_i \tag{6}$$

which shows that the only measured value is the object position, with added noise ν_i. The prediction of target position in given by the equation:

$$\begin{bmatrix} \hat{x}_{i+1} \\ \hat{\dot{x}}_{i+1} \\ \hat{\ddot{x}}_{i+1} \end{bmatrix} = \begin{bmatrix} 1 & T & \frac{T^2}{2} \\ 0 & 1 & T \\ 0 & 0 & 1 \end{bmatrix} \begin{bmatrix} \hat{x}_i \\ \hat{\dot{x}}_i \\ \hat{\ddot{x}}_i \end{bmatrix} + \begin{bmatrix} 0 \\ 0 \\ {}^{\ddot{x}}u_i \end{bmatrix} \tag{7}$$

Figure 3 shows the data flow diagram of the ${}^{c}\mathcal{B}_{\mathrm{rs},1}$ behavior. The control subsystem c_{rs} receives from the virtual receptor $r_{\mathrm{rs,c}}$, through the ${}^{r}_{x}c_{\mathrm{rs,c}}$ buffer, a homogeneous transformation matrix ${}^{C}_{G}T[i]$ specifying the object G (goal) pose with respect to (wrt) the camera coordinate frame in the current (ith) control cycle. Based on the constant camera pose wrt the base coordinate frame ${}^{0}_{C}T$, the current goal pose wrt base coordinates ${}^{0}_{G}T[i]$ is calculated. The object orientation is not taken into account and is assumed constant.

$$\begin{bmatrix} 1 & 0 & 0 & x_i \\ 0 & 1 & 0 & y_i \\ 0 & 0 & 1 & z_i \\ 0 & 0 & 0 & 1 \end{bmatrix} = {}^{0}_{G}T[i] = {}^{0}_{C}T\,{}^{C}_{G}T[i] \tag{8}$$

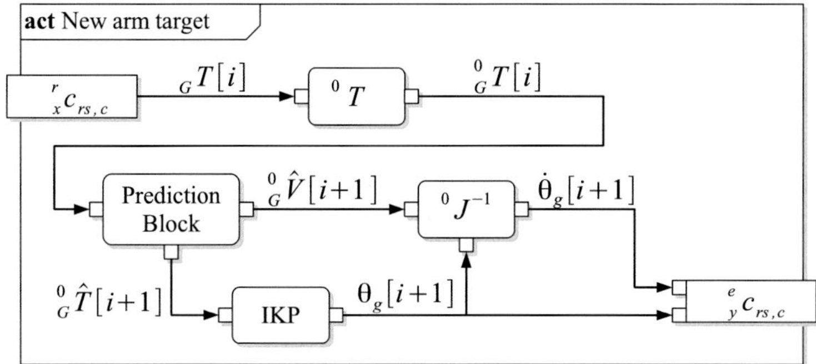

Fig. 3 Data flow diagram of the effector control transition function defining the *new arm target* behavior of the control subsystem c_{rs}

This value is passed on to the Kalman filter in a prediction block, which estimates the object's predicted pose ${}^{0}_{G}\hat{T}[i+1]$ and velocity ${}^{0}_{G}\hat{V}[i+1]$ in the next control cycle (this should appear in the next observation).

$$ {}^{0}_{G}\hat{T}[i+1] = \begin{bmatrix} 1 & 0 & 0 & \hat{x}_{i+1} \\ 0 & 1 & 0 & \hat{y}_{i+1} \\ 0 & 0 & 1 & \hat{z}_{i+1} \\ 0 & 0 & 0 & 1 \end{bmatrix}, \quad {}^{0}_{G}\hat{V}[i+1] = \begin{bmatrix} \hat{\dot{x}}_{i+1} \\ \hat{\dot{y}}_{i+1} \\ \hat{\dot{z}}_{i+1} \end{bmatrix} \tag{9} $$

Inverse kinematics problem solver calculates the manipulator joint positions $\Theta_g[i+1]$ for the manipulator tool center point located at the predicted pose ${}^{0}_{G}\hat{T}[i+1]$. This value and ${}^{0}_{G}\hat{V}[i+1]$ are used together with the inverse jacobian matrix ${}^{0}J^{-1}$ expressed wrt the base coordinate frame to calculate the joint velocities $\dot{\Theta}_g[i+1]$ of the manipulator TCP following the tracked object in the next control cycle.

$$ \Theta_g[i+1] = IKP({}^{0}_{G}\hat{T}[i+1]) \tag{10} $$

$$ \dot{\Theta}_g[i+1] = {}^{0}J(\Theta_g[i+1])^{-1}\, {}^{0}_{G}\hat{V}[i+1] \tag{11} $$

These values of the joint position and velocity vectors are sent to the virtual effector $e_{\mathrm{rs,m}}$ through the ${}^{e}_{y}c_{\mathrm{rs,m}}$ data buffer. They are treated as the desired position and velocity values that should be attained during the time period equal to one control cycle of the control subsystem. After dispatching the command to the virtual effector $e_{\mathrm{rs,m}}$, the ${}^{c}\mathcal{B}_{\mathrm{rs},1}$ behavior is terminated and the control subsystem FSM switches back to the ${}^{c}\mathcal{B}_{rs,0}$ behavior.

3.2 Virtual Effector

The goal of the virtual effector is to provide to the real effector a series of joint positions and velocities which will form a smooth configuration (joint) space trajectory between the poses commanded by the control subsystem. Its primary objective during trajectory generation is to satisfy the constraints imposed on joint positions, velocities and accelerations. Figure 4 shows the graph of the FSM governing the activities of the virtual effector $e_{\mathrm{rs,m}}$. Behavior ${}^{e}\mathcal{B}_{\mathrm{rs,m},0}$ keeps the manipulator immobile while the virtual effector waits for a command from the control subsystem c_{rs}, specifying the target pose and velocity for the manipulator motion. When the command arrives, the FSM switches to behavior ${}^{e}\mathcal{B}_{\mathrm{rs,m},1}$, which calculates a new manipulator joint trajectory, that is expected to attain the commanded joint pose $\Theta_g[i+1]$ and velocity $\dot{\Theta}_g[i+1]$ exactly in the time required for the control subsystem to provide the next command. In general, such trajectory is a series of intermediate poses in configuration space. It starts in the current joint position $\Theta_g[i]$ and with

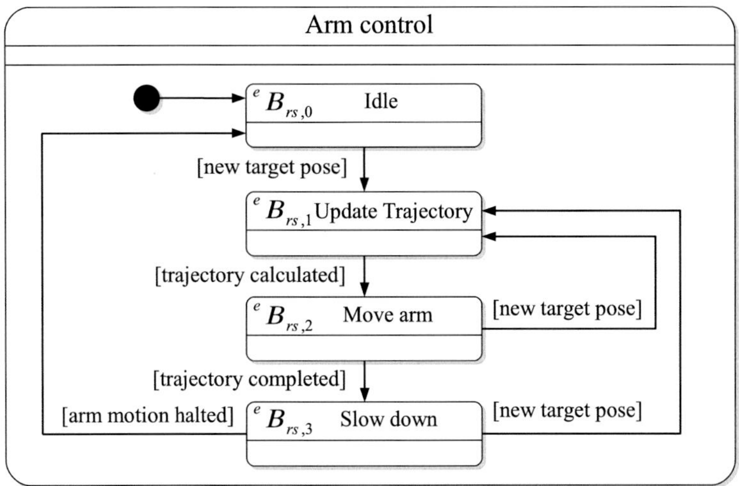

Fig. 4 FSM of the virtual effector $e_{\mathrm{rs,m}}$ executing *arm control*

the current joint velocity $\dot{\Theta}_g[i]$ at discrete time i and leads towards the final joint position $\Theta_g[i+1]$ and joint velocity $\dot{\Theta}_g[i+1]$ at time $i+1$, what ensures that the manipulator tool center point will follow the tracked object in the next control cycle of the control subsystem c_{rs}. Although the desired duration of the trajectory execution equals the period of the control subsystem control cycle, the resulting trajectory execution duration may be extended so that the velocity and acceleration limits are satisfied.

The joint trajectory generation in reactive control is performed at high frequency. Known methods of limited jerk trajectory generation composing the path from splines tend to have high computational load [13] which makes them not applicable. The most of simpler methods present in literature, including those of trapezoidal velocity profile trajectory, focus on the case where initial and final velocities are zero [14]. The method applied in presented application is similar to [15]. The trajectory of each joint has a generalized trapezoidal velocity profile, which is shown in Fig. 5a. In the general case, it consists of four phases, distributed between points 1 to 5. The motion begins at point 1 with an initial position and velocity. Phase 1–2 is of constant acceleration, then the velocity reaches a value which is kept constant through phase 2–3. Constant acceleration phase 3–4 leads to point 4, where the desired target position and velocity is reached. In the case when the next command from the control subsystem does not arrive until point 4 is reached, a deceleration phase 4–5 is planned to halt the manipulator motion (this extends the duration of trajectory execution beyond the control period of the control subsystem).

Figure 5 shows a three stage process of a single joint trajectory generation. For each joint, the trajectory generator is provided with: initial position θ_1, initial velocity $\dot{\theta}_1$, target position θ_4, target velocity $\dot{\theta}_4$, maximum velocity $\dot{\theta}_{max}$, maximum acceleration $\ddot{\theta}_{max}$ and desired trajectory duration t_{14d}.

The first step is to find a trajectory of a triangular velocity profile with minimum duration time, assuming maximum acceleration in 1–2 and 3–4 stages (Fig. 5b). This leads to the solution:

$$\begin{cases} \dot{\theta}_{23} = \dot{\theta}_1 + \ddot{\theta}_{12} t_{12} \\ \ddot{\theta}_{12} = -\ddot{\theta}_{34} = \pm\ddot{\theta}_{max} \\ t_{34} = \frac{-2\dot{\theta}_4 \pm \sqrt{2\dot{\theta}_1^2 + 2\dot{\theta}_4^2 + 4\ddot{\theta}_{12}(\theta_4 - \theta_1)}}{2\ddot{\theta}_{12}} \\ t_{12} = t_{34} + \frac{\dot{\theta}_4 - \dot{\theta}_1}{\ddot{\theta}_{12}} \end{cases} \tag{12}$$

For each proper solution (with calculated times $t_{12} > 0$ and $t_{34} > 0$), if velocity $|\dot{\theta}_{23}| > \dot{\theta}_{max}$, it is trimmed to the maximum value. As a result, a trapezoidal velocity profile trajectory with maximum accelerations and velocity is calculated (Fig. 5c):

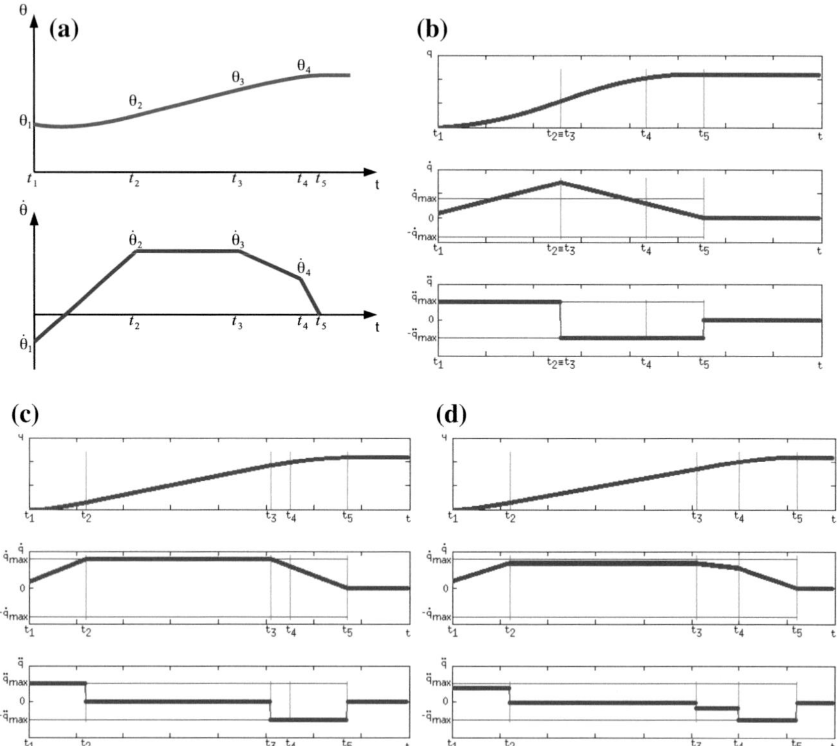

Fig. 5 Joint trajectory generation: general velocity profile and exemplary trajectory generation. *Thin horizontal lines* stand for velocity and acceleration limits, *vertical lines* delimit motion phases. **a** General velocity profile. **b** Profile generation: step 1. **c** Profile generation: step 2. **d** Profile generation: step 3

$$\begin{cases} \dot{\theta}_{23} = \pm\dot{\theta}_{max} \\ \ddot{\theta}_{12} = -\ddot{\theta}_{34} = \pm\ddot{\theta}_{max} \\ t_{12} = \frac{\dot{\theta}_{23}-\dot{\theta}_1}{\ddot{\theta}_{12}} \\ t_{34} = \frac{\dot{\theta}_4-\dot{\theta}_{23}}{\ddot{\theta}_{34}} \\ t_{23} = \frac{\theta_4-(\theta_1+\frac{(\dot{\theta}_1+\dot{\theta}_{23})t_{12}}{2}+\frac{(\dot{\theta}_{23}+\dot{\theta}_4)t_{34}}{2})}{\dot{\theta}_{23}} \end{cases} \tag{13}$$

The obtained trajectory parameter solutions are then sorted in ascending order of durations t_{14} and the one with the smallest duration is selected. If the calculated duration is greater or equal to the desired trajectory duration $t_{14} \geq t_{14d}$, the motion phases durations are extended to meet the desired trajectory duration (Fig. 5d):

- the smaller of t_{12} and t_{34} is increased until $t_{12} = t_{34}$ or $t_{14} \geq t_{14d}$,
- t_{12}, t_{23}, t_{34} are evenly increased until $t_{12} = t_{23} = t_{34} = 2\frac{\dot{\theta}_{max}}{\ddot{\theta}_{max}}$ or $t_{14} \geq t_{14d}$,
- if still $t_{14} < t_{14d}$, t_{23} is increased: $t_{23} = t_{14d} - (t_{12} + t_{34})$.

Then, the remaining parameters are calculated:

$$\begin{cases} \dot{\theta}_{23} = \frac{2(\theta_4-\theta_1)-\dot{\theta}_1 t_{12}-\dot{\theta}_4 t_{34}}{t_{12}+2t_{23}+t_{34}} \\ \ddot{\theta}_{12} = \frac{\dot{\theta}_{23}-\dot{\theta}_1}{t_{12}} \\ \ddot{\theta}_{34} = \frac{\dot{\theta}_4-\dot{\theta}_{23}}{t_{34}} \end{cases} \tag{14}$$

If $\dot{\theta}_{23}$, $\ddot{\theta}_{12}$ or $\ddot{\theta}_{34}$ exceed maximum values, the procedure is repeated for the next solution. Otherwise, the calculated parameters define the desired trajectory. In the last step, the 4–5 motion phase is calculated.

$$\begin{cases} \theta_5 = \theta_4 + \frac{\ddot{\theta}_{45}t_{45}^2}{2} \\ \dot{\theta}_5 = 0 \\ \ddot{\theta}_{45} = -\text{sgn}(\dot{\theta}_4)\ddot{\theta}_{max} \\ t_{45} = \frac{\dot{\theta}_4}{\ddot{\theta}_{45}} \end{cases} \tag{15}$$

When the single trajectories for separate joints are calculated, their durations t_{14} are compared. If they are not equal, the whole process of trajectory generation is repeated, assuming t_{14d} as the longest duration of previously calculated motions. This results in smooth interpolation of the whole motion towards target point in joint space.

Once the trajectory is calculated, the phases 1–4 are executed by the ${}^e\mathcal{B}_{\text{rs,m,2}}$ behavior. If the target manipulator pose and velocity is reached and no new command from the control subsystem is obtained, the FSM of the virtual effector switches to the ${}^e\mathcal{B}_{\text{rs,m,3}}$ behavior and executes the 4–5 trajectory phase, in which the manipulator motion slows down to a halt.

3.3 Virtual Receptor

The task of the virtual receptor is to provide an estimated pose of the tracked object, when requested by the control subsystem. The presented general approach to visual servo control does not specify any particular method of object pose detection. In service robotics, methods based on RGB-D images gain popularity, e.g. [16], however methods based on RGB camera images are also used [17]. In both cases image processing takes significant time, what results in large latency, reaching 100 ms. Such a delay is much higher than the period of the virtual effector behavior control cycle, dictated by ${}^{e}\mathcal{B}_{\mathrm{rs},2}$ (*Move arm*) behavior, which calculates the desired effector position once every 1ms.

An exemplary object detection algorithm, used in the experiments, utilizes data from real receptor being a stand alone RGB camera, detecting a chessboard pattern. The methods provided by OpenCV camera calibration and 3D reconstruction module find the coordinates of internal corners of the chessboard (`findChessboardCorners()`). Positions of these points on the chessboard and their projections onto the image plane are used to estimate a transformation matrix between the camera and the chessboard, given the specific camera model parameters (`solvePnP()`). Virtual receptor was implemented in component-oriented system, designed especially for sensory data processing [18].

4 Experiments

The presented general structure of the robot control system capable of object motion prediction, calculation and execution of a smooth trajectory that follows an object required verification. The most appropriate test bench for examining the algorithm consists of a vision system and a robot manipulator. At the initial stage, the manipulator was replaced with a Matlab simulation, which provides quicker way of verifying the control system operation.

The object to be followed was a chessboard held by the operator. Its motion was recorded by an RGB camera. Once the chessboard motion was complete, its trajectory was saved to a file and used in numerous experiments to obtain optimal parameters of the Kalman filter. The predicted position and velocity of the chessboard were used to calculate the target pose for the manipulator trajectory generation in Matlab

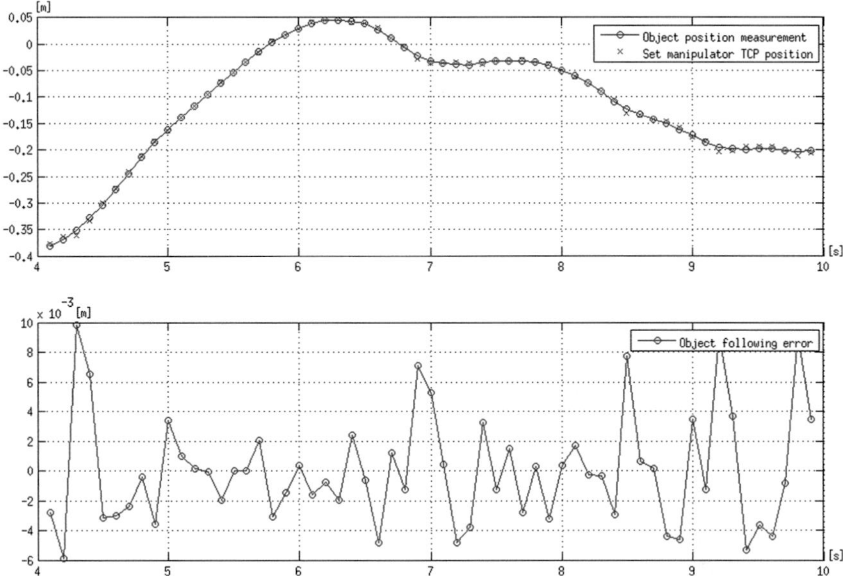

Fig. 6 Object position following

simulation. Figure 6 shows the results of following the chessboard position and the disparity between the simulated manipulator's TCP and the measured chessboard positions.

5 Conclusions

The experiments show that the approach utilizing the Kalman filter for object pose prediction in Cartesian space and a joint trajectory generator enables the development of a reliable visual servo controller that takes into account joint motion limits imposed on velocities and accelerations. Due to the application of a constant acceleration motion model in the Kalman filter, the algorithm successfully deals with low frequency measurements. The object position tracking deviation results from the error of object pose prediction rather than trajectory generation.

In the experiment the joint trajectory generator was used to calculate the trajectory towards the predicted object position every 100 ms. However, due to its moderate complexity, it could be run in every control cycle of the virtual effector with its role restricted to limiting joint velocities and accelerations. That enables the application of a simple cubic spline trajectory generator in Cartesian space resulting in a smoother movement in the task space.

Acknowledgments The authors gratefully acknowledge the support of this work by The National Centre for Research and Development grant no. PBS1/A3/8/2012.

References

1. Staniak, M., Winiarski, T., Zieliński, C.: Parallel visual-force control. In: Proceedings of the IEEE/RSJ International Conference on Intelligent Robots and Systems, IROS '08 (2008)
2. Sanderson, A., Weiss, L.: Adaptive visual servo control of robots. In: Pugh, A. (ed.) Robot Vision, International Trends in Manufacturing Technology. pp. 107–116. Springer, Berlin (1983)
3. Janabi-Sharifi, F., Marey, M.: A kalman-filter-based method for pose estimation in visual servoing. IEEE Trans. Robot. **26**(5), 939–947 (2010)
4. Gortcheva, E., Garrido, R., Gonzalez, E., Carvallo, A.: Predicting a moving object position for visual servoing: theory and experiments. Int. J. Adapt. Control Signal Process. **15**(4), 377–392 (2001)
5. Zieliński, C., Winiarski, T.: Motion generation in the MRROC++ robot programming framework. Int. J. Robot. Res. **29**(4), 386–413 (2010)
6. Friedenthal, S., Moore, A., Steiner, R.: A Practical Guide to SysML: The Systems Modeling Language. Morgan Kaufmann (2014)
7. Zieliński, C., Kornuta, T., Winiarski, T.: A systematic method of designing control systems for service and field robots. In: 19th IEEE International Conference on Methods and Models in Automation and Robotics, MMAR'2014, IEEE, pp. 1–14
8. Forsyth, D.A., Ponce, J.: Computer Vision: A Modern Approach. Prentice Hall (2002)
9. Cuevas, E.V., Zaldivar, D., Rojas, R., et al.: Kalman filter for vision tracking (2005)
10. Moose, R., Gholson, N.: Adaptive tracking of abruptly maneuvering targets. In: 1976 IEEE Conference on Decision and Control Including the 15th Symposium on Adaptive Processes, pp. 804–808 (1976)
11. Kiruluta, A., Eizenman, E., Pasupathy, S.: Predictive head movement tracking using a kalman filter. IEEE Trans. Syst. Man Cybern. Part B Cybern. **27**(2), 326–331 (1997)
12. Gutman, P.O., Velger, M.: Tracking targets using adaptive kalman filtering. IEEE Trans. Aerosp. Electron. Syst. **26**(5), 691–699 (1990)
13. Erkorkmaz, K., Altintas, Y.: High speed cnc system design. part i: jerk limited trajectory generation and quintic spline interpolation. Int. J. Mach. Tools Manuf. **41**(9), 1323–1345 (2001)
14. Kroger, T., Tomiczek, A., Wahl, F.M.: Towards on-line trajectory computation. In: 2006 IEEE/RSJ International Conference on Intelligent Robots and Systems, pp. 736–741. IEEE (2006)
15. Ramos, F., Gajamohan, M., Huebel, N., D'Andrea, R.: Time-optimal Online Trajectory Generator for Robotic Manipulators. Eidgenössische Technische Hochschule Zürich, Institute for Dynamic Systems and Control (2013)
16. Wilkowski, A., Kornuta, T., Kasprzak, W.: Point-Based Object Recognition in RGB-D Images. In: Filev, D., Jabłkowski, J., Kacprzyk, J., Krawczak, M., Popchev, I., Rutkowski, L., Sgurev, V., Sotirova, E., Szynkarczyk, P., Zadrozny, S. (eds.) Proceedings of the 7th IEEE International Conference Intelligent Systems IS'2014 of Advances in Intelligent Systems and Computing (AISC), vol. 323, pp. 593–604. Springer (2015)
17. Stefańczyk, M., Walçcki, M.: Localization of essential door features for mobile manipulation. In: Szewczyk, R., Zieliński, C., Kaliczyńska, M.,(eds.) Recent Advances in Automation, Robotics and Measuring Techniques of Advances in Intelligent Systems and Computing (AISC), vol. 267, pp. 487–496. Springer (2014)
18. Stefańczyk, M., Kornuta, T.: Handling of asynchronous data flow in robot perception subsystems. In: Simulation, Modeling, and Programming for Autonomous Robots, vol. 8810 of Lecture Notes in Computer Science, pp. 509–520. Springer (2014)

Low-Cost Canoe Counting System for Application in a Natural Environment

Artur Wilkowski and Marcin Luckner

Abstract This paper presents low-cost system for counting canoes and canoeists to control cannoning tourist routes. The created system was implemented on Raspberry Pi 2 and the total cost of the tracking device is less than 200$. The proposed algorithm uses background subtraction and Support Vector Machines to track vessels and recognize canoes among them. The obtained results are rewarding as for low-cost solution. Depending on considered group of objects the accuracy of the algorithm reaches 84, 89.5, and 96 % for canoes, vessels, and all objects respectively.

Keywords Computer vision · Pattern recognition · Classification with rejection

1 Introduction

One of challenging computer vision issues is tracking and recognition of moving objects. The tracking of vessels is especially interesting because of unstable background, varying illumination conditions and diversity of observed objects.

In this paper we propose the low-budget automatic canoe counting system. The system is a part of a tourist canoeing monitoring project. The information on a river route usage is useful during planning and managing of tourist infrastructure by provincial administration. However, to create a system that covers the whole voivodeship it is necessary to place hundreds of tracking devices. Therefore, the tracking device cost should be limited as much as it is possible.

The aims of the system are not constrained to tracking vessels as in [1, 2]. Canoe counting involves also such subproblems as: separation of the vessels from natural

A. Wilkowski (✉)
Faculty of Geodesy and Cartography, Warsaw University of Technology,
Pl. Politechniki 1, 00-661 Warsaw, Poland
e-mail: a.wilkowski@gik.pw.edu.pl

M. Luckner
Faculty of Mathematics and Information Sciences, Warsaw University of Technology,
Koszykowa 75, 00-662 Warsaw, Poland
e-mail: mluckner@mini.pw.edu.pl

R. Szewczyk et al. (eds.), *Challenges in Automation, Robotics and Measurement Techniques*, Advances in Intelligent Systems and Computing 440, DOI 10.1007/978-3-319-29357-8_61

occurrences such as whirlpools, waves, or ripples; isolation of canoes from other vessels; estimation of the number of canoeists.

Similar problems are covered in the literature. In paper [4] a vessels recognition problem was presented, but the vessels were recognised against stable objects such as rocks and islands. Paper [3] discussed classification of various sport photos including rowing and sailing. However, the experiment described in that paper was performed on pictures not on video as in our work.

The presented solution was implemented on low-cost equipment. As a computing device a Raspberry Pi 2 was used. The device is equipped with a 900 MHz quad-core ARM Cortex-A7 CPU processor and 1GB LPDDR2 SDRAM memory. The connected USB IP camera GEMINI-610B-3 has 1024×768 resolution. In the Pi design of Pi device the processing speed was consented on behalf of low energy consumption, passive cooling and low costs. The whole cost of the tracking device is less than 200$.

The experiments showed that our system working on Pi device is able to process up to seven frames per second, which was enough to recognise nearly 90 % of vessels and over 96 % of registered objects.

The rest of the paper is structured as follows. Section 2 presents algorithms and methods that were used to track and recognize vessels. Section 3 describes experiments that have been performed to evaluate the system. Finally, Sect. 4 sums up the results.

2 Methods and Algorithms Used

The algorithm proposed in this paper uses background subtraction to binarize the scene. After that foreground objects are tracked. Images of tracked objects are extracted from the original picture and fed to the classification system. The classification system rejects objects that are not vessels and recognizes canoes among all the vessels. For successfully recognized canoes the number of canoeists in each canoe is estimated. Details are given in the following sections.

2.1 Application of Background Subtraction

Background Subtraction In order to perform image segmentation the Adaptive Gaussian Mixture Model (AGMM) (described in [7, 8]) was utilized. In the model each of the pixels is treated independently. The probability distribution of the color for each pixel is given by the following Gaussian Mixture:

$$p(x|BG + FG) = \sum_{m=1}^{m} \hat{\pi}_m \mathcal{N}(x; \hat{\mu}_m, \hat{\sigma}_m^2 I)$$

where $\hat{\mu}_m$ and $\hat{\sigma}_m^2$ are estimates of mean and variance of Gaussian components, and $\hat{\pi}_m$ are non-negative mixing weights (adding up to 1).

When a new data sample $x^{(t)}$ arrives at time t the estimates are updated according to the following rules

$$\begin{aligned}
\hat{\pi}_m &\leftarrow \hat{\pi}_m + \alpha(o_m^{(t)} - \hat{\pi}_m) - \alpha c_T \\
\hat{\mu}_m &\leftarrow \hat{\mu}_m + o_m^{(t)}(\alpha/\hat{\pi}_m)\delta_m \\
\hat{\sigma}_m &\leftarrow \hat{\sigma}_m + o_m^{(t)}(\alpha/\hat{\pi}_m)(\delta_m^T\delta_m - \hat{\sigma}_m^2)
\end{aligned}$$

where α is the adaptation coefficient (it is typically assumed that $\alpha = 1/T$ and the T is the predefined time horizon), $\delta_m = x^{(t)} - \hat{\mu}_m$ and $o_m^{(t)}$ is set to 1 for the component 'close' to $x^{(t)}$ with the largest $\hat{\pi}_m$ and to 0 for other components. The 'close' component is the component closer to $x^{(t)}$ than 3 standard deviations (in the sense of Mahalanobis distance). c_T is a fixed constant and the term αc_T is responsible for 'smoothing' the distribution of $\hat{\pi}_m$ according to the Dirichlet prior [7]. The values $\hat{\pi}_m$ need to be normalized after update.

If no 'close' existing component could be found for the new sample $x^{(t)}$ a new component is created for this sample with parameters $\hat{\pi}_{M+1} = \alpha$, $\hat{\mu}_{M+1} = x^{(t)}$ and $\sigma_{M+1} = \sigma_0$ (σ_0 is a fixed initial variance). In case of reaching the maximum number of components, the one with the smallest $\hat{\pi}_m$ is discarded.

The model described above represents simultaneously foreground and background distributions. In order to separate foreground from background it is assumed that the background model is constituted by the largest Gaussian clusters (those with the highest values of $\hat{\pi}_m$). If components are ordered by descending $\hat{\pi}_m$ values, the number of components actually constituting background model B is given by

$$B = \arg\min_b \left(\sum_{m=1}^{b} \hat{\pi}_m > (1 - c_f) \right)$$

where c_f is a fixed constant describing the maximum proportion of the data that can belong to foreground objects. Consequently the background model is given by the following mixture

$$p(x|BG) \sim \sum_{m=1}^{B} \hat{\pi}_m \mathcal{N}(x; \hat{\mu}_m, \hat{\sigma}_m^2 I)$$

Then the decision concerning whether a pixel at time t (namely $x^{(t)}$) is to be classified as a background is performed by a simple thresholding of $p(x^{(t)}|BG)$

$$p(x^{(t)}|BG) > c_{thr}$$

and c_{thr} is the threshold value.

AGMM for Vessels' Detection In our application the vector x is a three-element color vector. There were performed several preliminary experiments concerning the selection of a color-space the most appropriate for the given problem. It turned out that the basic RGB space is not sufficient for the application. Eventually the Hue-Saturation-Value (HSV) color space was selected as the one giving the best relation between true and false foreground detections.

Binary Image Processing The output of the background subtraction module is a binary image with foreground pixels marked as 1 and background pixels marked as 0. In order to remove erroneous detections the n-times morphological erosion is applied to the image. Afterwards, in order to join nearly-connected image patches m-times morphological dilation is consequently applied. In our system typically $m \geq n$ is utilized.

After preprocessing step there is extracted a list of non-connected image patches. Internal holes in patches are filled. In order to do so an effective method described in [5] is applied. The extracted list is the list of image tracking candidates.

Prefiltering For each foreground patch (tracking candidate) there are computed two features: the minimum bounding box r, the center of the bounding box c and the size of the image patch s (effectively—the number of pixels). Then the patch is rejected if any of the conditions hold

$$\begin{cases} s < minArea \\ r.height > maxHeight \\ r.width > maxWidth \\ IWaterMask(c.x, c.y) = 0 \end{cases}$$

where *IWaterMask* is the mask containing area covered by water. The resulting list forms a set of observation for object recognizer and object tracker.

The output list of observation for frame t can be described as a list of pairs $o_i^t = (o_i^t.x, o_i^t.y)$, where (x, y) stand for the bounding box center coordinates for each patch.

2.2 Recognition of Water Objects

Recognition process consists of two main steps. In the first step, the system recognises canoes among all vessels and natural occurrences. In the second step, a type of the recognised canoe is detected. There were two types of canoes: an one-person canoe and a two-person canoe.

The implementation of both steps is based on Support Vector Machines (SVM) [6].

For two classes labelled -1 and $+1$ and the space of features included in R^d we define vectors of features of training elements $x_i \in R^d$ labelled as y_i, where y_i is either −1 or +1, for $i = 1, 2, \dots, N$ and N being the cardinality of the learning set. For a non linear separation of the two classes, the space R^d is mapped into a space of higher dimension using kernel function $K(x, x')$. Then an SVM decision function is implemented as:

$$f(x) = \operatorname{sgn}\left(\sum_{i=1}^{N} y_i * \alpha_i * K(x, x_i) + b\right) \tag{1}$$

where coefficients α_i and b are computed by maximization of the following convex quadratic programming (QP):

$$\sum_{i=1}^{N} \alpha_i - \frac{1}{2}\sum_{i=1}^{N}\sum_{j=1}^{N} \alpha_i * \alpha_j * y_i * y_j * K(x_j, x_i) \tag{2}$$

subject to the following constrains:

$$\bigwedge_{i \in \{1,2,\dots,N\}} 0 \leqslant \alpha_i \leqslant C \ \wedge \ \sum_{j=1}^{N} \alpha_i * y_i = 0 \tag{3}$$

The regularization coefficient C in Eq. 3 controls trade-off between margins and misclassification errors. In this work the regularization coefficient C is set to 1. As the kernel function $K(x, x')$ the Gaussian kernel is taken

$$\exp\left(-\frac{1}{d}||x - x'||^2\right), \tag{4}$$

where d is the number of features.

In both steps, recognised objects are described by a vector of 18 features. The features vector begins with a feature representing proportion between width and height of the object. The remaining features are calculated separately for each of color components: red, green, blue. The values of components are given in the range [0, 255]. The first feature calculated for each component is the 'darkness' factor calculated as the average value of the component. Next four features are horizontal and vertical histograms of the channel. Finally, the proportions between components are calculated for each pair of components.

2.3 Tracking and Counting of Water Objects

Object Tracking In the application of canoe tracking we propose a simple but fast tracking systems. The observations obtained from the image are associated with the *tracked objects* by a minimum distance criterion. Tracked objects that are too old are excluded from the matching. Then the list of tracked objects is updated using new associations. Let us now describe the process in more detail.

The output set of observation for frame t is described as a set of pairs $o_i^t = (o_i^t.x, o_i^t.y)$. Let us denote this set as O^t. Next, the set of *tracked object entries* A^t will contain elements in the form $a_j^t = (a_j^t.x, a_j^t.y, a_j^t.k)$, where $a_j^t.y$ and $a_j^t.k$ is the last known position of the tracked object, and $a_j^t.k$ is the last frame when it was observed.

In each frame a Cartesian product of sets $A^t \times O^t$ is produced. In the first filtering step there are left only pairs (a_j^t, o_i^t) for which the following conditions hold

$$\begin{cases} a_j^t.k < \varepsilon_{age} \\ dist(a_j^t, o_i^t) < \varepsilon_{dist} \end{cases}$$

where *dist* is a Euclidean distance function defined as

$$dist(a, o) = \sqrt{(a.x - o.x)^2 + (a.y - o.y)^2)}$$

Then the second filtering step is applied. The resulting set of pairs (a_j^t, o_i^t) is sorted according to the value of the *dist* function. Then the list is traversed starting from the smallest values of the *dist* function. For each pair (a_j^t, o_i^t) that is processed there are removed from the list all subsequent pairs that duplicate either a_j^t or o_i^t. In results we obtain one-to-one *tracking associations* (a_j^t, o_i^t) between objects currently tracked and the new observations.

The new set A^{t+1} of tracked object entries is initially set to A^t then updated in the following way:

- If for the observation o_i^t there exist a valid tracking association in the form (a_j^t, o_i^t) then the tracked object entry a_j^t is only updated according to o_i^t
- If for the observation o_i^t there do not exist any valid tracking association (a_j^t, o_i^t) it is first verified if tracking can be started at the given image location by checking if $ITrackStartMask(o_i^t.x, o_i^t.y) = 1$. If the result is positive then a new tracked object entry (say a_l^{t+1}) is added to A^{t+1} and then its values are set according to o_i^t.

Attribute update rules of a_j^{t+1} or a_l^{t+1} are the same in both cases

$$\begin{cases} a_\bullet^{t+1}.x \leftarrow o_i^t.x \\ a_\bullet^{t+1}.y \leftarrow o_i^t.y \\ a_\bullet^{t+1}.k \leftarrow t \end{cases}$$

where $^\bullet$ stands for either j or l.

Canoe Counting Parallelly to the set of tracked object entries A^t there is maintained a set of statistics used for object counting C^t. For each entry of A^t there is exactly one entry in C^t. Each entry in C^t is an $m+1$-tuple $c_i^t = (c_i^t.n, c_i^t.n_1, \dots c_i^t.n_m)$. $c_i^t.n$ -

denotes the total number of frames in which the given tracked object was observed, while $c_i^t.n_s$ denotes the number of frames in which the tracked object i was classified as a canoe of type s.

The set C^{t+1} is initially set to C^t and then updated simultaneously with the update of A^t. When an entry a_j^t is updated in A_t (so there was established some tracking association), then $c_j^{t+1}.n$ is incremented $c_j^{t+1} \leftarrow c_j^t.n + 1$. Otherwise a new entry c_l^{t+1} is added to C^{t+1}, with initial values $c_l^{t+1}.n \leftarrow 1$ and $c_l^{t+1}.n_s \leftarrow 0$ for all s.

Then if the observation associated with the tracked object is classified as a canoe of type p, the corresponding C^{t+1} entry is updated so that $c_\bullet^{t+1}.n_p \leftarrow c_\bullet^{t+1}.n_p + 1$, where $^\bullet$ stands for either i or l.

Canoe Recognition Decision For each frame t, and for each C^t entry there is made a decision whether a tracked object associated with some entry c_j could be recognized as a Canoe of type p. If the following conditions hold

$$\begin{cases} c_j^t.n > \varepsilon_{fcount} \\ \frac{c_j^t.n_p}{c_j^t.n} > \varepsilon_{fprop} \end{cases}$$

then the tracked object is recognized as a canoe of type p. If the object was recognized once, then the recognition process is not repeated any more.

3 Experiments

3.1 Water Objects Recognition

Rejection of Non-canoe Objects In the first experiment, registered natural phenomenons and vessels other that canoes were separated from canoes. During film registration 22980 pictures of objects were taken. Among them 16287 were pictures of canoes and 6693 were pictures of other objects. Figure 1 shows observed vessels, when Figs. 2, 3 shows observed canoes.

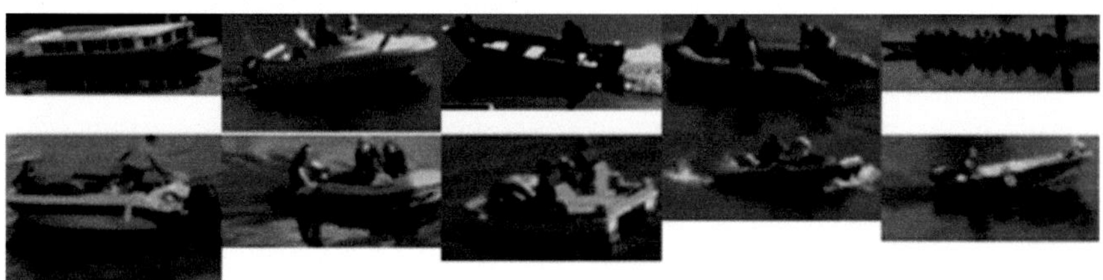

Fig. 1 Observed vessels

Fig. 2 Observed one-person canoes

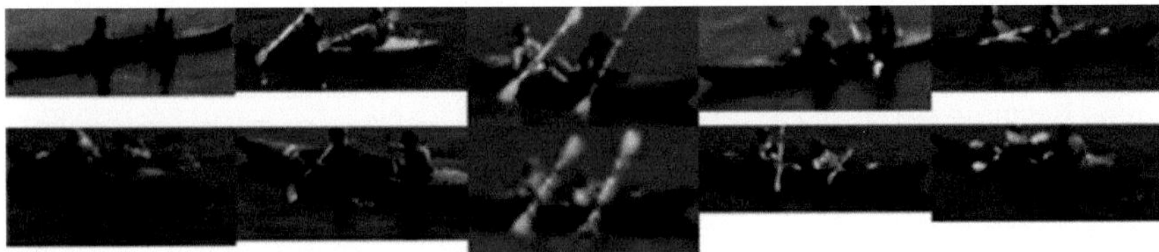

Fig. 3 Observed two-person canoes

The separation was done using an SVM with the Gaussian kernel. Parameters of the SVM were selected as $C = m * 10^n$ and $\gamma = m * 10^n$ where $m \in [1, \dots, 9]$ and $n \in [-3, \dots, 3]$. The best results were obtained for $C = 0.9$ and $\gamma = 90$.

The result obtained on the learning set were very high. The total accuracy was 99.64 %. The accuracy for canoes was 99.93 % and the accuracy for other objects was 99.95 %.

The SVM trained using the selected parameters was tested on the testing set that was collected separately. The testing set included 12154 pictures. In this number 8956 pictures were the pictures of canoes and 3198 pictures of other objects.

The result obtained on the testing set were worse that the results obtained on the learning set. The total accuracy was 85.33 %. The accuracy for canoes was 87.17 % and the accuracy for other objects was 80.17 %.

The results obtained on the testing set are good enough to create a recognizer based on a voting method described in the previous section. In the tracking system each object is described as a series of snapshots. The object is positively recognized as a canoe of some type only if more than $\varepsilon_{fprop} = 0.6$ of object snapshots are adequately recognized. Therefore, the accuracy over 80 % is rewarding.

Estimation of the Number of Canoeists In the second experiment, recognized canoes were separated into one-person and two-persons types. Among 16287 pictures of canoes 13772 pictures were pictures of the two-persons canoe and only 2515 pictures were pictures of the one-person canoe. Figure 2 shows observed one-person canoes and Fig. 3 shows two-persons canoes.

The separation was done using an SVM with the Gaussian kernel. Parameters of the SVM were selected as $C = m * 10^n$ and $\gamma = m * 10^n$ where $m \in [1, \dots, 9]$ and

$n \in [-3, \ldots, 2]$. The best results were obtained for $C = 0.9$ and $\gamma = 90$. To reduce the difference between the cardinal numbers of both sets various weights were used. The one-person canoes class was taken with weight 100 when two-person canoes class was taken with weight 1.

The result obtained on the learning set were once again very high. The total accuracy was 99.83 %. The accuracy for one-person canoes was 100 % and the accuracy for two-person canoes was 99.80 %.

The SVM trained using the selected parameters was tested on the testing set that was collected separately. The testing set included 8956 pictures. In this number 1156 pictures were the pictures of one-person canoes and 7800 pictures of two-persons canoes.

The total accuracy obtained in the test was 86.73. However, when the accuracy for two-persons canoes is still high—95.94 %— the accuracy for one-person canoes was very small—24.58 %.

The low results of one-person canoes recognition can bring overestimation of the canoeists number. However, if the number of one-person canoes will be as small as in collected data the total estimation error should not be gross.

We can compare our result from experimental results from [3, 4]. In work [3] the recognition of the pictures that presented rowing and sailing was at the level of 80 %. The results of recognition of vessels and stable objects presented in [4] were at the level of 81–92.3 % for various sets of features. That shows that our results can probably be improved by better selection of features, but the obtained accuracy is acceptable.

3.2 Water Objects Tracking

The tests of the whole system were performed on the material gathered during a single day 2015-09-19. The registered video was 8 h length. The video started at 8 am and finished at 4 pm. The video was cut in order to exclude long empty subsequences. A multi-threads video processing on RaspBerry Pi 2 allowed to process the video with the framerate of about 7 frames per second and the resolution of 640×360.

Fig. 4 Tracking scene

Figure 4 shows tracking scene with analyzed objects marked. A green rectangle denotes a positively recognized object, a yellow rectangle denotes an object still analyzed, and red rectangle denotes a rejected object. On the right the binary result of background subtraction step is given.

During the experiment 90 canoes were observed. Among them 12 were one-person canoes and 78 were two-person canoes. Additionally 43 other vessels were observed.

The algorithm underestimated the number of canoes to 76. The system missed 3 one-person canoes and 11 two-persons canoes. That give us accuracy of recognized canoes at the level of 84 %. However, when we take under consideration also other vessels that were not misclassified with canoes then the accuracy of vessels recognition grows to 89.5 %. We should also remember that vessels are not all of the moving objects that can be registered by the camera. During the experiment 265 moving object were registered. This number includes vessels and natural phenomenons. The accuracy of canoes recognition calculated with respect to all registered objects scores as much as 96 %.

The number of canoeists was estimated at 143 persons. The real number was higher slightly higher (168). The algorithm missed 25 persons. That give us the accuracy of canoeists recognition at the level of 85 %.

The system ARGOS [1] for vessels counting that is installed over the Grand Canal in Venice uses a group of IR/VIS cameras to obtain the accuracy at the level of 92–96 %. Therefore, our results at the level of 89.5 % are rewarding for the low-cost solution.

4 Conclusions

In this paper we proposed low-cost system for monitoring tourist canoe traffic. The created system was implemented and evaluated on Raspberry Pi 2 and the total cost of the tracking device is less than 200$.

The proposed algorithm succeeded in tracking moving objects and recognizing canoes among other vessels and natural phenomenons. The obtained results—accuracy of 84, 89.5, and 96 % for canoes, vessels, and all objects respectively—are rewarding for the low-cost solution. The comparison of the results with other works [1, 3, 4] showed that an alternative set of features may improve the results, but the obtained accuracy is acceptable.

Acknowledgments The project was carried out in cooperation with NFC Vision.

The research was partially supported by the National Science Center, grant No 2012/07/B/ST6/01501, decision no UMO–2012/07/B/ST6/01501.

References

1. Bloisi, D.D., Iocchi, L., Leone, G.R., Pigliacampo, R., Tombolini, L., Novelli, L.: A distributed vision system for boat traffic monitoring in the venice grand canal. In: Proceedings of 2nd International Conference on Computer Vision Theory and Applications (VISAPP-2007), pp. 549–556 (2007)
2. Broggi, A., Cerri, P., Grisleri, P., Paterlini, M.: Boat speed monitoring using artificial vision. In: Foggia, P., Sansone, C., Vento, M. (eds.) Image Analysis and Processing ICIAP 2009. Lecture Notes in Computer Science, vol. 5716, pp. 327–336. Springer, Berlin (2009)
3. Li, L.J., Fei-Fei, L.: What, where and who? classifying events by scene and object recognition. In: IEEE 11th International Conference on Computer Vision, 2007. ICCV 2007, pp. 1–8 (2007)
4. Ma, Z., Wen, J., Hao, L., Wang, X.: Multi-targets recognition for surface moving platform vision system based on combined features. In: 2014 IEEE International Conference on Mechatronics and Automation (ICMA), pp. 1833–1838 (2014)
5. Suzuki, S., be, K.: Topological structural analysis of digitized binary images by border following. Comput. Vis. Graph. Image Process. **30**(1), 32–46 (1985). http://www.sciencedirect.com/science/article/pii/0734189X85900167
6. Vapnik, V.: Statistical Learning Theory. Wiley, New York (1998)
7. Zivkovic, Z.: Improved adaptive gaussian mixture model for background subtraction. In: Proceedings of the 17th International Conference on Pattern Recognition, 2004. ICPR 2004, vol. 2, pp. 28–31 (2004)
8. Zivkovic, Z., van der Heijden, F.: Efficient adaptive density estimation per image pixel for the task of background subtraction. Pattern Recogn. Lett. **27**(7), 773–780 (2006). http://dx.doi.org/10.1016/j.patrec.2005.11.005

Hypothesis Generation in Generic, Model-Based Object Recognition System

Maciej Stefańczyk and Rafał Pietruch

Abstract One of the key problems in robotics is ability to search and recognize objects, as well as their reliable localization. Existing object-detection solutions relies either on instance-models, making them unable to generalize, or can't estimate object pose if the model is not known at hand. We aim at making system able to work with generic models, described using set of simple parts and relations between them, as well as localize them on scene. Article presents hypothesis generation part of the designed system.

Keywords Hypothesis generation · Virtual receptor · Object detection · Edge detection · Constraint satisfaction

1 Introduction

In robotics, ability to search and recognize objects, as well as their reliable localization, is one of the key problems. Problem of localization known objects is rather well studied, with multiple options possible. Pose estimation of 3D object in 2D image can be done based on e.g. linear segments [7] or feature points [8]. If input data is three dimensional (e.g. RGB-D from depth camera) other methods can be employed—matching object shape with input point cloud [3], which allows recognition of texture-less objects. All those methods, however, have similar drawbacks, of which requirement to deliver exact model of the object seems to be most limiting.

M. Stefańczyk (✉)
Institute of Control and Computation Engineering,
Nowowiejska 15/19, 06-665 Warsaw, Poland
email: M.Stefanczyk@elka.pw.edu.pl
URL: http://robotyka.ia.pw.edu.pl/

R. Pietruch
Przemyslowy Instytut Automatyki I Pomiarow PIAP,
Al. Jerozolimskie 202, 02-486 Warsaw, Poland
email: rpietruch@piap.pl
URL: http://www.antiterrorism.eu/

R. Szewczyk et al. (eds.), *Challenges in Automation, Robotics and Measurement Techniques*, Advances in Intelligent Systems and Computing 440, DOI 10.1007/978-3-319-29357-8_62

It is rather impossible to find object by giving only rough information about its type (e.g. find the bike on current scene). That's why more generic solutions were developed [9]. Objects are described by set of parts and relations between them, which gives the algorithm ability to generalize [1]. Part-based methods are great for scene type recognition and searching for new objects from known category, but (in general) are unable to estimate proper object pose and parameters.

We propose integrated, complex object recognition system, based on object description given in semantic form [6]. This allows us to recognize and describe basic object parts in generic way, and then build whole object from those smaller, basic blocks. System is able to work with different types of initial segments, e.g. edges and feature points from RGB image or shapes extracted from 3D information. This article presents first step in the process, which is aggregation of segments extracted from the image to the form of basic concepts to prepare set of initial object hypotheses.

Article is structured as follows. Section 2 shows basic concepts, that are connected with described part of the system. Example implementation of edge concepts is presented in Sect. 3, which is then used to test system quality (Sect. 4). Article ends with discussion regarding current and future system development (Sect. 5).

2 Inference System Description

2.1 System Structure

Overall structure of the inference module in our system is presented on Fig. 1. As the system is part of virtual receptor [5] and agent system as a whole, it communicates with other elements, in our case with control subsystem and segmentation module.

During whole process, system is fed with concept instances detected on scene (those are kept in internal memory). Two more repositories must be filled beforehand, during system configuration: models of known objects and used constraints. When a query from the control subsystem arrives, inference module loads appropriate model along with constraint calculation functions and processes all known instances to find ones best matching the model. Next section will describe those elements in detail.

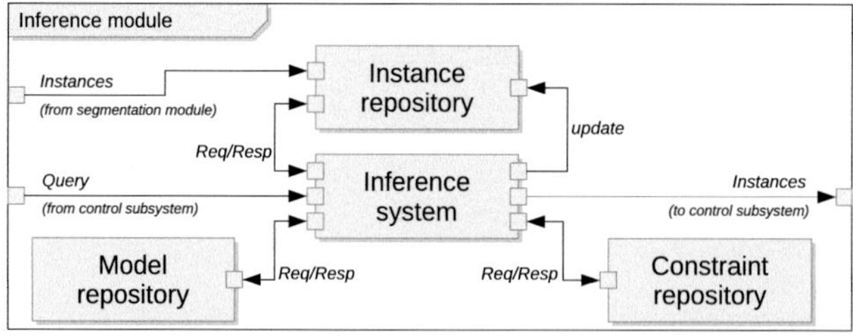

Fig. 1 Structure of the inference subsystem

2.2 Concept of Concepts

Model Structure Model M is built from parts P and constraints C defined between them. Each part must have assigned class (*class*(P)) and unique role in model (*role*(P)). Thus, model is a tuple consisting of two sets:

$$M = \left\{P = \left\{p_1 \ldots p_n\right\}, C = \left\{c_1 \ldots c_m\right\}\right\} \tag{1}$$

Constraints Constraints between the parts are defined by telling which predicate is to be used on which subset of parts. Predicates in our system returns extended information, i.e. from the $[0 \ldots 1]$ range, instead of classical $\{True, False\}$, where 1 means full constraint satisfaction and 0 total inconsistency of given set of parts with examined predicate. It enables us to treat constraint satisfaction result as an intermediate score in further processing steps, giving finally overall score for current model hypothesis.

$$pred(p_1, \ldots, p_k) \in [0 \ldots 1] \tag{2}$$

Hypotheses Each assignment of existing instances to model parts, made in a way that for each part of the model exactly one instance of proper class is selected, is called hypothesis $H(M)$. It can be defined as:

$$H(M) : \forall i \in \{1..n\} \exists j \in \{1..k\} : p_i \leftarrow inst_j \wedge class(p_i) = class(inst_j) \tag{3}$$

where n is the number of parts in the model and k is the number of already recognized instances ($inst_k$, e.g. segments). Overall hypothesis score is calculated by taking the product of all constraints for given assignment.

$$score\,(H\,(M)) = \prod_{m} eval(c_m, H) \tag{4}$$

2.3 Hypothesis Generation

Exhaustive Search The most naive way of generating hypotheses is exhaustive search of whole space of possibilities, i.e. generating all variations of existing instances that match model structure. This way, number of generated hypotheses, that must be further checked for score, is rather big:

$$|O| = \prod_{d \in D} |d| \tag{5}$$

where O is set of object hypotheses and $D = \{d_i\}$ is a domain for particular part (i.e. set of all instances of the same class as given part). For each hypothesis score is remembered and the best ones are returned.

Classic CSP with Backtracking To avoid full expanding of hypothesis before its verification (which is computationally very expensive), problem can be treated as constrain satisfaction (CSP), with modification on predicate score (mentioned earlier). Basic backtracking algorithm is used to build hypothesis step by step. After assigning new variable (in our case assigning existing instance to yet unassigned model part) hypothesis score is calculated and, if score is lower than some set threshold, current branch is pruned and algorithm goes backward to search for other, better possibilities.

2.4 *Modifications of CSP Problem*

Checking All Solutions Classic CSP works until it finds first solution satisfying all constraints. As the score in our system can be anywhere between 0 and 1, we can compare two hypotheses and select the best ones. CSP implementation is thus modified as follows. We keep track of N best hypotheses found so far (list is empty when algorithm starts). At each step, hypothesis score is calculated and, if it is lower than the worst from the list, branch is pruned. If hypothesis is fully generated and its score is high enough, it is assigned to the list. This way we achieve the same effect as in exhaustive search, but with much better performance—a lot of hypotheses are rejected at early stage.

Most Constrained Variable First In domain D, set of possible assignments for each model part is held. At each step of the algorithm, next yet unassigned part is selected and then value for this part is selected from its domain. Order, in which parts are selected, has great influence on overall algorithm performance. We apply "most constrained variable first" CSP optimization here, i.e. we are choosing part which has the least number of possible assignments. This way, when tree is pruned (partial hypothesis is bad), remaining domain for unassigned parts is bigger than while using other ordering.

Domain Ordering While previous optimization tells us which part is to be assigned next, domain ordering tells us which instance has to be assigned next for selected part. It works as follows. For each element in selected parts domain the score for partial instance is calculated and element giving the biggest score is selected as a next assignment. This way high-scored instances are generated relatively quick, which enables us to reject weaker instances on early stages of algorithm, further reducing processing time.

Domain Filtering After selecting next parts assignment, it can be used to filter remaining domain with respect to model constraints (forward checking, at the moment system can filter taking only binary constraints into account). For each constraint with one unassigned argument, every element in the domain of this argument is checked (score for this constraint is calculated). All elements giving overall score (constraint score multiplied by current partial assignment score) lower than the worst from already created full assignments are removed from the domain.

3 Exemplary 2D Concepts

3.1 Edge Segmentation

As a 2-dimensional concepts the two kinds of edges were defined in this work: *line segment* and *elliptical arc*.

For the edge detection step the *ELSD* [10] algorithm was used. It takes a gray-level image as input and returns a list of detected *line segments* and *elliptical arcs*. It is based on *a contrario* approach and provides an efficient technique for an automatic computation of detection thresholds based on Helmholtz's perception principle. The "non-accidentalness principle" states that there is no perception in white noise. Therefore the algorithm controls its own number of false detections. It depends on several parameters that determine its behavior. Their values were designed by the authors of ELSD algorithm, thus it doesn't require any parameter tuning by user. The procedures and its parameters were shortly presented in Table 1.

Because some of the straight lines in the image are detected as elliptical arcs in our approach detected arcs having small angle (less than $\pi/8$) were converted to line segments before recognition step.

Edge Attributes Every edge concept has its *start* and *end-point*. According to [2] line segments detected by *ELSD* are oriented and the order of their starting and ending points encodes the angle from the dark to light transition. The right angle from the orientation of line segment to gradient vector have positive sign. However an *elliptical arc* extracted using *ELSD* algorithm has orientation arbitrary to gradient. The arcs orientation is encoded as opposite to angle orientation on corresponding ellipse.

For simplification the length of the elliptical arc is calculated after converting corresponding ellipse to circle with the radius equivalent to semi-major axis of the ellipse. We made an assumption that most of ellipses presented in 2-D image are projections of circles in 3-D world. Attributes, that were used to calculate edge predicates, are length of corresponding arc on circle with radius equal to semi-major axis of ellipse containing the arc and ellipse major axis orientation angle.

Table 1 ELSD algorithm procedures and its parameters

Procedure	Params	Explanation
Image filtering	$S = 0.8$, $\Sigma = 0.6$	Image scaling with factor S using Gaussian subsampling with standard deviation of Gaussian kernel determined by $\sigma = \Sigma/S$
Gradient calculation	$W = 2$, $q = 2$	Gradient calculation with mask WxW, pixels with gradient magnitude $\rho < q/sin(\tau)$ are rejected from constructing *line-support regions*
Gradient segmentation	$\tau = \pi/8, N = 8, D = 0.7$	Connecting pixels with similar gradient angle up to a certain tolerance τ using N-neighbourhood, forming rectangle regions with gradient pixel density under D
Region chaining	$\Delta\Theta_t = \pi/2$	*Smoothness rule* for regions chaining to form arcs; maximum orientation difference of linked rectangular regions
Region validation	$\epsilon = 1$	ϵ threshold controls numbers of false detections in *a contrario* approach and the Helmholtz principle

3.2 Edge Concepts

Edge Predicates The scores of constraints satisfaction were calculated for pair of objects using presented Concepts' attributes according to Table 2. In this experiment we proposed linear functions for scores calculation shown on Fig. 2a. Because line segments are oriented the *Linked* predicate was divided into 4 sub-types according to Table 3.

Edge Models The conducted experiments deal with the recognition of two objects: a rectangular plane and a cylinder. Both models are kinds of composite instances, that correspond to 3-D world shapes. *A rectangular plane model* consists of four line segments $s_1, \ldots, s_4$. Those edges that are distant are also parallel while consecutive edges are perpendicular. All of them have similar sizes. A generic semantic network for 4-edge based models is presented on Fig. 2b.

Table 2 Specification of parameter $x = n/d$ for different predicate pairs (f_{close}, f_{far}), where for ith edge: l_i—length, α_i—angle, $\mathbf{p}_1^i$—starting or $\mathbf{p}_2^i$—ending point

f_{close} , f_{far}	n / d
Sizediffer , Sizematch	$min(l_1, l_2)$ / $max(l_1, l_2)$
Parallel , Perpendicular	$\lvert\alpha_1 - \alpha_2\rvert$ / $(\pi/2)$
Linked , Distant	$\lvert\lvert\mathbf{p}_i^1 - \mathbf{p}_j^2\rvert\rvert$ / $mean(l_1, l_2)$

Table 3 Four types of linking predicates were defined according to the distances between following *endpoints* $\mathbf{p}_i^1$ and $\mathbf{p}_j^2$

Predicate	Visualisation	$i\,j$
Consecutive	—— ——	* *
Continuous	→→	2 1
Converging	→←	2 2
Diverging	←→	1 1

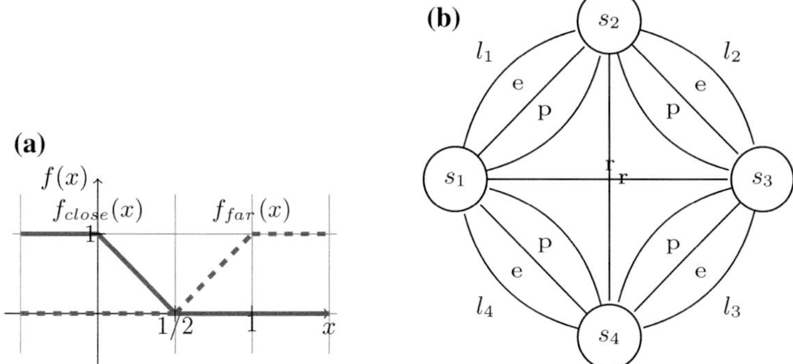

Fig. 2 **a** Score linear activation function, **b** semantic network for 4-*edge* based $(s_1, \ldots, s_4)$ model construction with predicates: *e* Sizematch, *p* Perpendicular, *r* Parallel, $[l_1 \ldots l_4]$ Linked (different types)

There were defined 4 submodels of rectangular plane according to edge directions presented on Fig. 3a. *A cylinder model* is build from two line segments and two elliptical arcs. The edges of the same type are parallel and distant. Figure 3b shows five types of edges orientation in cylinder models.

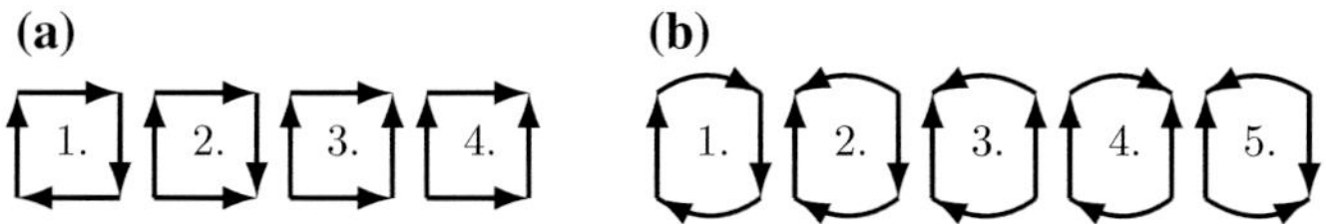

Fig. 3 Semantic network model construction according to edge orientation: **a** rectangular plane model types $(1, \ldots, 4)$, **b** cylinder model types $(1, \ldots, 5)$

4 Results

4.1 Material

The Datasets Images for testing computer vision algorithms consists of (13 objects, 12 views per objects) We used images aquired by the RGB camera, with resolution 1280×960px [11]. Objects positioned on a turntable against white background are observed from fixed viewpoint. For each object, the turntable was rotated 30° each.

4.2 2-D Results

For the presentation purposes all input images were cropped to 250×250 pixels regions including the object of interest before the detection process. The authors noticed different results for the same regions in original and cropped images as expected. The results of two models detection (plane and cylinder) are presented on Fig. 4.

4.3 Performance

Inference system, presented in previous sections, is meant to be used in object recognition tasks, also in applications that requires fast response (i.e. has to detect objects in seconds rather than minutes). Our approach of building object hypothesis from the low level segments, using CSP and constrained models, can be rather computationally expensive. Modifications and optimizations presented earlier were designed to cope with this problem. To check the influence of each of the modifications we conducted few experiments. Test scenes were created using a number of randomly generated, linear segment instances. Model, that was searched in those scenes, was 6-segment loop. Table 4 presents processing times for 10, 100 and 1000-element scenes for each algorithm version.

5 Discussion

5.1 Edge Fragmentation

In the presented solution detected object must satisfy the entire model. This is equivalent to logical conjunction of all predicates. Therefore the algorithm is sensitive for edge fragmentation and partial object occlusion. To minimize the fragmentation of

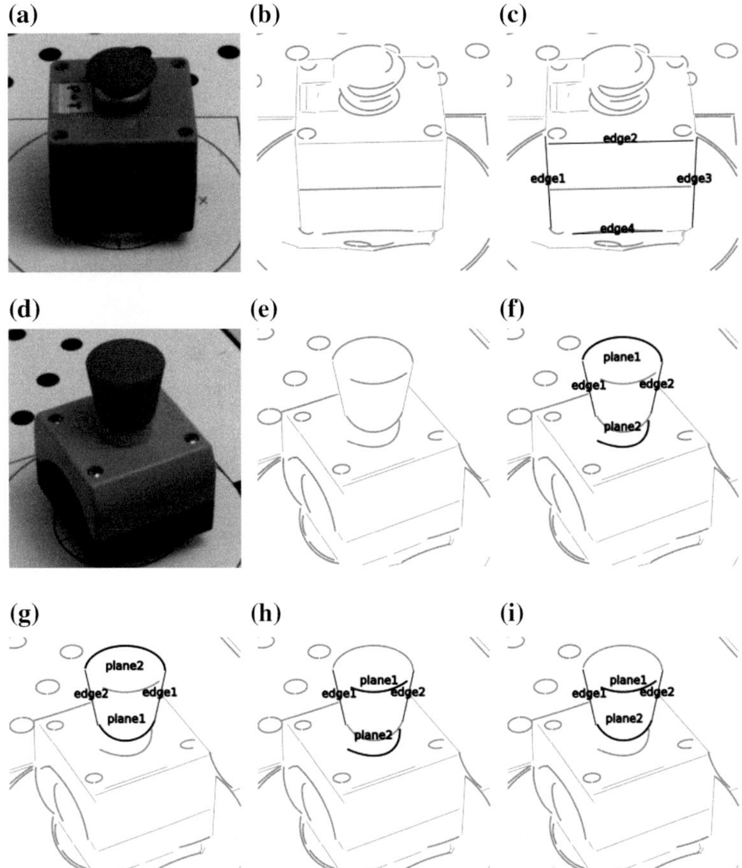

Fig. 4 Edge results displayed on images cropped to region of interest: **a** original image of button type-1, **b** result of ELSD detector for button 1, **c** result of square type-2 detection for button 1, **d** original image of button type-2, **e** result of ELSD for button 2, **f** cylinder type-1 detected for button 2 (score: $4.2 \cdot 10^{-5}$), **g** cylinder 1 ($6.2 \cdot 10^{-4}$), **h** cylinder 2 ($3.7 \cdot 10^{-6}$), **i** cylinder 2 ($5.8 \cdot 10^{-4}$)

Table 4 Processing times for generation of 10 hypotheses of 6-segment loop [s]

Number of segments	Exhaustive search	CSP with backtracking	Most constrained variable first	Domain ordering and filtering
10	25	<1	<1	<1
100	Timed out	2	2	<1
1000	Timed out	80	75	7

edges the scaling step is performed in *ELSD* algorithm. According to [2] it helps to cope with aliasing and quantization artifacts like staircase effect while the white noise statistics are preserved. However the authors shown that even if the image is filtered there are still fragmented lines.

The fragmentation problem could be further reduced at the edge segmentation level by redesigning *ELSD* algorithm parameters. The algorithm should be revised in terms of 2 x 2 window of gradient calculation and 8-pixels neighborhood of similar gradients forming the edge region. Hidden or not properly detected parts require partial solution of constraint satisfaction problem. The solution for this problem was presented in [4] and should be applied as a extension for our methods.

Multi-modal Models Because of missing logical disjunction operation in our inference model several sub-models were generated to cover all modalities of objects in terms of connected edges' orientation. The number of modalities may be reduced if we redesign the segmentation algorithm to make looser assumptions of straight lines, and to get rid of very slightly curved edges classified as arcs.

Skipping Color Information The *ELSD* algorithm processes only gray-level images. In current state the authors didn't extended the algorithm to cover all channels of RGB images. Most of significant edges in natural images were represented in brightness information. Moreover the color information is less reliable in terms of image compression in this experiment.

Image Cropping and Scaling The results of the *ELSD* algorithm differ when the image is analyzed at different scales or if algorithm is applied to small part of the image. According to [2] this is natural and corresponds to different details that one can see if an image is observed from a distance or if attention is paid to a specific part.

Acknowledgments The authors gratefully acknowledge the support of the National Centre for Research and Development (Poland), grant no. PBS1/A3/8/2012.

References

1. Felzenszwalb, P.F., Girshick, R.B., McAllester, D., Ramanan, D.: Object detection with discriminatively trained part-based models. IEEE Trans. Pattern Anal. Mach. Intell. **32**(9), 1627–1645 (2010)
2. Grompone von Gioi, R., Jakubowicz, J., Morel, J.M., Randall, G.: LSD: a line segment detector. Image Process. Line **2**, 35–55 (2012)
3. Hinterstoisser, S., Lepetit, V., Ilic, S., Holzer, S., Bradski, G., Konolige, K., Navab, N.: Model based training, detection and pose estimation of texture-less 3d objects in heavily cluttered scenes. In: Computer Vision-ACCV 2012, pp. 548–562. Springer (2013)
4. Kasprzak, W., Czajka, L., Wilkowski, A.: A constraint satisfaction framework with bayesian inference for model-based object recognition. Computer Vision and Graphics. Lecture Notes in Computer Science, vol. 6375, pp. 1–8. Springer, Berlin (2010)

5. Kasprzak, W., Kornuta, T., Zieliński, C.: A virtual receptor in a robot control framework. In: Recent Advances in Automation, Robotics and Measuring Techniques. Advances in Intelligent Systems and Computing (AISC), vol. 267, pp. 399–408. Springer (2014)
6. Kasprzak, W., Pietruch, R., Bojar, K., Wilkowski, A., Kornuta, T.: Integrating data—and model-driven analysis of rgb-d images. In: Intelligent Systems'2014, Advances in Intelligent Systems and Computing, vol. 323, pp. 605–616. Springer International Publishing (2015)
7. Lowe, D.G.: Three-dimensional object recognition from single two-dimensional images. Artif. Intell. **31**(3), 355–395 (1987)
8. Lowe, D.G.: Object recognition from local scale-invariant features. In: The Proceedings of the Seventh IEEE International Conference on Computer Vision, 1999, vol. 2, pp. 1150–1157. IEEE (1999)
9. Opelt, A., Pinz, A., Fussenegger, M., Auer, P.: Generic object recognition with boosting. IEEE Trans. Pattern Anal. Mach. Intell. **28**(3), 416–431 (2006)
10. Pătrăucean, V., Gurdjos, P., von Gioi, R.G.: A parameterless line segment and elliptical arc detector with enhanced ellipse fitting. In: Computer Vision—ECCV 2012, pp. 572–585. Lecture Notes in Computer Science, Springer, Berlin (2012)
11. Stefańczyk, M., Laszkowski, M., Kornuta, T.: WUT visual perception dataset—a dataset for registration and recognition of objects. In: Advances in Intelligent Systems and Computing (AISC) Progress in Automation, Robotics and Measuring Techniques. Springer (2016)

New Trends in the Use of Robotic Devices in Motor Rehabilitation of Upper Limbs

Marcin Bonikowski, Mikulas Hajduk, Lucia Koukolova, Wojciech J. Klimasara, Patrycja Mrozek and Zbigniew Pilat

Abstract In the years to come, robotic systems assisting physical rehabilitation will be used mainly by elderly, disabled, as well as children and adults after accidents and disorders limiting their physical capabilities. As the population is getting older, the issue becomes more and more critical. A growing number of people requiring rehabilitation generates significant costs, of which personal expenses are a major component. Providing the human personnel with appropriate mechatronic devices or replacing at least some rehabilitation medicine specialists with robots could reduce physical and mental workload of physicians. Broader application of such devices will also require among others new solutions in mechanic, control and human-robot communication. This paper presents overall vision of the development of rehabilitation robots, with consideration of the observed trends in this area, as well as expected achievements in electronics, material's engineering, ICT and other related fields of science and technology.

Keywords Robotics in rehabilitation · Motor rehabilitation · Disabled people · Elderly people · Development trends

---

M. Bonikowski (✉) · P. Mrozek
Mazovian Neuropsychiatry Center, Zagórze n., Warsaw, Poland
e-mail: martinbonikowski@gmail.com

P. Mrozek
e-mail: mrozek.patrycja@gmail.com

M. Hajduk · L. Koukolova
Technical University of Košice, Kosice, Slovakia
e-mail: mikulas.hajduk@tuke.sk

L. Koukolova
e-mail: lucia.koukolova@tuke.sk

W.J. Klimasara · Z. Pilat
Industrial Research Institute for Automation and Measurements PIAP, Warsaw, Poland
e-mail: wklimasara@piap.pl

Z. Pilat
e-mail: zpilat@piap.pl

R. Szewczyk et al. (eds.), *Challenges in Automation, Robotics and Measurement Techniques*, Advances in Intelligent Systems and Computing 440, DOI 10.1007/978-3-319-29357-8_63

1 Introduction

At the end of the 20th century in many R&D centers around the world they were carried out various projects related to the use of robotics solutions to aid and assist motor rehabilitation. Currently, these works have entered the phase of commercialization. There are many companies offering ready-made devices. There is a fairly widespread belief that the effectiveness of rehabilitation grows thanks to the use of modern robotics or broadly mechatronics systems. Despite this fact the widespread dissemination of them still encounters obstacles. The reason is certainly the price of these devices. They are simply expensive. There is also many to do in the field of construction. This applies to both the mechanics, control as well as of the widely understood human communication (patient or therapist) with the robot. Currently ongoing R&D works are concentrated on these problems.

2 Arguments for the Application of Mechatronic Devices in Supporting Rehabilitation

2.1 Current Medical Studies and Future Challenge to Robotized Technology

In the last decades we saw a remarkable change in the neuro-rehabilitation approach. New theories based on plasticity shoved property of the adult human brain to remap and create new neural pathways. The intensity of these processes depends on many factors. In the areas of the brain responsible for control and supervision of the motorics, a significant impact on the creation of new synaptic connections, is physical activity. These exercises can be performed by patient himself or with assistance of a therapist. The individual therapeutic movements must be repeated many times, over and over again. For this reason, the work of therapist is very hard physically and tedious. Understanding the opportunities offered by plasticity and awareness of the hardships and difficulties carrying out rehabilitation, led to innovative concepts to use in these therapeutic activities the robotic and mechatronic devices. A precursor was the MIT-Manus robot, which was developed in 1989, and in 1994 it was installed in the clinic, where the first tests were successful conducted [1].

Since then multitude of robotic devices were developed and tested for neuro-rehabilitation. A majority of these technically advanced devices are designed for clinical settings. However, there is still need to improve efficiency and reduce cost. Future development of accessible home-based devices for therapy and Activities of Daily Living (ADL) assistance is essential. Robot Assisted Therapy (RAT) efficacy has been studied in stroke patients. A recent meta-analysis concluded that RAT could improve upper limb (UL) structure and function and the ADL of these patients [2]. But still the effectiveness of robotic over conventional therapy is arguable and the best therapy strategy is still not clear. Recent recommendations based on motor learning theories state that intensive rehabilitation is necessary for improving motor function in children with Cerebral Palsy CP [3]. Robotic therapy which is: goal-oriented, assisted if needed, associated with sensory feedback may promote reorganization of the neuronal networks and restoration of normal function after brain lesion. RAT of the upper limb has been applied in children with CP with promising results [4]. However, there are no currently published randomized controlled trials, and such studies are needed to confirm the usefulness of RAT in children with CP.

Up to now, robotics and virtual reality, have proven to be applicable in the area of neurorehabilitation but not in the orthopedic and rheumatology. Potential applications of robotics in orthopedic and rheumatologic rehabilitation of upper limb are enormous. As with stroke patients a spread of home based inexpensive RAT systems may have huge influence on health care system.

Recent search into the US Clinical Trials database (http://clinicaltrials.gov/) from October 2015 using a combination of keywords: robot, hand, upper limb and rehabilitation identified 8 new trials recruiting patients. All studies focus on stroke survivors. Main topics are combination of RAT and electrical stimulation or pharmacotherapy, and efficacy of different robots designs and treatment strategies. Robots are: ReHapticKnob, InMotion3(IMT), Bi-Manu-Track (BMT), Armeo, Hand of Hope (HOH), Active and Sham tDCS. There is still a need for well-designed large scale multicentre studies to evaluate usefulness of RAT in UL rehabilitation. Further research should also investigate potential of RAT in ADL function recovery. Also in combination with neuro-imaging techniques might be useful to provide the evidence of neural remodeling.

2.2 *Robotic Devices for Motor Rehabilitation Currently Used in Therapeutic Practice*

Considering UL rehabilitation, we can talk about variety of devices, with different type of assistance, mechanical structure, control inputs, supported movements and fields of application. Basic division of robotic rehabilitation devices consist of the **type of assistance** provided by the actuators [5].

Active devices (e.g.Myomo, PowerGrip, ReoGo) provide active motion assistance. This devices possess active actuator(s) able to produce movement of the upper-extremity. Such assistance of movements is preferred for patients with significant movement disorders at early stage of rehabilitation.

Passive devices (e.g.Armeo Spring), are unable to move the limbs, but may resist the movement e.g. when exerted in the wrong direction. Thus, this type of devices is appropriate for rehabilitation of patients with preserved ability to move their limbs. Usually they are lighter and cheaper than those with active assistance [6].

Nowadays, most popular, commercial rehabilitation robots used in Poland, created for upper extremity are Armeo Power, Armeo Spring (Hocoma,

Switzerland) and ReoGo (Motorika Medical Ltd, Israel) [7]. All of them are exoskeleton-based devices, that enhances patient motivation through a variety of interactive and stimulating games that imitate natural hand movements.

Armeo Power enables intensive arm rehabilitation for early-stage patients even before they develop active movement. In the multicenter study the ArmeoPower was shown to lead to faster and better recovery after stroke compared to conventional therapy. However, the absolute difference between effects of robotic and conventional therapy was small and of weak significance, which leaves the clinical relevance in question [8].

Armeo Spring is suitable for the patients from severely to moderately affected. It is specifically designed for patients who are beginning to regain active movement of the arm and hand. Data from study comparing the effects of equal doses of robotic and conventional therapy in individuals with chronic stroke suggest, that Armeo Spring therapy can elicit improvements in arm function that are distinct from conventional therapy and supplements conventional methods to improve outcomes [9].

ReoGo is the robotic system designed to facilitate three-dimensional repetitive arm movements through the use of an advanced, fully motorized robotic arm. The system is suited for patients in all stages of stroke recovery, offering five modes of operation; ranging from passive support for highly functional patients to guided motion for cases with little or no arm mobility. Bovolent et al. [10] showed a positive evolution of the limitation of activity and functionality for all subjects treated with ReoGo.

3 Initial Development of Rehabilitation Robotics

The positive impact of exercises on both physical and mental health was known even in the ancient times. Back in those days, exercises were meant to help improve and maintain general fitness, as well as regain such fitness by people who had been injured. Physical exercises were systematized according to their therapeutic effects on specific body parts by Ling [11], which laid the groundwork for the emergence of so-called medical gymnastics. In the 19th century, it was widely promoted and developed in many centre's all over the world and often applied in the treatment of orthopedic disorders. The first therapeutic facilities were opened in places with appropriate climate, where patients received comprehensive recuperation treatment. In such centre's, where numerous patients would undergo medical gymnastic exercises at the same time, exercise-facilitating devices were becoming more and more popular. Those included mainly mechanical devices. One of the most renowned creator of devices used in medical gymnastics was a Swedish doctor named Jonas Gustaw Wilhelm Zander, who developed a method of treatment and regaining fitness through exercises performed on the apparatuses he designed [12].

Devices assisting physical rehabilitation were developed through upgrading their construction and introducing new materials. The natural consequence of this development was emergence of a new field of study—rehabilitation robotics. The first R&D works on the application of robotised technologies in supporting physical rehabilitation on record were carried out in USA in the early 1960s. Rancho Los Amigos National Rehabilitation Center (Rancho) created an electrically-powered orthosis with seven degrees of freedom. This device called Golden Arm [13] was initially designed for patients with the post-polio syndrome. At the same time, scientists from the Case Institute of Technology (Cleveland, Ohio, currently: Case Western Reserve University) created a pneumatic orthosis with four degrees of freedom. In both cases, practical application of the inventions was difficult due to insufficiently effective control systems and the lack of sensors that would ensure feedback depending on the position, speed and force. Further development of advanced sensor and computer technologies and their subsequent application in the field of robotics encouraged works on rehabilitation robots. In the 1980s, those efforts took place mainly on American and Western European universities and their research & development centres. It coincided with a breakthrough in medical research over the organisation of brain functioning and the nervous tissue structure, which significantly expanded the knowledge about brain and its highly flexible internal construction. Scientists coined a term of neuroplasticity meaning the ability of nerve cells in the brain to regenerate and create new networks with other neurons. As a result, the healthy nervous tissue can take over those functions of the brain, which have been impaired as a result of a local irreversible damage, e.g. caused by a stroke. It means that, through effective physical training requiring regular and long-term exercises, patients can teach their brains again to perform certain activities (such as walking, grabbing, etc.). The results of the medical research encouraged further research and development works over new and advanced rehabilitation devices using the solutions hitherto applied in robotics.

One of the first mechatronic rehabilitation devices, which have been positively evaluated by the global medical community, is the Manus robot developed in MIT [14] to assist the rehabilitation of upper limbs. Its mechanical part is composed of a manipulator having the kinematic structure of the robot named SCARA. The control system integrates the sensors of force and location with the complex patient-robot communication interface. The positive impact of the Manus robot application on the rehabilitation process was confirmed by research results. The response of patients to the new device and exercise method was highly positive. However, kinematic properties of the Manus robot manipulator enable only a single-plane motion, which somewhat limits its application.

Another concept was implemented in the University of Stanford. The solution was based on a classic industrial robot called PUMA (Staubli Unimation Inc.), which using a special mechanical interface, leads patient's arm along the programmed trajectory. This system named MIME (Mirror Image Movement Enabler) enables movement of the rehabilitated upper limbs along multi-plane trajectories [15].

Similar approach was applied in one of the first European projects in the field of robot-assisted arm rehabilitation. Reharob was created under the 5th EU

Framework Program [6]. Its main objective was to develop an arm rehabilitation system using standard robots. The project was coordinated by the University of Budapest and the consortium included the ABB company, which supplied two industrial robots. Owing to proper situation (one robot supports the arm near the elbow and another robot moves the patient's wrist) and control of those robots, the system enables the movement of the patient's arm on all anatomical planes of motion.

Research on devices supporting motor rehabilitation of UL has been performed for many years by the PIAP institute in Warsaw. One of their most interesting projects is Renus-1, mechatronic active rehabilitation assistance system. This device provides capability of performing complex, controlled spatial movements of patient's UL [12].

The solutions, developed and successful tested within R&D projects, were the basis for commercialization of rehabilitation robotics, through applied many similar constructions in devices designed for the market.

4 Mechanical Structure of Robotic Devices for Movement Rehabilitation

When comparing the **mechanical structure** of robotic devices for movement rehabilitation, two categories of devices are considered. These are end-effector-based and exoskeleton-based devices. The difference between those two is how the movement is transferred from the device to the patient's limb.

4.1 End-Effector Based Devices

This devices (e.g. InMotion ARM) contact the patient's limb only at its most distal part what simplifies the structure of the device. Although, this feature can reduce the control of the position of particular joints because of many possible degrees of freedom.

Movements of the end effector change the position of attached upper limb. Segments of upper limb create a mechanical chain, thus movements of the end effector also indirectly change the position of other segments of the patient's body.

The advantage of end-effector based systems is their simpler structure and less complicated control algorithms. However, it is difficult to isolate specific movements of a particular joint because these systems produce complex movements. The manipulator allows up to 6 unique movements (3 rotations, 3 translations). Control of the movements of the patients upper limb is possible only if the sum of possible anatomical movements of patient arm in all assisted joints is limited to 6.

The typical end-effector based systems are presented in Fig. 1.

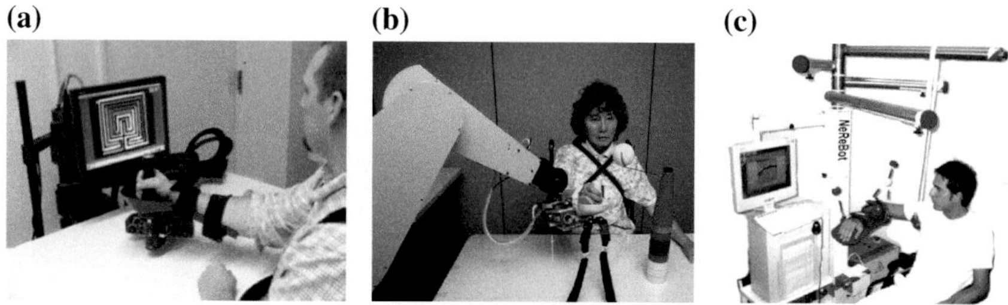

Fig. 1 **a** MIT-MANUS [14], **b** MIME [15], **c** NeReBot [19]

4.2 Exoskeleton-Based Devices

This devices (e.g. Armeo Power) are an external mounted mechanisms which have a mechanical structure that mirrors skeletal structure of patient's limb. Their joints match those of the human body. Movement in particular joint of the device directly produces a movement of the specific joint of the limb. It adapts to a person so that physical contact between the operator and the exoskeleton enables a direct transfer of mechanical power and information signals [16]. It's more complex than an end-effector based device, however this design allows independent, precise and isolated movements in a joint(s).

Application of the exoskeleton-based approach allows independent and concurrent control of particular movement of patient's arm in many joints, even if the overall number of assisted movements is higher than 6.

In order to avoid patient injury, it is necessary to adjust lengths of particular segments of the manipulator to the lengths of patient´s arm segments. Therefore setting-up such device for a particular patient may take a significant amount of time.

Furthermore, the position of the center of rotation of many joints of human body may change significantly during movement, therefore special mechanisms are necessary to ensure patient safety during the movements of these joints. For this reason, the mechanical and control algorithm complexity of such devices is usually significantly higher than of the end effector-based devices. The complexity escalates as the number of DOF increases. The typical exoskeleton based devices are presented in Fig. 2.

Although exoskeleton based devices are now more popular, both directions are being developed. End-effector based devices better placed to exercise typical utility movements of the upper limb. In addition, they are much cheaper than the exoskeleton based devices.

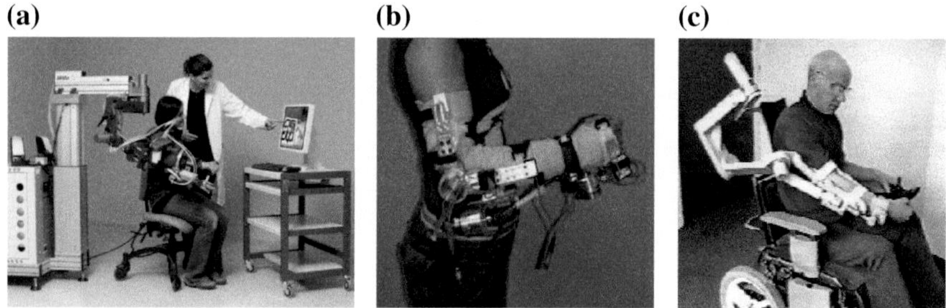

Fig. 2 **a** ARMin [5], **b** WOTAS [20], **c** MULOS [5]

5 New Solutions in Control

Nowadays majority of rehabilitation robots, used in practice, have very similar structure of control. The main controller is responsible for the trajectory realization. The function of master computer, usually PC type, contains human robot communication, planning and collecting of exercises, as well as supervision of their execution. On this computer data, the basis of patients and exercises is stored. This computer also contains different kinds of animations and games, for ease of patient's motivation to exercise. The R&D in area of control can be presented in 3 directions.

5.1 *Virtual Reality*

Rehabilitation based on VR is characterized by the integration of cognitive and physical tasks within a multimodal sensory environment [17]. The game based VR systems can be successfully used to the restoration and rehabilitation of hand function of after stroke persons. Possibility and attractiveness of VR in the post stroke individuals rehabilitation is evident and conformed by clinically performed tests.

5.2 *Haptics Interfaces*

Important limitation of mechatronic rehabilitation devices is the lack of function of feeling. When a therapist practices an exercise, he feels resistance of the patient and can change the movement according to this reactions. In order to implement such kinds of functions in rehabilitation robots, the first step is to equip it with special sensors. The solution is to use so called haptic interfaces, which enables measure the back forces between patient and robot. Two possibilities are analyzed and researched today. The first one is installation of special sensors in joints of the robot (measurement of motion resistance). The second method is installing these sensors in the point where robot has direct contact with patient (tactile sense of touch).

5.3 Human-Robot Communication

Nowadays usually two ways of communication of the rehabilitation robots are used: with patient and therapist. Screens or monitors are used as a standard solution [18]. Various information, such as mechanism for exercise planning, options, supervision as well as information about the progress of the exercise are displayed on the therapist screen. Animations and games are presented on the patient monitor. The next step should be introducing a voice communication. Especially it is important to implement such a functionality, where robot will recognize specific messages from patient, e.g. about pain that occurs during exercise. In relation to the perspective appearance of home rehabilitation robots, it is necessary to develop remote access that enables the therapist, doctor, caregiver to obtain information about the progress of rehabilitation, patient's condition, etc.

6 Summary

Robot aided neurorehabilitation is rapidly developing field, based on robotics and mechatronics, virtual reality, haptic interfaces, neuroscience and conventional rehabilitation. It can be characterized by challenges both in engineering and in clinical practice. There is a need for more integrated solutions to perform a therapy in a safe environment and with a good compliance from the patients. Robotic rehabilitation is certainly undergoing a period of rapid growth. Notwithstanding, many issues remain still open. Robotic rehabilitation may offer benefits not only in terms of cost reduction but also allows quantifying objectively the patient outputs and outcomes (for example strength improvement, coordination, time of movements). Present researches are applied to improve the efficiency of robotic manipulation, the strengths, trajectories, and multi-sensorial inputs in order to provide the quality and efficiency of rehabilitation.

In the near future robotic rehabilitation devices for UL will be available to clinicians, both in the hospitals and in homes. For this purpose the main barrier is the price, these devices must be considerably cheaper. In the case, where the robot-type rehabilitation apparatus, will not be operated by a trained therapist, but by the patient himself, a new security problems will appears.

Acknowledgements The paper presents results of researches supported by EU within the project RoboReha "Robotics in Rehabilitation", LdV—TOI no. 13310 0530, under the Lifelong Learning Programme. This publication represents only author´s opinion and neither the European Commission nor the National Agency is not responsible for any of the information contained in it.

References

1. Krebs, H., Hogan, N., Volpe, B.T., Alsen, M.L., Edelstein, L., Diels, C.: Overview of clinical trials with MIT-MANUS: a robot-aided neurorehabilitation facility. Technol. Health Care **7**(6), 419–423 (1999)
2. Mehrholz, J., Hädrich, A., Platz, T., Kugler, J., Pohl, M.: Electromechanical and robot-assisted arm training for improving generic activities of daily living, arm function, and arm muscle strength after stroke. Cochrane Database Syst Rev. **6** (2012)
3. Sakzewski, L., Ziviani, J., Boyd, R.: Systematic review and meta-analysis of therapeutic management of upper-limb dysfunction in children with congenital hemiplegia. Pediatrics **123**, 1111–1122 (2009)
4. Frascarelli, F., Masia, L., Di Rosa, G., et al.: The impact of robotic rehabilitation in children with acquired or congenital movement disorders. Eur. J. Phys. Rehabil. Med. **45**, 135–141 (2009)
5. Koukolová, L.: Overview of the robotic rehabilitation systems for upper limb rehabilitation, Transfer inovácií 30/2014
6. Maciejasz, et al.: A survey on robotic devices for upper limb rehabilitation. J. Neuroeng. Rehabil. **11**, 3 (2014)
7. Mikołajewska, E., Mikołajewski, D.: Usage of robotic rehabilitation for improvement, Niepełnosprawność - zagadnienia, problemy, rozwiązania. Nr IV/2013(9) (in Polish)
8. Klamroth-Marganska, V., Blanco, J., Campen, K., Curt, A., Dietz, V., Ettlin, T., Felder, M., Fellinghauer, B., Guidali, M., Kollmar, A., Luft, A., Nef, T., Schuster-Amft, C., Stahel, W., Riener, R.: Three-dimensional, task-specific robot therapy of are after stroke: a multicentre, parallel-group randomised trial. Lancet Neurol. **13**(2), 159–166 (2014)
9. Brokaw, E.B., Nichols, D., Holley, R.J., Lum, P.S.: Robotic therapy provides a stimulus for upper limb motor recovery after stroke that is complementary to and distinct from conventional therapy. Neurorehabil. Neural Repair. **28**(4), 367–376 (2014)
10. Bovolenta, F., Sale, P., Dall'Armi, V., Clerici, P., Frances, M.: Robot-aided therapy for upper limbs in patients with stroke-related lesions. Brief report of a clinical experience. J. Neuroeng. Rehabil. **8**, 18 (2011)
11. Brodin, H.: Per Henrik Ling and his impact on gymnastic. Sven Med Tidskr **12**(1), 61–68 (2008). (in Swedish)
12. Pilat, Z., Klimasara, W.J., Juszyński, Ł., Michnik, A.: Research and development of rehabilitation robotics in Poland. ROBTEP 2014. Applied Mechanics and Materials, vol. 613, pp. 196–207 (2014). © (2014) Trans Tech Publications, Switzerland
13. Lansberger, S., Leung, P., Vargas, V., Shaperman, J., Baumgarden, J., Yasuda Lynn, Y., Sumi, E., McNeal, D., Waters, R.: Mobile arm supports: history, application, and work in progress. Top Spiral Cord Inj. Rehabil. **11**(2), 74–94 (2005) (Thomas Land Publishers, Inc.)
14. Krebs, H.I., et al.: Robot–aided neurorehabilitation: from evidence-based to science-based rehabilitation. Top Stroke Rehabil. **8**, 54–70 (2002)
15. Lum, P.S., et al.: MIME robotic device for upper-limb neurorehabilitation in subacute stroke subjects: a follow-up study. J. Rehabil. Res. Dev. **43**(5), 631–642 (2006)
16. Pons, J.L., Rocon, E., Ruiz, A.F. Moreno, J.C.: Upper-limb robotic rehabilitation exoskeleton: tremor suppression. In: Kommu, S.S. (ed.) Rehabilitation Robotics. InTech (2007). ISBN: 978-3-902613-04-2

17. Wade, E., Winstein, C.J.: Virtual reality and robotics for stroke rehabilitation: where do we go from here. Top Stroke Rehabil. **18**(6), 685–700 (2011). (Thomas Land Publishers, Inc.)
18. Dunaj, J., Klimasara, W.J., Pilat, Z., Rycerski, W.: Human-robot communication in rehabilitation devices. JAMRIS J. Autom. Mobile Robot. Intell. Syst. **9**(2), 9–19 (2015). ISSN 1897-8649
19. Masiero, S., Celia, A., Rosati, G., Armani, M.: Robotic-assisted rehabilitation of the upper limb after acute stroke. Arch. Phys. Med. Rehabil. **88**(2), 142–149 (2007)
20. Rehabilitation Robotics. Chapter 25 Upper-Limb Robotic Rehabilitation. In: Kommu, S.S. (ed.) Exoskeleton: Tremor Suppression, I-Tech Education and Publishing (2007). ISBN 978-3-902613-04-2

Part III
Measuring Techniques and Systems

Multi-camera Vision System for the Inspection of Metal Shafts

Piotr Garbacz and Tomasz Giesko

Abstract To address the needs of the manufacturing industry, the automation of quality inspection processes is often performed. Ensuring the high reliability and accuracy of measurement usually requires the development of specialized systems dedicated for specific applications. In the paper, a multi-camera vision system developed for the automatic inspection of metal shafts in the automotive industry is presented. The novel solution of the inspection system is based on the application of a set of three cameras that allow the inspection of shafts in separated areas. The system enables the detection and recognition of the following features: thread, knurls, grooves, and measurements of the shaft length. The image processing is performed with the use of filtering, morphology operations (dilation and erosion), and edge detection. The two-dimensional Look-up Table has been created in order to calculate a correct shaft length considering perspective errors of lenses.

Keywords Shaft defects · Automatic optical inspection · Multi-camera vision system · Image analysis

1 Introduction

In many factories operating in the automotive sector, the quality inspection of products, consisting of visual observation and the use of basic gauges, is often performed by trained workers using only naked eyes. However, the complex quality inspection of more technologically complicated products includes a number of tasks,

P. Garbacz (✉) · T. Giesko
Instytut Technologii Eksploatacji – Państwowy Instytut Badawczy, Radom, Poland
e-mail: piotr.garbacz@itee.radom.pl

T. Giesko
e-mail: tomasz.giesko@itee.radom.pl

R. Szewczyk et al. (eds.), *Challenges in Automation, Robotics and Measurement Techniques*, Advances in Intelligent Systems and Computing 440, DOI 10.1007/978-3-319-29357-8_64

e.g., checking of product features, detecting various surface and shape defects, and geometrical measurements. In such cases, the planned inspection process is executed by a worker or workers, and includes individual tasks that are often performed using specialized measuring equipment at inspection stations. For example, high accuracy measurements of dimensions require contact or contactless precise measurement tools. Manufacturers need to increase production efficiency to enhance competition in markets, and the systematically increased demands to raise the level of quality requirements are the reasons to apply the automation in quality inspection. The dynamic growth of optomechatronics technologies enables new possibilities in developing the advanced automatic inspection systems with the use of machine vision methods [1]. Automation and robotics application in manufacturing processes is of key importance for increasing the innovation and competitiveness in industry, according to prospective goals defined in the "Industry of the Future" concept [2]. Mechatronics and optomechatronics are leading innovative technologies for the development of automated manufacturing systems. The current state of the art and a critical overview of real-world applications of machine vision methods for quality inspection in industry are presented in [3]. Furthermore, Neogi et al. [4] presented a wide review of optical inspection systems for defect detection and classification of steel surfaces. Innovative optomechatronics and advanced inspection systems develop a systematic approach to designing processes and the adoption of more effective design methods present a new paradigm. The development problems of design methodology dedicated to machine vision systems are the subject of a few studies [5]. The original design and implementation methodology for optomechatronic systems, including automatic optical inspection systems (AOI), is presented in monograph [6]. When designing systems for quality inspection in industry, an inspection of metal parts is one of most often-determined tasks. Special problems concerning the development of effective inspection methods of metal surfaces arise from specific phenomena of light interaction with the object's surface. A number of interferences during measurements and vision inspection can be caused by a high surface reflectivity and reflectivity irregularity at different areas on object's surface. Lighting plays a critical role in vision inspection as one of key factors that influence image quality and the effectiveness of inspection process. Therefore, problems of modelling the illumination for inspection processes are the subject of numerous research studies [7–11]. Apart from the detection of defects, the inspection process very often includes precision measurements of machine parts carried out in industrial environments. Increasing the reliability of measurements is possible by using hybrid methods involving mechanical contact and optical non-contact measurement techniques [12]. The inspection of metal products, particularly machine parts for the automotive industry includes the detection and recognition of various types of surface defects and shape defects. The most commonly occurring types of defects are the following: geometric defects, the lack of machining operation, forging defects, grinding defects, turning defects, cracks, scratches, material loss, and corrosion.

In computer vision systems, non-contact measurements are based on the analysis of information coded in the image registered by a camera. In order to measure geometric dimensions of the object presented in the image, edge detection and shape recognition is necessary. The edge detection and shape recognition methods are described in a number of papers [13, 14]. Among methods used for edge detection, gradient methods that use Robert's cross operator, the Sobel operator, the Prewitt operator, and the Laplace operator are the most popular [15, 16]. To obtain proper separation of the object from the background in the image, the analysis of the grey-level image, and comparing every pixel intensity with global intensity should be performed. Basic bi-level binary thresholding that converts a grey-level image to the set of points with values of 0 or 1 is often used in image pre-processing. Bi-level binary thresholding is defined by:

$$I'(x,y) = \begin{cases} 0 & \textit{if } I(x,y) \le B \\ 1 & \textit{if } I(x,y) > B \end{cases} \tag{1}$$

where

$I'(x,y)$ binary value at location (x, y) belonging to the binarized image,
$I(x,y)$ intensity pixel at location (x, y) belonging to the original image,
B threshold value.

Apart from bi-level thresholding, the multi-level thresholding techniques are used for segmenting the image into several separated regions. In 2015, Jiang et al. [17] proposed the optical inspection system for the detection of defects on shaft surfaces. A line-scan camera, and front light and back light illuminators were used in the system. The method to measure the shaft diameter and the application of algorithms for the correction of measurement errors caused by optical distortion are reported in [18]. Paper [19] presents the vision system for the inspection of shaft edges with the use of two cameras and parallel lighting. The application involving a high-speed camera for detection of the ellipticity of shafts during rotation is described in [20]. Nowadays, automatic systems for precision measurement of shafts [21, 22] are offered on the market. However, the manual clamping device is a significant limitation that reduces applications of these systems to workshop conditions. Furthermore applied solution of a single camera mounted on a linear drive requires a few seconds for the acquisition and analysis full part image not including installation time of the product in measuring socket. Therefore these kind of systems are dedicated rather for statistical control of products.

The paper presents the developed automatic optical inspection system for quality inspection of shafts in industry. The general concept of the solution and system structure is fundamentally influenced by customer requirements, as well as technical and economic limitations. The elements of original design and implementation methodology for optomechatronic systems were applied for developing the inspection system [6].

2 Inspection Method

The method is designed for the inspection of shafts mounted in motor wipers. According to customer requirements, the vision inspection system ought to ensure the detection and recognition of features: thread, conical knurl, cylindrical knurl, groove, and measurements of shaft length in the range from approx. 86 mm to approx. 140 mm, with an accuracy of ±0.1 mm (Fig. 1).

The vision system structure for the inspection of shafts is presented in Fig. 2. The modular vision system consists of three area-scan cameras with lenses and LED illuminators. To ensure the high contrast of images while measuring the shaft length, the line back-light LED illuminator of high brightness has been developed. The proper adjusting of cameras in the inspection area ensures the inspection of the entire shaft using three separated images. Dividing the global inspection area into smaller regions allows achieving high measurement accuracy in images by applying high-resolution cameras with sensors of 2592 × 1944 pixels. For the field of view of a dimension of 50 mm along shaft axis, the optical resolution is approx. 0.02 mm/pixel. To achieve the same optical resolution using a single camera vision

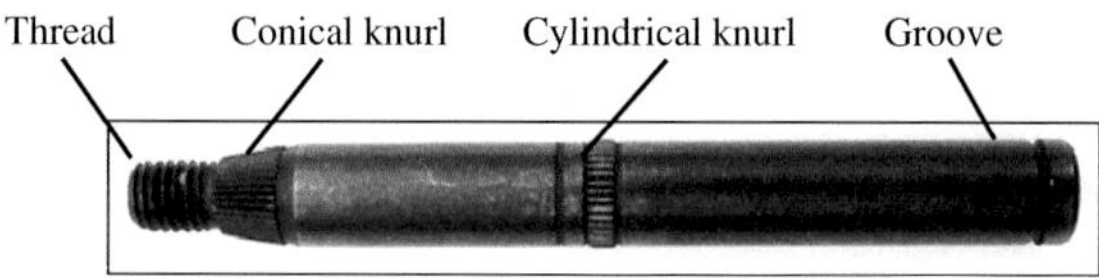

Fig. 1 Wiper motor shaft

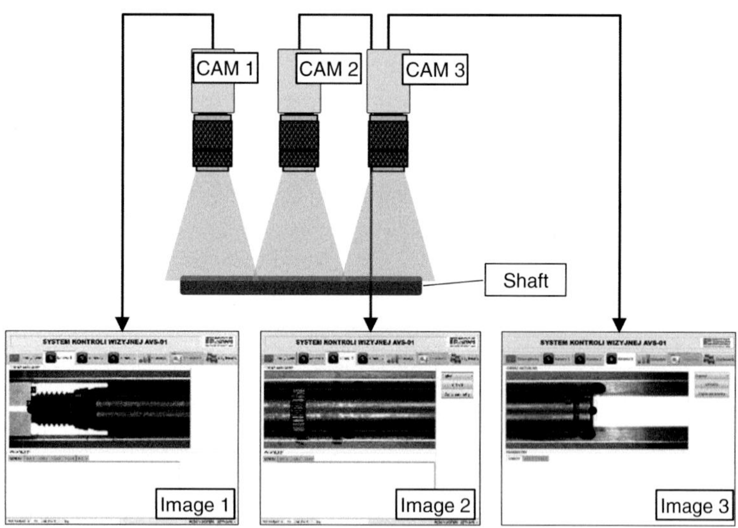

Fig. 2 Scheme of the multi camera vision system for the inspection of shafts

system, the sensor resolution would have to be not less 7500 pixels along shaft axis. However, currently available industrial cameras of maximum resolution 29 megapixels are equipped with sensors of 6576 × 4384 pixels in size. Furthermore, when using a single camera vision system with conventional lens to measure the length of an object, optical distortion, and perspective errors generated by conventional lens become a critical problem. At the initial phase of the design process, the possibility of applying telecentric lens instead of conventional lens in the vision system was analysed. Taking into account the high cost of the telecentric lens allowing the observation in the field of view of diameter approx. 150 mm, the solution concept of a single camera system has been dropped. After problem analysis, the concept of a multi-camera vision system has been accepted. Positions of cameras along the shaft axis are adjusted to sizes of shafts that belong to the determined series. When the inspection of longer shafts is needed, it is possible to reconfigure the structure of vision system. Such reconfiguration includes the modification of the position of cameras and the installation of an additional camera as well. According to customer's stated requirements and technical limitations, no drive unit for rotating the shaft during inspection is applied. It was assumed that the inspection of a shaft in any position is sufficient.

The image processing is performed by software with the use of following operations: filtering, morphology operations (dilation and erosion), and edge identification utilizing the Shen-Castan detector (Fig. 3). After this phase, the shaft length is calculated and features of the shaft are detected (thread, knurls, grooves) in the ROI's (Regions of Interest). When the shaft length is not equal to the nominal value or one of the features is lacked, the given shaft under inspection is classified as defective.

The shaft length is calculated based on the position of the shaft presented in pixels in images. In the first step, positions of shaft ends are detected and calculated in pixels. Values presented in pixels are converted into millimetres with use of the

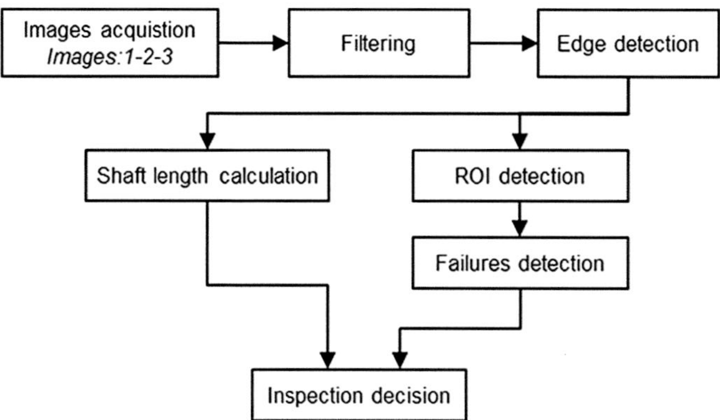

Fig. 3 Scheme of inspection procedure

Look-up Table. The two-dimensional Look-up Table was created in order to calculate a correct shaft length considering perspective errors. The developed table contains pairs of length values of standard shafts in millimetres and corresponding values in pixels. A set of pairs has been created for the sequence of different shaft positions under the vision system. Subsequent positions of a standard shaft in the camera window were adjusted by means of a micrometre screw gauge. The accuracy of the gauge is 0.01 mm. For positions that are placed between the reference points, a linear interpolation is used. Minimization of interpolation errors is possible by reducing intervals between reference points in the table during the calibration procedure. The shaft length is calculated using the following equation:

$$l_s = l_n + k \cdot \Delta_{px} \tag{2}$$

where

l_s shaft length (mm),
l_n nearest value of the reference point from the Look-up Table (mm),
k interpolation factor (mm/px),
Δ_{px} position of shaft's end in the image (px).

The measurement uncertainty budget includes a calibration error of the measurement set using a micrometre screw gauge and the detection error of edges in image. Influences of other factors are assumed as negligible, e.g. temperature. According to the GUM (Guide Uncertainty Measurement) [23], the measurement uncertainty is calculated. The standard uncertainty of a measurement is defined by the following:

$$u_c = \sqrt{u_{cal}^2 + u_{p1}^2 + u_{p2}^2} \tag{3}$$

where

u_{cal} calibration uncertainty,
u_{p1} resolution of the image, where the first shaft end is presented,
u_{p2} resolution of the image, where the second shaft end is presented.

Analysed measurement uncertainty depends on the used calibration method, where the error of micrometre screw gauge reading and the error of edge detection in image are dominant. A calibration uncertainty is expressed by the following equation:

$$u_{cal} = \sqrt{\left(\frac{\frac{1}{2}\Delta_m}{\sqrt{3}}\right)^2 + u_{p1}^2 + u_{p2}^2} \tag{4}$$

where

Δ_m scale division in the rotating head ($\Delta_m = 0.01$ mm),
u_{p1} resolution of the image presenting the first shaft end ($u_{p1} = 0.02$ mm),
u_{p2} resolution of the image presenting the second shaft end ($u_{p2} = 0.02$ mm).

Image resolution is determined by dividing the image dimension of 50 mm by the sensor resolution of 2594 pixels giving the result of approx. 0.02 mm/px. The lens used in the vision system is designed for a camera of resolution 10 Mpx then the optical resolution of the lens is not a critical limitation.

After calculating, the obtained measurement uncertainty is $u_c = 0.04$ mm. Expanded uncertainty U_c (coverage factor $k = 2$) is calculated using the following equation:

$$U_c = k \cdot u_c = 0.08\,\text{mm} \tag{5}$$

The group of tasks in the process of shaft inspection includes the detection and recognition of the following features: thread, conical knurl, cylindrical knurl, and grooves. Feature detection is based on the detection of "edge markers" in ROIs that are determined in images. For thread detection the back-light illumination is applied (Fig. 4a). The proper image contrast obtained using such a technique enables the high effectiveness of thread detection (Fig. 4b).

For detection of other shaft features, images recorded with the use of directional front-lighting technique are analysed (Fig. 5a). Inspection of conical knurl by means of edge detection between crest and root is presented in Fig. 5b. Quality assessment of the shaft is based on the analysis of edge strength and the correctness of a knurl pitch.

Results of individual tasks carried out in the inspection process are the basis for the final judgment and quality selection. The total time of the inspection, including image acquisition, exposure time, processing and analysis of images is less than 350 ms which is a significant difference to presented alternative solutions which are based on a single camera mounted on a linear drive.

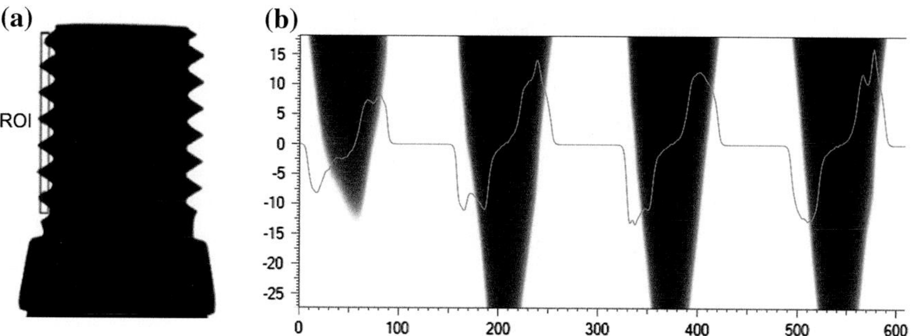

Fig. 4 Thread detection: **a** ROI determined in image, **b** edge strength—derivative filter used to locate edge candidates

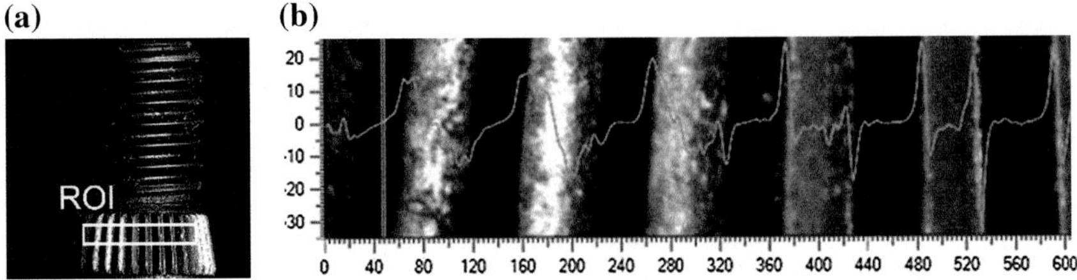

Fig. 5 Knurl detection: **a** ROI determined in image, **b** edge strength—derivative filter used to locate edge candidates

3 Developing the Automatic Optical Inspection System

The developed automatic optical system for quality inspection of shafts is presented in Fig. 6. The system consists of the following modules: vision module, computer based control system, operator panel, and the shaft feeder with container. The fully automated system enables the quality inspection and selection of shafts with efficiency up to 1,200 parts per hour. The operator's tasks are limited to loading shafts into the container and unloading a certifying the shafts after inspection. This performance corresponds to a production rate of the dedicated product and fully fills the requirements of the manufacturer.

The vision module consists of three cameras with lenses, two front-light LED illuminators, and the V-block holder with line back-light LED illuminator (Fig. 7). After leaving the container, shafts are placed in the V-block holder for the duration of quality inspection.

The system enables the inspection of 20 types of shafts that have their characteristics stored in a computer database. Tests of the system performed in industrial

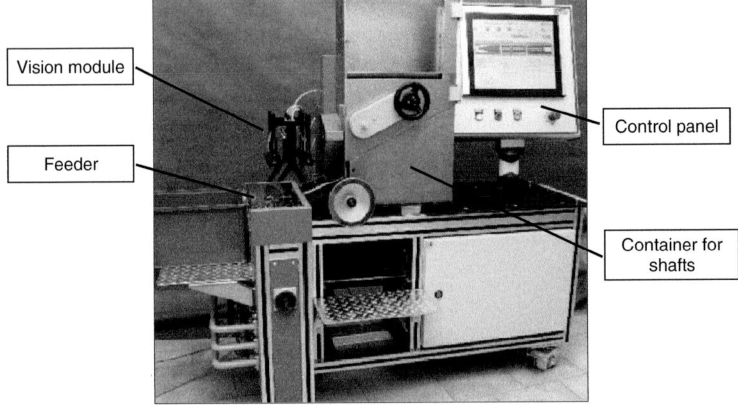

Fig. 6 View of the developed automated inspection system

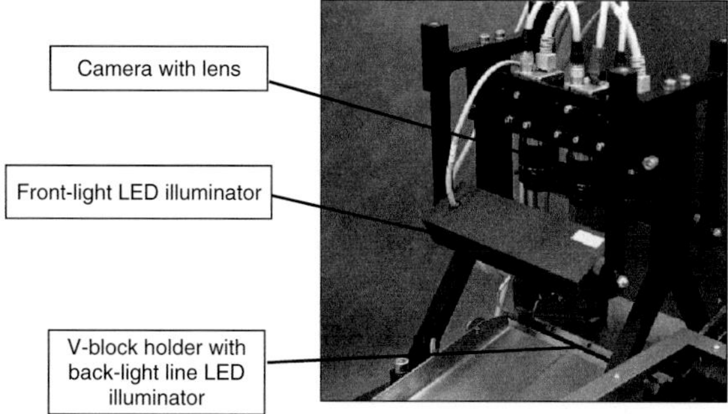

Fig. 7 Vision module

conditions have been confirmed its advantages. The developed system has been introduced into a manufacturing plant that makes shafts for leading companies in the automotive industry.

4 Conclusions

Vision inspection in industry while manufacturing slender parts needs to develop specialized dedicated vision systems that enable a high efficiency automatic quality inspection with appropriate measurement accuracy. In a conventional single-camera vision inspection systems, ensuring high image resolution, optical distortion, and perspective errors are critical problems. The developed multi-camera optical inspection system enables vision inspection of metal shafts in industrial conditions and includes measurements of shaft length and the detection and recognition of shaft features. The novel solution of the inspection system is based on the application of a set of three cameras that allow the inspection of shaft in separated areas with the required high measurement accuracy. Minimization of measurement errors enables the calibration method with the use of the created Look-up Table. Furthermore, the possibility to reconfigure the vision system and adapt it to the inspection of other series of shafts is a significant merit, since production lines can change. Further work will be focused on the development of a multi-camera vision system that applies very high-resolution cameras and new algorithms for the detection of other shaft defects.

References

1. Cho, H.: Opto-Mechatronic Systems Handbook. CRC Press Taylor & Francis Group (2003)
2. Daschenko, A.I.: Manufacturing Technologies for Machines of the Future. Springer, Berlin Heidelberg (2003)
3. Malamas, E.N.E., Petrakis, G.M., Zervakis, M., Petit, L., Legat, J.D.: A survey on industrial vision systems, applications and tools. Image Vis. Comput. **21**, 171–188 (2003)

4. Neogi, N., Mohantal, D.K., Dutta, P.K.: Review of vision-based steel surface inspection systems. URASIP. J. Image Video Process. 1–19 (2014)
5. Kumar, D.P., Kannan, K.: Roadmap for designing an automated visual inspection system. Int. J. Comput. Appl. **1**(19), 0975–8887 (2010)
6. Giesko, T.: Metodyka projektowania i implementacji innowacyjnych systemów optomechatronicznych. Wydawnictwo Naukowe Instytutu Technologii Eksploatacji – PIB, Radom (2013)
7. Zheng, H., Kong, L.X., Nahavandi, S.: Automatic inspection of metallic surface defects using genetic algorithms. J. Mater. Process. Technol. **125-126**, 427–433 (2002)
8. Medina, R., Gayubo, F., González-Rodrigo, L.M., Olmedo, D., Bermejo, Gómez-García-Pernkopf, F., O'Leary, P.: Image acquisition techniques for automatic visual inspection of metallic surfaces. NDT&E Int. **36**, 609–617 (2003)
9. Rosati, G., Boschetti, G., Biondi, A., Rossi, A.: Real-time defect detection on highly reflective curved surfaces. Opt. Lasers Eng. **47**, 379–384 (2009)
10. Bermejo, J., Zalama, E., Perán, J.R.: Automated visual classification of frequent defects in flat steel coils. Int. J. Adv. Manuf. Technol. **57**, 1087–1097 (2011)
11. Reiner, J.: Identyfikacja i modelowanie optyczne systemów wizyjnej kontroli jakości wytwarzania. Oficyna Wydawnicza Politechniki Wrocławskiej, Wrocław (2013)
12. Sitnik, R., Sladek, J., Kupiec, M., Blaszczyk, P. M., Kujawinska, M.: New concept of fast hybrid contact and no-contact measurement for automotive industry. In: International Society for Optical Engineering, Bellingham WA, United States, Strasbourg, France, p. 619803 (2006)
13. Hocenski, Ž., Keser, T.: Failure Detection and Isolation in Ceramic Tile Edges Based on Contour Descriptor Analysis. In: Proceedings of the 15th Mediterranean Conference on Control and Automation, Athens, Greece, pp. 514–519 (2007)
14. Miyatake, T., Matsushima, H., Ejiri, M.: Contour representation of binary images using run-type direction codes. Mach. Vis. Appl. **9**(4), 193–200 (1997)
15. Lindeberg, T.: Edge detection and ridge detection with automatic scale selection. Int. J. Comput. Vision **30**(2), 117–154 (1998)
16. Tadeusiewicz, R., Korohoda, P.: Komputerowa analiza i przetwarzanie obrazów. Wydawnictwo Fundacji Postępu Telekomunikacji, Kraków (1997)
17. Jiang, L., Sun, K., Zhao, F., Hao, X.: Automatic detection system of shaft part surface defect based on machine vision. In: Proceedings SPIE, Automated Visual Inspection and Machine Vision, p. 9530–9518 (2015)
18. Wei, G., Tan, Q.: Measurement of shaft diameters by machine vision. Appl. Opt. **50**(19), 3246–3253 (2011)
19. Song, Q., Wu, D., Liu, J., Zhang, C., Huang, J.: Instrumentation design and precision analysis of the external diameter measurement system based on CCD parallel light projection method. In: Proceedings SPIE, pp. 715621–715625 (2008)
20. Ayub, M.A., Mohamed, A.B., Esa, A.H.: In-line inspection of roundness using machine vision. Procedia Technol. **15**, 808–817 (2014)
21. Optical Gaging Products. http://www.ogpnet.com/ogpViciVision.jsp
22. Automation Artisans. http://www.automationartisans.com/shaftinspectionsystem.html
23. JCGM. Evaluation of measurement data: guide to the expression of uncertainty in measurement (2008)

Piezoelectric Gravimeter of the Aviation Gravimetric System

Igor Korobiichuk, Olena Bezvesilna, Andrii Tkachuk, Michał Nowicki and Roman Szewczyk

Abstract In the article the functional schema of the aviation gravimetric system is reviewed, its structure and the possibility of using the piezoelectric sensor as the gravimeter is analyzed, the options for installing and fixing the piezoelectric gravimeter of the aviation gravimetric system are presented.

Keywords Sensor element · Piezoelectric gravimeter · Gravity · Aviation gravity system

1 Introduction

Accurate knowledge of gravitational acceleration and anomalies of the gravitational field of the Earth is needed in geology and geodesy (study of the Earth's surface shape and the world ocean, search for mineral deposits), in geodynamics (study of

---

I. Korobiichuk (✉)
Industrial Research Institute for Automation and Measurements PIAP, Warsaw, Poland
e-mail: ikorobiichuk@piap.pl

O. Bezvesilna
National Technical University of Ukraine "Kyiv Polytechnic Institute", Kiev, Ukraine
e-mail: bezvesilna@mail.ru

A. Tkachuk
Zhytomyr State Technological University, Zhytomyr, Ukraine
e-mail: andrew_tkachuk@i.ua

M. Nowicki · R. Szewczyk
Institute of Metrology and Biomedical Engineering, Warsaw University of Technology, Warsaw, Poland
e-mail: nowicki@mchtr.pw.edu.pl

R. Szewczyk
e-mail: r.szewczyk@mchtr.pw.edu.pl

R. Szewczyk et al. (eds.), *Challenges in Automation, Robotics and Measurement Techniques*, Advances in Intelligent Systems and Computing 440, DOI 10.1007/978-3-319-29357-8_65

the Earth crust behavior, the movement of tectonic plates of the Earth), in aviation and space technology (correction of inertial navigation systems of aircraft, missiles), and also in the other fields of science and technology. To determine the characteristics of the Earth's gravitational field we can construct an aviation gravimetric system (AGS), the sensing element of which is the gravimeter. With the help of AGS we can get gravimetric information in remote areas of the globe much faster and at less cost than by sea or land gravity means [1–3].

The effectiveness of the AGS is largely provided by a choice of a system's sensitive element—gravimeter. To date, the most famous are such aviation gravimeters as a string (GS) [4–6], quartz (GAL-S) [7] and gyro (GG) [8, 9].

Measurement results of the acceleration of gravity obtained with the help of the above mentioned gravimeters contain large measurement errors caused by the influence of various factors: cross-angular velocity of the base and angular velocity of the Earth ($584 \times 10^{-5} m/s^2$), changes in temperature, atmospheric pressure, the emergence of noises of various origin and vibrations at the installation site of the device. For the high-precision measurements of the Earth's gravitational field the presence of the above indicated errors is unacceptable. Therefore, the problem of increasing of the aviation gravimetric measurements accuracy is important. Today one of the most accurate gravimeters AGS is piezoelectric gravimeter [10], which contributes to measuring the acceleration of gravity anomalies with a precision close to $0.1 \times 10^{-5} m/s^2$.

2 Analysis of the Scientific Researches

Studies have shown that a large contribution to the theory and practice of gravimetric measurements in the CIS was made by a number of prominent scientists: V.O. Bahromyants, A.M. Lozinska, V.V. Fedynskyi, N.P. Grushinsky, E.I. Popov and others. No less important role was played by foreign scientists: A. Graf, V. Torhe, M. Holvani, D. Harrison. In Ukraine, the well-known studies of various types of gyro gravimeters were led by Honored Scientist of Ukraine, Ph.D., professor O.M. Bezvesilna.

Development of the theory of piezoelectric transducers is associated with the names and scientific works of A.A. Andreyev, V.V. Malov, N.A. Shulga, V.V. Lavrinenko, S.I. Puhachov, A.E. Kolesnikov, P.O. Gribovskiy, V.M. Sharapov and others.

The aim of the article is the analysis of possible ways of placing the sensitive element in the structure of piezoelectric gravimeter of aviation gravimetric system and the choice of the most optimal one.

3 The Equation of Motion AGS

In the work [1] an equation of motion AGS was received:

$$f_z = g_z - \frac{v^2}{r} + 2e\frac{v^2}{r}\left[1 - 2\cos^2\varphi\cdot(1 - \frac{\sin^2 k}{2})\right] - 2\omega_3 v\sin k\cos\varphi \\ + 2\dot{h}\frac{e}{r}v\cos k\sin 2\varphi - 2\frac{\gamma_0 h}{r} - \omega_3^2 h\cos^2\varphi + \ddot{h} \tag{1}$$

where f_z—output signal of piezoelectric gravimeter (PG); g_z—gravity acceleration (GA) along the sensitivity axis of PG; v—velocity of the aircraft (AC); r—radius of the AC location; e—ellipsoid compression; φ—geographic latitude; k—AC path; ω_3—angular velocity of the Earth; h—AC height above ellipsoid; $\dot{h}$—AC vertical speed; $\ddot{h}$—AC vertical acceleration; γ_0—referential acceleration of gravity.

All members of Eq. (1), except g_z—are obstacles.

Let's rewrite the Eq. (1) in another form:

$$g_z = f_z + \frac{v^2}{r}\left\{1 - 2e\cdot\left[1 - \cos^2\varphi\cdot(1 - \frac{\sin^2 k}{2})\right]\right\} + 2\omega_3 v\sin k\cos\varphi \\ - 2\dot{h}\frac{e}{r}v\cos k\sin 2\varphi + 2\frac{\gamma_0 h}{r} + \omega_3^2 h\cos^2\varphi - \ddot{h} \tag{2}$$

AGS equations of motion to determine the GA Δg anomaly considering that the GA anomaly is equal to the difference of GA along the axis of PG sensitivity and the reference PST values, becomes:

$$\Delta g = f_z + \frac{v^2}{r}\left\{1 - 2e\cdot\left[1 - \cos^2\varphi\cdot(1 - \frac{\sin^2 k}{2})\right]\right\} + 2\omega_3 v\sin k\cos\varphi \\ - 2\dot{h}\frac{e}{r}v\cos k\sin 2\varphi + 2\frac{\gamma_0 h}{r} + \omega_3^2 h\cos^2\varphi - \ddot{h} - \gamma_0 \tag{3}$$

Let's rewrite the equation of AGS motion in the form of

$$\Delta g = f_z + E + A - \ddot{h} - \gamma_0, \tag{4}$$

where f_z—AGS gravimeter output signal; $E = \frac{v^2}{r}\left\{1 - 2e\cdot\left[1 - \cos^2\varphi\cdot(1 - \frac{\sin^2 k}{2})\right]\right\} + 2\omega_3 v\sin k\cos\varphi - 2\dot{h}\frac{e}{r}v\cos k\sin 2\varphi$—Etvesh amendment; $A = 2\frac{\gamma_0 h}{r} + \omega_3^2 h\cos^2\varphi$—correction to the height; $\gamma_0 = \gamma_{0e}(1 + 0.0052884\sin^2\varphi - 0.0000059\sin^2 2\varphi)$—the reference value of the gravity acceleration; $\gamma_{0e} = 9.78049$ m/s^2—the reference equatorial value of the gravity acceleration.

Analysis of the equations of motion of AGS shows that AGS should consist of subsystems that perform the following functions: measuring the gravity

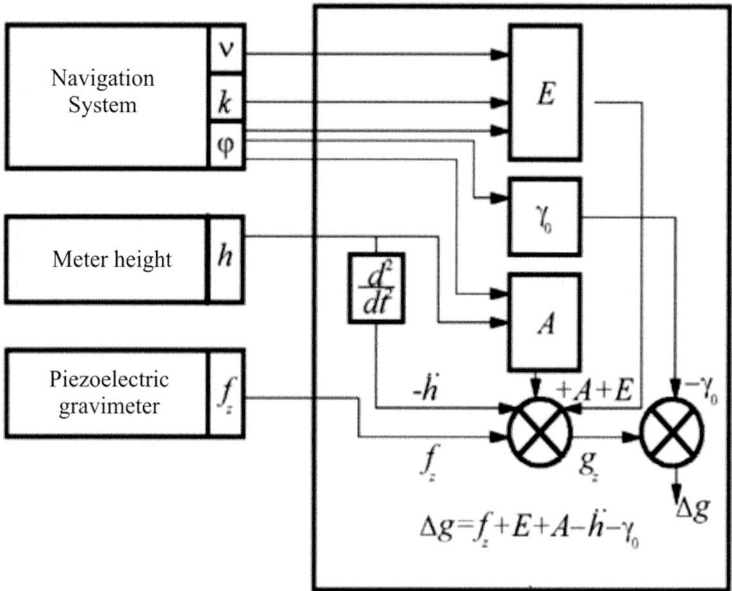

Fig. 1 The functional diagram of AGS

acceleration; stabilizing the sensitivity axis of piezoelectric gravimeter into vertical position; determining the location and speed coordinates; measuring the height; performing computing operations.

Expression (4) can be presented as the functional circuit shown in Fig. 1.

Piezoelectric gravimeter AGS in its composition contains sensitive element in the form of piezoelement (PE), which contains an inertial mass (MI). The principle of the gravimeter functioning is as follows. Under the influence of the gravity acceleration g_z onto the gravimeter MI, the gravity G arises, as a result of which the PE is bent (deformed) at an angle α:

$$\alpha \equiv f(g_z). \tag{5}$$

As a result of such deformation the phenomenon of direct piezoelectric effect is observed—formation on the PE surface the electric charge Q, which is directly proportional to g_z. In the gravimeter the output voltage value is U, but not the charge Q:

$$U \equiv \frac{Q(g_z)}{C_{PE}}, \tag{6}$$

where C_{PE}—electric capacity of PE.

Analytical expression for the voltage output gravimeter calculation taking into account its physical parameters is as follows:

$$U = \frac{d_{ij} \cdot m \cdot g_z}{C_{PE}}, \quad (7)$$

where d_{ij}—piezoelectric module (PE parameter, which depends on the material of manufacture); m—mass of PE.

4 Design of Piezoelectric Gravimeter

There are three basic types of designs of piezoelectric acceleration transducers (Fig. 2). Each of these types has its advantages and disadvantages. Compression converters (Fig. 2a) possess the quality of the high stability against vibration and are used under the conditions of the high vibration and temperature influence. The design that works on bending (Fig. 2b), provides plenty of high sensitivity in the required direction and eliminates extraneous sensitivity, such as transverse. PE on shift (Fig. 2c) has a complicated structure and is sensitive to transverse vibrations, but is used in the high-precision systems.

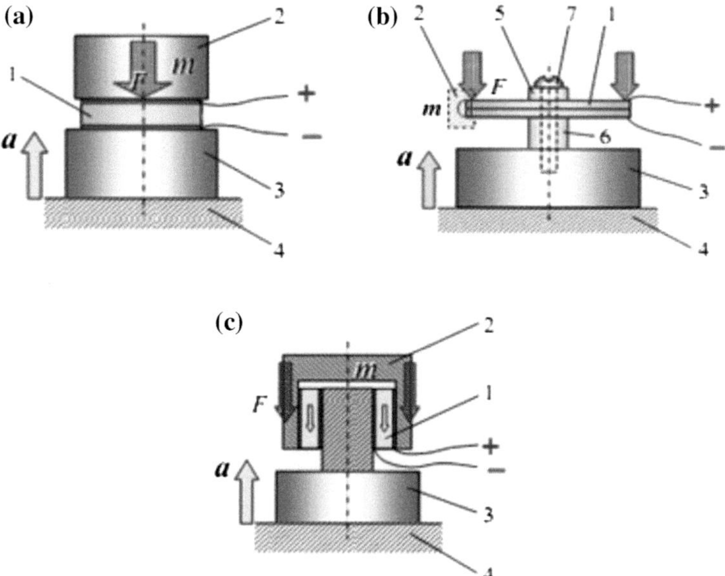

Fig. 2 Main types of piezoelectric accelerate transducers design: **a** tension-compression; **b** bend; **c** shift: *1* PE; *2* MI; *3* basis; *4* object of control; *5*, *6* support; *7* screw

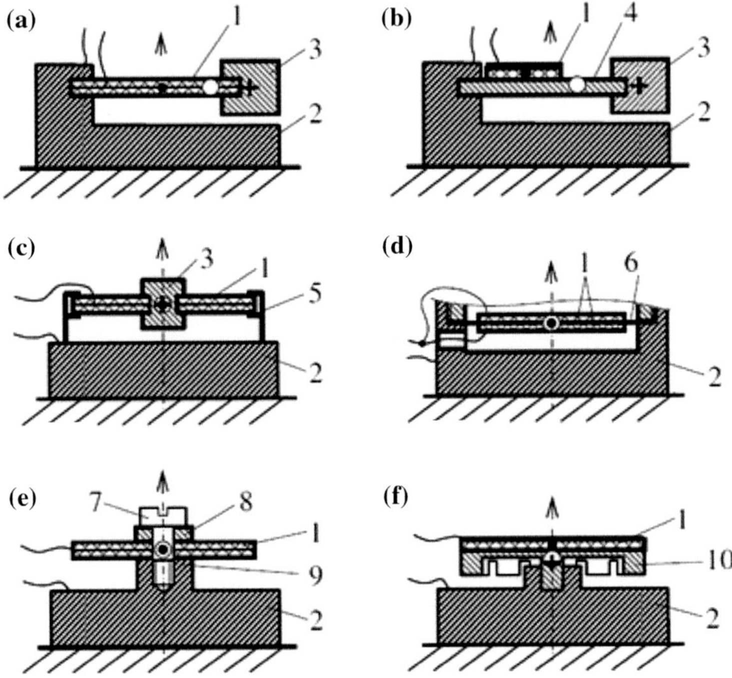

Fig. 3 SE layouts in the design of the piezoelectric AGS gravimeter: *1* PE; *2* basis; *3* MI; *4* beam; *5* stand; *6* plate; *7* screw; *8*, *9* support; *10* plate with inertial elements

After analyzing the advantages and disadvantages of the main types of structures of the piezoelectric acceleration transducers, we can conclude that for the measuring of the gravity acceleration (of the high accuracy and sensitivity) we should use the PE, the principle of which is the bending strain.

There are also several ways to place PE in the SE constructions which operate bending (Fig. 3). The simplest design of SE is—not-centered (Fig. 3a), which consists of bi- or uni-morph beam PE 1, which is cantilever fixed on the basis 2. Inertial mass 3 is located at the end of PE. Somewhat similar to the previous one is the design (Fig. 3b) with PE1 glued to beam 4. In SE centered design (Fig. 3c) the peripheral fastening of girder or disk PE 1 is performed using elastic bars 5, which provide the ability to bend free PE in the axial direction. In this construction the total charge is created through mechanical stress, operating both radial and circular in cross-section.

Another option of peripheral fixing of PE (Fig. 3d) is implemented with a thin metal plate 6, glued between the two disc plates PE 1. The system of peripheral PE fixing affects the character of its strain. On the one hand it negatively affects the performance of PE, but on the other—provides the stability of its metrological characteristics. This structure is suitable to use for measuring slowly changing physical quantities.

Table 1 Characteristics of the SE placement schemes in the AGS's PE design

	Type of structure	Advantages	Disadvantages
1	Not-centered	High sensitivity; temperature resistance	Small measuring range
2	Not-centered with the glued PE	Stability of metrological characteristics	Low accuracy and sensitivity; low rate of conversion; small measuring range
3	Centered with the side fixing and MI in the center	High conversion efficiency, high sensitivity	Exposed to transverse vibrations; temperature instability
4	Centered with the side fixing and the membrane (plate)	Stability of metrological characteristics; high sensitivity	Dependence of the plate properties
5	Centered by a screw	Isolated output; resistance to external magnetic fields	Low conversion factor; Dependence on the properties of the screw
6	Mushroom scheme	High conversion efficiency	Temperature instability; exposed to transverse vibrations

Piezoelement (Fig. 3g) is fixed in the center by the screw 7 between the supports 8 and 9. This design is centered. The usage of the additional insulators between the PE and supports allows us to implement the symmetrical output isolated from the housing.

In the variant of the so-called mushroom scheme (Fig. 3e) the disc PE 1 is adhered to a metal plate 10 with the MI disposed on the periphery. A significant drawback is—the temperature instability. The comparative characteristics of the mentioned above structural schemes is shown in the Table 1.

As you can see from the Table 1, each structural scheme has its advantages and disadvantages. Therefore, the proposed design of the piezoelectric AGS gravimeter use the following layout of SE (Fig. 4).

As can be seen from Fig. 4, the construction of the AGS piezoelectric gravimeter is not-centered. The sensing element is placed in a sealed package and is designed as a cantilevered fixed on the rod piezo element, at the free end of which an inertial mass is placed.

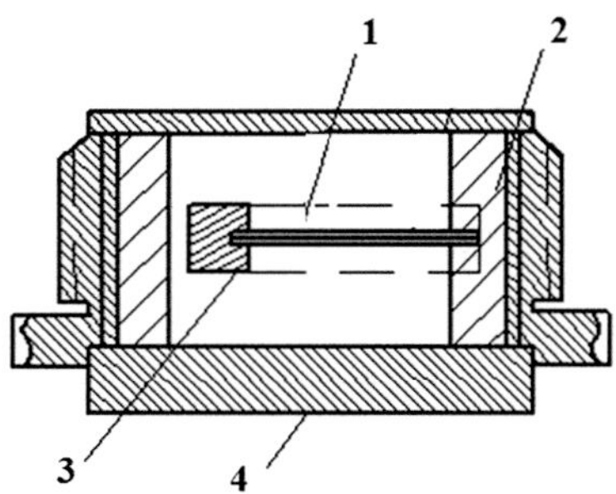

Fig. 4 The design of the piezoelectric AGS gravimeter: *1* PE; *2* rod; *3* IM; *4* air-tight casing

5 Results and Discussion

The design of the piezoelectric AGS gravimeter was chosen out of the following considerations. Whereas it is necessary to maximize the sensitivity of piezoelectric AGS gravimeter, the construction should not be centered (Table 1). Measuring of the gravity acceleration does not include a large working range, so the lack of the scheme is not significant. Since, as the SE, the piezo package [2] rather than the bimorph PE will be used, it must be fixed cantilever and with maximum elasticity. This fixing provides resistance to overloads and the maximum elasticity of PE, which also increases the reliability of gravimeter.

The recommended material for PE is quartz. This SE provides a measurement of the gravity acceleration in the only one direction—along the axis perpendicular to the plane of installation of the piezoelectric (O_z), which reduces the gravimeter sensitivity to transverse vibrations and cross velocities and therefore increases the measurement accuracy of the gravity acceleration.

Return of the piezo element to the original position, that is the force compensation, occurs through the introduction of a scheme of direct conversion of the gravitational acceleration signal onto the voltage of electromechanical negative feedback, which is based on the phenomenon of reverse piezoelectric effect. The circuit scheme implementing this power compensation involves coating the PE with three electrodes: the first—connected to the amplifier input voltage, the second—to the general scheme and the third wire—to the amplifier output voltage. As you can see, PE is also the power compensator, which allows it to provide the maximum resiliency, stability over time and as a result, high reliability.

6 Conclusions

Considered the aviation gravimetric system with piezoelectric gravimeter to measure the acceleration of gravity anomalies with greater accuracy (0.1×10^{-5} m/s^2) and speed (fully automated) in comparison with the currently known ($1–10 \times 10^{-5}$ m/s^2). The principle of the piezoelectric AGS gravimeter functioning is described. The possible SE layout in the construction of the piezoelectric AGS gravimeter is described and the most appropriate of them—not centered is selected.

References

1. Bezvesilna, O.M.: Aviation gravimetric systems and gravimeters: monograph. Zhitomir: ZDTU, p. 604 (2007) (in Ukraine)
2. Korobiichuk, I., Bezvesilna, O., Tkachuk, A., Nowicki, M., Szewczyk, R., Shadura, V.: Aviation gravimetric system. Int. J. Sci. Eng. Res. **6**(7), 1122–1127 (2015)
3. Bezvesilna, O.M.: Study of a new precision inertial navigation system at aviation gravimetric measurements: monograph. Zhytomyr: ZSTU, p. 476 (2012) (in Ukraine)

4. Jentzsch, G., Schulz, R., Weise, A.: A well-known principle in a new gravimeter: the automated Burris Gravity Meter. AVN Allgemeine Vermessungs-Nachrichten **122**(5), 168–175 (2015)
5. Liard, J., Gagnon, C.: The new A-10 absolute gravimeter at the 2001 international comparison of absolute gravimeters. Metrologia **39**(5), 477–483 (2002)
6. Baumann, H., Klingel´e, E.E., Marson, I.: Absolute airborne gravimetry: a feasibility study. Geophys. Prospect. **60**, 361–372 (2012)
7. Sharapov, V.M., Musyenko, M.P., Sharapova, E.V.: Piezoelectric sensors/pod red. In: Sharapov, V.M. (ed.) Tehnosfera, p. 632s (2006) (in Russian)
8. Korobiichuk, I., Nowicki, M., Szewczyk, R.: Design of the novel double-ring dynamical gravimeter. J. Autom. Mobile Robot. Intell. Syst. **9**(3), 47–52 (2015) (Industrial Research Institute for Automation and Measurements "PIAP")
9. Mikan, S., John, C.: Application of the Digital Nautical Chart (DNC) database to help identify areas of vertical deflection in the Ring Laser Gyro Inertial Navigator (AN/WSN-7). Record—IEEE PLANS, Position Location and Navigation Symposium, pp. 299–303 (2000)
10. Bezvesilna, O.M., Podchashunkiy, U.O., Tkachuk, A.G.: Piezogravimetr. Pat. 9084 Ukraine, MPK G 01V 7/00. the applicant and the owner of the patent.—№ a201113894; appl. 25.11.2011; publ. 10.07.2012, vol. № 13

Methodology of Reduction of the Offset Voltage in Hall-Effect Sensors

Maciej Kachniarz, Oleg Petruk and Roman Szewczyk

Abstract The paper presents new, original methodology of reduction of the offset voltage in Hall-effect sensors. The outline of innovative Hall-effect sensor with five electrical pins for each electrode is presented. Such a construction creates a possibility to use different combinations of pins in order to find minimum value of the offset voltage. The procedure of searching for the best combination of pins connected for minimization of the value of offset voltage is also presented. For practical implementation of the methodology, special experimental system was developed, utilizing system of relays controlled by Data Acquisition Card to set different combinations of pins and search for the minimum value of the offset voltage. Obtained results indicate that there is a possibility to reduce value of the offset voltage even up to one hundred times.

Keywords Magnetic field measurement · Hall-effect sensor · Offset voltage reduction

1 Introduction

Magnetic field sensors are often utilized in many industrial applications. The dominant type of magnetic field sensors are Hall-effect sensors, which hold over 70 % of the market share in industrial measurements of magnetic field [1]. They are

---

M. Kachniarz (✉) · O. Petruk
Industrial Research Institute for Automation and Measurements,
al. Jerozolimskie 202, 02-486 Warsaw, Poland
e-mail: mkachniarz@piap.pl

O. Petruk
e-mail: opetruk@piap.pl

R. Szewczyk
Institute of Metrology and Biomedical Engineering, Warsaw University
of Technology, św. Andrzeja Boboli 8, 02-525 Warsaw, Poland
e-mail: r.szewczyk@mchtr.pw.edu.pl

R. Szewczyk et al. (eds.), *Challenges in Automation, Robotics and Measurement Techniques*, Advances in Intelligent Systems and Computing 440, DOI 10.1007/978-3-319-29357-8_66

used in a wide range of industrial applications as incremental counters, position detectors and sensing elements in DC current transformers [2].

Hall-effect sensors are utilizing physical phenomenon known as the Hall-effect. It involves generation of the electric potential difference (Hall voltage) across an electric conductor placed in external magnetic field of direction perpendicular to the direction of electric current flowing through the conductor [3].

One of the most important parameters of the Hall-effect sensor is the offset voltage. It is the potential difference in electric conductor with electric current flow in the absence of external magnetic field. Offset voltage is undesirable, because it limits the ability of the sensor to detect low magnetic fields. Offset voltage results from several reasons. Geometrical imperfections like asymmetrical alignment of the sensing electrodes or asymmetrical shape of the conductor structure are one of them. The other reason is a presence of inhomogeneities in the material of the sensor. Finally, the source of the offset voltage can be a mechanical distortion changing resistivity of the sensor's material [4].

In possible industrial applications, where high sensitivity is desired, offset voltage could be important problem. Many attempts were made to solve this issue. Most of them involve reduction of the offset voltage with external analog circuits [5] or resistance bridges [6]. This paper presents new, original methodology allowing to reduce value of the offset voltage within the structure of the Hall-effect sensor using special measuring and control system, and preserve the reduced value without any external devices.

2 Outline of the Multiple-Pin Hall-Effect Sensor

The proposed outline of the exemplary multiple-pin Hall-effect sensor is schematically presented in Fig. 1.

Hall-effect sensor is an electronic component with four electrodes. Two opposite electrodes (supplying electrodes) are utilized for powering the sensor with supply current I_s, like electrodes B and D in Fig. 1. Two other electrodes (measuring electrodes), A and C in Fig. 1, are used to measure the value of Hall voltage generated in the sensor, when external magnetic field is present. If there is no external magnetic field, measured voltage will be equal to the offset voltage value ΔV resulting from the equation:

$$V = \Delta V + V' \tag{1}$$

where V and V' are electric potentials of measuring electrodes.

As it can be seen in Fig. 1, each of four electrodes of the sensor have five electrical pins, which can be connected in different configurations. If measuring electrodes have five pins of individual electric potentials, the Eq. 1 will take form:

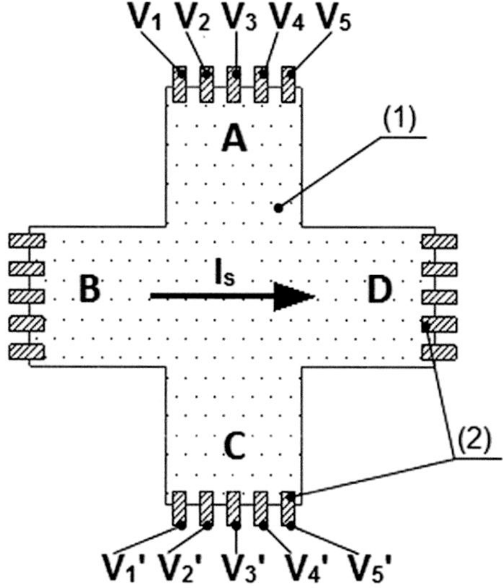

Fig. 1 Exemplary outline of the multiple-pin Hall-effect sensor: *1* Hall-effect structure, *2* pins of the electrodes, I_s supply current, A–D electrodes, V_i, V_i' electric potential (i = 1, 2, …, 5)

$$V_1 + V_2 + V_3 + V_4 + V_5 = \Delta V + V_1' + V_2' + V_3' + V_4' + V_5' \qquad (2)$$

Through multiple pins of the measuring electrodes it is possible to find the configuration of pins, which allow to significantly reduce the value of ΔV part in the right side of the Eq. (2), which results in significant reduction of the offset voltage in the Hall-effect sensor.

The exemplary Hall-effect sensor with multiple pins for each electrode presented in Fig. 1 has the form of symmetrical, equal-arm cross, but it can have any different shape, symmetrical or asymmetrical.

3 Procedure of Searching for Optimal Pin Connection of Hall-Effect Sensor

According to previously presented outline of the multiple-pin Hall-effect sensor, special procedure was developed to search for optimal connection of pins of the sensor providing the minimum value of the offset voltage.

In the first step the A–C bar of the Hall-effect sensor presented in Fig. 1 is connected as a measuring bar while supply current I_s is flowing through B–D bar. In the supply bar (B–D in Fig. 1) all pins are connected to provide uniform current flow through the bar. In the measuring bar, pins are connected in different configurations in order to find the minimum value of the offset voltage and for each configuration offset voltage is measured. Each of the pins can have two states: 1, when pin is connected and 0, when pin is disconnected. So the state of each measuring electrode can be presented as the string of five binary values, which can be converted into decimal number of value from 0 to 31, where each number from this range corresponds to the given configuration of the pins. This indicates that there are 32 possible configurations for each electrode A and C. The configurations of the electrode pins are checked in ascending order from 1 (when only first pin of the electrode is connected) to 31 (when all pins of the electrode are connected). The 0 state is ignored because in this state all pins of the electrode are disconnected. For each state of electrode A, all 31 states of electrode C are checked, which results in 961 possible configurations of the measuring bar A–C that have to be checked. From all this configurations the one characterized by the lowest offset voltage is chosen as the optimal.

In the second step, the electrodes are switched. Electrodes A and C are connected as the supply electrodes, while B and D electrodes are measuring electrodes. Then procedure described previously is repeated to find the optimal pin configuration for B–D bar.

The presented procedure is based on the assumption that Hall-effect sensor is powered with constant supply current I_s during the whole time of the searching procedure. In case of voltage supply, results could be influenced by the fluctuations of current resulting from changes of the resistance due to different configurations of pins connected.

Finally the offset voltage values for optimal configuration of A–C and B–D bars are compared and the bar with lower resulting offset voltage is chosen to be the measuring bar of the magnetic field sensor in normal operation.

4 Experimental Setup for Reducing Offset Voltage of the Hall-Effect Sensor

On the basis of outline of the multiple-pin Hall-effect sensor and the search procedure, special computer controlled system was developed to implement the presented methodology, schematically shown in Fig. 2.

The Hall-effect sensor is powered from a precise DC current supply. There are two sets of relays controlling the pins of multiple-pin Hall-effect sensor. Each set counts 10 normally open relays. One set is used to control the pins of power electrodes of the Hall-effect sensor and all relays in this set should be closed during the operation of the system. Second set of relays controls the pins of measuring electrodes, and this relays are set in different configurations in order to search for the best connection of the pins for the value of the offset voltage minimization. All relays are controlled with the National Instruments NI USB-6361 Data Acquisition Card (DAQ) integrated into the system. DAQ is used also for measuring the offset voltage in all tested configurations of the measuring electrodes pins. Data Acquisition Card is connected to the Personal Computer with special control application developed in National Instruments LabVIEW environment. Developed experimental system is presented in the Fig. 3.

Data Acquisition Card is connected to the system of relays with multiple wires. Each wire connects one digital output of the Card with one relay, so 20 digital outputs of the Card are used. The investigated multiple-pin Hall-effect sensor is mounted in the special test socket connected with the system of relays by the IDC ribbon. Analog input of the Data Acquisition Card is also used for measuring the offset voltage of the Hall-effect sensor for each configuration of the pins.

Important part of the experimental system is the application developed in LabVIEW environment. The user interface of the application is presented in Fig. 4. Application graphically presents current state of the pins connection of multiple-pin Hall-effect sensor using arrays of diode indicators. Developed program can operate in two modes switched with bistable toggle switch. In MANUAL mode application gives possibility to manually connect and disconnect each pin of the Hall-effect

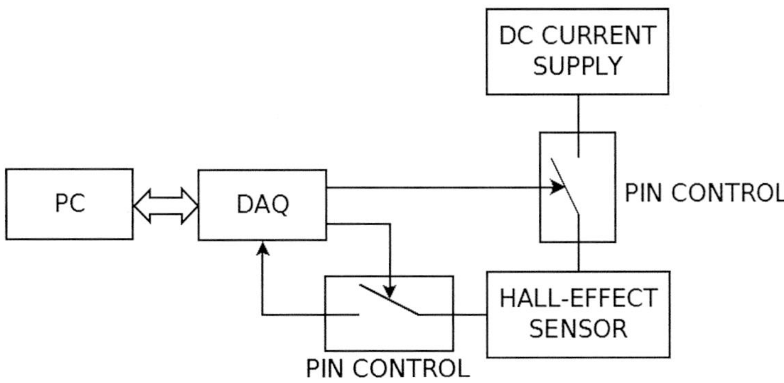

Fig. 2 Schematic block diagram of the developed experimental system

Fig. 3 View of the developed experimental system: *1* NI USB-6361Data Acquisition Card, *2* system of relays controlling the pins connection of the Hall-effect sensor, *3* IDC ribbon for connecting system of relays with Hall-effect sensor, *4* socket for mounting the multiple-pin Hall-effect sensor

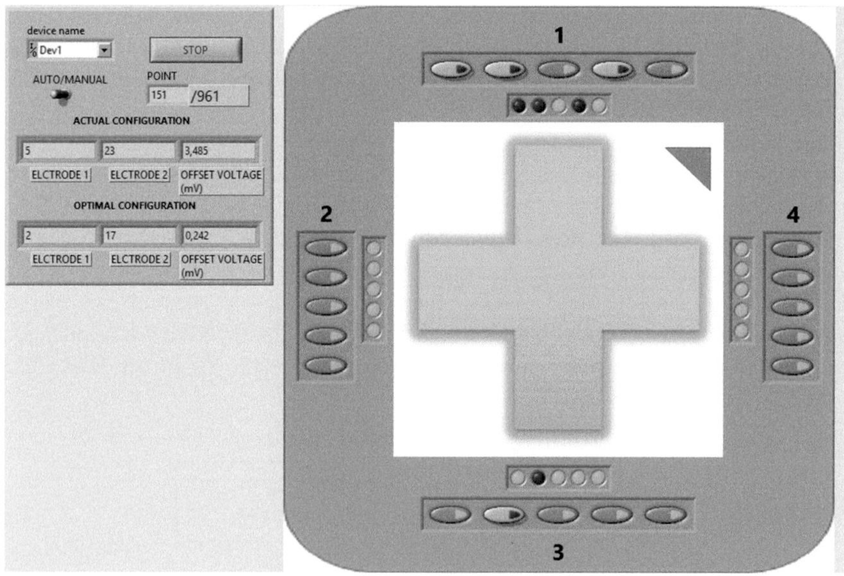

Fig. 4 User interface of the developed control application during operation of the system

sensor with four arrays of switches, each array controlling one electrode of the sensor. In AUTO mode application automatically perform search procedure described previously to search for the optimal configuration of measuring electrodes pins. Procedure is implemented in the application with two nested FOR loops. In the outer loop the first measuring electrode is controlled and the inner loop controls second electrode. In each iteration of the outer loop, which corresponds to the configuration of the pins of the first measuring electrode given with the number of iteration, all configurations of the second measuring electrode are searched within inner loop. The application displays the value of offset voltage for actual connection of the pins as well as the value for the best connection found so far with configurations of the electrodes corresponding to this value. After the searching is finished, in the field OPTIMAL CONFIGURATION, the configurations of electrodes and offset voltage value are displayed for the lowest found value of the offset voltage.

5 Experimental Results

In order to verify the correctness of the discussed methodology and developed experimental system, several tests were carried out on previously prepared multiple-pin Hall-effect sensors. All tests were performed for supply current $I_s = 1.0$ mA. The obtained results are presented in Table 1.

Table 1 Results of the searching for optimal configuration for 4 tested Hall-effect sensors

Number of tested sensor	Offset voltage for full connection (mV)	Offset voltage after optimization of the configuration (mV)	Ratio
1	40.59	0.85	47.75
2	12.76	0.24	53.17
3	99.78	4.14	24.10
4	30.37	0.32	94.91

Reduction of the offset voltage obtained by searching for optimal configuration of the pins is significant. Especially for the fourth investigated Hall-effect sensor the results are very good as the ratio of the initial offset voltage to the offset voltage for optimal connection is nearly 100. This indicates, that presented methodology and experimental setup are working correctly.

6 Conclusion

New and original methodology of reduction of the offset voltage in Hall-effect sensors was developed and tested. The outline of the multiple-pin Hall effect sensor was presented. On the basis of the described methodology experimental system for reducing the offset voltage value of the Hall-effect sensors was prepared. Obtained results are very good and indicate, that presented methodology is correct and allows to reduce the value of the offset voltage in Hall-effect sensor even up to one hundred times.

Developed solution can be possibly applied to optimize the offset voltage of Hall-effect sensors during the manufacturing process. After finding the optimal configuration of the pins for the lowest offset voltage, Hall-effect sensor could be mounted into the package permanently connected in this configuration. This would significantly improve the functional properties of Hall-effect sensors for industrial applications.

Acknowledgements This work was partially supported by The National Centre of Research and Development (Poland) within GARF-TECH program.

References

1. Magnetic Field Sensors Market by Type (Hall Effect, Magnetoresistive, SQUID, Others), Technology, Applications (Automotive, Consumer Electronics, Industrial & Infrastructure, Medical, Aerospace & Defense) & Geography—Forecasts & Analysis to 2013–2020. MarketsandMarkets (2014). http://www.marketsandmarkets.com
2. Ai, X., Bao, H., Song, Y.H.: Novel method of error current compensation for Hall-effect based high-accuracy current transformer. IEEE Trans. Power Deliv. **20**(1), 11–14 (2005)

3. Ripka, P.: Magnetic Sensors and Magnetometers. Artech, Boston (2001)
4. Ramsden, E.: Hall-Effect Sensors: Theory and Application. Newnes (2006)
5. Liu, C. Liu, J.G.: Offset error reduction in open loop effect current sensor powered with single voltage source. In: IEEE International Workshop on Applied Measurements for Power Systems Proceedings, pp. 1–6 (2014)
6. Warsza, Z.L.: New approach to the accuracy description of unbalanced bridge circuits with the example of pt sensor resistance bridges. J. Autom. Mobile Robot. Intell. Syst. **4**(2), 8–15 (2010)

Propagation of Distributions Versus Law of Uncertainty Propagation

Paweł Fotowicz

Abstract Basic methods for calculating a measurement uncertainty are presented. One method is based on the approach called the propagation of distributions, and the second method is based on the approach called the law of uncertainty propagation. The methods give not the same calculation result in evaluation of standard uncertainty associated with the measurand. The reasons for these discrepancies are explained.

Keywords Measurement uncertainty · Propagation of distributions · Law of uncertainty propagation

1 Introduction

The propagation of distributions and the law of uncertainty propagation are the methods for calculating a measurement uncertainty recommended in documents [1, 2], respectively. The first method involves calculation of measurand through the measurement model using a Monte Carlo procedure. This method is a numerical, but the second method is an analytical based on the formula called the law of uncertainty propagation. The propagation of distributions relies on the calculation of output quantity distribution, as the measurand, from the distributions associated with input quantities defined in measurement model. Usually the measurement model is a linear. In this case the alternative calculation may be done with the use of the parameters of distributions using analysis of variance, as the law of uncertainty propagation. But the results of those calculations are not equal.

P. Fotowicz (✉)
Central Office of Measures, Warsaw, Poland
e-mail: uncert@gum.gov.pl

R. Szewczyk et al. (eds.), *Challenges in Automation, Robotics and Measurement Techniques*, Advances in Intelligent Systems and Computing 440, DOI 10.1007/978-3-319-29357-8_67

2 Measurement Model

Measurement model determines the quantity to be measured. This quantity is called the mensurand. The mensurand is usually treated as an output quantity y represented by a measurement function. The arguments for this function are the input values x_i constituting the components of measurement uncertainty

$$y = f(x_1, \ldots, x_N) \tag{1}$$

Measurement function is a mathematical explicit function presents the relationship between the input quantities and output quantity. Usually, the measurement function is approximated by linear measurement equation based on a first-order Taylor series approximation. Any quantity is a random variable having a probability distribution and is characterized by two parameters: location and dispersion. The location parameter represents the most probable value of the quantity. In probability this parameter called the expectation. In metrology the expectation is represented by average of repeated observations, as the estimation of the quantity, and called the best estimate. The dispersion parameter is a standard deviation, in metrology called the standard uncertainty. So, the measurement uncertainty is defined as a parameter associated with the result of a measurement characterizing the dispersion of the values attributed to the measurand [2].

3 Calculation of Standard Uncertainty

The standard uncertainty is calculated with the use of two method, type A and type B. In the case of type A method the standard uncertainty u_{A} is expressed as an experimental standard deviation of the mean s for series n of observations q_k

$$u_{\mathrm{A}}(x_i) = s(\bar{q}) = \sqrt{\frac{\sum_{k=1}^{n} (q_k - \bar{q})^2}{(n-1)\,n}} \tag{2}$$

The best estimate for the quantity x_i, calculated by A method, is a mean of observations having Student's distribution.

In the case of the B method, the standard uncertainty u_{B} is a standard deviation σ of probability distribution for the input quantity x_i, created on available information associated with the measurement, deriving from calibration certifications, handbooks, specifications or previous measurement data

$$u_{\mathrm{B}}(x_i) = \sigma \tag{3}$$

In this case, the standard deviation is usually associated with a rectangular, triangular or trapezoidal distributions.

4 Law of Uncertainty Propagation

The law of uncertainty propagation is based on analysis of variance, and for the independent input quantities x_i is expressed by formula

$$u_c^2(y) = \sum_{i=1}^{N} \left(\frac{\partial f}{\partial x_i}\right)^2 u^2(x_i) \tag{4}$$

where u_c is a combined standard uncertainty associated with output quantity y. The expanded uncertainty of this quantity is defined as follows

$$U = k \cdot u_c(y) \tag{5}$$

where k is a coverage factor associated with output quantity distribution, usually for 95 % coverage probability.

5 Propagation of Distributions

Propagation of distributions is relied on determination of the measurand distribution from distributions for the input quantities through the measurement model. The recommended method for propagation of distributions is a Monte Carlo method basing on random sampling from probability distributions [1]. The use of Monte Carlo procedure creates the set of M possible values y_r associated with the output quantity as the numerical representation of the distribution function for measurand, which covers the coverage interval with low y_{low} and high y_{high} endpoints, for stated coverage probability, usually 95 %. The half of the coverage interval length is the expanded uncertainty

$$U = \frac{y_{\text{high}} - y_{\text{low}}}{2} \tag{6}$$

The standard uncertainty associated with the measurand, as an equivalent of combined standard uncertainty, is calculated from equation

$$u^2(y) = \frac{1}{M-1} \sum_{r=1}^{M} (y_r - y)^2 \tag{7}$$

6 Comparison of the Method

The methods for calculating a measurement uncertainty are different. The method based on the law of uncertainty propagation is an analytical method, but the method based on the propagation of distributions is a numerical method. Both methods utilize the measurement model in evaluation of measurement uncertainty, but in a different way. The analytical method uses the simple formula deriving from the analysis of variance for independent input quantities in measurement model. The numerical method uses the random sampling from distributions for input quantities through measurement model by Monte Carlo procedure. The summarize of these methods is presented in Table 1.

When we use the type A method for calculating the standard uncertainty, then the results are different. The calculation realized by Monte Carlo method gives the standard uncertainty associated with Student's distribution

$$u(x_i) = \sigma = \sqrt{\frac{v}{v-2}}\, s(\bar{q}) \tag{8}$$

where $v = n - 1$ is a degrees of freedom attributed to input quantity x_i, and σ is a standard deviation of Student's distribution. The relative difference δ between calculations realized both methods is presented in Fig. 1.

The standard uncertainty calculated according to approach recommended in [1] is bigger than defined by [2]. But the calculation of expanded uncertainty, defined as the half-width of the coverage interval, gives the same result when we use the recommended approaches in [1, 2], respectively. Then, the coverage factor k is different, as shown in Fig. 2. The coverage factor determines the relation between the expanded and standard uncertainties, and usually takes approximately the value $k = 2$ for 95 % coverage probability. This value is bigger than $k = 2$ when we define the standard uncertainty as a standard experimental uncertainty of the mean s, and

Table 1 Summarizing the methods for measurement uncertainty calculation

Framework	Law of uncertainty propagation	Propagation of distributions
Measurement result	$y \pm U$ Estimate ± expanded uncertainty	y—estimate $u(y)$—standard uncertainty $y_{\text{low}}, y_{\text{high}}$—coverage interval limits
Estimate	$y = f(x_1, \ldots, x_N)$	$y = \frac{1}{M}\sum_{r=1}^{M} y_r$
Standard uncertainty	$u_c^2(y) = \sum_{i=1}^{N} \left(\frac{\partial f}{\partial x_i}\right)^2 u^2(x_i)$	$u^2(y) = \frac{1}{M-1}\sum_{r=1}^{M} (y_r - y)^2$
Uncertainty expression	Expanded uncertainty $U = k \cdot u_c(y)$	Coverage interval $[y_{\text{low}}, y_{\text{high}}]$ $y_{\text{low}} = y_{M(1-p)/2}$ $y_{\text{high}} = y_{M(1+p)/2}$

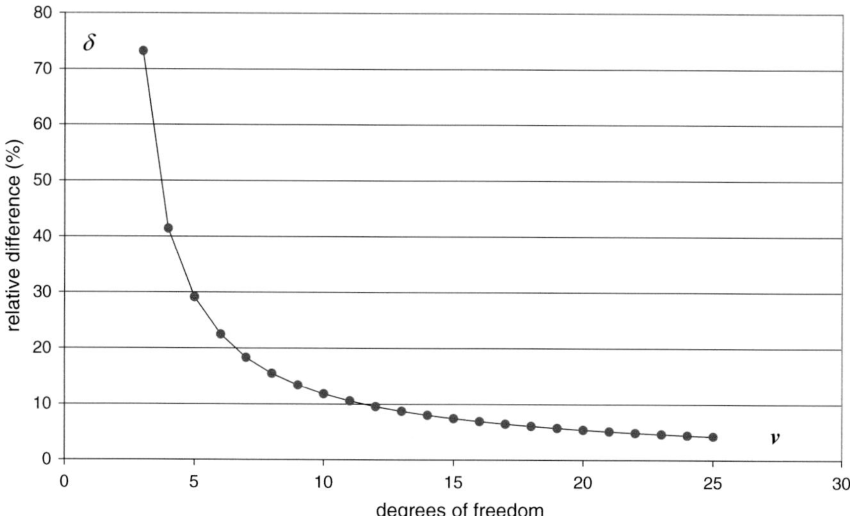

Fig. 1 Relative difference between the calculations of standard uncertainty of type a method with the use of two approach associated with the law of uncertainty propagation and the propagation of distributions

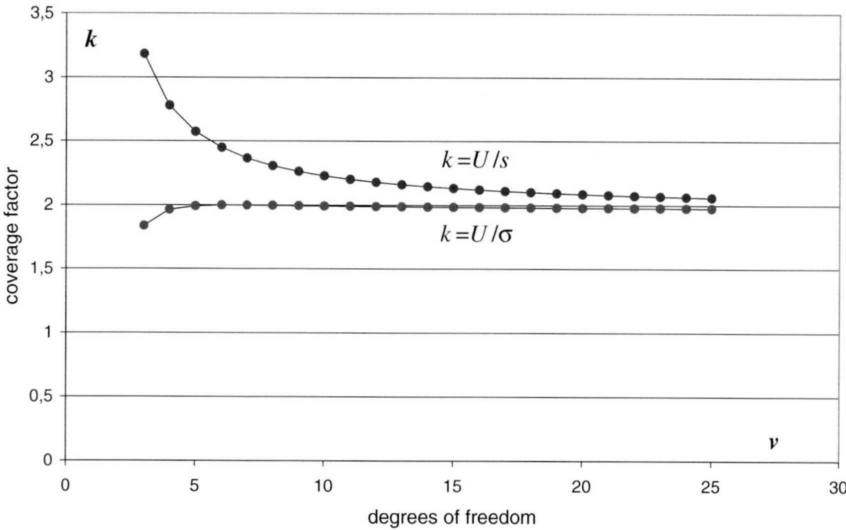

Fig. 2 The coverage factor for 95 % coverage probability when the standard uncertainty is defined as the standard experimental uncertainty of the mean, or the standard deviation of Student's distribution

the coverage factor is smaller than $k = 2$ when we define the standard uncertainty as a standard deviation of Student's distribution σ.

For example, the differences also occurs when we calculate the uncertainty budget associated with calibration of measuring instrument.

7 Practical Example of Calculation

We can use the presented methods for calculating the uncertainty associated with the calibration of typical industrial measuring instrument such as a manometer. The manometer is calibrated with the use of deadweight piston gauge. The measurement error of instrument e_p, as the measurand, is defined by equation

$$e_p = p + \delta p - p_s \tag{9}$$

where p is a reading of manometer pressure, δp is a resolution of manometer indication and p_s is a pressure forced by a deadweight piston gauge, as the standard reference. There are three input quantities in measurement model (9).

The estimate of the first input quantity p is a average of series of $n = 6$ observations of manometer reading p_k, presented in Table 2.

The experimental standard deviation of the mean is as follows

$$s(\bar{p}) = \sqrt{\frac{\sum_{k=1}^{n} (p_k - \bar{p})^2}{(n-1)\, n}} = 0.026 \text{ MPa} \tag{10}$$

The Student's distribution with degrees of freedom $\nu = n - 1 = 5$ is associated with this quantity. The standard deviation of Student's distribution is as follows

$$\sigma = \sqrt{\frac{\nu}{\nu - 2}}\, s(\bar{p}) = 0.033 \text{ MPa} \tag{11}$$

Table 2 Measurement data in calibration of manometer

p_k	19.9 MPa
	19.8 MPa
	19.9 MPa
	19.9 MPa
	20.0 MPa
	19.9 MPa
$\bar{p}$	19.9 MPa

The second quantity δp is a resolution of manometer indication. The rectangular distribution is associated with this quantity. The resolution $R = 0.1$ MPa, as the one tenth of a scale interval of manometer. Then, the standard uncertainty is as follows

$$u(\delta p) = \frac{R}{2\sqrt{3}} = 0.029 \text{ MPa} \tag{12}$$

The last quantity p_s is a standard pressure forced by deadweight piston gauge. The accuracy class of gauge tester is 0.05. Because the standard pressure $p_s = 20$ MPa, then the standard uncertainty associated with this pressure is as follows

$$u(p_s) = \frac{0.05 \cdot p_s}{100\sqrt{3}} = 0.0058 \text{ MPa} \tag{13}$$

When we use the approach recommended in [2], all the quantities may be summarized in the uncertainty budget, as in Table 3. The standard uncertainty associated with the output quantity e_p is a combined standard uncertainty, as follows

$$u_c(e_p) = \sqrt{s^2(\bar{p}) + u^2(\delta p) + u^2(p_s)} = 0.039 \text{ MPa} \tag{14}$$

When we use the approach recommended in [1], all the quantities may be summarized in the uncertainty budget, as in Table 4. The standard uncertainty associated with the output quantity e_p is as follows

Table 3 Uncertainty budget of calibrated manometer using a law of uncertainty propagation

Quantity	Estimate (MPa)	Standard uncertainty (MPa)	Probability distribution	Sensitivity coefficient	Uncertainty contribution (MPa)
p	19.9	0.026	Student	1	0.026
δp	0	0.29	Rectangular	1	0,29
p_s	20	0.0058	Rectangular	−1	−0.0058
e_p	−0.1	–	–	–	0.039

Table 4 Uncertainty budget of calibrated manometer using a propagation of distributions

Quantity	Estimate (MPa)	Standard uncertainty (MPa)	Probability distribution	Sensitivity coefficient	Uncertainty contribution (MPa)
p	19.9	0.033	Student	1	0.033
δp	0	0.29	Rectangular	1	0.29
p_s	20	0.0058	Rectangular	−1	−0.0058
e_p	−0.1	–	–	–	0.044

$$u(e_p) = \sqrt{\sigma^2 + u^2(\delta p) + u^2(p_s)} = 0.044 \text{ MPa} \tag{15}$$

The same result may be obtained using a Monte Carlo procedure with $M = 10^4$ trials, for calculating a values of the output quantity e_p through the measurement Eq. (9), and its standard uncertainty from formula (7).

The standard uncertainty of measurand calculated with the use of Monte Carlo method for propagation of distributions is bigger than the combined standard uncertainty when the uncertainty budget contains the input quantity having Student's distribution.

Regardless of the approach, when we use the method presented in [3, 4], based on the convolution of rectangular and Gaussian distribution [5] or the Flatten-Gaussian distribution [6], the expanded uncertainty is $U = 0.086$ MPa, for the 95 % coverage probability, so we obtain the same measurement result of instrument error. But the standard uncertainty associated with the measurand is not the same, and the coverage factor is also not the same. In the first case the coverage factor is $k = 2.19$, and in the latter case the coverage factor is $k = 1.93$.

8 Conclusion

The methods associated with the law of uncertainty propagation and the propagation of distributions give not the same results in calculation of the standard uncertainty attributed to the measurand. The combined standard uncertainty is smaller than the standard uncertainty of output quantity calculated by Monte Carlo procedure. The Monte Carlo procedure gives the standard deviations associated with any probability distribution, especially with Student's distribution. The standard deviation of Student's distribution is bigger than the experimental standard deviation of the mean for the same sample, regardless of the degrees of freedom. This affects the final result of standard uncertainty calculation, and the standard uncertainty attributed to the measurand in the first case is usually smaller than in the latter case (in practical example smaller about 12 %). But the calculation results of the expanded uncertainty are the same in any approach, because the coverage factors are not the same. In the first case the coverage factor is bigger than in the latter case (in practical example bigger about 12 %). For 95 % coverage probability k is usually bigger than 2, in the first case, and always smaller than 2, in the latter case.

References

1. Supplement 1 to the "Guide to the expression of uncertainty in measurement"—Propagation of distributions using a Monte Carlo method. JCGM 101:2008 (2008)
2. Guide to the expression of uncertainty in measurement. ISO (1995)
3. Fotowicz, P.: An analytical method for calculating a coverage interval. Metrologia **43**, 42–45 (2006)
4. Fotowicz, P.: Methods for calculating the coverage interval based on the Flatten-Gaussian distribution. Measurement **55**, 272–275 (2014)
5. Moszczyński, L., Bielski, T.: Development of analytical method for calculation the expanded uncertainty in convolution of rectangular and Gaussian distribution. Measurement **46**, 1896–1903 (2013)
6. Warsza, Z.L., Kubisa, S.: Midrange as estimator of measured value for samples from population of uniform and Flatten-Gaussian distributions. In: 20th IMEKO TC4 Symposium on Measurements of Electrical Quantities, pp. 720–725 (2014)

Assessment of Graphene Coatings Influence on Tribological Properties of Surfaces

Tadeusz Missala, Roman Szewczyk, Wojciech Winiarski, Marek Hamela, Marcin Kamiński, Andrzej Juś, Jan Tomasik, Michał Nowicki and Iwona Pasternak

Abstract This paper presents results of experiments carried out to determine influence of graphene coating on tribological properties of surfaces of friction pairs. Within these experiments series of 24 h tribological tests of sliding friction between examined surfaces on specially designed measuring stand were conducted. Subject of study in these tests were electrolytic copper plated steel samples, additionally covered with graphene and, as a reference, identically prepared samples without the graphene layer. As a result of these experiments characteristics of coefficient of friction and temperature between surfaces, as well as changes in mass and rough-

T. Missala · R. Szewczyk (✉) · W. Winiarski · M. Hamela · M. Kamiński · A. Juś
Industrial Research Institute for Automation and Measurements PIAP,
Al. Jerozolimskie 202, 02-486, Warsaw, Poland
e-mail: rszewczyk@piap.pl

T. Missala
e-mail: tmissala@piap.pl

W. Winiarski
e-mail: wwiniarski@piap.pl

M. Hamela
e-mail: mhamela@piap.pl

M. Kamiński
e-mail: mkaminski@piap.pl

A. Juś
e-mail: ajus@piap.pl

J. Tomasik · M. Nowicki
Institute of Metrology and Biomedical Engineering, Warsaw University of Technology,
Św. Andrzeja Boboli 8, 02-525 Warsaw, Poland
e-mail: j.tomasik@mchtr.pw.edu.pl

M. Nowicki
e-mail: Nowicki@mchtr.pw.edu.pl

I. Pasternak
Institute of Electronic Materials Technology, Wólczyńska 133, 01-919 Warsaw, Poland
e-mail: iwona.pasternak@itme.edu.pl

R. Szewczyk et al. (eds.), *Challenges in Automation, Robotics and Measurement Techniques*, Advances in Intelligent Systems and Computing 440, DOI 10.1007/978-3-319-29357-8_68

ness before and after tests were obtained. Based on these results improvement of tribological properties of sliding surfaces was shown.

Keywords Graphene · Tribology · Electrolytic copper plated steel

1 Introduction

Issues related to friction, wear and lubrication have been intensively researched for at least last 50 years [1–3]. Interest in this area is caused by its economic importance. It is estimated that over 1/3 of generated energy is wasted as a result of friction of interacting surfaces. In addition, wear of friction pair surfaces causes loss of their functionality and expensive repairs.

The most widespread way to decrease friction and wear of surfaces is the application of lubricating substance between them. Such substance creates a film which separates these surfaces, and replaces their external friction with internal friction of this substance. Since thickness of film depends on parameters like relative speed and force between surfaces, and on viscosity of base oil, it is not always possible to have fully separated surfaces. In areas where contact occurs, durability of boundary layer as well as tribological properties of interacting surfaces are important.

One of the new materials with potential to decrease coefficient of friction and wear in kinematic pairs is graphene—one atomic thickness carbon layer with hexagonal layout, which is known for very good mechanical properties as well as thermal and electrical conductivity. Various studies on use of graphene for tribological parameters enhancement are nowadays conducted [4, 5]. Researchers concentrate on application of graphene as an additive to grease or oil, or as self-contained solid grease or layer. They conduct experiments as well as numerical simulations [6]. For example in [7] results of study on coefficient of friction and wear of sliding bronze specimens lubricated with lithium grease with additive of graphene nanoplatelets are shown. Researches in this area are also described in [8, 9] wherein second one refers to application of graphene oxide. There are also known trials of application of graphene between sliding surfaces by use of mixture of graphene platelets and ethanol in humid air [10] and dry nitrogen [11] (after the evaporation of ethanol platelets of graphene are from technical point of view the only lubricating material in friction pairs in these tests). Studies on tribological properties of graphene as a layer are also conducted but they are concentrated on nano- or microscale [12, 13]. Results of all specified studies indicate improvement of tribological properties of kinematic pairs after application of graphene.

2 Experimental

Since tribological properties of graphene as a layer in macroscale are hardly known, the main goal of studies presented in this paper was to determine if graphene coatings decrease friction and wear of surfaces in this scale. For this purpose a measurement stand which provides relative movement between surfaces forming a friction pair, as well as measurement of friction force between this pair was built. Main part of the test stand is shown in Fig. 1.

Friction pairs in this stand (Fig. 1) consist of a ring-shaped replaceable track (1) and three samples (2) to be tested. During tests track (1) rotation in vertical axis the specimens (2) are kept in place by holders (3), which means there is a relative movement between surfaces of track (1) and specimens (2). Specimens (2) are pressed to track (1) by the gravity with use of disc (4), pin (5) and weights (6). In order to achieve parallel working surfaces (which is essential to provide uniform stress distribution) there are two rotational degrees of freedom (in horizontal axes) between holders (3) and shield (4), and two between shield (4) and pin (5). There is also one degree of freedom (rotation, in vertical axis) between pin (5) and bushing (7) which is blocked by the force sensor (8). Force measured by this device is reaction to friction forces between track (1) and samples (2). Since this force is recorded with numbers of pulses from rotation sensor (9) the results of experiments can be shown as characteristics of friction coefficient versus distance.

Parameters of experiments conducted on the above described stand are: type of grease, pressing force and linear velocity between surfaces of friction pairs, sliding distance, materials of friction pairs (with or without coating) and process of their machining. In order to determine influence of graphene on tribological properties of surfaces two series of tests were carried out. For this purpose two sets of $20 \times 20 \times 5$ mm samples of 41Cr4 steel were made. They were milled, polished and

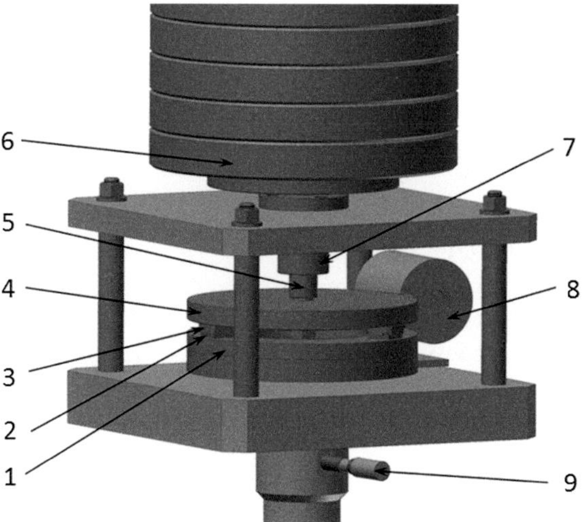

Fig. 1 The schematic of the stand for tribological properties of graphene layer measurement

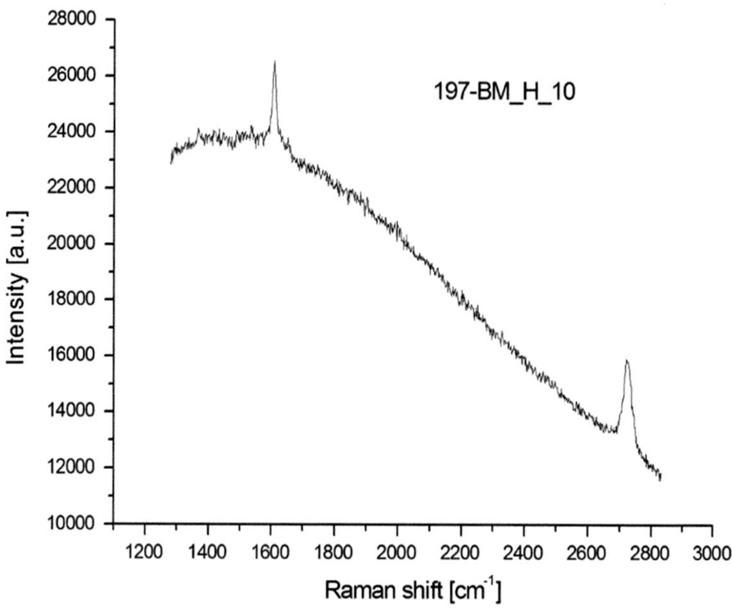

Fig. 2 Raman spectrum of electrolytic copper plated steel with graphene coating

copper plated. Difference between them was presence of graphene on their surfaces —half of them were graphene coated in result of a chemical vapor deposition process. Second half were reference specimens without graphene. Figure 2 presents Raman spectrum of electrolytic coppered steel. In this spectrum peaks characteristic for graphene—G (1600 cm^{-1}) and 2D (2700 cm^{-1})—are present.

Tensile strength of graphene is ca. 130 GPa and it is therefore two orders of magnitude higher than tensile strength of quenched and tempered steel. However, graphene consists of single atomic layer, which means it is not able to carry forces which occur in industrial mechanical parts. In addition, during coating graphene duplicates roughness of substrate, which in technical applications is three or even four orders of magnitude higher then thickness of graphene. These factors cause that it is impossible for graphene to stay undamaged during sliding friction. In case of no lubrication (dry friction) high wear appears, mainly abrasive and adhesive. Influence of graphene on decrease of wear is in this case not significant and therefore tests described in this paper were conducted in mixed friction mode (thanks to graphene presence facilitation of shear in places where no fluid friction occurs was expected). To achieve this kind of friction in each test lithium grease was applied, as well as geometry of interacting surfaces, their linear speed and press were properly selected.

Values of the parameters of the conducted tests are shown in Table 1.

Characteristics of friction coefficient versus distance, and temperature versus distance determined with use of the measuring stand (Fig. 1) do not give straight

Table 1 Values of the tests parameters

Parameter	Value
Type of grease	lithium soap (white lithium, lithium grease), Kinematic viscosity: 100 mm^2/s (at 40 °C), NLGI class: 2
Press (MPa)	0.55
Velocity (m/s)	0.52
Distance (m)	40,000
Specimen material	Steel 41Cr4: (a) Electrolytic copper plated and graphenized (b) Electrolytic copper plated
Track material	Steel 41Cr4
Specimen surface machining	Polishing
Track surface machining	Polishing

information about wear of specimens. Quantitative description of this wear was determined by measurement of mass of specimens before and after experiment with use of RADWAG PS 600.R2 scales. Roughness of surfaces before and after experiment was measured with Taylor-Hobson Talysurf PGI830 profilometer. Exemplary profilogram is shown in Fig. 3.

As a result of conducted tests it was determined if graphene enhances tribological properties of copper plated steel.

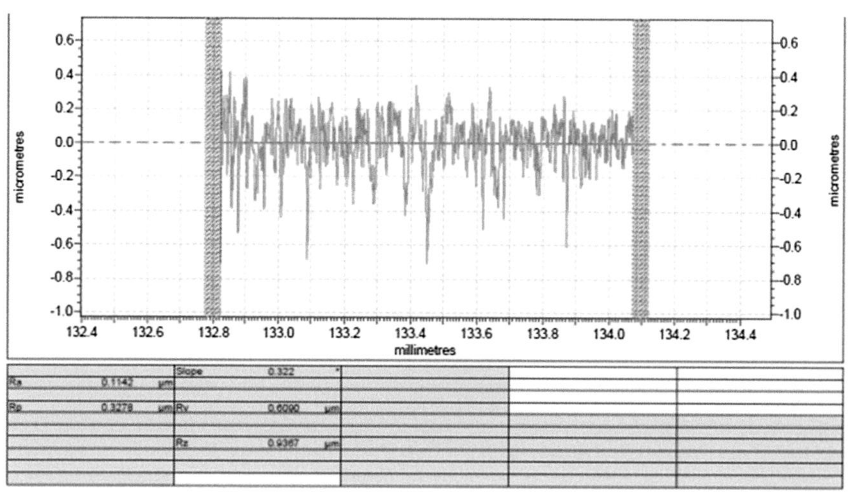

Fig. 3 Exemplary profilogram of electrolytic copper plated and graphenized specimen before experiment

3 Results

Results of experiments carried out on the test stand shown in Fig. 1 are presented as characteristics of friction coefficient versus distance and of temperature versus distance. In case of specimens with graphene (Fig. 4, blue diagram) after initial run coefficient of friction stays on level of 0.045 and is very stable, which means compression of graphene and as a result facilitation of sliding between surfaces of samples and track in places with little lubricating film. In experiment carried out on specimens without graphene (Fig. 4, orange diagram) coefficient of friction is over 0.06 and it is not as stable as for graphenized specimens. Reason for this can be lack of layer that prevents tearing off particles in areas where no lubricating film occurs. Reflection of lower coefficient of friction of graphenized specimens is also lower temperature of contact (Fig. 5)—below 40 °C for graphenized specimens and over 45 °C for not graphenized specimens.

Results of measurement of mass loss confirm results of measurement of friction coefficient (see Table 2). The higher friction between surfaces of specimens and track is, the bigger wear occurs. The lower mass loss (which means the lower wear) in graphenized specimens is thank to compressed graphene layer which facilitates shear in areas where lubricating film is too thin, and therefore decreases intensity of tearing off particles of surfaces in contact. These tests were conducted on high accuracy (verification scale interval e = 0.001 g, actual scale interval d = 0.00001 g), calibrated laboratory scale AS 60/220/C/2—it confirm reliability of presented results.

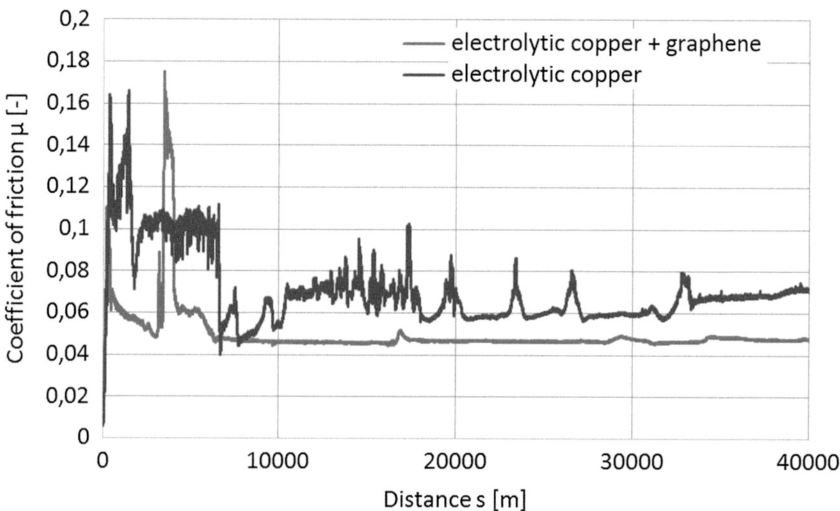

Fig. 4 Diagrams of coefficients of friction μ versus distance s of electrolytic coppered samples

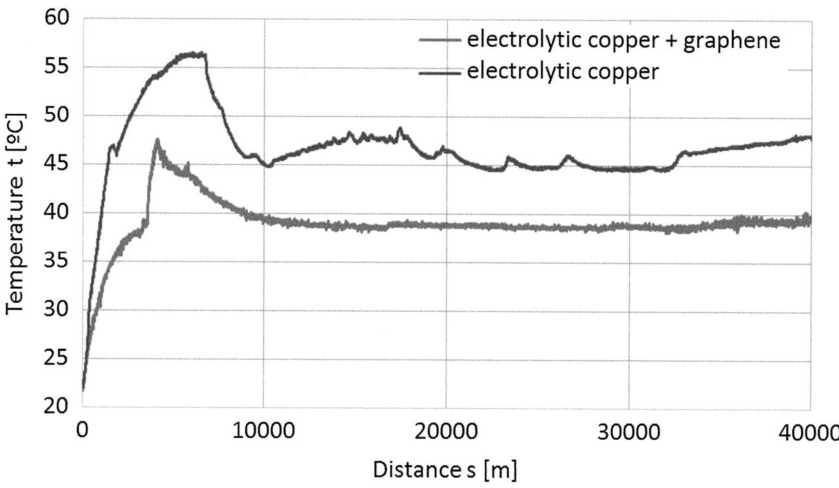

Fig. 5 Diagrams of temperatures t versus distance s of electrolytic coppered samples

Table 2 Mean mass loss of specimens as a result of tribological experiments

Type of layer	Mean mass (g)		Mean mass loss (g)
	Before experiment	After experiment	
Electrolytic copper + graphene	14.96687	14.95940	0.00747
Electrolytic copper	15.45613	15.43313	0.02300

Table 3 Changes in roughness of surfaces as a result of tribological experiments

Type of layer	Specimens		Track	
	Mean roughness R_a before experiment (μm)	Mean roughness R_a after experiment (μm)	Mean roughness R_a before experiment (μm)	Mean roughness R_a after experiment (μm)
Electrolytic copper + graphene	0.1171	0.0541	0.2212	0.2136
Electrolytic copper	0.1109	0.0544	0.2338	0.2181

Results of roughness measurements indicate that roughness gets lower in every case (Table 3). It is because of running in process and as a result removal of highest peaks. Since differences between graphenized and not graphenized specimens are lower than dispersion of results they are statistically not significant.

4 Conclusion

Experiments on graphene layers on electrolytic copper plated steel confirm improvement of tribological properties in case of mixed friction in macroscale. Graphene layer in this case caused 25 % decrease of coefficient of friction, and improvement of wear resistance.

Acknowledgement This work has been supported by the National Centre for Research and Development (NCBiR) within the GRAF-TECH programme (no. GRAF-TECH/NCBR/05/13/2012).

References

1. Hebda, M., Wachal, A.: Trybologia. Wydawnictwa Naukowo-Techniczne, Warszawa (1980)
2. Płaza, S., Margielewski, L., Celichowski, G.: Wstęp do Tribologii i Tribochemia. Wydawnictwo Uniwersytetu Łódzkiego, Łódź (2005)
3. Booser, E.R.: CRC Handbook of Lubrication. Theory and Design, vol. II. CRC Press LLC (1983)
4. Berman, D., Erdemir, A., Sumant, A.: Graphene: a new emerging lubricant. Mater. Today **17**, 31–42 (2014)
5. Penkov, O., Kim, H.-J., Kim, H.-J., Kim, D.: Tribology of graphene: a review. Int. J. Precis. Eng. Manuf. **15**, 577–586 (2014)
6. Bonelli, F., Manini, N., Cadelano, E., Colombo, L.: Atomistic simulation of the sliding friction of graphene flakes. Eur. Phys. J. B **70**, 449–459 (2009)
7. Missala, T., Szewczyk, R., Winiarski, W., Hamela, M., Kamiński, M., Dąbrowski, S., Pogorzelski, D., Jakubowska, M., Tomasik, J.: Study on tribological properties of lubricating grease with additive of graphene. Progress in automation, robotics and measuring techniques. Adv. Intell. Syst. Comput. **352**, 181–187 (2015)
8. Cheng, Z., Qin, X.: Study on friction performance of graphene-based semi-solid grease. Chin. Chem. Lett. **25**, 1305–1307 (2014)
9. Lin, J., Wang, L., Chen, G.: Modification of graphene platelets and their tribological properties as a lubricant additive. Tribol. Lett. **41**, 209–215 (2011)
10. Berman, D., Erdemir, A., Sumant, A.: Few layer graphene to reduce wear and friction on sliding steel surfaces. Carbon **54**, 454–459 (2013)
11. Berman, D., Erdemir, A., Sumant, A.: Reduced wear and friction enabled by graphene layers on sliding steel surfaces in dry nitrogen. Carbon **59**, 167–175 (2013)
12. Berman, D., Erdemir, A., Zinovev, A., Sumant, A.: Nanoscale friction properties of graphene and graphene oxide. Diam. Relat. Mater. **54**, 91–96 (2015)
13. Marchetto, D., Held, C., Hausen, F., Wahlisch, F., Dienwiebel, M., Bennewitz, R.: Friction and wear on single-layer epitaxial graphene in multi-asperity contacts. Tribol. Lett. **48**, 77–82 (2012)

Setup for Stereovision Simulation for Mutual Navigation of Satellites Formation

Paweł Nowak, Robert Ugodziński, Roman Szewczyk, Marcin Kamiński, Michał Nowicki and Maciej Kachniarz

Abstract Paper presents foundation of newly developed setup for stereovision simulation. Described system will be used for verification of different algorithms of mutual navigation for satellites formation. Designed setup contains image display system, remote vision system and image processing system. In paper hardware and software parts of all systems are described. Current state of setup development as well as future actions are presented. Setup utilization for its own calibration is described.

Keywords Stereovision simulation · Space navigation · Satellites vision navigation

1 Introduction

Mutual navigation of satellites is an important issue in a space industry. Simplify it is used to ensure proper relative movement of one satellite towards another. Thus it is crucial during remote satellites service without the need of bringing damaged satellite to earth. Reliable mutual navigation is required during satellite transport to the graveyard orbit as well. Proper navigation diminishes accident possibility, which reduces satellite average operating costs as well as number of space debris.

P. Nowak (✉) · R. Ugodziński · M. Kamiński
Industrial Research Institute for Automation and Measurements PIAP,
Al. Jerozolimskie 202, 02-486 Warsaw, Poland
e-mail: pnowak@piap.pl

R. Ugodziński
e-mail: rugodzinski@piap.pl

R. Szewczyk · M. Nowicki · M. Kachniarz
Institute of Metrology and Biomedical Engineering, Warsaw University of Technology,
Sw. A. Boboli 8, 02-525 Warsaw, Poland

R. Szewczyk et al. (eds.), *Challenges in Automation, Robotics and Measurement Techniques*, Advances in Intelligent Systems and Computing 440, DOI 10.1007/978-3-319-29357-8_69

Vision navigation for diverse types of systems [1] was significantly boosted by utilization of stereovision [2, 3]. Thanks too that depth of object can be calculated from single image acquisition [4]. Additionally stereovision solves problem of map scale, which occurs in monocular observations [5]. Thus, stereovision navigation was selected to achieve mutual navigation of satellites formation. In order to develop optimal algorithms many offline tests are required. To provide suitable environment for algorithms testing presented setup was developed.

2 General Idea of Setup for Stereovision Simulation

Designed setup for stereovision simulation was based on a standard stereovision setups, where two spaced apart cameras are observing the scene from different angles. Based on the distance between cameras data such as object distance and relative position can be calculated. Described setup has to simulate view in outer space with big variety of objects distances. Therefore, in order to achieve proper stereovision simulation, another solution was chosen.

As presented in Fig. 1, developed test stand is based on two cameras, which are observing two separately controlled monitors. Both cameras are part of remote vision system, which also contains their controller and system for initial image processing. Both monitors are controlled individually and are suitable for rendering images provided by control software. Data from remote vision system after optional preprocessing are transmitted to external image processing system.

2.1 Image Display System

Hardware part of image display system consists two monitors with a 4K2K resolution each. They are used to display properly prepared images simulating stereovision. In order to simultaneously display high resolution images utilization of efficient graphic card was required and AMD Radeon R9200 with two Display Port outputs was selected. This setup allows to simultaneously display two images of 4K2K resolution with a 30 Hz refresh frequency. This setup exceeds the capabilities of the vision system (which is described below), both in display resolution and refresh frequency. Thus it can be assumed that developed system is capable for simulating outer space from the satellite viewpoint.

Software part of image display system will change during the progress of project. Currently developed software is used to display simple images utilized during cameras calibration. Future software versions will properly modify given image in order to simulate different relative positions and distances between satellites and will influence optic distortion caused by parallelism error between monitors and cameras optical surfaces.

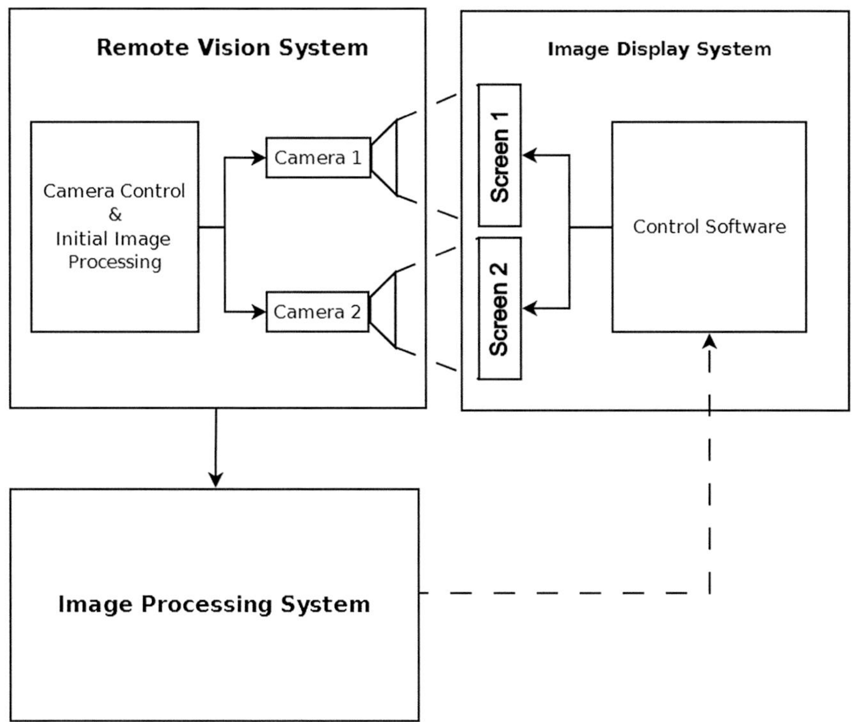

Fig. 1 Schematic block diagram of designed setup

2.2 Remote Vision System

Remote vision system is based on NI CVS-1457 RT (a compact vision system produced by National Instruments) and two Ethernet cameras. Utilized compact vision system is independent, fully programmable real time system, containing Intel Atom processor as well as reconfigurable FPGA, based on Spartan-6 LX25. Its basic functionality contains camera control and simple image processing. For camera control system utilizes NI IMAQ drivers, which are supporting most of the USB and Ethernet cameras on the market. Due to utilization of advanced vision system initial image processing (such as threshold or edge detection) can be applied in real time. Thus it can be programmed to extract only necessary data from camera images which greatly increases speed of whole system operation by eliminating necessity to transfer large amounts of data.

Hardware part of remote vision system contains described compact vision system as well as two Basler acA2500-14gc cameras. They allow continuous image acquisition with 5 MP resolution and 14 frames per second.

Software part of remote vision system will change during the progress of project. Currently two programs are developed. Both are based on continuous frames acquisition. First program was utilized during setup calibration and utilizes image processing functions. On acquired image automatic threshold is applied. Afterwards elements on image are segmented. Coordinates of mass center of each segment as well as their areas are calculated. Those data are broadcasted to local network by UDP protocol. Second software is for general use by external image processing system. Frames from cameras are continuously acquired. Afterwards each image is transformed to array of pixels intensities which is broadcasted with TCP protocol.

2.3 Image Processing System

Image processing system will be developed during the progress of the project. Currently simple image processing system was developed in order to proceed with camera calibration. More details about are presented in Sect. 4.

3 Setup Hardware Implementation

Setup was locked due to strict requirements for relative position of both cameras, as well as their position with respect to monitors. Remote vision system (cameras and compact vision system) was placed in casing presented on Fig. 2. Casing provides constant distance between both cameras and protects from external conditions.

To assure constant relative position between remote vision system and image display system both were fixed rigidly. As presented on Fig. 3, monitors from

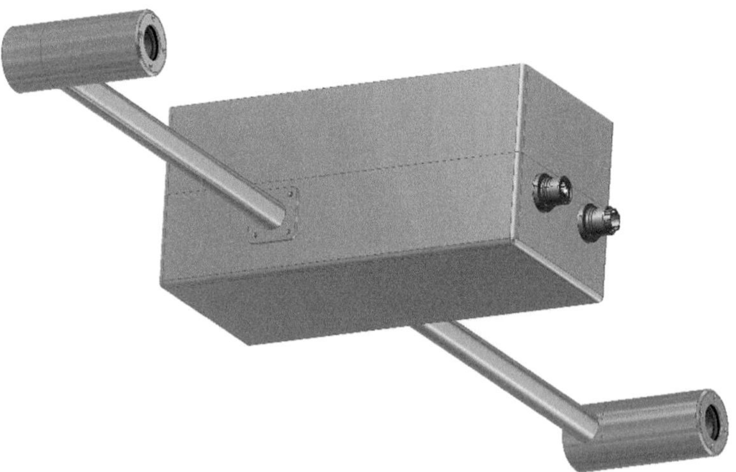

Fig. 2 Rendering of remote vision system housing

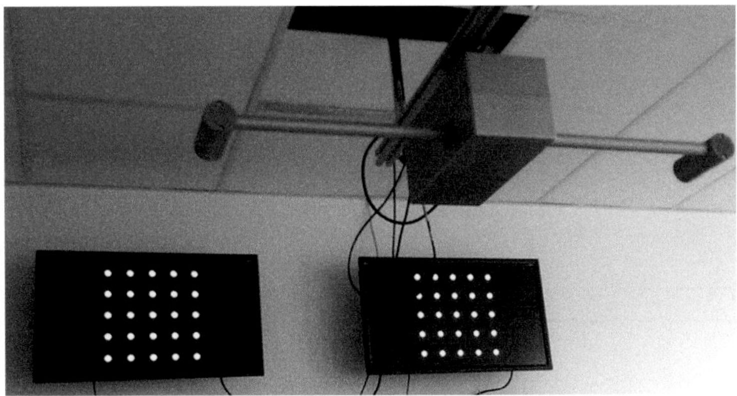

Fig. 3 View of system prototype placement

image display system are fixed to the wall, and remote image system is suspended from the ceiling. This solution assured constant distance between the crucial elements of described setup.

4 Setup Test—Camera Calibration

In order to assure proper operation of developed system during future research, whole setup had to be calibrated. During that process perspective correction matrixes [6] were determined. Those matrixes will be utilized in image rendering, in order to minimalize influence of optical distortions. Calibration procedure consists three stages described below and represented in Fig. 4.

- Stage 1—point rendering
  This stage is proceeded on image display system. Nettings of user-defined points are rendered on both monitors by specially developed software. User can modify the number of displayed rows and columns of points, spread between them as well as radius of points. Exemplary point nettings are visible in Fig. 3.
- Stage 2—Image acquisition and initial processing on remote system
  This stage is proceeded on remote vision system and utilizes software described in Sect. 2.2. Images from both cameras are continuously acquired and analyzed, in order to extract data about coordinates of each displayed point. Image proceeding is conducted on remote vision system and calculated coordinates of points are transmitted to calibration software.
- Stage 3—Calculation of perspective correction matrix
  Coordinates of acquired points as well as coordinates of displayed points are proceeded in MATLAB software. Based on coordinates of points in both reference systems (connected with monitors and cameras) perspective correction

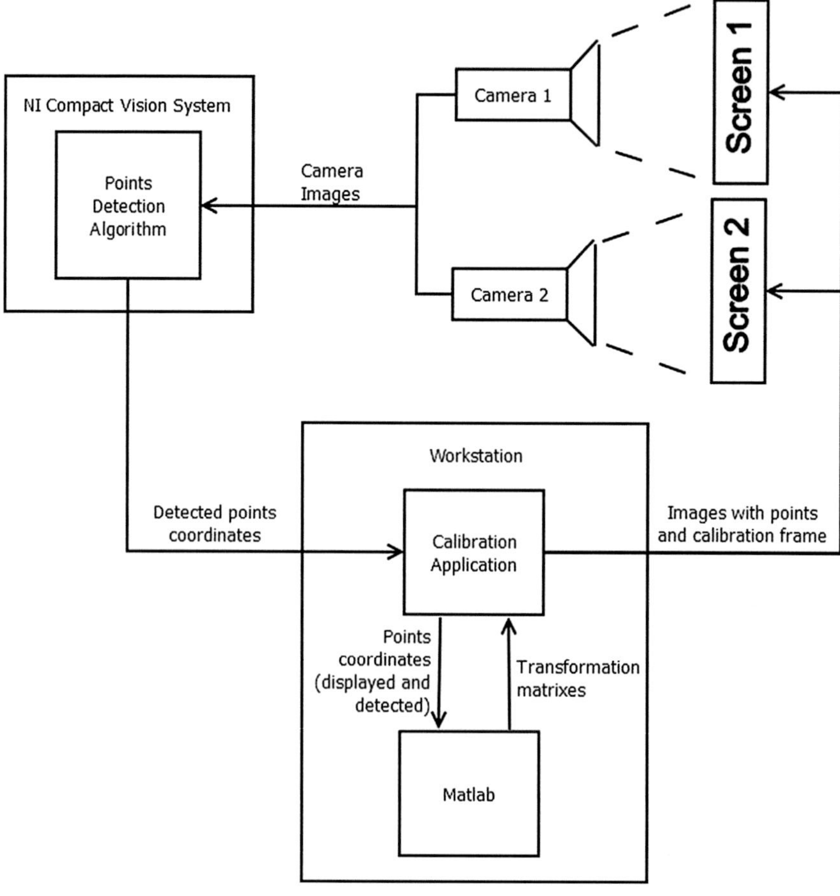

Fig. 4 Scheme of camera calibration procedure

matrix can be calculated for both cameras separately. Optimization problem for each camera can be formulated as (1) presents:

$$\min_{a_{11},a_{12},a_{21},a_{22},b_1,b_2} \sum_{i=1,2,..,n_x} \sum_{j=1,2,..,n_y} \left(\begin{bmatrix} x_{ij}^{disp} \\ y_{ij}^{disp} \end{bmatrix} \begin{bmatrix} a_{11} & a_{12} \\ a_{21} & a_{22} \end{bmatrix} + \begin{bmatrix} b_1 \\ b_2 \end{bmatrix} = \begin{bmatrix} x_{ij}^{cam} \\ y_{ij}^{cam} \end{bmatrix} \right) \quad (1)$$

where:

$a_{11}, a_{12}, a_{21}, a_{22}, b_1, b_2$	elements of perspective correction matrix
$n_x,\ n_y$	numbers of rows and columns of rendered points
$x_{ij}^{disp},\ y_{ij}^{disp}$	coordinates of displayed points
$x_{ij}^{cam},\ y_{ij}^{cam}$	coordinates of acquired points identified during image processing

Calculated matrixes are transferred to calibration software. Based on them information about relative lurch between camera and monitor planes are displayed. Additionally, visual feedback representing camera point of view is displayed on both monitors. Calibration procedure was based on simple optimization, in which the objective function was to maximize the area of the screens observed by each camera, and to minimize the optical distortions represented by the elements in perspective correction matrix. It was obtained by adjusting the zoom of both cameras as well as by assuring proper parallel position of the monitors and cameras.

5 Conclusion

Presented setup for stereovision simulation was developed. It was based on two separately controlled monitors in image display system, which allows simulating different stereovision situations. Based on that advanced mutual navigation algorithms for satellite formation will be developed. Currently setup contains fully developed remote vision system as well as examples of image display system and image processing system. Those systems were developed in order to conduct system vison calibration. Physical structure of setup, due to significant influence of relative position of cameras and monitors, was fixed after calibration. Developed system will be utilized in further research of mutual vision navigation algorithms.

Acknowledgements This work was partially supported by The National Center for Research and Development with PBS 2 Program—Grant no. PBS2/B3/17/2013.

References

1. Nowicki, M.: WiFi-guided visual loop closure for indoor navigation using mobile devices. J. Autom. Mobile Robot. Intell. Syst. **8**(3), 10–18 (2014)
2. Jesus, F., Ventura, R.: Simultaneous localization and mapping for tracked wheel robots combining monocular and stereo vision. J. Autom. Mobile Robot. Intell. Syst. **7** (2013)
3. Murray, D., Little, J.J.: Using real-time stereo vision for mobile robot navigation. Auton. Robots **8**(2), 161–171 (2000)
4. Hartley, R., Zisserman, A.: Multiple View Geometry in Computer Vision. Cambridge University Press (2003)
5. Piniés, P., et al.: Inertial aiding of inverse depth SLAM using a monocular camera. In: 2007 IEEE International Conference on Robotics and Automation, pp. 2797–2802. IEEE (2007)
6. Jagannathan, L., Jawahar, C.V. Perspective correction methods for camera based document analysis. In: Proceedings of First International Workshop on Camera-Based Document Analysis and Recognition, pp. 148–154 (2005)

Issues and Problems with Measuring the Temperature in the Hive

Jerzy Niewiatowski, Wojciech Winiarski, Paweł Nowak, Grażyna Topolska, Urszula Grzęda and Michał Gałek

Abstract Currently developed setups for measurement of the parameters of the inside of the hive are based on Internet network, which allows simple data acquisition from the distributed measurement points. Those setups does not focus on measurement methodology and commonly utilize temperature sensors hanging inside bee space inside the hive or sensors boxes placed on frames. This results with disruption inside the nest and does influence bees behavior. Acquired data are not reliable, because the measured phenomena was significantly influenced by measurement process. There are weak links to experts knowledge, which connects measurement results with the cause of the phenomenon, which is required to diagnose state of the bee colony and thus facilitate beekeepers work. To fulfill that demand in Industrial Research Institute for Automation and Measurements PIAP measurements system of hive temperature was developed. Described system will increase efficiency of beekeepers work by utilization of the expertise database and due to development of measurement device which will not influence bee behavior.

J. Niewiatowski (✉) · W. Winiarski · P. Nowak
Industrial Research Institute for Automation and Measurements PIAP, Al. Jerozolimskie 202, 02-486 Warsaw, Poland
e-mail: jniewiatowski@piap.pl

W. Winiarski
e-mail: wwiniarski@piap.pl

P. Nowak
e-mail: pnowak@piap.pl

G. Topolska · U. Grzęda
Warsaw University of Life Sciences-SGGW (WULS-SGGW), Nowoursynowska 166, 02-787 Warsaw, Poland
e-mail: grazyna_topolska@sggw.pl

U. Grzęda
e-mail: urszula.grzeda@gmail.com

M. Gałek
Beekeeping Farm, Ludwikow 109, 27-350 Sienno, Poland
e-mail: michal.galek1@wp.pl

R. Szewczyk et al. (eds.), *Challenges in Automation, Robotics and Measurement Techniques*, Advances in Intelligent Systems and Computing 440, DOI 10.1007/978-3-319-29357-8_70

Keywords Bee hive • Bee hive measurements • NTC resistor • Measurement frame • Temperature measurements

1 Introduction

Electronic industry is one of most developed sector of world economy. Electronisation progresses at the agriculture as well. Some examples are commonly used controllers for automated milking and milk cooling in bigger cowsheds as well as security systems and devices for energy management. Utilization of electronics is required in big agricultural holdings as an database systems [1]. Control systems are commonly used in greenhouse systems where real time measurements of soil and atmosphere allow to maintain their parameters. In the agricultural field parameters of soil are determined in laboratory studies and atmosphere parameters are obtained from weather station. Attempts of the measurements of animal welfare and thus reducing the human workload are not advanced. Instance is previously reported control of windows size in cowshed [2].

Beekeeping, despite being non-electrified, presents high possibility of electronic utilization. Beekeepers work involves multiple inspections of bee hive in order to determine breeding parameters, such as size of bee family, amount of collected feed (honey, pollen), number of offspring (maggots), family state (queenless, swarm mood, sickness etc.). Beekeepers has to connect those data with season and the state of environment (amount of feed) [3, 4]. Conducting those inspections, besides significant workload of beekeepers, influences development of bee family—each nets opening causes disruption after which hive need week or two to regenerate [5, 6]. Number of required overviews will be significantly reduced by acquisition of data from inside of bee hive. Basing on temperature, humidity, CO_2 saturation and hive mass state of bee family can be reasoned.

To utilize measurement data development of information management system is required. Such system should be based on experts knowledge obtained due years of study. The biggest recognized problem in development of system supporting the beekeepers work is caused by need of hiring highly qualified experts. Other significant issue is conduction of the measurements inside the hive with minimal disruption within the bee colony.

Research centers around the world developed measurement systems to monitor the state of bee hive and acquire data to control server [7–9]. Those setups have simple measuring systems and does not fulfill requirement of minimal influence on bee colony. Also no expert system were developed, due no linking of measurements results with bee behavior. WMA (Wireless Monitoring in Apiary) system created in 2010 [10] was mainly focused on data distribution system and not on development of measurement setup invisible for bees.

2 Development of Temperature Measurement System

Temperature were selected on the first stage of the project as a parameter relatively simple to measure but providing significant amount of data about bee family. In order to minimize influence of measurement system on bee family, temperature sensors were deluged in a beeswax and electronic PCB (Printed Circuit Board) was placed in drilled frame. Wooden hive frame supports honeycomb, which, for a wielkopolska frame, can weight more than 2 kg. Drilling an original supper frame side will result with a significant loss of frame strength. New supporting beam was made from aluminum which assured right mechanical properties. On the other hand, usage of thermal isolation was required in order to eliminate thermal bridge which would otherwise result with increased nest moisture. In order to achieve thermal isolation from the top of the frame, top wooden beam was left. Metal beam with control PCB was placed in a way which allowed utilization of a drilled holes in the top beam as a cables outlet. Scheme of frame is presented in Fig. 1.

As a temperature sensors NTC (Negative Temperature Coefficient) resistors in 0204 casing were selected because of small dimensions, acceptable measurement accuracy and high availability. In order to provide electric connection between sensors and control PCB, measurements strips were made. They contained conductive tracks deposited on the group of connected plastic films with mounted NTC temperature sensors. Similar technology is utilized for membrane keyboards. Schematic view of single measurement strip is presented in Fig. 2.

Each sensor is utilizing 4 wire connection in order to delete influence of connections resistance [11] which are significant and unrepeatable. Single PCB mounted in metal beam is suitable for controlling 3 measurement stripes with 4 temperature

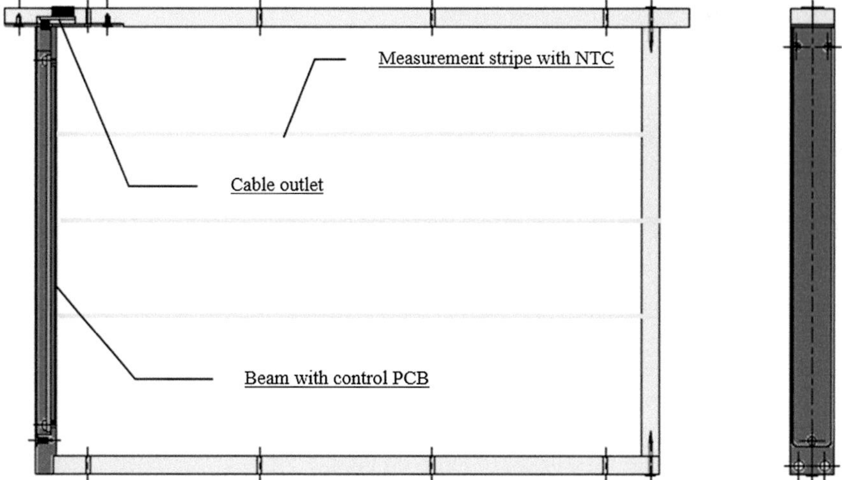

Fig. 1 The profile of the measuring frame

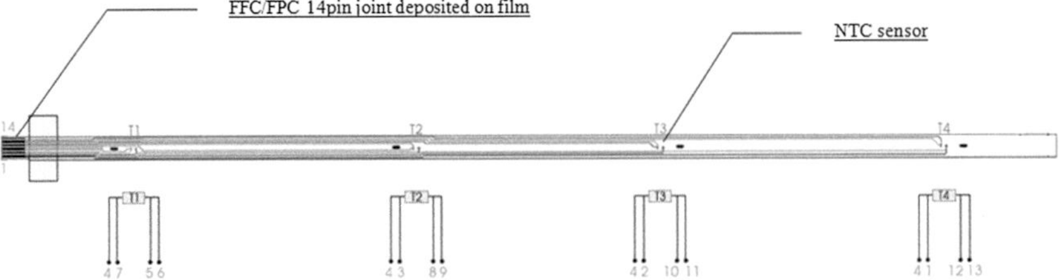

Fig. 2 Construction of measurement stripe

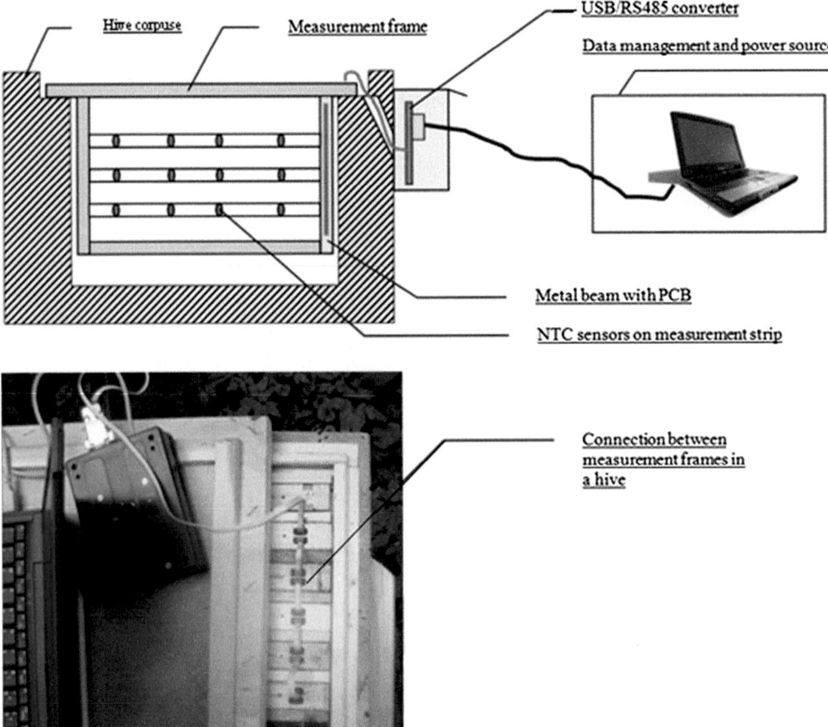

Fig. 3 Scheme of hive temperature measurement setup

sensors on each. Thus single measurement frame provides temperature data from 12 measurement points. Due to that temperature distribution in the hive can be measured. Scheme of temperature measurement setup for single hive is presented in Fig. 3. Tests in climatic chamber confirmed acceptable measurement error (±0.5 °C). Afterwards environmental studies on bee colony were conducted, in order to determine influence of measurement setup on bees.

3 Environmental Studies—Measurement Stripes in Wax Foundation

Innovative idea of hive temperature measurements utilizing sensors embedded in a honeycomb required tests of its influence on bees behavior. After mounting measurement stripes in a wax foundation sheet and putting it back to the hive observations were conducted whether bees would accept new frame and work in the presence of the measurement stripes. Additionally, tests were conducted with two measurement stripes arrangement—parallel and perpendicular to wax foundation sheet plane. Also influence of number of stripes and distance between them were examined.

15 measurement frames models were prepared. In some frames full measurement stripes were not used. They were depleted by sensors and conductive layers. Models of utilized frames are presented in Fig. 4. Wax foundation sheet was cut and placed between measurement stripes.

Measurement stripes were made of plastic with bonding layers method. Both plastics and glue had initially peculiar smell which could significantly influence bees behavior. All measurement stripes were exposed to the sun for 3 weeks, which highly reduced their smell.

Set of frames presented in Fig. 4 was put into hives in May 2014. After 6 weeks, on June 14, 2014, observations of bees work on new frames were conducted. Exemplary results are presented on Figs. 5, 6. Frame no. 11 (brood box frame) presented in Fig. 5 contained depleted measurement stripes placed parallel to wax

Frames 1-5 with PCB. Original measurements stripes placed perpendicular to wax foundation.

Frames 6-10 without PCB. Depleted stripes placed perpendicular to wax foundation.

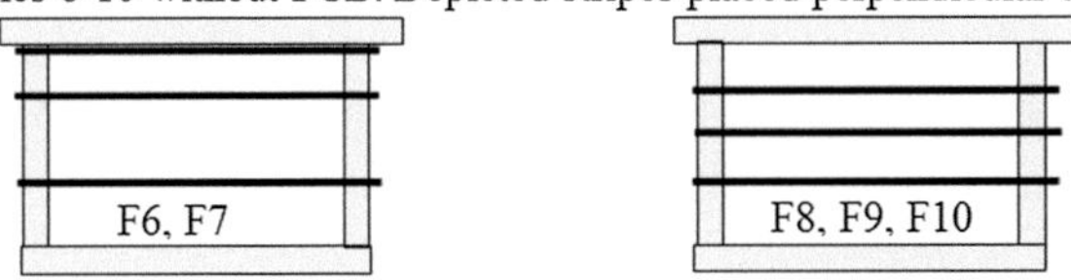

Frames 11-15 without PCB. Depleted stripes placed parallel to wax foundation.

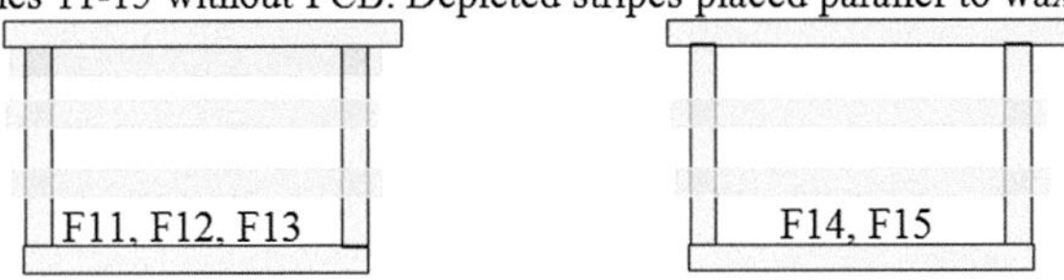

Fig. 4 Models of different frames configurations utilized for testing bees tolerance and work

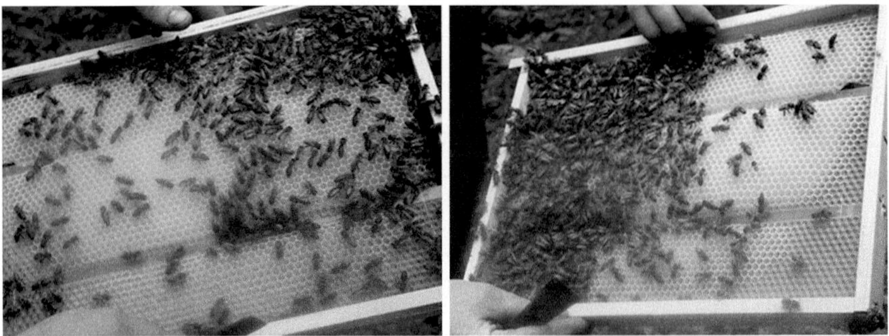

Fig. 5 Photograph of Frame no. 11

Fig. 6 Photograph of Frame no. 5

foundation sheet. Bees work is significantly uneven—stripes are partially covered with wax. Despite that bees started breeding—new eggs and larvae are visible. Nectar was gathered at the top of the frame.

Frame no. 5 (honey supper frame) presented in Fig. 6 contained original measurement stripes placed perpendicular to wax foundation sheet and was placed in the middle of the hive. Wax foundation sheet was fully reconstructed and nectar was gathered. Comb is characteristically uneven near the measurement stripes, where cells shapes were not fitted properly. Bees are avoiding cells nearby the measurement stripes during nectar gathering.

The research have shown, that bees accepted the modified frame unexpectedly well. Honeycomb was rebuild when stripes were placed perpendicularly and partially rebuilt when stripes were place parallel.

4 Environmental Studies—Experimental Setup on Hive with Measurement Frames

After confirming bees acceptance to modified frames with measurement stripes, experimental measurement setup was built in a new hive according to scheme presented in Fig. 3. Fully rebuilt frames from previous stages were utilized to assemble this system. As presented in Fig. 7 measurement stripes were fully covered with wax. Experimental hive was left in natural conditions for the winter. Regular temperature measurements were conducted. Exemplary results are presented in Fig. 8. Black dots represent hive bottoms and square indicates the beehive entrance. On left image measurement results acquired at 9th December, 2014 are presented. Red points, representing higher temperature, indicate place where bee family formed winter cluster. Right image represents data acquired on 28th April, 2015—temperature distribution is more uniform because of bees activity.

Fig. 7 Fully wax-covered measurement frame utilized for experimental setup

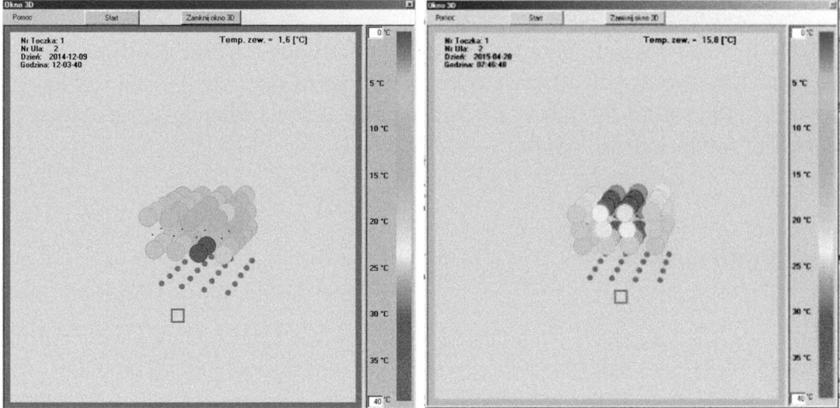

Fig. 8 View of exemplary measurement results

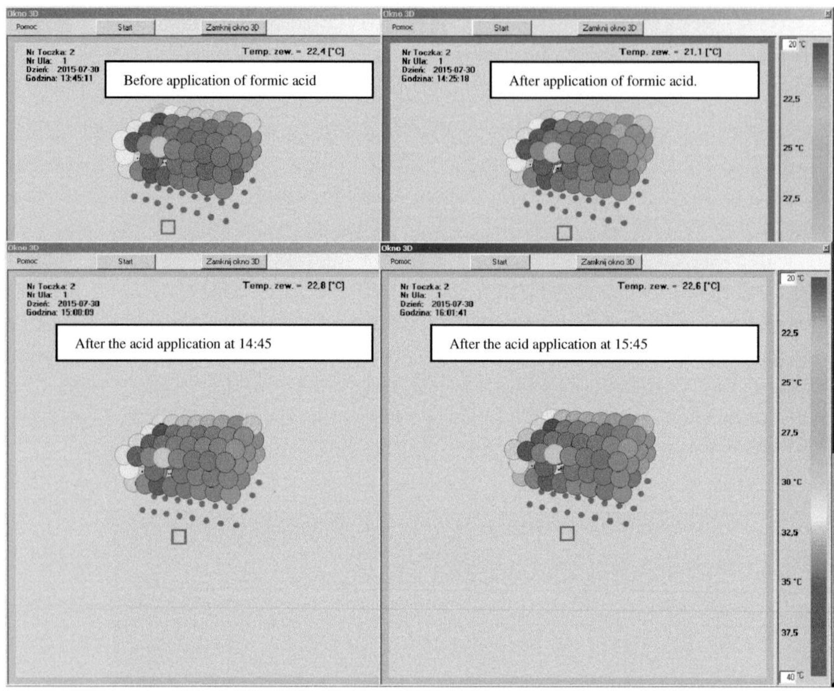

Fig. 9 Exemplary measurements conducted in expanded experimental hive at the Warsaw University of Life Sciences

Currently experimental hive is utilized for research in Warsaw University of Life Sciences on Faculty of Veterinary Medicine. Research concerns influence of formic acid on bees behavior. Experimental setup was expanded with additional measurement frames. Exemplary results of measurements conducted on 30th July, 2015, are presented in Fig. 9. Those measurements were conducted during tests of bees reaction to formic acid application. Formic acid was applied to control the infestation of the parasitic mite *Varroa destructor*. Bees respond poorly or did not react with the temperature change on the given dose of acid. On the 2nd and the 4th frame (counting from the left side) visible temperature variation is caused by the damage to the measuring point.

5 Conclusions

Observation of bees work on prepared measurement frames shown that, despite increased effort, bees were working on modified frames. Bees had no problems with cut wax foundation sheet and rebuilt honeycomb even on measurement stripes placed parallel. Increased workload of bees was required during fitting new honeycomb cells to cut and distanced foundation sheet. New cells differ with size and shape and are initially skipped during food storage. No influence of measurement stripes distance on rebuilt process was observed. Bees rebuilt measurement stripe placed tangentially to top beam. Rebuilt honeycomb was similar to comb without measurement stripes.

Frames with uncovered parallel stripes required more workload of bees than when stripes were placed perpendicular to wax foundation. When parallel oriented stripes were initially covered with a thin layer of wax rebuilt process was significantly faster. Also queen bee was not avoiding measurement frames and was breeding there as well.

Other issue concerned peculiar smell of measurement stripes including plastic and glue. Initial smell was significantly reduced by 3 week sun exposure but still could influence bees tolerance. Conducted observations confirmed, that bees fully accepted new frames even with metal beam with PCB inside. Honeycombs were fully rebuilt and measurement stripes were fully covered with wax.

It has been shown, that with current state of the measurement art, mass-produced elements can be used for development of hive measurement system, which is invisible to bees. Research concerning expert system development has barely begun, and without the necessary financing will be stopped at the end of the year.

References

1. Oduwole, A.A., Sowole, A.O.: Utilization and impact of The Essential Electronic Agricultural Database (TEEAL) on library services in a Nigerian university of agriculture. Program **40**(2), 157–167 (2006)
2. Polikarpus, A., Kaart, T., Kokin, et al.: Automatic monitoring of milking order in an large loose housing cowshed. In: Animal hygiene and sustainable livestock production. Proceedings of the XVth International Congress of the International Society for Animal Hygiene, Vienna, Austria, 3–7 July 2011, vol. 1, pp. 329–332, Tribun EU (2011)
3. Wilde, J., Prabucki, J. (Eds.): Hodowla pszczół. PWRiL (2008)
4. Hartwig, A., Topolska, G.: Honey Bee Diseases. Tutorial materials SGGW Warszawa (1996)
5. Rinderer, T.E. (Ed.): Bee Genetics and Breeding. Academic Press (2013)
6. Ostrowska, W.: Gospodarka pasieczna. PWRiL (2013)
7. Internet Monitoring of Beehives. www.e-ruche.fr
8. Open Energy Monitor-Bee Hive Monitor. www.beemonitor.org/
9. Nova Labs Helps Local Beekeepers Collect Hive Data. www.nova-labs.org/blog/
10. Howis, M., Berezowski, M., Nowakowski, P.: Monitorowanie mikroklimatu rodziny pszczelej system WMA. Wyd. Pasieka 2010/1s.41, 2010/2s.43, 2010/3s.41, 2010/4s.48, 2010/5s.50, 2010/6s.42
11. Warsza, Z.L.: New approach to the accuracy description of unbalanced bridge circuits with the example of Pt sensor resistance bridges. J. Autom. Mobile Robot. Intell. Syst. **4**, 8–15 (2010)

Research of Metal Film Resistor's Temperature Stability According to Their Nominal Wattage

Andrzej Juś, Paweł Nowak, Roman Szewczyk and Weronika Radzikowska-Juś

Abstract Paper presents the results of tests of temperature characteristics of resistors according to their nominal wattage. Object of the study are cheap, popular metal film resistors of the same type and producer in the wattage range 0.4–2 W. There is a possibility of selecting some of those resistors and using them as a current-voltage converters. In such application resistors are supposed to work in a wide range of temperatures, which can significantly change their values. So the influence of nominal wattage on temperature stability of resistors was checked. The study was conducted on high resolution test stand.

Keywords Metal film resistors · Temperature coefficient of resistance · Nominal power rating

1 Introduction

Resistors are the most common passive electrical components [1]. The reason of such situation is a huge variety of that elements applications. One of them is precision current-voltage conversion [2]. Such application have high importance in

A. Juś (✉) · P. Nowak
Industrial Research Institute for Automation and Measurements,
Al. Jerozolimskie 202, 02-486 Warsaw, Poland
e-mail: ajus@piap.pl

P. Nowak
e-mail: pnowak@piap.pl

R. Szewczyk
Faculty of Mechatronics, Warsaw University of Technology, sw. A. Boboli 8,
02-525 Warsaw, Poland

W. Radzikowska-Juś
Faculty of Civil Engineering and Geodesy, Military University of Technology,
gen. Sylwestra Kaliskiego 2, 00-908 Warsaw, Poland

R. Szewczyk et al. (eds.), *Challenges in Automation, Robotics and Measurement Techniques*, Advances in Intelligent Systems and Computing 440, DOI 10.1007/978-3-319-29357-8_71

measurement techniques, due to the highest accuracy of voltage measurement. Often other physical values (like strain or temperature) are measured after transforming them to voltage. However to ensure accurate voltage–current transforming usage of high stability resistors is necessary. Dedicated components are usually expensive, so it is desirable to select stable resistors from typical, cheaper elements.

Selection of the most stable elements is a laborious process, so finding parameters which would allow pre-selection before carrying out the verification tests is important.

Among many parameters characterizing commercially available resistors nominal power rating was selected. It is one of the most important parameters of resistors defined as the maximum permissible power flow through the resistor at a temperature below some specified value (typically 70 °C) [3]. Difference between resistors of various nominal wattages (but of the same type and manufacturer) is represented by its size. Due to the comparable prices of resistors, conducted tests attempted to determine the relationship between the resistors temperature stability, and its nominal power rating.

2 Object of Study

As an object of study typical metal film ROYALOHM's resistors of various values of nominal power dissipation were chosen. They are characterized by temperature coefficient of resistance (*TCR*) up to ±50 ppm/°C (declared by the manufacturer) and low price. Main parameters of tested resistors are presented Table 1 [4]. The appearance and the approximate size of resistors of different nominal power ratings is presented in Fig. 1. The resistors with the following manufacturer symbols were chosen to the testes: (a) MFF04FF2000A5, (b) MF006FF2000A50, (c) MF01SFF2000A10, (d) MF02SFF2000A10.

Due to the desire of using tested resistors as a resistance reference their properties was compared with the properties of the most popular reference metal foil resistors from Vishay. Vishay's resistors are made from one of the three alloys: K, C or Z. Each alloy has 2nd degree polynomial characteristic of changing resistance value in the temperature function with an extreme point at room temperature (Fig. 2) [5].

Table 1 Datasheet parameters of the ROYALOHM's precision metal film resistors [4]

Parameter	Value
Nominal power rating, P_{70}	(0.4, 0.6, 1.0, 2.0) W
Maximum working voltage, U_{max}	(200, 250, 350, 500) V
Tolerance	±1 %
Temperature coefficient of resistance, *TCR*	±50 ppm/°C

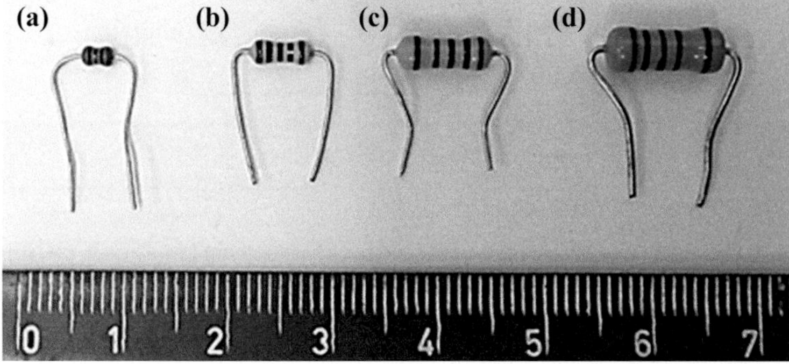

Fig. 1 Tested ROYALOHM's resistors of nominal wattages of: **a** 0.4 W, **b** 0.6 W, **c** 1.0 W, **d** 2.0 W

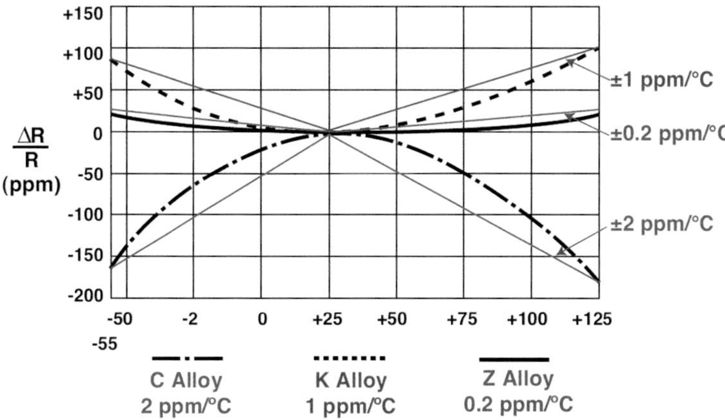

Fig. 2 Relative changes in resistance of the reference resistor made of type K alloy, C alloy and Z alloy [5]

3 Test Stand

Tests were carried out on a developed test stand for temperature characteristics of ultra-precise resistors [6, 7]. Schematic block diagram of utilized test stand is presented on Fig. 3.

Operation of test stand is based on differential resistance measurement. Through the resistors (tested R_1 and reference R_2), connected in series, flows extremely stable DC current [8, 9]. The voltage drop on the tested resistor (U_1) and on the reference resistor (U_2) are subtracted from each other and amplified 100 times by the

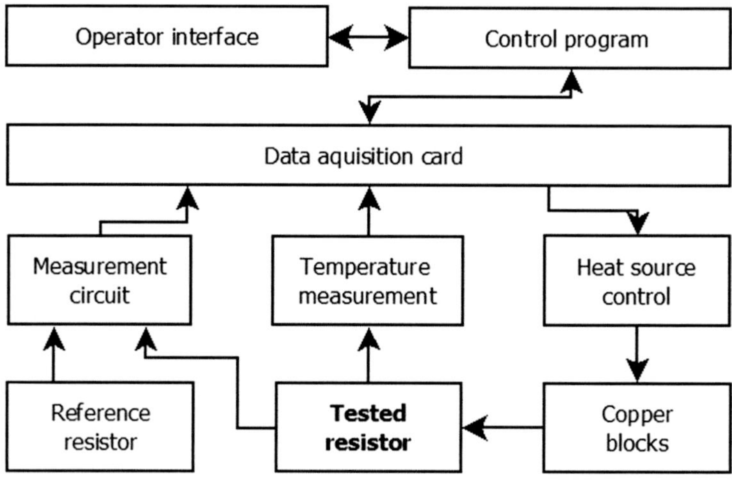

Fig. 3 Schematic block diagram of the utilized test stand [6, 7]

measurement circuit based on differential amplifier [2]. Voltage held on the output of the subsystem is given as (1) presents:

$$V_{out} = (U_1 - U_2) \cdot k = I_1 \cdot (R_1 - R_2) \cdot 100. \quad (1)$$

Output signal is measured by a data acquisition card. Detailed description of system operation was previously reported in [6, 7, 10].

In order to measure the *TCR* value, tested resistor is placed between two copper blocks (Fig. 4) and heated, while the reference resistor remains in constant

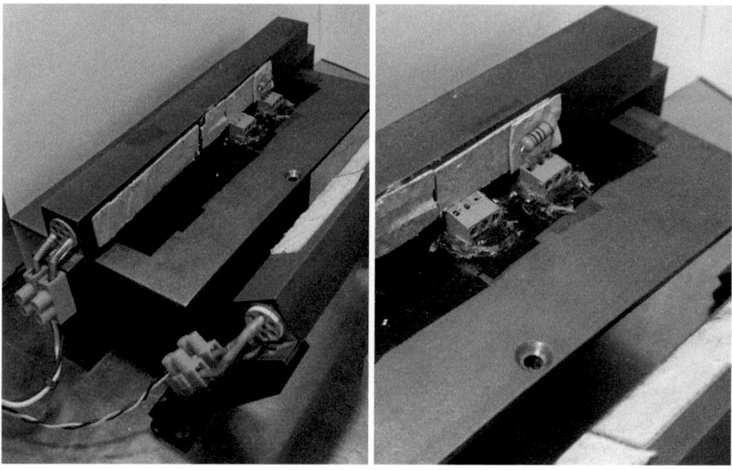

Fig. 4 Heat sources (copper blocks) with tested resistor placed between them

temperature. As a result of that process difference between resistors values is changing as well as the output signal from the measurement circuit. The temperature variation is measured by thermocouples placed between the heaters and software controlled. Test stand integration, user interface and data acquisition was implemented in LabVIEW software [7].

4 Measurements Results

Three specimens of resistors (marked for tests as A, B, C) of each analyzed power (0.4, 0.6, 1.0, 2.0 W) were tested. Measurements was based on controlled, repeatable heating of resistors from room temperature to 35 °C. During each test temperature (example waveform is presented in Fig. 5) and resistance variation (example waveform is presented in Fig. 6) were measured. On their basis characteristic of resistance variation in the function of temperature were obtained (example characteristic is presented in Fig. 7). Values of the *TCR* of tested resistors were obtained, as the value of the slope of $\Delta R(T)$ characteristics. *TCR* values obtained from all tests are summarized in Table 2.

Based on the Vishay's resistors $\Delta R(T)$ characteristics (Fig. 2) the proposition of a 2nd degree polynomial characteristic of the researched resistors was formulated and verified. To verify that theory 2nd degree curves were fitted to the obtained $\Delta R(T)$ characteristics (example results are presented on Fig. 8). To check how well data fits to the established statistical model coefficient of determination R^2 was calculated [11]. Furthermore, the abscissa values of the extremes (T_{extr}) of the fitted functions were calculated. The results of analysis are presented in Table 3.

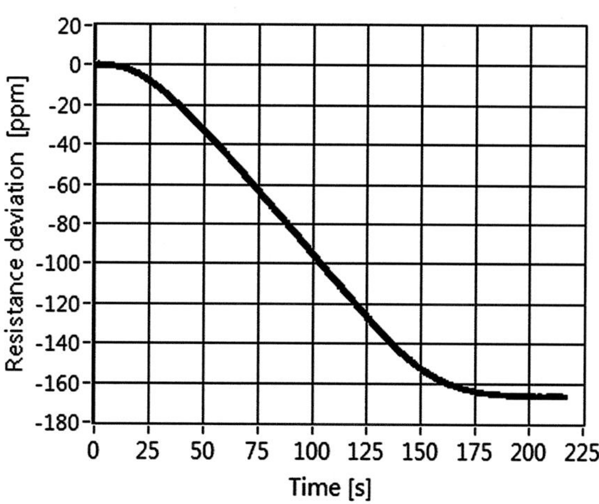

Fig. 5 Change of resistance deviation in time $\Delta R(t)$—resistor 0.4 W, B

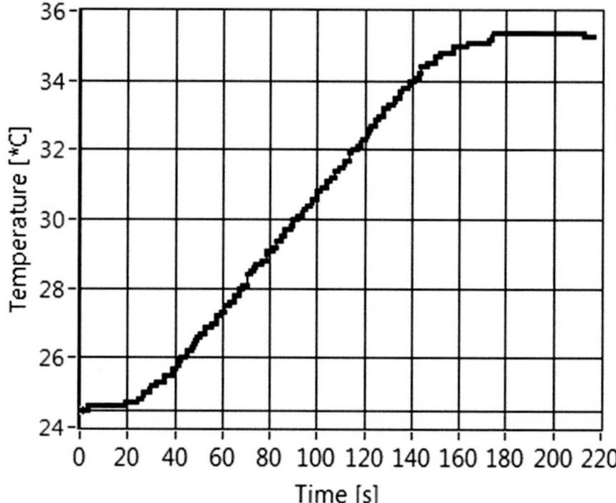

Fig. 6 Change of temperature in time *T*(*t*)—resistor 0.4 W, B

Table 2 Obtained *TCR* values of studied ROYALOHM's resistors

Nominal power rating, P_{70} (W)	Temperature coefficient of resistance, *TCR* (ppm/°C)		
	A	B	C
0.4	−18.6	−15.1	−17.8
0.6	0.8	−5.2	4.7
1.0	−10.2	−16.5	−15.5
2.0	−19.1	−17.5	−18.9

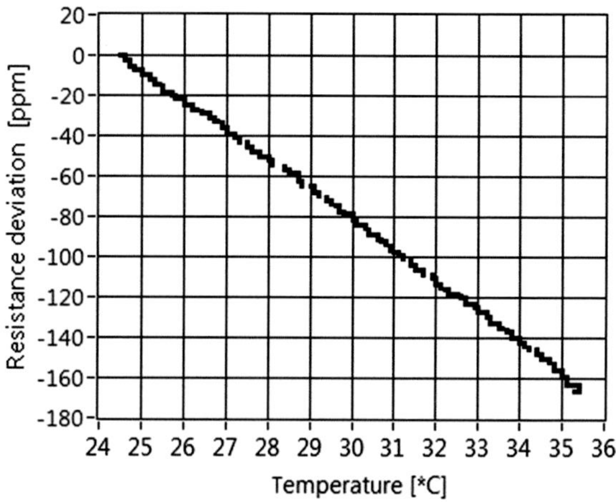

Fig. 7 Change of resistance deviation in the function of temperature $\Delta R(T)$—resistor 0.4 W, B

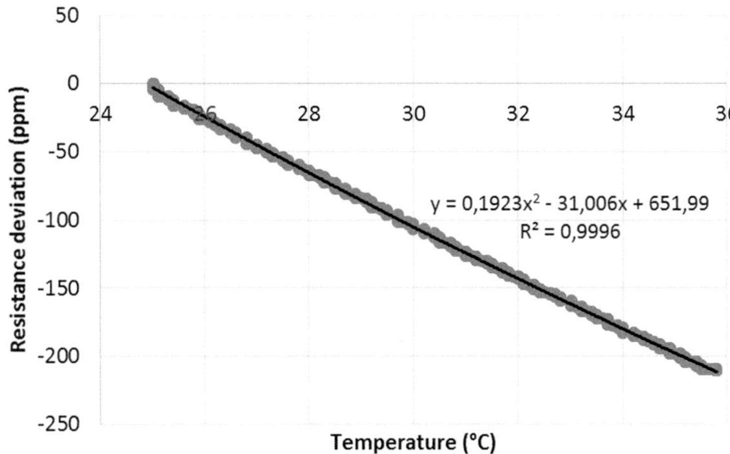

Fig. 8 Example $\Delta R(T)$ characteristic (resistor 2.0 W, A) with equation of matched 2nd degree polynomial and the value of coefficient of determination R^2

Table 3 Summary of the results of polynomial matching to the measured $\Delta R(T)$ characteristics of tested resistors

Resistor		R^2 (1st degree)	R^2 (2nd degree)	T_{extr} (°C)	*TCR* (ppm/°C) (in the range from 20 to 35 °C)
0.4 W	A	0.9988	0.9991	−25.6	−18.6
	B	0.9991	0.9994	−46.2	−15.1
	C	0.9994	0.9995	−84.1	−17.8
0.6 W	A	0.9966	0.9994	5.8	0.8
	B	0.9995	0.9995	247.3	−5.2
	C	0.9993	0.9995	−59.5	4.7
1.0 W	A	0.9995	0.9997	156.0	−10.2
	B	0.9957	0.9992	52.7	−16.5
	C	0.9981	0.9992	73.2	−15.5
2.0 W	A	0.9988	0.9996	80.6	−19.1
	B	0.9988	0.9996	80.3	−17.5
	C	0.9976	0.9994	61.6	−18.9

Analyzing results summary in presented above it can be stated, that:

- In all cases adopted 1st degree model describes the changes of $\Delta R(T)$ characteristics.
- In all cases adopted 2nd degree model describes the changes of $\Delta R(T)$ characteristics.
- Extremes of matching 2nd degree functions occur at temperatures in range from −84.1 to 247.3 °C—the temperature of the extreme of fitted function was highly unrepeatable for the tested resistors.

- Values of *TCR*'s of resistors of nominal wattages 0.4, 1.0, 2.0 W were in the range from −20 to −10 ppm/°C.
- Values of *TCR*'s of resistors of nominal wattage 0.6 W were in the range from −4.7 to 5.2 ppm/°C.

5 Conclusion

In the paper analyses of the dependency between the nominal power dissipation of a resistor and its temperature coefficient of resistance (*TCR*) were conducted. As an object of study popular metal film resistors were selected. Conducted tests had to conclude whether the nominal power dissipation can be a parameter which will facilitate the selection of cheap elements as a precise current-voltage converters.

Based on the $\Delta R(T)$ characteristics of precise reference resistors manufactured by Vishay, 2nd order model of the $\Delta R(T)$ characteristic of the tested resistors was assumed. This model was properly verified and the temperature values of extremes of fitted functions were calculated.

After analyzing the results, no significant correlation between the resistors nominal power dissipation and its thermal coefficient was observed. The lowest absolute *TCR* was measured for resistors with 0.6 W nominal power dissipation. On the other hand the test group consisted of only three resistors from the same production batch.

The 2nd order curve fitting resulted with very good fitting coefficients ($R^2 > 0.99$). On the other hand temperature the temperature of the extreme of fitted function was highly unrepeatable. Thus local value of resistors *TCR* cannot be minimized by setting their operating temperature near the extreme point.

The research confirmed, that from the tested cheap resistors some specimen of very low *TCR* can be selected. Also even after measuring only three resistors from each production batch range of *TCR* of all resistors can be partially determined. Thus the initial selection of proper production batch should be the basis for selection of the reference resistors among them.

References

1. Stepowicz, W.J., Górecki, K.: Materiały i elementy elektroniczne. Wyd.2. Wydawnictwo Akademii Morskiej w Gdyni, Gdynia (2008)
2. Horowitz, P., Hill, P.: The Art of Electronics, 2nd edn. Cambridge University Press, Cambridge (1989)
3. Górecki, P.: Listy od Piotra – Rezystory część 1. "Elektronika dla wszystkich" 1996, nr 1, 57–58
4. Precision Metal Film Fixed Resistors. ROYALOHM (datasheet)
5. Design and selector guide for high precision resistors, Vishay (2011)

6. Juś, A., Nowak, P., Szewczyk, R. et al.: Assessment of temperature coefficient of extremely stable resistors for industrial applications. In: Awrejcewicz, J., Szewczyk, R., Trojnacki, M., Kaliczyńska, M. (eds.) Mechatronics: Ideas for Industrial Applications, pp. 297–306. Springer International Publishing (2015)
7. Nowak, P., Juś, A., Szewczyk, R., et al.: Test stand for temperature characteristics of ultraprecise resistors. In: Awrejcewicz, J., Szewczyk, R., Trojnacki, M., Kaliczyńska, M. (eds.) Mechatronics: Ideas for Industrial Applications, pp. 345–352. Springer International Publishing (2015)
8. Korytkowski, J.: Układ elektroniczny cyfrowej syntezy rezystancji do dokładnej symulacji rezystancyjnych czujników temperatury. Pomiary Automatyka Robotyka **5**, 86–92 (2013)
9. Idzkowski, A., Makal, J., Warsza, Z.L.: Simultaneous measurement of two parameters by double current supplied bridge. J. Autom. Mobile Robot. Intell. Syst. **6**, 26–31 (2012)
10. Juś, A., Nowak, P., Szewczyk, R.: Automatic system for identification of temperature parameters of resistors based on self-heating phenomena. In: Szewczyk, R., Zieliński, C., Kaliczyńska, M. (eds.) Progress in Automation, Robotics and Measuring Techniques, vol. 3, pp. 91–100. Springer International Publishing (2015)
11. Cowan, G.: Statistical Data Analysis. Oxford University Press (1988)

Automated System for Testing Ferromagnetic Materials

Michał Urbański, Tomasz Charubin, Paweł Rozum, Michał Nowicki and Roman Szewczyk

Abstract The aim of this paper is to present an automated measuring station for ferromagnetic materials testing. The system is able to obtain full characteristics for magnetoelastic crystal magnetics, ceramic magnetics and amorphous magnetics. The paper presents the setup for measuring station and the process of obtaining measurement data for sample material. The results were analyzed and compared with material manufacturer's datasheet.

Keywords Ferromagnetic measurement · Non-destructive testing · Hysteresis graph

1 Introduction

The growing needs for non-destructive testing leads to development of various measurement methods [1, 2]. In case of ferromagnetic materials, which are present in many electronical devices as instances of magnetic cores in filters, transformers, chokes and other inductive elements [3], the investigation of magnetic parameters is supported by hysteresis graph. The hysteresis loops acquired during measurements provides almost complete characteristic of tested material. It also allows further investigation like testing material under stress [4]. For proper research the material

M. Urbański (✉) · M. Nowicki
Industrial Research Institute for Automation and Measurement, Warsaw, Poland
e-mail: murbanski@piap.pl

M. Nowicki
e-mail: mnowicki@piap.pl

T. Charubin · P. Rozum · R. Szewczyk
Institute of Metrology and Biomedical Engineering, Warsaw University of Technology, Warsaw, Poland
e-mail: r.szewczyk@mchtr.pw.edu.pl

R. Szewczyk et al. (eds.), *Challenges in Automation, Robotics and Measurement Techniques*, Advances in Intelligent Systems and Computing 440, DOI 10.1007/978-3-319-29357-8_72

needs to be tested under various input parameters such as frequency and amplitude of magnetization field.

Thorough analysis of given sample required building an automated testing stand for magnetic materials that would allow setting up a series of autonomous measurements to be conducted overtime in various configurations. The measurement station was designed in order to allow easy parameters input, as well as human friendly data presentation, analysis, and recording.

The paper presents configuration of designed testing station, the methodology of measurement and sample data of conducted research.

2 System Setup

The system main component is a PC class computer with LabVIEW based software. LabVIEW (Laboratory Virtual Instrument Engineering Workbench) is a system-design platform and development environment for a visual programming language from National Instruments. The designed software controls National Instruments Data Acquisition Device, which generates magnetization signal, that is processed by high resolution U/I converter. The magnetization current is driven through the magnetization winding, causing changes of voltage induced in the sensing winding of the researched sample (Fig. 1).

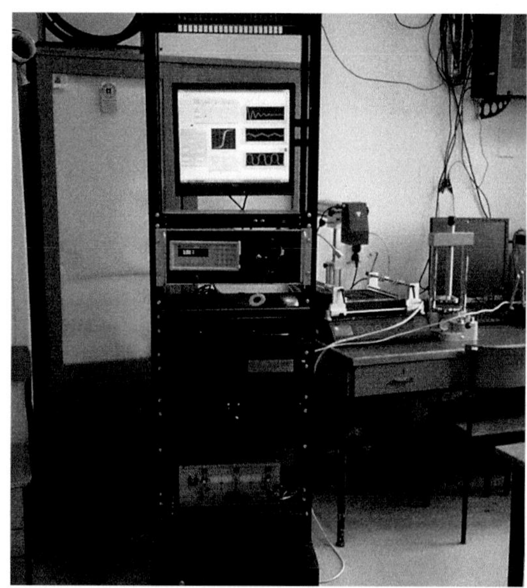

Fig. 1 Designed testing stand

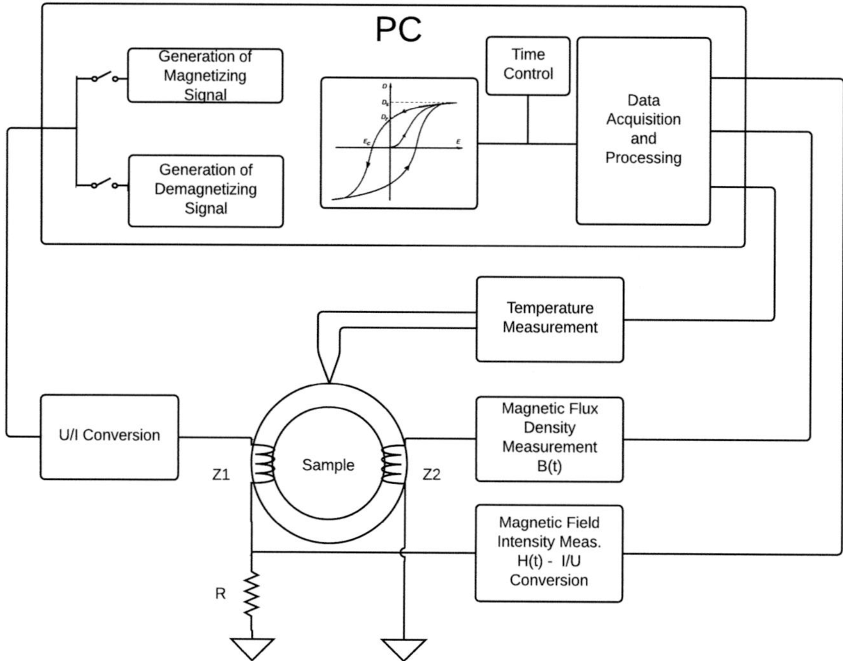

Fig. 2 Block diagram for designed system

The voltage signal from sensing winding is measured with Lakeshore 480 fluxmeter. The signal is integrated to a flux density value and back to NI Data Acquisition Device.

The LabVIEW based software is responsible for all the required calculations and control over the generating and sensing modules of the testing stand.

The temperature is monitored with a 'K' type thermocouple, which is natively supported by a National Instruments Data Acquisition Device (Fig. 2).

Based on these inputs, along with the setting of U/I conversion ratio, the maximal value of magnetic field intensity is being calculated. In this step the communication over RS-232 interface with Lakshore 480 fluxmeter is being initialized, and the necessary input values are fed to the device, to ensure proper voltage signal output for magnetic flux density values.

The modules of the testing stand are summarized in Table 1.

The interface for the designed software is presented in Fig. 3.

Table 1 List of devices used in the testing stand

Module	Device Name	Additional Info
U/I converter	Kepco BOP 36-6 M	Max. Current: ±6 A
Data acquisition device	NI PCI 6221	Max. Voltage: ±10 V
Fluxmeter	Lakeshore Model 480	N/A

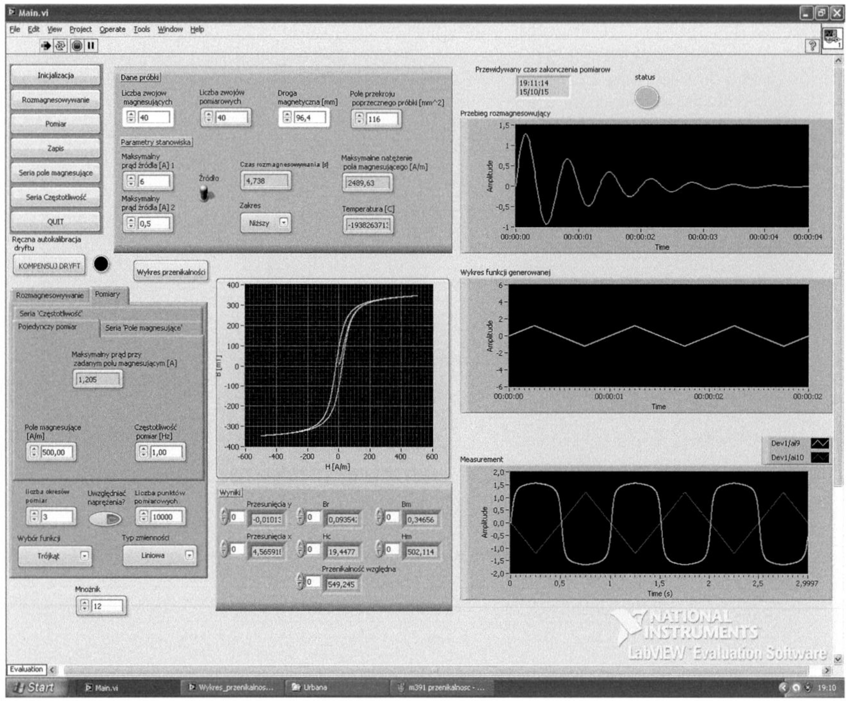

Fig. 3 Interface for the designed testing stand

3 Measurement Process

The complete measurement process flowchart is shown in Fig. 4. It represents the procedure of acquiring measurements step-by-step, along with additional operations required for the proper representation of testing results.

In the Initialization step of the software is conducting calculations based on several input parameters, such as:

- No. of magnetizing windings
- No. of sensing windings
- Sample's magnetic length
- Sample's cross-sectional area

Based on these inputs, along with the setting of U/I conversion ratio, the maximal value of magnetic field intensity is being calculated. In this step the communication over RS-232 interface with Lakeshore 480 fluxmeter is being initialized, and the necessary input values are fed to the device, to ensure proper voltage signal output for magnetic flux density values.

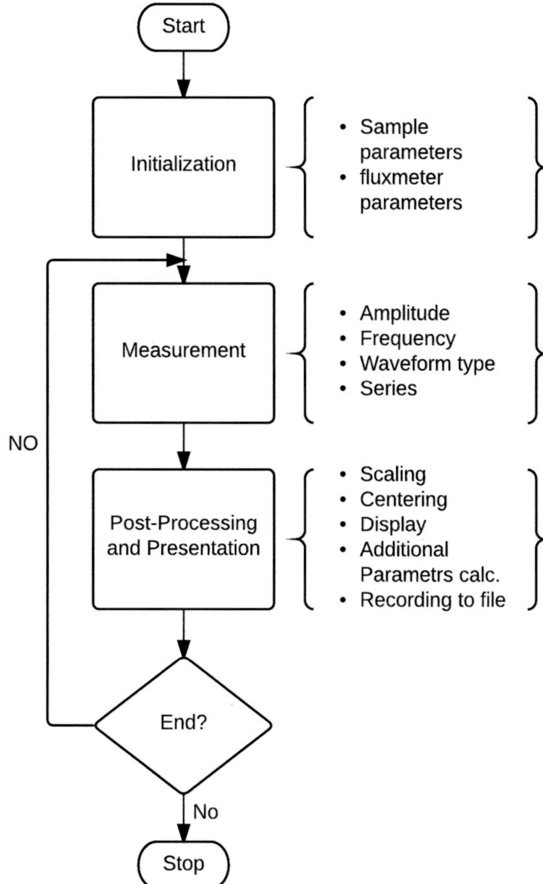

Fig. 4 Flowchart for measurement process

During the measurement there is an optional step of demagnetization of sample, in order to remove residual magnetism, which is the magnetization left behind in a ferromagnetic material after an external magnetic field is removed. This is provided by generating proper current signal that follows the formula 1.

$$I = I_0 \cdot \sin(2\pi ft) \cdot e^{\left(-\frac{t}{\tau}\right)} \quad (1)$$

where I—current value, I_0—max. current value, f—frequency, t—time, τ—time constant.

The values presented in formula above can be changed between measurements.

The measurement starts with computation of magnetizing waveform, with accordance to the desired magnetic field intensity, and frequency. There are three available options for magnetizing waveform:

Table 2 Parameters for measured samples

Sample	Magnetic length (cm)	Cross-sectional Area (cm^2)	No. of magnetizing windings	No. of sensing windings
F-3001	9.64	1.16	40	40
M-391	8.6	0.12	60	200

Table 3 Magnetic parameters of materials

Sample	Saturation flux density (mT)	Remnant flux density (mT)	Coercivity Hc (A/m)	Relative permeability
F-3001	~350	87	14	3000
M-391	~1200	–	<3	1000–100000

- Sinusoidal waveform,
- Sawtooth waveform
- Custom waveform (generated by user with LabVIEW's waveform tools)

In addition, the designed software allows to program a series of measurements with variable frequency or amplitude of magnetizing field intensity. The driving

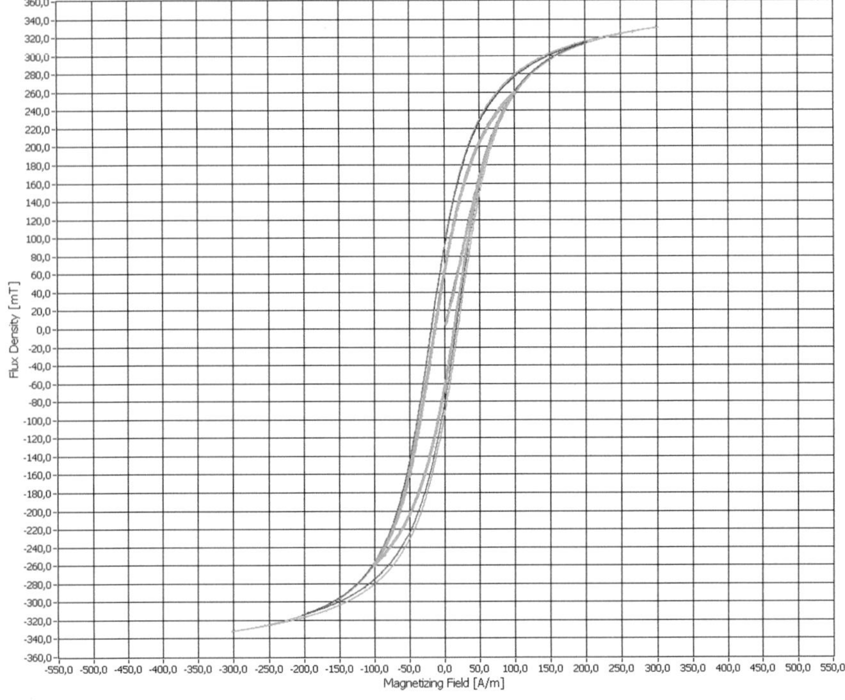

Fig. 5 F-3001—magnetizing field(s) H: (100, 200, 300) A/m

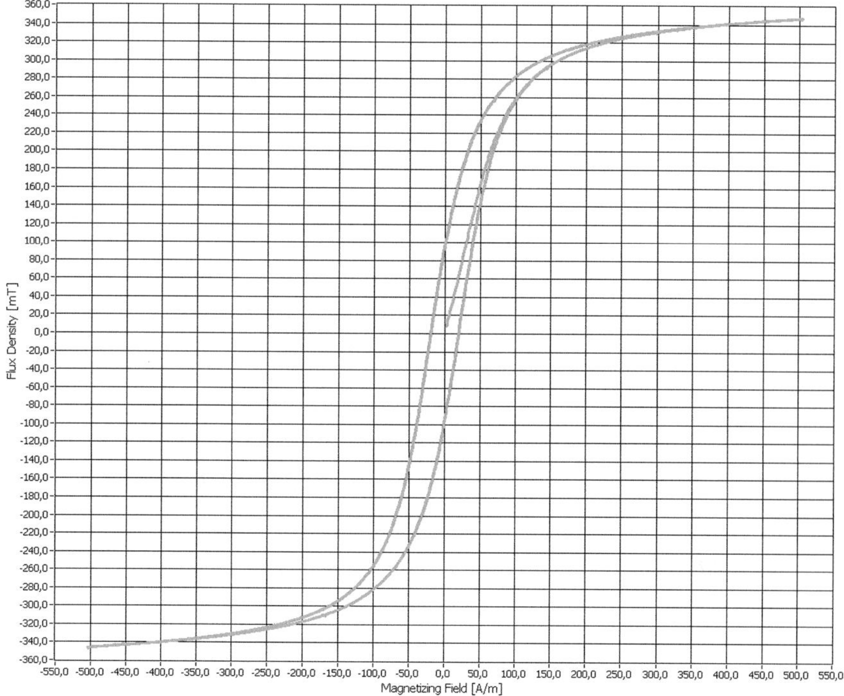

Fig. 6 F-3001—magnetizing field H 500 A/m

signal is generated as a voltage signal by National Instruments Data Acquisition Device, and then converted to current signal by U/I converter.

After proper initialization of the Lakeshore 480 fluxmeter, a voltage signal corresponding to magnetic flux density value is measured by NI DAQ Device. In the next step, the software communicates over RS-232 interface with fluxmeter, in order to capture peak values of the magnetic flux density, to provide parameters for post-processing module of the software. In the post-processing step, the hysteresis loop is being centered and scaled according to peak values. This step is important due to measured magnetic flux density value, as its indication on fluxmeter could change over time.

The last step of measurement process is displaying the hysteresis loop, calculating basic parameters of single hysteresis loop, and saving the results to a formatted text file. In case of series of measurements additional parameters are displayed, e.g. relative permeability or coercivity and saturation remanence.

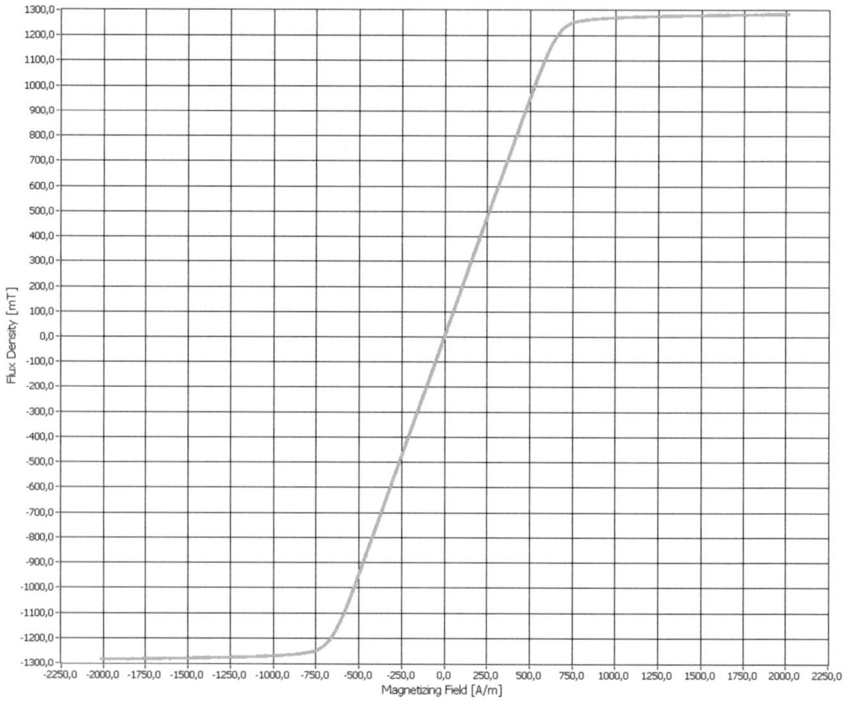

Fig. 7 M-391—magnetizing field H 2000 A/m

4 Results

To demonstrate the capabilities of the system a selection of ferromagnetic material samples was measured. The parameters of measured samples—material, cross-sectional area, magnetic length and number of windings—are presented in Table 2.

F-3001 is a soft ferrite produced by Zmm Polfer under trade-mark ferroxyd [5, 6]. The M-391 is a transformer core made from strongly anisotropic amorphous alloy NANOPERM LM, produced by Magnatec Company [7]. The magnetic parameters of these materials are presented in Table 3.

The results of measuring of given samples are presented in Figs. 5, 6 and 7. The x axis presents the value of magnetizing field (Amperes per Meter), while the y axis presents the magnetic flux density value (10^{-3} Teslas). The charts shown are exported directly from LabVIEW application.

The results of the measurements find confirmation in manufacturer datasheets for given materials, and previous measurements of researched materials [5, 6].

5 Conclusion

The results of measurement for given samples proves the possibility of utilizing the designed testing station for magnetic measurement of ferromagnetic materials. The data acquired during measurement for hysteresis loop correspond to official values provided by manufacturers or measured with professional devices for magnetic measurement.

The functionality of acquiring series of preprogrammed measurements in automated process and the human friendly graphical user interface are main advantages of this system. The modular design allows for quick modification of utilized devices, for example to ensure higher maximum value for generated current with interchangeable U/I converters.

Testing stand presented in this paper allowed for Validation of the Anhysteretic Magnetization Model for Soft Magnetic Materials with Perpendicular Anisotropy [7].

References

1. Blitz, J.: Electrical and Magnetic Methods of Non-Destructive Testing, vol. 3. Springer Science & Business Media (1997)
2. Runkiewicz, L.: Application of non-destructive testing methods to assess properties of construction materials in building diagnostics. Arch. Civil Eng. Environ. **2**, 79–86 (2009)
3. O'handley, R.C.: Modern Magnetic Materials: Principles and Applications, vol. 830622677. Wiley, New York (2000)
4. Jackiewicz, D., Szewczyk, R., Bieńkowski, A., Kachniarz, M.: New methodology of testing the stress dependence of magnetic hysteresis loop of the L17HMF heat resistant steel casting. J. Autom. Mobile Robot. Intell. Syst. **9**(2) (2015)
5. Catalogue R-9, Ferrite Materials and Cores, Soft-ferrites FERROXYD, WEMA (1971) (in Polish)
6. Izydorczyk, J.: A convenient method to compute parameters of Jiles and Atherton model for ferrite materials. J. Magnet. Soc. Jpn. **30**(5), 481–487 (2006)
7. Szewczyk, R.: Validation of the anhysteretic magnetization model for soft magnetic materials with perpendicular anisotropy. Materials **7**(7), 5109–5116 (2014)

Anisotropic Vector Preisach Model of Domain Rotation and Domain Wall Movement

Piotr Frydrych, Roman Szewczyk and Michał Nowicki

Abstract Developed two dimensional Preisach model enables simulation of material magnetization vector for different angles. Model includes anisotropy and simulates magnetization value in every direction, not only the mean magnetization vector. Model is based on collection of Preisach planes which describe material state in different angles. Implementation of two magnetization mechanisms: domain wall movement and domain rotation opens possibility to complete description of distribution of magnetization values and angles. Model shows changes in direction of material magnetization for different magnetic field, material magnetization and angle between magnetic field and easy axis.

Keywords Magnetic anisotropy · Vector preisach model · Amorphous alloys

1 Introduction

New high permeability amorphous and nanocrystalline materials exhibit high anisotropy. Core shapes used in fluxgate sensors and power conversion devices cannot ensure field homogenous direction. That causes inhomogeneous angles between material easy axis and its magnetization. For that reason magnetization characteristic of material, measured by producer can differ from characteristic of particular core. That can strongly change energetic parameters of the core [1]. Recently developed models [2, 3] enable efficient mean vector magnetic characteristics simulation for anisotropic materials. These models do not take into account domain rotation mechanism which seems to be crucial for magnetization vector modeling in

---

P. Frydrych (✉) · R. Szewczyk
Institute of Metrology and Biomedical Engineering,
Warsaw University of Technology, Warsaw, Poland
e-mail: p.frydrych@mchtr.pw.edu.pl

M. Nowicki
Industrial Research Institute for Automation and Measurements, Warsaw, Poland

R. Szewczyk et al. (eds.), *Challenges in Automation, Robotics and Measurement Techniques*, Advances in Intelligent Systems and Computing 440, DOI 10.1007/978-3-319-29357-8_73

different directions. On the other hand, rotational plane concept [4] is not appropriate for anisotropic materials. Therefore there is a need for efficient anisotropic vector model, which would be able to simulate not only mean vector value. Presented model is able to describe two dimensional angular magnetization distribution. Amorphous alloys are mainly produced in ribbon shape. They are 25 μm thin, therefore they can be treated as two dimensional objects. Thus modeling in three directions is not necessary.

2 Superposition of Preisach Planes

Preisach plane describes distribution of switching operators for increasing β and decreasing α magnetic field (Fig. 1). Distribution can be calculated from set of hysteresis loops for different field amplitudes. Distribution can be also fitted using two dimensional Gaussian or other probability density function. Best results were achieved for Lorenzian curve [5]. In that research Gaussian Probability Density Function was used. The reason to do this was its simplicity and direct correlation between distribution parameters and hysteresis curve shape. Gaussian distribution is sufficient for amorphous alloys [6]. It was assumed that hysteresis curve has to be symmetrical which means that center of distribution has to be on axis $\alpha = -\beta$. Distribution can be described by four parameters: amplitude A, mean value m, distribution width σ_c and σ_i.

One Preisach plane is sufficient for simulation of scalar hysteresis. To model distribution of magnetization vectors for different angles more Preisach planes are needed, depending on expected angular resolution. Mean magnetization is a sum of magnetization vectors for each angle (1). Integral of Preisach plane can have positive or negative values. Thus vector –m for angle θ is the same as vector m for angle $\theta + \pi$ (2).

$$\vec{M} = \int_{0}^{2\pi} \vec{m}(\theta) cos\theta d\theta \tag{1}$$

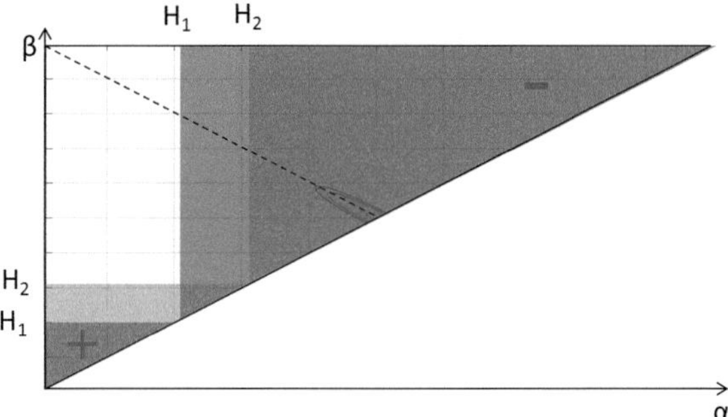

Fig. 1 Switching operator for different field values

$$\vec{m}(\theta) = -\vec{m}(\theta + \pi) \tag{2}$$

Nevertheless, reduction of angular range to $\langle 0, \pi)$ is not justified. The reason is that every mean magnetization vector can be the sum of infinite combination of vectors for each angle.

For every angular homogeneous plane two opposite Preisach planes exist. They represent domains oriented in the same angle but with opposite values.

3 Magnetization Mechanisms

For magnetic materials three kind of magnetization mechanisms can be observed: domain wall flexing, domain wall motion and domain rotation. Domain wall flexing is treated in this model as reversible kind of domain wall motion. Magnetization for each magnetic field value and angle can be calculated as integral (3):

$$m(H, \theta) = \int \int_{\beta \geq \alpha} p(\alpha, \beta, \theta) \cdot \gamma(\alpha, \beta, \theta, H) \cdot \chi(\alpha, \beta, \theta, H) d\alpha d\beta \tag{3}$$

where $\gamma(\alpha, \beta, \theta, H)$ is switching operator and $\chi(\alpha, \beta, \theta, H)$ is rotational operator.

3.1 Domain Wall Movement

When magnetic field increases domains oriented in magnetic field direction grow and domain in opposite direction shrink proportionally. Spins are switching to opposite directions. Domain movement occurs when field is parallel to domain direction. Thus projection of magnetic field on Preisach plane is responsible for switching process. Switching magnetic field can be calculated as (4):

$$H_{\theta\gamma} = Hcos\theta \tag{4}$$

Only spins which have lower β value or α value greater than projection of magnetic field switch to opposite value. Other spins remain in their previous state γ_{prev} (5) (Fig. 1).

$$\gamma = \begin{cases} +1 & \beta < Hcos\theta \cap \beta > \alpha \\ \gamma_{prev} & \alpha < Hcos\theta > \beta \\ -1 & \alpha > Hcos\theta \cap \beta > \alpha \end{cases} \tag{5}$$

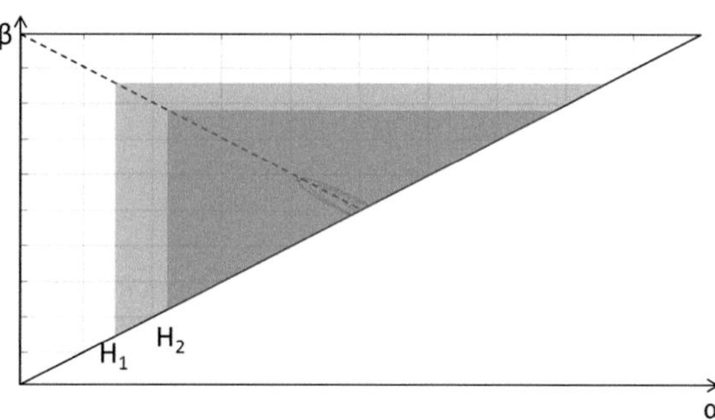

Fig. 2 Rotational operator for different field values

3.2 Domain Rotation

Domain rotation is a symmetrical process. It means that the same amount of positive and negative magnetic spins change their direction. Thus they rotate to another cross-section plane, so rotational operator can have 1 or 0 value. Rotation occurs when magnetic field is orthogonal to spin direction and higher than its switching vales. Therefore rotational magnetic field for each Preisach plane can be calculated as (6) (Fig. 2):

$$H_{\theta\chi} = H_{sat} - |Hsin\theta| \tag{6}$$

Domain rotational operator can defined be the formula (7):

$$\chi = \begin{cases} 1\alpha \le H_{sat} - |Hsin\theta| \ \cap \beta \le H_{sat} - |Hsin\theta| \\ 0\alpha > H_{sat} - |Hsin\theta| \ \cup \beta > H_{sat} - |Hsin\theta| \end{cases} \tag{7}$$

Consequence of that rotational operator description is that for $sin\theta = 0$ rotation mechanism has no influence on Preisach plane. For $H = 0$ every domain direction has the same probability.

Both motion and rotation mechanisms operates on the same Preisach planes. They have no influence on distribution shape.

4 Anisotropy

Magnetic properties anisotropy has influence on hysteresis shape for different directions of magnetization. Differences can be observed in different slope of hysteresis but the same magnetization saturation and coercivity. Thus in presented model only distribution width σ_c and σ_i changes for different angles. Elliptical anisotropy of distribution parameters was assumed. Therefore anisotropy coefficient can be expressed as (8):

$$a_{\varphi}^{2} = \frac{a^{2}}{a^{2} - (a^{2} - 1)cos^{2}\varphi} \tag{8}$$

where a is anisotropy strength, φ is distribution angle. a_{φ} reaches values between 0 and 1. Easy axis is for angle $\varphi = 0$, at which a_{φ} has the lowest value. For a = 0 there is only one magnetization direction possible, for a = 1 material is isotropic. Actual distribution width for given angle is therefore (9) and (10):

$$\sigma_{c}(\varphi) = a_{\varphi}\sigma_{c} \tag{9}$$

$$\sigma_{i}(\varphi) = a_{\varphi}\sigma_{i} \tag{10}$$

This approach enables modelling weak and strong anisotropy, which cannot be achieved by *w* parameter described in [7].

5 Magnetization Characteristics

Presented model has five parameters. For isotropic material *a* = *1*, thus only four parameters are needed. To verify the model, its ability to model isotropic material was investigated. Mn-Zn power ferrite was sufficiently simulated. Pearson's coefficient r^2 = 0.998 was achieved (Fig. 3).

For major hysteresis loop and three field values angular magnetization distribution was simulated. Two kinds of material were analyzed: isotropic Mn-Zn power ferrite and highly anisotropic material for easy axis, hard axis and for 45° to easy axis (Fig. 4).

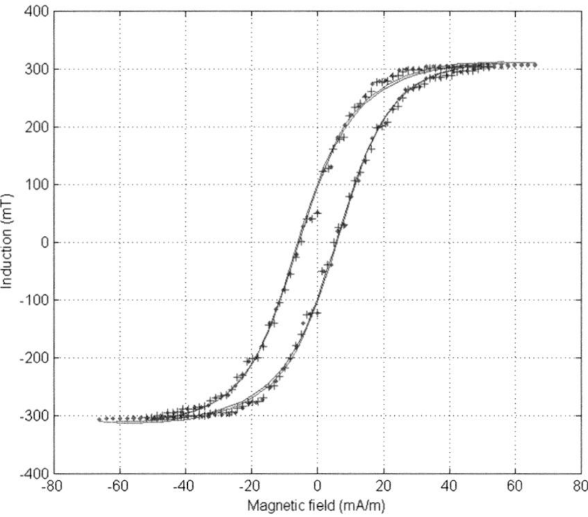

Fig. 3 Modeling results for Mn-Zn power ferrite. *Dots*—model, *line*—measured data

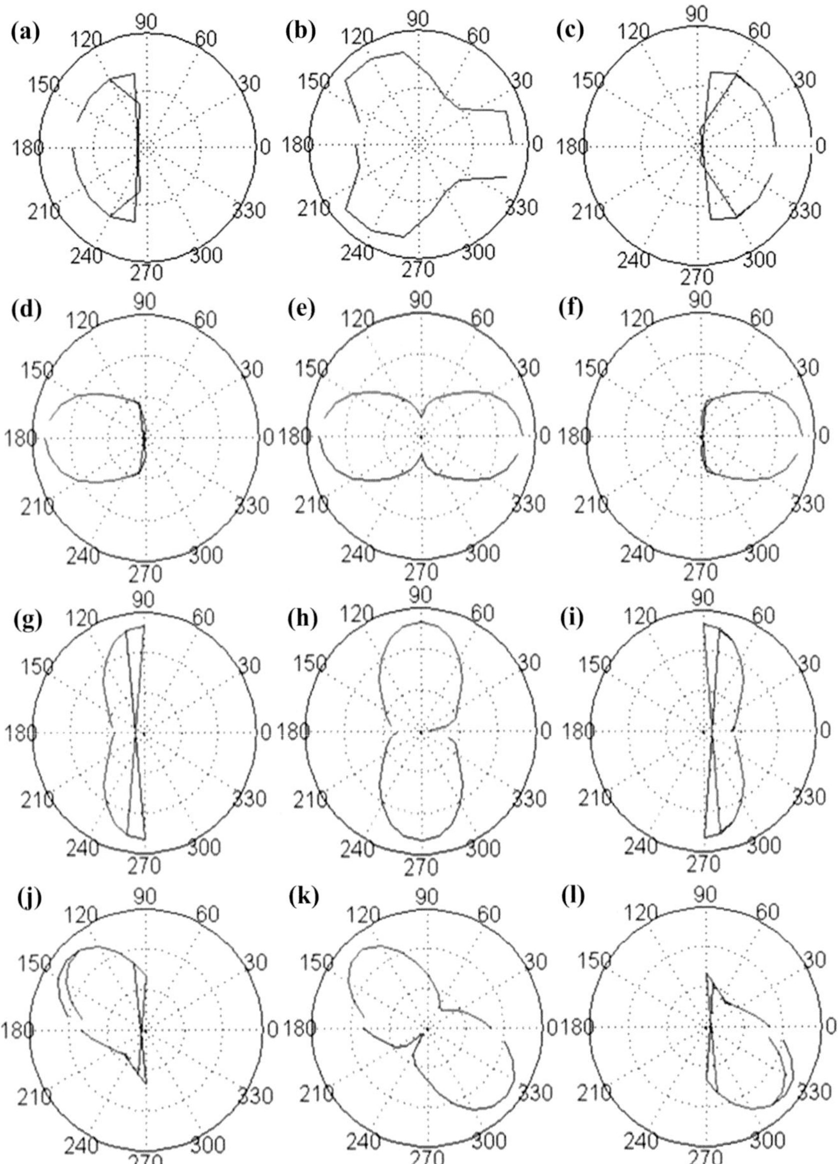

Fig. 4 Angular magnetization distribution for major hysteresis loop: Mn-Zn Power Ferrite: **a** $H = -H_{sat}$, **b** $H = 0$, **c** $H = H_{sat}$; highly anisotropic material for easy axis: **d** $H = -H_{sat}$, **e** $H = 0$, **f** $H = H_{sat}$, for hard axis: **g** $H = -H_{sat}$, **h** $H = 0$, **i** $H = H_{sat}$, for 45° to easy axis: **j** $H = -H_{sat}$, **k** $H = 0$, **l** $H = H_{sat}$

6 Results

Model shows good conformity with magnetic characteristics measured for Mn-Zn power ferrite. It should be emphasized that for isotropic material only four parameters are needed do describe material magnetic characteristics. Simulation shows, that zero magnetization can be achieved by different angular distribution of magnetic moments. For anisotropic materials mean magnetization is not parallel do field, when it is 45° to easy axis. This observation can be important when parallel magnetization of given object cannot be guaranteed. Thus presented model enables modeling of demagnetization process of objects with complicated shape made of anisotropic materials.

In further research model should be verified with vector hysteresis measurements. Unfortunately complete verification of angular magnetization distribution seems to be impossible. The reason is that for given direction only mean magnetization can be measured.

References

1. Stensland, T., Fuchs, E.F.: IEEE Trans. Power Delivery **12**(2) (1997). doi:10.1109/61.584366
2. Handgruber, P., Stermecki, A.: IEEE Trans. Magn. **51**(3) (2015). doi:10.1109/TMAG.2014.2353691
3. Kuczmann, M., et al.: J. Adv. Res. Phys. **1**(1) (2010)
4. Sutor, A., Kallwies, J., Lerch, R.: **111** (2012). doi:10.1063/1.3672069
5. Basso, V., Bertotti, G., Lnfortuna, A., Pasquale, M.: IEEE Trans. Magn. **31**(6) (1995)
6. Frydrych, P., Szewczyk, R.: Recent Adv. Autom. Robot. Measuring Tech. **267**, 591-596. doi:10.1007/978-3-319-05353-0_55
7. Enokizono, M.: JSAEM Stud. Appl. Electromagn. **1** (1992)

Validation of the Model of Anhysteretic Magnetisation Curve of Isotropic Soft Magnetic Materials

Roman Szewczyk and Michał Nowicki

Abstract Anhysteretic magnetization curve is the fundamental concept for Jiles-Atheton model of magnetic hysteresis loop. However, this connection between anhysteretic magnetization curve parameters and physical parameters of soft magnetic materials is not obvious. Paper presents the attempt to validate the parameters of anhysteretic magnetization curve for soft magnetic materials—soft Mn-Zn ferrites. On the base of experimental measurements of saturated magnetic hysteresis loops, the parameters of Jiles-Atherton model of anhysteretic magnetization curve were calculated. Parameters were also analysed from the point of view of connection with microscopic parameters of Mn-Zn ferrite structure.

Keywords Jiles-Atherton model · Magnetic hysteresis · Optimization

1 Introduction

Jiles-Atherton model is one of the most popular models of magnetic hysteresis loop of soft and semi-hard magnetic materials [1–3]. This model is especially useful due to the fact of possibility of calculation costs effective implementation in SPICE software [4] for modelling the magnetic circuits in electronics components.

One of the most important problems connected with Jiles-Atherton model was the method of determination of model's parameters [5, 6]. This problem was recently successfully overcome by application of cognitive optimisation methods not sensitive to local minima [7]. Especially differential evolutionary strategies were found to be effective in determination of model's parameters for steels [8], which

R. Szewczyk (✉) · M. Nowicki
Industrial Research Institute for Automation and Measurements,
Al. Jerozolimskie 202, PL-02-486 Warsaw, Poland
e-mail: szewczyk@mchtr.pw.edu.pl

R. Szewczyk et al. (eds.), *Challenges in Automation, Robotics and Measurement Techniques*, Advances in Intelligent Systems and Computing 440, DOI 10.1007/978-3-319-29357-8_74

created possibility of application o Jiles-Atherton model in non-destructive testing of ferromagnetic elements of civil engineering constructions.

On the other hand, one of the most controversial problems connected with Jiles-Atherton model is the fact, that it is based on anhysteretic magnetization curve of magnetic materials. It is very difficult to measure [9–11] such curve due to the fact, that such measurements requires continuous measurements of magnetic flux during the demagnetization process under the constant magnetic fields. However, anhysteretic magnetization curve can be easily observed in soft magnetic materials with negligible hysteresis loop. As a result, the model of anhysteretic magnetization curve of strongly anisotropic amorphous alloys was successfully validated previously [12]. However, there is still very limited information available about the connection between parameters of isotropic magnetization curve and microscopic parameters of soft magnetic material.

Presented paper is going to fill this gap on the base of experimental results. High permeability Mn-Zn ferrites seem to be very interesting materials for this purpose, due to stable production parameters, known microstructure and magnetic hysteresis loop suitable for accurate measurements using hysteresisgraphs.

2 Anhysteretic Magnetization Curve in Jiles-Atherton Model

In the Jiles-Atherton model, the anhysteretic magnetization curve is based on paramagnetic magnetization curve. In the case of paramagnetic materials, magnetization process is anhysteretic. Magnetization of paramagnetic material can be determined on the base of Boltzmann distribution by the following equation [2]:

$$M_{para} = M_s \frac{\int_0^{\pi} e^{\frac{-E_m(\theta)}{k_B \cdot T}} \sin\theta \cdot \cos\theta \cdot d\theta}{\int_0^{\pi} e^{\frac{-E_m(\theta)}{k_B \cdot T}} \sin\theta \cdot d\theta} \tag{1}$$

where M_s is saturation magnetization of material, k_B is Boltzmann constant, T is the temperature of magnetic materials and θ is the angle between the direction of atomic magnetic moment and direction of magnetizing field H. Energy $E_m(\theta)$ of atomic moment is given as [2]:

$$E_m(\theta) = -\mu_0 \cdot m_{at} \cdot H \cdot \cos\theta \tag{2}$$

where μ_0 is the magnetic constant. Considering the Eqs. (1) and (2), anhysteretic magnetization curve of isotropic (and only isotropic) ferromagnetic material may be given as [2]:

$$M_{anhysteretic} = M_s \frac{\int_0^{\pi} e^{\frac{-E_d(\theta)}{k_B \cdot T}} \sin\theta \cdot \cos\theta \cdot d\theta}{\int_0^{\pi} e^{\frac{-E_d(\theta)}{k_B \cdot T}} \sin\theta \cdot d\theta} \tag{3}$$

where θ is an angle between the direction of magnetization m_d of single magnetic domain and the direction of effective magnetizing field H_e. In such a case the energy of the single domain $E_d(t)$ is given as [2]:

$$E_d(\theta) = -\mu_0 m_d \cdot H_e \cos\theta \tag{4}$$

Effective magnetizing field H_e is given by the following equation [1]:

$$H_e = H + \alpha \cdot M \tag{5}$$

where α is interdomain coupling. Average magnetization of single domain m_d may be determined from the following equation [2]:

$$m_d = \frac{M_s}{N} \tag{6}$$

where N is the number of domains per cubic meter. Due to the fact, that integrals in Eq. (3) have antiderivatives, the set of Eqs. (3), (4) and (6) can be reduced to the following form determining anhysteretic magnetization of isotropic magnetic material $M_{anhysteretic}$ [2]:

$$M_{anhysteretic} = M_s \left[\coth\left(\frac{H_e}{a}\right) - \left(\frac{a}{H_e}\right)\right] \tag{7}$$

$$a = \frac{N \cdot k_B \cdot T}{\mu_0 \cdot M_s} \tag{8}$$

As a result, the shape of anhysteretic magnetization curve is determined by the parameters M_s, a and α. However, using the same model for paramagnetic anhysteretic magnetization curve and ferromagnetic anhysteretic magnetization (considering domain instead of single atoms) is not obvious from physical point of view. On the other hand, parameter a quantifies domain density in magnetic material and can be determined from its micromagnetic properties on the base of other experimental investigations, which creates the possibility of validation of Jiles-Atherton model.

3 Tested Material

Experiments were carried out on ring samples made of high permeability Mn-Zn F-6001 ferrite produced by Polfer. The outside diameter of the ring was 15.75 mm, inside diameter was 9.46 mm, whereas height was 5.53 mm.

Microstructure of Mn-Zn F-6001 ferrite was investigated previously [13]. Average grain size of such ferrite was about 1 μm.

4 Magnetic Measurements

Quasi-static magnetic hysteresis loops were measured by digitally controlled hysteresisgraph. Schematic block diagram of measuring system is presented in Fig. 1.

Magnetic hysteresis loop was measured using 480 fluxmeter produced by Lakeshore, whereas sample was magnetizing by BOP36-6 high power voltage-current converter produced by Kepco. System was controlled by computer with data acquisition card and software developed in National Instruments LabView environment. Measurements were carried out in the room temperature $T = 294$ K.

Results of measurements of magnetic hysteresis loop up to saturation are presented in Fig. 2 (black line). It can be calculated, that initial permeability of Mn-Zn ferrite is about 6000, which is in line with the data provided by the producer. Moreover, the area of hysteresis loop is limited, so it is possible to fit anhysteretic magnetization curve with reasonable accuracy.

5 Results of Modelling of Anhysteretic Magnetization Curve

Software for implementation of Jiles-Atherton model of anhysteretic magnetization curve was implemented in open source Octave 4.0. Parameters of Jiles-Atherton model were determined during the optimisation process. The target function F for optimization process was given by the following equation:

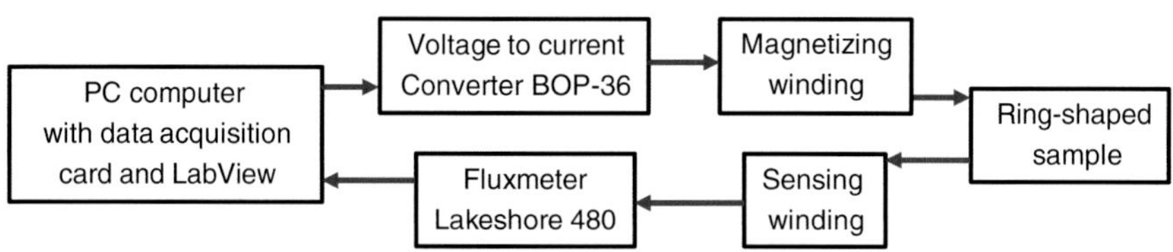

Fig. 1 Schematic block diagram of digitally controlled hysteresisgraph

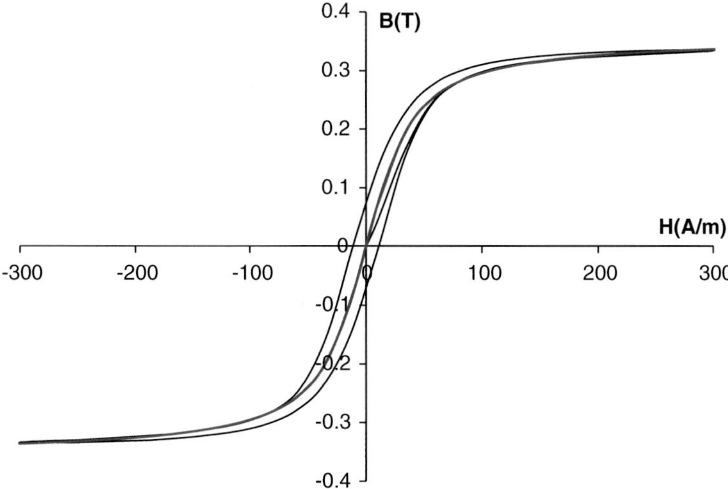

Fig. 2 Measured $B(H)$ magnetic hysteresis loop of high permeability F6001 Mn-Zn ferrite (*black line*), modelling the anhysteretic magnetization curve (*red line*)

$$F = \sum_{i=1}^{n} \left(M_{J-A}(H_i) - M_{meas}(H_i)\right)^2 \tag{9}$$

where $M_{J\text{-}A}$ were the results of the modelling and M_{meas} were the results of the experimental measurements, both for the value H_i of magnetizing field.

Optimisation process utilized gradientless Powell's method [14] for three dimensional minimization of a function F given by Eq. (9). Results of fitting anhysteretic curve to magnetic hysteresis loop of Mn-Zn ferrite are presented in Fig. 2.

Parameters of Jiles-Atherton model estimated during the optimisation process are presented in Table 1.

Saturation magnetization M_s and interdomain coupling α are in line with producer's data and expectations. From Eq. 8, number of domains per cubic meter may be calculated as:

$$N = \frac{\mu_0 \cdot M_s \cdot a}{k_B \cdot T} \tag{10}$$

In the presented case, N is equal $7.3 \cdot 10^{20}$, which means, that volume of average domain diameter in Mn-Zn ferrite is about $1 \cdot 10^{-7}$. In previous publication, the size of grain in high permeability Mn-Zn ferrite was estimated as 1 μm [13]. This means, that in one grain of ferrite about 1000 domains can be expected. This expectation is in line with typical, non-uniform shape of grain in ferrite as well as with the presence of significant number of domain wall pining points in such a grain.

Table 1 Estimated parameters of Jiles-Atherton model of anhysteretic magnetization curve for high permeability F6001 Mn-Zn ferrite

Parameter	Description	Value
M_s	Saturation magnetization	$2.84 \cdot 10^5$ (A/m)
a	Quantifies domain density (Eq. 8)	18.14 (A/m)
α	Interdomain coupling	$2.2 \cdot 10^{-5}$

6 Conclusions

Presented results confirm, that anhysteretic magnetization curve may be successfully modelled in the case of isotropic materials, such as high permeability Mn-Zn ferrites. Moreover, such modelling may be carried out with the use of open source software, such as Octave 4.0, which is especially important for industrial applications implemented by small or medium enterprises.

Results of modelling indicate, that one grain of high permeability Mn-Zn ferrite consist of about 1000 magnetic domains. This result is physically justified, especially in the case, when non-uniform grains with developed boundaries are observed. However, further detailed research on domain size utilizing e.g. Kerr effect may be necessary to confirm real domain size during the magnetization process.

Acknowledgment Calculations for the modelling were partially made in the Interdisciplinary Centre for Mathematical and Computational Modelling of Warsaw University, within grant G36-10.

References

1. Jiles, D.C., Atherton, D.L.: Theory of ferromagnetic hysteresis. J. Appl. Phys. **55**, 2115–2120 (1984)
2. Jiles, D.C., Atherton, D.L.: Theory of ferromagnetic hysteresis. J. Magn. Magn. Mater. **61**, 48–60 (1986)
3. Ramesh, A., Jiles, D.C., Roderik, J.: A model of anisotropic anhysteretic magnetization. IEEE Trans. Magn. **32**, 4234–4236 (1999)
4. Cundeva, S.: A transformer model based on the Jiles-Atherton theory of ferromagnetic hysteresis. Serbian J. Electr. Eng. **5**, 21 (2008)
5. Jiles, D.C., Thoelke, J.B., Devine, M.K.: Numerical determination of hysteresis parameters for the modeling of magnetic properties using the theory of ferromagnetic hysteresis. IEEE Trans. Magn. **28**, 27 (1992)
6. Lederer, D., Igarashi, H., Kost, A., Honmat, T.: On the parameter identification and application of the Jiles-Atherton hysteresis model for numerical modelling of measured characteristics. IEEE Trans. Magn. **35**, 1211–1214 (1999)
7. Chwastek, K., Szczyglowski, J.: Identification of a hysteresis model parameters with genetic algorithms. Math. Comput. Simul. **71**, 206–211 (2006)

8. Biedrzycki, R., Jackiewicz, D., Szewczyk, R.: Reliability and efficiency of differential evolution based method of determination of Jiles-Atherton model parameters for X30Cr13 corrosion resisting martensitic steel. J. Autom. Mobile Robot. Intell. Syst. **8**, 63 (2014)
9. Mousavi, S.A., Krings, A., Engdahl, G., Bissal, A.: Measurement and modeling of anhysteretic curves. In: Proceedings of 19th COMPUMAG Conference on the Computation of Electromagnetic Fields, Budapest (2013)
10. Leite, J.V., Sadowski, N., Kuo-Peng, P., Batistela, N.J., Bastos, J.: The in-verse Jiles-Atherton model parameters identification. IEEE Trans. Magn. **39**, 1397 (2003)
11. Kvasnica, B., Kundracík, F.: Fitting experimental anhysteretic curves of ferromagnetic materials and investigation of the effect of temperature and tensile stress. J. Magn. Magn. Mater. **162**, 43 (1996)
12. Szewczyk, R.: Validation of the anhysteretic magnetization model for soft magnetic materials with perpendicular anisotropy. Materials **7**, 5109–5116 (2014)
13. Jiang, K., Lil, K., Zhu, Y.: Effect of multi-additives on the microstructure and magnetic properties of high permeability Mn–Zn ferrite. J. Alloy. Compd. **541**, 472 (2012)
14. Powell, M.: An efficient method for finding the minimum of a function of several variables without calculating derivatives. Comput. J. **7**, 155 (1964)

Test Stand for Measuring Magnetostriction Phenomena Under External Mechanical Stress with Foil Strain Gauges

Oskar Gińko, Andrzej Juś and Roman Szewczyk

Abstract In this paper developed test stand for measuring magnetostriction deformation under external mechanical stress is presented. Magnetostriction phenomena is usually undesired, but there are applications which uses this phenomena in purpose for example to generate acoustic wave. In both cases complex approach and knowledge of magnetostriction factor λ in different material tensions are important. Usage of precision strain gauge, 24-bit analog to digital converter and digital signal processing allowed to achieve high resolution and accurate measurements of magnetostriction. All this also gives opportunity to measure small values of magnetostriction factors which are in materials like construction steels.

Keywords Magnetostriction · Magnetostriction under external stress · Strain gauge deformation measurement · Test stand

1 Introduction

Magnetostriction is a phenomenon which appears in change of geometrical dimensions on influence of magnetic field. Usually it is undesired, like in transformers or coils, but there are some applications which uses this phenomena in purpose. These applications are position detecting sensors, devices changing

O. Gińko · A. Juś (✉)
Industrial Research Institute for Automation and Measurements, Al. Jerozolimskie 202, 02-486 Warsaw, Poland
e-mail: ajus@piap.pl

O. Gińko
e-mail: oginko@piap.pl

R. Szewczyk
Faculty of Mechatronics, Warsaw University of Technology, sw. A. Boboli 8, 02-525 Warsaw, Poland

R. Szewczyk et al. (eds.), *Challenges in Automation, Robotics and Measurement Techniques*, Advances in Intelligent Systems and Computing 440, DOI 10.1007/978-3-319-29357-8_75

electrical oscillations into mechanical in frequency band of ultrasounds or perceptible by humans etc.

Characteristics of sample's deformation in magnetic field depends for example on sample material, strength of magnetic field and also tensions under which sample is expose to [1].

In purpose of measuring characteristic of deformation in magnetic field in function of mention factor test stand was developed, which is presented in this paper.

Benefit of this complex approach to research of magnetostriction, which is possible because of designed test stand, is possibility of material selection in inductive components with consideration of mechanical tension, which can appear in such elements. This approach will allow to optimize parameters of elements and reduce energy losses.

2 Magnetostriction Effect—Overview

2.1 Magnetostriction Effect

Magnetostriction (forced magnetostriction) appears in ferromagnetic. It is phenomena which consist in change of dimensions in magnetic material under influence of magnetic field. Inverse effect to magnetostriction is Villari effect [1].

Magnetostriction effect can appears in two different types: linear and volumetric. Linear magnetostriction means, that there are dimension changes in material in magnetic field without changes of its volume. Volumetric magnetostriction, like linear, makes dimensions change and also there are changes in material volume. In this case material bulge or contraction. Shape of the sample will be preserved. Magnetostriction is described by magnetostriction saturation factor λS. This factor describe relative change in dimension of ferromagnetic material in magnetic field ($\lambda_S = \frac{\Delta l}{l}$) [1].

Magnetostriction saturation factor can be positive for materials which extend in magnetic field and negative for those materials, which shrinks. Value of λS is in order of μm/m. For most ferromagnetics value of this factor is negative and it is in range of 0 to –200 μm/m. In construction steels value order of magnetostriction is several μm/m. On the other hand maximum value of magnetostriction appears in materials GMM (Giant Magnetostrictive Materials) up to 2000 μm/m. In this group of materials most popular is terfenol-D. Magnetostriction factor also depend on tensions in material [2].

Some sample characteristics of magnetostriction under tension are shown in Fig. 1. At the beginning negative deformation, while adding more and more

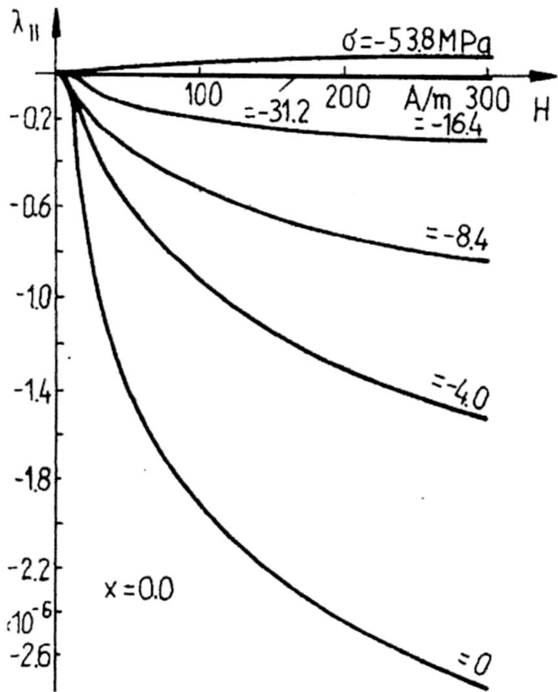

Fig. 1 Magnetostriction $\lambda_{||}(H)$ $Ni_{0.36}Zn_{0.64}Fe_2O_4$ under influence of external mechanical stress [2]

tension, were becoming smaller up to change of direction of deformation after applying significant tension [2].

2.2 Measurements of Magnetostriction

Ongoing research of magnetostriction mostly uses semiconductor strain gauge in tests stands. Signal from foil test stands is weak, which additionally make measurements more difficult to do [3]. Choosing semiconductor strain gauge was result of high gauge factor. After amplifying signal was strong enough to measure magnetostriction phenomena. Disadvantage of those strain gauges are low stability of parameters and nonlinear characteristic.

In last years progress in electronics allowed to measure magnetostriction deformation with foil strain gauge [4]. Results obtained with this method are burden with less error due to lack of previous mention disadvantages.

In developed test stand foil strain gauges are used. In contrast with other researches on magnetostriction under external mechanical stress this measurements were made on sample which was stretched, not squeezed [4].

3 Test Stand Idea and Implementation

3.1 General Idea

Schematic block diagram of test stand is shown on Fig. 1. Main element of test stand is tested sample of material with specially designed shape (3.2). Sample is put into hydraulic press (3.4). To achieve it sample is placed into reversing system, which is placed on force sensor (3.4, Fig. 6). Windings (3.2) on the sample are connected to voltage control current source (3.4). To know actual magnetic flux, there is fluxmeter connected to sample. Foil strain gauges connected into half-bridge are being used for measuring deformation in sample caused by magnetostriction (3.3). Signal from half-bridge is detected by Analog-to-digital converter (ADC), then in digital form it goes to microcontroller. All parts of test stand are integrated with LabView software (Fig. 2).

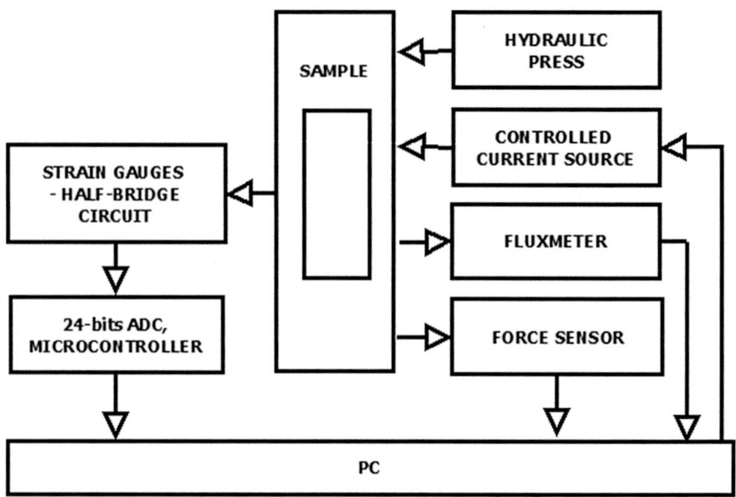

Fig. 2 Schematic block diagram of test stand

3.2 Investigated Sample

Measured sample is made of a material with ferromagnetic properties. For first tests construction steel 15HR was chosen. Shape of the sample is result of necessity of closing magnetic field loop inside the sample. Accurate drawing of sample with dimensions is presented on Fig. 3. Same size samples were also used in measurements of magnetoelastic effect presented in [5, 6].

On the one arm of the sample two strain gauges are stuck on in way that is shown on Fig. 5. Strain gauges are cover inside of the case (Fig. 4), on which coil was winded. Purpose of this case was to protect strain gauges from contact with winding.

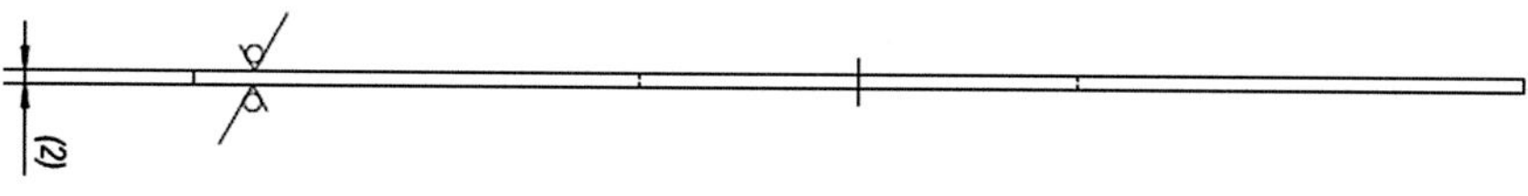

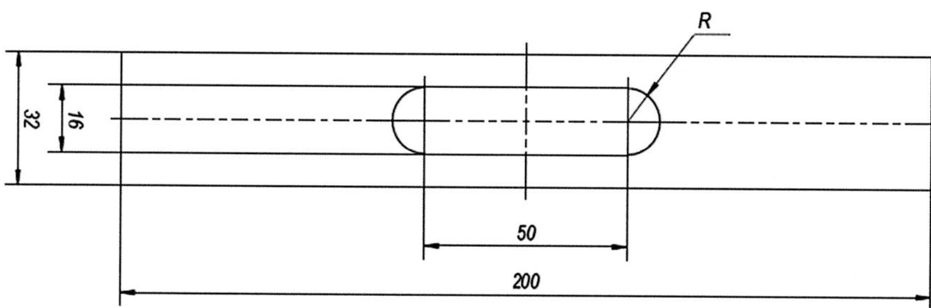

Fig. 3 Investigated sample geometry

Fig. 4 Visualization of closed case on one arm of the sample

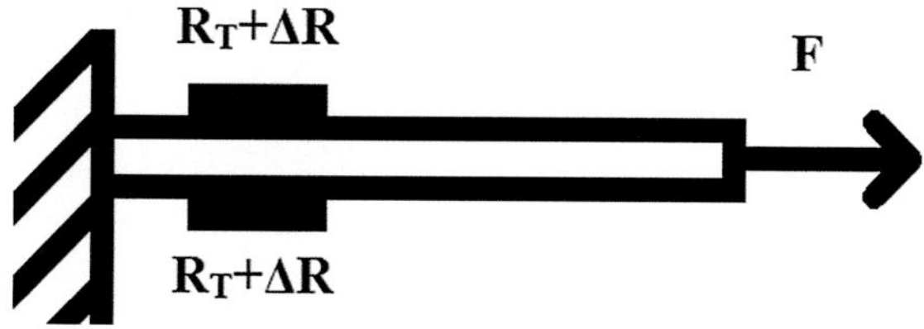

Fig. 5 Placement of strain gauges in used half-bridge

List of parameters which characterize tested sample and winding are below:

- magnetic path length—l = 176 (mm)
- number of magnetizing coils—N (–)
- magnetizing current—I (A)

Knowledge of this parameters allows to calculate value of magnetizing field H with Eq. (1) [7]:

$$H = N \cdot I/l \tag{1}$$

Length of magnetic path depend on geometry of sample (which can change in small range), while number of coils and magnetizing current must be chosen, that test stand will achieve planed value of magnetizing field. It must be considered, that in result of current flow through winding it will heat up—this effect is limiting maximum value of magnetizing current.

During tests sample was stretched using reversing system while at the same time applied by hydraulic press force was measured (3.4). Applied force shall not damage sample. Considering constant geometrics of sample and knowledge, that material is steel, applied tensions (σ) can be calculated from applied force (F) with Eq. (2).

$$F = S \cdot \sigma \tag{2}$$

where:

- S—surface area of sample cross-section (in this case 32 mm^2)
- R_s—yield strength which tell how much tension can be applied to sample without destroying it.

Value of yield strength for steel, which sample is made of, is 350 MPa [8]. Calculated force from Eq. (2) is equal to 11.2 kN. It shows, that during measurements tensions didn't reach yield point and sample was deforming elastically.

3.3 Strain Gauge Measurement of Deformation

Strain gain measurement of dimension change in sample was achieved with foil strain gauges CEA-06-240UZ-120 made by Vishay [9]. This strain gauges have thermal expansion coefficients similar to steel which allows easy solution to cancel thermal errors.

Strain gauges were connected into half-bridge [10]—two strain gauges on active sample, and the other two on other, not tested sample. This way of connecting strain gauges allow to measure extraction or shortening of the sample without impact of diffraction or contortion.

Half-bridge is powered up with 5 V voltage source from circuit measuring voltage across the bridge.

After measurement is done microcontroller reads data from analog to digital converter [11]. Next it change 24 bits into 6 hexadecimal digits, and send it through RS-232 interface to computer. Data are received by computer with software made in LabView.

3.4 Other Components of the Test Stand

Other elements of the test stand were standalone devices whose operation has been integrated during its implementation.

Magnetizing windings are supplied by power supplier KEPCO BOP 36-6 M. That device allows to generate current in range of ±6 A and voltage in range of ±36 V. Controlling of power supplier were made in National Instrument LabView environment—control voltage-signal ±10 V [12] derived from data acquisition card NI USB-6351.

Test stand allows to measurement value of voltage induced in the sensing winding, too. To realize that function fluxmeter LakeShore 480 is used. That device measured voltage induced in the sensing winding and convert it to the flux density value [13]. Signal from the fluxmeter is transmitted to the LabView software using RS232 interface.

Knowledge about supplied current parameters in magnetizing winding and about flux density in sensing winding allows to determine tested sample hysteresis loop.

To conduct research of magnetostriction of the samples under stretching loads used reversing system (Fig. 6), which was made to conduct a study presented in [5, 6]. Reversion system allow to attached the frame-shaped sample in the test stand and convert compressive to stretching loads.

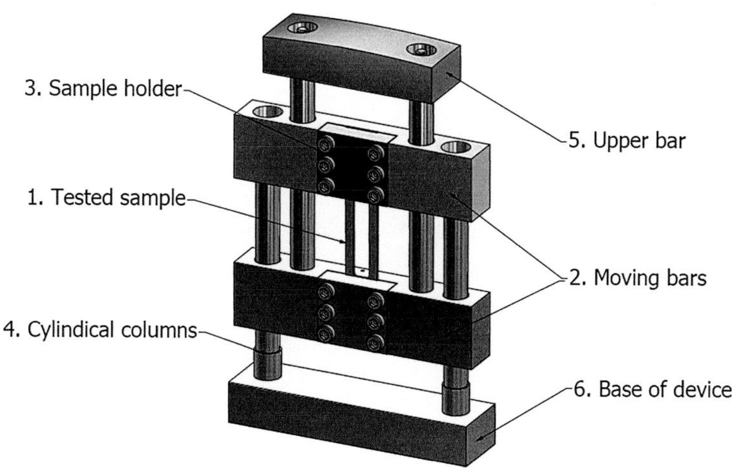

Fig. 6 Reversing system—mechanical device for attaching of the frame-shaped sample and for conversion compressive to stretching load [5, 6]

To load reversing system with the sample applied oil hydraulic press. The value of the load were manually controlled. To its measurement used strain gauge force sensor CL16U with a range of measured force values ±100 kN [14]. The measuring signal from the sensor hit on dedicated converter CL72-3U, and next it is read by data acquisition card.

3.5 Complete Test Stand

Using devices mention in 3.4: hydraulic press, reversing system, force sensor, voltage controlled current source, fluxmeter, and designed circuit for measuring voltage across the bridge with strain gauge half-bridge stuck on sample (3.3) fully functional system for measuring magnetostriction under external mechanical stress in sample (3.2) was created. Due to acquisition card and software made in LabView testing is partly automated.

4 Example Results

Results of measuring magnetostriction in sample made of 15 HR steel in magnetic field up to 3000 A/m under external mechanical stress 0 N and 7.5 kN (234 MPa) are shown below. For improving quality of received data voltage across the bridge was amplified 128 times by amplifier inside ADC chip [11]. This allows to change scale interval (achieved by 24-bit analog to digital converter) to 0.00045 μm/m.

Figure 7 shows measurements of magnetostriction under external mechanical stress. This data have digitally compensated change in applied force by hydraulic press during time (linear compensation).

The uncertainty in the measurement of magnetostrictive strain was calculated from equation describing strain gauge half-bridge [10].

$$U = \frac{E}{2}\left(k\frac{\Delta L}{L_0}\right) \quad (3)$$

where U is voltage across the bridge, E is bridge supply voltage (5 V), k is gauge factor and $\frac{\Delta L}{L_0}$ is reletive change of dimension.

According to producers datasheets uncertainty of measurement [9, 11]:

- u_U while amplifying signal 128 times is 15 nV,
- k has 0.5 %. Gauge factor is equal to 2.075, uncertainty u_k is 0.010375,
- u_E for load less than 50 mA is 25 mV.

In formula U was theoretical value of least significant bit, which was 1.16 nV.

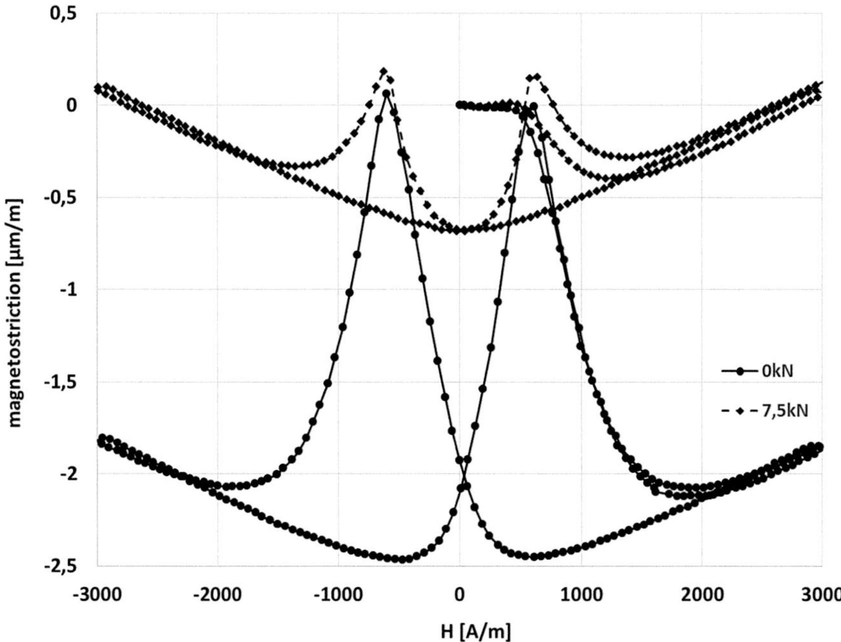

Fig. 7 Magnetostriction dependency of external mechanical stress in magnetic field in tested sample with signal amplified 128 times, and digitally compensated change in applied force by hydraulic press during time

In all datasheets distribution is not specified, therefore rectangular distribution was assumed. In this case all uncertainties were divided by $\sqrt{3}$. Uncertainty in measurement of magnetostriction is equal to:

$$u_c\left(\frac{\Delta L}{L}\right) = \sqrt{\left(\frac{d\frac{\Delta L}{L}}{dU}\right)^2 \cdot \left(\frac{u_U}{\sqrt{3}}\right)^2 + \left(\frac{d\frac{\Delta L}{L}}{dE}\right)^2 \cdot \left(\frac{u_E}{\sqrt{3}}\right)^2 + \left(\frac{d\frac{\Delta L}{L}}{dk}\right)^2 \cdot \left(\frac{u_k}{\sqrt{3}}\right)^2}$$
$$= \sqrt{\left(\frac{2}{E \cdot k}\right)^2 \cdot \left(\frac{u_U}{\sqrt{3}}\right)^2 + \left(-\frac{2U}{E^2 \cdot k}\right)^2 \cdot \left(\frac{u_E}{\sqrt{3}}\right)^2 + \left(-\frac{2U}{E \cdot k^2}\right)^2 \cdot \left(\frac{u_k}{\sqrt{3}}\right)^2} = 0.01669\,\frac{\mu\mathrm{m}}{\mathrm{m}} \tag{4}$$

In measurement of magnetostrictive strain influence of changing in applied, by hydraulic press, force was digitally compensated and it isn't taken into account in measurement uncertainty calculation.

Moreover in the measurements occurs uncertainties associated with supplying of magnetizing winding, measuring of voltage induced in sensing winding and measuring of value of the external load. But all this measurements was realized by standalone precise devices and values of its uncertainties are very low. So they are not included in measurement uncertainty calculation.

5 Conclusion

Obtained results (Fig. 7) shows, that test stand is operational. Uncertainty in measurement of magnetostriction is negligibly low (4). First changes of sample dimension under magnetic field have the same value and shape like in other researches in same subject (Fig. 1) [2, 4]. Further changes and ensuing hysteresis loops are analogous to results obtained in [2]. Nonetheless there are no hysteresis loops for this material. In this case there is need for more research for material with known hysteresis loop like shown in Fig. 7 for comparison.

During test a number of effects were observed, which accommodation will allow to upgrade build and obtained better performance:

- Sampling rate in analog to digital converter is to low (10 Hz)—it is necessary to change ADC for another integrated circuit with higher sampling rate. In case of lower resolution it will be important to use low-noise analog amplifier. This solution will allow to higher precision of measurement in different materials (changing gain in amplifier),
- During measurements it is necessary to stabilize temperature of the sample. Example of solution is test stand [15] where sample is submerge into oil, which temperature was stable and kept at 25 °C. In case of applying this solution strain gauge change won't be necessary.
- In current test stand only magnetostriction along the axis is being measure. Plans for upgrading test stands include adding additional possibility of measuring extension across the axis, too. This option will allow to immediate verify measurements correctness due to theoretical constant proportion between this two deformations [2], and also will make possible to measure volumetric magnetostriction.
- During measurements change in applied force by hydraulic press during time was seen. For more precise results more stable applied force is needed with for example usage of dead weight tester.

References

1. Problemy metrologii elektronicznej i fotonicznej 7, Mroczka, J. (ed.) Oficyna Wydawnicza Politechniki Wrocławskiej, Wrocław (2014)
2. Bieńkowski, A., Kulikowski, J.: Effect of stresses on the magnetostriction of Ni-Zn(Co) ferrites. J. Magn. Magn. Mater. **101**(1–3), 122–124 (1991)
3. Daou, R., Weickert, F., Nicklas, M., Steglich, F., Haase, A., Doerr, M.: High resolution magnetostriction measurements in pulsed magnetic fields using Fibre Bragg Gratings. Rev. Sci. Instrum. **81**, 033909 (2010)
4. ElBidweihy, H., Burgy, C.D., Della Torre, E., Wun-Fogle, M.: Modeling and experimental analysis of magnetostriction in high strength steels. In: EPJ Web of Conferences, vol. 40 (2013)

5. Jackiewicz, D., Szewczyk, R., Bienkowski, A., Kachniarz, M.: New methodology of testing the stress dependence of magnetic hysteresis loop of the L17HMF heat resistant steel casting. J. Autom. Mob. Rob. Intell. Syst. **9**(2), s.52–55 (2015)
6. Szewczyk, R., Svec Sr., P., Svec, P., Salach, J., Jackiewicz, D., Bienkowski, A., Hosko, J., Kaminski, M., Winiarski, W.: Thermal annealing of soft magnetic materials and measurements of its magnetoelastic properties Pomiary Automatyka Robotyka **2**, s.513–518 (2013)
7. Jiles, D.C.: Introduction to Magnetism and Magnetic Materials 2nd edn. CRC Press (1998)
8. Konstrukcja przyrządów i urządzeń precyzyjnych. Oleksiuk, W. (ed.) WNT, Warszawa (1996)
9. Vishay's strain gauges CEA-06–240UZ-120 datasheet. http://www.mae.ncsu.edu/ssml/Materials/MAE473/strain%20gage%20%28Student%20Guide%20309%29.pdf
10. Styburski, W.: Przetworniki tensometryczne Konstrukcja projektowanie użytkowe. Wydawnictwo Naukowo-Techniczne (1971)
11. ADC AD7191, Analog Devices. http://www.analog.com/media/en/technical-documentation/data-sheets/AD7191.pdf. Accessed 15 Oct 2015
12. Operator's manual BOP (M) (D) 100 W, 200 W, 400 W bipolar power supply, KEPCO INC 2011
13. LakeShore Model 480 Fluxmeter—User's Manual, Lake Shore Cryotronics, Inc 2004
14. Force sensor CL16U datasheet, ZEPWN. http://cms.zepwn.com.pl/zepwn/_media/products/pdf-pl/karta_cl16u_2013_09_24.pdf
15. Bieńkowski, A. Some problems of measurement of magnetostriction in ferrites under stresses. J. Magn. Magn. Mater. **112**(1–3), 143–145 (1992)

Method to Improve Accuracy of the Chromatography Mass Spectrometry Analysis

Adam Idźkowski, Eugenij Volodarsky, Zygmunt L. Warsza and Larysa A. Kosheva

Abstract The structural-algorithmic methods for improving the accuracy of quantitative chemical analysis are discussed. It is realized with the use of chromatography mass spectrometry measuring system. The idea is presented to exclude the influence of uncertainty at the stage of sample preparation. That allows, for the directly calibrated mass spectrometer, to assess the dispersion and to determine the overall uncertainty of analysis. Considerations are illustrated by numerical example with calibration at five points in a measuring range.

Keywords Calibration · Uncertainty · Chromatography mass spectrometry · Influence of sample preparation

1 Introduction

Quality control of industrial products i.e. many of foodstuffs and medicines in environmental monitoring, forensic, medical laboratory research needs quantitative analysis of multi-component fluids and gases. Their constituent components have to

E. Volodarsky
Department of Automation of Experimental Studies, National Technical University of Ukraine "KPI", Kiev, Ukraine
e-mail: vet-1@ukr.net

Z.L. Warsza
Industrial Research Institute of Automation and Measurement (PIAP), Warsaw, Poland
e-mail: zlw@op.pl

L.A. Kosheva
Department of Biocybernetics and Aerospace Medicine, National Aviation University of Ukraine, Kiev, Ukraine
e-mail: l.kosh@ukr.net

A. Idźkowski (✉)
Faculty of Electrical Engineering, Bialystok University of Technology, Bialystok, Poland
e-mail: a.idzkowski@pb.edu.pl

R. Szewczyk et al. (eds.), *Challenges in Automation, Robotics and Measurement Techniques*, Advances in Intelligent Systems and Computing 440, DOI 10.1007/978-3-319-29357-8_76

be identified and their concentration determined. To this aim a chromatography quantitative analysis may be applied. The primary information contained in the chromatogram is a sequence of chromatographic peaks, whose position on the time axis identifies a component. The area of the peaks is proportional to its relative content. This chromatogram is obtained by passing a sample of analyte through the sorbent inside a long capillary tube. The different components are delayed at different times due to their sorption capacity. The chromatogram is recorded and processed to determine the area of chromatographic peaks proportional to the number of relevant components. The presence of a test matrix of substances similar in chemical composition to the tested component and the inaccuracy of previous stages in preparation of tested probes cause uncertainties [1] and reduce the reliability of results. In view of their specific symptoms, the special techniques are required to reduce this influence and no computational methods can solve the problem [2, 3].

Recently the mass spectrometer becomes more widely available detector for a large number of chromatography studies. The combination of chromatography with mass spectrometry allows for the higher specificity of analysis including lower limits of detection and better quantification of analyte. Such measurement system is a useful tool in analysis. It allows the use of a simplified and shortened procedure for purification of extracts which are isolated from original matrix. That ultimately increases the percentage of certain analytes. Mass spectrometry coupled with chromatography increases the rapidity and accuracy of analysis [2, 4]. However, it is necessary to develop a method for reducing the influence of sample preparation on the result of analysis. That can be achieved by constructing the calibration characteristics of the entire measuring system, not only of chromatography mass spectrometer.

The aim of this paper is to review and study the possibility of applying structural methods to increase the precision of quantitative chemical analysis. Chromatography mass spectrometer as a measurement system is used.

2 Construction of Calibration Characteristics

Before measuring a chromatography-mass-spectrometry system should be calibrated. Measuring scale, that takes into account individual metrological characteristics of the system resulting from its construction, should be created. For this purpose the standard solutions of specified concentrations c_i, which are sometimes called "standards", are applied to the input of measurement system. Then the areas $\hat{S}_i^2$ which correspond to these concentrations are recorded. Number of standard solutions used for calibration depends on the measurement range and also on a type of mass-spectrometer characteristics.

When the calibration characteristics of the mass spectrometer is constructed, then in most cases it is assumed that it is a linear function $S = S_0 + bc$. Estimates $\hat{S}_0$

and $\hat{b}$ of coefficients S_0 and b are calculated using the method of least squares. If the calibration characteristics is known, then from the results of measurements S_k a concentration of analyte can be determined as

$$\hat{c}_k = \left(S - \hat{S}_0\right)/\hat{b}. \tag{1}$$

3 Influence of Sample Preparation on Uncertainty

An allocation of specific component from matrix is made in biomedical research before direct measurement of analyte concentration. The composition of matrix may be quite complex. This phase consists of many stages. The percentage of extraction of components from the matrix, i.e. "return", depends on the purity of reagents, dosing accuracy, thoroughness of execution in accordance with the procedure of cleaning mode, extraction etc. This introduces uncertainty into the result analysis. To reduce this uncertainty, experimental procedure should be calibrated including sample preparation.

During calibration, as described above, there are needed a set of standard solutions with values of concentrations covering the range of measurements and specially created blank samples of matrix. This is the same matrix as the test piece, but without the presence of the analyte, which is intended to be separated out, to determine its concentration. The number of such auxiliary matrices corresponding to the number of points is used to create the calibration curve. Quantities of standard solutions are added to these matrices. One should be absolutely sure that they do not contain the analyte. It is one of the difficulties of quantitative chemical analysis.

Another difficulty is the need to ensure equal uncertainties of multi-step sample preparation in a predetermined measuring range. It should be ensured for calibration of the measurement system as well as for analysis. Thus, reproducibility of sample preparation process should be complied.

With these considerations, parallel sample preparation with calibration (i.e. multi-track procedure) makes impossible to eliminate the uncertainty arising at this stage of the analysis process and sometimes may even increase it.

4 Structural-Algorithmic Method of Improving Accuracy

According to theory of invariance of B.N. Petrov the implementation of the measurement system function, which eliminates an unwanted influence of quantities, requires the existence of at least two channels of information processing. These channels can be realized not only separately but also by time division. In this second case the processing can be made in the same channel. Reference sample x_0 and analyzed sample x_k are transmitted over the same link by allocating a different time interval. If the level of influence quantities on the reference and measured values is small then non-ideal properties of the processing channel will in practice be identical. In channel output is the same relationship between two output values $y_0' = kx_0$ and $y_k' = kx_k$. Output values remain the same as the input for that channel. Then the measured value is based on the ratio y_0' and y_k'

$$x_k = x_0 \cdot y_k' / y_0'. \tag{2}$$

An imperfection of channel characteristics has no influence.

When carrying out quantitative chemical analysis this procedure cannot be used directly. The reason is that the stages of sample preparation procedures take place in a time sequence with a transition from one state to another. This is impossible without special methods to implement the principle of a single channel with the division in time.

However, in the quantitative chemical analysis the use of chromatography mass spectrometer makes possible to divide the "standard" and the measured analyte if the composition of any (but not on the main) basis is different. If this problem is solved directly by selecting "standard" with the other properties, the principle of invariance is violated. This can be avoided if the investigated matrix previously introduced as isotopically-labelled standard solution, which by definition, has the same properties as that of the analyte whose concentration is determined but it has a different weight. This makes possible, basing on the time of appearance of the peak in the chromatogram, separately assess the concentration of tested analyte and isotopically-labelled standard introduced to the matrix. Thus the uncertainty introduced in the stages of sample preparation, such as cleaning, evaporation, extraction etc., in accordance with (1) is excluded, since both the analyte and the "standard" will simultaneously undergo the same sample preparation steps.

With this approach, calibration can be performed for chromatograph mass spectrometer only, not for the whole system, since the uncertainty introduced on the stage of sample preparation, will be virtually eliminated.

5 Determination of Analyte Concentration

As already noted a solute concentration at the inlet of system is proportional to the peak area S recorded in chromatogram. In accordance with analysis of a measured component concentration c_k with the introduction of isotopically-labelled standard solution with a concentration c_h in the system of matrices will be proportional to the ratio of peak areas S_k/S_h in chromatogram. Then for the linear characteristic of the mass spectrometer and the absence of an additive bias

$$c_k = c_h \cdot S_k / S_h. \tag{3}$$

Thus, by determining the ratio of peak areas of chromatograms with known concentrations of "standard", one can find the desired analyte concentration c_k. This will be in an ideal situation when the characteristic is not biased. In a real situation (3) is

$$\frac{S_{ki}}{S_{hi}} = \frac{S_0 + bc_{ki}}{S_0 + bc_{hi}}, \tag{4}$$

where i—number of points in a measurement range which create calibration characteristics, s_{ki} and c_{hi}—known concentrations of studied analytes and isotopically-labelled prototype in ith standard solution.

Additive shift of conversion characteristics of a mass spectrometer introduces uncertainty in the result. At different concentrations c_{ki} and c_{hi} contribution is different. In addition the same ratio of peak areas S_k/S_h can be obtained for different ratios c_{ki}/c_{hi} which introduces additional uncertainty in the result.

To achieve uniqueness in the calibration and to avoid the dependence of results on the concentrations of the isotope-labeled standard solution, the grading should be carried out when c_h value is constant. It is necessary to have set of standard solutions in which the concentration of isotopically-labelled component remains constant while the concentrations of the analyte are represented by number of values c_{ki}. The values c_{ki} are within measuring range of matrix. Then, by changing the ratio

$$A_{ki} = \frac{S_{ki}(S_0 + bc_h)}{S_h(S_0 + bc_{ki})} = \frac{S_{ki}c_h\left(\frac{S_0}{c_h}\right) + b}{S_hc_{ki}\left(\frac{S_0}{c_{ki}} + b\right)} \tag{5}$$

the actual characteristic of mass spectrometer depending on c_{ki} can be created.

Assuming that $S_{ki}c_h = S_hc_{ki}$ expression (5) becomes

$$A_{ki} = \left(\frac{S_0}{c_h} + b\right) / \left(\frac{S_0}{c_{ki}} + b\right). \tag{5a}$$

Deviation of the real rate of A_{ki} from unity must be taken into account by introducing a correction factor to obtain the current result.

Making calibration for each ith point in the measuring range one can determine a set of calibration coefficients, which differ from unity in general

$$A_{gri} = \frac{S_h}{S_{\mathrm{ki}}} \cdot \frac{c_{\mathrm{ki}}}{c_h}. \tag{6}$$

Using the method of least squares [3, 5] one can find the coefficients of regression equation

$$\hat{A}_{gri} = \hat{a} + \hat{d}c_{ki}. \tag{7}$$

Values of these coefficients minimize the sum of squared deviations of calibration coefficients calculated on the basis of experimental data according to expression (6) and found from the Eq. (7) for *i*th component concentrations. It is done for all experimental points in the measuring range.

6 Checking the Adequacy of Calibration Characteristics

Adequacy of presented characteristics with experimental data can be assessed on the basis of variance [3, 5]

$$s_{ad}^2 = \frac{1}{N-2}\sum_{i=1}^{N}\left(\hat{A}_{gr\,i} - A_{gri}\right)^2, \tag{8}$$

where $\hat{A}_{gr\,i}$ and A_{gr_i}—concentrations, respectively—calculated on the basis of regression equation and found on the basis of experimental data in the *i*th point, $i = \overline{1, N}$, *N*—number of points.

According to the theory of mathematical statistics for a single experiment and for each concentration ratio c_{ki}/c_h, which takes place in this case, on the basis of variance s_{ad}^2 the hypothesis H_0 can be tested: the horizontal line $\bar{A} = \frac{1}{N}\sum_{i=1}^{N} A_{gr\,i}$ corresponds to the experimental data and dispersion of the values with respect to this line depends only on the impact of random variables. Under the alternative hypothesis H_1: the experimental data corresponds to the sloping line.

Hypothesis is tested using the Fisher F-test. The estimated ratio is defined as

$$F = \frac{s^2(A)}{s_{ad}^2}. \tag{9}$$

Variance, which characterizes the dispersion of experimentally determined values A_{gri} in relation to their average $\bar{A}$, is calculated as

$$s^2(A) = \frac{1}{N-1}\sum_{i=1}^{N}\left(A_{gr\,i} - \bar{A}\right)^2. \tag{10}$$

The estimated ratio *F* is compared with the critical value F_{kr} at the selected level of statistical significance α and with the number of degrees of freedom for the numerator (*N* − 1) and for the denominator (*N* − 2).

If $F > F_{kr}$ then on the basis of available data it can be accepted that the linear relationship with probability $(1 - \alpha)$ corresponds with the experimental data. That indicates a change in the mass spectrometer characteristics in a function of measured concentration.

However, using a fixed value c_h during calibration in the relative values allows statistical estimation of repeatability. It results from constant concentration of the isotopically-labelled standard solution c_h during calibration of mass spectrometer characteristics. Therefore, the area of the corresponding peaks S_{hi} have also to be constant. In practice this situation is not observed. At different points of the calibration curve and for the same concentration of "standard" there are differences in the area of chromatographic peaks S_{hi}. The reason is the influence of random variables including noises. It can be estimated as

$$\hat{s}^2(S_{hi}) = \frac{1}{n-1}\sum_{j=1}^{n}\left(S_{hij} - \bar{S}_{hi}\right)^2, \tag{11}$$

where $\bar{S}_{h\,i} = \frac{1}{n}\sum_{j=1}^{n} S_{hi\,j}$—average value of the area of chromatographic peaks at the ith point of calibration curve; $\hat{s}^2(S_{hi})$—variance of results—n-time observations at the ith point.

On the basis of $\hat{s}^2(S_{hi})$ value, the coefficient F_1 is calculated

$$F_1 = \frac{s_{ad}^2}{\sum_{i=1}^{N} k_i\hat{s}^2(S_{hi})/N}, \tag{12}$$

where k_i—coefficient taking into account the different scales of components.

The coefficient F_1 allows to test the hypothesis H_0: A linear dependence is adequate to experimental data. In this case F_1 is compared with the critical value F_{kr1}. It is chosen for a given statistical significance α and a given number of degrees of freedom for the numerator $(N - 2)$ and for the denominator $N(N - 1)$. If $F_1 < F_{kr1}$ then the hypothesis H_0 can be accepted. This allows to improve the statistical reliability of results.

Estimation of variance $\sum_{i=1}^{N} k_i\hat{s}^2(S_{hi})/N$ for the average value of results characterizes their repeatability [6]. Expanded uncertainty of single analysis can be calculated on the basis of the value $\sum_{i=1}^{N} k_i\hat{s}^2(S_{hi})/N$, the uncertainty of preparation of isotopically-labelled standard solution and the uncertainty of volume of analyzed matrix (type B uncertainty [1, 7]). According to the traditional approach it is not possible.

The sources of uncertainty of correction factor $\hat{A}_{gr}$ in constructing the calibration characteristics are following: inaccuracy of determining the coefficient $\hat{a}$, inaccuracy of determining the coefficient $\hat{d}$, error in determining the peak areas corresponding to the analyzed solution and isotopically-labelled standard solution.

Standard uncertainty of coefficient is estimated as

$$u(\hat{a}) = s_{ad}\sqrt{\frac{1}{N} + \frac{\bar{c}_k}{\sum\limits_{i=1}^{N}(c_{ki} - \bar{c}_k)^2}}, \tag{13}$$

where $\bar{c}_k = \frac{1}{N}\sum\limits_{i=1}^{N} c_{ki}$—the average concentration of standard solutions used in the construction of calibration characteristics.

Standard uncertainty of correction factor $\hat{d}$, including the change of correction factor with the value of measured concentration, is defined as

$$u(\hat{d}) = s_{ad}\frac{1}{\sqrt{\sum\limits_{i=1}^{N}(c_{ki} - \bar{c}_k)^2}}. \tag{14}$$

Then the combined standard uncertainty is

$$u_c(\hat{A}_{gr}) = \sqrt{u^2(\hat{a}) + u^2(\hat{d})\,c_k^2}. \tag{15}$$

The value $\hat{a}$ is approximately equal to one. Then, in the measurements of low concentrations, the second element in the quotation (15) can be neglected. However, the second element becomes significant at high concentrations measured. In relation to it, the error definition of the area of peaks in a chromatogram is a value of the second order. It is due to the 24-bit analog-to-digital converters in a mass spectrometer are used.

7 Numerical Example

Let us consider a chromatograph calibration carried out in five points of the operating range by introducing 2 ml of calibration standard solutions CS1–CS5 (Table 1).

The least-squares approach was utilized to find coefficients $\hat{a}$ and $\hat{d}$ of the straight line

Table 1 The concentration of components in calibration solution (ng/ml)

Components	CS1	CS2	CS3	CS4	CS5
Analyzed	0.5	2	10	40	200
Isotopically labelled	100	100	100	100	100

$$\hat{a} = \hat{A}_{gr} - \hat{d}\,\bar{c}_k, \quad \hat{d} = \frac{\sum_{i=1}^{5} (c_{ik} - \bar{c}_k) \cdot A_{gr\,i}}{\sum_{i=1}^{5} (c_{ik} - \bar{c}_k)^2}. \tag{16}$$

where $\bar{c}_k = \frac{1}{5}\sum_{i=1}^{5} c_{ik}$—mean value of the analyzed component in the calibration solutions CS1–CS5.

Then $\bar{c}_k = \frac{0.5+2+10+40+200}{5} = 50.5\,\frac{\text{ng}}{\text{ml}}$, $\hat{d} = \frac{7.77}{28953} = 0.00027\,\frac{\text{ml}}{\text{ng}}$, $\bar{A}_{gr} = \frac{1}{5}\sum_{i=1}^{5} A_{gr\,i} = 1.1895$,

$$\hat{a} = 1.1895 - 0.00027 \cdot 50.5 = 1.1759.$$

The values $A_{gr\,i}$, calculated in accordance with (16), are presented in Table 2.

Linear regression equation, with the correction factor $\hat{A}_{gr}$ of the same value at any point in the measuring range, has the form

$$\hat{A}_{gr} = 1.1759 + 0.00027\,c_k \tag{17}$$

On the basis of (17), the calibration points corresponding to CS1–CS5 need to be calculated to determine the variance s^2_{ad}. The calculations are presented in Table 3.

The variance of adequacy is equal to

$$s^2_{\text{ad}} = \frac{1}{5-2} \cdot 0.00572 = 0.00191.$$

Table 2 Calibration coefficients

Solutions	CS1	CS2	CS3	CS4	CS5
$A_{gr\,i}$	1.2296	1.2296	1.2296	1.2296	1.2296

Table 3 Calibration coefficients (6), (7), absolute value of difference, squared difference

Standard solutions	$A_{gr\,i}$	$\hat{A}_{gr\,i}$	$\lvert A_{gr\,i} - \hat{A}_{gr\,i} \rvert$	$(A_{gr\,i} - \hat{A}_{gr\,i})^2$
CS1	1.2296	0.1760	0.0536	0.00287
CS2	1.1346	1.1764	0.0418	0.00175
CS3	1.1917	1.1786	0.0131	0.00017
CS4	1.1565	1.1867	0.0302	0.00091
CS5	1.2353	1.2309	0.0044	0.00002

Standard uncertainties of the regression coefficients in Eq. (7) were calculated in accordance with (13), (14)

$$u^2(\hat{a}) = 0.00191\left(\frac{1}{5} + \frac{(50.5)^2}{28953}\right) = 0.00055, \quad u^2(\hat{d}) = 0.00191\left(\frac{1}{28953}\right)$$
$$= 6.6 \cdot 10^{-8}\frac{\mathrm{ml}^2}{\mathrm{ng}^2}.$$

The combined standard uncertainty of calibration equation was calculated in accordance with (15)

$$u_c(\hat{A}_{gr}) = \sqrt{5.5 \cdot 10^{-4} + 6.6 \cdot 10^{-8} c_k^2}.$$

In this formula for lower range value (c_k = 0.5 ng/ml) the second element in the quotation can be neglected. Then we obtain

$$u_c(\hat{A}_{gr}) = 2.34 \cdot 10^{-2}$$

However, for upper range value (c_k = 200 ng/ml) the second element in the quotation is

$$u^2(\hat{d}) = 6.6 \cdot 10^{-8} \cdot 4 \cdot 10^4 = 26.4 \cdot 10^{-4}\frac{\mathrm{ml}^2}{\mathrm{ng}^2}$$

The contribution of the second element in the quotation is larger. Then finally we obtain

$$u_c(\hat{A}_{gr}) = \sqrt{5.5 \cdot 10^{-4} + 26.4 \cdot 10^{-2}} = 5.65 \cdot 10^{-2}$$

which corresponds to 5.65 %.

To reduce the uncertainty of measurements taken in wide range (0.5 ÷ 200) ng/ml, it is necessary to increase the number of experimental points (concentration of standard solutions).

8 Conclusions

Introducing a test matrix of isotopically-labelled standard solution which has the same physicochemical properties as an analyzed solution excludes the impact of uncertainty at the stage of sample preparation.

This allows to calibrate the mass spectrometer only. Conducting the calibration in relative terms at a fixed concentration of "standard" makes possible to evaluate the repeatability and the expanded uncertainty of the result.

References

1. Guide to the Expression of Uncertainty in Measurement First edition. ISO Switzerland. Revised in 2008 as JCGM-100-2008 (1993)
2. Ettre, L.S.: Nomenclature for chromatography. Pure Appl. Chem. **65**(4), 819–872 (1993)
3. Freund, J.E., Walpole, R.E.: Mathematical Statistics, 4th edn. Prentice-Hall, Englewood Cliffs (1987)
4. Fenton, J.: Toxicology: a case oriental approach. CRC Press LLC, Florida (2002)
5. Volodarsky, E.T., Kosheva, L.O.: Statistical data processing Tutorial. K.: NAU (2008)
6. Watson, J.T., Sparkman, O.D.: Introduction to Mass Spectrometry: Instrumentation, Applications, and Strategies for Data Interpretation, 4th edn. Wiley, Chichester (2007)
7. Volodarsky, E.T., Kosheva, L.A., Warsza, Z.L.: Niepewność jako miara poziomu zaufania do wyników procedur doświadczalnych - Uncertainty as a measure of the confidence level to results of certain experimental procedures. Pomiary Automatyka Kontrola **5**, 483–485 (2011)

Primary Standard of Electrolytic Conductivity Based on the AC Four Electrode Cell

Alexander A. Mikhal, Zygmunt L. Warsza and Vladimir G. Gavrylkin

Abstract The principle of operation and metrological model of the Ukrainian primary standard of electrolytic conductivity (EC) is presented. Its original construction is based on the four electrode conductivity cell. The equations for calculating the cell constant from its dimensions, and the budget of the type B uncertainty from estimated unknown systematic errors are given. Methods: how to minimize these errors and the uncertainty are proposed and implemented. Obtained results are given and successfully verified in International Key Comparisons.

Keywords Conductivity • Primary cell • Geometric errors • Uncertainty

1 Introduction

The electrolytic conductivity (EC) of solutions is measured in many fields as electro-technology industry, chemistry, biology, medicine, oceanology, environment monitoring, and other fields. Reliable and comparable results of EC measurements are obtained due creation the national metrological system with the primary standard and also due traceability of EC unit, calibration regulations and international Key Comparisons. In the World practice a different ways to implement standard of EC are used. However, the principle of its operation is almost the same—

A.A. Mikhal (✉)
National Academy of Science of Ukraine (NANU), Institute of Electrodynamics, Peremogy av. 56, 03680 Kiev, Ukraine
e-mail: a_mikhal@ukr.net

Z.L. Warsza
Industrial Research Institute of Automation and Measurements, PIAP, Al. Jerozolimskie 202, 02-486 Warsaw, Poland
e-mail: zlw@op.pl

V.G. Gavrylkin
Ukrmetrteststandard, Metrologichna Str. 4, 03143 Kiev, Ukraine
e-mail: vgavrilkin@ukrcsm.kiev.ua

R. Szewczyk et al. (eds.), *Challenges in Automation, Robotics and Measurement Techniques*, Advances in Intelligent Systems and Computing 440, DOI 10.1007/978-3-319-29357-8_77

the “absolute”, i.e. direct method of reproduction of a unit of electrolytic conductivity [1–3]. This method is based on the measurement of the liquid column resistance in the conductivity cell and on the calculation of the EC for the known length and cross-sectional area of this column.

Established in Ukraine the national primary standard of the EC is designed due to the original conception. This EC standard as basic components includes the four-electrode cell with calculated constant K, special high precision AC bridge for conductance measurements, thermostat with temperature control at 25 °C and precision digital temperature meter.

Value of the liquid conductivity k is determined as

$$k = GK(1 + \alpha t_{25}) \quad (1)$$

where: G—the conductance of the liquid column in the cell of known shape and dimensions; K—the constant calculated from cell dimensions; α—the temperature coefficient for conductivity of the solution; t_{25}—is the temperature deviation from 25 °C.

On the example of this EC standard the methods of estimation and minimizing the uncertainty type B of the electric conductivity EC of electrolytes measurements is presented.

2 Measurement System for the EC Unit Reproduction

The block scheme of the instrumentation system of the laboratory stand of the Ukrainian primary EC standard is shown in Fig. 1.

This measurement and control system of primary EC standard consists several subsystems: subsystem of the preparation of solutions, subsystem of EC reproduction, subsystem of EC value transmission, subsystem of the thermal and climate stabilization and measurement, and also computer PC subsystem for control functions, processing and storage the final results of measurements.

The main part of the EC standard stand is the subsystem of EC standard value reproduction. Its simplified functional diagram is shown in Fig. 2. Its basic components are: the four-electrode cylindrical cell with calculated constant K and special high precision AC resistance bridge 1 specially designed for conductance $G = 1/R$ measurements.

The main element of the conductivity sensor is the tube with internal diameter D, which is filled with electrolyte solution. Typically, this is a solution of potassium chloride. The tube is used to fix the geometry of liquid conductor and has three parts.

The central part 1 of the tube has length L and two side portions 2 have also the same length l. The ends of the central portion of the tube 1 are coated with circular potential electrodes 4. Their width corresponds to the tube wall thickness. At the edges of the tube two discs 3 are fixed. Inner surface of the discs is coated with

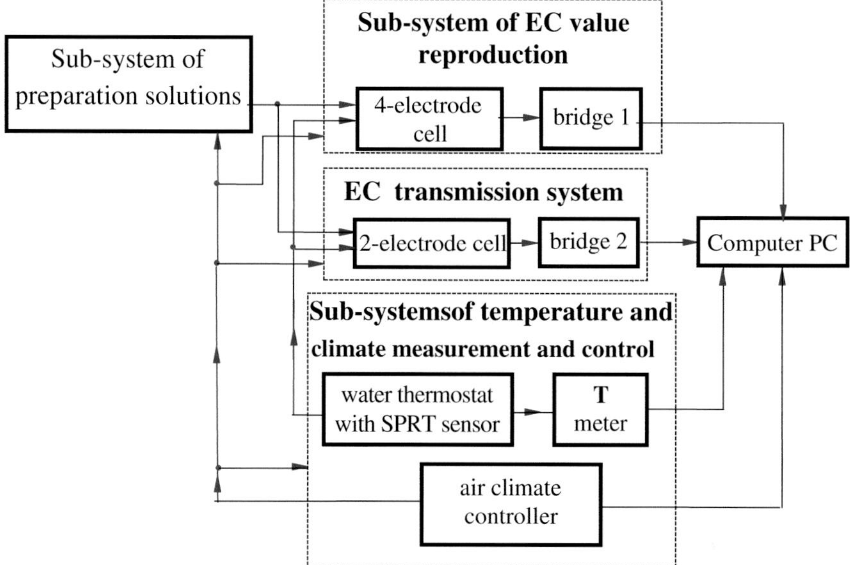

Fig. 1 Scheme of the instrumental system of Ukrainian primary 4-electrode cell *EC* standard: bridge *1*—special AC conductance bridge KM-4E, bridge *2*—AC bridge KM-2E, *SPRT*—stand and temperature platinum sensor, *T* meter—precision digital thermometer, computer *PC*—with software of all needed data processing

metallic film 5 and are current electrodes. Discs 3 have central holes 6 of the diameter *d,* which serve for filling liquid. Inner disc surface has the form of cone with an angle α. Such configuration is intended to facilitate the removal of air bubbles when filling cell with the liquid. The tube and the discs are made of quartz

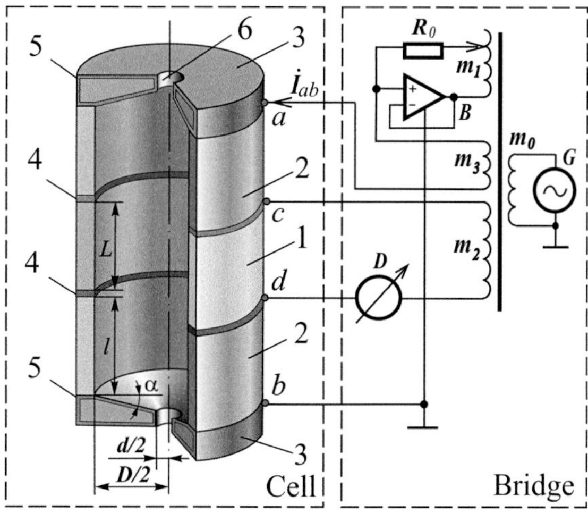

Fig. 2 Sketch of the primary conductivity cell and its connections to the AC bridge

glass which has good insulating properties, temporal stability and minimum coefficient of thermal expansion. The cell electrodes are made from platinum which has a minimum polarization effect. Four points, a, b, c and d, of these electrodes, are connected to the AC bridge (left side of Fig. 2), which measures the conductance G of liquid column between potential electrodes c, d. This is the automatic AC bridge designed especially for this EC standard. The admittance components of the parallel two-element equivalent circuit of the conductivity cell are measured by this bridge. The voltage comparison circuit is also used in it. That allows to eliminate the influence of impedances which occur on the border of electrolyte and current and potential electrodes of the conductivity cell.

This AC bridge works in the following way. The AC voltage of generator G is transferred from the winding m_0 with fixed number of winding sections, into winding m_1 with changing number of sections. Resistor R_0, windings m_1 and m_3, voltage follower (buffer) B jointly present an active node which performs the function of power supply source controlled by voltage. Thus, current $\dot{I}_{ab}$, the value of which depends on the voltage of generator G and the number of winding sections m_1, goes through the circuit of current electrodes of the cell. This current produces voltage drop U_{cd} on the potential electrodes c d which is proportional to the resistance $R = 1/G$ of the liquid column between them. On the other side in the circuit of potential electrodes we observe differential-mode voltage across the winding with fixed number of winding sections m_2. By changing the number of winding sections m_1 the value of current $\dot{I}_{ab}$ is adjusted until the difference of voltages on the input of detector D reaches zero for balance. From the condition of the bridge circuit balance the result of conductance G measurement is defined as

$$G = \frac{m_1}{m_2} \frac{1}{R_0}. \tag{2}$$

Very important for obtaining the enough high accuracy of measurement is also thermostat WT for temperature control (at 25 °C) and precision digital temperature meter TM with the platinum temperature sensor SPRT.

3 Metrological Model of the Four-Electrode EC Standard

One of the most important between metrological parameters of EC standard is the uncertainty of the conductivity unit reproduction. Upon absence of correlation between the parameters of the Eq. (1), the combined standard uncertainty u_k is calculated in accordance with Guide GUM [4] as

$$u_k = \sqrt{\left(\frac{\partial k}{\partial G} u_G\right)^2 + \left(\frac{\partial k}{\partial K} u_K\right)^2 + \left(\frac{\partial k}{\partial t} u_t\right)^2}. \tag{3}$$

where: u_G, u_t are estimated standard uncertainties of the conductance G and temperature t measurements correspondingly and u_K is the calculated uncertainty of the cell constant K (here u_k is referred only as type B uncertainty).

Partial derivatives as influence coefficients can be find from Eq. (1) and are expressed as

$$\frac{\partial k}{\partial G} = K(1+\alpha t_{25}),\ \frac{\partial k}{\partial K} = G(1+\alpha t_{25}),\ \frac{\partial k}{\partial t} = GK\alpha. \tag{4a–c}$$

Thus, the expanded uncertainty $U_k = c_P u_k$ and its relative value $U_k^* \equiv \frac{U_k}{k}$ is

$$U_k^* = c_P\sqrt{\left(\frac{u_G}{G}\right)^2 + \left(\frac{u_K}{K}\right)^2 + \left(\frac{u_t\alpha}{1+\alpha t_{25}}\right)^2}, \tag{5}$$

where: c_P—coefficient depended from desired probability P and type of the final distribution of uncertainty (For Normal distribution and P = 0.95 coefficient c_P = 2).

Estimated value of U_k^* is rounded to two significant figures. Therefore, we can assume $(1+\alpha t_{25})\approx 1$.

Uncertainties u_G, u_K, and u_t will depend on errors from various factors. The major factors are: metrological characteristics of measuring devices and instruments, errors of calibration standards, methodological errors of calculation of model parameters, manufacturing accuracy of calculated elements, environmental conditions, parameters of power supply, etc. Usage of relative values on the stage of preliminary evaluation gives us possibility to compare contribution of each error component and to estimate a reasonable amount of errors upon budget formation. The uncertainty u_k^* is estimated from the budget of the relative limited errors: $\delta_{iG} = \Delta G_i/G$—of the conductance G measurement, $\delta_{jK} = \Delta K_i/K$—limited error of the constant K calculations, and δ_{nt}—limited error of the maximum temperature increment $\Delta t\alpha$. Thus, for expanded relative uncertainty U_k^* the following formula is obtained

$$U_k^* = c_P\sqrt{\sum_{i=1}^{N}(A_{iG}\delta_{iG})^2 + \sum_{j=1}^{M}\left(A_{jK}\delta_{jK}\right)^2 + \sum_{n=1}^{H}(A_{nt}\delta_{nt})^2}. \tag{6}$$

The influence coefficients of limited errors δ_{iG}, δ_{jK}, δ_{nt} from factors $x \in (x_i, x_j, x_n)$ are

$$C_{iG}\frac{\partial G(x_i)}{\partial x_i} \equiv A_{iG},\ C_{jK}\frac{\partial K(x_j)}{\partial x_j} \equiv A_{jK},\ C_{nt}\frac{\partial t(x_n)}{\partial x_n} \equiv A_{nt}. \tag{7a–c}$$

Coefficients C_{iG}, C_{jK}, C_{nt} depend on the probability distributions of the relevant interacting errors [5].

From above it follows that the main metrological task lies in the optimization or reduction to the minimum U_k^* in (6). For the Ukrainian EC standard the solution of such task has few independent strategies (direction of research). The first strategy implies minimization of the function (6) by choosing solutions which influence coefficients close to zero $A(x) \rightarrow 0$. The second strategy involves minimization of the objective Eq. (6) by minimizing each error component $\delta(x) \rightarrow \min$. The third strategy is connected with estimation of the unknown systematic error components

$$\delta(x) = \delta_C(const) \tag{8}$$

where: $C \in \{i \rightarrow N, j \rightarrow M, n \rightarrow H\}$.

C is considered as some constant and as correction factor upon estimation. Error δ_C can be obtained experimentally by the external standard calibration or as a result of theoretical calculation of the real physical model of measurement cell. Thus, the realization of the third strategy lies in determining of the maximum number C and, in its turn, in minimizing of the argument as well as of the whole function $\Phi[\delta(x) - \delta_C]$. The above mentioned strategies can be combined in one expression as follows:

$$U_k^* = \min_{A(x) \rightarrow 0} \left\{ \min_{\delta(x) \rightarrow \min} \langle \min_{C \rightarrow N \wedge M \wedge H} \{\Phi[A(x), \delta(x) - \delta_C]\} \rangle \right\} \tag{9}$$

For realization of the first strategy such structure of standard is used, which covers construction principles of measuring converters, measurement techniques, calibration and processing of the conversion results. Then most of influence coefficients A_{iG}, A_{jK}, and A_{nt} are below the sensitivity threshold of measuring devices. For example, upon construction of the measurement cell we have chosen such material as quartz glass. This material is well process-able, chemically-stable, and has stable physical properties. Therefore, after evaluating influence coefficients A_{jK} in the cell for all influencing in practice climatic factors x_j, including coefficient of temperature expansion is find that is possible to ignore them. Similar approach is used to the construction principles of secondary measuring converters. For instance, conductivity AC bridge (model KM-4E [in Rus. «КМ-4Э»]) is a specialized device with fully balanced measuring circuit. It allows ignoring the coefficients A_{iG}, which are related with parameters of supply voltage (amplitude and frequency of AC voltage) and power generator supplying the measuring circuit. Measuring circuit of the bridge is built on the basis of transformers with close inductive coupling. Therefore, the result of conversion does not depend on any (except temperature) parameter of the environment.

Temperature meter has similar properties. It is attributed to the fact that the construction principle of its measuring circuit is built on ratio metric conversion of two alternating voltages with use the sigma-delta analog-to-digital (AD) converter. Measuring temperature by AC current circuit enables additionally to ignore the coefficient $A_{nt,}$ which is related with thermal and electrochemical EMFs.

The choice of elements for the standard structure is significantly affected by the temperature of surrounding air. This temperature influences in two ways. Firstly, it affects the resistance of the solution inside the cell. Temperature coefficient of most solutions ranges within (0.02–0.03) 1/°C. It is rather considerable destabilizing factor. Thus, it becomes necessary to stabilize the cell's temperature. Therefore, the structure of the standard contains improved water thermostat WT of TWP-6M type [in Rus. «ТВП-6М»] which enables to maintain stable temperature for 8 h just as well as ±0.002 °C. Secondly, the temperature also affects the base resistors located inside the bridge and inside the temperature meter. Their temperature changes are in the range (2–10) ppm. That is why it is important to maintain indoor temperature inside the range (20 ± 1) °C. Such requirements are met with the help of air-conditioning system and special construction of the room as passive thermostat. The room is separated with special gates from windows and entrance doors.

For the realization of the second and the third strategies, first and foremost it is needed to restrict {N}, {M}, {H} numbers of influencing factors in the Eq. (6). Imposition of restrictions is connected with the choice of expert evaluation for the error value δ_e below which all the other errors will be considered irrelevant and will not be taken in calculations. Based on the long-term experience and expertise in precise measurements, the following value $\delta_e = 5$ ppm is taken. It is by 20 times less than the value of residual systematic error (100 ppm) which is written in technical specifications of the standard. Further implementation of the strategies lies in usage of the procedures such as calibration of measuring instruments and estimation of the systematic error as a calculated value for correction.

Type B uncertainty of the measurement system includes components from the uncertainty of several instruments, almost entirely dependent on estimation of unknown systematic errors from their limited values. Therefore, a complete set or budget of uncertainties of the EC standard is basic to assess the quality and adequacy of the mathematical model describing the accuracy of that standard. The uncertainty full budget "tree" of Ukrainian EC standard is shown in Fig. 3.

Two parameters from the Eq. (1), conductance G and temperature increment t_{25}, only are measured directly with instruments. In this case the main methods of reducing uncertainty are well-known technique of calibration and temperature stabilization of these instruments. Uncertainty components from the errors that have been almost completely eliminated are denoted in Fig. 3 by (calibr.) and (t stab.).

The third parameter (cell conductivity constant K) is obtained from calculations. Therefore, the estimation of its uncertainty needs to pay the increased attention. The cell constant is determined by calculating the ratio of the length of the tube 1 to the cross-section area. However, this definition is true for an idealized object of measurements with uniform distribution of current flow lines. Distortion of such lines will be due to: the presence of air holes in the solution filling; the form of current electrodes and the presence of potential electrodes; non-ideal profile of the inner tube 1 in Fig. 2. Therefore, the calculations constant K will have errors.

The error budget for the constant calculation of the cell from Fig. 2 can be written as a set of unknown systematic relative limited errors:

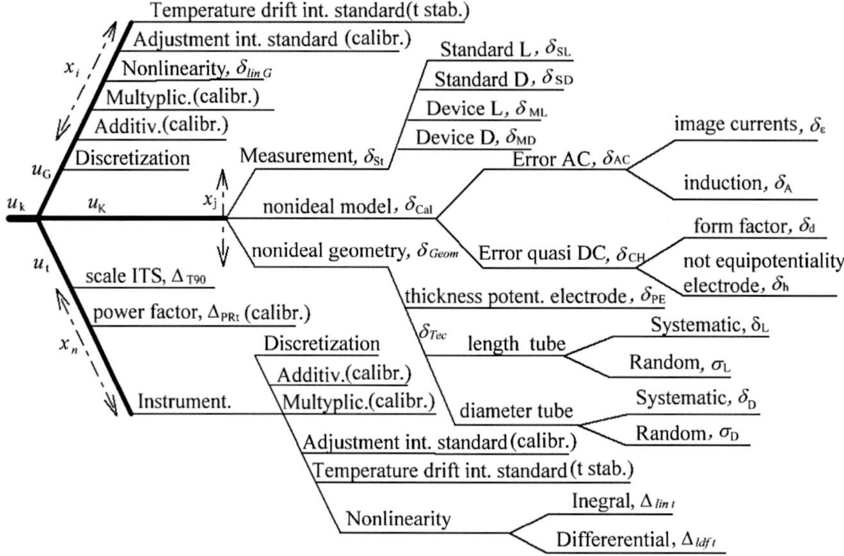

Fig. 3 "Tree" of error sources for the estimation of the uncertainty u_k of the value of electrolytic conductivity k

$$U_K^* \equiv \frac{u_K}{K} \in \{\delta_{St}, \delta_{Cal}, \delta_{Geom}(\delta_{Tec}, \delta_{PE})\} \tag{10}$$

where: δ_{St}—an error due to the accuracy of measurement standards and measuring instruments to determine the length and diameter of the tube; δ_{Cal} is an error due to the deviation of the calculation model for the cell constant in real conditions relative to the idealized model; δ_{Geom} is an error in assessment of geometrical dimensions.

Each argument of set (10) has several components. Let us consider them in detail. Minimizing of error δ_{St} is limited by the level of metrological assurance for measurements of tube length and diameter. It is defined by the metrological parameters of standards and instruments for the length and internal diameter measurements.

An error δ_{Cal} has two components: an error δ_{AC} due to alternating current measurement and an error δ_{CH} due to discontinuity of an electric field in the cell because of its finite dimensions and design features. Analysis of all δ_{Cal} components is described in detail in [6, 7].

Geometric error δ_{Geom} is an error dependent on manufacturing technology of the tube sections and their assemblage. The error δ_{Geom} is due to the deviation of actual profile of the inner surface of the tube 1 from the ideal profile of that tube, which is presented as rectangle along the longitudinal section and as circle in cross-section.

It should be mentioned that the cost of tube production from a monolithic quartz crystal is extremely high. As a rule, tubes are manufactured from work pieces (preform) which undergo precision machining. If precision machining of the inner surface is too deep, the mechanical resistance of the tube will be reduced significantly. Tubes of less than 1 mm in thickness will crack (fracture) under elastic forces (adhesive polymerization, temperature differences). Therefore, grinding of the work piece inner profile should be of minimum depth. On the other side, the work piece inner surface can have wedge-like cracks which are parallel to the axis of the work piece. These cracks are due to manufacturing techniques of work piece production and depend on the quality of nozzles through which the work piece is pulled itself. Therefore, due to the lack of deep machining of tubes, we can observe deviation from circle in cross-section along the entire profile. The second reason of non-ideal profile may be the precession of the grinding tool. During processing, quality control of the tube is practically impossible. After final grinding, the tube profile may differ from the ideal rectangle. As a result there is a systematic error that cannot be completely excluded and its component of uncertainty type B is estimated.

Two methods for correction of the non-ideal profile of the inner surface of the tube are proposed and results of corrections to the cell constant K are calculated.

The method of linear interpolation reduced the bias (systematic error) to 0.027 % and allows to reduce 10 times the uncertainty as random component.

The method of equivalent triangles reduced these same parameters respectively to 0.015 % and at 2.5 times. Main results of this research were presented at the IMEKO ISMQC 2013 Symposium in Krakow-Kielce and published in [7, 8].

Two designs of the conductivity cell, (a) with protection tube and (b) without tube, and (c) the general view of primary EC standard of Ukraine are presented in Fig. 4.

Dynamic properties of two types of cells from Fig. 4a, b are described in paper [9].

For use in the measurements of EC standard of Ukraine the several units of both types of the conductivity cell were made.

4 Experimental Results

Correctness, sufficiency and adequacy of the selected models of correction of systematic errors are confirmed by international Pilot and Key Comparisons (P22, P47, K36), in which the primary standard of Ukraine (laboratory UkrCSM) was involved. Each participated laboratory is using the Eq. (1) and its own metrological resources to determine the precise value k_{lab} of EC and value of uncertainty $U(k_{lab})$ for measured conductivity samples. The rules of Pilot and Key Comparison specify that a Reference Value of any of them must be derived and should be compared with the participants' results, and a Degree of Equivalence of each laboratory must be inferred. The Degree of Equivalence is given as:

$$D(S/m) = k_{lab} - k_{ref} \qquad (11)$$

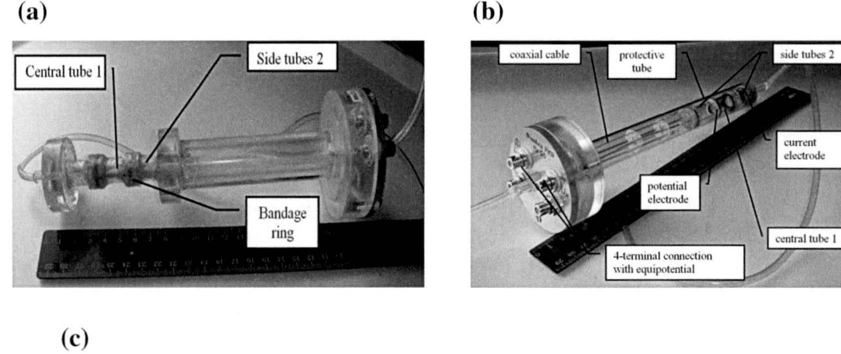

(c)

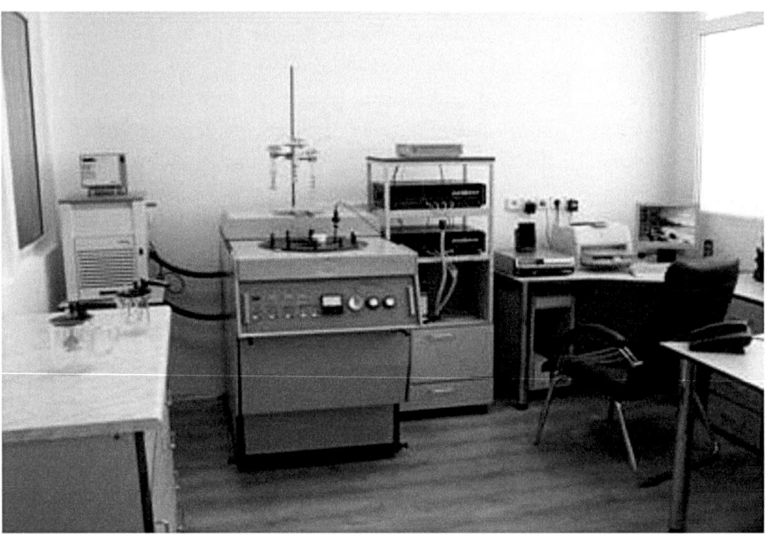

Fig. 4 Two cells of the primary EC standard of Ukraine: **a** without and **b** with protective tube; **c** the general view of the stand of primary EC standard of Ukraine

Figure 5 shows the results of the Pilot Comparison P 47 for a solution with a conductivity of 5 mS/m. The result was very close to the median (line 0.00 is line ref.) and has minimal uncertainty.

The best results were obtained in international comparisons CCQM-K36. In these comparisons 14 laboratories of the leading NMI of the world were participated, i.e.: USA (NIST), Germany (PTB), Israel (INPL), Slovakia (SMU), Denmark (DFM), and others. All participants received samples of potassium chloride solutions with nominal electrolytic conductivity 0.5 S/m and 5 mS/m. In report [10] the model of Reference Value k_{ref} of K36 was presented. Figure 6 shows the results of K36.

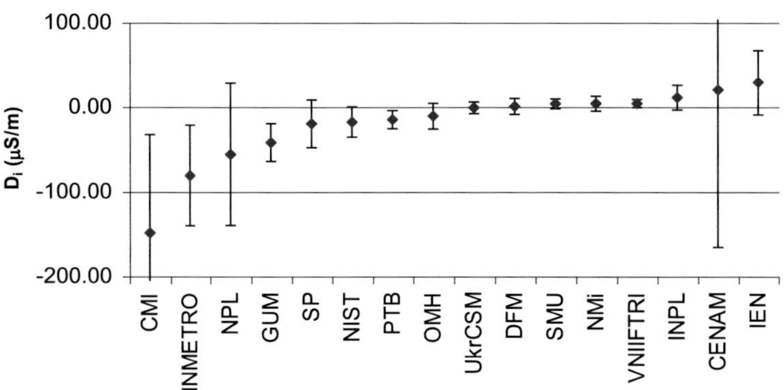

Fig. 5 Results of Pilot Comparisons P47

After calculation of all known corrections, obtained values of parameters are listed in the Table 1. Hence, the value of standard uncertainty u_k from (3) is equal to

$$u_k = \sqrt{(6.5 \times 10^{-5})^2 + (6.8 \times 10^{-5})^2 + (5.1 \times 10^{-5})^2} \approx 1.1 \times 10^{-4}, \quad \mathrm{S/m} \qquad (12)$$

For the second solution with electrolytic conductivity k = 5 mS/m value of uncertainty 0.0033 mS/m was obtained. These both results are put as final in the

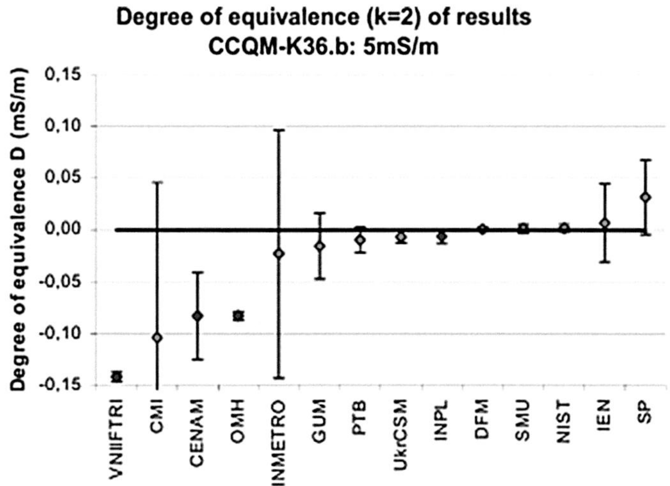

Fig. 6 Results of Key Comparisons K36

Table 1 Final parameters of uncertainty at the value 0.5 S/m

Source of uncertainty	Sensitivity coefficient, Eq. (4a–c), $\partial k/\partial i$	Standard uncertainty, u_i	Contribution to standard uncertainty
Conductivity	297 1/m	$u_G = 2.2 \times 10^{-7}$ S	6.5×10^{-5} S/m
Constant cell	1.7×10^{-3} S	$u_K = 4.0 \times 10^{-2}$ 1/m	6.8×10^{-5} S/m
Temperature	1.01×10^{-2} S/m°C	$u_t = 0.005$ °C	5.1×10^{-5} S/m

statement of international comparisons of CCQM-K36. According to the results of comparisons (K36) the CMC (calibration and measurement capabilities) of Ukraine in the field of EC measurements are added to the BIPM database.

5 Conclusions

Measuring system of the electrolytic conductivity (EC) primary standard of the Ukraine (Fig. 1) is based on the four-electrode conductivity cell (Fig. 2) and the high precision specialized AC bridge. The operating principle of this system has no analogues among the EC standards of NMI (National Metrology Institute) of other countries [10]. Given in Eqs. (5)–(7a–c) the metrological model of the accuracy description is applied. The minimization of the uncertainty of this EC standard is obtain due the implementation of three type of strategies (9): keeping coefficients of influencing variables close to zero, minimizing components of the standard EC uncertainty and proper estimation of the final standard EC uncertainty from unknown systematic errors. In Ukrainian EC standard two given below problems are solved.

Firstly, the value obtained for lab. UkrCSM k_{lab}, practically coincides with the value k_{ref}, which is the value of electrolytic conductivity reproduced by the NMI of the World leading countries. Results were achieved by minimizing systematic error.

Secondly, during a long period of time (P22, P47, K36), the minimum value of uncertainty $u(k_{lab})$ has been obtained. It is evident that obtained results of accuracy, repeatability and traceability of the reproduction of the unit of electrolytic conductivity are very good.

References

1. Shreiner, R.H., Pratt, K.W.: Standard Reference Materials: Primary Standards and Standard Reference Materials for Electrolytic Conductivity. NIST Special Publication, pp. 260–142 (2004)
2. Marriarssy, M., Pratt, K.W., Spitzer, P.: Major applications of electro-chemical techniques at National Metrology Institutes. Metrologia **46**, 199–213 (2009)

3. Brinkmann, F., et al.: Primary methods for the measurement of electrolytic conductivity. Accred Qual. Assur. **8**, 346–353 (2003)
4. Evaluation of measurement Data—Guide to the expression of uncertainty in measurement. ISO, Geneva, Switzerland (1995), last revised version: BIPM JCGM 100 (2008)
5. Mikhal, A.A., Warsza, Z.L.: Impact of AC electric field non-uniformity on impedance of the conductivity cell. In: XXI IMEKO World Congress Measurement in Research and Industry, Prague, Czech Rep (2015)
6. Gavrylkin, V.G., Glukhenkiy, A.I., Mikhal, A.A.: An analysis of the error when determining the constant of the primary standard conductometric cell. Meas. Tech. **56**(8), 935–941 (2013)
7. Mikhal, A.A., Warsza. Z.L.: Geometric part of uncertainties in the calculation constant of the primary four electrode conductivity cell. Acta IMEKO **4**(2), 18–22 (2015)
8. Mikhal, A.A., Warsza, Z.L., Gavrylkin, V.G.: Correction of the influence of not ideal geometric profile on the conductivity of reference cell. In monograph: Jabłoński, R., Brezina, T. (eds.) Advanced Mechatronics Solutions vol. 393, pp. 385–396. Springer (2016). doi:10.1007/978-3319-23923-1_59
9. Mikhal, A.A., Gavrylkin, V.G., Warsza, Z.L.: Experimental evaluation of dynamic properties of the four electrode reference conductivity cells. Measurement Automation Monitoring MAM 61 (12), (2015) (in print)
10. Jensen, H.D.: Final Report of Key Comparison CCQM-K36. 15 August 2006. http://kcdb.bipm.org/AppendixB/appbresults/ccqm-k36/ccqm-36_final_report.pdf

Experimental Definition of Compressive Stiffness of Cotton Flock

Vytautas Bucinskas, Nikolaj Sesok, Andzela Sesok, Igor Iljin, Ernestas Sutinys, Rimantas Subacius, Gintautas Bureika and Zygmunt Lech Warsza

Abstract Definition of Young's modulus for majority of materials is the routine operation. In case of materials with low stiffness this task requires special procedure and sophisticated machinery. Cotton flock is the case, where compressive stiffness of cotton flock is so small for direct measurement, so one of possible solution is proposed indirect method of its definition. Recently cotton processing industry uses dynamic methods of cotton cleaning, therefore it is required to know compressive stiffness of cotton flock in order to optimize vibration parameters of device and adjust them to current status of cleaning material. Real compression stiffness of cotton depends on many factors, so in this paper is made an attempt to create simple method of compressive stiffness definition, which can be applied in simple workshop. Paper contains initial assumptions, mathematical derivations of final formula, methodology of experimental research and results.

V. Bucinskas (✉) · N. Sesok · A. Sesok · I. Iljin · E. Sutinys · R. Subacius · G. Bureika
Vilnius Gediminas Technical University, Vilnius, Lithuania
e-mail: vytautas.bucinskas@vgtu.lt

N. Sesok
e-mail: nikolaj.sesok@vgtu.lt

A. Sesok
e-mail: andzela.sesok@vgtu.lt

I. Iljin
e-mail: igor.iljin@vgtu.lt

E. Sutinys
e-mail: ernestas.sutinys@vgtu.lt

R. Subacius
e-mail: rimantas.subacius@vgtu.lt

G. Bureika
e-mail: gintautas.bureika@vgtu.lt

Z.L. Warsza
Industrial Research Institute for Automation and Measurements, Warsaw, Poland
e-mail: zlw@op.pl

R. Szewczyk et al. (eds.), *Challenges in Automation, Robotics and Measurement Techniques*, Advances in Intelligent Systems and Computing 440, DOI 10.1007/978-3-319-29357-8_78

Keywords Young modulus of cotton • Compressive stiffness • Cotton-cleaning methods • Dynamic method

1 Introduction

Despite competition from artificial counterparts cotton is still the best selling fiber material in the world, mostly in textile industries [1]. These natural cellulosic fibers do not create many harmful effects as the artificial fibers do [2].

Cotton fibers are product of cotton plant processing, therefore they contain some organic inserts—seeds, leafs and other parts of plant. During harvesting process flocks are contaminated by some non-organic mechanic particles like sand, small stones or metal. Quality of commercial cotton depends on a wide variety of factors like production technology, biologic variety of plants, quality of plant environment, harvesting procedures and the ginning process [3]. The term *ginning* refers to the process of separating the cotton fiber from cottonseeds.

There are well-known methods of cotton ginning and available equipment for it [4–6]. Before the fiber is separated from seed, all non-organic includes have to be separated. Cleaning equipment of cotton is used in combination with ginning machines. Therefore mechanical properties of cotton flock became important.

Siddaiah et al. [7] analyzed cotton quality indicators, classified all includes of cotton and proposed method of cleaning automating. Tensile stiffness of cotton knit is well known and not requires commenting, but compressive stiffness of cotton flock is a key for vibration cleaning procedure.

Mechanical properties of cotton were researched in the papers [1, 8, 9, 11]. Basu [1] measured mechanical characteristic of cotton using Agilent UTM T150 tensile test machine. S. Basu defined tensile Young's modulus of cotton as 30 ± 4 GPa, tensile strength 1066 ± 41 MPa, strain at failure 5 ± 1 %. Although the cellulose crystals in the mercerized cotton fibers exhibit high modulus and strength, they are also the least ductile comparing to the other fibers. Moreover, the mechanical properties of cotton also vary with the length of the fiber and the chemical treatment it undergoes before application [1, 2].

Cotton fruit size depends on a kind of cotton and can vary in size from 2 to 10 cm in diameter with weight correspondingly from 2 to 10 g [10].

There are a few research papers on vibration transportation and cleaning of cotton [12, 13]. Vibration transport machines are intended for transporting and/or separating variable bulk materials [16]. In paper [12] static and dynamic analysis are performed for the main working head of a saw cylinder of a fiber extracting machine for the cotton cleaning industry of Uzbekistan. Paper [13] is dedicated to research of dynamic behavior of saw-gin in the ginning machines with special reference to separator units.

Authors in the special research [14] analyzed operation of separation of seeds using layer of the mixture model which executes vibration motions on a plane with friction. The processes of vibrational-transport and vibrational-separation are based on the dynamic displacements, with sliding of the mixture layer on the surface of flat sieves what executes vibrations on the same direction [14]. The working process of vibratory conveying device is a result of the aggregate effects of individual pulses with greater frequency following one after another. These pulses are the result of movement of these devices—vibrations, i.e. fluctuations with high frequency and small amplitude. In vibrational transport devices has been widely used the harmonic law of motion [15]. There is presented vibrational method of hazelnut cleaning in the paper [17]. A system has been developed to separate in-shell pistachio nuts from kernel by processing streams based on the vibration induced on a plate from an impacting nut.

Vibrational transportation method is relatively easy applicable and widely used for transportation of materials [18].

Mass of transported material during transportation is small in comparison with weight of material, stiffness and other properties of transported materials became important for proper process of sorting and cleaning during transportation. Evaluation of real stiffness of cotton flock helps a lot to adjust vibration parameters of the cleaning and transporting equipment.

Aim of this paper is to create experimental method to define coefficient of compressive stiffness of cotton flock, necessary for optimization of vibrational method of cotton cleaning.

2 Main Assumptions and Model of Cotton Flock for Experimental Analysis

Body of the cotton flock is treated as ideal body with mass in the center of mass and with spring, attached in the bottom of the body with linear stiffness, which represents compression stiffness of cotton flock. This dynamic model, which is presented in Fig. 1, used to define stiffness of whole piece of cotton and spring stiffness is assumed to be linear.

In this case, position of the body is defined by coordinate x, excitation to the body applied cinematically, along coordinate x_0. At the moment of time $t = 0$ body is moved with initial velocity v_0. Height h we found experimentally by methodology, described below.

According energy conservation law for researched system is possible to state:

$$\frac{mv_0^2}{2} = mgh + A_{pas} + \frac{k(\Delta x)^2}{2}, \tag{1}$$

where m—mass of the body (cotton flock), m = 7.68 g, v_0—velocity due to kinematic excitation, h—height, to which our body will fly after excitation applied (defined experimentally), A_{pas}—work of aerodynamic drag force; k—requested

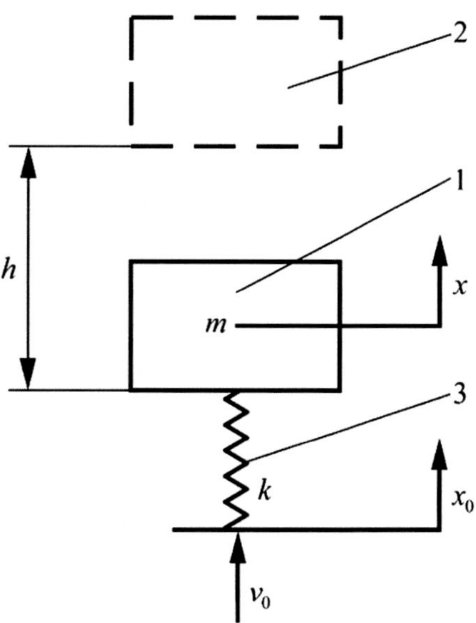

Fig. 1 Dynamical model of cotton flock: m—mass of the body, v_0—initial speed of the body, k—stiffness of the spring, x_0—initial displacement of body

stiffness, here represented by mass-less spring, Δx—absolute deformation of the body (spring) (defined experimentally).

Initially kinetic energy $\frac{mv_0^2}{2}$, is supplied to the body ant then body achieves height h (thus gained potential energy $m \cdot g \cdot h$), compress the body (spring) (and gains potential energy equal $\frac{k(\Delta x)^2}{2}$) and overcomes air drag force (air drag force performs work A_{pas}). Then coefficient of stiffness k of the cotton flock, which is represented as single mass-less spring is found from (2):

$$k = \frac{mv_0^2 - 2mgh - 2A_{pas}}{(\Delta x)^2}. \tag{2}$$

In this expression, there are two unknown members—A_{pas} and v_0.

Work of air drag force A_{pas} is found from formula (3):

$$A_{pas} = \int {}_0^h F_{pas} dx. \tag{3}$$

For definition of air drag force, there is assumed [19] that air drag force is proportional to square of velocity:

$$F_{pas} = k_2 v^2, \tag{4}$$

where k_2—coefficient of air drag, which is defined experimentally by procedure given below.

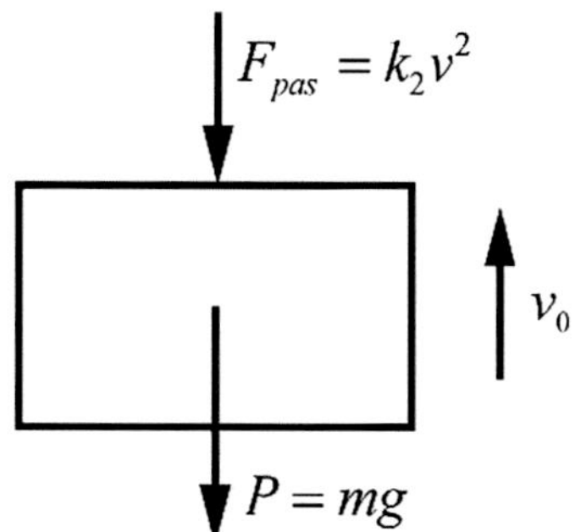

Fig. 2 Forces, applied to researched body during free-fall: F_{pas} aerodynamic drag force; P force of gravity; v_0 direction of speed vector

Forces, acting to cotton flock during free fall, shown in the Fig. 2.

Then we can draw differential equation of linear movement of body when velocity is function of coordinate $v = f(x)$ and keeping force balance:

$$mv\frac{dv}{dx} = -P - k_2 v^2, \tag{5}$$

where v—velocity of body, $P = mg$—gravity force, $k_2 v^2$—air drag force.

Rearranging Eq. (5) in order to obtain desired form of equation:

$$v\frac{dv}{dx} = -g\left(1 + \frac{k_2}{P}v^2\right). \tag{6}$$

Then we apply substitution:

$$a^2 = \frac{P}{k_2}. \tag{7}$$

Thus, equation takes form:

$$v\frac{dv}{dx} = -g\left(1 + \frac{v^2}{a^2}\right). \tag{8}$$

and continue rearranging we get:

$$\frac{vdv}{a^2 + v^2} = -\frac{g}{a^2}dx. \tag{9}$$

Further Eq. (9) is integrated and we obtaining:

$$\ln\left(a^2 + v^v\right) = -2\frac{g}{a^2} \cdot x + C_1. \tag{10}$$

Coefficient C_1 is found from initial condition. In case when $x = 0$, then $v = v_0$.

Substituting these values to Eq. (10), we get:

$$C_1 = \ln\left(a^2 + v_0^2\right). \tag{11}$$

Then after rearrangement and substituting, we get desired equation:

$$\ln\left(a^2 + v^2\right) = \frac{2g}{a^2} \cdot x + \ln\left(a^2 + v_0^2\right). \tag{12}$$

Solution of (12) delivers expression of velocity:

$$v = \sqrt{\left(a^2 + v_0^2\right) \cdot e^{-\frac{2g}{a^2}x} - a^2}. \tag{13}$$

By substituting (13) into (3) work of air drag force is obtained:

$$A_{pas} = \int_0^h k_2\left(\left(a^2 + v_0^2\right) \cdot e^{-\frac{2gx}{a^2}} - a^2\right) \cdot dx. \tag{14}$$

Finally, expression takes form:

$$A_{pas} = -k_2\left(a^2 \cdot h + \frac{e^{-\frac{2gh}{a^2}} - 1}{2g}\left(a^4 + a^2 \cdot v_0^2\right)\right). \tag{15}$$

This equation will be used in experimental research to define work of aerodynamic drag.

3 Methodology of Experimental Research

Experimental research is performed using lever as device for generating impact of known value to cotton flock. Initially, lever is in position, when its side with cotton flock is placed down, as shown in Fig. 3.

Load of known mass m lay on lever and its position fixed. Then load 5 is released and it moves down. Force, created by load is equal $F = m \cdot g = 28.9395$ N and generates torque equal

$$M = F \cdot L = 6.36669\,\text{Nm}.$$

During the experimental research is obtained that mentioned moment is applied to lever 0.166 s, and it continues until other arm of lever reaches support (3). Using experimentally obtained time of lever movement, velocity v_0 is calculated using procedure, described below.

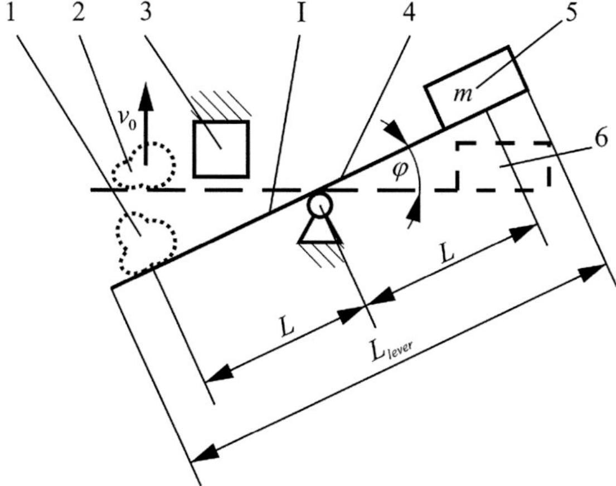

Fig. 3 Diagram of experimental research of cotton flock using lever: *1* cotton flock at rest position on the lever; *2* cotton flock leaving lever at the stopping point of lever; *3* support of lever; *4* lever; *5* load; *6* load at the stopping point of lever; φ angle of lever rotation

In order to evaluate influence of moment of inertia of lever with attached mass and nest for test piece of cotton, Eq. (16) is built as:

$$I \cdot \ddot{\varphi} = M, \quad (16)$$

where *I*—moment of inertia of lever with attached mass and cotton, $\ddot{\varphi}$—angular acceleration of lever, *M*—force moment on the lever.

To solve Eq. (16) by MATLAB we transform into operator form (using $p = \frac{d}{dt}$):

$$I \cdot \varphi \cdot p^2 = M. \quad (17)$$

Value of moment of inertia:

$$I = \frac{1}{12} m_{lever} \cdot L_{lever}^2 + m_{weight} \cdot L^2 = 0.153483 \text{ kg} \cdot \text{m}^2 \quad (18)$$

where $m_{lever} = 0.475$ kg, $L_{lever} = 0.52$ m, $m_{weight} = 2.950$ kg, $L = 0.22$ m.

Solution of the (17) is obtained in the form $\varphi = \frac{M}{Ip^2}$. For solution of the Eq. (17) we build SIMULINK model, which is presented in Fig. 4.

By entering experimentally obtained time value $t = 0.166$ s to SIMULINK model, we get values of φ and ω, correspondingly $w = 6.635$ rad/s. Initial velocity v_0 we obtain from angular velocity as $v_0 = w\,L = 1.426$ m/s.

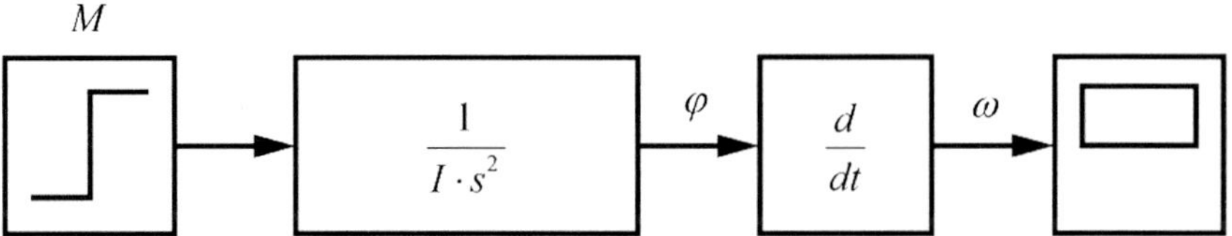

Fig. 4 Simulink model to define lever rotation angle φ and rotational speed ω

4 Definition of Coefficient of Aerodynamic Drag

There are known from theory [19] that during body free fall its velocity increases in the beginning, until reaches velocity limit and fall with constant velocity. This effect appears, when gravity forces became equal to aerodynamic drag force. In order to use this phenomenon, the simple experimental test is done (Fig. 5).

During the experimental research time t of body, falling was measured in conditions, when velocity of falling is practically constant. Time registering was performed using video camera. After experimental research, the mean falling time of body was $t = 0.304$ s. Height of falling is known and set to 1 m, we get average falling velocity $v_{drop} = 3.289$ m/s.

In free fall of the body, only two forces are applied—aerodynamic drag and gravity. As it observed during test, body falls linearly and then air drag can be defined as:

$$k_2 = \frac{mg}{v_{drop}^2} = \frac{6.79 \times 10^{-3} \times 9.81}{3.289^2} = 0.006147.$$

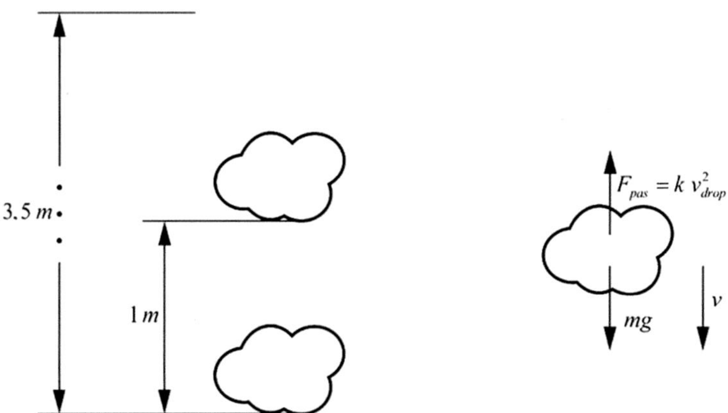

Fig. 5 Diagram of cotton flock flow during experimental research and forces, applied to falling cotton flock: F_{pas} air drag force; mg gravity force; v direction of velocity of movement

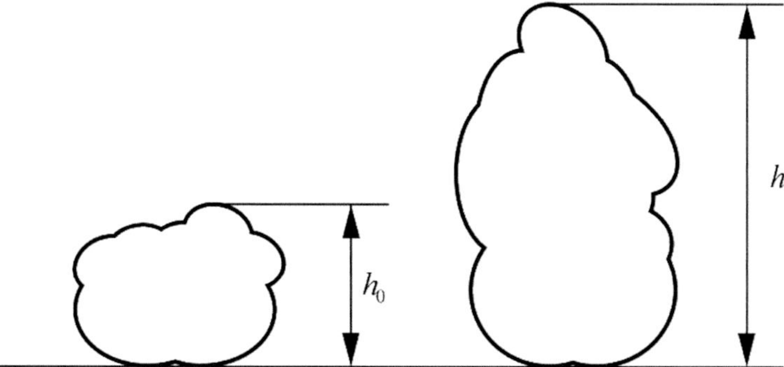

Fig. 6 Change of cotton flock height under action of inertia: h^* height under rest or free fall; h_0 height of cotton clock during acceleration

Deformation of the body Δx is defined by procedure described below. Researched body (cotton flock) is under action of moment M and is compressed from inertia, thus its dimension is diminished from h^* to h_0, as shown in the Fig. 6.

At the rest of during free fall dimension of cotton flock is restored to h^*.

Compression of the body is $\Delta x = h^* - h_0$ and is found from experimental research.

Mean value of experimentally defined height $\Delta x = 24.4$ mm.

5 Results of Experimental Research

During the experimental research was obtained height of rise of cotton flock and control piece of foam with the same mass (Fig. 7). Control piece of foam has already known coefficient of compressive stiffness, and results of indirect definition of the stiffness fits to real numbers of foam stiffness.

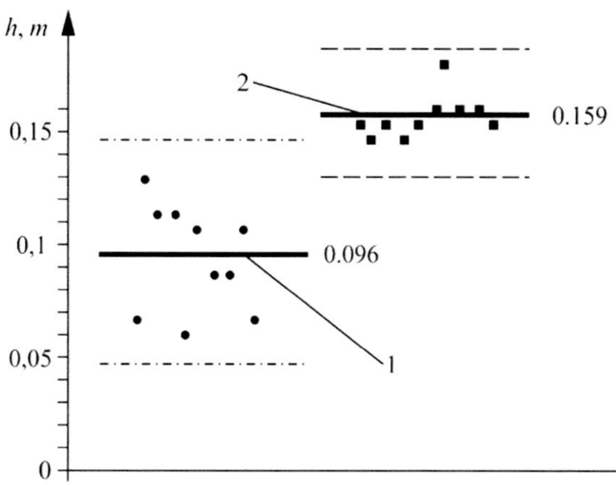

Fig. 7 Height of rise of test piece from lever impact: *1* cotton flock; *2* control piece of foam

This experiment shows correctness of assumptions on possibility to detect coefficient of stiffness of cotton. Coefficient of stiffness is given when data from experimental research is entered to Eq. 2. Then defined stiffness of cotton flock is $k = 6.54 \pm 0.36$ N/m.

In case of different size of flock, compressed area can be excluded from calculation by dividing given coefficient of stiffness by compressed area.

6 Discussion and Conclusions

Definition of cotton compression stiffness is important task in defining parameters of machinery of inertial cleaning. This kind of cleaning does not damage fiber, keeps cotton fiber structure and cost is low in comparison with mechanical processes. Therefore, defining of compression stiffness of cotton becomes important task. Tuning of machinery brings efficiency in cleaning process and diminishes energy consumption.

There were many attempts to create methodology for cotton flock stiffness using compression machines, but proposed method is cheap, fast and requires minimal efforts to build specialized machine for such purpose.

Authors described the process and methodology of definition of cotton flock compression stiffness and provided new methodology of definition of such value with simple and easy accessed instrumentation. Finally, authors bring three main contributions. Firstly, there are given methodology to define air drag force from free fall of cotton flock in equilibrium with permanent velocity. This means, that gravity force and aerodynamic drag force are equal and this allows finding desired values from experimental measurement. The second contribution is definition of lever angular speed and initial velocity of cotton flock free fly as terminal velocity of lever nest from applied mass influence. This allows creating simple system with simple, well-defined and accurate system. Third contribution is definition of height of cotton flock fly using video camera and ruler in the same view. Such method allows finding height of test piece and defining energy of deformation of cotton flock from initial impact. In order to test initial assumptions that test piece of the same mass and cross-section with different coefficient of stiffness will fly to different height is proven during experimental research. Such result illustrates success of initial assumptions.

Future research works will focus mainly on justification of such method and implementing it to definition of compressive stiffness of similar substances. This can be very soft foam, heat insulation materials, in some case composite materials, where other methods are not capable.

References

1. Basu, S.: Tensile Deformation of Fibers Used in Textile Industry. Agilent Technologies Application Note, pp. 1–7 (2012)
2. Eichhorn, S., Hearle, J., Jaffe, M., Kikutani, T.: Handbook of Textile Fibre Structure Natural, Regenerated, inorganic and Specialist Fibres. Woodhead Publishing (2009)
3. Wegerich, K.: Natural drought or human-made water scarcity in Uzbekistan. Central Asia Caucasus **2**, 154–162 (2002)
4. Agblevor, F.A., Batz, S., Trumbo, J.: Composition and ethanol production potential of cotton gin residues. Appl. Biochem. Biotechnol. **105–108**, 219–230 (2003)
5. Underhill, P.: The cotton gin, oil, robots and the store of 2020. Disp. Des. Ideas **20**(10) (2008)
6. Impact of Ginning on Fiber Quality: The Best Ginning Practices, International Cotton Advisory Committee (2001)
7. Siddaiah, M., Lieberman, M., Prasad, N., Hughs, S.E.: Automation in cotton ginning. Int. J. Intell. Syst. **19**, 111–129 (2004)
8. Kim, S., Park, S.K., Daugherty, K.E.: Some physical characteristics and heavy metal analyses of cotton gin waste for potential use as an alternative fuel. Korean J. Chem. Eng. **21**(3), 640–646 (2004)
9. Kim, S., Moon, J., Kim, G., Ha, C.: Mechanical properties of polypropylene/natural fiber composites: comparison of wood fiber and cotton fiber. Polym. Test. **27**, 801–806 (2008)
10. Greenberg, S., Sappington, T., Tamou, M., Coleman, R.: Influence of different cotton fruit sizes on boll weevil oviposition and survival to adulthood. Environ. Entomol. **33**, 443–449 (2004)
11. Lee, J.T., Kim, M.W., Song, Y.S., Kang, T.J., Youn, J.R.: Mechanical properties of denim fabric reinforced poly(lactic acid). Fibers Polym. **11**(1), 60–66 (2010)
12. Mukhammadiev, D., Rakhmatkariev, Sh, Arifdzhanov, A.: Analysis of static and dynamic characteristics of a saw cylinder of a gin. J. Mach. Manuf. Reliab. **38**(2), 120–123 (2009)
13. Erdem, K., Oğuz, D.: Theoretical investigation of separator units in saw-gin machines part ii. dynamic behaviour. Fibres Text. Eastern Europe **14**(3), 70–76 (2006)
14. Ilea, R., Tonea, C., Popa, D., Drăgoi, Gh: Mechanical models of the mixture layer from the cleaning system of harvesting combines. J. Hortic. For. Biotechnol. **15**, 138–142 (2011)
15. Bratoev, K., Mitev, G., Todorov, T.: Theoretical interpretation of the working process on machine form exact fixing of seeds into biodegradable band. Mach. Technol. Mater. **6**, 60–66 (2013)
16. Rumyantsev, S., Tarasov, D.: Numerical simulation of non-linear dynamics of vibration transport machines in case of three independently rotating vibration exciters. Recent Adv. Appl. Math. 190–194 (2010)
17. Ron, P., Haff, E., Pearson, C.: Separating in-shell pistachio nuts from kernels using impact vibration analysis. Sens. Instrum. Food Qual. **1**, 188–192 (2007)
18. Blechman, I.: Vibrational Mechanics: Nonlinear Dynamic Effects, General Approach, Applications. World Scientific Publishing Co. Pte. Ltd. (2003)
19. Bertin, J., Cummings R.: Aerodynamics for Engineers, International Edition. Pearson, vol. 6, (2013)

Method of Improving Accuracy of Measurement of the Acoustic Pulses Time-of-Flight Based on Linear Modulation of Period

Mykhaylo Dorozhovets, Olha Zahurska and Zygmunt L. Warsza

Abstract In this paper the main problems that appear in measurement of time-of-flight intervals in acoustic tomography are analyzed. They are related to the influence of noise and with a limited slew rate of the signal amplitude envelope. The method based on linear modulation of the acoustic pulses period is proposed. Correction of directly measured time interval is based on the additional measurement of duration of the given number of half periods of received signal when the signal envelope is stabilized at a sufficient level. The efficiency of proposed method of correction is investigated by simulation.

Keywords Time-of-flight measurement · Acoustic signal · Period modulation · Correction

1 Introduction

Acoustic measurements are used in various fields of study. They are used in medicine, hydro-location, flow and level measurement instrumentation, tomography systems, security systems, car parking elements etc. Use of acoustic signal allows non-contact measurement of investigated mediums. Depending on the investigated object and purpose of the measurement, passed through an object signal, reflected signal, or dispersed acoustic signal are used [1]. The often used

M. Dorozhovets
Rzeszow University of Technology, Rzeszow, Poland
e-mail: michdor@prz.edu.pl

O. Zahurska
National University Lviv Polytechnic, Lviv, Ukraine
e-mail: zahurska.olha.m@gmail.com

Z.L. Warsza (✉)
Industrial Research Institute of Automation and Measurement (PIAP), Warsaw, Poland
e-mail: zlw@op.pl

R. Szewczyk et al. (eds.), *Challenges in Automation, Robotics and Measurement Techniques*, Advances in Intelligent Systems and Computing 440, DOI 10.1007/978-3-319-29357-8_79

method of implementation of acoustic measurement is based on measuring of time-of-flight of propagation of the acoustic signal at a known path signal propagation. To investigate objects using the acoustic method are known the relationship between the velocity of sound waves and parameters of the environment. Velocity v of an acoustic signal in an ideal gas is determined by the known dependency [2]:

$$v = \sqrt{\frac{\gamma RT}{M}}, \tag{1}$$

where γ—thermal capacity at a constant pressure and fixed volume, R—gas constant, T—absolute temperature, M—gas molar mass.

In gas mixtures the speed of propagation of sound depends on the concentration of gas.

Method of time-of-flight tomography of acoustic signal in gases with a known relationship between the speed of sound and medium temperature (1), allows to determine of a temperature spatial distribution (or temperature field) inside investigated object [3–9]. In tomographic measurements the acoustic transducers (transmitters (T_{ij}) and receivers (R_{ij})) are located on the perimeter of the measured object (Fig. 1) [3–6].

Along each path $l_{i,j}$ between transducers the time-of-flight $t_{m,i,j}$ are measured and their values are described by integral:

$$t_{m,ij} = \int_{l_{ij}} \frac{dl(x,y)}{v(x,y)} = \int_{l_{ij}} w(x,y)dl(x,y), \tag{2}$$

where $v(x,y)$ is velocity of the acoustic signal propagation dependent of coordinates (x, y); $w(x,y) = 1/\ v(x, y)$ is inverse velocity.

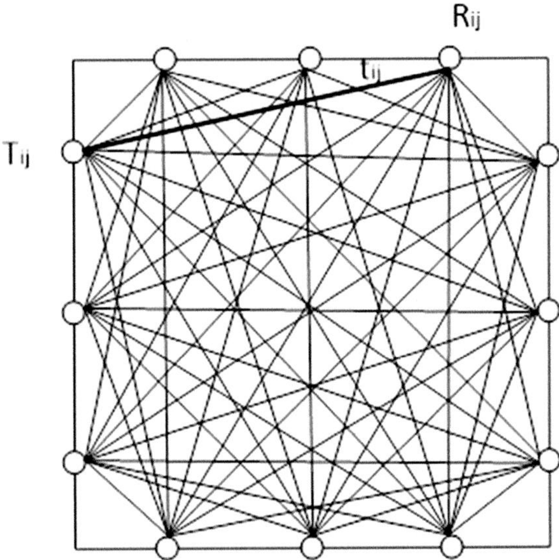

Fig. 1 Scheme of the all paths in tomography measurements using three receivers and transmitters in each side of study object

The sets of the measurement results (2) are used in tomography reconstruction technique which directly provide a spatial distribution of a inverse acoustic velocity distribution $w(x, y)$ inside the investigate object [3–9]. Next after simple functional transformation the acoustic velocity $v(x, y)$ and finally temperature distribution $T(x, y)$ can be calculated from Eq. (1).

Accuracy of the reconstructed temperature distribution depended on approximation of distribution of inverse acoustic velocity, number of measurement results (number of acoustic transducers), method of reconstruction but mainly depended from accuracy of measurement time-of-flight of acoustic pulses.

2 Measurement the Time-of-Flight of Acoustic Pulses

For generating and receiving acoustic signals transducers which convert electrical signals into acoustic and vice versa are commonly used [1]. For example electrodynamics, piezoelectric and others transducers are often used for this purpose. Measurement of acoustic signal time-of-flight intervals is realized by the typical scheme shown in Fig. 2.

The signal with specific frequency is generated by the transmitter of acoustic signal passes through the studied medium and obtained using a receiver (microphone). To measure the time-of-flight interval of acoustic signal propagation is used a counter, which counts the pulses of the reference frequency (clock). During passage through the studied medium the acoustic signal is partly dissipated, reflected from different types of obstacles and distorted from noise. Besides noise distortion the received signal envelope increased with limited speed depending on the properties of tested medium and measured time-of-flight [10–12].

A typical front of received acoustic pulse in gas medium (in the time window from 14 to 16 ms with carrier frequency of 10 kHz) is shown in Fig. 3a [12].

When pulse probing in a liquid medium received signal (center frequency of piezo-transducer ~5 MHz) in the time window from 16 to 21 μs (interval informative about 19 μs) is shown in Fig. 3b [12]. These effects causes difficulties in correct detection and measurement of received signals.

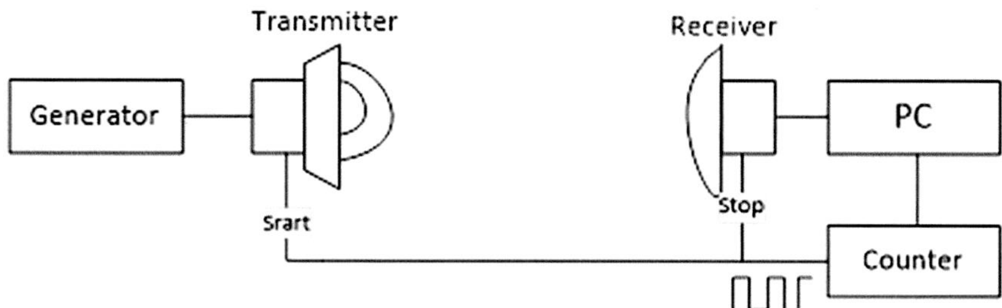

Fig. 2 Typical block diagram of time-of-flight intervals measurement

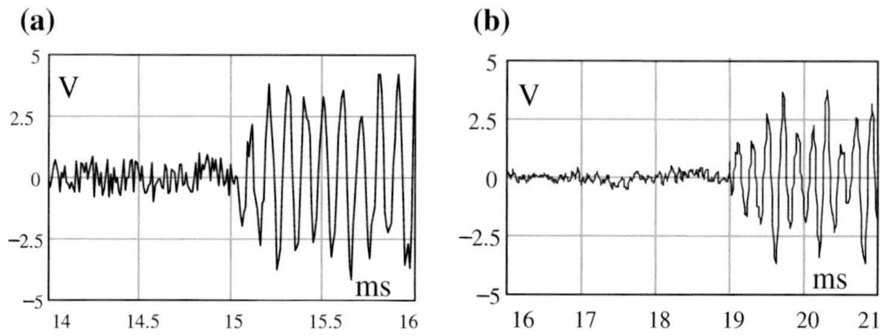

Fig. 3 The shape of received pulses: **a** in gas (time in ms), **b** in water (time is in μs)

In temperature distribution measurements the initial phase of the received signal depends not only of the distance between transmitter and receiver but also depended on acoustic speed. This in turn is dependent on the temperature distribution in the test object. On the moment of detection of the signal, which depended on the noise level and signal envelope achieving a certain level, it can be measured with error from a half a period and even up to several half-periods [12]. It is clear that required precision of measured time interval depended on precision of reconstruction of temperature distribution. For example the standard deviation of reconstructed the average temperature of 1 % required the accuracy of measured time-of-flight intervals is about 0.1–0.2 %. However, for the reconstruction local space temperature, accuracy of the measurement of time-of-flight intervals should be much higher, more than 10 times comparatively with accuracy of the measurement of time-of-flight intervals for the purpose reconstruction of the average temperature. This places stringent requirements on the accuracy of methods and means of measuring time intervals.

To ensure high measurement accuracy using the acoustic signal it is important to solve the problem of correct detecting the first period of the received acoustic pulses.

3 Improvement the Accuracy of Measurement of Time-of-Flight of Acoustic Pulses

To improve the accuracy of received signal detection, various methods can be used, in particular the chirp method [13], cross-correlation method and frequency modulation. One of the most promising methods based on the modulation offrequency of transmitted signal [12]. The neighboring periods of frequency modulated signal have varying duration. That makes it possible to the precisely determine time of arrival of the pulse, by determining the length of a certain number of periods in the least deformed part of the received pulses [12]. In the measurement method [12] received signal is compared with unipolar threshold and therefore it is possible to additionally measure certain number periods which requires that the duration of received pulse was big enough.

To improve this method we propose measure duration a certain number of half periods instead certain periods using comparison of received signal with bipolar thresholds $\pm U_{\mathrm{p}}$. The main idea of proposed measurement method is shown in Fig. 4.

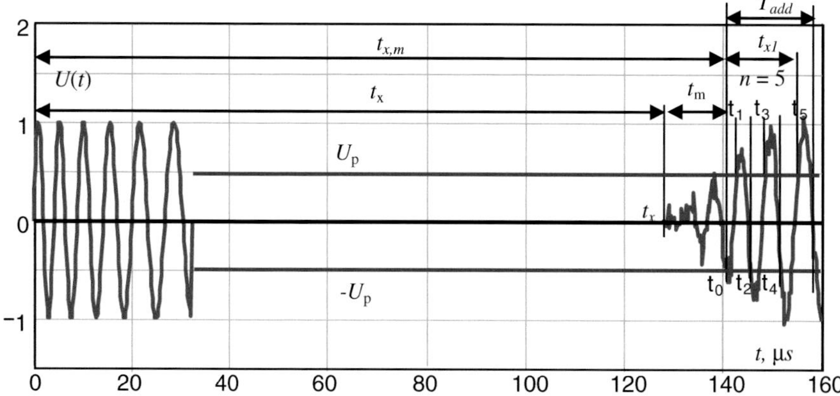

Fig. 4 The probing and receiving hyperbolically frequency modulated impulses and the time-of-flight intervals at its receiving: $\pm U_p$—noise cutoff bipolar thresholds; t_0—time of the first crossing of threshold of the received signal ($+U_p$ or $-U_p$), t_1, t_2, t_3, … times of crossing of threshold of the received signal; $t_{x,m}$—basic values of measured time-of-flight interval from starting the probing impulse to the first crossing of the received signal of threshold level (at time t_0); n—estimated number of crossings (except the first) of the received signal of threshold levels in a given time window T_{dod} ($n = 5$); t_{x1}—values of additional measured time-of-flight interval during detected n half periods of signal that fit a given time window T_{ad}

Using measured additional interval t_{x1} and number n of half-periods it is possible determine delay interval t_m and next determine the true interval t_x:

$$t_x = t_{x,m} - t_m. \tag{3}$$

Dependence between measured additional interval t_{x1}, number n of half-periods and parameters of frequency modulation can be determined by the next algorithm.

Period $T(t)$ of the line-period modulated signal with initial period T_0 and parameter of modulation ε linearly changes in time:

$$T(t) = T_0\left(1 + \varepsilon\frac{t}{T_0}\right). \tag{4}$$

Therefore the frequency of such signal changes hyperbolically over the time:

$$f(t) = \frac{1}{T(t)} = \frac{f_0}{1 + \varepsilon \cdot f_0 t}. \tag{5}$$

Received modulated signal with amplitude envelope $U_m(t)$ (for zero initial phase) approximately can be described by the following expression:

$$U_{in}(t) = U_m(t) \cdot \sin\left(\frac{2\pi f_0 \cdot t}{1+\varepsilon \cdot f_0 \cdot t}\right) + u_n(t), \tag{6}$$

where $u_n(t)$ is additive noise.

For the estimation of temporal delay t_m of pulse arriving the first of all is needed to estimate the moments t_k when the signal crossing of the bipolar threshold levels $\pm U_{\mathrm{p}}$. The moments t_k can be calculated from solving nonlinear equation:

$$U_m(t)\sin(\omega_0 t_k/(1+\varepsilon t_k/T_0)) + u_n(t) = \pm U_p \; (k=0,\ 1,\ 2,\ \ldots). \tag{7}$$

If the signal-to-noise ratio $SNR = U_m/\sqrt{2}\sigma_n$ (where U_m is a signal amplitude, σ_n—is a noise standard deviation) is more than 20 ÷ 30 (26 ÷ 30 dB) then the value of threshold can be set on level U_p < (0.2 – 0.5), · therefore in the first approximation moments t_k can by calculated from equation:

$$\sin(2\pi f_0 t_k/(1+\varepsilon t_k/T_0)) \approx 0, \quad 2\pi f_0 t_k/(1+\varepsilon t_k/T_0) \approx \pi k. \tag{8}$$

Therefore approximately:

$$t_k \approx T_0 k/(2-\varepsilon k). \tag{9}$$

Using dependence (9) it is possible calculate interval t_m: $k = m$:

$$t_m = T_0 m/(2-\varepsilon \cdot m); \tag{10}$$

and interval t_{m+n} $k = m + n$:

$$t_{m+n} = t_m + t_{x1} = T_0(m+n)/(2-\varepsilon \cdot (m+n)). \tag{11}$$

From (10) and (11) duration of additional measured time interval is:

$$t_{x1} = t_{m+n} - t_m = 2T_0 n/(2-\varepsilon \cdot (m+n))(2-\varepsilon \cdot m). \tag{12}$$

Because initial period T_0 and parameter of modulation ε are known and number n of additionally measured half periods is determined during measurement process therefore from the Eq. (12) can be determined the number m:

$$m = round\left[\frac{2}{\varepsilon} - \frac{n}{2}\left(1+\sqrt{1+\frac{8T_0}{t_{x1}\varepsilon^2 n}}\right)\right] \tag{13}$$

According to the m we can estimate the time t_m (10) and using the measured time interval $t_{x,m}$ it is possible to determinate searched time interval t_x (3):

$$t_{x,cor} = t_{x,m} - t_m = t_{x,m} - T_0 m/(2 - \varepsilon \cdot m). \tag{14}$$

4 Analysis of the Method Precision

The noise $u_n(t)$ may change the value of number m in (13). Let caused by noises standard uncertainty of measured additional interval t_{x1} is $u(t_{x1})$. The coefficient of sensitivity of changes of interval t_{x1} is equal to:

$$S_t = \frac{\partial m}{\partial t_{x1}} = \frac{2T_0}{\varepsilon \cdot t_{x1}\sqrt{(\varepsilon \cdot t_{x1})^2 + \frac{8t_{x1}T_0}{n}}} \approx \frac{1}{\varepsilon \cdot t_{x1}}\sqrt{\frac{nT_0}{2t_{x1}}}. \tag{15}$$

Because $t_{x1} \approx n \cdot T_0/2$, then with (15) the standard uncertainty $u(m)$ is:

$$u(m) = |S_t| \cdot u(t_{x1}) \approx \frac{u_{rel}(t_{x1})}{\varepsilon}, \tag{16}$$

where $u_{rel}(t_{x1}) = u(t_{x1})/t_{x1}$ is a relative standard uncertainty of the measurement result of additional interval t_{x1}.

Due to rounding the error not appeared when in (15) rounded argument changes is less then ±0.5. Therefore the stabled result in determination number m and next stable calculated interval t_m can be obtained if expanded uncertainty $U(m) = k \cdot u(m) < 0.5$, where k is coverage factor, typically $k = 2 \div 3$. Therefore from (16) the value of parameter of modulation should be corresponded to condition:

$$\varepsilon > \approx (4 \div 6) \cdot u_{rel}(t_{x1}). \tag{17}$$

For example, if relative standard uncertainty of the measured time interval t_{x1} is 0.5 % thus from (17) parameter of modulation should be more than 0.02÷0.03.

5 Simulated Results

Effectiveness of the proposed method has been carried out by simulation in Mathcad-11. Initial signal frequency is f_0 = 12.5 kHz; initial period is T_0 = 80 μs; signal amplitude U_m = 1 V, standard deviation of random Gaussian noise σ_n = 70 mV (SNR = 20 dB); modulation parameter ε = 0.05; bipolar thresholds ±0.5 V. The measurement of time intervals are obtained for the clock frequency f_{clc} = 1 MHz (period T_{clc} = 1 μs).

The part of the received signal which is used for measurement of time interval and time moments of crossing the threshold levels by received signal is presented in Fig. 5.

Calculations:

(1) base measured time interval: $t_{x,m} = Nt_{x,m}T_w = 1426.1\ \mu s = 1.4260$ ms;
(2) additional measure time interval: $t_{x1} = Nt_{x1}T_w = 141.1\ \mu s = 0.141$ ms
(3) number of half period from start of measurement:

$$m = round\left[\frac{2}{0.05} - \frac{4}{2}\left(1 + \sqrt{1 + \frac{8 \cdot 0.080\,\text{ms}}{0.244\,\text{ms} \cdot 0.05^2 \cdot 4}}\right)\right] = round[5.547] = 6;$$

(4) the value of time interval correction (12):

$$t_m = 0.080\,\text{ms} \cdot 6/(2 - 0.05 \cdot 6) = 0.1412\,\text{ms};$$

(5) the value of corrected time interval (16):

$$t_{x,cor} = t_{x,m} - t_m = 1.4260 - 0.1412\,\text{ms} \approx 1.285\,\text{ms};$$

(6) true value of measured time interval: $t_{x,true} = 1277.1\ \mu s = 1.277$ ms;
(7) error after correction of measured time interval:

$$\Delta t_{x,cor} = t_{x,cor} - t_{x,true} = 1.285 - 1.277\,\text{ms} = 0.008\,\text{ms}$$

(8) relative error after correction of measured time interval:

$$\delta_{t,cor}\frac{\Delta t_{x,cor}}{t_x}100\,\% = \frac{0.008\,\text{ms}}{1.277\,\text{ms}}100\,\% = 0.63\,\%;$$

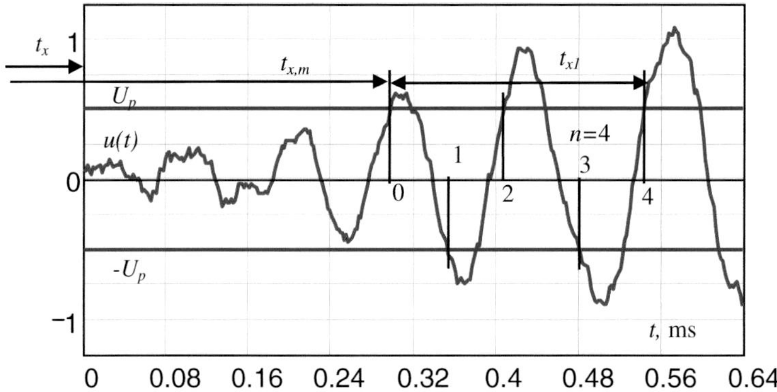

Fig. 5 Time moments of crossing the threshold levels by received signal (simulation)

(9) error without correction of measured time interval:

$$\Delta t_{x,\,meas} = t_{x,\,m} - t_{x,\,true} = 1.426 - 1.277\,\text{ms} = 0.149\,\text{ms};$$

(10) relative error without correction of the measured time interval:

$$\delta_{t\,x,\,meas} = \frac{\Delta t_{x,\,meas}}{t_x} 100\,\% = \frac{0.149\,\text{ms}}{1.277\,\text{ms}} 100\,\% \approx 11.7\,\%.$$

Efficiency of correction of measured time interval:

$$E_{cor} = \left|\frac{\Delta t_{x,\,meas}}{\Delta t_{x,\,cor}}\right| = \left|\frac{0.149\,\text{ms}}{0.008\,\text{ms}}\right| \approx 18.6\,\text{times}.$$

6 Conclusions

In the acoustic tomography of temperature distribution the requirement for precision of measurement of time intervals are more than ten times higher than required precision of reconstructed temperature.

The main problem of time-of-flight measurements of acoustic pulses is detecting of the first half period of received signal. The error of measured interval may be in range of half a period and even up to several of half periods dependently of noise intensity and slew rate of amplitude envelope of received signal.

An effective method to improve the accuracy of measurement of time-of-flight intervals is based on a linear modulation of period acoustic signal. Correction of directly measured time interval is based on measurement of additional interval of several half period of modulated signal in the least deformed part of the received pulses.

Results of theoretical research and simulation confirmed the effectiveness of the proposed method for increasing the accuracy of the measurement of time intervals.

The future step will be experimental tests connected with influences of the impact of noise, slew rate of amplitude envelope, modulation parameter and other parameters on the accuracy of the measurement time intervals.

References

1. Dobrucki, A.: Przetworniki elektroakustyczne. WNT, Warszawa (2007)
2. Mayer, A.V.: On acoustic pyrometer. Philos. Mag. **45**, 18–22 (1873)
3. Green, S.F., Woodham, A.V.: Rapid Furnace Temperature Distribution Measurement by Sonic Pyrometer. Central Electricity Generation Board. Matchwood Engineering Laboratories. Matchwood, Southampton (1983)

4. Dorozhovets, M., Baran, A., Kusij, A., Stadnyk, B.: Mathematical Aspects of Measurement of Temperature Fields in Gases and Flames. VDI/VDE-Gesselschaft Mess-und Automatisierungstechnik, "Temperatur '92", Dusseldorf, pp. 249–252 (1992)
5. Dorozhovets, M., Kouzii, A., Stadnyk, B.: The Measurement of Temperature Field by Ultrasonic Computer Tomography Methods. In: IMEKO TC-4, International Symposium on Intelligent Instrumentation for Remote and on-Site Measurements, Brussels, pp. 159–163 (1993)
6. Schwarz, A.: Acoustic measurement of temperature and velocity fields in furnaces. Tomographic techniques. In: Proceeding ECAPT-92 3, Manchester, UK (1992)
7. Codel International Ltd.: Stack Emission Monitoring. The Total Concept. Combustion Development Ltd. Bakewell, Derbyshire, UK (1994)
8. Dorozhovets, M., Basarab, H.I.: Measurement of the Temperature Distribution in Fluid Using Ultrasonic Tomography. Institute of Electrical and Electronic Engineers. IEEE Ultrasonics Symposium Proceedings vol. 3, pp. 1891–1894 (1994)
9. Barth, M.: Akustische Tomographie zur zeitgleichen Erfassung von Temperatur- und Strömungsfeldern, Leipzig (2009)
10. Honarvar, F., Sheikhzadeh, H., Moles, M., Sinclair, A.N.: Improving the time-resolution and signal-to-noise ratio of ultrasonic NDE signals. Ultrasonics **41**, 755–763 (2004)
11. Svilainis, L., Dumbrava, V., Chaziachmetovas, A., Jurkstas, P., Aleksandrovas, A.: Technique for the Performance Evaluation of the Ultrasonic Preamplifier Input Protection Circuits. In: Proceedings of the 2015 IEEE 8th International Conference on Intelligent Data Acquisition and Advanced Computing Systems: Technology and Applications (IDAACS) vol. 1, pp. 102–107 (2015)
12. Dorozhovets, M.: Application of hyperbolic-frequency modulated pulses to improve the accuracy of measurement of time-of-flight intervals in the acoustic tomography system. Automation, Measurement and Control, Lviv Polytechnic National University, N 356, Lviv, pp. 15–21 (1998) (in Ukrainian)
13. Heyser, R.C.: Acoustical measurement by time delay spectrometry. J. Audio Eng. Soc. **15**, (1967)

Semi-parametric Estimation of the Change-Point of Parameters of Non-gaussian Sequences by Polynomial Maximization Method

Serhii W. Zabolotnii and Zygmunt L. Warsza

Abstract This paper deals with application of the maximization method in the synthesis of polynomial adaptive algorithms for a posteriori estimation of the change-point of the mean value or variance of random non-Gaussian sequences. Statistical simulation shows a significant increase in the accuracy of polynomial estimates, which is achieved by taking into account the non-Gaussian character of statistical data.

Keywords Change-point estimation • Non-Gaussian sequence • Stochastic polynomial • Mean value • Variance • Cumulant coefficients

1 Introduction

One of the important tasks of technical diagnostics is the measurement of the position (in time or space) of point at which the statistical properties of the observed process are subject to a change. This point is called "the change-point". Statistical methods of detecting the change-point can be used either in real time or a posteriori, in the latter case called also the retrospective methods. A posteriori statistical estimation is based on the analysis of a fixed-volume sample of all information received from the diagnosed object. This approach requires a longer period of reaction, but it provides a more reliable and accurate estimation of the change-point [1].

A posteriori estimation of the time of changes of parameters of stochastic processes is needed in many practical applications, such as the diagnostics of some industrial processes, detection of climate changes [2], analysis of the genetic time

S.W. Zabolotnii (✉)
Cherkasy State Technological University, Cherkasy, Ukraine
e-mail: zabolotni@ukr.net

Z.L. Warsza
Industrial Research Institute for Automation and Measurements PIAP, Warsaw, Poland
e-mail: zlw1936@gmail.com

R. Szewczyk et al. (eds.), *Challenges in Automation, Robotics and Measurement Techniques*, Advances in Intelligent Systems and Computing 440, DOI 10.1007/978-3-319-29357-8_80

series [3], identification of intrusion in computer networks [4], segmentation of speech signals and messages of social networks [5]. For such a wide range of tasks the development of a large variety of mathematical models and statistical processing tools is required. It should be noted that most of the theoretical studies connected with the estimation of the change point are focused on the class of random processes described by the Gaussian probability distribution. However, the real statistical data are often different from the Gaussian model.

The classical methods based on the probability density are called the parametric methods. The main problems in parametric approaches (Bayesian and maximum-likelihood) are connected with the requirement of an a priori information about the form of distribution, as well as with the potentially high complexity of their implementation and analysis of the properties. Thus, a significant amount of the contemporary research is concerned with the construction of applied statistical methods which would allow to remove or minimize the required amount of the a priori information. Such methods are based on robust statistical processing procedures that are insensitive to "non-exactness" of probabilistic models, or on nonparametric criteria independent of specific types of distributions. The price for "omission" of probabilistic properties in the handled statistical data is the deterioration of quality characteristics in comparison with the optimal parametric methods [6].

The use of higher-order statistics (described by moments or cumulants is one of the alternative approaches in solving problems related to processing of non-Gaussian signals and data. This mathematical tool can find applications in various areas, where the estimation of the change-points is important, e.g., in: defining the moment of arrival of the acoustic emission signals [7], detection of video signal streams [8], intrusion detection in telecommunication networks [9]. It can also be applied for testing the performance of random components of a continuous industrial process.

In this paper the application of a new unconventional statistical method in solving problems of a posteriori type estimation of change points is considered. This method is called the polynomial maximization method (under acronym: PMM) and it was proposed by Kunchenko [10]. He applied the stochastic polynomials as the mathematical tool. The method, used in conjunction with the description of probability distribution by moments and cumulants, allows to simplify substantially the process of synthesis of adaptive statistical algorithms.

This method can be called adaptive as it allows to estimate certain probabilistic properties of non-Gaussian data in the form of values of the cumulant coefficients of order 3 and higher. If such data is simulated by a Gaussian distribution it is impossible to find these coefficients since they are obtained equal to zero. The account of this kind of probabilistic properties together with an additional non-linear data transformation allows to improve the accuracy, i.e. variances of estimates are smaller. The probability of erroneous decisions is also reduced.

The moment-cumulant description of the probability distribution is only the approximated statistical method. But applications based on that description allow to obtain asymptotically optimal results (i.e., its uncertainty increases with the order of the used statistics). Thus, the proposed methods can be classified as semi-parametric.

The aims of the first chapters are the following:

- application of the method of polynomial optimization to synthesis of algorithms of a posteriori estimation of the change-point of the mean-value or variance of the non-Gaussian random sequences,
- investigation of effectiveness of those algorithms using statistical modeling.

2 Mathematical Formulation of the Problem

Suppose there is a random sample $\bar{x} = \{x_1, x_2, \ldots, x_n\}$. Elements of this sample can be interpreted as a set of n independent random variables. The probabilistic nature of this sample can be described by the mean value θ, variance σ^2 and cumulant coefficients γ_l up to a given order $l = \overrightarrow{3, 2s}$. Up to some (a priori unknown) point of the discrete time τ, the mean value of the sample is equal to θ_0, and the variance is σ_0^2. Then, at the time $\tau + 1$ the value θ_0 jumps to θ_1 or variance jumps to σ_1^2. Our purpose is to estimate the value τ of the change-point of one of the above sequence parameters on the basis of the analysis of the data sample $\bar{x}$.

Various variants of the change point estimation may differ, depending on the availability of a priori information about values of a variable parameter (before and/or after the change), as well as on the knowledge of the probabilistic character of the other parameters of the random sequence model.

3 A Posteriori Estimation of the Change-Point of Mean Value by the Maximum Likelihood Method

One of the basic directions in investigations of a posteriori problems of the change point study is based on the idea of the maximization of the likelihood. It was elaborated in details by Hinckley [11]. He proposed a general asymptotic approach to obtain distributions of a priori change-point estimates by method of maximum likelihood (MML). Application of this approach requires a priori information about the distribution law of statistical data, before and after the change.

For a Gaussian distribution it is known that estimation of the mean value by MML method is the same as a linear estimation by method of moments (MM), i.e.

$$\theta = \frac{1}{n} \sum_{v=1}^{n} x_v \tag{1}$$

The estimate of the form (1) is consistent and not shifted. This MM estimator can be used for estimation of the mean value of random variables of any arbitrary distribution. However, this assessment is effective only in the few special cases, such as the Gaussian model. For this probabilistic model, the logarithm of the maximum likelihood function (MML) with known variance σ^2 is transformed [1] into statistics of the form:

$$T_r(\theta_0, \theta_1) = r \sum_{v=1}^{r} (x_v - \theta_0)^2 + (n - r) \sum_{v=r+1}^{n} (x_v - \theta_1)^2 \tag{2}$$

$T_r(\theta_0, \theta_1)$ has a maximum in a neighborhood of the true value of the change-point τ. Thus, the desired change-point estimate can be find by the following algorithm:

$$\hat{\tau} = \arg \max_{1 \le r \le n-1} T_r(\theta_0, \theta_1) \tag{2a}$$

Hinckley [11] considered also the case when parameters θ_0 and θ_1 of the Gaussian distribution are unknown. In this case, the MML estimate for the change-point of the mean value has the form:

$$\hat{\tau} = \arg \max_{1 \le r \le n-1} \left[r(\theta_{0,r} - \theta)^2 + (n - r)(\theta_{1,r} - \theta)^2 \right] \tag{3}$$

where

$$\theta_{0,r} = \frac{1}{r} \sum_{v=1}^{r} x_v, \; \theta_{1,r} = \frac{1}{n-r} \sum_{v=r+1}^{n} x_v. \tag{4}$$

Since statistics (2) and (3) do not depend on any other probabilistic parameters they can be used for nonparametric estimation of the change-point of the mean value of random sequences with an arbitrary distribution. However, in such situations, similar to the case where the mean is evaluated according to (1), the nonparametric algorithms lose their optimality. To overcome this difficulty, the nonlinear estimation algorithms based on the polynomial maximization method are described below. They allow to take into account, in a simple way, the degree of non-Gaussian character of the statistical data.

4 A Posteriori Estimation of the Change-Point of Variance by the Maximum Likelihood Method

The Gaussian distribution of variance estimate by PMM for a known value of the expectation θ is

$$\hat{\sigma}^2 = \frac{1}{n}\sum_{\nu=1}^{n}(x_\nu - \theta)^2. \tag{5}$$

Estimate $\hat{\sigma}^2$ of the form (5) is consistent and asymptotically not biased. As estimation of the method of moments (MM), expression (5) can be used to estimate the variance of the random variable with arbitrary distribution. For this probabilistic model the logarithm of the maximum likelihood function can be written as:

$$\ln L(\sigma_0^2, \sigma_1^2) = -\frac{n}{2}\ln(2\pi) - \frac{r}{2}\ln(\sigma_0^2) - \frac{n-r}{2}\ln(\sigma_1^2) - \frac{\sum_{\nu=1}^{r}(x_\nu - \theta)^2}{2\sigma_0^2} - \frac{\sum_{\nu=r+1}^{n}(x_\nu - \theta)^2}{2\sigma_1^2}.$$

For the situation, where the value θ is a priori known and invariable, the sample $\bar{x}$ can be centered by assigning it the mean value of zero. Thus it is possible to form the following statistics

$$T_r(\sigma_0^2, \sigma_1^2) = -n\ln(2\pi) - r\ln(\sigma_0^2) - (n-r)\ln(\sigma_1^2) - \frac{\sum_{\nu=1}^{r}x_\nu^2}{\sigma_0^2} - \frac{\sum_{\nu=r+1}^{n}x_\nu^2}{\sigma_1^2} \tag{6}$$

It has a maximum in the vicinity of the true value of the change point τ.

$$\hat{\sigma}_{0,r}^2 = \frac{1}{r}\sum_{\nu=1}^{r}x_\nu^2. \tag{7}$$

The resulting algorithm for estimating the change-point will be as follows [1]

$$\hat{\tau} = \arg\max_{1\le r\le n-1}\left[n\ln(\hat{\sigma}^2) - r\ln(\hat{\sigma}_{0,r}^2) - (n-r)\ln(\hat{\sigma}_{1,r}^2)\right] \tag{8}$$

where: the estimated value $\hat{\sigma}^2$ is calculated for the whole sample by the formula (5).

As the statistics (6) and (8) do not depend on any other parameters, these results can be used for the nonparametric estimation of change-point of the dispersion of random sequences from the arbitrary distribution. However, in situations similar to those described earlier for the case of mean value estimation, nonparametric algorithms lose their optimality, i.e. may provide precision that will be significantly less than for effective estimates. Because of that, the new polynomial algorithms for the retrospective evaluation of the change point of the dispersion of non-Gaussian sequences are described below.

5 General Algorithm of Polynomial Change-Point Estimation

Let $\vec{x}$ be evenly distributed sampled elements. Consider the algorithm presented in [10] and denoted by PMM. In that paper it is shown that the estimate of an arbitrary parameter ϑ can be found by solving such stochastic equations with respect to ϑ:

$$\sum_{i=1}^{s} h_i(\vartheta)\left[\frac{1}{n}\sum_{v=1}^{n} x_v^i - \alpha_i(\vartheta)\right]\Bigg|_{\vartheta=\vartheta} = 0,$$

where: s—is the order of the polynomial used for parameter estimation, $\alpha_i(\vartheta)$—are the theoretical initial moments of the ith order.

Coefficients $h_i(\vartheta)$ (for $i=\overrightarrow{1,s}$) can be found by solving the system of linear algebraic equations, given by conditions of minimization of variance (with the appropriate order s) of the estimate of the parameter ϑ, namely:

$$\sum_{i=1}^{s} h_i(\vartheta)F_{i,j}(\vartheta) = \frac{d}{d\vartheta}\alpha_j(\vartheta),\ j=\overline{1,s} \tag{9}$$

where $F_{i,j}(\vartheta)=\alpha_{i+j}(\vartheta)-\alpha_i(\vartheta)\alpha_j(\vartheta)$—is the centered correlants of i,j dimensions.

Equation (9) can be solved analytically using the Kramer method, i.e.,

$$h_i(\vartheta) = \frac{\Delta_{is}}{\Delta_s},\ i=\overline{1,s},$$

where $\Delta_s = \det\left\|F_{i,j}\right\|$; $(i,j=\overline{1,\ s})$—volume of the stochastic polynomial of dimension s, Δ_{is}—is the determinant obtained from Δ_s by replacing the ith column by the column of free terms of Eq. (9).

A new approach for finding the posteriori estimates of change-point, proposed in this paper, is based on application of the PMM method. In this approach a property of the following stochastic polynomials [10] is used:

$$l_{sn}(\bar{x}/\vartheta) = nk_0(\vartheta) + \sum_{i=1}^{s} k_i(\vartheta)\sum_{v=1}^{n} x_v^i,\ i=\overrightarrow{1,s} \tag{10}$$

where

$$k_0(\vartheta) = \int_a^{\vartheta} \sum_{i=1}^{s} [h_i(\vartheta)\,\alpha_i(\vartheta)]\,d\vartheta,\ k_i(\vartheta) = \int_a^{\vartheta} h_i(\vartheta)\,d\vartheta. \tag{11}$$

The mathematical expectation $E\{l_{sn}(\bar{x}/\vartheta)\}$, treated as a function of ϑ assumes the maximum at the true value of the change point of ϑ.

Values of the parameter ϑ belong to some interval (a, b). If the stochastic polynomial of the form (10) will be maximized with use of a parameter ϑ which has a change-point (step change from value ϑ_0 to value ϑ_1), then we can build a polynomial form statistics:

$$P_r^{(s)}(\vartheta_0, \vartheta_1) = rk_0(\vartheta_0) + \sum_{i=1}^{s} k_i(\vartheta_0) \sum_{v=1}^{r} x_v^i + (n-r)k_0(\vartheta_1) + \sum_{i=1}^{s} k_i(\vartheta_1) \sum_{v=r+1}^{n} x_v^i \tag{12}$$

which will reach the maximum in a neighborhood of the true value of the change-point τ. Thus, the general algorithm of applying PMM method for estimation of the change point τ can be formulated as follows:

$$\hat{\tau} = \arg \max_{1 \le r \le n-1} P_r^{(s)}(\vartheta_0, \vartheta_1). \tag{13}$$

6 Polynomial Estimation of the Change-Point of Mean Value

It is known from [10] that the estimate of the mean value θ obtained by PMM method using a polynomial of degree $s = 1$ coincides with the form (1) of the linear estimate MM. Hence, the synthesis of polynomial algorithms for estimating the change-point of this parameter is justified only for degrees $s \ge 2$.

At a degree $s = 2$ polynomial estimate of the mean value can be calculated by solving the following quadratic equation [10]

$$\gamma_3\theta^2 - \left[2\gamma_3 \frac{1}{n}\sum_{v=1}^{n} x_v - \sigma(2+\gamma_4)\right]\theta - \sigma(2+\gamma_4)\frac{1}{n}\sum_{v=1}^{n} x_v + \gamma_3\left[\frac{1}{n}\sum_{v=1}^{n} (x_v)^2 - \sigma^2\right]\Bigg|_{\theta=\theta} = 0 \tag{14}$$

The analysis of Eq. (14) shows that the estimated value of $\theta_{s=2}$ depends additionally on coefficients of skewness γ_3 and kurtosis γ_4. If the values of these parameters are equal to zero, then the distribution is the normal (Gaussian) one. In this case the polynomial estimate (14) reduces to the classical estimate of the form (1). It is shown in [10] that the use of Eq. (14) ensures higher accuracy (decrease of the variance) than the estimate (1).

The asymptotic value of this estimate (for $n \to \infty$) is given by the following formula:

$$g_2(\gamma_3, \gamma_4) = 1 - \frac{\gamma_3^2}{2 + \gamma_4} \tag{15}$$

Using the analytical expressions (9) and (11) one can easily find that, for order $s = 2$, the coefficients maximizing the selected stochastic polynomial of the form (10) in a neighborhood of the true value of the parameter θ are the following:

$$k_0(\theta) = \frac{\sigma^3}{6\Delta_2}\left[2\gamma_3\theta^3 + 3(2+\gamma_4)\sigma\theta^2 - 6\gamma_3\sigma^2\theta\right] \tag{16a}$$

$$k_1(\theta) = \frac{\sigma^3}{\Delta_2}\left[\gamma_3\theta^2 + (2+\gamma_4)\sigma\theta\right],\ k_2(\theta) = -\frac{\sigma^3}{\Delta_2}\gamma_3\theta \tag{16b, c}$$

where $\Delta_2 = \sigma^6\left(2 + \gamma_4 - \gamma_3^2\right)$.

In the presence of an a priori information about the mean values of θ_0 before and θ_1 after the change-point, and on condition $\theta_1 > \theta_0$, the polynomial (12) for the order $s = 2$ can be expressed as follows

$$\begin{aligned} P_r^{(2)}(\theta_0, \theta_1) = (n-r)&\left[\frac{1}{3}\gamma_3\left(\theta_1^3 - \theta_0^3\right) + \frac{1}{2}\sigma(2+\gamma_4)\left(\theta_1^2 - \theta_0^2\right) - \sigma^2\gamma_3(\theta_1 - \theta_0)\right] \\ &+ \left[\gamma_3\left(\theta_1^2 - \theta_0^2\right) + \sigma(2+\gamma_4)(\theta_1 - \theta_0)\right]\sum_{v=r+1}^{n} x_v - \gamma_3(\theta_1 - \theta_0)\sum_{v=r+1}^{n} x_v^2. \end{aligned} \tag{17}$$

7 Adaptive Polynomial Estimation of Change-Point for the Partial a Priori Information

As noted above, different formulations of the problem of a posteriori estimation of the time of change point occurrence may vary depending on the amount of available a priori information about the meaning of variant parameters (before and after the change point) as well as about other probabilistic characteristics of the random processes. In particular, the use of moment-cumulant description and application of stochastic polynomial of degree s for constructing the polynomial statistics estimating the change-point of mean value θ, information about the true value of the dispersion σ^2 and about cumulant coefficients γ_l up to 2 s order is required. In many practical situations, such information before the study is fully or partially unavailable. In the formulation of the real problems of the retrospective evaluation of the change point time two main types of lack of a priori knowledge can be identified:

(1) there is no information about the magnitude of the informative parameter of a random process model, the value of which is controlled and changed after the change point;
(2) lack of information about other parameters of the random process model that determine its probabilistic properties.

The classical approach to overcome the a priori uncertainty of the first type is based on applying the values of informative parameters of a posteriori estimates, which are found from the data of tested sample, in the algorithm of the corresponding statistics formation. In particular, when the change-point of mean value θ is estimated, as a posteriori estimates of the current value of the parameter (before and after the change point) the value of sample mean of the form (4) can be used. For the Gaussian distribution model this leads to the formation of statistics of the form (3). It, like statistics (2), has a maximum value in the vicinity of the change point, but will be characterized by lower accuracy (the greater variance of the change-point estimate). It is caused by a greater degree of a priori lack of information.

This approach can also be applied to find estimates of the change point of non-Gaussian sequences with polynomial statistics of the general form (8). In this case the true values of variable parameter ϑ have to be substituted by the respective posterior estimates. For example in the case, where an a priori information about the mean values θ_0 and θ_1 is unknown, polynomial estimation of the change-point can be found (as in the classical case) by replacing the unknown values of these parameters by their a posterior estimates of the form (4) . These estimates are formed for each potential change-point. Thus, for $s = 2$ the adaptive algorithm for estimating the time of change-point $\hat{\tau}$, based on PMM method can be formulated as follows:

$$\hat{\tau} = \arg \max_{1 \le r \le n-1} \left\{ r \left[\begin{array}{l} \frac{4}{3}\gamma_3 \theta_{0,r}^3 + \frac{3}{2}\sigma(2+\gamma_4)\theta_{0,r}^2 + \\ - \left[\sigma^2 + \sum_{v=1}^{r} (x_v)^2 \right] \gamma_3 \theta_{0,r} \end{array} \right] + (n-r) \left[\begin{array}{l} \frac{4}{3}\gamma_3 \theta_{1,r}^3 + \frac{3}{2}\sigma(2+\gamma_4)\theta_{1,r}^2 + \\ - \left(\sigma^2 + \sum_{v=r+1}^{n} (x_v)^2 \right) \gamma_3 \theta_{1,r} \end{array} \right] \right\} \tag{18}$$

The analysis of the structure of the polynomial statistics (17) and (18) confirms again the fact that, for s = 2, the use of PMM is justified only in the case of an asymmetry ($\gamma_3 \neq 0$) of the distribution of the statistical data.

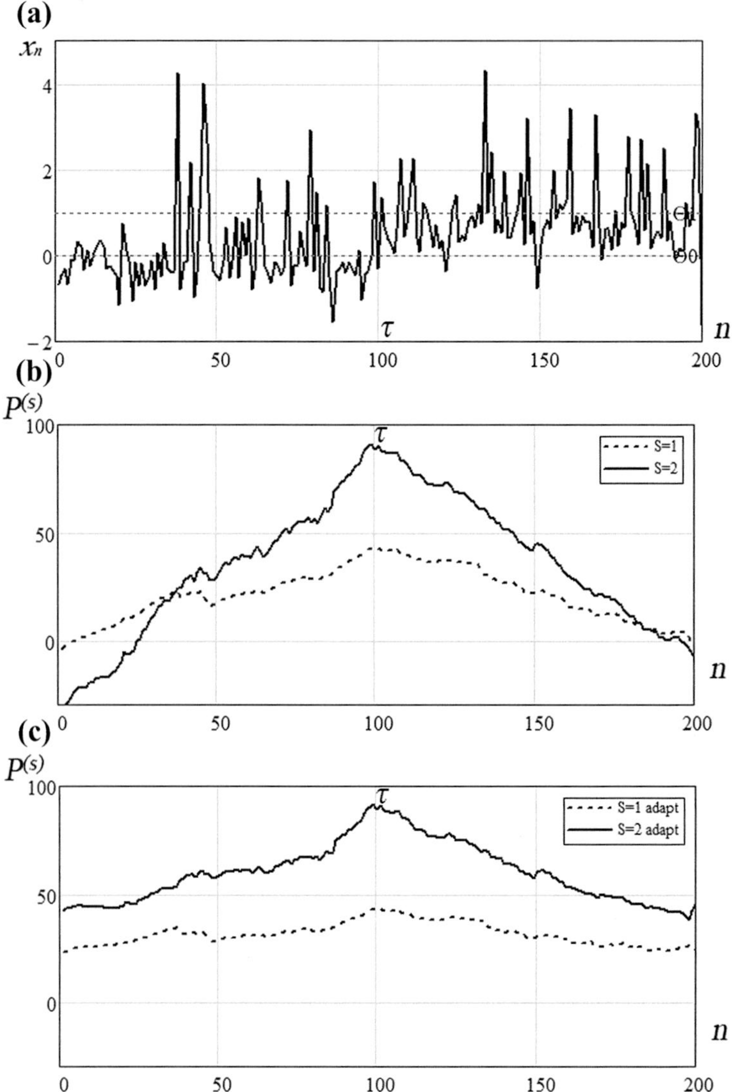

Fig. 1 Example of a posteriori estimation of the mean value of the change-point

In Fig. 1 the results are presented for a numerical example obtained by estimation procedures for the mean value of the change-point of mean values $\theta_0 = 0$ and $\theta_1 = 1$ of the non-Gaussian sequence (Fig. 1a), where $\sigma = 1$, $\gamma_3 = 2$ and $\gamma_4 = 5$.

Creation of $P^{(S)}$ statistics was performed using the classical versions (2) and (3) of the algorithm of a posteriori estimation by MML method (which coincides with PMM if $s = 1$) as well by the polynomial algorithms (12) and (13) of PMM for

$s = 2$. The graphs are plotted for the cases when a priori information about the parameter values before and after the change point are available (Fig. 1b) or are unknown (Fig. 1c).

The results presented in Fig. 1 clearly confirm the potentially higher precision obtained by polynomial statistics for $s = 2$, since the maximum of corresponding function is strongly marked, as compared with the smoothed form of the statistic for $s = 1$.

The lack of information about the values of variable parameters generally leads to a deterioration of the selective properties of the resulting statistics.

They become less useful in comparison with the relevant statistics, which are formed with the availability of a priori information. The increase in the degree of the polynomial provides growth of the accuracy of estimation. The relative accuracy of estimators depends primarily on the values of the higher-order cumulant coefficients and negligibly on the amount of a priori information. That is confirmed by the results of statistical modeling.

Overcoming the second type of a priori lack of information is especially important when parametric methods are used. In particular, it is important for the method of the maximum likelihood (MML). Application of this method requires a description of the statistical model in the form of the distribution of the probability density. If such a priori information is absent, then the complex approximation problems including the selection of the relevant law of distribution, statistical estimation of its parameters and testing the adequacy based on some criteria, have to be solved.

An alternative to this computationally expensive plan approach, is the use of non-parametric methods which usually are near to optimal in terms of accuracy. In such situation, the use of moment-cumulant description greatly simplifies the process of overcoming a priori uncertainty about the probabilistic properties of the model of the investigated processes. This is due to the fact that the problem of estimating the unknown parameters of the model (cumulants or cumulant coefficients) can be solved simply by using the method of moments. For example, estimates of the skewness and kurtosis coefficients, necessary to compile statistics (12) or (13), may be prepared using the statistics [12]:

$$\hat{\gamma}_3 = \frac{\frac{1}{n}\sum_{v=1}^{N}(x_v-\theta)^3}{\left[\frac{1}{n-1}\sum_{v=1}^{N}(x_v-\theta)^2\right]^{\frac{3}{2}}}; \quad \hat{\gamma}_4 = \frac{\frac{1}{n}\sum_{v=1}^{N}(x_v-\theta)^4}{\left[\frac{1}{n-1}\sum_{v=1}^{N}(x_v-\theta)^2\right]^{2}} - 3. \tag{19}$$

The only significant obstacle to finding estimates of the form (13) is that the original data sample, which contains the change point, is inhomogeneous. This requires its preliminary segmentation. An important feature of the problem of retrospective evaluation of the change point of the mean value is the fact that the relevant statistics (2) and (3), obtained on the basis of maximum likelihood method (assuming the adequacy of the Gaussian model), can be used as non-parametric statistics. These statistics are not optimal, but it is possible to determine the time of disorder in parameters for random sequences with the arbitrary distribution. This fact allows us to offer the following three-step procedure of adaptive polynomial retrospective estimation of the times of change point of parameters of non-Gaussian random sequences:

(1) the estimate of change-point $\hat{\tau}$ inside the sample $\vec{x}$ of n values can be approximate using an appropriate non-parametric statistics;
(2) the original sample is divided into individual segments $\bar{x}'$ and $\bar{x}''$ of volumes $\hat{\tau}$ and $n-\hat{\tau}$ values respectively. For each of them the required parameters of the model are estimated, e.g. the cumulant coefficients $\hat{\gamma}_l'$ and $\hat{\gamma}_l''$, $l=\overrightarrow{3,2s}$;
(3) based on an analysis of the original sample $\vec{x}$ the adaptive estimation of change-point $\hat{\tau}^{(S)}$ is carried out. It is based on the application of the corresponding polynomial statistics of degree s and on average of estimated values of parameters

$$\hat{\gamma}_l = \left[(\hat{\tau})\hat{\gamma}_l' + (n-\hat{\tau})\hat{\gamma}_l''\right]/n,\ l=\overrightarrow{3,2s}.$$

8 Polynomial Estimation of the Change-Point of Variance

It is known [10] that the estimate of variance σ^2 by PMM method can be found only for the degree of polynomial $s \geq 2$. When $s=2$ polynomial estimate of this parameter at $\theta=0$ can be written as:

$$\hat{\sigma}_{s=2}^2 = \frac{1}{n}\sum_{v=1}^{n} x_v^2 + \frac{1}{2}\gamma_3^2\left(\frac{1}{n}\sum_{v=1}^{n} x_v\right)^2 - \gamma_3\left(\frac{1}{n}\sum_{v=1}^{n} x_v\right)\left\{\frac{1}{4}\gamma_3^2\left(\frac{1}{n}\sum_{v=1}^{n} x_v\right)^2 + \frac{1}{n}\sum_{v=1}^{n} x_v^2\right\}^{1/2} \tag{20}$$

Analysis of Eq. (20) shows that the only additional parameter that affects the determination of $\hat{\sigma}_{s=2}^2$ is the coefficient of asymmetry γ_3. For its value equal to zero (which corresponds to the symmetry of the distribution) polynomial (20) is degenerated into the classical estimate of the form (5). It was shown [10] that in the asymptotic case (when $n \to \infty$) the use of (20) provides an increase in accuracy (decrease of variance of the estimate) comparative with estimation (5). Value of $g_2(\gamma_3, \gamma_4)$ when $n \to \infty$ is described by (15).

By using (9) and (11) it is easy to find the coefficients that maximize the sampled stochastic polynomial of the order $s=2$ of the form (10) in the vicinity of the true values of the parameter σ^2. Their values are as follows:

$$k_0(\sigma) = \frac{-\ln(\sigma^2)}{(2+\gamma_4-\gamma_3^2)}, \quad k_1(\sigma) = \frac{2\gamma_3}{\sigma(2+\gamma_4-\gamma_3^2)}, \quad k_2(\sigma) = \frac{-1}{\sigma^2(2+\gamma_4-\gamma_3^2)}. \tag{21a–c}$$

In the denominator of each component (21a–c) the expression $(2+\gamma_4-\gamma_3^2)$ is present. It does not depend on σ^2. Then in the construction of the final statistics it can be neglected. Thus, when a priori information about the dispersion σ_0^2 before and σ_1^2 after the change point is given, the polynomial statistics of degree $s=2$ can be written as:

$$P_r^{(2)}(\sigma_0^2, \sigma_1^2) = -r\ln(\sigma_0^2) + 2\gamma_3 \frac{\sum_{v=1}^{r} x_v}{\sigma_0} - \frac{\sum_{v=1}^{r} x_v^2}{\sigma_0^2} - (n-r)\ln(\sigma_1^2) + 2\gamma_3 \frac{\sum_{v=r+1}^{n} x_v}{\sigma_1} - \frac{\sum_{v=r+1}^{n} x_v^2}{\sigma_1^2}. \tag{22}$$

It can be shown that for the a priori known values of the variance of the random sequence before and after the change, the statistic of the following form is equivalent to (22) but more convenient from the point of view of computation:

$$R_r^{(2)}(\sigma_0^2, \sigma_1^2) = -(n-r)\left[\ln(\sigma_1^2) - \ln(\sigma_0^2)\right] + 2\gamma_3 \left[\frac{1}{\sigma_1} - \frac{1}{\sigma_0}\right] \sum_{v=r+1}^{n} x_v - \left[\frac{1}{\sigma_1^2} - \frac{1}{\sigma_0^2}\right] \sum_{v=r+1}^{n} x_v^2. \tag{23}$$

When a priori values of the variance of a random sequence are unknown, polynomial evaluation of the change point can be found by replacing the unknown quantities of the dispersion by their posterior estimates of the form (5). They have to be formed for each potential change-point.

If a priori unknown is only a value of variance after the change point, the algorithm for finding the estimation $\hat{\tau}$ of the change point can be represented as:

$$\hat{\tau} = \arg \max_{1 \le r \le n-1} - \left\{ (n-r)\left(\ln(\hat{\sigma}_{1,r}^2) - \ln(\sigma_0^2) + 1 - \frac{\hat{\sigma}_{1,r}^2}{\sigma_0^2} \right) - 2\gamma_3 \left(\frac{1}{\hat{\sigma}_{1,r}} - \frac{1}{\sigma_0} \right) \sum_{v=r+1}^{n} x_v \right\} \tag{24}$$

When values of the parameters before and after the change-point are a priori unknown then the algorithm of estimation takes the following form:

$$\hat{\tau} = \arg \max_{1 \le r \le n-1} \left[2\gamma_3 \left(\frac{\sum_{v=1}^{r} x_v}{\hat{\sigma}_{0,r}} + \frac{\sum_{v=r+1}^{n} x_v}{\hat{\sigma}_{1,r}} \right) - r \ln\left(\hat{\sigma}_{0,r}^2\right) - (n-r) \ln\left(\hat{\sigma}_{1,r}^2\right) - n \right] \tag{25}$$

Analysis of the structure of polynomial statistics (22)–(25) once again confirms the fact that the use of PMM method with polynomial degree $s = 2$ is appropriate only for the asymmetry ($\gamma_3 \neq 0$) of the statistical data distribution.

9 Statistical Modeling of a Posteriori Estimate of Change-Point

Based on results of above considerations, a software package in the software environment MATLAB has been developed. It allows to perform the statistical modeling of the proposed semi-parametric estimation procedures, applied to the estimation of the change-points of the mean value or variance of non-Gaussian random sequences. Both, single and multi- experiments (in the sense of the Monte Carlo method) can be simulated.

The accuracy obtained by the classical and proposed polynomial algorithms for the experimental data can be also compared.

Results of single experiments, as shown on Fig. 1, do not allow to compare adequately the accuracy of the statistical estimation algorithms. As a comparative criterion of efficiency, the ratio of variances of the estimates of the change-point is used. That can be obtained by a series of experiments with the same initial values of the model parameters.

It should be noted that results of the simulation by statistical algorithms of a posteriori estimation of the change-point can depend on a large number of various factors, including e.g. the relative value of the step jump at the change-point, the probabilistic nature (values of coefficients of higher order cumulants) of non-Gaussian random sequences, the presence of an a priori information about values of variable parameters. Furthermore, the accuracy of estimations of the change-point depends on the chosen number n of the sample elements and on the accuracy of determination of the variance of estimates, i.e., on the number m of experiments with the same initial conditions.

As the example, results of statistical modeling for $n = 200$ and $m = 2000$ are shown on Figs. 2 and 3.

Coefficient G_2 is the ratio of variances of the change-point estimates obtained by PMM method with the polynomial order $s = 2$ and for $s = 1$ statistics, respectively. The value of G_2 characterizes the relative increase of accuracy.

Figure 2a shows the dependence of G_2 on the relative values of the jump $q = (\theta_1 - \theta_0)/\sigma$ at the change-point, obtained with different coefficients of skewness

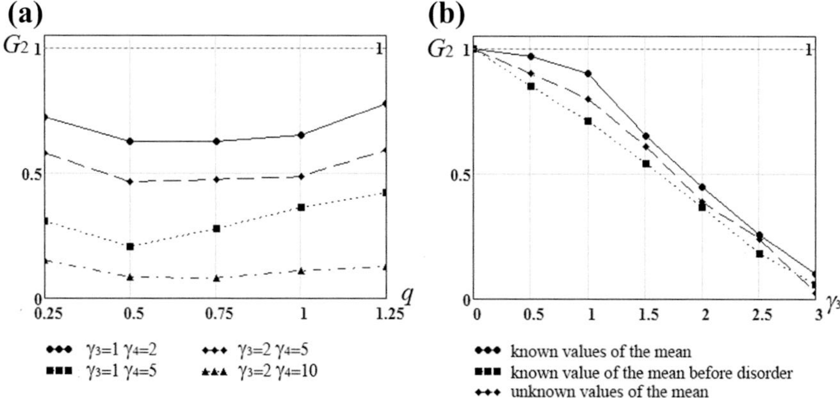

Fig. 2 Experimental values of coefficients G_2 of the reduction of estimates of variance of the mean value change-point

γ_3 and kurtosis γ_4. Figure 2b shows the dependence of G_2 on the coefficients of skewness γ_3.

Figure 3a shows the dependence G_2 on the ratio of variance magnitudes $D = \sigma_1^2/\sigma_0^2$ for various values of the coefficients of skewness γ_3 and kurtosis γ_4. In Fig. 3b G_2 is given as function of γ_3 (when $\gamma_4 = 10$ and $D = 4$) obtained under the conditions of various sizes of a priori information and values of variable parameters.

Analysis of these and many other experimental results confirm the theoretical results concerning the effectiveness of the polynomial method in the change-point estimation. It turns out that the relative growth of accuracy is roughly the same for different formulations of the problem, related to the presence or absence of an a priori information about the values of the variable parameter.

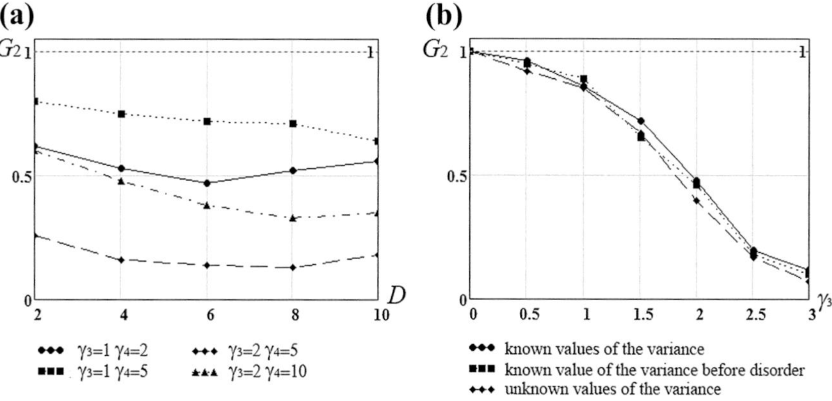

Fig. 3 Experimental values of coefficients of the variance reduction of estimates of the change-point of dispersion

The improvement does not depend significantly on the relative magnitude of the jump at the change-point. It is determined primarily by the degree of "non-Gaussian-ness" of the process, which numerically is expressed as the absolute values of the higher-order cumulant coefficients.

10 Conclusions

Presented partly earlier on conferences [13, 14] and summarized in this paper results of the research lead to the general conclusion about the potentially high efficiency of the implementation of the polynomial maximization method (PMM) to the synthesis of simple adaptive algorithms for estimation of change-points of parameters of stochastic processes of non-Gaussian character of statistical data.

Described in this paper theoretical results allowed to develop a new approach to the construction of semi-parametric algorithms for a posteriori estimation of the change-point. The approach is based on application of the stochastic polynomials proposed by Kunchenko [10].

Among possible directions of further research one should mention the following:

- increase of the degree of the stochastic polynomial, which is necessary to get more effective solutions, especially for non-Gaussian sequences with symmetrical distributions;
- analysis of the dependence of the accuracy of determination of the parameters of non-Gaussian model (higher-order statistics) on the stability of polynomial algorithms for a posteriori estimation of the change-points;
- synthesis of polynomial algorithms for estimating the change-point with respect to other parameters (e.g., dispersion or correlation and regression coefficients), or in the case where the values of several parameters are changed simultaneously (e.g., the mean value and the variance, etc.).

Both described methods of a posteriori detection of the change point of stochastic process parameter and results of theoretical studies on real-time methods [15, 16] can be applied in many fields. As examples are: the software for analysis of data recorded by the virtual measurement system for on line monitoring of random component of continuous processes or software for the estimation type A uncertainty of automatic measurements of sampled signals [17].

References

1. Chen, J., Gupta, A.K.: Parametric Statistical Change Point Analysis, p. 273. Birkhaeuser (2012)
2. Reeves, J., Chen, J., Wang, X.L., Lund, R., Lu, Q.: A review and comparison of change point detection techniques for climate data. J. Appl. Meteorol. Climatol. **46**(6), 900–915 (2007)

3. Wang, Y., Wu, C., Ji, Z., Wang, B., Liang, Y.: Non-parametric change-point method for differential gene expression detection. PLoS One **6**(5), e20060 (2011)
4. Yamanishi, K., Takeuchi, J., Williams, G., Milne, P.: On-line unsupervised outlier detection using finite mixtures with discounting learning algorithms. In: Proceedings of the Sixth ACM SIGKDD International Conference on Knowledge Discovery and Data Mining, pp. 320–324 (2000)
5. Liu, S., Yamada, M., Collier, N., Sugiyama, M.: Change-point detection in time-series data by relative density-ratio estimation. Neural Netw. **43**, 72–83 (2013)
6. Brodsky, B., Darkhovsky, B.: Nonparametric Methods in Change-Point Problems. Kluwer Academic Publishers, Dordrecht (1993)
7. Lokajicek, T., Klima, K.: A first arrival identification system of acoustic emission (AE) signals by means of a higher-order statistics approach. Meas. Sci. Technol. **17**, 2461–2466 (2006)
8. Wang, Y.R.: The signal change-point detection using the high-order statistics of log-likelihood difference functions. In: Proceedings of IEEE Inter-national Conference on, Acoustics, Speech and Signal Processing ICASSP, pp. 4381–4384 (2008)
9. Hilas, C.S., Rekanos, I.T., Mastorocostas P.A.: Change point detection in time series using higher-order statistics: a heuristic approach. mathematical problems in engineering. Article ID 317613 (2013)
10. Kunchenko, Y.: Polynomial Parameter Estimations of Close to Gaussian Random Variables. Shaker Verlag, Aachen (2002)
11. Hinkley, D.: Inference about the change-point in a sequence of random variables. Biometrika **57**(1), 1–17 (1970)
12. Cramér, H.: Mathematical Methods of Statistics, vol. 9. Princeton University Press (1999)
13. Zabolotnii, S.W., Warsza, Z.L.: Semi-parametric polynomial method for retrospective estimation of the change-point of parameters of non-Gaussian sequences. In: Pavese, F., et al. (eds.) Monograph Advanced Mathematical and Computational Tools in Metrology and Testing X (AMCTM X), vol. 10, Series on Advances in Mathematics for Applied Sciences vol. 86, pp. 400–408. World Scientific, Singapore (2015)
14. Zabolotnii, S.W., Warsza, Z.L.: Semi-parametric estimation of the change-point of mean value of non-Gaussian random sequences by polynomial maximization method. In: Proceedings of 13th IMEKO TC10 Workshop on Technical Diagnostics Advanced Measurement Tools in Technical Diagnostics for Systems' Reliability and Safety, Warsaw, Poland (2014)
15. Nosek, K.: Methods of the change point detection with limitation of the form of alternatives. PhD Thesis. AGH, Krakow (2012) (in Polish)
16. Zabolotnii, S., Warsza, Z.L.: Semi-parametric polynomial modification of CUSUM algorithms for change-point detection of non-Gaussian sequences. In: Proceedings of XXI IMEKO World Congress Measurement in Research and Industry. Prague, Czech Republic, pp. 2088–2091 (2015)
17. Warsza, Z.L., Korczynski, M.J.: A new instrument enriched by type A uncertainty evaluation. In: Proceedings of 16th IMEKO TC4 Symposium in Florence. Paper no. 1181 (2008)

Author Index

R. Szewczyk et al. (eds.), *Challenges in Automation, Robotics and Measurement Techniques*, Advances in Intelligent Systems and Computing 440, DOI 10.1007/978-3-319-29357-8

Printed in the United States
By Bookmasters